METAL IONS IN BIOLOGICAL SYSTEMS

METAL IONS IN BIOLOGICAL SYSTEMS

Edited by

Astrid Sigel
and Helmut Sigel

Institute of Inorganic Chemistry
University of Basel
CH-4056 Basel, Switzerland

VOLUME 35

Iron Transport and Storage in Microorganisms, Plants, and Animals

MARCEL DEKKER, INC. NEW YORK • BASEL • HONG KONG

Library of Congress Cataloging-in-Publication Data

Iron transport and storage in microorganisms, plants, and animals / edited by Astrid Sigel and Helmut Sigel.
p. cm. — (Metal ions in biological systems ; v.35)
Includes bibliographical references and index.
ISBN 0-8247-9984-4 (alk. paper)
1. Iron—Physiological transport. 2. Iron—Metabolism. 3. Microbial metabolism. 4. Plants—Metabolism. I. Sigel, Astrid. II. Sigel, Helmut. III. Series.
QP535.F4I765 1998
572'.517—dc21 97-46947
CIP

The publisher offers discounts on this book when ordered in bulk quantities. For more information, write to Special Sales/Professional Marketing at the address below.

This book is printed on acid-free paper.

MARCEL DEKKER, INC.
270 Madison Avenue, New York, New York 10016
http://www.dekker.com

Current printing (last digit):
10 9 8 7 6 5 4 3 2 1

PRINTED IN THE UNITED STATES OF AMERICA

Preface to the Series

Recently, the importance of metal ions to the vital functions of living organisms, hence their health and well-being, has become increasingly apparent. As a result, the long-neglected field of "bioinorganic chemistry" is now developing at a rapid pace. The research centers on the synthesis, stability, formation, structure, and reactivity of biological metal ion-containing compounds of low and high molecular weight. The metabolism and transport of metal ions and their complexes is being studied, and new models for complicated natural structures and processes are being devised and tested. The focal point of our attention is the connection between the chemistry of metal ions and their role for life.

No doubt, we are only at the brink of this process. Thus, it is with the intention of linking coordination chemistry and biochemistry in their widest sense that the Metal Ions in Biological Systems series reflects the growing field of "bioinorganic chemistry". We hope, also, that this series will help to break down the barriers between the historically separate spheres of chemistry, biochemistry, biology, medicine, and physics, with the expectation that a good deal of future outstanding discoveries will be made in the interdisciplinary areas of science.

Should this series prove a stimulus for new activities in this fascinating "field", it would well serve its purpose and would be a satisfactory result for the efforts spent by the authors.

Fall 1973

Helmut Sigel
Institute of Inorganic Chemistry
University of Basel
CH-4056 Basel, Switzerland

Preface to Volume 35

Iron is the fourth most abundant element in the earth's crust and, after aluminum, the second most abundant metal. Thus, when the earth was young, and its atmosphere benignly anaerobic, iron-dependent organisms met no difficulty in satisfying their needs for this versatile element, since the soluble ferrous state of iron (Fe^{2+}) predominated. However, on a bright day when blue-green algae (cyanobacteria) started to synthesize chlorophyll to capture the energy of the sunlight, problems began. With the appearance of dioxygen in the atmosphere due to photosynthesis, ferrous iron was oxidized to ferric iron (Fe^{3+}), which has a solubility product in water of only about 10^{-38} M^4, and consequently, aquated Fe^{3+} reaches an equilibrium concentration of only about 10^{-17} M at neutral pH. Indeed, as discussed in the opening chapter of this volume the supply of iron may limit biological productivity in some areas of the sea, as, for example, in parts of the equatorial Pacific Ocean where iron is the limiting element in the ecosystem.

Since most living systems require iron for their growth and development, they had to devise strategies to solubilize and acquire this essential metal. The production of siderophores, which are low-molecular-weight, high-affinity Fe(III) chelators, serves this purpose. These compounds are common not only in prokaryotic bacteria, but also in eukaryotic fungi and some plants. The processes involved in siderophore-mediated (microbial) Fe transport are solubilization, transport to and across the cell membrane, and deposition at an appropriate site within the cell. With the advent of multicellular organisms and their circulating systems, more complex molecules were used for seizing and transporting iron, namely the transferrins, which comprise a class of iron-binding proteins once believed to originate with the evolutionary emergence of vertebrates or prevertebrates, but now known to be at least as ancient as insects.

Iron is not only difficult to acquire, it must constantly be "disciplined," monitored and appropriately shielded when stored. The problem is that the same property underlying the versatility of iron in biological electron transfer reactions, that is, the easy one-electron interconvertibility of the ferrous and ferric states in many iron complexes, enables the generation of oxygen-derived radicals and related oxidizing species, capable of attacking almost all biomolecules, and these most reactive species pose a threat to the viability of cells in which they arise. Hence, in order to protect cells from the potentially toxic effects of iron, special means for storing are needed. These are found in the ferritins, which are widely distributed among plants, animals, and microbes, and which store iron in biologically accessible forms as a crystalline, hydrous ferric oxide mineral core covered by a protein shell.

The complicated, interwoven, and intriguing events indicated above are detailed and discussed in the first fifteen chapters of this book. In the remaining three contributions, the effect of other metals on iron transport is considered, and especially the mechanisms underlying homeostasis in bacteria, yeast, plants, and animals are outlined with an emphasis on humans. In the latter, iron metabolism is highly conservative, that is, it is essentially controlled at the level of dietary iron absorption. There is no physiological mechanism in humans for excretion of excess iron! Thus, homeostasis is wholly reliant on dietary absorption and storage regulation. Aberrations in the homeostatic control mechanisms will lead therefore to an elevation in the total body burden of iron, and result ultimately in tissue/organ damage and eventual mortality. Consequently, iron removal is required under such conditions, and this challenge is discussed in the final chapter dealing with iron chelating agents for clinical use.

Astrid Sigel
Helmut Sigel

Contents

PREFACE TO THE SERIES iii
PREFACE TO VOLUME 35 v
CONTRIBUTORS xv
CONTENTS OF PREVIOUS VOLUMES xix
HANDBOOK ON TOXICITY OF INORGANIC COMPOUNDS xli
HANDBOOK ON METALS IN CLINICAL AND ANALYTICAL CHEMISTRY xlii

Chapter 1
BIOLOGICAL CYCLING OF IRON IN THE OCEAN 1
Neil M. Price and François M. M. Morel
1. Introduction 2
2. Distribution, Sources, and Sinks 2
3. Iron Chemistry in Seawater 5
4. Biological Cycling in Surface Waters 9
5. Kinetics and Mechanisms of Iron Acquisition 13
6. Physiological Adaptation to Fe Limitation 15
7. Siderophores 19
8. Evidence for Fe Limitation in the Ocean 20
9. Global and Paleooceanographic Perspective 25
Acknowledgments 29
Abbreviations 29
References 29

Chapter 2
MICROBIAL IRON TRANSPORT: IRON ACQUISITION BY PATHOGENIC MICROORGANISMS 37
B. Rowe Byers and Jean E. L. Arceneaux
1. Introduction 38
2. Why Iron Is an Elusive but Hazardous Necessity 39

3. Why Iron Has a Decisive Role in Infectious Diseases 45
4. Summary and Future Prospects 61
Acknowledgments 62
Abbreviations 62
References 62

Chapter 3
BACTERIAL IRON TRANSPORT: MECHANISMS, GENETICS, AND REGULATION 67
Volkmar Braun, Klaus Hantke, and Wolfgang Köster
1. Introduction 68
2. Transport Across the Outer Membrane 73
3. Transport Across the Cytoplasmic Membrane 84
4. Intracellular Iron Metabolism 96
5. Genetic Organization of Iron Transport Systems 101
6. Iron as a Central Regulatory Element 113
7. Outlook 127
Acknowledgments 128
Abbreviations 129
References 131

Chapter 4
MOLECULAR BIOLOGY OF IRON TRANSPORT IN FUNGI 147
Sally A. Leong and Günther Winkelmann
1. Introduction 148
2. Fungal Siderophores 149
3. Biochemistry of Siderophore Transport 153
4. Molecular Biology of Siderophore Biosynthesis and Transport 156
5. Molecular Biology of Iron Transport in Yeast 171
6. Conclusions 176
Acknowledgments 178
Abbreviations 178
References 179

Chapter 5
SOIL MICROORGANISMS AND IRON UPTAKE BY HIGHER PLANTS 187
Janette Palma Fett, Kristin LeVier, and Mary Lou Guerinot
1. Introduction 188

2. Iron Uptake Systems of Selected Plant-Associated Microorganisms 189
3. Iron Uptake Systems of Plants 200
4. Utilization of Microbial Siderophores by Plants 203
5. Utilization of Phytosiderophores by Microbes 205
6. Conclusions 206
Acknowledgments 207
Abbreviations 207
References 208

Chapter 6
IRON TRANSPORT IN GRAMINACEOUS PLANTS 215
Satoshi Mori
1. Introduction 216
2. Iron Transport from Root to Shoot 219
3. Iron Transport from Shoot to Root 222
4. Effect of Iron Status on Iron Transport in Plants 224
5. Light Effect on Iron Transport in Plants in Comparison with Phosphate Transport 227
6. Light-Dependent Iron Transport into Chloroplasts 231
7. Conclusions 233
Acknowledgments 235
Abbreviations 235
References 236

Chapter 7
COORDINATION CHEMISTRY OF SIDEROPHORES: THERMODYNAMICS AND KINETICS OF IRON CHELATION AND RELEASE 239
Anne-Marie Albrecht-Gary and Alvin L. Crumbliss
1. Introduction 240
2. Siderophore Ligands: Structural Features 241
3. Fe(III) and Fe(II) Complexes: Solution Structures 248
4. Fe(III) and Fe(II) Complexes: Stability 268
5. Iron(III) Uptake and Release Mechanism: Kinetics in Solution 296
6. Perspectives 313
Acknowledgments 314
Abbreviations 314
References 316

Chapter 8
BIOMIMETIC SIDEROPHORES: FROM STRUCTURAL PROBES TO DIAGNOSTIC TOOLS 329
Abraham Shanzer and Jacqueline Libman†
1. Introduction 330
2. Methodology 333
3. Enterobactin: The Minimalistic Approach toward Perfection 335
4. Hydrophilic Ferrichrome Analogues as Species-Specific Microbial Growth Promoters 338
5. Hydrophilic Ferrioxamine Analogues as Variable Range Microbial Growth Promoters 340
6. Fluorescent Labeled Siderophore Analogues: Probes for Iron Trafficking in Real Time 343
7. Lipophilic Ferrichrome Analogues or Reversed Siderophores as Iron Scavengers 346
8. Summary and Conclusion 350
Acknowledgments 350
Abbreviations 351
References 351

Chapter 9
THE PHYSICAL CHEMISTRY OF BACTERIAL OUTER-MEMBRANE SIDEROPHORE RECEPTOR PROTEINS 355
Dick van der Helm
1. Introduction 356
2. Receptor Proteins with Known Sequences 357
3. Homology, Models, and Channels 364
4. Interactions and Specificity 372
5. Mechanism 386
6. Conclusions 391
Acknowledgments 391
Abbreviations 391
References 392

Chapter 10
THE IRON RESPONSIVE ELEMENT (IRE) FAMILY OF mRNA REGULATORS 403
Elizabeth C. Theil
1. Introduction: mRNA Regulation and IREs 404
2. Specific Features of Iso-IREs: mRNA Translation Regulators 413

3. Specific Features of Iso-IREs: mRNA-Stability/Turnover Regulators 416
4. Proteins Recognized by IREs 417
5. Regulation of Iron Uptake and Storage in Plants and Microorganisms 421
6. Perspective 424
Acknowledgments 426
Abbreviations 426
References 427

Chapter 11
STRUCTURE-FUNCTION RELATIONSHIPS IN THE FERRITINS 435
Pauline M. Harrison, Paul C. Hempstead, Peter J. Artymiuk and Simon C. Andrews
1. Introduction 436
2. Diversity and Evolution of Ferritins 438
3. Three-Dimensional Structures of Ferritins 444
4. Structure-Function Relationships in Ferritins 459
5. Concluding Remarks 471
Acknowledgments 471
Abbreviations 471
References 472

Chapter 12
FERRITIN. UPTAKE, STORAGE, AND RELEASE OF IRON 479
N. Dennis Chasteen
1. Introduction 480
2. Iron Deposition 481
3. Iron Mobilization 498
4. Conclusions 503
Acknowledgments 504
Abbreviations 504
References 505

Chapter 13
FERRITIN. ITS MINERALIZATION 515
Annie K. Powell
1. Introduction 516
2. Biomineral Formation as a Means of Controlling Iron Homeostasis 517
3. Mineralization 521

4. Core Structures 534
5. A Comparison of the Use of Mineral and Cluster Aggregate Models for Elucidating the Core Mineral Structure in Ferritins 543
6. Conclusions 555
Abbreviations 556
References 556

Chapter 14
IRON STORAGE AND FERRITIN IN PLANTS 563
Jean-François Briat and Stéphane Lobréaux
1. Introduction 564
2. Iron Storage in the Apoplast and Plant Vacuoles 565
3. Structure-Function of Plant Ferritins 566
4. Ferritin and Plant Development 572
5. Ferritin and Recovery from Iron Starvation 575
6. Ferritin and Iron-Mediated Oxidative Stress 577
7. Outlook 578
Acknowledgments 579
Abbreviations 579
References 579

Chapter 15
TRANSFERRIN, THE TRANSFERRIN RECEPTOR, AND THE UPTAKE OF IRON BY CELLS 585
Philip Aisen
1. Introduction: Evolution and the Need for Iron-Binding Molecules 586
2. Transferrins 587
3. The Transferrin Receptor 606
4. Uptake of Iron by Cells 610
5. Perspectives and Prospects 618
Acknowledgments 618
Abbreviations 619
References 619

Chapter 16
IRON HOMEOSTASIS 633
Robert R. Crichton and Roberta J. Ward
1. Introduction 634
2. Iron Homeostasis in Bacteria 638

3. Iron Homeostasis in Plants and Yeast 639
4. Cellular Iron Homeostasis in Humans 643
5. Iron Transport, Storage, and Homeostasis in Humans 650
6. Conclusions and Unresolved Problems 657
Acknowledgment 660
Abbreviations 660
References 661

Chapter 17
THE ROLE OF OTHER METALS IN IRON TRANSPORT 667
Barry Chiswell, Kelvin O'Halloran, and Jarrod Wall
1. Introduction 668
2. Other Metals and Iron-Binding Proteins in Prokaryotes and Lower Eukaryotes 672
3. Other Metals and Iron-Binding Proteins in Higher Eukaryotes 675
4. Nitrogenase and Iron Transport 682
5. Conclusions 683
Abbreviations 684
References 684

Chapter 18
IRON CHELATORS FOR CLINICAL USE 691
Gary S. Tilbrook and Robert C. Hider
1. Introduction 692
2. Design Features of Fe(III) Ligands 695
3. Iron Chelator Toxicity 704
4. Clinically Useful Bidentate and Tridentate Ligands 709
5. Clinically Useful Hexadentate Ligands 714
6. Therapeutic Applications of Fe(III)-Selective Ligands 719
7. Conclusions 725
Abbreviations 726
References 727

SUBJECT INDEX 731

Contributors

Numbers in parentheses indicate the pages on which the authors' contributions begin.

Philip Aisen Departments of Physiology and Biophysics, and Medicine, Albert Einstein College of Medicine, Yeshiva University, 1300 Morris Park Avenue, Bronx, NY 10461, USA (FAX: ++1-718-430-8819; <aisen@aecom.yu.edu>) (585)

Anne-Marie Albrecht-Gary CNRS-URA 405-EHICS, Université Louis Pasteur, 1 rue Blaise Pascal, F-67008 Strasbourg Cedex, France (FAX: ++33-88-60-4392; <amalbre@chimie.u-strasbg.fr>) (239)

Simon C. Andrews The Krebs Institute for Biomolecular Research and Department of Molecular Biology and Biotechnology, University of Sheffield, Firth Court, Western Bank, Sheffield, S10 2TN, UK (FAX: ++44-114-272-8697) (435)

Jean E. L. Arceneaux Department of Microbiology, The University of Mississippi Medical Center, 2500 North State Street, Jackson, MS 39216-4505, USA (FAX: ++1-601-984-1708) (37)

Peter J. Artymiuk The Krebs Institute for Biomolecular Research and Department of Molecular Biology and Biotechnology, University of Sheffield, Firth Court, Western Bank, Sheffield, S10 2TN, UK (FAX: ++44-114-272-8697) (435)

Volkmar Braun Department of Microbiology/Membrane Physiology, University of Tübingen, Auf der Morgenstelle 28, D-72076 Tübingen, Germany (FAX: ++49-7071-294634; <volkmar.braun@mikrobio.uni-tuebingen.de>) (67)

Jean-François Briat CNRS/Unité de Recherche 573, ENSA-INRA, Biochimie et Physiologie Vegétales, Place Viala, F-34060 Montpellier, France (FAX: ++33-67-525737; <briat@msdos.ensam.inra.fr>) (563)

B. Rowe Byers Department of Microbiology, The University of Mississippi Medical Center, 2500 North State Street, Jackson, MS 39216-4505, USA (FAX: ++1-601-984-1708; <byers@fiona.umsmed.edu>) (37)

N. Dennis Chasteen Department of Chemistry, University of New Hampshire, Parsons Hall, Durham, NH 03824-3598, USA (FAX: ++1-603-862-4278; <dennis.chasteen@unh.edu>) (479)

Barry Chiswell Department of Chemistry, University of Queensland, Brisbane, QLD 4072, Australia (FAX: ++61-7-3365-4299) (667)

Robert R. Crichton Unité de Biochimie, Université Catholique de Louvain, Bâtiment Lavoisier, Place Louis Pasteur, 1, B-1348 Louvain-la-Neuve, Belgium (FAX: ++32-10-472796; <crichton@bioc.ucl.ac.be>) (633)

Alvin L. Crumbliss Department of Chemistry, Duke University, Box 90346, Durham, NC 27708-0346, USA (FAX: ++1-919-660-1605; <alc@chem.duke.edu>) (239)

Janette Palma Fett Department of Biological Sciences, Dartmouth College, 6044 Gilman, Hanover, NH 03755, USA (FAX: ++1-603-646-1347) Current address: Departamento de Botanica, Centro Biotecnologia, UFRGS, Porto Alegre, 90046-900, Brasil (187)

Mary Lou Guerinot Department of Biological Sciences, Dartmouth College, 6044 Gilman, Hanover, NH 03755, USA (FAX: ++1-603-646-1347; <mary.lou.guerinot@dartmouth.edu>) (187)

Klaus Hantke Department of Microbiology/Membrane Physiology, University of Tübingen, Auf der Morgenstelle 28, D-72076 Tübingen, Germany (FAX: ++49-7071-294634; <klaus.hantke@mikrobio.uni-tuebingen.de>) (67)

Pauline M. Harrison The Krebs Institute for Biomolecular Research and Department of Molecular Biology and Biotechnology, University of Sheffield, Firth Court, Western Bank, Sheffield, S10 2TN, UK (FAX: ++44-114-272-8697; <p.harrison@sheffield.ac.uk>) (435)

Paul C. Hempstead The Krebs Institute for Biomolecular Research and Department of Molecular Biology and Biotechnology, University of Sheffield, Firth Court, Western Bank, Sheffield, S10 2TN, UK (435)

Robert C. Hider Department of Pharmacy, King's College London, University of London, Manresa Road, London, SW3 6LX, UK (FAX: ++44-171-351-5307; <r.hider@bay.cc.kcl.ac.uk>) (691)

Wolfgang Köster Department of Microbiology/Membrane Physiology, University of Tübingen, Auf der Morgenstelle 28, D-72076 Tübingen, Germany (FAX: ++49-7071-294634) (67)

Sally A. Leong USDA/ARS, and Department of Plant Pathology, University of Wisconsin—Madison, 1630 Linden Drive, Madison, WI 53706-1598, USA (147)

Kristin LeVier Department of Biological Sciences, Dartmouth College, 6044 Gilman, Hanover, NH 03755, USA (FAX: ++1-603-646-1347). Current address: Department of Biology, Building 68-633, Massachusetts Institute of Technology, Cambridge, MA 02139-4391, USA (187)

Jacqueline Libman[†] Department of Organic Chemistry, The Weizmann Institute of Science, 76100 Rehovot, Israel (329)

Stéphane Lobréaux CNRS/Unité de Recherche 573, ENSA-INRA, Biochimie et Physiologie Vegétales, Place Viala, F-34060 Montpellier, France (FAX: ++33-67-525737) (563)

François M. M. Morel Department of Geosciences, Princeton University, Princeton, NJ 08540, USA (<morel@geo.princeton.edu>) (1)

Satoshi Mori Laboratory of Plant Molecular Physiology, Department of Applied Biological Chemistry, Division of Agriculture and Agricultural Life Sciences, The University of Tokyo, Yayoi 1-1, Bunkyo-ku, Tokyo 113, Japan (FAX: ++81-3-3812-0544; <asmori@hongo.ecc.u-tokyo.ac.jp>) (215)

Kelvin O'Halloran Centre for Water and Waste Technology, University of New South Wales, Sydney, NSW 2052, Australia (667)

Annie K. Powell Centre for Metalloprotein Spectroscopy and Biology, Schools of Chemical Sciences and Biological Sciences, University of East Anglia, Norwich, NR4 7TJ, UK (FAX: ++44-1603-593140; <a.powell @uea.ac.uk>) (515)

Neil M. Price Department of Biology, McGill University, 1205 Penfield Avenue, Montréal, PQ, H3A 1B1, Canada (FAX: ++1-514-398-5069; <nprice@biol.lan.mcgill.ca>) (1)

Abraham Shanzer Department of Organic Chemistry, The Weizmann Institute of Science, 76100 Rehovot, Israel (FAX: ++972-8-344142; <coshanzr@wiccmail.weizmann.ac.il>) (329)

Elizabeth C. Theil Department of Biochemistry, North Carolina State University, Box 7622, Raleigh, NC 27695-7622, USA (FAX: ++1-919-515-5805; <theil@bchserver.bch.ncsu.edu>) (403)

Gary S. Tilbrook Department of Pharmacy, King's College London, University of London, Manresa Road, London, SW3 6LX, UK (FAX: ++44-171-351-5307; <gary.tilbrook@kcl.ac.uk>) (691)

Dick van der Helm Department of Chemistry and Biochemistry, University of Oklahoma, 620 Parrington Oval, Room 208, Norman, Oklahoma 73019-0370, USA (FAX: ++1-405-325-6111) (355)

Jarrod Wall Department of Chemistry, University of Queensland, Brisbane, QLD 4072, Australia (FAX: ++61-7-3365-4299) (667)

Roberta J. Ward Unité de Biologie Animale, Université Catholique de Louvain, Place Croix du Sud 1, B-1348 Louvain-la-Neuve, Belgium (633)

Günther Winkelmann Department of Mikrobiology & Biotechnology, University of Tübingen, Auf der Morgenstelle 1, D-72076 Tübingen, Germany (FAX: ++49-7071-293094; <winkelmann@uni-tuebingen.de>) (147)

Contents of Previous Volumes

Volume 1. Simple Complexes*

Volume 2. Mixed-Ligand Complexes*

Volume 3. High Molecular Complexes*

Volume 4. Metal Ions as Probes*

Volume 5. Reactivity of Coordination Compounds*

Volume 6. Biological Action of Metal Ions*

Volume 7. Iron in Model and Natural Compounds*

Volume 8. Nucleotides and Derivatives: Their Ligating Ambivalency*

Volume 9. Amino Acids and Derivatives as Ambivalent Ligands

1. Complexes of α-Amino Acids with Chelatable Side Chain Donor Atoms
R. Bruce Martin
2. Metal Complexes of Aspartic Acid and Glutamic Acid
Christopher A. Evans, Roger Guevremont, and Dallas L. Rabenstein
3. The Coordination Chemistry of L-Cysteine and D-Penicillamine
Arthur Gergely and Imre Sóvágó
4. Glutathione and Its Metal Complexes
Dallas L. Rabenstein, Roger Guevremont, and Christopher A. Evans
5. Coordination Chemistry of *L*-Dopa and Related Ligands
Arthur Gergely and Tamás Kiss

---

*Out of print.

6. Stereoselectivity in the Metal Complexes of Amino Acids and Dipeptides
Leslie D. Pettit and Robert J. W. Hefford
7. Protonation and Complexation of Macromolecular Polypeptides: Corticotropin Fragments and Basic Trypsin Inhibitor (Kunitz Base)
Kálmán Burger

Author Index-Subject Index

Volume 10. Carcinogenicity and Metal Ions

1. The Function of Metal Ions in Genetic Regulation
Gunther L. Eichhorn
2. A Comparison of Carcinogenic Metals
C. Peter Flessel, Arthur Furst, and Shirley B. Radding
3. The Role of Metals in Tumor Development and Inhibition
Haleem J. Issaq
4. Paramagnetic Metal Ions in Tissue During Malignant Development
Nicholas J. F. Dodd
5. Ceruloplasmin and Iron Transferrin in Human Malignant Disease
Margaret A. Foster, Trevor Pocklington, and Audrey A. Dawson
6. Human Leukemia and Trace Elements
E. L. Andronikashvili and L. M. Mosulishvili
7. Zinc and Tumor Growth
Andre M. van Rij and Walter J. Pories
8. Cyanocobalamin and Tumor Growth
Sofija Kanopkaitė and Gediminas Bražėnas
9. The Role of Selenium as a Cancer-Protecting Trace Element
Birger Jansson
10. Tumor Diagnosis Using Radioactive Metal Ions and Their Complexes
Akira Yokoyama and Hideo Saji

Author Index-Subject Index

Volume 11. Metal Complexes as Anticancer Agents*

Volume 12. Properties of Copper

1. The Coordination Chemistry of Copper with Regard to Biological Systems
R. F. Jameson

---

*Out of print.

2. Copper(II) as Probe in Substituted Metalloproteins
Ivano Bertini and A. Scozzafava
3. Copper(III) Complexes and Their Reactions
Dale W. Margerum and Grover D. Owens
4. Copper Catalyzed Oxidation and Oxygenation
Harold Gampp and Andreas D. Zuberbühler
5. Copper and the Oxidation of Hemoglobins
Joseph M. Rifkind
6. Transport of Copper
Bibudhendra Sarkar
7. The Role of Low-Molecular-Weight Copper Complexes in the Control of Rheumatoid Arthritis
Peter M. May and David R. Williams

Author Index-Subject Index

Volume 13. Copper Proteins*

Volume 14. Inorganic Drugs in Deficiency and Disease

1. Drug-Metal Ion Interaction in the Gut
P. F. D'Arcy and J. C. McElnay
2. Zinc Deficiency and Its Therapy
Ananda S. Prasad
3. The Pharmacological Use of Zinc
George J. Brewer
4. The Anti-inflammatory Activities of Copper Complexes
John R. J. Sorenson
5. Iron-Containing Drugs
David A. Brown and M. V. Chidambaram
6. Gold Complexes as Metallo-Drugs
Kailash C. Dash and Hubert Schmidbaur
7. Metal Ions and Chelating Agents in Antiviral Chemotherapy
D. D. Perrin and Hans Stünzi
8. Complexes of Hallucinogenic Drugs
Wolfram Hänsel
9. Lithium in Psychiatry
Nicholas J. Birch

Author Index-Subject Index

*Out of print.

Volume 15. Zinc and Its Role in Biology and Nutrition

1. Categories of Zinc Metalloenzymes
Alphonse Galdes and Bert L. Vallee

2. Models for Zn(II) Binding Sites in Enzymes
Robert S. Brown, Joan Huguet, and Neville J. Curtis

3. An Insight on the Active Site of Zinc Enzymes Through Metal Substitution
Ivano Bertini and Claudio Luchinat

4. The Role of Zinc in DNA and RNA Polymerases
Felicia Ying-Hsiueh Wu and Cheng-Wen Wu

5. The Role of Zinc in Snake Toxins
Anthony T. Tu

6. Spectroscopic Properties of Metallothionein
Milan Vašák and Jeremias H. R. Kägi

7. Interaction of Zinc with Erythrocytes
Joseph M. Rifkind

8. Zinc Absorption and Excretion in Relation to Nutrition
Manfred Kirchgessner and Edgar Weigand

9. Nutritional Influence of Zinc on the Activity of Enzymes and Hormones
Manfred Kirchgessner and Hans-Peter Roth

10. Zinc Deficiency Syndrome During Parenteral Nutrition
Karin Ladefoged and Stig Jarnum

Author Index-Subject Index

Volume 16. Methods Involving Metal Ions and Complexes in Clinical Chemistry

1. Some Aspects of Nutritional Trace Element Research
Clare E. Casey and Marion F. Robinson

2. Metals and Immunity
Lucy Treagan

3. Therapeutic Chelating Agents
Mark M. Jones

4. Computer-Directed Chelate Therapy of Renal Stone Disease
Martin Rubin and Arthur E. Martell

5. Determination of Trace Metals in Biological Materials by Stable Isotope Dilution
Claude Veillon and Robert Alvarez

6. Trace Elements in Clinical Chemistry Determined by Neutron Activation Analysis
Kaj Heydorn

7. Determination of Lithium, Sodium, and Potassium in Clinical Chemistry
Adam Uldall and Arne Jensen

8. Determination of Magnesium and Calcium in Serum
Arne Jensen and Erik Riber

9. Determination of Manganese, Iron, Cobalt, Nickel, Copper, and Zinc in Clinical Chemistry
Arne Jensen, Erik Riber, Poul Persson, and Kaj Heydorn

10. Determination of Lead, Cadmium, and Mercury in Clinical Chemistry
Arne Jensen, Jytte Molin Christensen, and Poul Persson

11. Determination of Chromium in Urine and Blood
Ole Jøns, Arne Jensen, and Poul Persson

12. Determination of Aluminum in Clinical Chemistry
Arne Jensen, Erik Riber, and Poul Persson

13. Determination of Gold in Clinical Chemistry
Arne Jensen, Erik Riber, Poul Persson, and Kaj Heydorn

14. Determination of Phosphates in Clinical Chemistry
Arne Jensen and Adam Uldall

15. Identification and Quantification of Some Drugs in Body Fluids by Metal Chelate Formation
R. Bourdon, M. Galliot, and J. Hoffelt

16. Metal Complexes of Sulfanilamides in Pharmaceutical Analysis and Therapy
Auke Bult

17. Basis for the Clinical Use of Gallium and Indium Radionuclides
Raymond L. Hayes and Karl F. Hübner

18. Aspects of Technetium Chemistry as Related to Nuclear Medicine
Hans G. Seiler

Author Index-Subject Index

Volume 17. Calcium and Its Role in Biology

1. Bioinorganic Chemistry of Calcium
R. Bruce Martin

2. Crystal Structure Studies of Calcium Complexes and Implications for Biological Systems
H. Einspahr and C. E. Bugg

3. Intestinal and Renal Absorption of Calcium
Piotr Gmaj and Heini Murer

4. Calcium Transport Across Biological Membranes
Ernesto Carafoli, Giuseppe Inesi, and Barry Rosen

5. Physiological Aspects of Mitochondrial Calcium Transport
Gary Fiskum
6. Mode of Action of the Regulatory Protein Calmodulin
Jos A. Cox, Michelle Comte, Armand Malnoë, Danielle Burger, and Eric A. Stein
7. Calcium and Brain Proteins
S. Alamà
8. The Roles of Ca^{2+} in the Regulation and Mechanism of Exocytosis
Carl E. Creutz
9. Calcium Function in Blood Coagulation
Gary L. Nelsestuen
10. The Role of Calcium in the Regulation of the Skeletal Muscle Contraction-Relaxation Cycle
Henry G. Zot and James D. Potter
11. Calcification of Vertebrate Hard Tissues
Roy E. Wuthier

Author Index-Subject Index

Volume 18. Circulation of Metals in the Environment

1. Introduction to "Circulation of Metals in the Environment"
Peter Baccini
2. Analytical Chemistry Applied to Metal Ions in the Environment
Arne Jensen and Sven Erik Jørgensen
3. Processes of Metal Ions in the Environment
Sven Erik Jørgensen and Arne Jensen
4. Surface Complexation
Paul W. Schindler
5. Relationships Between Biological Availability and Chemical Measurements
David R. Turner
6. Natural Organic Matter and Metal-Organic Interactions in Aquatic Systems
Jacques Buffle
7. Evolutionary Aspects of Metal Ion Transport Through Cell Membranes
John M. Wood
8. Regulation of Trace Metal Concentrations in Fresh Water Systems
Peter Baccini
9. Cycling of Metal Ions in the Soil Environment
Garrison Sposito and Albert L. Page
10. Microbiological Strategies in Resistance to Metal Ion Toxicity
John M. Wood

11. Conclusions and Outlook
Peter Baccini

Author Index-Subject Index

Volume 19. Antibiotics and Their Complexes

1. The Discovery of Ionophores: An Historical Account
Berton C. Pressman
2. Tetracyclines and Daunorubicin
R. Bruce Martin
3. Interaction of Metal Ions with Streptonigrin and Biological Properties of the Complexes
Joseph Hajdu
4. Bleomycin Antibiotics: Metal Complexes and Their Biological Action
Yukio Sugiura, Tomohisa Takita, and Hamao Umezawa
5. Interaction Between Valinomycin and Metal Ions
K. R. K. Easwaran
6. Beauvericin and the Other Enniatins
Larry K. Steinrauf
7. Complexing Properties of Gramicidins
James F. Hinton and Roger E. Koeppe II
8. Nactins: Their Complexes and Biological Properties
Yoshiharu Nawata, Kunio Ando, and Yoichi Iitaka
9. Cation Complexes of the Monovalent and Polyvalent Carboxylic Ionophores: Lasalocid (X-537A), Monensin, A-23187 (Calcimycin) and Related Antibiotics
George R. Painter and Berton C. Pressman
10. Complexes of *D*-Cycloserine and Related Amino Acids with Antibiotic Properties
Paul O'Brien
11. Iron-Containing Antibiotics
J. B. Neilands and J. R. Valenta
12. Cation-Ionophore Interactions: Quantification of the Factors Underlying Selective Complexation by Means of Theoretical Computations
Nohad Gresh and Alberte Pullman

Author Index-Subject Index

Volume 20. Concepts on Metal Ion Toxicity

1. Distribution of Potentially Hazardous Trace Metals
Garrison Sposito

2. Bioinorganic Chemistry of Metal Ion Toxicity
R. Bruce Martin

3. The Interrelation Between Essentiality and Toxicity of Metals in the Aquatic Ecosystem
Elie Eichenberger

4. Metal Ion Speciation and Toxicity in Aquatic Systems
Gordon K. Pagenkopf

5. Metal Toxicity to Agricultural Crops
Frank T. Bingham, Frank J. Peryea, and Wesley M. Jarrell

6. Metal Ion Toxicity in Man and Animals
Paul B. Hammond and Ernest C. Foulkes

7. Human Nutrition and Metal Ion Toxicity
M. R. Spivey Fox and Richard M. Jacobs

8. Chromosome Damage in Individuals Exposed to Heavy Metals
Alain Léonard

9. Metal Ion Carcinogenesis: Mechanistic Aspects
Max Costa and J. Daniel Heck

10. Methods for the In Vitro Assessment of Metal Ion Toxicity
J. Daniel Heck and Max Costa

11. Some Problems Encountered in the Analysis of Biological Materials for Toxic Trace Elements
Hans G. Seiler

Author Index-Subject Index

Volume 21. Applications of Nuclear Magnetic Resonance to Paramagnetic Species

1. Nuclear Relaxation Times as a Source of Structural Information
Gil Navon and Gianni Valensin

2. Nuclear Relaxation in NMR of Paramagnetic Systems
Ivano Bertini, Claudio Luchinat, and Luigi Messori

3. NMR Studies of Magnetically Coupled Metalloproteins
Lawrence Que, Jr., and Michael J. Maroney

4. Proton NMR Studies of Biological Problems Involving Paramagnetic Heme Proteins
James D. Satterlee

5. Metal-Porphyrin Induced NMR Dipolar Shifts and Their Use in Conformational Analysis
Nigel J. Clayden, Geoffrey R. Moore, and Glyn Williams

6. Relaxometry of Paramagnetic Ions in Tissue
Seymour H. Koenig and Rodney D. Brown III

Author Index-Subject Index

Volume 22. ENDOR, EPR, and Electron Spin Echo for Probing Coordination Spheres

1. ENDOR: Probing the Coordination Environment in Metalloproteins
Jürgen Hüttermann and Reinhard Kappl
2. Identification of Oxygen Ligands in Metal-Nucleotide-Protein Complexes by Observation of the Mn(II)-^{17}O Superhyperfine Coupling
Hans R. Kalbitzer
3. Use of EPR Spectroscopy for Studying Solution Equilibria
Harald Gampp
4. Application of EPR Saturation Methods to Paramagnetic Metal Ions in Proteins
Marvin W. Makinen and Gregg B. Wells
5. Electron Spin Echo: Applications to Biological Systems
Yuri D. Tsvetkov and Sergei A. Dikanov

Author Index-Subject Index

Volume 23. Nickel and Its Role in Biology

1. Nickel in the Natural Environment
Robert W. Boyle and Heather W. Robinson
2. Nickel in Aquatic Systems
Pamela Stokes
3. Nickel and Plants
Margaret E. Farago and Monica M. Cole
4. Nickel Metabolism in Man and Animals
Evert Nieboer, Rickey T. Tom, and W. (Bill) E. Sanford
5. Nickel Ion Binding to Amino Acids and Peptides
R. Bruce Martin
6. Nickel in Proteins and Enzymes
Robert K. Andrews, Robert L. Blakeley, and Burt Zerner
7. Nickel-Containing Hydrogenases
José J. G. Moura, Isabel Moura, Miguel Teixeira, António V. Xavier, Guy D. Fauque, and Jean LeGall
8. Nickel Ion Binding to Nucleosides and Nucleotides
R. Bruce Martin
9. Interactions Between Nickel and DNA: Considerations About the Role of Nickel in Carcinogenesis
E. L. Andronikashvili, V. G. Bregadze, and J. R. Monaselidze
10. Toxicology of Nickel Compounds
Evert Nieboer, Franco E. Rossetto, and C. Rajeshwari Menon

11. Analysis of Nickel in Biological Materials
Hans G. Seiler

Author Index-Subject Index

Volume 24. Aluminum and Its Role in Biology

1. Bioinorganic Chemistry of Aluminum
R. Bruce Martin
2. Aluminum in the Environment
Charles T. Driscoll and William D. Schecher
3. The Physiology of Aluminum Phytotoxicity
Gregory J. Taylor
4. The Physiology of Aluminum Tolerance
Gregory J. Taylor
5. Aluminum in the Diet and Mineral Metabolism
Janet L. Greger
6. Aluminum Intoxication: History of Its Clinical Recognition and Management
David N. S. Kerr and M. K. Ward
7. Aluminum and Alzheimer's Disease, Methodologic Approaches
Daniel P. Perl
8. Mechanisms of Aluminum Neurotoxicity—Relevance to Human Disease
Theo P. A. Kruck and Donald R. McLachlan
9. Aluminum Toxicity and Chronic Renal Failure
Michael R. Wills and John Savory
10. Analysis of Aluminum in Biological Materials
John Savory and Michael R. Wills

Author Index-Subject Index

Volume 25. Interrelations Among Metal Ions, Enzymes, and Gene Expression

1. Metal Ion-Induced Mutagenesis In Vitro: Molecular Mechanisms
Kathleen M. Downey and Antero G. So
2. Metallonucleases: Real and Artificial
Lena A. Basile and Jacqueline K. Barton
3. Metalloregulatory Proteins: Metal-Responsive Molecular Switches Governing Gene Expression
Thomas V. O'Halloran
4. Yeast Metallothionein: Gene Function and Regulation by Metal Ions
David J. Ecker, Tauseef R. Butt, and Stanley T. Crooke

5. Zinc-Binding Proteins Involved in Nucleic Acid Replication
Joseph E. Coleman and David P. Giedroc

6. "Zinc Fingers": The Role of Zinc(II) in Transcription Factor IIIA and Related Proteins
Jeremy M. Berg

7. Site-Directed Mutagenesis and Structure-Function Relations in EF-Hand Ca^{2+}-Binding Proteins
Sture Forsén, Torbjörn Drakenberg, Sara Linse, Peter Brodin, Peter Sellers, Charlotta Johansson, Eva Thulin, and Thomas Grundström

8. Genetic Alteration of Active Site Residues of Staphylococcal Nuclease: Insights into the Enzyme Mechanism
Albert S. Mildvan and Engin H. Serpersu

9. Alcohol Dehydrogenase: Structure, Catalysis, and Site-Directed Mutagenesis
Y. Pocker

10. Probing the Mechanism of Action of Carboxypeptidase A by Inorganic, Organic, and Mutagenic Modifications
David S. Auld, James F. Riordan, and Bert L. Vallee

11. Site-Directed Mutagenesis of *E. coli* Alkaline Phosphatase: Probing the Active-Site Mechanism and the Signal Sequence-Mediated Transport of the Enzyme
John E. Butler-Ransohoff, Debra A. Kendall, and Emil Thomas Kaiser

12. Site-Directed Mutagenesis of Heme Proteins
Patrick R. Stayton, William M. Atkins, Barry A. Springer, and Stephen G. Sligar

13. Exploring Structure-Function Relationships in Yeast Cytochrome c Peroxidase Using Mutagenesis and Crystallography
J. Matthew Mauro, Mark A. Miller, Stephen L. Edwards, Jimin Wang, Laurence A. Fishel, and Joseph Kraut

Author Index-Subject Index

Volume 26. Compendium on Magnesium and Its Role in Biology, Nutrition, and Physiology

1. Bioinorganic Chemistry of Magnesium
R. Bruce Martin

2. Magnesium in the Environment
Raili Jokinen

3. Magnesium in Plants: Uptake, Distribution, Function, and Utilization by Man and Animals
Stanley R. Wilkinson, Ross M. Welch, Henry F. Mayland, and David L. Grunes

4. Magnesium in Animal Nutrition
H. Meyer and J. Zentak

5. Dietary Magnesium and Drinking Water: Effects on Human Health Status
John R. Marier

6. Magnesium in Biology and Medicine: An Overview
Nicholas J. Birch

7. Role of Magnesium in Enzyme Systems
Frank W. Heaton

8. Magnesium: A Regulated and Regulatory Cation
Michael E. Maguire

9. Magnesium Transport in Prokaryotic Cells
Marshall D. Snavely

10. Hormonal Regulation of Magnesium Homeostasis in Cultured Mammalian Cells
Robert D. Grubbs

11. Functional Compartmentation of Intracellular Magnesium
T. Günther

12. Membrane Transport of Magnesium
T. Günther and H. Ebel

13. Intestinal Magnesium Absorption
H. Ebel

14. The Renal Handling of Magnesium
Michael P. Ryan

15. Magnesium and Liver Cirrhosis
Leon Cohen

16. Hypomagnesemia and Hypermagnesemia
Nachman Brautbar, Atul T. Roy, Philip Hom, and David B. N. Lee

17. Systemic Stress and the Role of Magnesium
H. G. Classen

18. Magnesium and Lipid Metabolism
Yves Rayssiguier

19. Magnesium and the Cardiovascular System: Experimental and Clinical Aspects Updated
Burton M. Altura and Bella T. Altura

20. Magnesium and the Peripheral (Extradural) Nervous System: Metabolism, Neurophysiological Functions, and Clinical Disorders
Jerry G. Chutkow

21. Magnesium and the Central (Intradural) Nervous System: Metabolism, Neurophysiological Functions, and Clinical Disorders
Jerry G. Chutkow

22. The Role of Magnesium in the Regulation of Muscle Function
Christopher H. Fry and Sarah K. Hall
23. Magnesium in Bone and Tooth
Colin Robinson and John A. Weatherell
24. Magnesium and Osteoporosis
Leon Cohen
25. The Role of Magnesium in Pregnancy, for the Newborn, and in Children's Diseases
Ludwig Spätling
26. Magnesium and Its Role in Allergy
Nicole Hunziker
27. Magnesium and Its Relationship to Oncology
Jean Durlach, Michel Bara, and Andrée Guiet-Bara
28. The Assessment of Magnesium Status in Humans
Ronald J. Elin
29. Magnesium and Placebo Effects in Human Medicine
H. G. Classen
30. Determination of Magnesium in Biological Materials
Hans G. Seiler

Author Index-Subject Index

Volume 27. Electron Transfer Reactions in Metalloproteins

1. Mediation of Electron Transfer by Peptides and Proteins: Current Status
Stephan S. Isied
2. Electron and Atom Group Transfer Properties of Protein Systems
Hans E. M. Christensen, Lars S. Conrad, Jens Ulstrup, and Kurt V. Mikkelsen
3. Electron Tunneling Pathways in Proteins
David N. Beratan, José Nelson Onuchic, and Harry B. Gray
4. Diprotein Complexes and Their Electron Transfer Reactions
Nenad M. Kostić
5. Electron Transfer Between Bound Proteins
George McLendon
6. Long-Range Electron Transfer in Cytochrome c Derivatives with Covalently Attached Probe Complexes
Robert A. Scott, David W. Conrad, Marly K. Eidsness, Antonius C. F. Gorren, and Sten A. Wallin
7. Photoinduced Electron Transfer Reactions in Metalloprotein Complexes Labeled with Ruthenium Polypyridine
Francis Millett and Bill Durham

8. Stereoselective Effects in Electron Transfer Reactions Involving Synthetic Metal Complexes and Metalloproteins
Klaus Bernauer

9. Properties and Electron Transfer Reactivity of [2Fe-2S] Ferredoxins
A. Geoffrey Sykes

10. Electron Transfer in Photosynthetic Reaction Centers
Sethulakshmi Kartha, Ranjan Das, and James A. Norris

11. Modeling the Primary Electron Transfer Events of Photosynthesis
Michael R. Wasielewski

12. Electrochemical Investigation of Metalloproteins and Enzymes Using a Model Incorporating Microscopic Aspects of the Electrode-Solution Interface
Alan M. Bond and H. Allen O. Hill

Author Index-Subject Index

Volume 28. Degradation of Environmental Pollutants by Microorganisms and Their Metalloenzymes

1. General Strategies in the Biodegradation of Pollutants
Thomas W. Egli

2. Oxidation of Aromatic Pollutants by Lignin-Degrading Fungi and Their Extracellular Peroxidases
Kenneth E. Hammel

3. Biodegradation of Tannins
James A. Field and G. Lettinga

4. Aerobic Biodegradation of Aromatic Hydrocarbons by Bacteria
Shigeaki Harayama and Kenneth N. Timmis

5. Degradation of Halogenated Aromatics by Actinomycetes
Bruno Winter and Wolfgang Zimmermann

6. Enzymes Catalyzing Oxidative Coupling Reactions of Pollutants
Jean-Marc Bollag

7. Mechanism of Action of Peroxidases
Helen Anni and Takashi Yonetani

8. Mechanistic Aspects of Dihydroxybenzoate Dioxygenases
John D. Lipscomb and Allen M. Orville

9. Aerobic and Anaerobic Degradation of Halogenated Aliphatics
Dick B. Janssen and Bernard Witholt

10. Mechanisms of Reductive Dehalogenation by Transition Metal Cofactors Found in Anaerobic Bacteria
Lawrence P. Wackett and Craig A. Schanke

11. Bacterial Degradation of Hemicelluloses
Wolfgang Zimmermann

12. Degradation of Cellulose and Effects of Metal Ions on Cellulases
Anil Goyal and Douglas E. Eveleigh

13. Metalloproteases and Their Role in Biotechnology
Guido Grandi and Giuliano Galli

14. Metal-Dependent Conversion of Inorganic Nitrogen and Sulfur Compounds
Peter M. H. Kroneck, Joachim Beuerle, and Wolfram Schumacher

Author Index-Subject Index

Volume 29. Biological Properties of Metal Alkyl Derivatives

1. Global Bioalkylation of the Heavy Elements
John S. Thayer

2. Analysis of Organometallic Compounds in the Environment
Darren Mennie and Peter J. Craig

3. Biogeochemistry of Methylgermanium Species in Natural Waters
Brent L. Lewis and H. Peter Mayer

4. Biological Properties of Alkyltin Compounds
Yasuaki Arakawa and Osamu Wada

5. Biological Properties of Alkyl Derivatives of Lead
Yukio Yamamura and Fumio Arai

6. Metabolism of Alkyl Arsenic and Antimony Compounds
Marie Vahter and Erminio Marafante

7. Biological Alkylation of Selenium and Tellurium
Ulrich Karlson and William T. Frankenberger, Jr.

8. Making and Breaking the Co-Alkyl Bond in B_{12} Derivatives
John M. Pratt

9. Methane Formation by Methanogenic Bacteria: Redox Chemistry of Coenzyme F430
Bernhard Jaun

10. Synthesis and Degradation of Organomercurials by Bacteria —A Comment by the Editors
Helmut Sigel and Astrid Sigel

11. Biogenesis and Metabolic Role of Halomethanes in Fungi
David B. Harper

Author Index-Subject Index

Volume 30. Metalloenzymes Involving Amino Acid-Residue and Related Radicals

1. Free Radicals and Metalloenzymes: General Considerations
Ei-Ichiro Ochiai

2. Peroxidases: Structure, Function, and Engineering
Thomas L. Poulos and Roger E. Fenna

3. Photosystem II
Curtis W. Hoganson and Gerald T. Babcock

4. Ribonucleotide Reductase in Mammalian Systems
Lars Thelander and Astrid Gräslund

5. Manganese-Dependent Ribonucleotide Reduction and Overproduction of Nucleotides in Coryneform Bacteria
George Auling and Hartmut Follmann

6. Prostaglandin Endoperoxide Synthases
William L. Smith and Lawrence J. Marnett

7. Diol Dehydrase from *Clostridium glycolicum*: The Non-B_{12}-Dependent Enzyme
Maris G. N. Hartmanis

8. Diol Dehydrase and Glycerol Dehydrase, Coenzyme B_{12}-Dependent Isozymes
Tetsua Toraya

9. Adenosylcobalamin (Vitamin B_{12} Coenzyme)-Dependent Enzymes
Ei-Ichiro Ochiai

10. *S*-Adenosylmethionine-Dependent Radical Formation in Anaerobic Systems
Kenny K. Wong and John W. Kozarich

11. The Free Radical-Coupled Copper Active Site of Galactose Oxidase
James W. Whittaker

12. Amine Oxidases
Peter F. Knowles and David M. Dooley

13. Bacterial Transport of and Resistance to Copper
Nigel L. Brown, Barry T. O. Lee, and Simon Silver

Author Index-Subject Index

Volume 31. Vanadium and Its Role for Life

1. Inorganic Considerations on the Function of Vanadium in Biological Systems
Dieter Rehder

2. Solution Properties of Vanadium(III) with Regard to Biological Systems
Roland Meier, Martin Boddin, Steffi Mitzenheim, and Kan Kanamori

3. The Vanadyl Ion: Molecular Structure of Coordinating Ligands by Electron Paramagnetic Resonance and Electron Nuclear Double Resonance Spectroscopy
Marvin W. Makinen and Devkumar Mustafi

4. Vanadyl(IV) Complexes of Nucleotides
Enrique J. Baran

5. Interactions of Vanadates with Biogenic Ligands
Debbie C. Crans

6. Use of Vanadate-Induced Photocleavage for Detecting Phosphate Binding Sites in Proteins
Andras Muhlrad and Israel Ringel

7. Vanadium-Protein Interactions
N. Dennis Chasteen

8. Stimulation of Enzyme Activity by Oxovanadium Complexes
Paul J. Stankiewicz and Alan S. Tracy

9. Inhibition of Phosphate-Metabolizing Enzymes by Oxovanadium(V) Complexes
Paul J. Stankiewicz, Alan S. Tracy, and Debbie C. Crans

10. Vanadium-Dependent Haloperoxidases
Hans Vilter

11. Vanadium Nitrogenases of *Azotobacter*
Robert R. Eady

12. Amavadin, the Vanadium Compound of Amanitae
Ernst Bayer

13. Vanadium in Ascidians and the Chemistry of Tunichromes
Mitchell J. Smith, Daniel E. Ryan, Koji Nakanishi, Patrick Frank, and Keith O. Hodgson

14. Biochemical Significance of Vanadium in a Polychaete Worm
Toshiaki Ishii, Izumi Nakai, and Kenji Okoshi

15. Vanadium Transport in Animal Systems
Kenneth Kustin and William E. Robinson

16. Vanadium in Mammalian Physiology and Nutrition
Forrest H. Nielsen

17. Vanadium Compounds as Insulin Mimics
Chris Orvig, Katherine H. Thompson, Mary Battell, and John H. McNeill

18. Antitumor Activity of Vanadium Compounds
Cirila Djordjevic

19. Methods for the Spectroscopic Characterization of Vanadium Centers in Biological and Related Chemical Systems
C. David Garner, David Collison, and Frank E. Mabbs

20. Analytical Procedures for the Determination of Vanadium in Biological Materials
Hans G. Seiler

Author Index-Subject Index

Volume 32. Interactions of Metal Ions with Nucleotides, Nucleic Acids, and Their Constituents

1. Phosphate-Metal Ion Interactions of Nucleotides and Polynucleotides
Cindy Klevickis and Charles M. Grisham
2. Sugar-Metal Ion Interactions
Shigenobu Yano and Masami Otsuka
3. Dichotomy of Metal Ion Binding to N1 and N7 of Purines
R. Bruce Martin
4. General Conclusions from Solid State Studies of Nucleotide-Metal Ion Complexes
Katsuyuki Aoki
5. Solution Structures of Nucleotide Metal Ion Complexes. Isomeric Equilibria
Helmut Sigel and Bin Song
6. Stacking Interactions Involving Nucleotides and Metal Ion Complexes
Osamu Yamauchi, Akira Odani, Hideki Masuda, and Helmut Sigel
7. Effect of Metal Ions on the Hydrolytic Reactions of Nucleosides and Their Phosphoesters
Satu Kuusela and Harri Lönnberg
8. Metal Complexes of Sulfur-Containing Purine Derivatives
Erich Dubler
9. Mechanistic Insight from Kinetic Studies on the Interaction of Model Palladium(II) Complexes with Nucleic Acid Components
Tobias Rau and Rudi van Eldik
10. Platinum(II)-Nucleobase Interactions. A Kinetic Approach
Jorma Arpalahti
11. NMR Studies of Oligonucleotide-Metal Ion Interactions
Einar Sletten and Nils Åge Frøystein
12. Metal Ion Interactions with DNA: Considerations on Structure, Stability, and Effects from Metal Ion Binding
Vasil G. Bregadze
13. Electron Transfer Reactions Through the DNA Double Helix
Thomas J. Meade
14. The Role of Metal Ions in Ribozymes
Anna Marie Pyle

15. Ternary Metal Ion-Nucleic Acid Base-Protein Complexes
Michal Sabat

16. Metal-Responsive Gene Expression and the Zinc-Metalloregulatory Model
David A. Suhy and Thomas V. O'Halloran

17. The Role of Iron-Sulfur Proteins Involved in Gene Regulation
M. Claire Kennedy

18. Current Status of Structure-Activity Relationships of Platinum Anticancer Drugs: Activation of the *trans*-Geometry
Nicholas Farrell

19. Cisplatin and Derived Anticancer Drugs: Mechanism and Current Status of DNA Binding
Marieke J. Bloemink and Jan Reedijk

20. Proteins that Bind to and Mediate the Biologial Activity of Platinum Anticancer Drug-DNA Adducts
Joyce P. Whitehead and Stephen J. Lippard

21. Interactions of Metallopharmaceuticals with DNA
Michael J. Clarke and Michael Stubbs

Subject Index

Volume 33. Probing of Nucleic Acids by Metal Ion Complexes of Small Molecules

1. Molecular Modeling of Transition Metal Complexes with Nucleic Acids and Their Constituents
Jiří Kozelka

2. Zinc Complexes as Targeting Agents for Nucleic Acids
Eiichi Kimura and Mitsuhiko Shionoya

3. Metallocene Interactions with DNA and DNA Processing Enzymes
Louis Y. Kuo, Andrew H. Liu, and Tobin J. Marks

4. Evidences for a Catalytic Activity of the DNA Double Helix in the Reaction between DNA, Platinum(II), and Intercalators
Marc Boudvillain, Rozenn Dalbiès, and Marc Leng

5. *Trans*-Diammineplatinum(II)—What Makes it Different from *cis*-DDP? Coordination Chemistry of a Neglected Relative of Cisplatin and Its Interaction with Nucleic Acids
Bernhard Lippert

6. Metal Ions in Multiple-Stranded DNA
Michal Sabat and Bernhard Lippert

7. DNA Interactions with Substitution-Inert Transition Metal Ion Complexes
Bengt Nordén, Per Lincoln, Björn Åkerman, and Eimer Tuite

8. Effect of Metal Ions on the Fluorescence of Dyes Bound to DNA
Vasil G. Bregadze, Jemal G. Chkhaberidze, and Irine G. Khutsishvili

9. Photolytic Covalent Binding of Metal Complexes to DNA
Mark A. Billadeau and Harry Morrison

10. Electrochemically Activated Nucleic Acid Oxidation
Dean H. Johnston, Thomas W. Welch, and H. Holden Thorp

11. Electron Transfer between Metal Complexes Bound to DNA: Is DNA a Wire?
Eric D. A. Stemp and Jacqueline K. Barton

12. Porphyrin and Metalloporphyrin Interactions with Nucleic Acids
Robert F. Pasternack and Esther J. Gibbs

13. Selective DNA Cleavage by Metalloporphyrin Derivatives
Geneviève Pratviel, Jean Bernadou, and Bernard Meunier

14. Synthetic Metallopeptides as Probes of Protein-DNA Interactions
Eric C. Long, Paula Denney Eason, and Qi Liang

15. Targeting of Nucleic Acids by Iron Complexes
Alexandra Draganescu and Thomas D. Tullius

16. Nucleic Acid Chemistry of the Cuprous Complexes of 1,10-Phenanthroline and Derivatives
David S. Sigman, Ralf Landgraf, David M. Perrin, and Lori Pearson

17. Specific DNA Cleavage by Manganese(III) Complexes
Dennis J. Gravert and John H. Griffin

18. Nickel Complexes as Probes of Guanine Sites in Nucleic Acid Folding
Cynthia J. Burrows and Steven E. Rokita

19. Hydrolytic Cleavage of RNA Catalyzed by Metal Ion Complexes
Janet R. Morrow

20. RNA Recognition and Cleavage by Iron(II)-Bleomycin
Jean-Marc Battigello, Mei Cui, and Barbara J. Carter

21. Metallobleomycin-DNA Interactions: Structures and Reactions Related to Bleomycin-Induced DNA Damage
David H. Petering, Qunkai Mao, Wenbao Li, Eugene DeRose, and William E. Antholine

Subject Index

Volume 34. Mercury and Its Effects on Environment and Biology

1. Analytical Methods for the Determination of Mercury(II) and Methylmercury Compounds: The Problem of Speciation
James H. Weber

2. Mercury in Lakes and Rivers
Markus Meili

3. Biogeochemical Cycling of Mercury in the Marine Environment
William F. Fitzgerald and Robert P. Mason

4. Catchments as a Source of Mercury/Methylmercury in Boreal Surface Waters
Kevin H. Bishop and Ying-Hua Lee

5. Gold Rushes and Mercury Pollution
Jerome O. Nriagu and Henry K. T. Wong

6. Accumulation of Mercury in Soil and Effects on the Soil Biota
Lage Bringmark

7. Biogeochemistry of Mercury in the Air-Soil-Plant System
Ki-Hyun Kim, Paul J. Hanson, Mark O. Barnett, and Steven E. Lindberg

8. Microbial Transformation of Mercury Species and Their Importance in the Biogeochemical Cycle of Mercury
Franco Baldi

9. Bioaccumulation of Mercury in the Aquatic Food Chain in Newly Flooded Areas
R. A. (Drew) Bodaly, Vincent L. St. Louis, Michael J. Paterson, Robert J. P. Fudge, Britt D. Hall, David M. Rosenberg, and John W. M. Rudd

10. Mercury in the Food Web: Accumulation and Transfer Mechanisms
Alain Boudou and Francis Ribeyre

11. Physiology and Toxicology of Mercury
Lászlo Magos

12. Metabolism of Methylmercury in the Brain and Its Toxicological Significance
N. Karle Mottet, Marie E. Vahter, Jay S. Charleston, and Lars T. Friberg

13. Maternal-Fetal Mercury Transport and Fetal Methylmercury Poisoning
Rikuzo Hamada, Kimiyoshi Arimura, and Mitsuhiro Osame

14. Effects of Mercury on the Immune System
K. Michael Pollard and Per Hultman

15. The Impact of Mercury Released from Dental 'Silver' Fillings on Antibiotic Resistances in the Primate Oral and Intestinal Bacterial Flora
Cynthia A. Liebert, Joy Wireman, Tracy Smith, and Anne O. Summers

16. Inhibition of Brain Tubulin-Guanosine 5′-Triphosphate Interactions by Mercury: Similarity to Observations in Alzheimer's Diseased Brain
James C. Pendergrass and Boyd E. Haley

17. Interaction of Mercury with Nucleic Acids and Their Components
Einar Sletten and Willy Nerdal

18. Mercury Responsive Gene Regulation and Mercury-199 as a Probe of Protein Structure
David L. Huffman, Lisa M. Utschig, and Thomas V. O'Halloran

19. Bacterial Mercury Resistance Genes
Jon L. Hobman and Nigel L. Brown

Subject Index

Volume 36. Interrelations Between Free Radicals and Metal Ions in Life Processes (tentative)

1. The Mechanism of "Fenton-like" Reactions and Their Importance for Biological Systems
Stefan I. Liochev

2. Reactions of Aliphatic and Aliphatic-Peroxyl Radicals with Transition Metal Complexes as a Plausible Source for Biological Damage in Radical Processes
Dan Meyerstein

3. Free Radicals as a Result of Dioxygen Metabolism
Bruce P. Branchaud

4. Free Radicals as a Source of Uncommon Oxidation States of Transition Metals
George V. Buxton and Giuliano Mulazzani

5. Biological Chemistry of Copper–Zinc Superoxide Dismutase and Its Link to Amyotrophic Lateral Sclerosis
Edith Butler Gralla and Joan Selverstone Valentine

6. Free Radical Reactions of Cytochrome P_{450}
John T. Groves

7. DNA Damage Mediated by Metal Ions
José-Luis Sagripanti

8. Radical Migration Through the DNA Helix: Chemistry at a Distance
Shana O. Kelley and Jacqueline K. Barton

9. Involvement of Metal Ions in Lipid Peroxidation: Biological Implications
Odile Sergent

10. Formation of Methemoglobin and Free Radicals in Erythrocytes
Hans Nohl and Klaus Stolze

11. The Role of Free Radicals and Metal Ions in the Pathogenesis of Alzheimer's Disease
Craig S. Atwood, Xudong Huang, Robert D. Moir, Rudolph E. Tanzi, and Ashley I. Bush

12. Metal Binding and Radical Generation of Proteins in Human Neurologic Diseases and Aging
Gerd Multhaup

13. Thiyl Radicals in Glutathione and Other Biochemically Important Analogs in the Presence of Metal Ions
Hans-Jürgen Hartmann, Christian Sievers, and Ulrich Weser

14. Methylmercury-Induced Generation of Free Radicals. Biological Implications
Theodore A. Sarafian

15. The Role of Free Radicals in Metal-Induced Carcinogenesis
Joseph R. Landolph

16. pH-Dependent Organocobalt Sources of Active Radical Species: A New Type of Anticancer Agents
Mark E. Vol'pin, Ilia Ya. Levitin, and Sergei P. Osinsky

17. Detection of Chromatin-Associated Hydroxyl Radicals Generated by DNA-Bound Metal Compounds and Antitumor Antibiotics
G. Mike Makrigiorgos

18. Nitrogen Monoxide (NO). Its Formation and Biological Roles in Mammalian Systems
Jon M. Fukuto

19. The Chemistry of Peroxynitrite and Its Relevance for Biological Systems
Willem H. Koppenol

20. Novel NO-Liberating Heme Proteins in the Saliva of Bloodsucking Insects
F. Ann Walker, José M. C. Ribeiro, and William R. Montfort

21. Nitrogen Monoxide-Related Disease and NO Scavengers as Potential Drugs
Simon P. Fricker

22. Therapeutics of Nitric Oxide Donors
Ho-Leung Fung, Brian P. Booth, and Mohammad A. Tabrizi-Fard

Subject Index

Comments and suggestions with regard to contents, topics, and the like for future volumes of the series are welcome.

The following Marcel Dekker, Inc. books are also of interest for any reader dealing with metals or other inorganic compounds:

Handbook on Toxicity of Inorganic Compounds
edited by Hans G. Seiler and Helmut Sigel, with Astrid Sigel
In 74 chapters, written by 84 international authorities, this book covers the physiology, toxicity, and levels of tolerance, including prescriptions for detoxification, for all elements of the Periodic Table (up to atomic number 103). The book also contains short summary sections for each element,

dealing with the distribution of the elements, their chemistry, technological uses, and ecotoxicity as well as their analytical chemistry.

Handbook on Metals in Clinical and Analytical Chemistry
edited by Hans G. Seiler, Astrid Sigel, and Helmut Sigel
This book is written by 80 international authorities and covers over 3500 references. The first part (15 chapters) focuses on sample treatment, quality control, etc., and on the detailed description of the analytical procedures relevant for clinical chemistry. The second part (43 chapters) is devoted to a total of 61 metals and metalloids; all these contributions are identically organized covering the clinical relevance and analytical determination of each element as well as, in short summary sections, its chemistry, distribution, and technical uses.

METAL IONS IN BIOLOGICAL SYSTEMS

1

Biological Cycling of Iron in the Ocean

Neil M. Price[1] *and François M. M. Morel*[2]

[1]Department of Biology, McGill University, Montreal, QC, H3A 1B1, Canada

[2]Department of Geosciences, Princeton University, Princeton, NJ 08540, USA

1.	INTRODUCTION	2
2.	DISTRIBUTION, SOURCES, AND SINKS	2
3.	IRON CHEMISTRY IN SEAWATER	5
4.	BIOLOGICAL CYCLING IN SURFACE WATERS	9
5.	KINETICS AND MECHANISMS OF IRON ACQUISITION	13
6.	PHYSIOLOGICAL ADAPTATION TO Fe LIMITATION	15
7.	SIDEROPHORES	19
8.	EVIDENCE FOR Fe LIMITATION IN THE OCEAN	20
9.	GLOBAL AND PALEOOCEANOGRAPHIC PERSPECTIVE	25
	ACKNOWLEDGMENTS	29
	ABBREVIATIONS	29
	REFERENCES	29

1. INTRODUCTION

Over the past few years, the old idea [1,2] that the supply of iron may limit biological productivity of some parts of the ocean has been revived and has generated a great deal of research and speculation. Here we review what we know about Fe in the ocean, focusing on its interaction with the biota.

2. DISTRIBUTION, SOURCES, AND SINKS

Iron is the fourth most abundant element in the earth crust, yet it is present only at ultratrace concentrations in seawater. This difference between the iron content of continental material and oceanic samples creates ubiquitous contamination problems. It is only recently that we have begun to obtain trustworthy measurements of iron concentrations in seawater [3], and these measurements are so exacting that they are still not numerous [4–8]. A typical concentration profile (e.g., Fig. 1 for the North Pacific Ocean) shows dissolved iron concentrations of about 50 pM at the surface, increasing to a broad maximum of about 70 pM at 1000 m, and decreasing slightly with depth in deeper waters. The top 1000 m of this concentration profile is typical of "nutrients", elements such as nitrate or phosphate that are utilized by the biota at the surface, sink as biological particulate material, and are remineralized in the water column as the organic matter is respired. The iron maximum at 1000 m in the North Pacific corresponds to the depth of the oxygen minimum, where respiration is high. The slight decrease in iron concentrations below 1000 m indicates that the sediments are a sink for seawater Fe.

The vertical distribution of dissolved iron concentration in seawater is similar to that of many other "bioactive" trace metals such as zinc, cadmium, or nickel. These other metals, however, are present almost entirely in the dissolved phase; iron differs from them in having a particulate concentration of the same order as and often larger than its dissolved concentration [6,9] (see Fig. 1).

In coastal waters, the two major sources of iron are rivers and anoxic sediments, whereas atmospheric dust and upwelling of Fe-rich

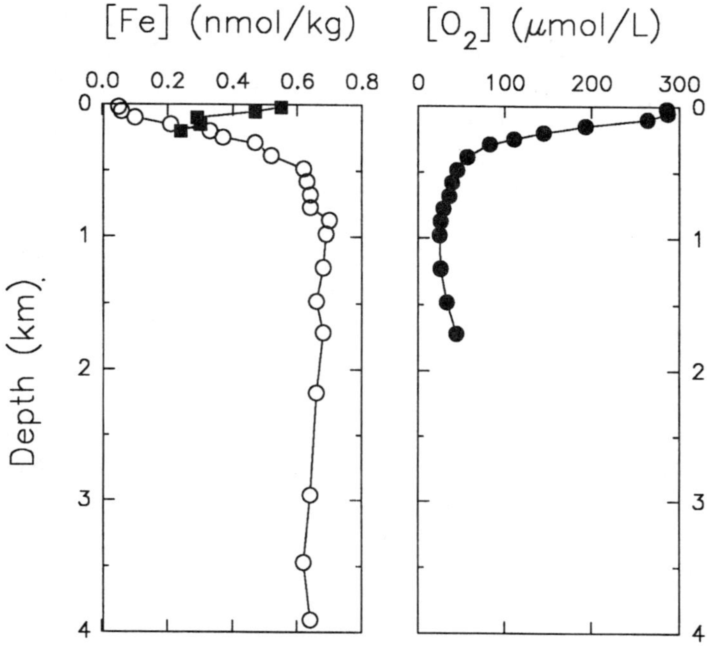

FIG. 1 Vertical distribution of dissolved (○) and particulate (■) Fe and dissolved oxygen at Ocean Station P, subarctic Pacific Ocean (OSP: 50°N, 145°W) (Redrawn by permission from [6].)

deep water are the major sources in the open oceans [10,11]. Compared with the oceans, rivers contain a huge concentration of iron, chiefly in colloidal and particulate form. Most of the riverine iron, however, is efficiently removed in estuaries by precipitation, coagulation, and sedimentation processes as river water mixes with seawater and the salinity increases [12,13]. As a result, elevated iron concentrations are only measured very near to the shore.

In productive coastal waters where the sediments are sometimes anoxic, iron can be resolubilized by reduction of iron oxyhydroxides and Fe(II) and then diffuses to the overlying waters. Because oxidation of Fe(II) and reprecipitation of Fe(III) is very effective, very little of this sedimentary iron reaches oxic surface waters. Nonetheless, both riverine and sedimentary inputs provide coastal waters with a higher Fe concentration than found in the open oceans, and advective processes can sometimes transport such iron-rich coastal water offshore along isopycnals and result in a layer of high dissolved Fe at some depth in the water column. Thus a concentration maximum at a given depth may reflect long range advection processes rather than in situ remineralization.

Eolian transport of continental dust particles is estimated to supply about three times as much dissolved Fe to the oceans as rivers [10]. This continental dust material contains about 5% Fe, which is brought down chiefly by wet deposition. During its atmospheric transport, this Fe is subjected to very acidic conditions in aerosols (due to SO_2 oxidation) and exposed to intense sunlight. The result is that a sizable fraction of this iron (up to 80%; [14–16]) is reduced and dissolved in the aerosol with only the most unreactive iron in silicates and aluminosilicates remaining particulate. Atmospheric dust inputs to the surface oceans are highly nonuniform spatially (Fig. 2), being on average eight times higher in the Northern than in the Southern Hemisphere. Atmospheric inputs are also episodic and, as discussed later, volcanic erup-

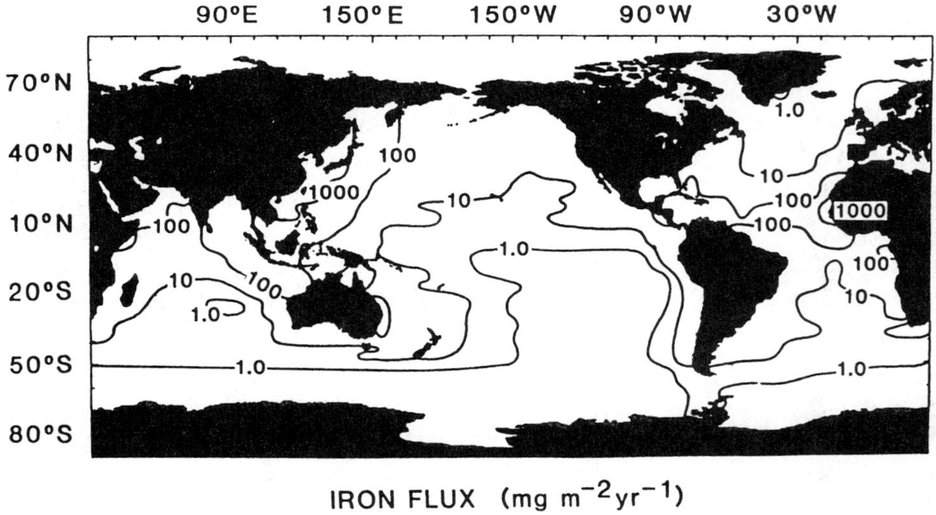

FIG. 2 Calculated flux of total Fe (particulate plus dissolved) from the atmosphere to the ocean. (Reproduced with permission from [10].)

tions, for example, may result in sudden Fe enrichment of surface waters [17].

As seen in Fig. 1, deep seawater contains high concentrations of iron compared with surface seawater as a result of remineralization of the iron settling as particulate organic matter. Thus, transport by eddy diffusion and upwelling of deep water supplies iron to the surface as it supplies major algal nutrients, nitrate, and phosphate. Depending on the hydrographic conditions that control the magnitude of eddy diffusion and upwelling velocities, the flux of Fe from deep waters to the surface may dominate all other inputs. For example, this upward Fe flux has been estimated to be 60 mol m^{-2} yr^{-1} in the Antarctic circumpolar current and in the Equatorial Pacific, about 10 times the atmospheric inputs to these areas [11,18]. In the North Pacific, the relative contributions from upwelling deep water and atmospheric dust are exactly the opposite (~1:10) for a similar Fe input rate to the euphotic zone.

A special situation may be obtained in circumpolar regions such as the Weddell Sea in the Antarctic. There the atmospheric dust input that is collected for much of the year on sea ice is released to the water in the spring when the ice melts [18,19]. In some cases, such inputs from melting sea ice may be very significant in providing Fe for spring blooms of phytoplankton at very high latitudes.

3. IRON CHEMISTRY IN SEAWATER

In view of the difficulty inherent in just measuring iron concentrations in the oceans, it is not surprising that our knowledge of the processes that control Fe chemistry in seawater is incomplete. What we know is largely based on experiments conducted at high concentrations, sometimes at lower pH than the ambient value of 8.2, and their extrapolation to natural conditions is somewhat speculative. In coastal as in open ocean areas, the major forms of iron entering surface seawater, are dissolved Fe(II) and Fe(III) and particulate Fe(III) in oxyhydroxides, silicates, and aluminosilicates. Iron silicates and aluminosilicates that enter the oceans from riverine or atmospheric inputs are very unreactive and they likely contribute a negligible fraction of the dissolved iron in seawater. Like that of all colloids and particles, their fate is governed by coagulation and settling processes which bring them to the sediments. Ingestion by suspension-feeding grazers may possibly prolong the water column residence time of Fe in these particles by increasing their solubilization, but this is likely a small effect compared with the total flux. Some iron oxides are also unreactive and like aluminosilicates simply contribute to the sedimentary iron input. Others, particularly the metastable hydrous ferric oxide, $Fe(OH)_3$, can be partly dissolved as Fe(III) species through thermal dissolution reactions or as Fe(II) species through direct or indirect photochemical and biological reductive processes. Except for some coastal basins and fjords and the Black Sea, seawater is well oxygenated (see Fig. 1) and Fe(III) is the thermodynamically stable form of iron. The Fe(II) produced photochemically and biologically, along with direct Fe(II) inputs, is thus reoxidized to dissolved Fe(III) and may be reprecipitated as $Fe(OH)_3$, involving iron in a classic redox cycle in surface seawater (Fig. 3).

Dissolution of freshly precipitated colloidal $(FeOH)_3$ in seawater in the absence of chelators or light is fast relative to that of other iron oxides, with a pseudo-first-order rate constant, $k_d \sim 10^{-2}\ h^{-1}$; but this rate must be increased by the presence of strong natural Fe(III) organic chelators (see below), as has been demonstrated for many model chelators. Light can increase the dissolution rate of $Fe(OH)_3$ 100-fold ($k_d \sim 1\ h^{-1}$ at typical noontime irradiance of 100 mW cm^{-2} in the tropics; [20–24]). The chromophores for this reaction are most likely surface Fe(III) complexes with humic substances, although the surface hydroxy complexes of Fe(III) are themselves photoreactive. Similar photoreduction is observed for dissolved Fe(III)-humate complexes, but it is not known if the strong Fe(III) complexes present in seawater (see below) are photoreactive. Once dissolved as Fe(II), inorganic iron is rapidly reoxidized in seawater by reactions with O_2 and H_2O_2 [25,26]. The reaction with O_2 results in a half-life of 2 min in oxic surface waters (pH = 8.1, $[O_2] = 2.1 \times 10^{-4}$ M). Hydrogen peroxide is formed by dismutation of superoxide, O_2^-, produced by the irradiation of dissolved humic substances in the presence of O_2. At noontime in the tropics, H_2O_2 may reach concentrations up to 10^{-7} M and accelerate the oxidation of Fe^{2+} with a half-life down to 1 min. Depending on the kinetics of the major dismutation mechanisms, it is estimated that O_2^- may reach concentrations up to 10^{-9} M (B. M. Völker, personal communication). In this case, O_2^- itself may become the dominant oxidant for Fe(II), but because it

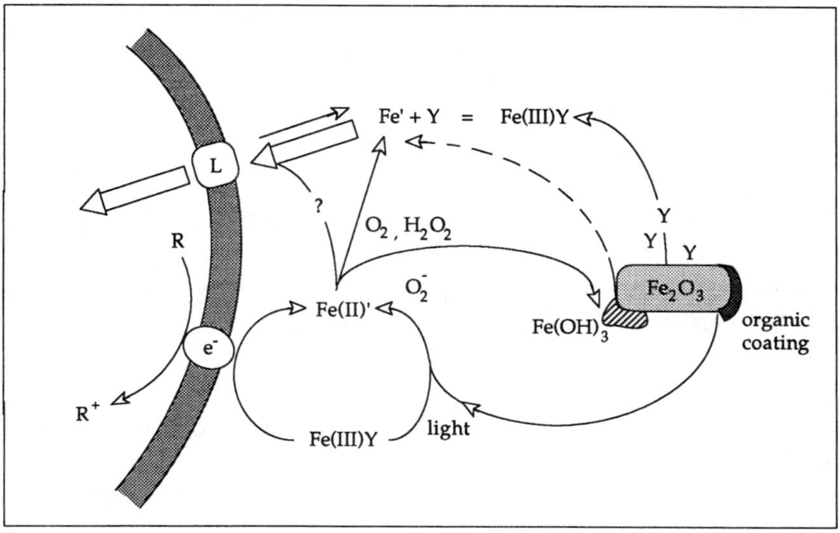

FIG. 3 Schematic representation of Fe transport by eukaryotic phytoplankton (for example *Thalassiosira* spp.), showing potential sources of Fe and chemical and biological transformations of dissolved and colloid Fe species. Colloid Fe is represented by hematite (Fe_2O_3). Thermal processes, chemical and photochemical reduction, and solubilization by organic ligands (Y) dissolve colloidal Fe. Phytoplankton plasmalemma is depicted by the shaded arc traversed by a square box and an oval which represent the Fe transporter (L) and ferric reductase (e^-), respectively. Description of abbreviations and symbols: Fe′, inorganic Fe(III) species; Fe(II)′, inorganic Fe(II) species; R, intracellular reductant (e.g., NADH).

is also an effective reductant for inorganic Fe(III), the result may be the maintenance of a steady-state balance of Fe(II) and Fe(III) [27].

The relative proportions of Fe(III) and Fe(II) in the dissolved Fe content of surface seawater is dependent on the relative rates of reduction and oxidation by the various mechanisms just described (which depend on the season and the time of day) and by the extent of stabilization of the oxidized and reduced iron by inorganic and organic complexation. Measurements of Fe(II) concentrations are extremely difficult and subject to artifacts, for most methods rely on the stabilization of Fe(II) by chelators and are likely to measure "reducible" as well as reduced Fe [21]. Nonetheless, Fe(II) measurements in surface seawater have shown that a sizable fraction of the dissolved Fe may be Fe(II) and that this fraction varies with sunlight [28]. Likewise, incubations of natural seawater with added Fe have shown a distinct diurnal variation in dissolved Fe concentration [22].

The principal inorganic species of Fe(II) in seawater at pH ~ 8.1 are the hydrated ferrous ion ($Fe^{2+} = Fe[H_2O]_6^{2+}$, 76%) and the carbonate ($FeCO_3$, 23%) and monohydroxy ($FeOH^+$, 1%) complexes [29]. In addition, Fe(II) can also be complexed by natural organic ligands, such as humic substances, but it is doubtful that these ligands are strong enough to be effective at natural concentrations. There are at present no data on the possible existence of strong specific Fe(II) chelators in seawater.

Although the importance of the neutral trihydroxy complex of Fe(III), $Fe(OH)_3$, has been debated [21,29,30], it appears that the major inorganic dissolved Fe(III) species in seawater are $Fe(OH)_2^+$ and $Fe(OH)_4^-$ (50% ± 20% each, depending on small pH variations). However, most of the Fe(III) in seawater appears to be complexed to unidentified strong organic ligands. The organic complexation of Fe(III) has been demonstrated by cathodic stripping voltammetry (CSV) [31–33]. The method consists of equilibrating a suitable concentration of a strong artificial complexing agent (e.g., *L*-salicylaldoxime) with natural seawater and quantifying the resulting FeL complex by adsorbing it on the cathode of an electrochemical cell and measuring the peak current obtained on reduction of the Fe when scanning the voltage to negative potentials. By varying the added Fe concentration, a titration of the natural Fe(III) ligand in seawater can be obtained. The CSV data have been interpreted as showing the presence of two ligands (or two ligand classes). The concentrations of these ligands and their conditional stability constants with respect to the total inorganic concentration of Fe (Fe′) are on the order of 0.4–1.0 nM $\times 10^{13}$ M^{-1} for L1 and 1.5 nM $\times 3 \times 10^{11}$ M^{-1} for L2. Thus, in waters containing a few tenths of a nanomolar total Fe, the concentration of Fe(III)′ predicted from equilibrium thermodynamic calculations is about 0.07 pM. It has been hypothesized that the natural Fe(III) chelators that are measured in this way may be partial breakdown products of Fe-proteins (hemes or Fe-sulfur centers) or Fe(III)-siderophores produced by marine bacteria. Indeed, some species of marine bacteria have been shown to produce siderophores under Fe-limiting conditions (see below). As pointed out by Rue and Bruland [31], a well-characterized siderophore, such as desferal, has a conditional stability constant in seawater of $10^{16.5}$ M^{-1}, similar to L1. The stronger ligand, L1, is apparent only in the surface waters, whereas L2 is found at depths of 2000 m and shows a subsurface maximum of 500 m. Localization of L1 in surface waters is consistent with a biological source, although the heterotrophic bacteria that are a possible source of siderophores are more or less uniformly distributed with depth [34].

4. BIOLOGICAL CYCLING IN SURFACE WATERS

To illustrate the role of biota in Fe cycling, we have constructed a simple mass balance model of reservoirs and fluxes of Fe (Fig. 4) in the subarctic Pacific (Ocean Station PAPA: 50°N, 145°W), a remote ocean region identified as Fe deficient [35]. This is an area for which we have the most complete inventory of the various forms of Fe in seawater. In surface waters at this station, the total concentration of all forms of Fe is 560 pM, of which 55 pM is dissolved and 165 pM is bound in aluminosilicates [6]. Iron input via atmospheric deposition is equivalent to 1–2 pM/d [10]. This value closely approximates the output rate (0.5 pM/d) calculated from measurements of new production (38 gC/m²/yr [36]) and the Fe:C content of Fe-limited phytoplankton (3.2 ± 2.0 μmol/mol [37]). For the purposes of this exercise, we have assumed an input and output rate of 0.5 pM/d. Accordingly, the system can be viewed as in steady state and the residence time of Fe in the euphotic zone is on the order of 1000 days. This time is short compared with the 55-year residence time of Fe in the world ocean [38].

Figure 4 illustrates that biota contain only a small fraction of the total Fe in the system. The majority of it is part of a detrital pool composed of the remains of living organisms. As defined here, the detrital Fe pool is not directly measured but calculated by difference, so its precise identification is somewhat uncertain. A large concentration of Fe in detritus, however, is not unlike that observed for N and C in low nutrient, oligotrophic gyres [39]. Producers contain the largest stocks of biogenic Fe [40]. These organisms, which directly assimilate Fe from solution, include phototrophic eukaryotes and prokaryotes and heterotrophic bacteria. Because of their high biomass and rich iron

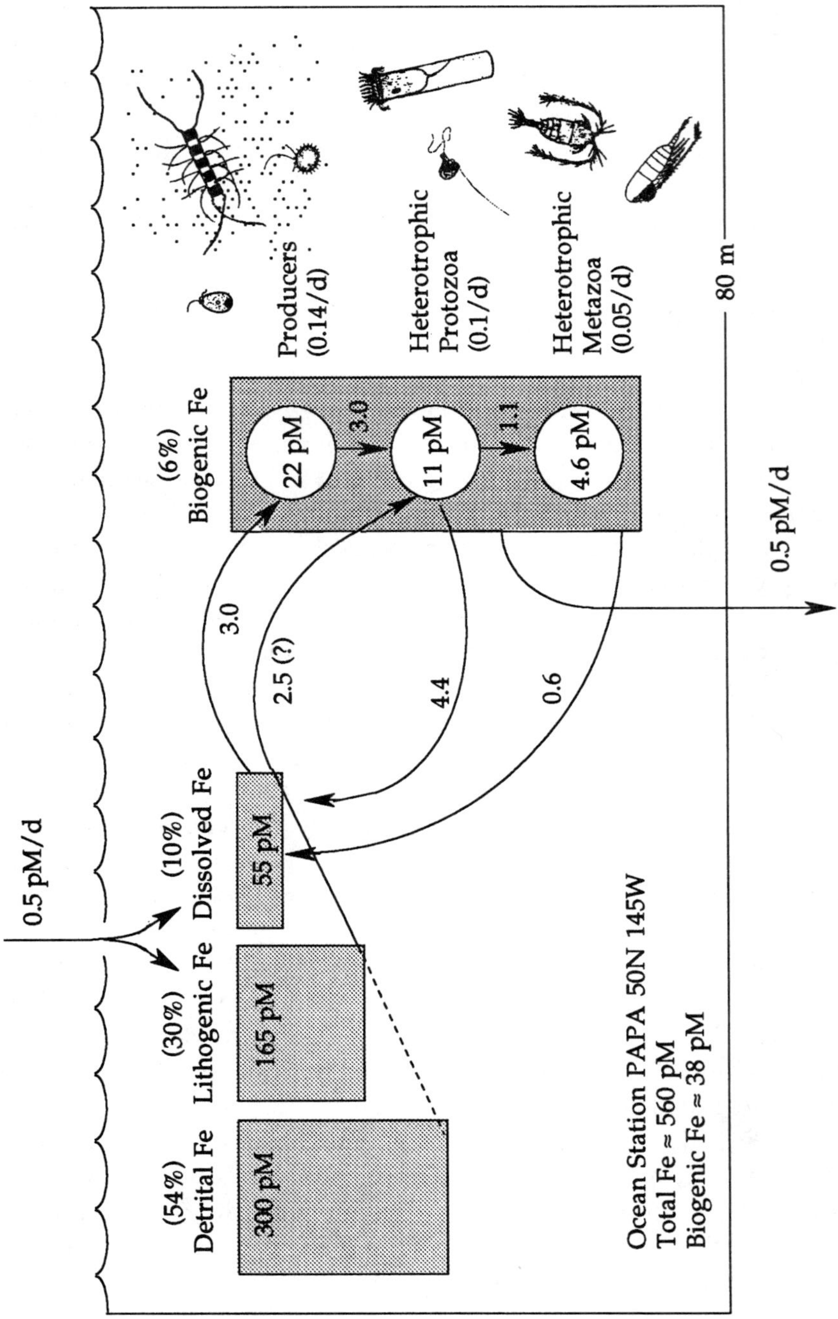

Producers
(0.14/d)
Heterotrophic
Protozoa
(0.1/d)
Heterotrophic
Metazoa
(0.05/d)
80 m
(6%)
Biogenic Fe
22 pM
11 pM
4.6 pM
3.0
1.1
0.5 pM/d
3.0
2.5 (?)
4.4
0.6
0.5 pM/d
(10%)
Dissolved Fe
55 pM
(30%)
Lithogenic Fe
165 pM
(54%)
Detrital Fe
300 pM
Ocean Station PAPA 50N 145W
Total Fe ≈ 560 pM
Biogenic Fe ≈ 38 pM

FIG. 4 Biological Fe cycle in the upper mixed layer (<80 m) of the subarctic Pacific Ocean at Ocean Station P (OSP: 50°N, 145°W). The cycle is portrayed in steady state with inputs balancing outputs. Iron stocks (P) are picomoles/liter (pM) and flows are pM/d. Biogenic Fe is estimated from laboratory and field measurements of Fe:C ratios of oceanic organisms [40,46,117] and annual averages of their C standing stocks [118,119]. Lithogenic Fe is calculated from the concentration of particulate Al and a Fe:Al crustal abundance ratio of 0.33 [6]. Detrital Fe is the difference between the total particulate Fe [6] and the lithogenic and biogenic components. Dissolved Fe is operationally defined as that fraction which passed through a 0.4-μm filter and undoubtedly includes colloidal (1–1000 nm particles) as well as truly dissolved species [32]. Percentages of the total Fe in each of the four pools are indicated in parentheses. Flows among the Fe pools were determined assuming that all daily Fe production was consumed by the next higher trophic level. Iron uptake of producers, which include heterotrophic bacteria, phototrophic cyanobacteria, and eukaryotic phytoplankton, is the product of their average rate of turnover (T, in parentheses) and Fe content [40]. Heterotrophic protozoan consumption of Fe (R) was calculated from measured rates of turnover [44,49], assuming a gross Fe growth efficiency (GFeGE) of 0.2 [46]; namely, R = T·P/0.2. Iron consumption by metazoa was set equal to protozoan production, and turnover rates were calculated using a GFeGE of 0.2 [120]. Consult the text for a more complete description of the cycle.

content, heterotrophic bacteria account for more than 50% of the Fe in the producer pool. The remaining 11 pM Fe is distributed equally between the cyanobacteria and the photosynthetic eukaryotes [40]. High concentration of Fe in bacteria (and in the protozoa) is intriguing in that photoautotrophs, which are demonstrably Fe limited in these waters [35,41,42], must compete with heterotrophs for this scarce resource. Since the photoautotrophs supply all the energy and C to these trophic levels, the heterotrophs effectively magnify the limiting effect of low Fe in the water by sequestering large quantities in their biomass.

Iron requirements of the producers are equivalent to 3 pM/d based on their measured turnover rates [40]. This supply is thought to come from the dissolved pool and is larger than Fe inputs. Thus most of the Fe supplied to producers must be regenerated; that is, released by higher trophic levels during assimilation and feeding, as proposed [43]. Our calculations show that 85% of the Fe requirements of the producers are indeed met in this fashion. The importance of alternative pathways of Fe recycling, such as viral lysis or bacterial degradation of dying phytoplankton, cannot be established at present. If the reactivity of the recycled Fe differs depending on the mechanism of solubilization (D. A. Hutchins, personal communication), the pathway of resupply could have important ramifications for the Fe nutrition of the producers.

The biomass of phytoplankton and heterotrophic bacteria is constant on average, so grazing must balance production, or in other words, losses must equal gains. The calculated rate constant for turnover of Fe in heterotrophic protozoa is thus 0.05/d if their only source for Fe is from producers. This rate is two to three times slower than that which is measured [44], suggesting that the flagellates and ciliates may use Fe more efficiently than we believe or that they may use other sources of Fe directly for growth. Recent observations [44] that flagellated protozoans ingest and solubilize Fe colloids lends credence to the latter proposal, particularly since these organisms can use Fe bound to prey surfaces as a nutritional supplement [46]. To achieve their measured turnover rate of 0.1/d [44], protozoa would have to utilize 30% of the lithogenic Fe reservoir. Alternatively, some fraction of the detrital pool could serve as an Fe source. These scenarios should be viewed speculatively, although they represent plausible pathways for Fe to enter the biological loop. The importance of protozoans in Fe regeneration is apparent from our model, as they provide 90% of the recycled supply. This observation is perhaps not too surprising, since they are the dominant phytoplankton consumers in these regions [47].

If metazoa consume all of the daily production of protozoan Fe (1.1 pM/d), their turnover rate would be 0.05/d. This rate is almost identical to the rate of growth of *Neocalanus plumchrus*, the dominant copepod in these waters [48]. The Fe ration of metazoa can be estimated independently using ingestion rate and biomass data [49] (i.e., 4 μgC/copepod/d × 0.28 copepods/L × Fe : C prey (12 μmol/mol) = 1.2 pM Fe/day). Metazoan consumption of phytoplankton represents about 10% of their total production [50] and would thus contribute a further 0.15 pM/d to their overall Fe acquisition rate. Total ingested Fe calculated in this way (1.3 pM/d) is thus similar to that proposed in our mass balance model. Apparently, the biological system can be portrayed in terms of Fe as well as C. The source of Fe exiting the surface layer is unknown, but most likely is in the form of fecal pellets and intact phytoplankton cells.

If the flow of Fe from the dissolved compartment into the biota occurs through soluble Fe species (ignoring direct consumption of Fe colloids by mixotrophic species), turnover of the dissolved pool is 0.10/d, similar to the rate constant for the biota. This calculation of course does not imply that all Fe in the dissolved pool turns over slowly. Indeed, the inorganically complexed species that represent about 0.03% of this reservoir would be turning over every 3 min if they provided all of the Fe requirements of the producers. This concentration of Fe, however, is so low that its diffusion rate to a 1-μm diameter cell could only support a growth rate of 0.05 day^{-1}. Since the eukaryotic phytoplankton that reside in these waters are considerably larger than 1 μm, they must be obtaining Fe from the organic complexes. Although cyanobacteria are roughly this size, they too could not obtain sufficient Fe from this pool, because their Fe requirements are about five times higher than the eukaryotes. Photochemical reduction could increase the steady-state concentration of Fe' if the irradiance were high enough and the Fe organic complexes photolabile.

5. KINETICS AND MECHANISMS OF IRON ACQUISITION

Exceedingly low concentrations of Fe in surface waters of the open sea have imposed unique constraints on Fe transport and acquisition by phytoplankton. From what we know of studies of a few coastal species of eukaryotes, uptake of Fe is believed to occur via binding to a surface ligand and subsequent internalization by transfer across the plasmalemma [51,52] (see Fig. 3). The kinetics of this reaction are well described by the Michaelis-Menten equation, relating uptake rate to external Fe concentration through a hyperbolic rate law. Measurements of the rate constants for dissociation of Fe from the transporter and Fe internalization, however, show that the Fe-transport complex is not at equilibrium with the free ferric ion concentration [52]. Iron is internalized from surface-bound sites at a rate of 0.1 h^{-1} and does not readily exchange with external Fe once it is attached to the cell surface [52]. The rate-limiting step of transport is thus binding of Fe to the surface ligand, a reaction which is determined by the concentrations of labile inorganic ferric species (Fe′) and their kinetics of water loss. Low iron concentrations and slow reaction kinetics in seawater thus bring about a requirement for high ligand densities on the phytoplankton membranes. Indeed, measurements of surface-bound Fe formed during transient uptake experiments show roughly 10 amol (10^{-17} mol) of sites per cell, corresponding to about 0.1 amol of sites/μm^2 of membrane. By necessity, these ligands must be small or they would otherwise occupy a large fraction of the plasmalemma [53]. Although not depicted in our diagram (see Fig. 3), the Fe binding ligands are likely to be independent of the transporter but closely associated with it. Cell surface labeling experiments have tentatively identified a membranous, low molecular weight compound that binds Fe [54], but further characterization of Fe transport proteins in phytoplankton has not been attempted.

The importance of ferric ion species in controlling Fe transport is well established. In the sea, however, the concentrations of these species are apparently vanishingly small and it is difficult to envisage how phytoplankton could obtain the Fe they need for growth (see above). Biologically mediated reduction of Fe at the cell's surface may be a way to increase the rate of Fe exchange from organic complexes. A large body of literature has documented reduction of Fe(III) by higher plants [55] and phytoplankton [56,57] (not to mention unicellular fungi for which the molecular genetics have been elucidated [58–60]). Plasmalemma redox enzymes, spanning the membrane, are thought to be responsible for mediating the reduction by transferring electrons from intracellular donors to extracellular Fe organic complexes. In phytoplankton, details of this pathway are just now being worked out. Reductase activity has been detected in a number of diatoms and green algae. Copper(II) and Fe(III) complexes, including ferricyanide, FeEDTA, FeDTPA, and desferrioxamine B, are reduced directly by the redox enzyme [57,61]. The reductase activity is inhibited by antibodies to cytosolic nitrate reductase (NR) suggesting that a NR-like enzyme is mediating the reduction [62]. Using a Fe(II) chemiluminescence assay [63], we have recently obtained direct proof of Fe(III) reduction by a centric marine diatom, *Thalassiosira oceanica* (M. T. Maldonado and N. M. Price, unpublished observations). Reduction rates are enhanced under Fe deficiency and are faster in nitrate than in ammonium grown cells, consistent with the hypothesized involvement of nitrate reductase. High concentrations of Fe (10 μM) are used in reductase experiments, so the significance of biological Fe reduction in the sea cannot be reliably predicted. The relationship between Fe reductase activity and Fe uptake is presently unknown, but some preliminary evidence suggests that changes in the redox state of Fe may occur during transport [51,56].

Availability of particulate forms of iron to phytoplankton like *Thalassiosira* spp. is limited by their rates of dissolution, since colloids themselves cannot be utilized directly. Highly crystalline forms of Fe oxide like α-FeOOH (goethite), β-FeOOH (akaganite), and Fe_2O_3 (hematite) are thus not capable of supporting typical rates of phytoplankton growth, because they are so stable [64,65]. Even the more labile forms of colloidal Fe dissolve slowly and can limit growth rates and maximum cell yields [66].

6. PHYSIOLOGICAL ADAPTATION TO Fe LIMITATION

Differences in supply of Fe to nearshore and oceanic waters creates a gradient in Fe concentration that has influenced plankton during the course of their evolution. Species isolated from oceanic waters, for example, are renowned for their ability to grow well with little Fe, whereas coastal species cannot [67,68]. As discussed above, kinetic analysis demonstrates that Fe-limited coastal phytoplankton can obtain Fe at a rate that approximates the maximum rate that Fe can diffuse to the cell surface or react with cellular ligands [52,53]. These organisms thus possess an extremely high-efficiency Fe transport system that operates near the limits of what is physically and chemically possible, and phytoplankton from oceanic waters cannot be made more efficient. Cell surface-normalized uptake rates of taxa isolated from oceanic and coastal waters are indeed identical over a range of low, rate-limiting Fe concentrations [69]. The only possible adaptive response of oceanic phytoplankton to low Fe is thus to decrease their size or their Fe requirements. Size reduction of phytoplankton in low Fe waters increases Fe acquisition rates relative to growth (as transport decreases in proportion to radius^2 and requirements decrease in proportion to radius^3) and is undoubtedly important. This response, however, cannot account for the greater fitness of oceanic phytoplankton, since coastal species also decrease in size under Fe-limiting conditions. As shown in Table 1, Fe requirements (or so-called Fe quotas) of open ocean phytoplankton are significantly lower than coastal species. By itself this is

TABLE 1

Iron Quotas (μmol Fe/mol C), Growth Rates (d^{-1}), and Iron Use Efficiencies (IUE: 10^5 mol C/mol Fe/d) of Eukaryotic Phytoplankton, Cyanobacteria, and Heterotrophic Protozoa.

Taxon	Habitat	Fe:C	Growth Rate	IUE	n
Phytoplankton	Coastal	11.0 ± 3.7	0.62 ± 0.3	0.59 ± 0.2	11
Phytoplankton	Oceanic	3.1 ± 2.0	1.2 ± 0.3	5.0 ± 3.0	13
Cyanobacteria	Coastal	21.2 ± 8.8	0.9 ± 0.05	0.46 ± 0.17	1
Cyanobacteria[a]	Coastal	150	0.94	0.06	1
Heterotrophic protozoa	Coastal/ oceanic	10.4 ± 3.0	1.2 ± 0.4	6.9 ± 4.0	2

Data for phytoplankton and cyanobacteria (*Synechococcus* DC-2) are average values of cells growing with nitrate or ammonium as N source [37,100]. Growth rates and Fe quotas of the cyanobacterium[a] *Synechococcus* PC7002 from [116]. Two species of heterotrophic protozoa (*Paraphysomonas*) were fed Fe-limited bacteria [46].

not a remarkable feat, since Fe-starved coastal phytoplankton can reduce Fe quotas to these levels. What differentiates species from these two habitats, however, is the ability of oceanic phytoplankton to maintain near maximal rates of growth at such low quotas. Such fast rates of growth are possible because iron use efficiencies (mol C/mol Fe/d: IUE) of oceanic phytoplankton are very high (Table 1).

The quotas reported here are determined in the laboratory in chemically well-defined seawater media which attempt to mimic the extraordinarily low concentrations of Fe in the field by employing high concentrations of artificial chelators (i.e., 100 μM EDTA) to buffer the Fe in solution. In this context, the laboratory experiments have been remarkably successful judging from the concordance of these quotas with those measured in situ (4.4 μmol/mol [40]). Since the distribution of Fe in the sea is controlled by biological uptake and remineralization (see above), the molar ratio of Fe and any other algal nutrient in the nutricline should be equivalent to the ratio of these elements in plankton (the quotas) settling to depth [70]. Sunda and Huntsman [69] applied this reasoning in their analysis of Martin's data from the subarctic Pacific Ocean [5,6]. The Fe : C ratios of the sinking plankton determined in this way were equivalent to 3.6 ± 0.2 μmol/mol, assuming a C : P of 70 for Fe-limited phytoplankton (N.M. Price, unpublished observations), remarkably similar to the other independent measurements discussed above.

Very little information is available on the Fe requirements and growth of marine cyanobacteria. Their Fe quotas are considerably higher than those of eukaryotic phytoplankton [71,72], which is a result that is thought to be due to the abundance of the Fe-rich photosystem I in cyanobacteria [73,74] and the Fe-poor cytoplasm in eukaryotes [71]. Yet cyanobacteria are among the most abundant photoautotrophs in Fe-poor parts of the sea; evidently their high Fe quotas do not prevent them from inhabiting these waters. An examination of the two marine cyanobacteria for which we have complete data shows a 100-fold difference in their Fe quotas and IUEs. *Synechococcus* DC-2, a strain not known to produce siderophores, is surprisingly similar to other coastal eukaryotic phytoplankton. The siderophore producer PC7002 has the lowest IUE of any marine microbe yet examined. From an evolutionary perspective, these species may represent two alternative strategies for dealing with low Fe. On one hand, the organism's ability to scavenge Fe from solution is maximized by the production of siderophores, while on the other hand, the organism has evolved to economize what little Fe is available.

Table 1 also includes the first measurements of iron use efficiencies for marine heterotrophs. Because the iron use efficiency represents the C assimilation rate per unit Fe, calculations for heterotrophs must take into account the portion of the C ration respired, which for these protists amounts to 80%. The IUEs of a coastal and oceanic representative of *Paraphysomonas* are apparently very similar and are among the highest measured for any plankton. Like oceanic phytoplankton, these heterotrophs have maximized their use of Fe in metabolic pathways.

The high Fe use efficiencies of oceanic phytoplankton exceed theoretical predictions based on Fe content of known catalysts and their turnover rates [73]. Part of this discrepancy arises from the replacement of Fe-containing redox proteins by functionally equivalent proteins that contain no metal cofactor. A striking example of this involves the redox protein flavodoxin which replaces the Fe-S protein ferredoxin in Fe-deficient phytoplankton cultures [75,76] and in natural populations [77,78]. Other substitutions, including replacement of cytochrome *c*-553 by the copper-containing redox protein, plastocyanin [79], may further decrease the cells' dependence on Fe and lessen the impact of low Fe on fitness.

Although substitutions reduce phytoplankton requirements for Fe, they cannot completely supplant the need for this element, because Fe plays such an important catalytic role in photosynthesis and respiration and in a number of enzymes involved N and O cycling [80]. Low concentrations of Fe are thus observed to limit these reactions and to impair phytoplankton growth. One well-documented effect of Fe limitation is a reduction in photosynthetic energy conversion caused by a decrease in the number of functional photosystem II (PSII) reaction centers and electron transport [81]. The result is a decrease in the C fixation rate of Fe-limited phytoplankton. In heterotrophic bacteria and protozoa, Fe deficiency also affects C metabolism, but in these cases, it is the respiratory chain that is impaired. Slow rates of electron transport are thought to limit adenosine triphosphate (ATP) production during C catabolism, effectively depriving the cells of energy for growth [40,46]. Less of the ingested C is converted into biomass and more is recycled through CO_2. Iron limitation of heterotroph metabolism could make C transfer through food webs less efficient.

7. SIDEROPHORES

Although the Fe transport kinetics of eukaryotic phytoplankton are well established, virtually nothing is known in this regard for marine phototrophic and heterotrophic bacteria. Most of the bacterial work has focused on production of siderophores, soluble Fe-binding ligands which are synthesized in response to Fe stress [82,83]. The Fe-reactive moieties of these siderophores are hydroxamates, catechols, and mixed-function ligands (containing carboxyl, keto, and alkyl moieties). All of these types of siderophores are synthesized by *Synechococcus* species, and most of the species produce more than one type [84]. Three siderophores of aquatic bacteria have been characterized structurally (Fig. 5): shizokinen, a hydroxamate siderophore from *Anabaena*, and aerobactin A and B, catechols isolated from *Alteromonas*, a heterotrophic marine bacterium [85]. The ligand atoms of the siderophore are predominantly oxygens and they coordinate Fe(III), forming extremely stable complexes. By analogy with the Fe transport systems of enteric bacteria [86], siderophores are released to the external milieu where they bind Fe and the entire ferrisiderophore complex is internalized through a specific outer membrane receptor. In some microbes, iron is reduced off the siderophore outside the cell and the Fe is then taken up [61]. Ferrisiderophore receptors have not yet been identified in marine microbes, but some cyanobacteria and a heterotrophic bacterium are known to produce unique membrane proteins under low Fe conditions [87–89].

The recent discovery of organically complexed Fe has rekindled speculation that bacterial siderophores may be controlling Fe speciation in the sea. This would imply that bacteria experience Fe limitation at certain times, and indeed bioassay experiments suggest that this may be true for heterotrophic species [90]. Photosynthetic efficiencies of the equatorial plankton community are suboptimal [91], so cyanobacteria are apparently experiencing some kind of Fe stress as well, although their growth rates are not severely depressed [92]. During the second Fe mesoscale fertilization experiment (see below), Rue and Bruland [93] monitored the speciation of Fe before and after the infusion of Fe to low Fe waters of the equatorial Pacific Ocean. They observed that within 1 day after Fe addition, the concentration of the strong chelator L1 increased from 0.3 to 1.3 nM, so that the added Fe was almost immediately complexed. Since siderophore synthesis is typically re-

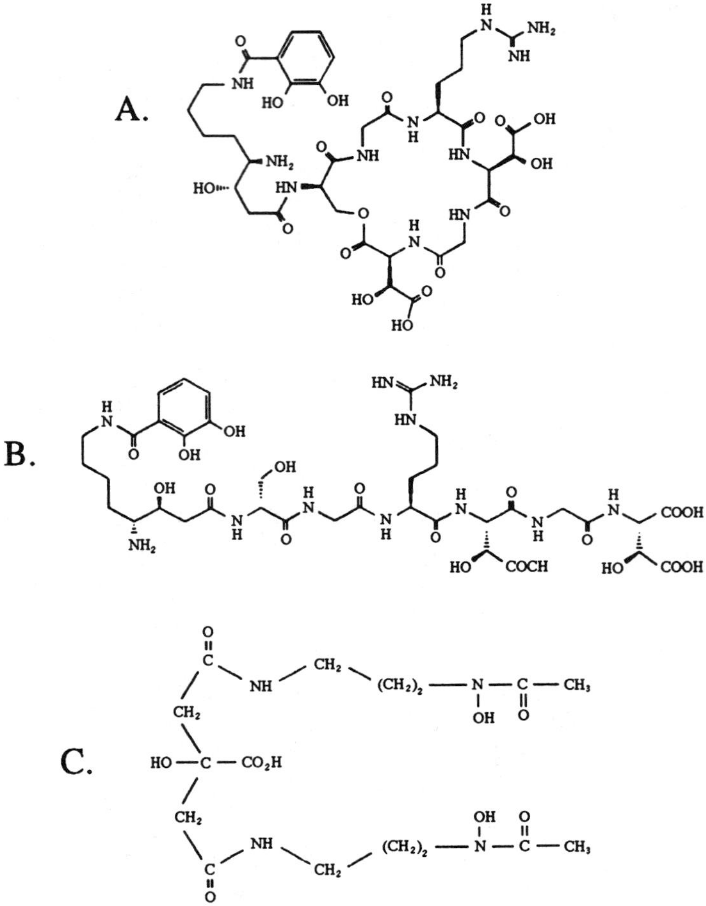

FIG. 5 Chemical structures of siderophores produced by the free-living heterotrophic marine bacterium, *Alteromonas luteoviolacea* [85]: (A) alterobactin A; (B) alterobactin B; and cyanobacterium, *Anabaena* sp. [121]: (C) shizokinen.

pressed by Fe, the response of the "natural ligand" L1 argues against this compound being a siderophore.

8. EVIDENCE FOR Fe LIMITATION IN THE OCEAN

Seasonal cycles of phytoplankton abundance are visible features in many parts of the ocean. Increased solar radiation in the spring creates ideal growth conditions for phytoplankton by increasing the amount and rate of photosynthesis and by decreasing the mixing depth. Nutrients brought to the surface by winter mixing are rapidly consumed as the phytoplankton grow and their biomass increases. Blooming of the phytoplankton represents an increase in biomass of anywhere from 10 to 100 μg chlorophyll *a*/L. Secondary producers lag behind the phytoplankton, but eventually these grazing organisms consume them and the system enters an oligotrophic stage until the cycle is repeated the following year or later in the season. In a few regions of the sea, particularly the Equatorial Pacific, the Subarctic Pacific, and the Southern Ocean around Antarctica, nitrate and phosphate remain at measurable concentrations throughout the year (Fig. 6) and standing stocks of phytoplankton are always low. These three regions are known collectively by the abbreviation HNLC (high nutrient, low chlorophyll) and are hypothesized to be Fe limited. The surface distribution of nitrate and phosphate is not too surprising, as they are all regions of upwelling where nutrient rich water is brought to the surface as a result of Ekman transport [94]. But by comparison with coastal regions containing similar amounts of nitrate and phosphate, the levels of phytoplankton biomass are very low (between 0.3–0.4 μg chlorophyll *a*/L). The issue is thus: Why are phytoplankton levels so low and major nutrients not consumed (the HNLC enigma [95])?

Ambient phytoplankton stocks represent a balance between production due to growth and loss due to grazing (and to a lesser extent sinking and other factors) [96]. For this reason, low levels of biomass in HNLC regions could represent high rates of loss or slow rates of production or some combination of the two. This issue of regulation of phytoplankton biomass by resource limitation (Fe limitation) versus top-down control (grazing) is at the heart of the controversy that developed following Martin's and Fitzwater's 1988 paper [35] describing phytoplankton biomass increase in the subarctic Pacific following Fe enrichment.

Bottle incubation experiments [7,35,97,98] with water sampled from all three of the HNLC regions indeed showed increased algal growth with Fe additions; thus iron appears to be a limiting nutrient in HNLC waters as is consistent with the low atmospheric Fe input to these regions (see Fig. 2). Physiological stimulation of phytoplankton by Fe was evident from changes in the N uptake rates as the community switched from using ammonium as the primary nitrogen source to nitrate (Fig. 7). Normalized to phytoplankton biomass, the rates of

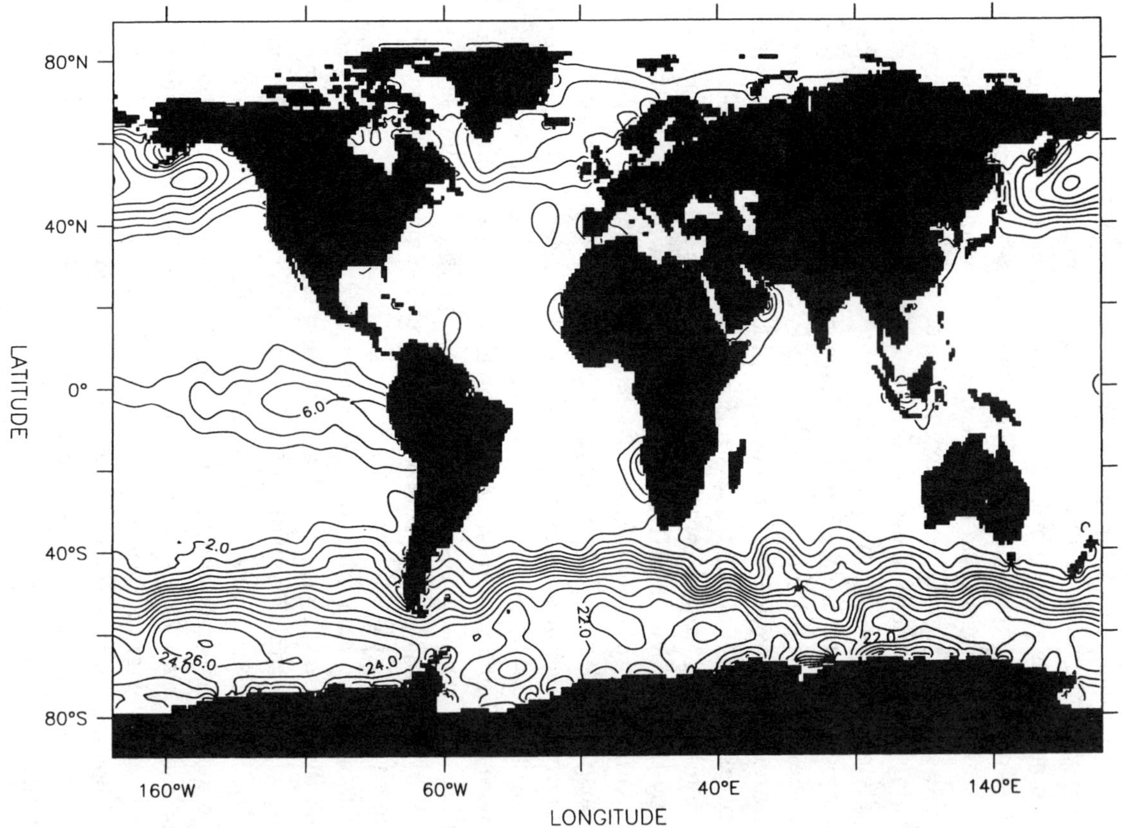

FIG. 6 Average annual nitrate concentration in surface waters of the ocean. Contour lines represent 2 μM increments. (Data from the World Ocean Atlas 1994 (Conkright et al. [122].)

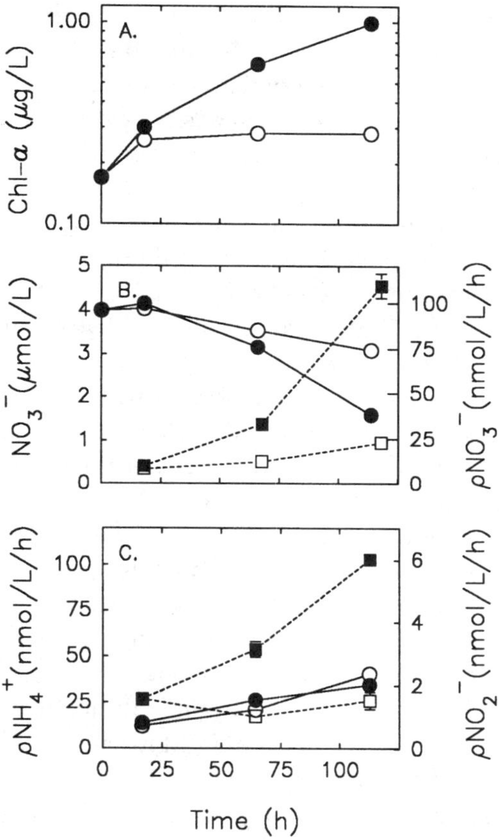

FIG. 7 Response of equatorial Pacific phytoplankton community (0°, 140°W) to iron enrichment. Ambient dissolved Fe concentration was about 0.06 nM: solid symbols represent samples taken from experimental bottles enriched with 1 nM Fe(III); open symbols are from control bottles containing unamended seawater. Changes in (A) phytoplankton biomass (chlorophyll *a*), (B) dissolved nitrate concentration (●, ○) and nitrate uptake rate (■, □), and (C) ammonium (●, ○) and nitrite (■, □) uptake rates. Uptake rates (ρN: nmol N/L/h) were measured during 4-hr incubations in subsamples taken from experimental and control bottles 16, 64, and 113 h after the start of the experiment. Error bars are 1 SD of the mean; where absent, the deviation was less than the symbol width (Reproduced with permission from [123].)

nitrate uptake were on average three times faster in Fe-amended samples than in controls after 2 days of incubation [99]. These results are in agreement with theoretical predictions that nitrate assimilation is more Fe dependent than ammonium assimilation [73] and with measurements of Fe quotas of nitrate and ammonium grown phytoplankton [37,100].

Ecosystem scale fertilization experiments have recently provided dramatic and unambiguous tests of the Fe-limitation hypothesis in the sea. Two such experiments have been conducted in the equatorial Pacific Ocean west of the Galapagos Islands [101,102]. In both of these, large patches of surface water (>60 km^2) were enriched with 100's of kilograms of ferrous sulfate mixed with an inert chemical tracer to monitor the location of the enriched water. In the first experiment, IRONEX I, a patch of water containing 11 μM nitrate and little phytoplankton was enriched to an average final Fe concentration of 4 nM. Biomass, chlorophyll *a*, and C fixation rates were all significantly increased relative to waters outside the patch, but very little nitrate was consumed compared with the response observed in bottle experiments. Measurements of the quantum efficiency of photosystem II determined from the variable to maximum fluorescence (Fv/Fm) ratio showed a dramatic and rapid response of the phytoplankton, demonstrating that they were Fe deficient in situ and that Fe additions alleviated this stress [103]. The second fertilization experiment, IRONEX II, which employed sequential enrichments of Fe, was conducted in May, 1995 [102]. The results of this experiment were definitive. Within 24 h, Fv/Fm increased twofold from 0.25 to 0.5 and photosynthetic efficiency of the abundant picophytoplankton was enhanced [91]. Chlorophyll *a* concentrations increased from 0.15 to 0.2 μg/L to 4 μg/L and nitrate decreased from 10.5 to about 7.0 μM as pennate diatoms bloomed. Short-term nitrate uptake rates were also elevated and N-specific uptake rates were highest in the large phytoplankton size class, similar to previous results [99]. Only a modest increase was observed in biomass of the small picophytoplankton, as they were quickly consumed by herbivorous microzooplankton. Net growth of the large diatoms was not constrained by grazing, as they were too large to be effectively removed by microzooplankton and were not kept in check by mesozooplankton. Thus, as predicted by the "ecumenical hypothesis" [104], the equatorial Pacific Ocean represents a community of grazer-controlled phytoplankton populations in an iron-limited ecosystem [99].

9. GLOBAL AND PALEOOCEANOGRAPHIC PERSPECTIVE

Much of the present interest in the biological cycling of iron in the oceans centers on its possible role as a global limiting nutrient: Are the inputs and concentrations of iron in the oceans limiting oceanic primary production in a way that affects the global cycle of carbon, or more precisely, the CO_2 concentration in the atmosphere? To see the import of this question, we must first outline how oceanic productivity is tied to atmospheric CO_2.

The concentration of CO_2 in surface seawater is about 10 μM; roughly at equilibrium with the 360 ppm in the atmosphere. In contrast, the partial pressure of CO_2 in the deep oceans is about three times its surface value, such that deep water upwelled to the surface should release CO_2 to the atmosphere. This deep water also contains high concentrations of algal nutrients such as phosphate and nitrate, however, so that the excess CO_2 is fixed photosynthetically and sent back as settling particulate organic matter to the deep ocean where it is remineralized. This process by which the oceanic biota keeps a thin lid of low CO_2 surface water on top of a high CO_2 ocean is known as "the biological CO_2 pump" and is a subject of great interest at a time when we are worried about increasing levels of CO_2 in the atmosphere. The pertinent question is what controls the efficiency of this biological pump?

In most of the oceans, particularly in the large gyres that occupy the central mid latitude regions between the continents, nitrate and phosphate are effectively maintained at near zero concentrations at the surface by phytoplankton uptake. In these regions, the biological pump functions at maximum efficiency and removes excess CO_2 in the approximate fixed stoichiometry of the biota: 106 C:16 N:1 P. Nothing short of a massive input of N and P to the oceans could foster a greater biological removal of surface CO_2. In the HNLC regions where iron has been hypothesized to be a limiting nutrient, variations in Fe concentrations may affect the efficiency of the biological pump. The recent IRONEX II experiment in the equatorial Pacific was designed to test precisely this hypothesis and the measured draw-down of CO_2 from 510 to 420 ppm in the "patch" following iron addition was seen as a crucial part of the test [105].

On a spatial scale of kilometers and over a time scale of days, iron clearly limits the growth of phytoplankton in some regions of the oceans. But to conclude that iron controls the efficiency of the biological pump, we need to demonstrate that it is a limiting nutrient on a global scale. One of the more interesting ways to approach the question of the global role of iron in primary production is by studying the history of the atmosphere and oceans as recorded in polar ice and sediments. As seen in Fig. 8, the air bubbles trapped in polar ice show that atmospheric CO_2 was much lower than present during the last ice age. There is on average a difference of about 70 ppm CO_2 between interglacial and glacial times. Also in Fig. 8, we see that atmospheric inputs of dust (as measured by aluminum concentrations in ice) to the far regions of the earth are much increased during glacial times. This is explained

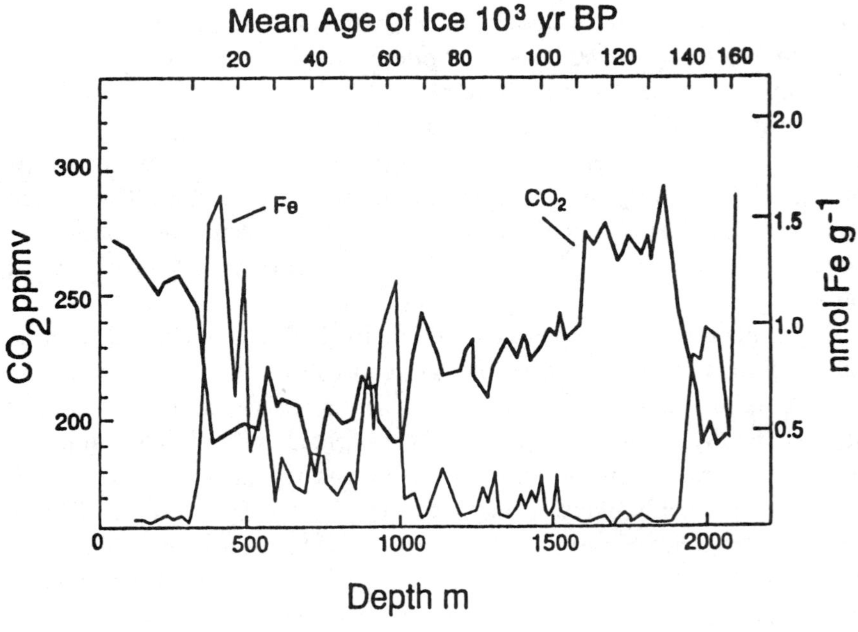

FIG. 8 Iron concentration as a function of depth in the Antarctic Vostok ice core and CO_2 concentration of air trapped in the ice as a function of ice age, determined from mean ice air age. Iron concentrations were calculated from Al measurements using a crustal Fe:Al ratio of 0.33. (Reproduced with permission from [107].)

by the much dryer and stormier weather that is thought to occur during ice ages. This strong anticorrelation between CO_2 and dust content in the ice is the basis of a "paleo-iron hypothesis" according to which increased atmospheric inputs of iron to the oceans (particularly to the HNLC regions) resulted in higher oceanic primary production and a greater efficiency of the biological CO_2 pump during glacial times [106,107]. Indeed, the sedimentary record in the equatorial Pacific shows that the rate of barium inputs from surface waters was much higher than now during the last ice age [108]. Barium is taken up by phytoplankton and its rate of sedimentation can be used as a proxy for net "export" production. In parts of the southern oceans (north of the Antarctic polar front 50–60°S) other sedimentary proxies of export production (e.g., $^{231}Pa/^{230}Th$ and $^{10}Be/^{230}Th$ [109]) also show strong correlations with the ^{18}O content of the sediments which records the extent of glaciation (because of the preferential incorporation of ^{16}O in ice). In addition, the Fe content of the sediments in the same regions also show a strong correlation with ^{18}O. Thus both oceanic primary production and iron inputs were much higher during the last ice age and the "paleo-iron hypothesis" is at least plausible.

There is, however, a major difficulty with this hypothesis: Simple and complex model calculations reveal that iron fertilization of the HNLC regions could only have reduced atmospheric CO_2 by about half of what is observed in the ice record during the last ice age [109,110]. Processes other than an increase in oceanic primary production due to increased iron inputs are then necessary to explain the data. A possible explanation comes from the direct and indirect effects of the marine biota on the precipitation and sedimentation of $CaCO_3$. Coccolithophores (like some nonphotosynthetic organisms) precipitate $CaCO_3$ at the same time that they fix inorganic carbon into biomass. At the pH of seawater, precipitation of $CaCO_3$ results paradoxically in an increase in CO_2 concentration. Thus if there was a relative decrease in the formation and export of $CaCO_3$-carbon compared with organic matter during glacial times, the result would be a decrease in P_{CO_2} on a time scale of 10–100 years. On a time scale of thousands of years, this effect would be amplified if there was a greater dissolution of $CaCO_3$ at depth, resulting from higher downward fluxes of decomposing organic matter [111]. Such processes caused both by a switch in dominant algal species and an indirect effect of increased productivity could result in a much higher efficiency of the biological pump; iron fertilization could then account fully for the 70-ppm decrease in atmospheric CO_2 during the last ice age. The question of what environmental parameters determine the dominant species of primary producers is a very complex oceanographic and ecological question. Although iron may well be one of the dominant factors, other trace elements such as zinc which co-occur in atmospheric dust may also be important [112].

Only paleooceanographic data can resolve the question of whether changes in the iron supply to the oceans may affect global primary production on geological time scales. But the question of whether changes in iron inputs may affect global primary production in the short term might be addressable by examining natural experiments such as large volcanic eruptions. For example, the eruption of Mount Pinatubo in June 1991 injected about 10^{13} kg of continental material into the atmosphere. With crustal abundance of 5% by weight, the global iron input from Pinatubo was about 10^9 mol Fe. Coincident with the Pinatubo eruption was an observed slowing down of the increase in atmospheric CO_2, unprecedented in the 35-year record at Mauna Loa, corresponding to a "deficit" of a 1.4×10^{14} mol C [113]. At the same time, an anomalous increase in atmospheric O_2 corresponding roughly to twice the CO_2 "deficit" was observed, which is on the order of what would be expected if the CO_2 deficit had been caused by increased oceanic primary production [114]. It is tempting to relate those coincidal events to each other and hypothesize that iron fertilization from the Pinatubo eruption caused the global atmospheric CO_2 and O_2 anomalies through ocean fertilization. Rough calculations show that such a mechanism is plausible: about one tenth of the volcanic material may have been deposited in the Southern Ocean [115], corresponding to a potential 10^{14}–10^{15} mol of CO_2 fixed depending on the actual Fe:C stoichiometry in phytoplankton [17]. It is noteworthy that iron addition experiments conducted in the spring of 1992 in the equatorial Pacific showed relatively high phytoplankton growth rates in the control (no Fe added) incubation bottles [22]. Of course, the eruption fertilized the ocean with all the trace elements found in crustal material and the causality of the Pinatubo anomalies cannot be deduced from their coincidence. Nonetheless natural experiments such as large volcanic eruptions may allow us to examine the complex issue of iron limitation of seawater productivity on the global scale.

ACKNOWLEDGMENTS

This work was supported in part by grants from the Natural Sciences and Engineering Research Council of Canada and the Fonds pour la Formation de Chercheurs et l'Aide à la Recherche du Quebec to N.M.P., and also in part by grants from the U. S. National Science Foundation and the U. S. Office of Naval Research to F. M. M. M.

ABBREVIATIONS

ATP	adenosine 5′-triphosphate
CSV	cathodic stripping voltammetry
DTPA	diethylenetrinitrilopentaacetic acid = 1,4,7-triazaheptane-1,1,4,7,7-pentaacetic acid
EDTA	ethylenediamine-*N,N,N′,N′*-tetraacetate
Fe′	inorganic Fe species
GFeGE	gross iron growth efficiency
HNLC	high nutrient, low chlorophyll
IUE	iron use efficiency
NADH	nicotinamide adenine dinucleotide (reduced)
NR	nitrate reductase
ppmv	parts per million by volume
PSII	photosystem II

REFERENCES

1. H. H. Gran, *Rapp. Proc. Verb. Cons. Int. Explor. Mer.*, *75*, 37–46 (1931).
2. T. J. Hart, *Discovery Reports*, *8*, 1–268 (1934).
3. R. M. Gordon, J. H. Martin, and G. A. Knauer, *Nature*, *299*, 611–612 (1982).
4. W. M. Landing and K. W. Bruland, *Geochim. Cosmochim. Acta*, *51*, 29–43 (1987).

5. J. H. Martin and R. M. Gordon, *Deep-Sea Res.*, *35*, 177–196 (1988).
6. J. H. Martin, R. M. Gordon, S. Fitzwater, and W. W. Broenkow, *Deep-Sea Res.*, *36*, 649–680 (1989).
7. J. H. Martin, R. M. Gordon, and S. E. Fitzwater, *Nature*, *345*, 156–158 (1990).
8. R. T. Powell, D. W. King, and W. M. Landing, *Mar. Chem.*, *50*, 13–20 (1995).
9. J. Wu and G. W. Luther III, *Limnol. Oceanogr.*, *39*, 1119–1129 (1994).
10. R. A. Duce and N. W. Tindale, *Limnol. Oceanogr.*, *36*, 1715–1726 (1991).
11. K. H. Coale, S. E. Fitzwater, R. M. Gordon, K. S. Johnson, and R. T. Barber, *Nature*, *379*, 621–624 (1996).
12. E. R. Sholkovitz, *Geochim. Cosmochim. Acta*, *40*, 831–845 (1976).
13. A. M. L. Kraepiel, J.-F. Chiffoleau, J.-M. Martin, and F. M. M. Morel, *Geochim. Cosmochim. Acta*, *61*, 1421–1436 (1997).
14. G. Zhuang, Z. Yi, R. A. Duce, and P. R. Brown, *Nature*, *355*, 537–539 (1992).
15. P. Behra and L. Sigg, *Nature*, *344*, 419–421 (1990).
16. L. J. Spokes and T. D. Jickells, *Aquatic Geochem.*, *1*, 355–374 (1996).
17. A. J. Watson, *Nature*, *385*, 587–588 (1997).
18. H. J. W. De Baar, J. T. M. De Jong, D. C. E. Bakker, B. M. Loscher, C. Veth, U. Barthmann, and V. Smetacek, *Nature*, *373*, 412–415 (1995).
19. P. N. Sedwick and G. R. DiTullio, American Society of Limnology and Oceanography Conference, Santa Fe, New Mexico, February 10–14, 1997 (abstr.).
20. T. D. Waite and F. M. M. Morel, *Environ. Sci. Tech.*, *18*, 860–868 (1984).
21. R. J. M. Hudson, D. T. Covault, and F. M. M. Morel, *Mar. Chem.*, *38*, 209–235 (1992).
22. K. S. Johnson, K. H. Coale, V. A. Elrod, and N. W. Tindale, *Mar. Chem.*, *46*, 319–334 (1994).
23. M. L. Wells and L. M. Mayer, *Deep-Sea Res.*, *38*, 1379–1395 (1991).

24. M. L. Wells, L. M. Mayer, O. F. X. Donard, M. M. de Souza Sierra, and S. G. Ackelson, *Nature*, *353*, 248–250 (1991).
25. F. J. Millero, S. Stolongo, and M. Izaguirre, *Geochim. Cosmochim. Acta*, *51*, 793–801 (1987).
26. J. W. Moffett and R. G. Zika, *Environ. Sci. Technol.*, *21*, 804–810 (1987).
27. B. M. Völker and D. L. Sedlak, *Marine Chem.*, *50*, 93–102 (1995).
28. D. W. O'Sullivan, A. K. Hanson, W. L. Miller, and D. R. Kester, *Limnol. Oceanogr.*, *36*, 1727–1741 (1991).
29. F. J. Millero, W. Yao, and J. Aicher, *Mar. Chem.*, *50*, 21–39 (1995).
30. R. H. Byrne and D. R. Kester, *Mar. Chem.*, *4*, 255–274 (1976).
31. E. L. Rue and K. W. Bruland, *Mar. Chem.*, *50*, 117–138 (1995).
32. J. Wu and G. W. Luther, *Mar. Chem.*, *50*, 159–177 (1995).
33. M. Gledhill and C. M. G. van den Berg, *Mar. Chem.*, *47*, 41–54 (1994).
34. B. C. Cho and F. Azam, *Nature*, *332*, 441–443 (1988).
35. J. H. Martin and S. E. Fitzwater, *Nature*, *331*, 341–343 (1988).
36. R. N. Sambrotto and C. J. Lorenzen, in *The Gulf of Alaska: Physical Environment and Biological Resources*, (D. W. Hood and S. T. Zimmerman, eds.), Ocean Assessments Division, Alaska Office, NOAA, US Minerals Management Service, OCS study, 1986, pp. 249–282.
37. M. T. Maldonado and N. M. Price, *Mar. Ecol. Prog. Ser.*, *141*, 161–172 (1996).
38. W. S. Broecker and T.-H. Peng, *Tracers in the Sea*, Lamont-Doherty Geological Observatory, Columbia University, Palisade, NY, 1982.
39. I. Valiela, *Marine Ecological Processes*, Springer-Verlag, New York, 1984.
40. P. D. Tortell, M. T. Maldonado, and N. M. Price, *Nature*, *383*, 330–332 (1996).
41. K. H. Coale, *Limnol. Oceanogr.*, *36*, 1851–1864 (1991).
42. P. W. Boyd, D. L. Muggli, D. E. Varela, R. H. Goldblatt, R. Chretien, K. J. Orians, and P. J. Harrison, *Mar. Ecol. Prog. Ser.*, *136*, 179–193 (1996).

43. D. A. Hutchins, G. R. DiTullio, and K. W. Bruland, *Limnol. Oceanogr.*, *38*, 1242–1255 (1993).
44. M. R. Landry, D. J. Gifford, D. L. Kirchman, P. A. Wheeler, and B. C. Monger, *Prog. Oceanog.*, *32*, 239–258 (1993).
45. K. Barbeau, J. W. Moffett, D. A. Caron, P. L. Croot, and D. L. Erdner, *Nature*, *380*, 61–64 (1996).
46. Z. Chase and N. M. Price, *Limnol. Oceanogr.* In press.
47. C. B. Miller, B. W. Frost, P. A. Wheeler, M. R. Landry, N. Welschmeyer, and T. M. Powell, *Limnol. Oceanogr.*, *36*, 1600–1615 (1991).
48. T. R. Parsons, R. J. Le Brasseur, J. D. Fulton, and O. D. Kennedy, *J. Exp. Mar. Biol. Ecol.*, *3*, 39–50 (1969).
49. D. J. Gifford, *Prog. Oceanog.*, *32*, 223–237 (1993).
50. M. Dagg, *Prog. Oceanog.*, *32*, 163–183 (1993).
51. M. A. Anderson and F. M. M. Morel, *Limnol. Oceanogr.*, *27*, 789–813 (1982).
52. R. J. Hudson and F. M. M. Morel, *Limnol. Oceanogr.*, *35*, 1002–1020 (1990).
53. R. J. Hudson and F. M. M. Morel, *Deep-Sea Res.*, *40*, 129–150 (1993).
54. R. J. Hudson, The chemical kinetics of iron uptake by phytoplankton, PhD Thesis, MIT, Cambridge, MA, 1989.
55. F. Bienfait, in *Iron Transport in Microbes, Plants and Animals* (G. Winkelmann, D. van der Helm, and J. B. Neilands, eds.), VCH Press, Weinheim, Germany, 1987, pp. 339–349.
56. M. A. Anderson and F. M. M. Morel, *Mar. Biol. Lett.*, *1*, 263–268 (1980).
57. G. J. Jones, B. P. Palenik, and F. M. M. Morel, *J. Phycol.*, *23*, 237–244 (1987).
58. A. Dancis, D. G. Roman, G. J. Anderson, A. G. Hinnebusch, and R. D. Klausner, *Proc. Natl. Acad. Sci. USA*, *89*, 3869–3873 (1992).
59. D. M. DeSilva, C. C. Askwith, D. Eide, and J. Kaplan, *J. Biol. Chem.*, *270*, 1098–1101 (1995).
60. R. Stearman, D. S. Yuan, Y. Yamaguchi-Iwai, R. D. Klausner, and A. Dancis, *Science*, *271*, 1552–1557 (1996).
61. F. C. T. Allnutt and W. D. Bonner, *Plant Physiol.*, *85*, 751–756 (1987).

62. G. J. Jones and F. M. M. Morel, *Plant Physiol.*, *87*, 143–147 (1988).
63. D. W. King, H. A. Lounsbury, and F. J. Millero, *Environ. Sci. Technol.*, *29*, 818–824 (1995).
64. M. L. Wells, N. G. Zorkin, and A. G. Lewis, *J. Mar. Res.*, *41*, 731–746 (1983).
65. H. W. Rich and F. M. M. Morel, *Limnol. Oceanogr.*, *35*, 652–662 (1990).
66. K. Kuma and K. Matsunaga, *Mar. Biol.*, *122*, 1–11 (1995).
67. J. H. Ryther and D. D. Kramer, *Ecology*, *42*, 444–446 (1961).
68. W. G. Sunda, D. G. Swift, and S. A. Huntsman, *Nature*, *351*, 55–57 (1991).
69. W. G. Sunda and S. A. Huntsman, *Mar. Chem.*, *50*, 189–206 (1995).
70. A. C. Redfield, in *James Johnstone Memorial Volume*, Liverpool University Press, Liverpool, UK, 1934, pp. 176–192.
71. L. E. Brand, *Limnol. Oceanogr.*, *36*, 1756–1771 (1991).
72. N. M. Price and F. M. M. Morel, in *Algae and Water Pollution*, (Advances in Limnology; Heft 42), (L. C. Rai, J. P. Gaur, and C. J. Soeder, eds.), E. Schweizebbartsche Verlagsbuchhandlung, Stuttgart, 1994, pp. 79–97.
73. J. A. Raven, *New Phytol.*, *109*, 279–287 (1988).
74. J. A. Raven, *New Phytol.*, *116*, 1–18 (1990).
75. J. LaRoche, R. J. Geider, L. M. Graziano, H. Murray, and K. Lewis, *J. Phycol.*, *29*, 767–777 (1993).
76. G. J. Doucette, D. L. Erdner, M. L. Peleato, J. J. Hartman, and D. M. Anderson, *Mar. Ecol. Prog. Ser.*, *130*, 269–276 (1996).
77. B. Entsch, R. G. Sim, and B. G. Hatcher, *Mar. Biol.*, *73*, 17–30 (1983).
78. J. LaRoche, P. W. Boyd, R. M. L. McKay, and R. J. Geider, *Nature*, *382*, 802–805 (1996).
79. M. Nakamura, M. Yamagishi, F. Yoshizaki, and Y. Sugimura, *J. Biochem.*, *111*, 219–224 (1992).
80. R. J. Geider and J. LaRoche, *Photosyn. Res.*, *39*, 275–301 (1994).
81. R. M. Greene, R. J. Geider, Z. Kolber, and P. G. Falkowski, *Plant Physiol.*, *100*, 565–575 (1992).

82. J. E. Armstrong and C. Van Baalen, *J. Gen. Microbiol.*, *111*, 253–262 (1979).
83. C. G. Trick, *Curr. Microbiol.*, *18*, 373–378 (1989).
84. S. W. Wilhelm and C. G. Trick, *Limnol. Oceanogr.*, *39*, 1979–1984 (1994).
85. R. T. Reid, D. H. Live, D. J. Falkner, and A. Butler, *Nature*, *366*, 455–458 (1993).
86. J. B. Neilands, *Adv. Chem. Ser.*, *162*, 1–32 (1977).
87. C. G. Trick, S. W. Wilhelm, and C. M. Brown, *Can. J. Microbiol.*, *41*, 1117–1123 (1995).
88. H. E. Chadd, I. R. Joint, N. H. Mann, and N. G. Carr, *FEMS Microbiol. Ecol.*, *21*, 69–76 (1996).
89. R. T. Reid and A. Butler, *Limnol. Oceanogr.*, *36*, 1783–1792 (1991).
90. J. D. Pakulski, R. B. Coffin, C. A. Kelley, S. L. Holder, R. Downer, P. Aas, M. M. Lyons, and W. H. Jeffrey, *Nature*, *383*, 133–134 (1996).
91. M. J. Behrenfeld, A. J. Bale, Z. S. Kolber, J. Aiken, and P. G. Falkowski, *Nature*, *383*, 508–511 (1996).
92. M. L. Wells, N. M. Price, and K. H. Bruland, *Limnol. Oceanogr.*, *39*, 1481–1486 (1994).
93. E. L. Rue and K. W. Bruland, *Limnol. Oceanogr.* In press.
94. A. E. Gargett, *Limnol. Oceanogr.*, *36*, 1527–1545 (1991).
95. S. W. Chisholm and F. M. M. Morel, *Limnol. Oceanogr.*, *36*, 1507–1970 (1991).
96. B. W. Frost, *Limnol. Oceanogr.*, *36*, 1616–1630 (1991).
97. J. H. Martin, R. M. Gordon, and S. E. Fitzwater, *Limnol. Oceanogr.*, *36*, 1793–1802 (1991).
98. S. E. Fitzwater, K. H. Coale, R. M. Gordon, K. S. Johnson, and M. E. Ondrusek, *Deep-Sea Res.*, *43*, 995–1015 (1996).
99. N. M. Price, B. A. Ahner, and F. M. M. Morel, *Limnol. Oceanogr.*, *39*, 520–534 (1994).
100. I. Kudo and P. J. Harrison, *J. Phycol.*, *33*, 232–240 (1997).
101. J. H. Martin, K. H. Coale, K. S. Johnson, S. E. Fitzwater, R. M. Gordon, S. J. Tanner, C. N. Hunter, V. A. Elrod, J. L. Nowicki, T. L. Coley, R. T. Barber, S. Lindley, A. J. Watson, K. Van Scoy, C. S.

Law, M. I. Liddicoat, R. Ling, T. Stanton, J. Stockel, C. Collins, A. Anderson, R. Bidigare, M. Ondrusek, M. Latasa, F. J. Millero, K. Lee, W. Yao, J. Z. Zhang, G. Friederich, C. Sakamoto, F. Chavez, K. Buck, Z. Kolber, R. Green, P. Falkowski, S. W. Chisholm, F. Hoge, R. Swift, J. Yungel, S. Turner, P. Nightingale, A. Hatton, P. Liss, and N. W. Tindale, *Nature*, *371*, 123–129 (1994).

102. K. H. Coale, K. S. Johnson, S. E. Fitzwater, R. M. Gordon, S. Tanner, F. P. Chavez, L. Ferioli, C. Sakamoto, P. Rogers, F. Millero, P. Steinberg, P. Nightingale, D. Cooper, W. P. Cochlane, M. R. Landry, J. Constantinou, G. Rollwagen, A. Trasvina, and R. Kudela, *Nature*, *383*, 495–501 (1996).

103. Z. S. Kolber, R. T. Barber, K. H. Coale, S. E. Fitzwater, R. M. Greene, K. S. Johnson, S. Lindley, and P. J. Falkowski, *Nature*, *371*, 145–149 (1994).

104. F. M. M. Morel, J. G. Rueter, and N. M. Price, *Oceanography*, *4*, 56–61 (1991).

105. D. J. Cooper, A. J. Watson, and P. D. Nightingale, *Nature*, *383*, 511–513 (1996).

106. J. H. Martin, S. E. Fitzwater, and R. M. Gordon, *Global Biogeochem. Cycles*, *4*, 5–12 (1990).

107. J. H. Martin, *Paleoceanogr.*, *5*, 1–13 (1990).

108. A. Paytan, M. Kastner, and F. P. Chavez, *Science*, *274*, 1355–1357 (1996).

109. N. Kumar, R. F. Anderson, R. A. Mortlock, P. N. Froelich, P. Kubik, B. Dittrich-Hannen, and M. Suter, *Nature*, *378*, 675–680 (1995).

110. J. L. Sarmiento and J. C. Orr, *Limnol. Oceanogr.*, *36*, 1928–1950 (1991).

111. W. S. Broecker and T.-H. Peng, *Global Biogeochem. Cycles*, *1*, 15–29 (1987).

112. F. M. M. Morel, J. R. Reinfelder, S. B. Roberts, C. P. Chamberlain, J. G. Lee, and D. Yee, *Nature*, *369*, 740–742 (1994).

113. J. L. Sarmiento, *Nature*, *365*, 697–698 (1993).

114. R. F. Keeling, S. C. Piper, and M. Heimann, *Nature*, *381*, 218–221 (1996).

115. V. K. Saxena, J. Anderson, and N.-H. Lin, *J. Geophys. Res.*, *100*, 16,735–16,751 (1995).

116. S. W. Wilhelm, D. P. Maxwell, and C. G. Trick, *Limnol. Oceanogr.*, *41*, 89–97 (1996).

117. J. H. Martin and G. A. Knauer, *Geochim. Cosmochim. Acta*, *37*, 1639–1653 (1973).

118. B. C. Booth, J. C. Lewin, and J. R. Postel, *Prog. Oceanog.*, *32*, 57–99 (1993).

119. T. R. Parsons and C. M. Lalli, *Oceanogr. Mar. Biol. Annu. Rev.*, *26*, 317–359 (1988).

120. D. A. Hutchins and K. W. Bruland, *Mar. Ecol. Prog. Ser.*, *110*, 259–269 (1994).

121. F. B. Simpson and J. R. Neilands, *J. Phycol.*, *12*, 44–48 (1976).

122. M. E. Conkright, S. Levitus, and T. P. Boyer, *World Ocean Atlas*, Vol. 1: *Nutrients*; NOAA Atlas NESDIS 1., U.S. Department of Commerce, Washington, D.C., 1994.

123. N. M. Price, L. F. Andersen, and F. M. M. Morel, *Deep-Sea Res.*, *38*, 1361–1378 (1991).

2

Microbial Iron Transport: Iron Acquisition by Pathogenic Microorganisms*

B. Rowe Byers and Jean E. L. Arceneaux

Department of Microbiology, The University of Mississippi Medical Center, Jackson, MS 39216, USA

1. INTRODUCTION 38
2. WHY IRON IS AN ELUSIVE BUT HAZARDOUS NECESSITY 39
   2.1. Some Salient Properties of Iron 39
   2.2. Some Iron Principles 40
   2.3. Oxidative Risk of Too Much or Too Little Iron 42
3. WHY IRON HAS A DECISIVE ROLE IN INFECTIOUS DISEASES 45
   3.1. Microbial Iron-Transport Systems 45
      3.1.1. Reductive (Ferrous) Uptake 48
      3.1.2. Siderophore-Based Systems 49
      3.1.3. Commonality of Siderophore Biosynthesis 50

---

*This chapter is dedicated to Eugene D. Weinberg in recognition of his many contributions to this field.

3.1.4. Iron Containing Macromolecules as Iron Sources 52
3.1.5. Regulation of Iron Gene Expression 53
3.2. The "Iron Biofulcrum" 56
3.2.1. Host Iron-Withholding Defenses 56
3.2.2. Pathogen Iron-Acquisition Strategies 57
3.3. Rationale for an Iron Biofulcrum 60
4. SUMMARY AND FUTURE PROSPECTS 61
ACKNOWLEDGMENTS 62
ABBREVIATIONS 62
REFERENCES 62

1. INTRODUCTION

Although a stimulating review by J. B. Neilands in 1957 [1] correctly identified the physiological function of certain microbial ferric chelators, it was in 1973 that Charles Lankford suggested the term *siderophore* for these low molecular mass, iron transport cofactors [2]. A computer search of a U. S. national library database for the year 1973 retrieved less than 20 papers concerned specifically with microbial iron acquisition. A similar search of the database for the year 1996 yielded more than 100 publications with the term *siderophore* in the titles or abstracts, whereas a database on microbial iron transport maintained by the authors contains about 300 citations for 1996. Scrutiny of the recent literature shows that the number of papers on bacterial iron metabolism has almost doubled between 1990 and 1996, with a nearly steady increase each year. These data support the recent suggestion [3] by another of the pioneers in this field, E. D. Weinberg, that hundreds of studies on bacterial iron acquisition have been published in the past three decades. Many of these publications examine the decisive role of microbial iron-acquisition systems in virulence of human (and other vertebrate) and certain plant pathogens. For obvious reasons, many of the critical studies cannot be reviewed here. We will discuss some of the important aspects of microbial iron metabolism and survey the microbial iron-acquisition systems and their roles in virulence. We will suggest a reason why hosts may have chosen iron withholding as one of the prime defensive postures. Other chapters in this book discuss additional aspects of the chemistry, genetics, and functions of microbial iron-acquisition systems. In-depth discussions of the many aspects of this extensive subject may be found in two excellent monographs [4,5]. Examples of recent reviews of the role of iron in infection and uptake of iron by vertebrate pathogens include those of Byers and Arceneaux [6], Wooldridge and Williams [7], Payne [8], Litwin and Calderwood [9], Weinberg [3], Cornelissen and Sparling [10], Askwith et al. [11], Trivier and Courcol [12], and Lee [13]. Guerinot [14] has prepared a detailed review of iron acquisition by pathogenic and nonpathogenic bacteria and fungi.

2. WHY IRON IS AN ELUSIVE BUT HAZARDOUS NECESSITY

2.1. Some Salient Properties of Iron

The value of iron to biological systems is largely due to the two stable valencies that impart considerable range to the reactivity of iron. When inserted into appropriate organic ligands, the metal's reactivity can be manipulated, allowing it to catalyze a wide range of chemical reactions essential to the cell. Iron also is the active component of the oxygen carrier proteins. The necessity of sufficient iron for microbial metabolism is reflected by the almost universal requirement for the metal; only members of the genus *Lactobacillus* may have replaced iron with other cofactors [15].

The reactivity which makes iron so useful to most cells also renders the metal hazardous to handle. In several ways, the dangers coupled to the use of iron in biology parallel the problems faced by human technology in use of long-lived radioisotopes as fuel. Like radioactive fuel, iron may be difficult to acquire, must be constantly disciplined and monitored, and must be appropriately shielded when stored. It is likely true that an accidental release of only a few iron atoms from the intracellular iron-handling mechanism can unleash a cascade of rapidly lethal reactions not unlike those produced by radiation. The basis for this danger is described as "Fenton chemistry". Iron acts as a central catalyst yielding the highly reactive hydroxyl radical ($OH^\bullet$) from hydrogen peroxide [7,16]. The $OH^\bullet$ is most notorious for its damage to DNA [17]. Ferrous iron catalyzes $OH^\bullet$ production from H_2O_2:

$$H_2O_2 + Fe^{2+} \longrightarrow Fe^{3+} + OH^- + OH^\bullet$$

The hydrogen peroxide can be generated from the superoxide anion (O_2^-):

$$2O_2^- + 2H^+ \longrightarrow H_2O_2 + O_2$$

To perpetuate $OH^\bullet$ production, cellular reductants may promote reutilization of iron in $OH^\bullet$ synthesis:

$$Fe^{3+} + NADH \longrightarrow Fe^{2+} + [NADH^+]$$

Not only is iron difficult to handle, its acquisition can be elusive. In an oxidized environment at near pH 7, the ferric iron precipitates from solution. The propensity of iron to disappear from solution can be enhanced by the presence of ordinary ions like phosphate and aluminum which can coprecipitate with iron. These considerations reveal that, to biological systems, iron is essential, but it is expensive to acquire and costly to handle.

2.2. Some Iron Principles

Effective use of iron as a biocatalyst requires stringent compliance with the constraints dictated by the solubility and reactivity properties of iron. Disregard of these principles will sooner or later mean disaster for a biological cell. Some of the most important of the rules for safe use and handling of iron by a biological organism are briefly discussed.

First, a high-affinity iron-capture mechanism is required. The propensity of Fe^{3+} to precipitate from solution means that aerobic microorganisms must either build complete iron-acquisition systems able to solubilize, transport, and assimilate iron, or (in the words of Cornelissen and Sparling [10]), become an "iron pirate" and steal iron from another microorganism's iron-accumulating pathway. The most competitive cells may do both. Moreover, multiple iron-uptake systems may be required to meet different iron-acquisition problems in the environment. The various known iron-uptake systems are listed in Table 1 and are discussed in a later section.

Second, iron uptake must be regulated and coordinated with need. Control of uptake processes must be coupled with a system for controlled intracellular trafficking of iron. Iron-catalyzed production of free radicals means that loose or free intracellular iron cannot be tolerated. An excess of iron or iron that escapes the intracellular iron-handling molecules and becomes misplaced within the cell can be lethal. On the other hand, too little iron will have the obvious consequence of restricted growth and possibly cell death. The correct amount of iron for a cell is "just enough" to satisfy the needs of the iron-requiring systems with some iron available to synthesize new cell material as a prelude to cell division. The underlying rule must be that iron is moved into the cell and within the cell only when and to where needed. Iron in transit is likely to be most capable of causing oxidative damage, especially when the metal is passed from one system to another. Accordingly, the cell may need to keep the amount of iron in transit at a low level.

A third principle is that cells must have a means to detoxify excess iron which might accumulate and a way safely to store iron for use during periods of environmental iron restriction. Defense against the toxic effect of iron is accomplished in part by shielding excess or stored iron within the hollow shell of ferritin where iron is maintained as Fe^{3+} and is unable to catalyze free radical production unless mobilized from the shell. Ferritin is ubiquitous and of essentially the same architecture, although differences (the bacterial form may contain heme groups) are evident in the structure of bacterial and vertebrate ferritins. The extensive research efforts on ferritin synthesis (including regulation of synthesis) and on the roles of this substance as an iron reserve and as an iron-detoxifying agent have been recently reviewed [16]. It has been suggested that detoxification of iron is the major role of ferritin in bacteria [16].

An antioxidant defense system is a fourth critical necessity. It is possible that the normal flow of iron within the cell will periodically expose the metal in its catalytic form; hence, an important function of intracellular antioxidant defense enzymes is to destroy O_2^- and H_2O_2 before they can come into contact with iron. It is somewhat ironic that the catalytic properties of iron have been used to form important enzymes of the antioxidant defense. Iron and heme are essential cofactors for the H_2O_2-cleaving catalases, various peroxidases, and the iron-containing superoxide dismutases [18]. The mitochondrial and some bacterial superoxide dismutases use manganese as a cofactor, but many other bacterial superoxide dismutases have an iron cofactor [18]. An organism typically may produce either an iron or manganese superoxide dismutase, or both. In general, the iron and manganese cofactors are not interchangeable in these enzymes, although the proteins have significant amino acid sequence homologies. A small group of superoxide dismutase enzymes (produced mainly by anaerobic or facultatively aerobic organisms) accept either manganese or iron as cofactors. This group has been designated as the cambialistic superoxide dismutases, and it was suggested that they might be evolutionary ancestors of the modern manganese and iron enzymes [19].

2.3. Oxidative Risk of Too Much or Too Little Iron

Iron can have a dual role in the life of a cell both as a protector from oxidative damage and as a destroyer by promoting such damage. Both iron insufficiency (by depressing the oxidative defenses) and iron excess (by promoting the Fenton reaction) render cells at risk to damage from oxy-radicals. The following brief summary highlights the principle that the correct amount of iron for a cell is "just enough" to satisfy its needs and not much more.

The effects of excess iron on bacteria have been examined either during growth at high iron concentrations or by studying the physiology of mutants that are deregulated for iron uptake. Growth of *Staphylococcus aureus* at an elevated iron level produced cells that were very sensitive to the lethal action of H_2O_2 [20,21]. An *Escherichia coli* strain that was deregulated for iron assimilation (*fur* null mutant) became iron overloaded and showed increased sensitivity to oxidative damage [22]. A mutant of *Campylobacter jejuni* unable to detoxify iron by making ferritin showed increased sensitivity to H_2O_2 [23].

In several mycobacterial species, an *ideR* (*i*ron *de*pendent *r*egulation) gene encodes a regulatory protein that represses transcription of genes for iron uptake when sufficient iron is present [24]. The cloned *ideR* homologue of *Mycobacterium smegmatis* was inactivated by insertion of a transposon and an *ideR* mutant strain was generated by allelic exchange [24]. The *ideR* defective mutants showed increased sensitivity to H_2O_2 both on solid and in liquid media (Fig. 1). Although the phenotype of the *M. smegmatis ideR* mutants resembles certain *E. coli fur*

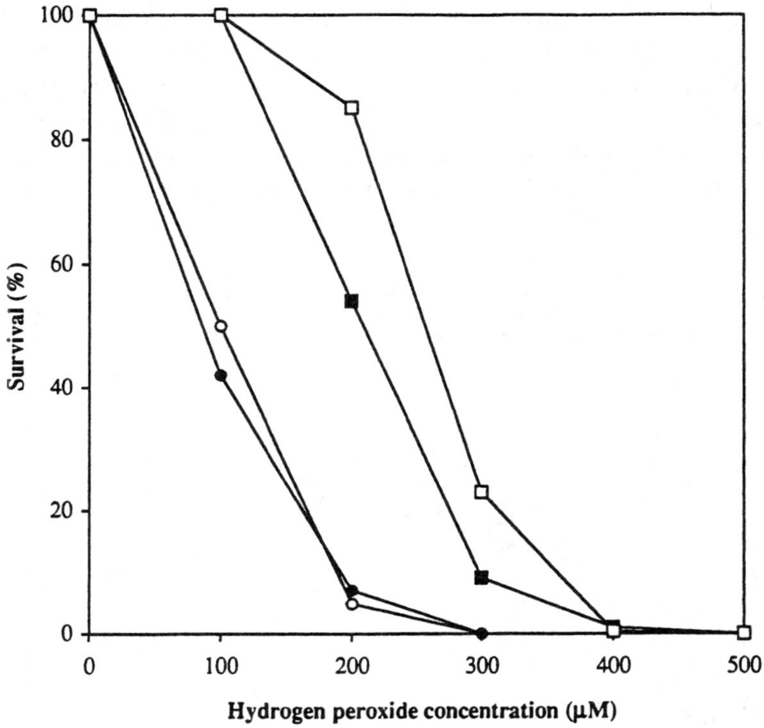

FIG. 1 Increased sensitivity of *Mycobacterium smegmatis* regulatory mutants to H_2O_2. The wild-type strain (closed square), two iron-dependent regulatory (*ideR*)-defective mutant strains (open circle and closed circle), and a mutant strain (open square) complemented with a wild-type copy of *ideR* were plated on Middlebrook 7H10 agar supplemented with 50 μM iron and various concentrations of H_2O_2; 100% survival approximates 10^8 colony-forming units per milliliter. (Reproduced with permission from [24].)

null mutants [22], it is not known whether the increased sensitivity to H_2O_2 in the *ideR* mutants is also the result of iron overload. The IdeR protein may be a pleiotropic regulator which also triggers components of the oxidative stress response [24]. Illustrating the complex role of the Fur- and IdeR-like regulatory proteins in iron metabolism and oxidative defenses, missense mutations within the *fur* locus of *Pseudomonas aeruginosa* yielded strains that were impaired in iron uptake and were highly sensitive to H_2O_2 [25].

Because iron is needed as cofactor in several of the enzymes of the antioxidant defense system, an insufficient intracellular iron level also may inflate oxy-radical damage by depressing the activities of the antioxidant defenses. When made iron deficient by growth in low-iron medium, *M. smegmatis* showed increased sensitivity to H_2O_2 [26]. The threshold levels of H_2O_2 necessary for killing this organism were in-

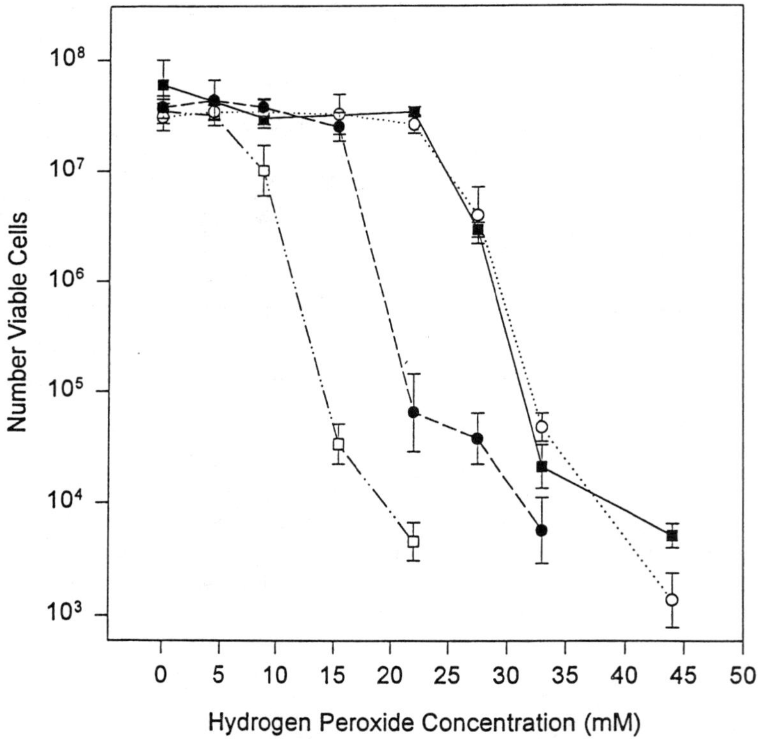

FIG. 2 Increased sensitivity of iron starved *Mycobacterium smegmatis* wild-type to H_2O_2. Exponential phase cultures growing in a glucose-mineral salts medium (treated with Chelex-100) that was supplemented with various iron concentrations (μM) of 0.05 (open square), 0.2 (closed circle), 1.0 (open circle), or 20.0 (closed square) were exposed to the indicated concentrations of H_2O_2 for 15 min and the surviving fractions determined by counting colony-forming units. The differences in effective H_2O_2 concentrations shown in Figs. 1 and 2 may be explained by the different H_2O_2 exposure protocols (continuous and only 15 min, respectively). (Reproduced with permission from [25].)

creased proportionally by increasing the iron concentration in the growth medium; however, greatest resistance to H_2O_2 was achieved by growth at the iron level which supported maximum growth and increasing the iron addition to the medium above this concentration did not increase resistance to H_2O_2 (Fig. 2). Although the vulnerability of iron-starved *M. smegmatis* to H_2O_2 killing may be due in part to a deficiency in the antioxidant defenses, iron-starved *M. smegmatis* also was unable to produce all of the oxidative stress proteins noted in iron-sufficient cells [26].

The need for iron to combat oxy-radicals can sometimes be made evident by switching a facultative organism from anaerobic to aerobic conditions. *Streptococcus mutans* requires an iron supplement for aerobic (but not for anaerobic) growth in a chemically defined medium that was treated to lower trace metal contamination [27]. Manganese could replace iron in supporting aerobic growth of *S. mutans*. It was shown that this organism has a cambialistic superoxide dismutase that is active with either iron or manganese as a cofactor, which may account for the aerobic requirement for iron or manganese of this organism [19].

3. WHY IRON HAS A DECISIVE ROLE IN INFECTIOUS DISEASES

3.1. Microbial Iron-Transport Systems

The constraints imposed by the solubility and reactivity properties of iron are satisfied in microorganisms by the syntheses of special iron-handling systems. Of these, the best understood are the membrane-embedded iron uptake processes. Because iron is as indispensable for the life of a cell as are nitrogen and carbon for which there are several assimilation systems, so also there are multiple mechanisms of iron assimilation (Table 1) even for a single organism [28]. Although few (or perhaps no) microbes may have all of the systems listed in Table 1, multiple iron-uptake systems would confer the ability to live in a changing environment; those microbes that are restricted to a single habitat may have fewer systems. Moreover, it would be arrogant to suggest that all of the microbial iron uptake systems have now been defined, and Table 1 includes an “Other” category to accommodate new discoveries in this rapidly moving field.

TABLE 1

Iron-Acquisition Systems of Microorganisms[a]

System	Major type	Iron sources used or subtype	Probable system components
Ferrous	Ferrous	Ferrous iron	Ferrous uptake system; possible surface reductase or excreted reductant
Siderophore-based	Endogenous siderophore	Ferric complexes of diverse types; Fe-containing macromolecules	Siderophore synthetic and excretion system, plus ferrisiderophore receptor and transport system; extracellular pathogens may need cytolysins and proteases for access to intracellular host iron
	Exogenous siderophore	Same as endogenous siderophore	Exogenous ferrisiderophore receptor and transport system

Iron containing macromolecule	Heme	Heme-proteins or heme-hemopexin and hemoglobin-haptoglobin complexes	Heme receptor and transport system, an alternative system may need heme oxygenase; extracellular pathogens may need cytolysins and proteases
	Fe-transferrin and Fe-lactoferrin	Siderophore based	See siderophore-based systems above
		Nonsiderophore based	Fe-transferrin (and Fe-lactoferrin) surface receptor (may be specific for transferrin type); mechanism for release of iron
	Ferritin	Ferritin	Reductase for release of iron
Other	Unknown	Host intracellular low- or high-molecular mass, non-heme, non-ferritin iron	Extracellular pathogens may require cytolysins (and a siderophore) to access host intracellular iron; intracellular pathogen's systems are uncertain, type may depend on infection site (e.g., within a vacuole or free in host cytosol)

[a]A comprehensive list of pertinent publications can be found in Refs. 4–14.

3.1.1. Reductive (Ferrous) Uptake

Although only speculation, the most ancient of the iron uptake systems may be transport of the more soluble ferrous state of the metal [28]. The ferrous transport system may persist in many organisms, although it may be useful only in special environmental niches (which might be created by the microorganism) where a supply of reduced iron is available. Exclusive use of ferrous transport for iron acquisition means that the microorganism must be assured of an adequate and dependable supply of reduced iron. Although it may be coincidental, several of the microorganisms in which ferrous transport systems were first identified live in oxygen-restricted environments where reduced iron may be prevalent [29]. Coupling an outward-facing membrane reductase with a ferrous transporter may represent a suitable combination that supports iron uptake in certain aerobic environments (Fig. 3). The aerotolerant bacterium *S. mutans* has only this type of iron transport system; a membrane-associated flavin reductase generates the appropriate form of iron for transport [30]. *Listeria monocytogenes* may excrete a reductant to satisfy a ferrous iron transporter [31]. *E. coli* does not have an

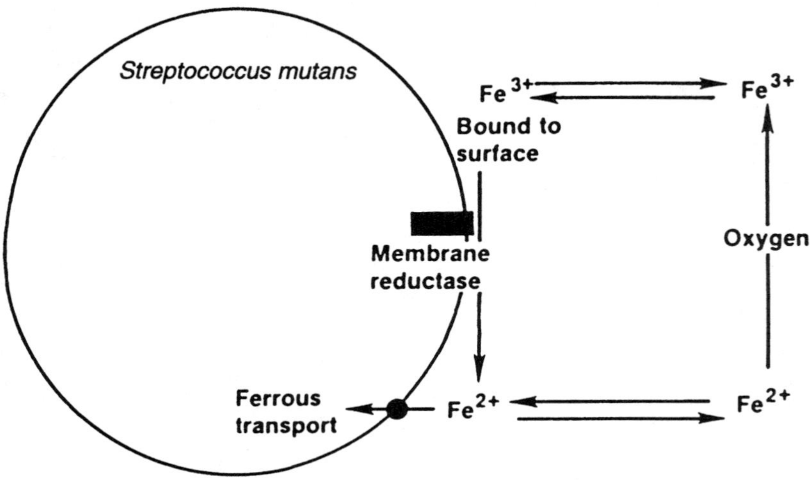

FIG. 3 Model for uptake of ferrous iron by *Streptococcus mutans* showing the role of a surface reductase. (Reproduced with permission from [30].)

external iron reductase system, but it maintains the ability to transport ferrous iron [29]. The two genes involved in this system have been cloned and sequenced; the predicted amino acid sequences of their products suggest that transport of ferrous iron is driven by ATP [29].

The eukaryotic yeast *Saccharomyces cerevisiae* also has a cell surface reductase that provides ferrous iron for uptake by the transport system [11,32]. The external ferric reductase is a cytochrome *b*, probably a flavocytochrome [33]. The iron transporter of this system is a two-component complex comprising a multicopper iron oxidase and a permease. The oxidase catalyzes the conversion of ferrous to ferric iron which is necessary for iron transport by the permease component. Therefore, iron transport involves a puzzling set of reactions at the cell surface. The iron reductase generates ferrous iron from the ferric form and the iron oxidase converts this to ferric iron for uptake. A baffling aspect of this system is that the apparent cytoplasmic product is highly insoluble ferric iron. This suggests the existence of another component, a cytoplasmic chelator, of newly transported ferric iron.

3.1.2. Siderophore-Based Systems

The most studied and perhaps best-understood systems of microbial iron acquisition are those based on the low molecular mass (500–1000 Da), iron-chelating molecules called siderophores [34,35]. Siderophores have high affinity ($K_{aff} > 10^{30}$) for the ferric atom, being able to solubilize iron from insoluble complexes of the metal and forming six-coordinate octrahedral complexes with ferric iron [4,5]. When chelated by a siderophore, iron also is protected from participation in oxy-radical production; therefore, the ferrisiderophore chelate is an exceptional solution to the solubility and reactivity constraints of iron utilization. There is considerable structural variation among siderophores; however, siderophores generally fall into two basic structural groups, the hydroxamates and the catecholates [35]. Siderophores usually are secreted and the externally formed ferrisiderophore is subsequently transported into the cell. The size of the siderophores exceeds the size which allows free movement across biological membranes and uptake of the ferrisiderophore requires a membrane-embedded transport system. A cognate surface receptor with high specificity for the ferrisiderophore is an essential component of the transport system. Energy is expended to actively transport the ferrisiderophore. A successful microorganism often is equipped with receptors that recognize its endogenous siderophore as well as receptors that bind and allow utilization of exogenous, structurally different siderophores produced by other microorganisms (see Table 1). Because siderophores generally have low affinity for ferrous iron, removal of the metal for metabolic use involves a reductase [36]. Expression of genes encoding syntheses of siderophores and their transport systems is triggered (or increased) by low iron conditions. The siderophore-based strategy for iron acquisition has been exploited by many types of both bacteria and fungi, and with some exceptions, it is likely true that an aerobic microorganism will produce a siderophore [35]. The genetics, transport mechanisms, and chemistry of microbial siderophores are discussed in several other chapters in this book and in two excellent monographs [4,5].

3.1.3. Commonality of Siderophore Biosyntheses

The structural diversity of siderophores may conceal a relatedness at the genetic and biosynthetic levels. There is an underlying structural commonality among siderophores of both groups. Many siderophores contain lysine or ornithine or they are built on polyamine backbones derived from these two amino acids [37]. In one sense, siderophores may be considered as biologically active derivatives of small peptides or peptide-like molecules and their biosynthesis might resemble biosynthesis of other small peptides. Alignment of the predicted amino acid sequences encoded by several siderophore biosynthetic genes indicates that some of these genes belong to a large family of adenylate-forming enzymes that activate and bind amino acids in the nonribosomal syntheses of small peptides, including peptide antibiotics [38]. Siderophore biosynthetic genes contained in this group include *angR* required for anguibactin synthesis in *Vibrio anguillarum*; *entE* and *entF* required for enterobactin synthesis in *E. coli*; and *irp2* required for yersiniabactin synthesis in *Yersinia pestis* and *Y. enterocolitica* [38–40]. This suggests that siderophore biosynthetic systems may be derived from an ancestral system related to the present nonribosomal peptide synthetases and that the biosynthetic steps in the siderophore pathways will be similar.

Although all of the biosynthetic genes for the amonabactins (novel catecholate siderophores from *Aeromonas hydrophila*) have not been cloned, physiological data suggest that these siderophores also may be assembled by enzymes resembling the nonribosomal peptide synthetases. The amonabactins are catecholate siderophores containing 2,3-dihydroxybenzoic acid (DHB) that are built on a backbone composed of glycine-lysine-lysine and either tryptophan (amonabactin T) or phenylalanine (amonabactin P) [41]. Tryptophan and phenylalanine in the amonabactins are in the D form. Both amonabactin T and amonabactin P usually are made simultaneously; however, synthesis of the tryptophan or the phenylalanine forms of the siderophores depends on availability of these amino acids. When iron-deficient cultures of *A. hydrophila* were supplemented with either *L*-tryptophan or *L*-phenylalanine, siderophore synthesis was markedly shifted to favor amonabactin T or amonabactin P, respectively [42]. Curiously, when the cultures were supplemented with *D*-tryptophan, syntheses of both amonabactin T and amonabactin P were inhibited. The reason for inhibition by *D*-tryptophan is unknown, but there may be a parallel with the syntheses of certain peptide antibiotics. Both tyrocidine synthetase and gramicidin synthetase activate phenylalanine, racemize it, and form an amino acid-adenylate-enzyme complex [43]. If the amonabactin biosynthetic enzymes are related to the nonribosomal peptide synthetases, then an amonabactin synthetase may follow this pattern of synthesis. Initial association of *D*-tryptophan (rather than the usual *L* form) with the enzyme prior to the racemization step might halt amonabactin biosynthesis.

Other evidence supporting the relatedness of siderophore biosynthetic genes is known. More than 20 structurally different siderophores (produced by at least that many genera) that contain the phenolate moiety DHB are known, and it was suggested that the microorganisms producing these siderophores harbor a related version of the operon encoding DHB biosynthesis [6,44]. The DHB component is synthesized and activated by the products of four genes [45]. The branch point compound (shikimic acid) in the aromatic biosynthetic pathway is the immediate precursor. Comparisons of restriction maps, hybridization patterns of gene probes, and complementation of siderophore mutants with cloned DHB genes [44,46–49] suggest that the various operons encoding DHB synthesis may have diverged from a single ancestral gene group [6,44]. Determinations of the nucleotide sequences of the

DHB operons might permit construction of the evolutionary history of the phenolate siderophores [6].

If there is an underlying relatedness of siderophore biosynthetic systems at the genetic and enzyme levels, then we should consider why siderophores are structurally different. Much of the differences may be accounted for by evolutionary diversion; however, because different siderophores require unique surface receptors for utilization by an organism, siderophore structural diversity may represent an effort to increase competitiveness of an organism's iron-acquisition system. Like the precious metal-laden sailing ships of the past, a very unique ferrisiderophore requiring a specific surface receptor for utilization might be a good defense against some iron pirates.

3.1.4. Iron Containing Macromolecules as Iron Sources

In addition to the siderophore-based uptake processes, many types of bacteria have other specific mechanisms for efficient acquisition of iron that is bound to macromolecules like heme-proteins, transferrin, and lactoferrin. At least 24 (and likely many more) different genera of pathogenic bacteria use heme [3]. Nonsiderophore-mediated acquisition of heme by gram-negative bacteria has been reviewed recently by Lee [13]. Most of the known heme-utilization systems of these bacteria appear to involve an outer membrane heme receptor and a transport process that incorporates the intact heme molecule for internal release of iron or for direct incorporation of heme into heme proteins. Some of the heme transport systems may resemble the ferrisiderophore uptake processes, using some of the same membrane components. Genetic loci encoding proteins required for transport of heme have been identified and cloned in several organisms. Heme utilization is often accompanied by production of iron-regulated hemolysins and cytolysins that release heme-proteins from cells. For over a century, it has been recognized that the ability of bacteria to lyse erythrocytes may be a virulence factor [3].

A possible second mechanism of iron acquisition from heme has been discovered in *Corynebacterium diphtheriae* [50]. Utilization of heme iron is dependent on a gene which encodes a protein with 33% identity and 70% similarity to the human heme oxygenase enzyme. Previously unknown in bacteria, the heme oxygenases of eukaryotes oxidize heme and subsequently release iron [50]. The corynebacterial ferrisiderophore and its uptake system were not required for acquisition of iron by the oxygenase-dependent system. The possible remaining components of this new system are yet to be described.

The ability of some bacteria to bind heme at the cell surface has long been recognized as a virulence trait [51,52]. The reason that heme-clad bacteria show increased virulence is presently unknown; it may not involve uptake of heme or iron from heme [52].

Although many bacteria can use their siderophores to capture iron from Fe-transferrin or Fe-lactoferrin complexes, several bacterial genera can access the iron bound to these glycoproteins without intervention of a siderophore. The mechanisms of nonsiderophore-mediated utilization of Fe-transferrin have been most studied in pathogenic *Neisseria* species and in *Haemophilus* species, although gram-negative pathogens, including *Moraxella catarrhalis*, *Actinobacillus pleuropneumoniae* (reviewed in Ref. 10), and *Porphyromonas gingivalis* [53], also have a transferrin receptor. A current model [10] of iron acquisition from Fe-transferrin by gram-negative bacteria suggests an outer membrane transferrin receptor which directly binds the glycoprotein. Iron is released from the complex and is then bound by a periplasmic protein for subsequent transport by a cytoplasmic membrane permease system. Cloning of genes encoding the system suggests that some of these may be homologues of the genes involved in uptake of ferrisiderophores [10]. Utilization of transferrin by bacteria differs from the process found in vertebrate cells; the bacterial transferrin receptor does not appear to be evolutionarily related to the eukaryotic transferrin receptor [10]. This evidence also suggests that transferrin receptors of bacterial pathogens may display specificity for the transferrin of the preferred host.

3.1.5. Regulation of Iron Gene Expression

Regulation of the syntheses and transactions of the various systems that are concerned with acquisition, distribution, and storage of iron is essential, not only for efficiency but because mistakes may be lethal. Our understanding of the control of the iron processes is fragmentary, but an elegant hierarchy of governing metabolic switches is coming into focus.

In almost every microorganism studied, iron restriction causes (or increases) expression of the siderophore biosynthetic and uptake systems. In many gram-negative and a few gram-positive bacteria, this expression is controlled by the *fur* (*f*erric *u*ptake *r*egulation) gene. Binding of ferrous iron configures the *fur* gene product (the Fur protein) so that it acts as a classic negative repressor of transcription of the siderophore operons [35,54]. The ferrous iron–Fur protein complex binds to a consensus nucleotide sequence (the "Fur box" or "iron box") in the operator region of the operons which Fur controls. In addition to the siderophore operons, a number of other genes are included within the Fur regulon. Many of these genes are unrelated to iron acquisition; some of them encode the iron-regulated virulence determinants of gram-negative bacteria [9,55,56]. These determinants may be included within the Fur regulon because iron restriction may be a strong signal to the microbe that the host environment has been encountered.

Although usually considered a negative repressor, the Fur protein may positively regulate the iron-superoxide dismutase gene in *E. coli* [57]. Other studies suggest a role for Fur as a positive regulator in *P. aeruginosa* [25].

A second iron-responsive, negative repressor has been characterized. It was initially designated as the *dtxR* (*d*iphtheria *t*oxin *r*egulation) gene because it was discovered in *C. diphtheriae* where it regulates both toxin production and the siderophore systems [58,59]. The product of *dtxR* (DtxR protein) shares little amino acid sequence homology with the Fur protein and the DtxR binding site on DNA has little similarity with the Fur box. The cloning of a functional homologue of *dtxR* from *Mycobacterium tuberculosis*, as well as evidence for similar homologues in other mycobacteria, in *Brevibacterium lactofermentum*, and in *Streptomyces pilosus* suggest a family of regulatory proteins. It was proposed that *ideR* (*i*ron-*de*pendent *r*egulation) might be a more appropriate designation for genes encoding the apparently global, iron-responsive, regulatory proteins homologous to DtxR [60].

Another facet of the control of iron genes was revealed by the discovery of other proteins that function, in general, to activate gene expression. In some instances, the Fur protein regulates expression of these proteins. Installing a second level of positive control (by activating transcription) above the negative control exerted by Fur- or IdeR-like regulatory proteins allows for specific gene expression only when certain conditions of environmental iron restriction are prevalent. An obvious system to be controlled in this fashion is expression of the receptor (and uptake system) for an exogenous siderophore. If the specific siderophore is not present, there is no reason to waste effort making its receptor [61,62]. *P. aeruginosa* uses a classic two-component signal transduction system to activate the expression of the genes encoding the receptor for the exogenous siderophore enterobactin [62]. Bacteria often detect and respond to environmental stimuli by using homologues of two proteins, a membrane-associated "sensor" which detects the signal and transfers the message to a cytoplasmic "response regulator" which in turn modifies expression of specific genes. Because the enterobactin receptor gene also is negatively regulated by Fur, at least two conditions must be met for its expression, iron restriction and the presence of enterobactin.

Several members of the genus *Pseudomonas* have coupled components of their Fur regulons to signals other than response proteins. Activation of transcription of some siderophore uptake genes may require an alternative sigma factor (which also may be regulated by Fur) and a protein that signals the presence of a siderophore (sometimes even the endogenous siderophore) in the environment [63–67]. In *E. coli*, iron may be acquired as the ferric dicitrate chelate. A special receptor is induced for uptake of this chelate but only if iron is scarce and if citrate is present externally [68]. Activation of expression of the ferric dicitrate receptor requires transmembrane signaling (of the presence of the chelate) by a protein which does not resemble the classic components of two-step signal transduction systems and a second cytoplasmic protein [68]. Regulation of the ferric dicitrate uptake system is discussed in Chapter 3 of this volume.

There are at least two examples (in *P. aeruginosa* and *Yersinia pestis*) of the regulation of siderophore receptors by proteins resembling the AraC family of transcriptional activators [69,70]. These regulatory proteins may inhibit expression of the siderophore receptor gene in the absence of its siderophore; when the appropriate siderophore is present, the AraC-like proteins activate transcription of the gene. In *Vibrio cholerae*, an iron-regulated (Fur-controlled) virulence gene also is activated by a protein which is homologous to the LysR family of transcriptional activators [71]. Like several other members of this family, the activator gene in *V. cholerae* is located upstream of, and in inverse orientation to, the virulence gene it activates. Other workers have reported evidence suggesting control of iron-regulated genes by an "antisense RNA" transcript located upstream from the gene it regulates in *V. anguillarum* [72].

Although the control systems mentioned above are complex and sophisticated (as well as incompletely understood), they govern expression of genetic information and tell us nothing about regulation of the operation of the transport systems. Unbridled iron uptake might overwhelm the iron-storage (detoxifying) system, rapidly leading to fatal destruction of DNA by iron-generated oxy-radicals. Little is known about the potential control exerted to prevent iron overload and almost nothing is understood about the signals and commands that oversee microbial intracellular iron trafficking. Knowledge of the putative operational regulation of the iron systems is essential for a complete picture of this exciting area.

3.2. The "Iron Biofulcrum"

3.2.1. Host Iron-Withholding Defenses

Vertebrates must comply with the same rules of iron management which are enforced on the microorganisms. In humans, iron is used in almost a closed circuit with most of the metal in hemoglobin, iron-containing enzymes, and ferritin; less than 1% of the total body iron is associated with circulating transferrin (human iron homeostasis is discussed in Chapter 16 of this volume). An invading pathogenic microorganism must interrupt this flow of iron, diverting the metal to the microbe. This struggle establishes a critical biofulcrum with iron as the contested precious nutrient (Table 2). If the balance tilts to favor the host, then the microbe may be eliminated. The major factors that comprise this "iron biofulcrum" have been detailed by Weinberg [3] and in other recent reviews [6–14]. Briefly, vertebrate compliance with the principle that iron must not be exposed inadvertently to oxygen means that the metal also is essentially secluded from invading microorganisms. Moreover, in response to an infection, the already scarce level of circulating iron is lowered further by the hypoferremic response in which the concentration of iron in transferrin is drastically reduced. Lactoferrin may be another major component of the iron withholding defenses. Lactoferrin is present in mucosal secretions and in the granules of polymorphonuclear leukocytes [73]. Unlike transferrin, lactofer-

TABLE 2

The Iron Biofulcrum in Bacterial and Fungal Infectious Diseases

PATHOGEN (iron-acquisition virulence factors[a])	← IRON →	HOST (iron-withholding defenses[b])
Siderophores		Hypoferremic response (ferritin iron storage)
Transferrin receptor		Transferrin
Lactoferrin receptor		Lactoferrin
Heme utilization (cytolysins, proteases, and heme receptor or heme oxygenase may be required)		Heme-proteins
		Heme- and hemoglobin-binding proteins (hemopexin, haptoglobin)
Other?		

[a]Comprehensive reviews of pathogenic iron acquisition appear in Refs. 3–14.
[b]Weinberg [3] discusses additional processes that may be involved in iron-withholding defenses.

rin will bind iron at pH below 4.5 and the antibacterial activity of lactoferrin has been related to nutritional deprivation of iron from microorganisms, which is an essentially bacteriostatic effect [73]. But lactoferrin also has a bactericidal effect [74], and this activity resides in a peptide fragment (designated lactoferricin) which is released by enzymatic digestion of lactoferrin [75,76]. This peptide damages the membranes of bacteria and yeast by a detergent-like action. A complete explanation of the unique defensive activity of lactoferrin is not yet available.

3.2.2. *Pathogen Iron-Acquisition Strategies*

Microorganisms that infect and cause diseases in humans are often considered to be a unique group of microbes. From a microbial standpoint, infecting a human may represent no more than adapting to and inhabiting yet another hostile environment. Pathogenic microorganisms that successfully establish an infection of a vertebrate host use one or more of the previously mentioned iron-acquisition systems to tilt the iron biofulcrum in favor of the microbe (see Tables 1 and 2). An understanding of the microbial strategies for deployment of the iron-capture processes is incomplete but some of the major approaches have been clarified [6–14]. A single microorganism may not possess all of the iron systems, although many microbes have more than one. A popular tactic of extracellular pathogens (those which do not inhabit host cells) is to derive iron from Fe-transferrin or from heme (after host cell disruption), and sometimes these two methods may be complementary, with each one being able to substitute for the other. The host environment contains numerous biological niches which may differ in sources and availability of iron [77]. During the normal course of infection, bacteria may pass through several of these host compartments before reaching their final site of infection. Whether or not iron must be obtained at any or all of the intermediate steps may depend on the infectious organism. For example, in mouse typhoid, the organism (*Salmonella typhimurium*) will pass from the lumen of the intestine through Peyer's patches, into the blood, and finally reach an intracellular niche in certain organs like the liver and spleen. Other microorganisms, like *M. tuberculosis* which usually moves by an airborne route from lung to lung, may have a more direct route to its preferred site. The microbial iron-acquisition system(s) used during an infection must be determined with each organism on a case-by-case basis, remembering that some iron uptake systems may loom critical in certain host niches. The most common experimental approach has been to generate mutant strains with a defect in one or more of the iron-acquisition systems and to determine the virulence of these mutants in a laboratory model infection. This protocol has inherent problems. Often, high doses of the microorganism are used to achieve statistically significant infection rates, but this may circumvent the usual requirement for growth of an initially small inoculum. Moreover, the inocula may bring sufficient iron for several divisions owing to growth in high-iron laboratory media. The route of infection (e.g., by intraperitoneal inoculation) may bypass many of the host compartments normally encountered by the bacteria.

Attempts have been made to overcome some of these experimental difficulties by studying the virulence of *S. typhimurium* strains with either a mutation in the ferrous uptake system (*feoB*) or by mutation of a factor (*tonB*) commonly required for several ferric iron uptake systems or both. Intragastric and intraperitoneal infection routes were used to delimit the effects of the mutations. The *feoB* mutation did not alter the LD_{50} doses, but the *tonB* mutation impaired the organism's ability to move beyond the Peyer's patches or the mesenteric lymph nodes, or both, suggesting that *tonB*-supported iron uptake might be essential in these host compartments. When given intraperitoneally, a *tonB feoB* double-mutant strain infected the liver and spleen at wild-type doses and replicated normally in tissue culture cells, supporting earlier studies [78] showing that production of the siderophore enterobactin was not required for intracellular replication of *S. typhimurium* and suggesting to these authors [77] that an undescribed mechanism of iron acquisition was involved during intracellular growth of *S. typhimurium*.

The iron sources used by intracellular pathogens are not clearly defined. There are several niches within a host cell and the types of iron available at these sites may be different. Some intracellular pathogens (e.g., *M. tuberculosis*, *Legionella pneumophila*) remain within phagosomes, whereas others (e.g., *Listeria monocytogenes*, *Shigella* species) escape into the host cytosol where sources like heme are available [31,79–81]. The microbial iron-acquisition strategies employed in these intracellular niches likely will be different. The intraphagosomal pathogen *L. pneumophila* may not produce siderophores [82]. Intracellular growth of this organism requires iron and monocytes may inhibit growth by limiting the availability of intracellular iron [80]. Moreover, loading the macrophage with iron stimulated intracellular growth of *L. pneumophila* and blocking macrophage iron uptake inhibited bacterial growth [83]. Illustrating the diversity of interactions between host cell and intracellular parasites, other studies [84] indicated that iron loading augmented killing of *Brucella abortus* by macrophages. In *L. pneumophila*, the Fur regulatory protein has been identified and a Fur-regulated gene (*frgA*) has been cloned [85]. Inactivation of *frgA* significantly impaired intracellular growth of *L. pneumophila*. The *frgA* product has homology with two siderophore (aerobactin) synthases of *E. coli*, suggesting to these investigators [85] that *L. pneumophila* may encode a siderophore which is required for intracellular growth. In addition to utilizing iron present within a host cell, intracellular parasites might direct incoming iron to themselves. External transferrin is delivered directly to the *M. tuberculosis* phagosome [79], suggesting that external

Fe-transferrin may be a major iron source for intraphagosomal *M. tuberculosis*. The iron acquisition mechanisms used by intracellular pathogens are both diverse and poorly understood.

3.3. Rationale for an Iron Biofulcrum

Why have vertebrate hosts selected iron as the contested nutrient for a major biofulcrum in infectious diseases? Iron withholding is a logical defense because the host's iron managing systems (that prevent unintended exposure of the metal to oxygen) could be enlisted and enhanced with other special systems to seclude iron from microorganisms. Simple nutritional iron deprivation would arrest or slow bacterial growth. Although this would be beneficial to the host, the discovery that iron restriction yields bacteria with increased sensitivity to killing by H_2O_2 (see Fig. 2) suggests that iron-withholding defenses may attempt to exploit this bacterial handicap. Indispensable components of the bacterial oxidative defense system require iron or heme as cofactors. Iron restriction might depress oxidative defenses, leading to vulnerability to oxy-radicals. The oxy-radicals play a key role in killing of bacteria by phagocytes [86]. Therefore, in addition to the bacteriostatic effect of iron deprivation, vertebrate iron withholding may lower the bacterial oxidative shields placing the bacteria at high risk to killing by oxy-radicals about to be unleashed by phagocytic cells. This bactericidal effect may be the major reason for vertebrate iron-withholding defenses.

Dependence on iron as a cofactor for the oxidative defense system may create a dilemma for the microbe during periods of iron deprivation. If too much of the scarce metal is diverted to metabolic systems, then the bacteria may not be prepared to cope with a sudden surge in oxy-radical concentration. If the intent of host iron-withholding defenses is to make bacteria vulnerable to the oxy-radicals of phagocytic cells, then many successful pathogens may not allow their oxidative defenses to decrease during iron restriction. *M. tuberculosis* may not allow its oxidative shields to fluctuate [87], suggesting that this obligate parasite requires full protection from oxy-radicals at all times. Maintenance by pathogenic microorganisms of a constant level of security against oxidative damage in low iron conditions might be a characteristic distinguishing them from nonpathogenic forms.

4. SUMMARY AND FUTURE PROSPECTS

The many studies of the role of iron in infectious diseases attest to the propositions that, to an invading microorganism, the environment of vertebrate hosts is iron deficient and that hosts attempt to restrict further iron availability during an infection. Although arresting microbial growth by depriving them of iron is beneficial to the host, another (and perhaps major) intent of host iron-withholding defenses might be to increase the vulnerability of the microorganism to killing by the oxy-radicals of phagocytic cells. Indispensable components of the microbial oxidative defenses require iron or heme as cofactors. Maintenance of protection from oxy-radicals during iron restriction might be a distinctive property of highly virulent microorganisms. Microbes have an array of iron-acquisition processes, and during an infection, a contest for iron ensues, with the direction of the tilt of this iron biofulcrum determining, in part, the outcome of the infection. The microbial iron-acquisition systems include those that gather soluble ferrous iron (sometimes first reducing iron to this state), the widespread siderophore-based ferric uptake mechanisms, systems that bind and remove iron from macromolecules like transferrin, and processes that utilize heme as an iron source. There may be additional microbial iron-uptake processes to be discovered; evidence suggests that some of the iron-acquisition mechanisms of intracellular pathogenic bacteria are yet to be identified. Complex regulatory circuits (including negative repressors, positive activators, and environmental sensing systems) by which expression of iron uptake genes is controlled are coming into focus. Still to be defined are the means by which microbes regulate the operation of the transport systems to prevent lethal iron overload. There may be a commonality at the genetic level among the iron-acquisition systems and their regulatory agents; some of the genes encoding biosyntheses of the structurally diverse siderophores may belong to a family of non-ribosomal peptide synthetases and components of various iron transport systems also may be shared.

The field has a bright future and much promise. Ultimately, the knowledge gained should permit design of new therapies to assist the host in tilting the iron biofulcrum to breach the microbes seemingly iron-clad systems for acquisition of iron.

ACKNOWLEDGMENTS

The authors are grateful to Michael D. Lundrigan for his helpful advice.

ABBREVIATIONS

ATP	adenosine 5′-triphosphate
DHB	2,3-dihydroxybenzoic acid
dtxR	diphtheria toxin regulation (gene)
fur	ferric uptake regulation (gene)
ideR	iron dependent regulation (gene)
LD_{50}	lethal dosis of a substance which leads, given in a single dosis, to the death of 50% of a population

REFERENCES

1. J. B. Neilands, *Bacteriol. Rev.*, *21*, 101–111 (1957).
2. C. E. Lankford, *CRC Crit. Rev. Microbiol.*, 273–331 (1973).
3. E. D. Weinberg, in *Mechanisms of Bacterial Virulence*, 2nd ed. (J. A. Roth, C. A. Bolin, K. A. Brogden, F. C. Minion, and M. J. Wannemeuhler, eds.), ASM Press, Washington, D.C., 1995, pp. 79–93.
4. G. Winkelmann, D. van der Helm, and J. B. Neilands (eds.), *Iron Transport in Microbes, Plants and Animals*, VCH Press, Weinheim, Germany, 1987.
5. G. Winkelmann (ed.), *Handbook of Microbial Iron Chelates*, CRC Press, Boca Raton, FL, 1991.
6. B. R. Byers and J. E. L. Arceneaux, in *Biochemistry of Metal Micronutrients in the Rhizosphere* (N. A. Manthey, D. E. Crowley, and L. D. G. Luster, eds.), CRC Press, Boca Raton, FL, 1994, pp. 29–40.
7. K. G. Wooldridge and P. H. Williams, *FEMS Microbial. Rev.*, *12*, 325–348 (1993).
8. S. M. Payne, *Crit. Rev. Microbiol.*, *16*, 81–111 (1988).
9. C. M. Litwin and S. B. Calderwood, *Clin. Microbiol. Rev.*, 6, 137–149 (1993).

10. C. N. Cornelissen and P. F. Sparling, *Mol. Microbiol.*, *14*, 843–850 (1994).
11. C. C. Askwith, D. de Silva, and J. Kaplan, *Mol. Microbiol.*, *20*, 27–34 (1996).
12. D. Trivier and R. J. Courcol, *FEMS Microbiol. Lett.*, *141*, 117–127 (1996).
13. B. C. Lee, *Mol. Microbiol.*, *18*, 383–390 (1995).
14. M. L. Guerinot, *Annu. Rev. Microbiol.*, *48*, 743–772 (1994).
15. F. S. Archibald, *FEMS Microbiol. Lett.*, *19*, 29–32 (1983).
16. P. M. Harrison and P. Arosio, *Biochim. Biophys. Acta*, *1275*, 161–203 (1996).
17. E. S. Henle, Y. Luo, W. Gassmann, and S. Linn, *J. Biol. Chem.*, *217*, 21177–21186 (1996).
18. I. Fridovich, *Science*, *201*, 875–880 (1978).
19. M. E. Martin, B. R. Byers, M. O. J. Olson, M. L. Salin, J. E. L. Arceneaux, and C. Tolbert, *J. Biol. Chem.*, *261*, 9361–9367 (1986).
20. I. M. Hoepelman, W. A. Bezemer, C. M. J. E. Vandenbroucke-Grauls, J. J. M. Marx, and J. Verhoef, *Infect. Immun.*, *58*, 26–31 (1990).
21. J. E. Repine, R. B. Fox, E. M. Berger, and R. N. Harada, *Infect. Immun.*, *32*, 407–410 (1981).
22. D. Touati, M. Jacques, B. Tardat, L. Bouchard, and S. Despied, *J. Bacteriol.*, *177*, 2305–2314 (1995).
23. S. N. Wai, K. Nakayama, K. Umene, T. Moriya, and K. Amako, *Mol. Microbiol.*, *20*, 1127–1134 (1996).
24. O. Dussurget, M. Rodriguez, and I. Smith, *Mol. Microbiol.*, *22*, 535–544 (1996).
25. D. J. Hasset, P. A. Sokol, M. L. Howell, J.-F. Ma, H. T. Schweizer, U. Ochsner, and M. L. Vasil, *J. Bacteriol.*, *178*, 3996–4003 (1996).
26. M. D. Lundrigan, J. E. L. Arceneaux, W. Zhu, and B. R. Byers, *Biometals*, *10*, 215–225 (1997).
27. M. E. Martin, R. C. Strachan, H. Aranha, S. L. Evans, M. L. Salin, B. Welch, J. E. L. Arceneaux, and B. R. Byers, *J. Bacteriol.*, *159*, 745–749 (1984).
28. T. Emery, in *Iron Transport in Microbes, Plants and Animals* (G. Winkelmann, D. van der Helm, and J. B. Neilands, eds.), VCH Press, Weinheim, Germany, 1987, pp. 235–250.

29. M. Kammler, C. Schön, and K. Hantke, *J. Bacteriol.*, *175*, 6212–6219 (1993).
30. S. L. Evans, J.E.L. Arceneaux, B. R. Byers, M. E. Martin, and H. Aranha, *J. Bacteriol.*, *168*, 1096–1099 (1986).
31. R. E. Cowart and B. G. Foster, *J. Infect. Dis.*, *151*, 721–730 (1985).
32. R. Stearman, D. S. Yuan, Y. Yamaguchi-Iwai, R. D. Klausner, and A. Dancis, *Science*, *271*, 1552–1557 (1996).
33. K. P. Shatwell, A. Dancis, A. R. Cross, R. D. Klausner, and A. W. Segal, *J. Biol. Chem.*, *271*, 14270–17244 (1996).
34. J. B. Neilands, *Annu. Rev. Biochem.*, *50*, 715–731 (1981).
35. J. B. Neilands, *J. Biol. Chem.*, *270*, 26723–26726 (1995).
36. M. Fontecave, J. Cove, and J.-L. Pierre, *Biometals*, *7*, 3–8 (1994).
37. R. J. Bergeron, in *Handbook of Microbial Iron Chelates* (G. Winkelmann, ed.), CRC Press, Boca Raton, FL, 1987, pp. 271–307.
38. I. Guilvout, O. Mercereau-Puijalon, S. Bonnefoy, A. P. Pugsley, and E. Carniel, *J. Bacteriol.*, *175*, 5488–5504 (1993).
39. J. D. Fetherston, J. W. Lillard, Jr., and R. D. Perry, *J. Bacteriol.*, *177*, 1824–1833 (1995).
40. J.-H. Zhang, N. B. Quigley, and D. C. Gross, *J. Bacteriol.*, *177*, 4009–4020 (1995).
41. J. R. Telford, J. A. Leary, L. M. G. Tunstad, B. R. Byers, and K. N. Raymond, *J. Am. Chem. Soc.*, *116*, 4499–4500 (1994).
42. S. Barghouthi, R. Young, J. E. L. Arceneaux, and B. R. Byers, *Biol. Metals*, *2*, 155–160 (1989).
43. H. Kleinkauf and H. von Döhren, *Eur. J. Biochem.*, *192*, 1–15 (1990).
44. G. Massad, J. E. L. Arceneaux, and B. R. Byers, *Biometals*, *7*, 227–236 (1994).
45. C. F. Earhart, in *Iron Transport in Microbes, Plants and Animals* (G. Winkelmann, D. van der Helm, and J. B. Neilands, eds.), VCH Press, Weinheim, Germany, 1987, pp. 67–84.
46. S. Barghouthi, S. M. Payne, J. E. L. Arceneaux, and B. R. Byers, *J. Bacteriol.*, *173*, 5121–5128 (1991).
47. M. S. Nahlik, T. J. Brickman, B. A. Ozenberger, and M. A. McIntosh, *J. Bacteriol.*, *169*, 784–790 (1989).

48. T. Franza, C. Enard, F. V. Gijsegem, and D. Expert, *Mol. Microbiol.*, *5*, 1319–1329 (1991).
49. B. M. Rowland, T. H. Grossman, M. S. Osburne, and H. W. Taber, *Gene*, *178*, 119–123 (1996).
50. M. P. Schmitt, *J. Bacteriol.*, *179*, 838–854 (1997).
51. S. M. Payne, *Mol. Microbiol.*, *3*, 1301–1306 (1989).
52. R. D. Perry, *Infect. Immun.*, *61*, 32–39 (1993).
53. K. Tazaki, E. Inoshita, A. Amano, T. Hanioka, H. Tamagawa, and S. Shizukuishi, *FEMS Microbiol. Lett.*, *131*, 161–166 (1995).
54. A. Bagg and J. B. Neilands, *Microbiol. Rev.*, *51*, 509–518 (1987).
55. K. Hantke, *Mol. Gen. Genet.*, *182*, 288–292 (1981).
56. K. Hantke, *Mol. Gen. Genet.*, *197*, 337–341 (1984).
57. E. C. Niederhoffer, L. C. M. Naanjo, K. L. Bradley, and J. L. Fee, *J. Bacteriol.*, *172*, 1930–1938 (1990).
58. J. Boyd, M. N. Oza, and J. R. Murphy, *Proc. Natl. Acad. Sci. USA*, *87*, 5968–5972 (1990).
59. M. P. Schmitt and R. K. Holmes, *Infect. Immun.*, *9*, 1899–1904 (1991).
60. M. P. Schmitt, M. Prdich, L. Doukhan, I. Smith, and R. K. Holmes, *Infect. Immun.*, *63*, 4284–4289 (1995).
61. P. Cornelis, N. Moguilevsky, J. F. Jacques, and P. L. Masson, in *Basic Research and Clinical Aspects of* Pseudomonas aeruginosa (G. Döring, I. A. Holder, and K. Botzenhart, eds.), S. Karger, Basel, 1987, pp. 290–306.
62. C. R. Dean and K. Poole, *Mol. Microbiol.*, *8*, 1095–1103 (1993).
63. M. Koster, W. van Klopenburg, W. Bitter, J. Leong, and P. Weisbeek, *EMBO J.*, *13*, 2805–2813 (1995).
64. H. E. Cunliffe, T. R. Merriman, and I. L. Lamont, *J. Bacteriol.*, *177*, 2744–2750 (1995).
65. H. Miyazaki, H. Kato, T. Nakazawa, and M. Tsuda, *Mol. Gen. Genet.*, *248*, 17–24 (1995).
66. R. Sexton, P. R. Gill, Jr., M. J. Callanan, D. J. O'Sullivan, D. N. Dowling, and F. O'Gara, *Mol. Microbiol.*, *15*, 297–306 (1995).
67. V. Venturi, P. Weisbeek, and M. Koster, *Mol. Microbiol.*, *17*, 603–610 (1995).

68. M. Ochs, S. Veitinger, I. Kim, D. Welz, A. Angerer, and V. Braun, *Mol. Microbiol.*, *15*, 119–132 (1995).
69. D. E. Heinrichs and K. Poole, *J. Bacteriol.*, *178*, 2586–2592 (1996).
70. J. D. Fetherston, S. W. Bearden, and R. D. Perry, *Mol. Microbiol.*, *22*, 315–325 (1996).
71. M. Goldberg, S. A. Boyko, and S. B. Calderwood, *Proc. Natl. Acad. Sci. USA*, *88*, 1125–1129 (1991).
72. P. C. Salinas, L. S. Waldbeser, and J. H. Crosa, *Gene*, *123*, 33–38 (1993).
73. R. A. Finkelstein, C. V. Sciortino, and M. A. McIntosh, *Rev. Infect. Dis.*, *5*, S759–S777 (1983).
74. R. R. Arnold, M. F. Cole, and J. R. McGhee, *Science*, *197*, 263–265 (1977).
75. W. Bellamy, M. Takase, K. Yamauchi, H. Wakabayashi, K. Kawase, and M. Tomita, *Biochim. Biophys. Acta*, *1121*, 130–136 (1992).
76. K. Yamauchi, M. Tomita, T. J. Giehl, and R. T. Ellison, III, *Infect. Immun.*, *61*, 719–728 (1993).
77. R. M. Tsolis, A. J. Bäumler, F. Hefron, and I. Stojiljkovic, *Infect. Immun.*, *64*, 4549–4556 (1996).
78. W. H. Benjamin, C. L. Turnbough, B. S. Posey, and D. Briles, *Infect. Immun.*, *50*, 392–397 (1985).
79. D. L. Clemens and M. A. Horwitz, *J. Exp. Med.*, *184*, 1349–1355 (1996).
80. T. Byrd and M. A. Horwitz, *J. Clin. Invest.*, *91*, 969–976 (1993).
81. K. M. Lawlor, P. A. Daskaleros, R. E. Robinson, and S. M. Payne, *Infect. Immun.*, *55*, 594–599 (1987).
82. M. W. Reeves, L. Pine, J. B. Neilands, and A. Ballows, *J. Bacteriol.*, *154*, 324–329 (1983).
83. S. J. Gebran, C. Newton, Y. Yamamoto, R. Widen, T. W. Klein, and H. Friedman, *Infect. Immun.*, *62*, 564–568 (1994).
84. X. S. Jiang and C. L. Baldwin, *Cell Immunol.*, *148*, 397–407 (1993).
85. E. K. Hickey and N. P. Cianciotto, *Infect. Immun.*, *65*, 133–143 (1997).
86. S. Okada, *Pathol. Int.*, *46*, 311–332 (1996).
87. D. R. Sherman, P. Sabo, T. Arain, A. Resconi, and C. K. Stover, *J. Cell Biochem.*, *19B(Suppl.)*, 89 (1995).

3

Bacterial Iron Transport: Mechanisms, Genetics, and Regulation

Volkmar Braun, Klaus Hantke, and Wolfgang Köster

Department of Microbiology/Membranephysiology,
University of Tübingen, D-72076 Tübingen, Germany

1. INTRODUCTION 68
   1.1. Questions to Be Answered 69
2. TRANSPORT ACROSS THE OUTER MEMBRANE 73
   2.1. The Need for Receptor Proteins 73
   2.2. The Receptor Proteins Form Regulated Channels 74
   2.3. Energy Requirement for Channel Opening 79
   2.4. Transduction of Energy from the Cytoplasmic Membrane into the Outer Membrane 81
3. TRANSPORT ACROSS THE CYTOPLASMIC MEMBRANE 84
   3.1. Ferrous Iron Transport Systems (Feo of *E. coli*) 86
   3.2. Transport of Ferric Siderophores and Heme 87
   3.3. Transport of Ferric Ions 94
4. INTRACELLULAR IRON METABOLISM 96
   4.1. Storage Proteins 96

4.2. Incorporation of Iron into Proteins 99
4.2.1. Incorporation of Iron-Sulfur Centers into Proteins 100

5. GENETIC ORGANIZATION OF IRON TRANSPORT SYSTEMS 101
5.1. Ferric Hydroxamate Transport Systems 101
5.2. Ferric Catecholate Transport Systems 104
5.3. Ferric Citrate Transport System 106
5.4. Ferric Carboxylate Transport Systems 106
5.5. Heme Transport Systems 107
5.6. Transport of Iron Delivered as Transferrin and Lactoferrin 111
5.7. Iron Transport of the Sfu Type 112

6. IRON AS A CENTRAL REGULATORY ELEMENT 113
6.1. Regulation of Gene Transcription by Fe^{2+}-Fur 114
6.2. Regulation of Gene Transcription by Fe^{2+}-DtxR 115
6.3. Regulation of Gene Transcription by Ferric Citrate and Ferric Pseudobactin from the Cell Surface 117
6.4. Indirect Regulation of Gene Transcription by Iron and Fur 121

7. OUTLOOK 127

ACKNOWLEDGMENTS 128

ABBREVIATIONS 129

REFERENCES 131

1. INTRODUCTION

To cover all that is known on bacterial iron transport would in itself fill a book. For this chapter, we have selected those transport systems that provide the most advanced insights. We refer to other systems to illustrate new aspects even if these aspects have not yet been fully studied. We cite recent reviews in which special aspects are covered in depth and in which most of the relevant literature is cited [1–13].

1.1. Questions to Be Answered

The Fe^{2+}/Fe^{3+} pair displays a wide range of redox potentials from −300 to +700 mV, depending on the iron ligands and the protein environment. All organisms, with the exception of certain lactobacilli, take advantage of this unusually wide range of electron transport capacity. However, despite its high abundance in nature, iron is difficult to acquire by most organisms. Under oxic conditions and at the physiological pH of 7, the concentration of free ferric ions in equilibrium with the ferric hydroxide polymer is in the order of 10^{-12} μM. The growth-promoting concentration of iron for microbes is approximately 10^{-1} μM. Bacteria contain about 10^5 iron ions per cell, of which Mössbauer spectroscopy indicates that less than 10% are incorporated into heme and iron sulfur centers of redox proteins [14]. The remaining iron is in a poorly defined state and awaits thorough investigation. Since bacteria frequently reach densities of 10^9 cells per mL, they require 10^{14} iron ions per generation, which sharply contrasts with the calculated 10^3 free ferric ions. Bacteria can only grow when they synthesize and secrete iron-chelating compounds, called siderophores, and take up the ferric siderophores via highly efficient transport systems, or when they take up iron chelated by compounds of their environment (Fig. 1). In the human body, iron is bound to proteins: transferrin in the serum, lactoferrin in secretory fluids and in granules of polymorphonuclear leukocytes, and in intracellular ferritin. The equilibrium concentration of transferrin iron and lactoferrin iron is in the order of 1 ion per liter. The siderophore-producing bacteria must be able to compete for iron with the host iron-binding proteins. Other bacteria are able to take up iron delivered by transferrin and lactoferrin.

Escherichia coli K-12, the most widely used laboratory strain, synthesizes six distinct siderophore-mediated Fe^{3+} transport systems, each of which recognizes a single type of ferric siderophore, and one Fe^{2+} transport system. However, *E. coli* K-12 synthesizes only one siderophore, enterobactin (cyclic trimer of 2,3-dihydroxybenzoylserine, also designated enterochelin). Two of the transport systems take up the biosynthetic precursors of enterobactin, 2,3-dihydroxybenzoic acid and 2,3-dihydroxybenzoylserine [169]. One transport system is specific for ferric citrate, although the concentration of citrate released from *E. coli*

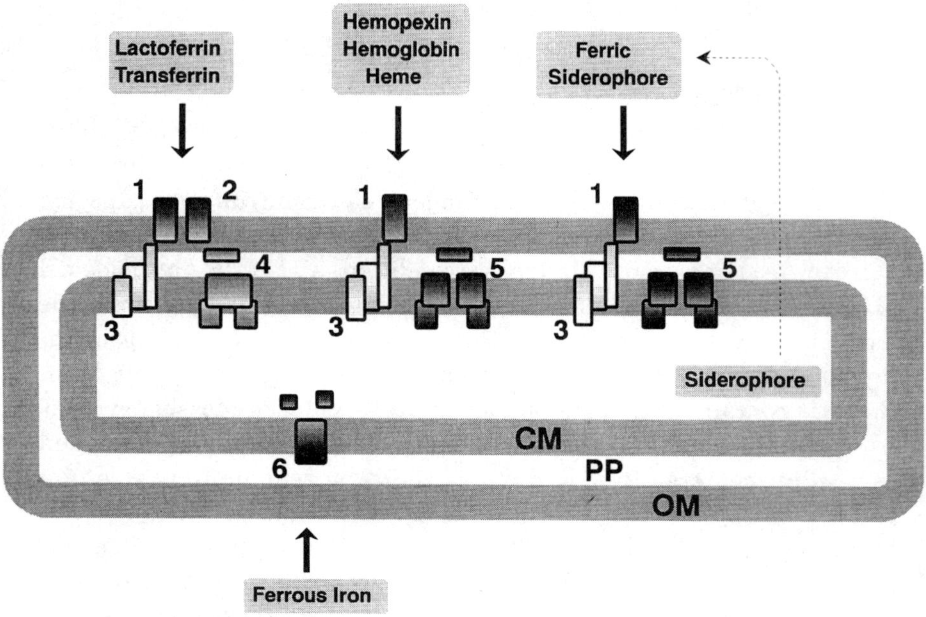

FIG. 1. Iron transport systems of gram-negative bacteria. Transport proteins located in the outer membrane (OM), periplasm (PP), and in the cytoplasmic membrane (CM) are symbolized by rectangles. Siderophores are produced by some species and released as indicated by a dotted line and an arrowhead. Acquisition of ferric iron from lactoferrin, transferrin, heme, hemoglobin, hemopexin, and ferric siderophores or as ferrous iron is indicated by solid arrows. Numbers depict single transport proteins or protein complexes: 1, TonB-dependent outer membrane receptor; 2, outer membrane protein involved in transferrin or lactoferrin utilization; 3, Ton complex composed of TonB, ExbB, ExbD; 4, binding protein-dependent transport (PBT) system for ferric iron across the CM; 5, PBT system for siderophores, or heme across the CM; 6, Feo system for ferrous iron.

K-12 into the growth medium under laboratory conditions (not more than 15 μM) is not sufficient to induce the system (threshold concentration 100 μM citrate). Citrate is an exogenous siderophore that is not formed by *E. coli* for iron uptake; ferrichrome and coprogen, which are of fungal origin, are also used as siderophores by *E. coli*. *E. coli* apparently is able to use heterologous siderophores. It is less likely that the "true" *E. coli* siderophores synthesized by *E. coli* and structurally similar to citrate, ferrichrome, and coprogen still have to be discovered. Pathogenic *E. coli* strains frequently synthesize aerobactin and a ferric aerobactin-specific transport protein, whose genes are encoded on plasmids. The question arises why there are so many iron transport systems in *E. coli* and in other bacteria. Studies under natural conditions have to be performed to determine the environments in which citrate, ferrichrome, and coprogen occur in sufficient concentrations to serve as siderophores and which ferric siderophore is preferentially used when several siderophores are present.

Gram-negative bacteria contain highly specific ferric siderophore receptor proteins in the outer membrane that discriminate between different ferric hydroxamates, whereas a single transport system transports all ferric hydroxamates across the cytoplasmic membrane (see Fig. 1). The same holds true for the ferric catecholates. Why is transport across the outer membrane more specific than transport across the cytoplasmic membrane? Exact complementary fit between the ferric siderophore and the receptor causes high specificity and usually also high affinity, which is advantageous if the outer membrane proteins serve the purpose of extracting the rare ferric siderophores from the culture medium and concentrating them at the bacterial cell surface. Outer membrane receptor proteins discriminate between the structurally similar transferrin and lactoferrin and between transferrin and lactoferrin of human and animal origin. It is unknown how Fe^{3+} is released from transferrin and lactoferrin and through which components iron is transported across the outer membrane.

Translocation of ferric siderophores across the outer membrane consumes energy, which is provided by the electrochemical potential of the cytoplasmic membrane. A protein complex consisting of the proteins TonB, ExbB, and ExbD is involved in the transfer of energy from the cytoplasmic membrane to the outer membrane. There are no experimental data that show how this occurs and for which molecular event energy is needed. There is good evidence that the outer membrane transport proteins form closed channels, but no direct data exists to show that energy is used to open the channels and how this might occur. It is also not clear whether transport of Fe^{3+} siderophores across the outer membrane is a unidirectional process from the receptor proteins into the periplasm or whether the Fe^{3+} siderophores are also released into the medium.

All ferric siderophore transport systems across the cytoplasmic membrane studied to date involve a periplasmic binding protein, one or two hydrophobic proteins in the cytoplasmic membrane, and an adenosine triphosphate (ATP) hydrolyzing protein associated with the inside of the cytoplasmic membrane. This is the most common transport mechanism found in bacteria through which amino acids, sugars, and anions are taken up. Despite many excellent studies, especially on the histidine and maltose transport, the molecular mechanism of substrate translocation through the membrane is not known [15]. For some ferric siderophores, uptake into the cytoplasm has been demonstrated; for others, such as ferric citrate, it is more likely that iron enters the cytoplasm without the chelator. It is assumed that Fe^{3+} is released from the siderophores in the cytoplasm by reduction to Fe^{2+}, since the binding constant for Fe^{2+} is orders of magnitudes below that of Fe^{3+} and ferric siderophore reductase activities have been detected in vitro. However, no mutants have been isolated that transport ferric siderophores but cannot use them for growth, which indicates the presence of several reductases that can replace each other or that such mutations are lethal and cells cannot grow without the reductases. There is also virtually nothing known about the intracellular iron metabolism once iron has been liberated from the siderophores—whether it is immediately incorporated into enzymes or whether it necessarily flows through storage compartments, including storage proteins. It is also not known how the siderophores are secreted after delivery of iron and how newly synthesized siderophores are exported.

Siderophores and the ferric siderophore transport systems are usually formed when they are needed under iron-restricted growth conditions. At a sufficient intracellular iron content, Fe^{2+} associates with the Fur protein in gram-negative bacteria and with the DtxR protein in most gram-positive bacteria and converts Fur (DtxR) to a repressor that binds to a control region, the so-called Fur (DtxR) box, upstream of iron-regulated genes. Binding of Fe^{2+} Fur inhibits transcription of the iron-regulated genes. Under iron-limiting growth conditions, Fe^{2+} dissociates from Fur (DtxR), and the iron-free form of Fur (DtxR) is inactive. However, there are Fur activities which have not been sufficiently studied to understand their role in cell metabolism. Fur controls genes with no apparent relationship to iron metabolism. Fur can induce transcription of genes, and iron metabolism can be controlled by genes other than Fur. Some cases are known in which the ferric siderophores exert a positive effect on the synthesis of the ferric siderophore transport systems, with different and in some cases novel mechanisms. Iron probably controls a central regulatory network that remains to be elucidated. Regulation of the cellular iron content by transport control also has to be considered under the aspect of iron toxicity based on iron-catalyzed oxygen radical formation.

2. TRANSPORT ACROSS THE OUTER MEMBRANE

2.1. The Need for Receptor Proteins

The need for receptor proteins in iron transport may be envisioned in contrast to the open porin channels through which most substrates diffuse through the outer membrane of gram-negative cells. Fe^{3+} is not transported as metal ion but as a complex with siderophores. The Fe^{3+} siderophores with molecular masses of 700–1000 Da are too large to diffuse through the porins, which have an exclusion limit of approximately 600 Da [16]. Cells may not tolerate permanently open channels of the size required for the diffusion of Fe^{3+} siderophores. With such large open channels, cells may not be as well protected against toxic compounds by the permeability barrier of the outer membrane. For example, bile salts in the gut would find access to the plasma membrane of *Escherichia coli* and dissolve it. The concentration of Fe^{3+} siderophores may also be too low for a sufficient iron supply if one considers the strong dilution of the exported siderophores in the environment of the cells. Competition of the siderophores for iron with other strong iron-chelating compounds, such as siderophores produced by bacteria and fungi which cannot be used by a particular strain, or proteins such as transferrin, lactoferrin, and ferritins produced by the infected hosts, further lowers the Fe^{3+} siderophore concentration. Fe^{3+} siderophores that make contact with a bacterial cell are captured by their cognate receptor proteins. By this means, the Fe^{3+} siderophores are extracted from the medium and concentrated at the bacterial cell surface. The low K_d of the receptors, in the order of 0.1 μM, favors binding of the Fe^{3+} siderophores. A function similar to that of the receptor proteins at the cell surface of gram-negative bacteria is displayed by the binding proteins which are anchored by a lipid moiety of the murein-lipoprotein type [17] to the outer face of the cytoplasmic membrane of gram-positive bacteria. In gram-positive bacteria, the Fe^{3+} siderophores diffuse through the multilayered murein and teichoic acids and gain direct access to the Fe^{3+} siderophore transport systems in the cytoplasmic membrane.

2.2. The Receptor Proteins Form Regulated Channels

Evidence for channels in the receptor proteins was obtained by excision of DNA fragments from the structural genes of two receptors which resulted in proteins that carry internal deletions and form permanently open channels. Deletion of residues 322–355 of FhuA (FhuAΔ322–355), the outer membrane receptor of ferrichrome, converted FhuA into an open channel through which ferrichrome enters cells at a rate proportional to the external ferrichrome concentration without showing saturation at higher ferrichrome concentrations. Transport across the cytoplasmic membrane is not rate limiting. FhuAΔ322–355 incorporated into artificial lipid bilayers (black lipid membranes) forms open channels with a single channel conductance more than three times as high as that of *E. coli* porins [18]. Segment 322–355 of FhuA is exposed to the cell surface, as revealed by proteolysis of FhuA by added proteases after insertion of a tetrapeptide and a hexadecapeptide after residue 321, which resulted in ferrichrome-transport-active FhuA derivatives [19]. A C3 viral reporter epitope inserted after residue 321 reacts with C3 antibodies [20], and a monoclonal anti-FhuA antibody reacted with this region in whole cells [21]. FhuA contains four cysteine residues that form disulfide bridges [22]. In living cells, the two cysteines at positions 318 and 329 react weakly with biotin-maleimide (after cleavage of the disulfide bridge with mercaptoethanol), but a newly introduced cysteine at position 336 is highly reactive, which demonstrates its exposure near the cell surface [22]. Deletion of a single amino acid, Asp-348 [23], and insertion of two amino acids after residue 321 inactivate ferrichrome transport activity of FhuA. Cells that synthesize FhuAΔ322–355 become sensitive to sodium dodecylsulfate (SDS) and bacitracin [18], presumably because these noxious agents enter cells via the open FhuA channels.

The permeability-determining region in the loop was narrowed down further to a segment of 21 residues (out of 741 in the entire FhuA protein) extending from residue 335 to residue 355. Cells that synthesize FhuAΔ335–355 display properties very similar to cells that synthesize FhuAΔ322–355 in that uptake of ferrichrome is proportional to the amount of added ferrichrome and cells are sensitive to SDS [24]. Cells containing FhuAΔ335–355 can grow on maltotetraose and maltopentaose as carbon sources in the absence of the LamB protein, which is required for the uptake of these sugars across the outer membrane. FhuAΔ335–355 forms an open, nonspecific channel through which ferrichrome, SDS, and the maltodextrins diffuse into the periplasm. At a low ferrichrome concentration (1 μM), both FhuA deletion derivatives support iron uptake much less than the FhuA wild-type. At higher ferrichrome concentrations (above 7 μM), iron uptake mediated by the FhuA deletion derivatives exceeds uptake via the FhuA wild-type, as one would expect for a diffusion-controlled process. Both FhuA deletion derivatives display virtually no binding of ferrichrome. Apparently, residues 335–355 mainly control the permeability of FhuA. The entire loop from residue 316 to residue 356 is designated as the "gating loop" [18] without implying that the permeability of FhuA is controlled sensu stricto by a gating mechanism (see Sections 2.3 and 2.4).

FhuA not only catalyzes transport of ferrichrome and of the structurally related antibiotic albomycin, but also serves as a receptor for the infection of a number of phages and for killing by toxic proteins (Fig. 2). The large variety of ligands that bind to the multifunctional FhuA protein makes FhuA a particularly attractive receptor to study structure-function relationships. Cells that synthesize FhuAΔ322–355 and FhuAΔ335–355 are resistant to the phages T1, T5, and φ80 (UC-1 was not determined) and to colicin M (microcin 25 was not determined). This shows that the gating loop is also involved in binding of the phages and in binding and uptake of colicin M.

The binding site of the phages in the gating loop were demonstrated by competitive peptide mapping [25]. This method avoids long-range conformational changes at regions outside the gating loop that may be involved in phage binding caused by mutations in the gating loop. Synthetic hexapeptides that covered the entire gating loop were employed, and those identical to sequences of three segments inhibited infection of phages T1, T5, and φ80. Similar results were obtained with

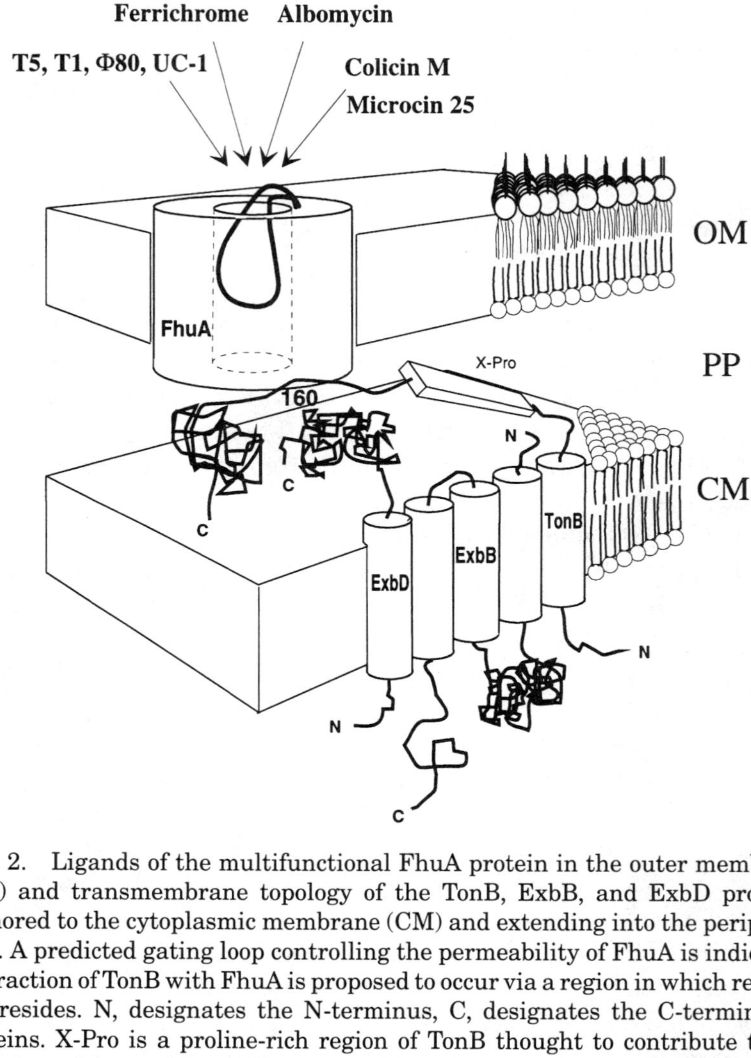

FIG. 2. Ligands of the multifunctional FhuA protein in the outer membrane (OM) and transmembrane topology of the TonB, ExbB, and ExbD proteins anchored to the cytoplasmic membrane (CM) and extending into the periplasm (PP). A predicted gating loop controlling the permeability of FhuA is indicated. Interaction of TonB with FhuA is proposed to occur via a region in which residue 160 resides. N, designates the N-terminus, C, designates the C-terminus of proteins. X-Pro is a proline-rich region of TonB thought to contribute to the spanning of the periplasm. T5, T1, ϕ80, UC-1 designate phages that use FhuA as a receptor to infect *E. coli* cells.

pentapeptides comprising residues 316–320, 332–337, and 348–351 and with selected tetrapeptides [26]. The hexapeptides not only interfered with phage binding but caused inactivation of the phages by triggering DNA release from the phage head [25]. The peptides mimicked binding of phage T5 to the entire isolated FhuA protein that causes DNA release from the phage as shown by electron microscopy. However, the peptides also trigger release of DNA from phage ϕ80, in contrast to the entire FhuA molecule, which does not cause DNA release from ϕ80 [25,27], or ϕ80 inactivation [25]. Presumably, the peptides assume a flexible conformation that enables them to adapt to the surface of the phage tail region with which the phages bind to FhuA. The entire FhuA protein assumes a defined conformation with less flexibility. These results indicate that the gating loop forms, or is part of the active center for all FhuA activities.

The kinetics of DNA release caused by isolated FhuA was determined with a fluorescent probe that intercalates DNA and does not penetrate into the phage head [27]. Increase in fluorescence starts 30 s after addition of FhuA and reaches a plateau after about 20 min. Under the conditions used, 1 μM ferrichrome completely inhibits DNA release from phage T5; this is approximately the inhibitory concentration observed in vivo [28]. The purified FhuA protein, which is active as T5 phage receptor, is monomeric [27], which suggests that FhuA is contained in the outer membrane as a monomer in contrast to the trimeric porins. This conclusion is supported by in vivo experiments that show a lack of negative complementation of active FhuA by a surplus of partially active FhuA molecules [29].

Sequence comparison of the ferrichrome receptors of *E. coli* K-12, *Salmonella paratyphi*, *Salmonella typhimurium*, and *Pantoea agglomerans* (formerly *Erwinia herbicola*) [30] supports the proposal that the second half of the gating loop determines ferrichrome binding and that the entire loop is involved in phage binding. *S. paratyphi* FhuA contains a region that has strong sequence similarity to the gating loop of *E. coli* (Fig. 3). *S. paratyphi* FhuA synthesized in *E. coli* confers the same activity with regard to ferrichrome uptake and phage T1, T5, and ϕ80 binding as the *E. coli* FhuA. *S. typhimurium* and *P. agglomerans* lack the first half of the *E. coli* FhuA gating loop; their FhuA proteins confer no sensitivity to the *E. coli* phages but transport ferrichrome [19].

A very important result was obtained when phage T5 was added to

```
              322          335                    355

FhuA E.c.  GVCSDPANAYSKQCAALAPADKGHYLARKYVVDDEKLQNFSVDTQ

FhuA S.p.  GVCSDPANGYSKQCAALAPADKGHYLARKYVVDDEKLQNFSVDTQ

FhuA S.t.  GIAAD-----------------GHTLNRGTIVDNERLQNFSVDTQ

FhuA P.a.  GF--------------LPATQE---ITRGSAVSEEKLNQLAVDTQ
```

FIG. 3. Comparison of the FhuA sequences of *E. coli* (E.c.), *S. parathyphi* (S.p.), *S. typhimurium* (S.t.) and *P. agglomerans* (P.a.) around the predicted gating loop. The dashes indicate the lack of amino acids as compared with the *E. coli* FhuA sequence.

isolated FhuA integrated in a planar lipid bilayer [31]. Phage T5 opens a channel in FhuA that presumably is required for translocation of the phage DNA across the outer membrane. The phage T5-induced channel has electrophysiological properties similar to the permanently open channel of FhuAΔ322–355 [18]. This is the first direct proof that FhuA forms a closed channel that is opened by interaction with phage T5 and demonstrates the physiological relevance of the data obtained with the FhuA deletion derivatives. FhuA incorporated into lipid vesicles mediates release of ferrichrome from the vesicles on addition of phage T5 [32,33].

FepA serves as a receptor for ferric enterobactin and the colicins B and D. Two very similar transmembrane topology models were derived from the insertion of the dipeptide Leu-Glu along the polypeptide [34], the addition of the foreign M2 epitope Tyr-Lys-Asp-Asp-X-Asp at the Leu-Glu insertion sites [35], use of monoclonal antibodies raised against FepA [36], and by site-directed spin-labeling at natural and introduced cysteine residues [37]. The models propose an amphiphilic protein that spans the outer membrane 29 times, starting with the N-terminus in the periplasm. Leu-Glu and Leu-Glu-M2 insertion derivatives after residue 324 are impaired in ferric enterobactin uptake and are resistant to colicins B and D. Monoclonal antibodies that interact with the regions 290–339 and 382–400 inhibit ferric enterobactin uptake and killing by colicins B and D. Replacement of the Glu residues at positions 280 and 310 by Cys results in FepA derivatives with the Cys residues exposed at the protein surface as suggested by spin-labeling studies. All these data agree with the proposal of a large FepA loop at the cell surface extending from residue 258 to residue 339. Excision of residues 202–340, which comprise the large surface loop, an additional surface loop, and the connecting transmembrane segments, convert FepA into a permanently open channel that no longer binds ferric enterobactin and colicins B and D [38]. The FepA deletion derivative facilitates entry of ferric enterobactin and of FepA-unrelated compounds such as ferrichrome, SDS, and the antibiotics erythromycin, rifampin, and bacitracin into cells. Uptake increases linearly with the concentration of these compounds and is independent of TonB. The membrane topology of the FepA deletion derivative agrees with that of wild-type FepA except that some epitopes of the deletion derivatives react with monoclonal antibodies which in wild-type FepA are presumably sterically shielded by the fragment that is excised in the mutant. The eliminated fragment is part of the binding site of the FepA ligands and occludes an underlying channel. Formation of an open channel by a FepA deletion derivative was demonstrated by its insertion into liposomes where it facilitated diffusion of stachyose (666 Da) and ferrichrome (740 Da) into liposomes. In living cells, the FepA deletion derivative facilitates entry of maltodextrins into cells that are too large to diffuse through the porins. The diameter of the FepA deletion channel is estimated to be at least twice as large as that of porin channels. These data agree with those obtained with the FhuA deletion derivatives. However, the segment that controls the permeability of the channels is more precisely defined in FhuA than in FepA. In FhuA, this segment has been confined to residues 335–356; in FepA, it is confined to residues 202–340. FhuA has the advantage that it is the receptor for many ligands. This allowed precise identification of subdomains in the gating loop for the binding of ferrichrome, albomycin, colicin M, and the various phages.

2.3. Energy Requirement for Channel Opening

The FhuA and FepA deletion derivatives render cells sensitive to SDS and antibiotics. The similar properties of the channel opened by phage T5 and the permanently open FhuAΔ322–355 channel indicates that the properties of FhuAΔ322–355 reflect closely the properties of the naturally open FhuA channel. Resistance of cells that synthesize wild-type FhuA and their failure to grow on maltodextrins when LamB is lacking indicate that FhuA and FepA are usually closed. How are they opened physiologically as opposed to the opening of FhuA by phage T5? The first indication how this may happen and still the best evidence obtained came from studies on the infection of phage T1 [39]. Phage-resistant cells were mutated in two genes, designated *tonA* and *tonB* (*ton* from T one). *tonA* was later renamed *fhuA* [40] to indicate its physiological activity in *f*erric *h*ydroxamate *u*ptake (ferrichrome). Early studies at the times when phage genetics opened a field now known as molecular biology revealed that T1 adsorbs to *tonB* mutants reversibly and that cellular energy is required for irreversible adsorption accompanied by infection. T1 does not bind to *tonA* mutants, suggesting that *tonA* determines the primary binding site and *tonB* defines a later step in infection (cited in [39]). The nature of the energy required for irreversible adsorption of T1 (and ϕ80) was then shown to be the electrochemical potential of the cytoplasmic membrane [39]. Correlation of the energy and TonB requirement suggests that TonB is somehow involved in coupling the energy conserved in the transmembrane potential of the cytoplasmic membrane to irreversible adsorption of T1 to the outer membrane. Later, the *tonA* gene product was shown to be a protein in the outer membrane [41] and that phage T5, which does not require energy and TonB for productive adsorption (infection), is inactivated by the isolated protein, in contrast to phages T1 and ϕ80, which do require TonB in energized cells [42]. Phage T5 inactivation is prevented by colicin M, which suggests a common binding site for phage T5 and colicin M [42]. This has been proven by recent experiments with point mutants and deletions in the gating loop and by competitive peptide mapping [25]. The requirement for energy and the TonB activity was related to the adsorption step by isolating T1 host mutants that infected *tonB* mutants. Infection by these mutants demonstrates that transfer of DNA across the outer membrane does not require TonB activity. The energized cytoplasmic membrane through the activity of TonB induces an infection-competent conformation in FhuA. This conformation is recognized by phage T1 and ϕ80, whereas phage T5 recognizes the unenergized conformation. Evidence for two FhuA conformations was also obtained by competition studies between phage T5 and ferrichrome at FhuA. Ferrichrome inhibits phage T5 adsorption to FhuA of unenergized cells (*tonB* mutants or energy-deprived *tonB*$^+$ cells) much stronger than to FhuA of energized cells [28]. It is likely that all Ton-dependent receptors receive similar conformational changes on energization, which results in translocation of the ferric siderophores across the outer membrane.

Uptake of ferrichrome into the periplasm of *E. coli* was determined with a *fhuB* mutant devoid of ferrichrome transport across the cytoplasmic membrane (see Sec. 5.1). The periplasmic FhuD-binding protein had to be overproduced to measure accumulation of radioactive [$^{55}Fe^{3+}$]ferrichrome in the periplasm [43]. About 8000 ferrichrome molecules bound to FhuD per cell and could be chased with nonradioactive ferrichrome. Under the same conditions, 100 000 ferrichrome molecules are taken up into the cytoplasm of a *fhuB*$^+$ strain, which no longer can be chased, because iron is released from ferrichrome and incorporated into heme and nonheme iron proteins and into the undefined iron pool of the cell.

Energization of ferrichrome transport across the outer membrane was determined by comparing uptake of ferrichrome into cells that synthesize wild-type FhuA and cells that form the deletion mutant FhuAΔ322–355 [26]. Dissipation of the electrochemical potential across the cytoplasmic membrane by 10 μM carbonylcyanide-*m*-chlorophenylhydrazone (CCCP) completely inhibits ferrichrome transport into FhuA cells but does not reduce uptake into FhuAΔ322–355 cells. When ferrichrome diffuses through the FhuAΔ322–355 channel, CCCP does not affect transport across the cytoplasmic membrane. The electrochemical potential is only required for active transport across the outer membrane and not for active transport across the cytoplasmic membrane. In both cell types, transport is inhibited by 100 mM arsenate, which interferes with the ATP-consuming transport across the cytoplasmic membrane (see Sec. 5.1). These experiments demonstrate an entirely different energization mechanism of transport across the outer membrane and transport across the cytoplasmic membrane.

2.4. Transduction of Energy from the Cytoplasmic Membrane into the Outer Membrane

Energization of outer membrane transport by the electrochemical potential of the cytoplasmic membrane raises the question of how the energy is transduced from the cytoplasmic membrane into the outer membrane. All that is actually known is that three proteins are required for activation of outer membrane receptors with regard to ferric siderophore transport, vitamin B_{12} transport, infection by certain phages, and uptake of group B colicins. The proteins are TonB, ExbB, and ExbD. *tonB* mutants are devoid of all these outer membrane receptor activities, whereas *exbB* and *exbD* mutants display residual activities. For this reason, an accessory role was ascribed to the ExbB and ExbD proteins until it was shown that two other proteins, TolQ and TolR, can partially substitute for ExbB and ExbD, respectively [44–46]. Double mutants in *exbB tolQ* and *exbD tolR* are completely inactive, as are *tonB* deletion mutants. *tonB* mutants do not accumulate ferrichrome in the periplasm of a *fhuB* mutant [43], and uptake in cells that synthesize FhuAΔ322–355 does not require TonB, which suggests that TonB is required for active transport across the outer membrane.

The transmembrane topology of TonB (see Fig. 2) further supports its role as an energy transducer between the cytoplasmic and the outer membrane. TonB is anchored by the N-terminal end to the cytoplasmic membrane and extends into the periplasm [47]. Interaction of TonB with outer membrane receptors was demonstrated by mutations close to the N-terminal end of receptors, in the so-called TonB box, which are suppressed by mutations at residue 160 of TonB [48–50]. Furthermore, overproduced FhuA prevents proteolytic degradation of overproduced TonB in cells only when both proteins functionally interact with each other [51].

Evidence for a structural complex between TonB, ExbB, and ExbD is based on the inhibition by ExbB of TonB and ExbD degradation by cellular proteases [47,52]. A fragment of TonB, consisting of only 44 N-terminal residues, is still stabilized by ExbB [53]. Since this fragment consists mainly of the portion of TonB that spans the cytoplasmic membrane, it is likely that interaction between the two proteins occurs within or close to the cytoplasmic membrane. Weak suppression of a TonB derivative lacking Val-17 by an Ala-39 to Glu mutation in the first transmembrane region of ExbB further indicates functional interaction of the two proteins in the cytoplasmic membrane [54]. Furthermore, TonB and ExbB could be chemically cross linked to each other in cells [54]. In vitro evidence for interaction of the three proteins was obtained by binding isolated ExbB which carried a His-tag at the C-terminal end to a nickel-nitrilotriacetate agarose column. His-tag ExbB specifically retains ExbD and TonB on the column [55].

Although the N-terminal transmembrane segment of TonB and the region around residue 160 are important functional regions of TonB, they are certainly not the only regions that determine TonB activity. Mutations in the C-terminal region beyond residue 160 (*E. coli* TonB consists of 239 residues) strongly affect uptake of ferrichrome, sensitivity to certain colicins and phages [53,56], and the activity of certain mutated FhuA proteins with phage T5 [24].

ExbD is, like TonB, anchored by the N-terminal region in the cytoplasmic membrane (see Fig. 2) and extends into the periplasm [57]. Replacement of Asp-25, the only charged amino acid in the transmembrane region, by asparagine results in an inactive ExbD [55]. It is conceivable that Asp-25 interacts with His-20 of TonB, which when replaced by arginine largely inactivates TonB [53]. Substitution of Leu-132 by Asn also completely inactivates ExbD [55]. As in TonB, ExbD contains important functional sites in the cytoplasmic membrane and in the periplasm. In contrast to TonB and ExbD, ExbB spans the cytoplasmic membrane three times (see Fig. 2), and most of the protein is located in the cytoplasm [58]. The ExbB protein may be the key of the device that senses the electrochemical potential of the cytoplasmic membrane by the TonB-ExbB-ExbD protein complex.

In *E. coli* K-12, the *tonB* (28.2 min) and *exbB exbD* (67.8 min) genes are located far apart on the chromosome. In *Pseudomonas putida* WCS358, the three genes are contiguous and read *exbB exbD tonB* [59]; in *Haemophilus influenzae* Rd and in *Neisseria meningitidis*, they read *tonB exbB exbD* [60]. This arrangement may control the coordinate expression of the three genes and supports their functional interdependence.

Transfer of energy from the cytoplasmic membrane into the outer membrane via the TonB-ExbB-ExbD protein complex can be envisioned to occur by an allosteric mechanism. TonB assumes an energized conformation through the action of ExbB and ExbD induced by the electrochemical potential of the cytoplasmic membrane. Interaction of energized TonB with outer membrane receptors induces a conformational change that opens the receptor channels and lowers the affinity of the receptors for the ferric siderophores. The ferric siderophores are released from the receptors and diffuse through the channels into the periplasm where they bind to the binding proteins. It is not known whether diffusion through the open receptor channels takes place vectorially, or whether the ferric siderophores can also be released into the culture medium, from where they have to adsorb again to the receptors to be transported across the outer membrane. Binding of the ferric siderophores to the periplasmic binding proteins imposes a concentration gradient outside to inside that facilitates diffusion from the cell surface into the periplasm. At low concentrations, the rate of energy-dependent ferrichrome and ferric enterobactin transport is much higher than the diffusion rate through the permanently open channels of the FhuA and FepA deletion derivatives. This could mean that the channels open only toward the periplasm and prevent escape of the ferric siderophores into the culture medium. The higher rate could also be caused by binding of the ferric siderophores to the wild-type receptors, which does not occur with the deletion derivatives.

Binding of ferrichrome induces a conformational change in FhuA, as evidenced by a change in the pattern of proteolytic degradation products [61], prevention of trypsin cleavage at Lys-67 of isolated FhuA, and inhibition of certain monoclonal antibody binding to FhuA [62]. It is conceivable that ferrichrome binding triggers energization via the Ton system. However, it is unlikely that this is an essential requirement for FhuA energization, because all the other Ton-dependent FhuA ligands would also have to induce the same or a very similar FhuA conformation capable of interacting with the Ton system.

3. TRANSPORT ACROSS THE CYTOPLASMIC MEMBRANE

Three types of transport systems responsible for the translocation of iron across the cytoplasmic membrane are found in bacteria. They mediate the import of ferrous iron, of ferric iron in the ionic form, and of ferric iron coupled to siderophores or heme. The latter two systems represent members of the periplasmic binding protein-dependent transport (PBT) (also called ABC transporter or traffic adenosine triphosphatase [ATPase]) composed of a periplasmic binding protein, one or two different integral membrane proteins, and one or two different ATPases that face the cytoplasm and supply the systems with energy. The almost identical design suggests a common origin of all PBT systems. During the evolutionary process, a stepwise change in substrate specificity may have led to the development of several families of PBT systems.

The periplasmic solute-binding proteins are the best characterized components of multicomponent PBT systems [for an overview, see Refs. 63 and 64]. They are soluble proteins and can be crystallized. X-ray analysis revealed two major lobes connected by a flexible hinge region [65]. Although they display little similarity in their primary structure, they all seem to assume a very similar three-dimensional structure. Substrate binding induces a conformational change from an open to a closed form and/or stabilizes the closed form. For example, the glucose/galactose-binding protein of *Salmonella typhimurium* has been crystallized with and without the bound substrate [66,67]. In gram-positive bacteria and also in some gram-negative bacteria (e.g., FatB of *Vibrio anguillarum* anguibactin uptake) [68], the binding proteins are anchored as lipoproteins to the cytoplasmic membrane.

The substrate-loaded solute-binding proteins interact with extremely hydrophobic integral membrane proteins and thereby trigger ATP hydrolysis by ATPases bound from the inside of the cytoplasmic membrane to the integral transport proteins. ATP is the energy source for transport across the cytoplasmic membrane. The histidine- and maltose-transport systems have been reconstituted in lipid vesicles, and it has been shown that HisMQ$(P)_2$ [69] and MalFG$(K)_2$ [66,70,71] form functional complexes in the cytoplasmic membrane. In these complexes, each of the hydrophobic components (MalFG, HisMQ) seems to be associated with one ATPase subunit (MalK, HisP). This stoichiometry might be the same for all PBT systems. No data are available from any ABC transporter that could explain the mechanism of substrate translocation at the molecular level.

The members of the two different iron transport families, the siderophore type and the Sfu type, are clearly distinguishable with respect to the primary structure of the binding proteins and the hydrophobic components. However, the mode of uptake of substrates into the cytosol might be similar regardless whether ferric iron is transported in its ionic form or as a ferric siderophore complex. Of the permeases involved in the transport of iron, the *E. coli* FhuCDB system, which accepts a variety of ferric hydroxamates and the structurally related antibiotic albomycin, is the best-studied system to date. Therefore, this system is taken as a model system to examine protein-substrate interaction(s) and membrane assembly of polytopic integral membrane proteins as well as intramolecular and intermolecular protein interactions (see Sec. 3.2).

3.1. Ferrous Iron Transport Systems (Feo of *E. coli*)

Enterobacteriaceae are facultative anaerobic bacteria. Under anoxic conditions, ferrous iron is stable, and for many bacteria, it has been shown that they reduce under anoxic conditions ferric iron to ferrous iron. Fe^{2+} is more soluble than Fe^{3+}, which allows its transport without being complexed by ligands. In *E. coli*, a ferrous iron transport system was found which could not be cloned on a plasmid, most likely due to the high toxicity of Fe^{2+} under oxic conditions. Three genes, *feoABC*, were identified [72,73]. *feoA* and *feoC* encode two small proteins with a molecular mass below 10 kDa with an unknown function. *feoB* encodes a 84-kDa protein which was localized in the cytoplasmic membrane. Analysis of the FeoB sequence showed a typical ATP-binding motif at the N-terminal end of the protein which led to the assumption that ferrous iron uptake is driven by ATP hydrolysis.

Mutants in *feoA* or *feoB* showed a strongly reduced ferrous iron uptake. In addition, *feo* mutants were derepressed for many Fur-regulated genes, which indicates that ferrous iron transport contributes also under oxic conditions to the iron supply of the cells [72]. *feo* genes were also identified in *S. typhimurium* [74] and a gene similar to *feoB* was found in *Methanococcus jannaschii* [75].

Mg^{2+} is taken up by the constitutively expressed CorA protein that is localized in the cytoplasmic membrane and which has been studied in *E. coli* and in *S. typhimurium* [76]. As the name of the gene *corA* (cobalt resistance) indicates, cobalt can be used for the selection of mutations in this gene. This reflects that also Co^{2+}, Mn^{2+}, and Ni^{2+} can be taken up by this transport system, which may be toxic for the cells when high concentrations of these metals are in the medium. Ferrous iron is also a substrate for this transporter [73] which may be the often mentioned "low-affinity" iron-uptake system of *E. coli*.

3.2. Transport of Ferric Siderophores and Heme

Uptake systems of the siderophore type that transport ferric siderophores, heme, or vitamin B_{12} are found in a variety of gram-positive and gram-negative bacteria, and archaebacteria, among them a number of pathogens in plants, animals, and humans. As shown for several ferric siderophores, and most recently for rhizoferrin [77] and heme [78], in these systems, the iron is transported in its liganded form into the cytoplasm. Distinctive signatures are found in all components of the systems belonging to the siderophore family. A model of the components involved in iron(III) hydroxamate uptake across the two *E. coli* membranes is shown in Fig. 4.

The binding proteins of the siderophore family are less conserved

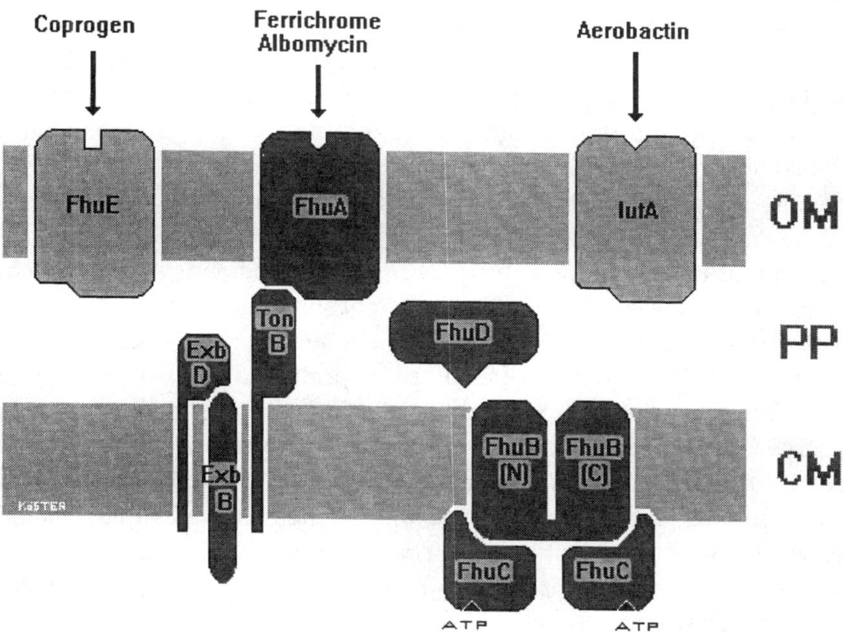

FIG. 4. Schematic model of the arrangement of proteins that mediate transport of ferric hydroxamates in *Escherichia coli*. OM, outer membrane; PP, periplasm; CM, cytoplasmic membrane. For details see text.

at the amino acid sequence level, and there is a remarkable variability in size (25–40 kDa). However, distinctive signatures typical of this family of proteins have been identified (Fig. 5). Sequence data and, in a number of cases, biochemical data provide evidence for the existence of siderophore- or heme-binding proteins in *Bacillus subtilis*, *B. intermedius*, *Listeria monocytogenes*, *E. coli*, *Morganella morganii*, *Yersinia pestis*, *Y. enterocolitica*, *Neisseria meningitidis*, *Vibrio anguillarum*, *Erwinia chrysanthemi*, *Campylobacter jejunii*, *C. coli*, *Mycobacterium tuberculosis*, and *Methanosarcina barkeri*. Moreover, although not shown by transport assays, the genomes of *Haemophilus influenzae*, *Methanococcus jannaschii*, and *Synechocystis* spp. contain genes encoding putative siderophore- or heme-binding proteins. Interestingly, no genes related to iron transport have been identified in the genomes of *Mycoplasma genitalium* [79] and *M. pneumoniae* [80].

The best-characterized siderophore-binding protein is FhuD of the *Escherichia coli* iron(III) hydroxamate uptake system, which accepts a number of structurally different siderophores of the hydroxamate type (e.g., ferrichrome, coprogen, aerobactin, ferrioxamine B, schizokinen, rhodotorulic acid) and the antibiotic albomycin. FhuD is synthesized as a precursor with a typical signal sequence and then is processed and exported into the periplasmic space [81–84]. Binding of iron(III) hydroxamates to the mature FhuD protein was identified by three types of experiments. First, accumulation of [$^{55}Fe^{3+}$]ferrichrome in the periplasm of intact cells was shown in a FhuD-overproducing strain, which, owing to a mutation in the integral membrane protein FhuB, is unable to translocate the substrate into the cytoplasm. In a second assay, radiolabeled FhuD from the periplasm is protected against proteolytic degradation by proteinase K and trypsin. This protection is exclusively observed in the presence of those ferric hydroxamates that support growth of the bacterial cells under iron-limiting conditions [43]. In a third approach, FhuD was isolated and purified as a "His-tag"-labeled derivative on a Ni-chelate resin. The dissociation constants of ferric hydroxamates were estimated from the concentration-dependent decrease in the intrinsic fluorescence intensity of "His-tag"-FhuD, and are 0.4 μM for ferric aerobactin, 1.0 μM for ferrichrome, 0.3 μM for ferric coprogen, and 5.4 μM for the antibiotic albomycin. Ferrichrome A, ferrioxamine B, and ferrioxamine E, which are taken up poorly via the Fhu system, display dissociation constants of 79, 36, and 42 μM, respectively. FhuD is essential for ferric hydroxamate transport. The anti-

Organism	Binding Protein	Size (aa)	Conserved regions			Accession #
Escherichia coli	FhuD	296	35 PNRIVALE 42	90 ELLTEMKPSFMVWS 103	223 SIDRLAAYKDVDVL 236	M12486
Escherichia coli	FecB	300	37 PQRIVVLE 44	93 EAIAALKPDLIIAD 106	220 GLEQLLAVNPAWLL 233	M26397
Escherichia coli	FepB	318	46 PQRIVSTS 53	109 EAVAAQMPDLILIS 122	249 GGENLAAGLNGESL 262	M2837
Yersinia enterocolitica	HemT	279	26 AERIVTIG 33	77 EGILAMKPTMLLVS 90	204 SQEGVIASAPDLLL 217	X77867
Yersinia pestis	HmuT	278	26 AERIVTIG 33	76 EGILAMKPTMLLVS 89	203 SQEGVIASAPDLLL 216	U60647
Morganella morganii	RumB	386	52 VSRIMLAD 59	120 EKALTYQPDLVIFD 133	264 NEEQIIAADPQFYL 277	I64030
Haemophilus influenzae	1472	351	39 INRAVVLQ 46	99 ESLLALKPDVVFVT 112	247 SLENVLEWNPAVIL 260	L46103
Erwinia chrysanthemi	CbrA	305	35 PQRVITLF 42	91 EDIVLLKPDAIVAS 104	223 NKESIPVVDADIFF 236	X87208
Vibrio anguillarum	FatB	322	49 PQRVAALD 56	103 EKIYALKPDLVLMT 116	244 SSEFINQANPDILY 257	M74068
Listeria monocytogenes	no name	309	58 PKRIVALQ 65	106 EKIANLKPDLIIIS 109	228 TKAISSEAIPEYAA 241	D55807
Bacillus subtilis	FhuD	315	58 PKRVVVMA 65	109 EKVIDLNPDLIIVW 122	239 YTSISLEKLPDFAG 252	M87283
Bacillus subtilis	FhuD*	319	56 PKRIVTDF 63	105 EKVMELQPDLIVVM 117	235 MKQLSQEVIPEYAA 248	D45912
Bacillus subtilis	FeuA	317	53 PTDKIAIT 60	109 EKILEMKPDVILAS 122	238 SLEKLSEMNPDHIF 251	L19945
Bacillus subtilis	YclP	317	54 PKKVVVFD 61	109 DKVAELDPDLIIIS 122	239 SYEYISKTNPDYLF 252	D50453
Campylobacter jejuni	no name	329	70 PSKVVILD 77	123 EAINALKPDLIIIS 136	251 NSEFILEKNPDYIF 264	1096254
Campylobacter coli	CeuE	328	69 PSKVVILD 76	122 EAINALKPDLIIIS 135	250 NSEFILEKNPDYLF 263	X88849
Mycobacterium tuberculosis	CY06A4.09cMt	330	56 PKRVVSAG 63	116 DRIAALKPDLIVAI 128	250 PRDHIKAVLDAADV 263	Z86089
Synechococcus spec.	slr1319	315	53 PQRIVVLT 60	109 EKIVALQPDLIIGS 122	235 SFEQIPQMEADALF 248	D90899
Synechococcus spec.	slr1491	330	59 PRRVVALD 66	110 ETLAVLQPDLIIGA 123	244 SRESMTEADGDVIF 257	D90899
Synechococcus spec.	slr1492	311	37 IDRLVTLD 44	92 EAILGTNPDLIVGL 105	230 SREVLEQADGDVIF 243	D90899
Synechococcus spec.	sll1202	347	57 PQRVVVLG 64	122 EGIFKAKPDLILSP 135	252 SLESLPKLNNANSI 265	D90899
Methanococcus jannaschii	0085	373	56 VNRIVCCG 63	121 EAIIQVKPDVIFVT 134	258 TKEQILKWNPDIIF 271	U67466
Methanococcus jannaschii	0872	405	116 VKRIVTDF 123	177 ELIVNLTPDIVFLG 190	339 TKEDLLKINPNFAE 352	U67531
Methanosarcina barkeri	Orf2	319	61 ISSTIGTS 68	119 ETVIDMHPDIFIES 132	259 SIEQVMNWNPEVII 272	X93084
Consensus			P-RIVTLD V V	EKI-ALKPDLII-S L	S-E-I-A-NPD-IF L E LL	

FIG. 5. Regions conserved among binding proteins related to siderophore and heme transport in the bacteria listed. The sizes of the binding proteins are given in numbers of amino acid residues. An alignment of regions showing significant similarity is shown above consensus sequences. Residues matching these motifs are printed in boldface type. Numbers preceding and following the aligned stretches indicate their positions in the polypeptide chain.

biotic albomycin, which is structurally similar to ferrichrome in the iron carrier portion, but completely different in the antibiotic portion, exhibits slower binding kinetics than the ferric siderophores. FhuD contributes to a large extent to the substrate specificity of transport through the cytoplasmic membrane. The dissociation constant for ferrichrome bound to FhuD is similar to the Michaelis constant determined for ferrichrome uptake into wild-type cells (K_D 0.1 μM, K_M 0.7 μM [85], 0.5 μM [16]). Purified "His-tag"-FhuD, introduced by calcium treatment into the periplasm of a chromosomal *fhuD* mutant, reconstitutes ferrichrome transport [86].

According to the PBT mechanism, it is assumed that FhuD delivers the ferric hydroxamates to the FhuB transport protein in the cytoplasmic membrane. Three experimental approaches indicate a physical interaction of FhuD with FhuB: (1) in spheroplasts, FhuD protects radioactively labeled overproduced FhuB from being degraded by trypsin and proteinase K [87], (2) "His-tag"-FhuD added to spheroplasts is chemically cross linked to overproduced [^{35}S]methionine-labeled FhuB, (3) peptides of 10- and 20-amino acid residues identical in sequence to the periplasmic loop PIII of FhuB[N] in Fig. 6 and to the connecting transmembrane sequence between the periplasmic loop PIII and the cytoplasmic loop CIV of FhuB[C] in Fig. 6) bind to FhuD and inhibit ferrichrome transport [88]. In all three experiments, FhuD free of substrate was at least as effective as FhuD loaded with ferrichrome or aerobactin. In the absence of ferrichrome, FhuD may fluctuate between an open and a closed conformation so that the ferrichrome-free closed form interacts with FhuB [87]. The competitive peptide binding experiment suggests that FhuD inserts deeply in a transmembrane channel of FhuB.

FhuB is unique among the integral membrane proteins in that it is about twice the size (70 kDa) of integral membrane proteins of other PBT systems. It consists of two major domains that display significant sequence similarity to each other. Both halves, FhuB(N) and FhuB(C), are essential for transport; no activity is observed with one individually expressed domain. The FhuB halves are connected by a linker region of a few amino acid residues. Connection by a flexible linker may favor interaction between the two halves, since they are synthesized at identical levels, and their spatial proximity may facilitate assembly into a transport-competent complex. However, restoration of ferric hydroxa-

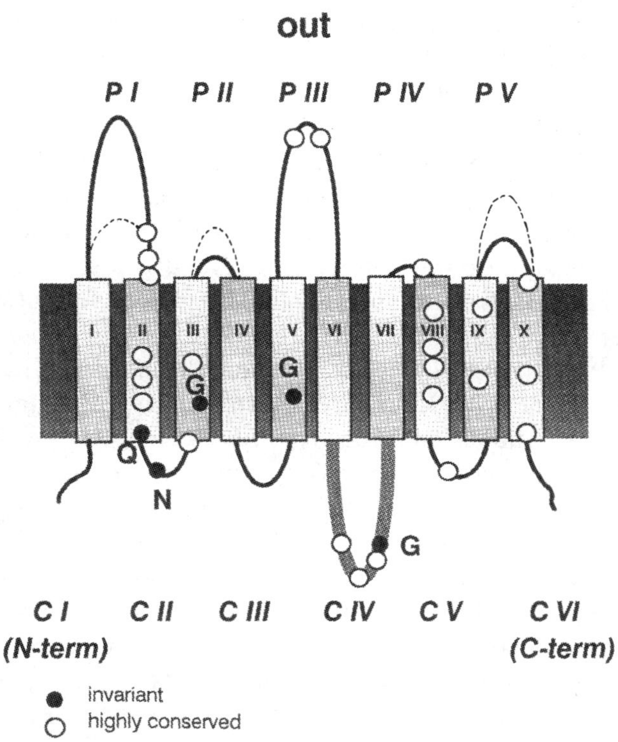

FIG. 6. Two-dimensional topology of an integral cytoplasmic membrane protein representing the hydrophobic components of ABC transporters involved in uptake of heme and ferric siderophores. The model relies on experimental data obtained from FhuB-β-lactamase fusions and the alignment of 35 proteins of the siderophore family. Two of these integral membrane proteins most likely form a unit interacting with each other, with the corresponding binding protein and ATPase, respectively, and with the appropriate substrates. Depending on the transport system, the integral membrane proteins assemble to homodimers (e.g., heme uptake), pseudoheterodimers (ferric hydroxamate transport), or heterodimers (most systems). Putative transmembrane helices (shown as boxes numbered I–X) are connected by loops P I–P V facing the periplasm (out) and loops C I–C VI oriented to the cytoplasm, respectively. In polypeptides of different systems P I, P II, and P V display a significant variability in length (as indicated by broken lines). The conserved region (CR) typical of hydrophobic components from PBT systems is printed in boldface type. The positions of the invariant amino acid residues (e.g., G 226 in FhuB[N] and G 559 in FhuB[C]) are indicated by a black dot. Positions of residues which are to at least 80% conserved are indicated by open circles.

mate uptake in a *fhuB* mutant by two distinct polypeptides demonstrates that FhuB[N] and FhuB[C] properly assemble without being connected covalently [89].

In order to determine the topology (cytosolic, periplasmic, or transmembrane) of important sites within the membrane-embedded polypeptide chain, FhuB′-β-lactamase chimeric proteins have been analyzed. The experimental data suggest that each half of FhuB is able to fold autonomously into the lipid bilayer; this is a prerequisite for the assembly of both halves into a transport-competent formation. Based on sequence alignments, it is concluded that the hydrophobic components from PBT systems involved in the uptake of siderophores, heme, and vitamin B_{12} define a subclass of polytopic integral membrane proteins. Their two-dimensional topology differs from that of equivalent components belonging to other PBT systems in that each polypeptide (and each half of FhuB) consists of 10 membrane-spanning regions. The location of N- and C-termini (and the linker region combining the two FhuB halves) is cytosolic. Based on the FhuB′-β-lactamase fusions and a multiple sequence alignment of 33 proteins, a schematic model of the two-dimensional arrangement of a typical integral membrane protein of the siderophore family has been developed [90] (see Fig. 6). In the functional transport unit, including the binding protein and ATPase molecules, two of these hydrophobic proteins constitute a membrane-embedded complex. In the heme and vitamin B_{12} transporters, two identical molecules may constitute such a complex. In most of the other systems, two different hydrophobic proteins are essential, whereas in the Fhu system, the two distinct domains are synthesized as a single polypeptide chain.

Complementation studies with FhuB dissected in two halves demonstrates that each component has the potential to insert independently into the lipid bilayer where it diffuses and associates stochastically with the complementary subunit. This may reflect the situation that is typical of PBT systems, which are usually composed of two integral membrane proteins. Little is known, however, about the interaction of the integral membrane proteins at the molecular level. The unique FhuB[N]-FhuB[C] arrangement enabled the identification of the mutant proteins FhuB(P60L) and FhuB(G426R), which are most likely impaired in the interaction of the two FhuB halves. In these two FhuB derivatives, the mutated half, still connected to the rest of the protein, can be substituted by the separately synthesized wild-type half (or parts thereof), resulting in a transport-competent complex. It is suggested that the inactive portion of the protein is displaced by the active wild-type portion within the cytoplasmic membrane [91]. Since "domain displacement" did not work with other point mutants in FhuB, there might be a tight interaction of FhuB[N] and FhuB[C] in wild-type FhuB and most of the derivatives. Therefore, it is highly probable that amino acid exchanges at positions 60 and 426 negatively affect the FhuB[N]-FhuB[C] "interdomain" interaction. The equivalent amino acid residues in the other membrane proteins of the ferric siderophore family might play a similar role.

Several areas of striking sequence similarity are found in the primary structures of hydrophobic components of PBT systems belonging to the siderophore family. One of these areas, the so-called "conserved region" (CR), located in the last third of the polypeptide chains, includes an invariant glycine residue at a distance of about 100 amino acids from the C-terminus (see Fig. 6). A region, comparable in sequence and location, has been identified within integral membrane proteins of all PBT systems [92,93]. The conserved Gly is identical with (or corresponds to) a conserved Gly, contained in the "E A A - - - G - - - - - - - - - I - L P" motif defined by Dassa and Hofnung [94]. Point mutations at two corresponding glycine residues located within FhuB[N] (at position 226) and FhuB[C] (at position 595) decrease transport activity. According to the topology model, G226 and G559 are located within cytosolic loops that connect transmembrane segments VI and VII, and VIa and VIIa, respectively. Interestingly, in the siderophore permease family, these loops are followed by four instead of two transmembrane spans, as reported for other PBT integral membrane components studied to date.

Since the analysis of the FhuC amino acid sequence suggested a function of FhuC as an ATP-binding component of the *fhu* system, highly conserved residues in the so-called "Walker A" and "Walker B" consensus motifs were altered. A total loss of function in all these FhuC derivatives suggests that FhuC indeed acts as an ATP-hydrolase, thereby energizing the transport process, most likely by inducing conformational changes in the components of the permease complex that might open a channel [95]. Interaction of FhuC with FhuB was demonstrated by dominant negative effects on the transport of FhuC derivatives with single amino acid replacements in the putative ATP-binding domains [96].

It is likely that FhuC binds to the conserved regions (CRs) of

FhuB. Overproduced FhuB derivatives (mutations at positions G226 or G559) with only one defective CR strain display a negative complementing phenotype in a chromosomal *fhu* wild-type strain, as shown by albomycin-sensitivity, siderophore growth tests, and transport rates. The derivatives are still able to interact with one FhuC unit, and as a result, iron(III) hydroxamate uptake is reduced, because there is not enough FhuC protein available to interact with wild-type FhuB encoded on the chromosome. In contrast, FhuB derivatives with two defective CRs sequester FhuC to a significantly lower extent (or not at all), thus causing only slight reduction in siderophore transport. ATP-binding proteins of all PBT systems share a significant sequence similarity along the entire length of the polypeptide chains. However, ATP-binding components of one system cannot be replaced by ATP-binding proteins of another system, indicating specific interactions between the ATP-binding proteins and the integral transport proteins.

3.3. Transport of Ferric Ions

A ferric iron transport system was first found and characterized in *Serratia marcescens* (Sfu) [97]. In iron-depleted medium, the plasmid-encoded *sfuABC* genes conferred growth to an *E. coli* K-12 strain that does not synthesize enterobactin, the only siderophore formed by wild-type *E. coli* K-12, and to *E. coli* mutants unable to transport ferric enterobactin [98]. The Sfu system also stimulates growth of *tonB* and *exbB* mutants, and no outer membrane protein could be related to the Sfu transport, thereby excluding an active ferric siderophore transport system across the outer membrane. Ferric iron solubilized in phosphate buffer or oxalate buffer is transported equally well, which indicates that there is no requirement for a specific ligand. At very low iron concentrations, ferric citrate stimulates iron uptake via the Sfu system provided that the FecA outer membrane protein and the Ton system are active. Under these conditions, ferric citrate is actively transported across the outer membrane by the ferric citrate transport system, and iron is released from citrate and is further transported by the Sfu system. Similar uptake systems for ferric iron that is mobilized from transferrin or lactoferrin at the cell surface or from citrate are found in *Neisseria gonorrhoeae* (Fbp) [99], *Neisseria meningitidis* (Fbp) [100],

Haemophilus influenzae (Hit) [101], *Yersinia enterocolitica* (Yfu) [102], and *Actinobacillus pleuropneumoniae* (Afu) [103]. Sequence similarities of these systems are found with PBT systems involved in the uptake of molybdenum, molybdopterin, and thiosulfate but not with siderophore import systems.

The binding proteins of the Sfu type, which are synthesized with a cleavable signal sequence, are released into the periplasm. The primary structures of the binding proteins are highly similar, thus indicating that substrate binding and interaction with the hydrophobic membrane protein functions in the same manner in all systems studied to date. Binding of ferric iron to SfuA has been demonstrated. Degradation of the periplasmic SfuA protein by the staphylococcal proteinase V8 is prevented by ferric chloride and much less by ferric citrate [97]. Similarly, binding of ferric iron to FbpA has been shown [104]. In vivo and in vitro studies suggest a reversible binding of ferric iron to FbpA involving two tyrosines, one histidine, and one bicarbonate anion as coordination sites [105]. A single FbpA molecule binds only one iron and, in agreement with its size, resembles one-half of the transferrin molecule. The periplasmic HitA protein of *H. influenzae* binds iron supplied as a complex with 2,2′-bipyridyl [101], which limits the available iron unless a siderophore-mediated ferric iron transport system or a transport system of the Sfu type is functioning. The process of iron mobilization from transferrin and lactoferrin depends on outer membrane receptors and a functional Ton system (see Sec. 2.1.–2.4.), but it is not known how iron is released from the proteins and in which form (free or complexed) it is translocated across the outer membrane. The binding protein of each system is thought to interact with the integral membrane protein, which mediates further translocation into the cytosol.

No detailed study has been performed on the integral membrane proteins of the Sfu type. In analogy to other ABC transporters, the integral membrane proteins (SfuB, FbpB, HitB, YfuB, AfuB) are thought to bind Fe^{3+}, which is delivered by the related periplasmic binding proteins (SfuA, FbpA, HitA, YfuA, AfuA). Further translocation into the cytoplasm is probably achieved at the expense of ATP hydrolysed by the corresponding ATPases (SfuC, FbpC, HitC, YfuC, AfuC). The amino acid sequences suggest two putative CR sites [106] (see Sec. 3.2) that would indicate two ATPase proteins interacting with one integral membrane protein.

4. INTRACELLULAR IRON METABOLISM

4.1. Storage Proteins

In vertebrates, the main iron-storage protein is ferritin, which has been known since 1937 when the first ferritin from horse spleen was purified by Laufberger [107]. For the storage of iron, 24 ferritin subunits assemble into a spherical protein shell that contains up to 4500 Fe(III) atoms. Two types of subunits, H and L, form the normally heterooligomeric ferritin shell. The ratio of H to L chains depends on the organ from which the ferritin is isolated. The H-type ferritins and the L-type ferritins are closely related and contain about 50% identical residues [108]. However, the H chains contain a ferroxidase center, which is lacking in the L chains. It is assumed that ferrous iron enters the shell through specific channels between the subunits and is oxidized at the ferroxidase centers to iron(III), which is deposited inside the shell as ferrihydrite [108]. The structure of this core is crystalline; however, cores with a high phosphate content are amorphous.

It was only in 1979 and later that the soluble cytochromes b_{557} from *Azotobacter* [109] and b_1 from *E. coli* [110] where recognized as bacterioferritins. It was surprising that these ferritins contain nearly 12 hemes per 24 subunits. The sequence of the bacterioferritin Bfr from *E. coli* [111] shows only 22–24% sequence identity to H chains and 16–18% identity to L chains of vertebrates, but its x-ray structure [112] reveals a high structural similarity to the ferritins. Bacterioferritin has been found in several other bacteria: several species of mycobacteria [113,114], *Rhodobacter capsulatus* [115], *Neisseria gonorrhoeae*, *Synechocystis* spp., *Brucella melitensis* [116], *Pseudomonas aeruginosa*, and *Nitrobacter winogradsky*. The list of topical known sequences may be found by a Blast search with a Bfr sequence against a nonredundant protein sequence database. At present, there are 10 complete sequences in the database. Interestingly, there are two Bfr-type sequences found in *Neisseria gonorrhoeae* that are not very similar.

Bacterioferritins are a subfamily of the ferritins defined by their heme content. The heme is found in a cleft between two subunits uniquely coordinated between the sulfur residues of two methionines [117] of α-helix B [112]. When one of these methionines (position 52 of *E. coli* bacterioferritin) is substituted by leucine or histidine, a bacterioferritin without heme is isolated [118]. The iron content of this ferritin is at least four times higher than that of the unchanged bacterioferritin. Heme probably functions in mobilizing the iron in the core for cell metabolism [118]. In addition to bacterioferritin, another type of ferritin that resembles the H chain ferritins is found in *E. coli* and other bacteria. These ferritins contain no heme. Five sequences are in the database: one from *Helicobacter pylorii*, two from *Haemophilus influenzae*, and two from *E. coli*. Unfortunately, the first gene for this ferritin type detected in *E. coli* [119] can be found under three names in the literature: *gene 165*, *rsgA*, and *ftn*. In this chapter, we will use *ftn* for ferritin [108]. A second ferritin-like protein in *E. coli* may be encoded by the open reading frame *yecI* only 1.4 kb upstream from the *ftn* gene. It is not known if this gene is expressed and if the protein has any function in *E. coli*.

Ferrous iron is the substrate of all ferritins. After oxidation of iron(II) to iron(III), it is deposited in the core as ferrihydrite. In vitro iron oxidation is faster in the presence of ferritin than in its absence, which leads to other forms of FeOOH (goethite or lepidocrocite) [108]. Oxidase centers are found in H-chain ferritin [108] and in both types of bacterial ferritins [112,120] but not in L-chain ferritin. The conserved residues in the ferroxidase center, mainly glutamate, histidine, and tyrosine, were changed by site-directed mutagenesis in *E. coli* Bfr and lead to ferritins with lowered oxidase activity [121]. The loss of heme does not influence the oxidation rate [118]. It seems that the di-iron ferroxidase centers have structural similarity to the di-iron centers of ribonucleotide reductase, methane monooxygenase hydroxylase, and hemerythrin [118]. However, not all residues are strictly conserved in the different ferritins, which indicates that the ferroxidase centers have different structures [108]. Inspired by bacterioferritin, researchers then studied the interaction of vertebrate ferritin with heme [122]. A strong interaction of platinum-loaded hematoporphyrin IX with L-chain ferritin was observed [123]. During crystallization of Pt-hematoporphyrin IX with L-chain ferritin, Pt was released and the protoporphyrin IX was embedded in a hydrophobic pocket in a position comparable to that of heme in bacterioferritin. In this case, it is not the sulfur of methionine that interacts with heme but hydrophobic interactions bury the porphyrin between the two subunits [123]. The extraction of the metal from Pt-hematoporphyrin is surprising and not understood. The physiological significance of the interaction of protoporphyrin IX with L-chain ferritin is not known. It is interesting to note that a ferritin similar to bacterioferritin was isolated from the fungus *Absidia spinosa* [124]. A functional specialization may have occurred in that the L chains may be important for mobilization of iron, possibly with protoporphyrin IX, analogous to the speculative function of heme in bacterioferritin, and H-chains, with a functional ferroxidase center, may be important for the detoxification of Fe(II) and the storage process.

In vivo functions of the different bacterial ferritins are not clear. The *E. coli* bacterioferritin is mainly produced during the stationary phase, and the stored iron is rapidly used in the early exponential phase of growth [125]. Upstream of *bfr*, a short open reading frame (64 amino acids) has been identified [111] with a functional Fur box [126], which indicates that this gene may be derepressed under conditions of low iron. The protein, named Bfd [127], is found to be an iron sulfur protein with a [2Fe-2S] center [127]. Iron sulfur proteins are often involved in redox processes, and in this case, one can speculate that under low iron conditions, as might occur during rapid growth, this protein may help in reductive mobilization of ferric iron from Bfr. No regulation of the Ftn ferritin in *E. coli* has been observed [125]. Mutants in *bfr* and/or *ftn* have no phenotype. However, in an *E. coli fur recA* mutant, which is unable to survive aerobically, the introduction of an *ftn* gene on a high-copy number plasmid allows growth [128]. Such a protection is not possible with overproduced Bfr. This indicates that Ftn lowers the high level of toxic ferrous iron in these cells through its ferroxidase activity. Further support for the protective role against oxidative stress by Ftn-type ferritins was found in *Campylobacter jejuni*. The gene *cft* encodes a protein homologous to Ftn, and a mutation in this gene renders the strain more sensitive to killing by H_2O_2 and paraquat [129]. In *E. coli*, the amount of Ftn is rather low, and it may be more important for metabolism and protection against Fe(II) than for storage of iron. Bfr seems to function mainly as an iron storage molecule during the stationary phase and as an iron donor during the early exponential phase.

In *E. coli* and in other bacteria, Dps proteins bind to DNA in stationary phase cells and protect the DNA from oxidative damage [130]. A DpsA protein of *Synechococcus* was isolated and found to be a DNA-binding heme-protein [131]. Sequencing and comparison to other proteins revealed an astonishing similarity of the carboxy-terminal domain of DpsA to Bfr [130]. Dps sequences of *E. coli* and other bacteria did not show this high similarity to Bfr. However, an unusual ferritin was isolated recently from *Listeria innocua*. The purified protein had only a molecular mass of about 240 kDa which contained up to 500 iron atoms [132]. It was composed of one type of subunit of 18 kDa, which suggests that the protein shell is composed of 12 subunits in contrast to the 24 subunits as it is found in all other ferritins. No significant homologies were observed to other ferritins, with the exception of some similarities to mammalian L-chain ferritin. However, a high similarity to Dps proteins was observed [132], which again demonstrates an evolutionary link between ferritins and Dps proteins.

4.2. Incorporation of Iron into Proteins

Internal iron metabolism in *E. coli* and in other bacteria has not been well studied. Even the distribution of the different iron species in the cell is ill defined. However, it seems certain that in *E. coli* the cytochromes contain less than 5%, and iron-sulfur proteins contain less than 10% of the total iron. Also ferritins contain less than 10% under low iron growth conditions or when *E. coli* is supplied with siderophore iron [14]. This indicates that the ferritins do not have the iron-storage function as observed in vertebrates. The bulk of iron seems to be in a badly defined "low molecular weight mobile iron pool," which, depending on the conditions, contains roughly 50% iron(II) and 50% iron(III). One component of this "mobile iron pool" has been isolated and identified as an acidic carbohydrate that is assumed to bind approximately 40% of the cytoplasmic iron as iron(II) [133].

When iron(III) is taken up with a siderophore, it is assumed that the iron is reduced to Fe^{2+} to lower its affinity to the siderophore. The empty siderophore is rapidly extruded from the cell [134]. Indeed, 30 min after uptake of iron(III) with the help of a siderophore, most of the iron is found in the iron(II) pool [14]. Different siderophore reductases have been found in cell extracts and have been purified [135,136]. Further studies indicate that these enzymes are flavin reductases for which in several cases other main functions have been demonstrated in vivo [136]. It has been suggested that the reduced flavins from these enzymes are the electron donors for ferric iron reduction [136]. However, these enzymes are not regulated by iron [135] and in no case has their involvement in the ferri-siderophore reduction been demonstrated in vivo.

4.2.1. Incorporation of Iron-Sulfur Centers into Proteins

More and more proteins with Fe-S clusters are being identified that have very diverse functions in cell metabolism such as electron transport, catalysis, and protein stabilization. In addition, the regulatory function of iron-sulfur proteins as sensors of iron and oxidants has found wide interest during the last years. The iron-responsive element in eukaryotes, which regulates the synthesis of ferritin, transferrin-receptors, and other iron-related proteins, has been identified as aconitase [137,138]. In the presence of iron, aconitase contains an iron-sulfur center and displays enzyme activity. At low iron conditions, the iron-sulfur center is destroyed and the protein binds to mRNAs of iron-related proteins and regulates the translation of these proteins.

The transcription factor Fnr, which activates the expression of many metabolic genes under anoxic conditions in *E. coli*, is an iron sulfur protein and only active with an intact reduced iron-sulfur cluster [139,140]. In contrast to Fnr, the SoxR protein is only active with an oxidized iron-sulfur center. SoxR as a redox sensor regulates the response to superoxide stress [141,142]. The Fe-S cluster of SoxR may be destroyed by superoxide [141]. Another possibility is that univalent oxidation of the Fe-S cluster and a lack of NAD(P)H to reduce it activates SoxR [143].

The first reports on the biosynthesis of Fe-S clusters appeared on the nitrogenase from *Azotobacter vinelandii*. Two components of the nitrogenase, the Fe-protein and the MoFe-protein, are known to be Fe-S proteins. A *nifS* mutant produces these two components of the nitrogenase complex with a lowered activity. The NifS protein uses pyridoxal phosphate as cofactor and liberates alanine and H_2S from cysteine in the presence of dithiothreitol [144]. An enzyme-bound persulfide was identified as an intermediate, and the activation of the apo-Fe-enzyme with iron and this sulfur donor was achieved [145]. By site-directed mutagenesis and other experiments, Cys-325 of NifS was identified as the active site.

Based on the above observations, similar activities were searched for in *E. coli*. Dihydroxyacid dehydratase is a [4Fe-4S] protein involved in the biosynthesis of isoleucine and valine. One iron in the cluster is not ligated to a cysteine residue, which presumably makes this cluster very sensitive to oxygen. The apo form of this protein was reconstituted in vitro with extracts from *E. coli* [146]. Four proteins with reconstituting activity were isolated. The three enzymes, *O*-acetylserine sulfhydrolase A (CysK) and B (CysM) and β-cystathionase (MetC), are able to mobilize sulfur from cysteine for the formation of the [4Fe-4S] center under in vitro conditions. It is assumed that the sulfur-mobilizing activity is only a side reaction and that the in vivo activities are those in intermediary metabolism, as indicated above [146]. The fourth enzyme isolated has a striking similarity to NifS of *A. vinelandii* [147]. The *E. coli* NifS-like protein has pyridoxal phosphate as cofactor and catalyzes the removal of sulfur from cysteine to form S^0 and alanine. In the presence of dithiothreitol, S^0 is reduced to S^{2-}. The NifS-like protein was isolated bound to the acyl-carrier protein via a disulfide link between the active cysteine residues. It is unknown whether the presence of the acyl carrier protein has any physiological significance. The isolated protein is less active than crude fractions, which indicates that additional factors were removed during purification which are necessary for an efficient synthesis of [Fe-S] clusters.

All the factors isolated to date are engaged in the supply of sulfur. Certainly, it is also necessary to present the iron complexed in the right valency and form. Candidates for the iron source may be found in the low molecular weight mobile iron pool as described by Böhnke and Matzanke [133]. Another possibility is that iron(II) is supplied by a protein like Fur, which binds iron(II).

5. GENETIC ORGANIZATION OF IRON TRANSPORT SYSTEMS

5.1. Ferric Hydroxamate Transport Systems

The *fhu* genes located at 4 min on the *Escherichia coli* chromosome are involved in the uptake of iron(III) hydroxamates into the bacterial cell. They constitute an operon, starting with the *fhuA* gene, which encodes the outer membrane receptor for ferrichrome and the antibiotic albo-

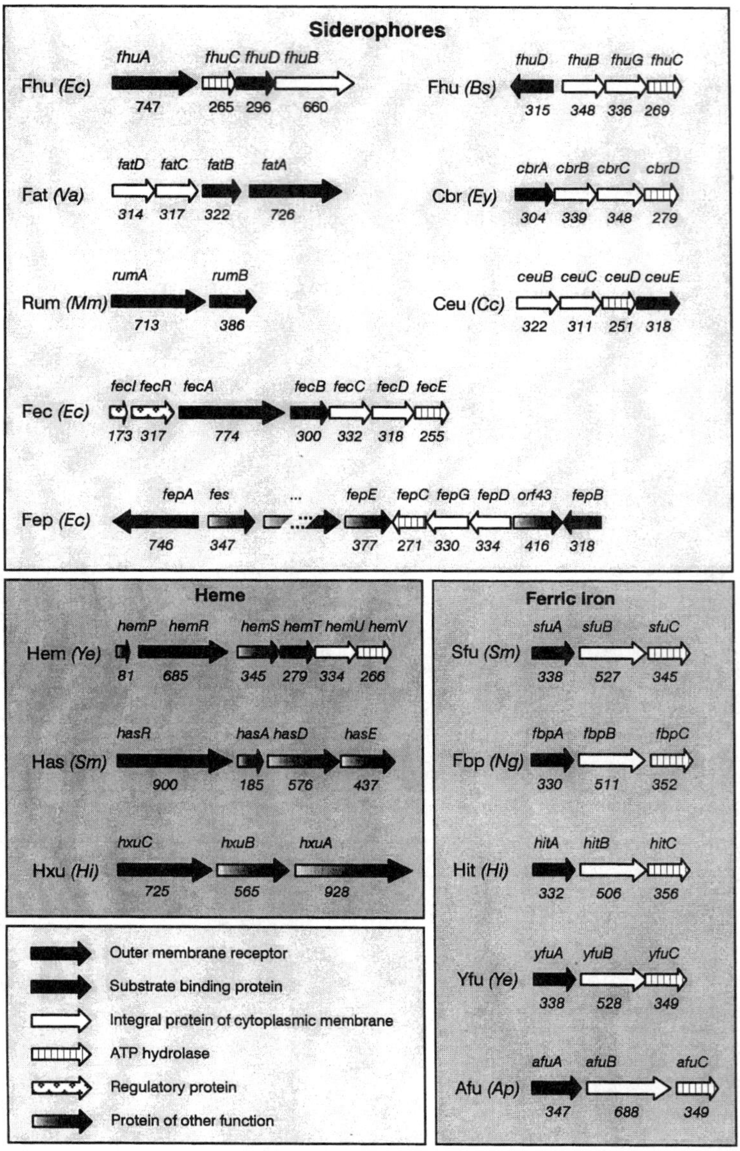

FIG. 7. Gene clusters related to bacterial iron transport. Examples are shown for systems involved in the acquisition of ferric siderophores, heme, and uncomplexed ferric iron. Direction of transcription of the transport genes is indicated by arrows, and the number of amino acids in the proteins is given. *Ec*, *Esche-*

mycin [81,148]. The *fhuC*, *fhuD*, and *fhuB* genes are essential for transport of albomycin, ferrichrome, and a number of other ferric hydroxamates such as coprogen, aerobactin, schizokinen, rhodotorulic acid, and ferrioxamine B across the cytoplasmic membrane (Fig. 7). The FhuC, D, B polypeptides constitute a typical periplasmic binding protein-dependent transport (PBT) system composed of an ATP-hydrolase (FhuC), a solute binding protein (FhuD), and an integral membrane protein (FhuB) [82–84,149]. *fhuC-fhuD* and *fhuD-fhuB* are translationally coupled with overlapping stop/start codons, as are other genes that are part of transport operons (e.g., *hisQ-hisM*, *pstC-pstA*, *btuE-btuD*, *fecB-fecC-fecD*, *fatD-fatC*) [150–154]. Translational coupling may regulate gene expression by influencing the frequency of translation initiation. This may also apply to the ferric hydroxamate uptake system where the stoichiometry of the Fhu components seems to be important for proper functioning. Fur and iron-regulated genes encoding two other outer membrane receptors involved in iron(III) hydroxamates are localized on the *E. coli* chromosome at 25 min (FhuE for coprogen, schizokinen, rhodotorulic acid) and on a number of ColV plasmids (IutA for aerobactin).

In *Bacillus subtilis*, the *fhu* genes for transport of ferrichrome are organized as two divergent transcription units (see Fig. 7) with overlapping promoter regions that are regulated by iron and Fur. *fhuD*, which encodes a binding protein with features typical of lipoproteins, is oriented in the opposite direction to *fhuB*, *fhuG*, and *fhuC*, which most likely form an operon. Unlike FhuB in *E. coli*, transport across the cyt oplasmic membrane is mediated by two integral membrane proteins, FhuB and FhuG, each of which is half the size of the *E. coli* FhuB. *fhuC*,

---

richia coli; *Bs*, *Bacillus subtilis*; *Va*, *Vibrio anguillarum*; *Ey*, *Erwinia chrysanthemi*; *Mm*, *Morganella morganii*; *Cc*, *Campylobacter coli*; *Ye*, *Yersinia enterocolitica*; *Sm*, *Serratia marcescens*; *Hi*, *Haemophilus influenzae*; *Ng*, *Neisseria gonorrhoeae*; *Ap*, *Actinobacillus pleuropneumoniae*; Fhu, ferric hydroxamate uptake; Fat, ferric anguibactin transport; Cbr, chrysobactin regulation; Rum, rhizoferrin uptake of *Morganella*; Ceu, Campylobacter enterobactin uptake; Fec, ferric citrate transport; Fep, ferric enterobactin uptake; Hem, heme uptake; Has, heme aquisition in *Serratia*; Hxu, hemopexin utilization; Sfu, *Serratia* ferric uptake; Fbp, ferric binding protein; Hit, *Haemophilus* iron transport; Yfu, *Yersinia* ferric uptake; Afu, *Actinobacillus* ferric uptake.

the last gene in the operon, encodes an ATPase [155]. A second putative siderophore-binding protein has been detected by sequencing. The high sequence similarity to FhuD suggests a function as a binding protein for an iron(III) hydroxamate (e.g., ferrioxamine B, or schizokinen). A transport system for anguibactin, a siderophore with structural characteristics of hydroxamates and phenolates [156], was described in the marine fish pathogen *Vibrio anguillarum*. Genes involved in ferric anguibactin synthesis and transport and the regulation of synthesis are located on the 65-kb plasmid pJM1 [157,158]. The *fatDCBA* genes, which facilitate uptake of the iron-anguibactin complex, constitute a polycistronic operon (see Fig. 7). The organization of the genes is unusual in that the two genes for the hydrophobic membrane proteins FatD and FatC are located upstream of *fatB* and *fatA* encoding the periplasmic binding protein and the outer membrane receptor [154]. Moreover, no gene for a putative ATPase has been found on pJM1. Presumably, the ATPase-encoding gene is located on the chromosome or the ATPase of another transport system is used.

5.2. Ferric Catecholate Transport Systems

Most *E. coli* strains and many Enterobacteriaceae produce the catecholate-type siderophore enterobactin (enterochelin), which is a cyclic trimer composed of 2,3-dihydroxy-N-benzoyl-serine. Seven transport genes (*fep*), six biosynthesis genes (*ent*), and two open reading frames of unknown function are clustered at 13.1 min of the *E. coli* linkage map (see Fig. 7) [159]. Three bidirectional Fur-regulated promoters have been characterized. Biosynthesis and transport genes are found in one transcriptional unit. FepA is the receptor protein for enterobactin, FepB is the periplasmic binding protein, FepD and G are the membrane components, and FepC is the ATPase that provides the energy (see Sec. 3.2). The *fepE* gene has been claimed to encode a component of the enterobactin transport system. However, multiple sequence alignments reveal a surprisingly high similarity of FepE to Cld proteins. The Cld (*c*hain-*l*ength *d*etermining) proteins control the length of the lipopolysaccharide O-chain [160,161], which raises the question about the role of FepE in ferric enterobactin transport [162].

Enterobactin uptake systems are found not only in enterobacteria, but also in *Pseudomonas aeruginosa* [163], *Aeromonas* [164], and *Campylobacter coli* [165]. Even in *Streptomyces*, production of enterobactin has been demonstrated, which indicates that even some gram-positive organisms can use enterobactin as an iron donor [166]. The outer membrane receptor PfeA for ferric enterobactin and its regulation has been studied in *P. aeruginosa* (see Sec. 6.4). The ABC transporter for enterobactin (see Fig. 7) has been characterized in *C. coli* [165].

The receptor proteins in the outer membrane, Cir and Fiu, were originally defined as iron-regulated colicin receptors in the outer membrane of *E. coli* [167,168]. Their in vivo function is the uptake of ferric catecholates and ferric dihydroxybenzoyl serine [169]. Both receptors recognize an astonishingly broad spectrum of catecholate derivatives, including catecholate-containing cephalosporin derivatives that are transported into the periplasm, the location of their action. Transport of the catecholate cephalosporins as opposed to diffusion of the cephalosporins reduces the minimal inhibitory concentration about 100-fold [170]. Receptors with a similar substrate specificity are also present in *Pseudomonas aeruginosa* strains, which makes these cephalosporin-resistant strains highly sensitive to catecholate-cephalosporins [171].

Vibriobactin, a catecholate siderophore produced by *Vibrio cholerae*, contains a spermidine backbone to which two oxazoline-2,3-dihydroxybenzoate residues and one 2,3-dihydroxybenzoate residue are bound. The receptor ViuA has been cloned and sequenced [172], and upstream an additional gene *viuB* has been identified which seemed to have a Fes-like (*f*erric enterobactin *es*terase) function and is necessary for vibriobactin-dependent iron utilization [173].

Chrysobactin, *N*-[*N*^{2}-(2,3-dihydroxybenzoyl)-D-lysyl]-L-serine, is produced by *Erwinia chrysanthemi*. The outer membrane receptor, Fct, has been cloned and sequenced [174]. In an operon structure, *fct* precedes the genes *cbsCEBA* which have been shown to be necessary for 2,3-dihydroxybenzoate biosynthesis and which seem to be similar to the *entCEBA* gene cluster of *E. coli* [175]. The *cbr* gene locus was originally identified as a regulator of chrysobactin biosynthesis. Sequence analysis identified the genes *cbrABCD* encoding an ABC-transporter (see Fig. 7) for the previously not recognized second siderophore achromobactin produced by *E. chrysanthemi* [176].

5.3. Ferric Citrate Transport System

A ferric citrate transport system has only been characterized in *E. coli* (see Fig. 7). Since a 20-fold surplus of citrate over iron is required to obtain a predominantly low molecular form of ferric citrate, the transport active composition of ferric citrate is not certain but most likely consists of ferric dicitrate [177,178]. Ferric citrate is transported across the outer membrane via the FecA protein at the expense of the electrochemical potential of the cytoplasmic membrane, mediated by the Ton system. Once in the periplasm, ferric citrate binds to the FecB protein. Only iron seems to be further transported across the cytoplasmic membrane, since 10 times more radioactive iron than radioactive citrate is firmly associated with the cells, and iron transport is inducible but citrate uptake is not inducible (see Sec. 6.3). It is not known where iron dissociates from citrate. This could occur at the FecB protein, because FecB also binds iron in the absence of citrate [179]. Iron is transported across the cytoplasmic membrane through the help of the two integral membrane proteins, FecC and FecD and the FecE protein, which is presumably bound to the inside surface of the cytoplasmic membrane. FecE contains the two Walker motifs of nucleotide-binding proteins and could be photoaffinity-labeled with [^{32}P]8-azido-ATP [180]. These results indicate that iron delivered as ferric citrate is transported across the cytoplasmic membrane by a mechanism that is typical for binding protein-dependent transport systems. FecB delivers ferric iron to the FecE,D transport proteins in the cytoplasmic membrane, and subsequent translocation of iron across the cytoplasmic membrane is energized by ATP hydrolysis catalyzed by FecE.

5.4. Ferric Carboxylate Transport Systems

Iron uptake mediated by the fungal siderophore rhizoferrin has been studied in *Morganella morganii*. Two genes essential for the utilization of this polyhydroxycarboxylate siderophore have been identified (see Fig. 7). They encode an outer membrane protein and a periplasmic protein named RumA and RumB, respectively (rhizoferrin uptake into *Morganella*). RumA displays striking sequence similarities to FecA from *E. coli* (Sec. 5.3) and to other TonB-dependent receptors, and RumB shows similarity to binding proteins of the siderophore family. *rumA* and *rumB* have the same transcription polarity and are probably cotranscribed from an iron-regulated promoter upstream of *rumA*. A predicted Fur regulatory sequence upstream of *rumA* was confirmed by the Fur titration assay. Analysis of a 10-kb sequence flanking *rumA* and *rumB* upstream and downstream revealed seven additional open reading frames for which no role in ferric rhizoferrin uptake could be discerned. Thus, *rumA* and *rumB* form an isolated operon, and additional genes required for the uptake of ferric rhizoferrin across the cytoplasmic membrane must map at sites distinct from *rumA* and *rumB* [77].

5.5. Heme Transport Systems

In vertebrates, heme is the most abundant iron carrier because of the high hemoglobin content of erythrocytes. In addition, heme is released by tissue damage and desquamation of epithelial cells, which is important for bacteria colonizing the mucosa. Heme can be used as an iron source by several bacteria [181]. The uptake systems have been studied mainly in pathogenic bacteria and have revealed an astonishing multitude of substrates and ways how bacteria extract heme from different sources. An overview is given in Table 1.

The first heme-uptake system characterized at a molecular level was in *Yersinia enterocolitica*. This system strongly resembles a typical siderophore-uptake system [78,182]. The six genes *hemPRSTUV* seem to constitute an operon (Fig. 7). HemP is a small protein that may have regulatory functions, and HemR is the outer membrane receptor for heme. The sequence of HemR shows similarities to the TonB-dependent siderophore receptors. In addition, it could be shown with a *Y. enterocolitica tonB* mutant that the heme uptake is TonB dependent [78]. HemS most likely helps to degrade heme and to liberate the iron inside the cell. The three genes *hemTUV* code for a binding protein-dependent transport system across the cytoplasmic membrane. HemT is the periplasmic binding protein, HemU is the integral membrane protein, and HemV provides the energy as an ATPase for the transport process [182].

Nearly identical TonB-dependent receptor proteins for heme were cloned and sequenced from *Shigella dysenteriae* serotype 1 (ShuA) and

TABLE 1

Bacterial Heme Transport Systems

TonB-dependent receptor (kDa)	Additional proteins (kDa)	Substrates	Organism	Ref.
HemR (78)	HemP (6), HemS (42), HemTUV (27, 35, 30)	Heme	*Y. enterocolitica*	78,182
HmuR (78)	HmuP, HmuS, HmuTUV	Heme	*Y. pestis*	275
ShuA, ChuA (69)		Heme	*S. dysenteriae*, *E. coli* O157	183,184
HutA (77)	HutB (26)	Heme	*V. cholerae*	185,186
HmbR (89)		Hemoglobin	*N. meningitidis*	187,188
HgbA [= HupA] (109)		Hemoglobin, hemoglobin-haptoglobin	*H. ducreyi*	189,190
HhuA (117)		Hemoglobin-haptoglobin	*H. influenzae*	191
HpuB (88)	HpuA (35)	Hemoglobin-haptoglobin	*N. meningitidis* *N. gonorrhoeae*	192,193
HxuC (78)	HxuA (100) HxuB (60)	Heme, hemopexin	*H. influenzae*	197
HasR[a] (92)	HasA (19) HasD (576) HasE (437) HasF (53)	Heme, hemoglobin	*S. marcescens*	199,200, 202
	HupA (44)	Heme	*B. fragilis*	203

[a]Homologue of TonB-type receptors but not a TonB-dependent receptor.

E. coli O157:H7 (ChuA) [183,184], which are very similar to HemR (68% identity) of *Y. enterocolitica* [184]. It is not known whether genes homologous to *hemSTUV* of *Y. enterocolitica* are present in *S. dysenteriae*. However, ShuA alone was sufficient for porphyrin and iron supply to an *E. coli* K-12 strain, indicating that a heme transport system through the cytoplasmic membrane is present in *E. coli* K-12. This has also been observed with HemR of *Y. enterocolitica* [78].

Another heme utilization system has been characterized in *Vibrio cholerae* [185]. The outer membrane receptor protein, HutA (77 kDa), shows sequence similarity to BtuB (22% identity), the vitamin B_{12} receptor of *E. coli*, and HemR (21% identity) from *Y. enterocolitica*. For heme uptake in *E. coli*, a second gene of *V. cholerae* is necessary; this gene may have a TonB-like function, since the presence or absence of *E. coli* TonB had no influence on heme utilization [186]. *V. cholerae* uses heme as well as hemoglobin as an iron source, whereas *E. coli* with the cloned Hut system is only able to use heme as an iron source. This may indicate that an additional factor responsible for utilization of hemoglobin remains to be identified in *V. cholerae*.

Several bacteria are able to extract heme from hemoglobin. A hemoglobin receptor, HmbR, has been cloned and characterized from *N. meningitidis* [187]. HmbR shows sequence similarities to transferrin (TbpA) and lactoferrin (LbpA) receptors and heme receptors of *Vibrio cholerae* (HutA) and *Y. enterocolitica* (HemR). HmbR shows the highest affinity to human hemoglobin; hemoglobins from animals bind less well which illustrates the high adaptation of *N. meningitidis* to its host [188]. Hemoglobin from lysed erythrocytes is rapidly complexed by haptoglobin in the serum. This prevents heme utilization for many bacteria such as *E. coli*. However, the specialized pathogens *Neisseria* and *Haemophilus* are able to use the hemoglobin-haptoglobin complex as a source for porphyrin and iron. In *H. ducreyi*, a hemoglobin-binding protein HgbA was identified [189] that showed similarities to TonB-dependent siderophore receptors. Later, HgbA was shown to be also important for the utilization of the hemoglobin-haptoglobin complex in *H. ducreyi* [190]. Another receptor, HhuA, with 117 kDa, necessary for hemoglobin-haptoglobin utilization, was identified in nontypeable *H. influenzae* strains. This protein showed remarkable 48% identity to HgbA from *H. ducreyi* [191].

A different system with substrate specificity for hemoglobin and hemoglobin-haptoglobin was identified in *N. meningitidis* [192]. Two genes, *hpuB* encoding a TonB type receptor, and *hpuA* encoding a lipoprotein, were found in an operon structure. This strongly resembles the two-component TbpBA receptor proteins found for transferrin-iron transport (see Sec 5.6) in Neisseriaceae and Pasteurellaceae (which include *Haemophilus* and *Actinobacillus*). The HpuB protein was also found in *N. gonorrhoeae* strains; however, in most strains, only a minority of the population was able to express this protein under selective conditions [193].

Depletion of haptoglobin in the serum leads to oxidation of the free hemoglobin and liberation of heme which is bound to albumin with a K_D of 10 nM and to hemopexin with a $K_D < 1$ pM. Only *Haemophilus* strains can use hemopexin as a heme source. Three proteins, HxuABC (see Fig. 7), are necessary for the efficient uptake of heme from hemopexin. The protein HxuA with 100 kDa is able to bind heme-hemopexin complexes. It was originally found at the cell surface of *H. influenzae* [194]. Later it was shown to be secreted into the medium [195] by the outer membrane protein HxuB. HxuB displays similarities to ShlB, a protein which specifically exports across the outer membrane the ShlA hemolysin (160 kDa) of *S. marcescens* [196]. A third protein, HxuC, localized in the outer membrane, showed similarities to the TonB-dependent receptors, and seems to interact with HxuA to transport heme through the outer membrane [197]. TonB dependence of the transport process has been shown [198], but further details of the mechanism are unknown.

Also in *S. marcescens*, a heme-binding protein, HasA (19 kDa), is secreted into the medium [199] which is much smaller than HxuA (100 kDa). However, in this case, an ABC-transporter composed of HasDEF [200,201], with similarities to the α-hemolysin export system of *E. coli*, secretes HasA into the medium. HasA has been shown to bind heme from hemoglobin. A gene similar to TonB-dependent receptors was identified upstream of *hasA* and cloned in *E. coli*. The expression of this gene, *hasR*, in *E. coli hemA* allows growth on heme or hemoglobin (see Fig. 7) [202]. Additional expression of *hasA* and secretion with the help of HasDE allows *E. coli hemA* to grow on a 100-fold lower hemoglobin concentration [202]. TolC from *E. coli* seems to substitute HasF quite well owing to its high similarity (74% identity) to HasF [201]. The complete system has a high affinity for heme. This indicates that HasA functions as a hemophore, which binds heme in the medium and delivers it to the membrane receptor HasR. Although HasR shows similarities to different TonB-dependent receptors, a TonB box was not identified. In addition, it is stated that heme uptake by HasR is TonB independent in *S. marcescens* and in *E. coli*, although the data are not shown [202]. It remains to be seen how heme transport by HasR across the outer membrane is energized.

In the outer membrane of *Bacteroides fragilis*, the heme-binding protein HupA with 44 kDa was found. The gene *hupA* was cloned in *E. coli* and found to encode a 44-kDa protein which was localized in *E. coli* in the cytoplasmic membrane. The sequence of this protein had no similarity with a TonB-dependent receptor protein, and it is assumed that *B. fragilis* uses a different strategy for the uptake of heme [203].

5.6. Transport of Iron Delivered as Transferrin and Lactoferrin

Bacteria adapted to live on or inside of animals have the problem of being in competition with their host for iron. Between animal cells, the iron is transported by transferrin, a protein that binds up to two ferric iron molecules with high affinity. In addition, in the secretions of vertebrates, the homologous protein lactoferrin is found; lactoferrin avidly binds ferric iron to withhold it from any parasites. This nutritional limitation is an effective defense against many bacteria [204]. However, some specialists use iron bound to transferrin or lactoferrin. The examples most intensely studied are the transferrin iron and lactoferrin iron uptake systems of *Neisseria* and *Haemophilus*.

It was shown already in 1981 that *Neisseria meningitidis* and *Neisseria gonorrhoeae* can use transferrin iron directly [205], and in 1982, the direct use of lactoferrin iron was demonstrated [206]. Receptor proteins of 70 and 100 kDa were identified by binding assays with transferrin labeled with horseradish peroxidase [207,208]. With the same technique, the high specificity of these receptors for transferrin from human and primate origin was demonstrated. This might explain why *N. gonorrhoeae* and *N. meningitidis* are only found in humans and not in other animals. The transferrin-binding protein TbpA (~100 kDa) was cloned from *N. gonorrhoeae*; TbpA shows some characteristic similarities to the TonB-dependent siderophore receptors. A second group of transferrin-binding proteins (TbpB) have molar masses between 68 and 86 kDa. The proteins are encoded in an operon that reads *tbpB tbpA*. TbpB is a lipoprotein [209] with a major part exposed on the surface of the cell. TbpA is necessary for uptake of transferrin, whereas TbpB seems not always to be necessary, but it facilitates the uptake of transferrin [209,210]. How these two proteins remove the iron from the transferrin bound to cells and forward it to FbpA, the ferric iron-binding protein in the periplasm (see Sec. 5.5), is not known. However, one would expect that TonB plays a role as an energy transducer in this ferric iron transport across the outer membrane.

A lactoferrin-binding protein, LbpA (102 kDa), was cloned and sequenced and found to be similar to the TbpA proteins [211,212]. Upstream of *lbpA*, an open reading frame similar to *tbpB* was found, and it is assumed that uptake of iron from lactoferrin is very similar to iron uptake from transferrin [213]. However, several clinical neisserial isolates are missing a functional lactoferrin iron-uptake system [206] even though one would expect that this is necessary for survival on mucosal surfaces where lactoferrin resides.

Haemophilus influenzae can only use human transferrin as an iron source but not any other vertebrate transferrin [214]. A similar high specialization for avian transferrin has been found in *H. paragallinarum*, which cannot grow on transferrins from other animals [215]. The *tbpA* and *tbpB* genes of *H. influenzae* show sequence similarities to the corresponding proteins of the Neisseriaceae [216]. In *H. influenzae*, the first experimental proof for the TonB dependence of the TbpA-type receptors was given [198]. Many different *tbpA* genes have been sequenced in *H. influenzae*. The operon structure *tbpB tbpA* was found, only in a few cases, whereas, in most cases, isolated *tbpA* genes are found.

5.7. Iron Transport of the Sfu Type

Sequencing of a 4.8-kb fragment of the *S. marcescens* chromosome revealed three genes, *sfuA*, *sfuB*, and *sfuC*, which are arranged in an operon [97] (see Fig. 7). The *sfuC* start codon overlaps the *sfuB* stop codon, indicating a regulation of protein expression by translational coupling. Upstream of the *sfu* operon, a putative Fur box overlaps a putative promoter region. The Sfu proteins constitute a typical PBT system of which the SfuA protein is localized in the periplasm and the very hydrophobic SfuB protein is localized in the cytoplasmic membrane; the SfuC protein is degraded too fast to be localized, but it contains the two Walker nucleotide-binding motifs of bacterial traffic ATPases.

Homologous genes organized in operons were sequenced from *Neisseria gonorrhoeae* (*fbp*) [99], *Haemophilus influenzae* (*hit*) [101], *Yersinia enterocolitica* (*yfu*) [102], and *Actinobacillus pleuropneumoniae* (*afu*) [103] (see Fig. 7). These systems transport ferric iron. In *Neisseria meningitidis*, DNase I footprinting experiments suggest the binding of two Fur repressor dimers upstream of *fbp*A at two putative Fur consensus sequences that overlap the potential −35 region [217].

6. IRON AS A CENTRAL REGULATORY ELEMENT

Empirically, it has been long known that the iron supply of bacteria may have a strong influence on their metabolism. One famous example is the observation in the 1930s that low iron stimulates the production of diphtheria toxin by *Corynebacterium diphtheriae* [218]. This type of regulation was later also observed for many other toxins and virulence factors of pathogenic bacteria [219]. Another early observation was that many microorganisms secrete under iron-limiting conditions chelating substances which produce red to brown iron complexes. When their function as Fe^{3+} carriers was recognized, they were termed siderophores [220]. How the cells sense their iron needs became more clear with the help of molecular biology techniques. The first mutation in the iron regulatory gene *fur* (*f*erric iron *u*ptake *r*egulation) was isolated in 1978 in *S. typhimurium* [221]. Later, a similar mutant was isolated in *E. coli* [222] which allowed the cloning and sequencing of the *fur* gene [223].

Important for the study of iron regulation was the development of the reporter gene technology by Casadaban and Cohen [224]. The *lac* gene expression is put under the control of the promoter of the gene of interest. The regulation of this gene under different growth conditions, for instance +/− iron, can be seen directly on MacConkey lactose plates or with X-gal as an indicator of β-galactosidase activity. The *lac* gene under control of a promoter of an iron-uptake gene responds to low iron growth conditions by strong expression—the colonies are red on MacConkey lactose plates and with high iron the colonies are white [168]. Most mutants which produce red colonies on high-iron MacConkey plates contain mutations in the regulatory gene *fur*.

6.1. Regulation of Gene Transcription by Fe^{2+}-Fur

The Fur protein of *E. coli* is a repressor with 148-amino acid residues [223]. The protein contains 12 histidines. The high content of histidines allows the isolation of the Fur protein by chelate affinity chromatography [225,226]. It is assumed that histidines and carboxylate groups from aspartate or glutamate bind the corepressor Fe^{2+} [227], mainly to the C-terminal domain of Fur. ^{1}H-NMR (nuclear magnetic resonance) studies with the isolated Fur protein indicated that metal binding will depend on both metal ion concentration and proton concentration [228]. This may be a link to the observed influence of Fur on the acid tolerance response, which was shown to be defect in Fur mutants of *S. typhimurium* [229]. The function of the four cysteines in the C-terminal domain is not clear. However, they seem not to bind Fe^{2+} directly [230,231].

By footprinting experiments with the promoter region of the aerobactin biosynthesis genes, it was shown that the binding of Fur to DNA was also mediated by other divalent ions, such as Co^{2+} and Mn^{2+} [232]. The protected region on the DNA was a 19-bp long degenerate palindrome. A similar sequence was found in the promoter region of *fur*, and autoregulation was demonstrated [233]. In the promoter regions of many other iron-regulated genes, similar sequences are found which are assumed to be binding sites for Fur, and which are termed Fur box or iron box. An overview on Fur-mediated regulation of gene transcription is given elsewhere [126]. When Fur-binding sites are cloned on a high-copy number plasmid, Fur is removed from its chromosomal binding sites, and derepression of a reporter gene with an iron-regulated promoter is observed.

This Fur titration effect is used to clone new genes with a Fur-box. In addition, it was also observed that genes coding for iron-binding proteins could lead to a derepression of the reporter gene by lowering the regulatory concentration of iron in the cell [126]. Since the Fur boxes from different bacteria are similar to the Fur box of *E. coli*, this system is also used to clone iron-regulated genes from other bacteria.

To study regulation by Fur, mutations in the *fur* gene were isolated which were able to inhibit repression of iron-regulated genes by the chromosomally encoded Fur protein. Such mutants were obtained and indicated for the first time that Fur acted as an oligomer. Defective Fur protein dimerizes with intact Fur protein encoded by the chromosome which leads to an inactive dimer unable to bind to the Fur box on the DNA [226]. Further proof for binding of a Fur oligomer to DNA was obtained by electron microscopy and atomic force microscopy of Fur-DNA complexes. At higher concentrations, Fur polymerizes to a multimer along the DNA [234], which is in accord with DNA footprinting experiments [232].

Many regulatory proteins have a typical helix-turn-helix motif at their DNA-binding site. Structure prediction programs do not identify such a motif for Fur. By construction of Fur-lambda cI hybrid proteins, it was shown that the N-terminal domain of Fur contains the DNA-binding site [235]. This result agrees with the predicted structural similarities of the N-terminal domain of Fur to the LexA helix-turn-helix DNA-binding motif [236].

Fur-like proteins have been found in many other bacteria; a homology search identified about 22 homologues at the time of this writing, from which a Fur-like function was demonstrated for 17 proteins. This shows that this principle of iron regulation is widely distributed among bacteria. In gram-positive bacteria with a low GC nucleotide content like *Bacillus subtilis*, two Fur-like genes were found in the genome sequencing project. From the promoter structure of several iron-related genes, it was postulated that a Fur-like regulator should also exist in *B. subtilis* [237]. It remains to be shown if one of these two genes has a Fur activity. The only exception seems to be gram-positive bacteria with a high GC content like *Corynebacterium* and *Streptomyces*. At present, no Fur-like proteins have been identified in these genera. DtxR-like proteins, another type of iron regulatory protein, fulfill the function of Fur.

6.2. Regulation of Gene Transcription by Fe^{2+}-DtxR

Knowledge on the iron-dependent regulation of diphtheria toxin production by *C. diphtheriae* stimulated many attempts to isolate and characterize the regulatory protein from this gram-positive organism. The regulatory *dtxR* gene was cloned and sequenced from *C. diphtheriae*, and the derived protein sequence turned out to be quite different from Fur [238]. A *dtxR* point mutation shows a phenotype similar to *fur* mutants in that it displays an iron-insensitive production of toxin and siderophore [239]. By footprinting experiments, 27 bp were identified as the DtxR-binding site in the promoter region of the *tox* gene [240]. Further studies revealed a 19-bp palindrome as a minimal DtxR-binding site [241].

An x-ray structure of the protein defined the second and third helix at the N-terminal end of the protein as the helix-turn-helix motif of the DNA-binding site [242]. Reevaluation of the metal-binding site defined His-106, Glu-105, Cys-102, and Met-10 as directly interacting with the metal, whereas Asp-6, Glu-9, Arg-13, and Glu-17 contributed indirectly to the stability of metal binding [243]. This is interesting, since Met-10 and most of the indirectly interacting amino acids are on the first helix at the N-terminal end very close to the DNA-binding site. The bridge formed by the metal is decisive for the proper conformation for DNA binding [243]. It is interesting to note that, in Fur, the second and third helices in the N-terminal domain are also the postulated DNA-binding site [236]. In addition, it was suggested that Fe^{2+} binds to His-32 and His-31 (which are in the first postulated helix of the N-terminal) and to His-131 in the C-terminal domain and to carboxylates of Asp and Glu [230]. This model derived from NMR studies is similar to the DtxR model. However, until now, crystallization of Fur has failed, possibly because of the oxygen sensitivity of the cysteines and the low solubility of the protein.

DtxR-like proteins were sequenced from other gram-positive organisms with a high GC base content. In *Streptomyces* spp., it was shown that a DtxR-like protein regulates the biosynthesis of the siderophore ferrioxamine B [244]. DtxR-like proteins were sequenced from *C. glutamicum*, *Mycobacterium smegmatis* (called IdeR; Table 2) and other mycobacteria. An *ideR* mutant of *M. smegmatis* constructed by allelic replacement showed a derepressed biosynthesis of siderophores, and a high peroxide sensitivity which was related to lower levels of Mn superoxide dismutase and catalase [245]. Like Fur, IdeR seems to be a global regulator in coupling iron metabolism to oxidative stress.

6.3. Regulation of Gene Transcription by Ferric Citrate and Ferric Pseudobactin from the Cell Surface

Ferric citrate is the only ferric siderophore of *E. coli* K-12 that induces transcription of the ferric siderophore transport genes. However, it is not intracellular ferric citrate but extracellular ferric citrate that serves as an inducer. This is not unexpected, since, as outlined in Sec. 5.3, iron, but not ferric citrate, is taken up into the cytoplasm. Mutants in *fec-BCDE* impaired in the transport of iron across the cytoplasmic membrane are fully inducible, but mutants in *fecA*, *tonB*, *exbB*, or *exbD* (the latter in combination with *tolQ* or *tolR* mutations), which are devoid of transport across the outer membrane, are not inducible [98,246]. The obvious conclusion that entry of ferric citrate into the periplasm is required for induction was ruled out by supplying to a *fecA* null mutant growth-promoting concentrations of ferric citrate that enter the periplasm through the porins. Ferric dicitrate, the most probable molecular entity for induction and transport, has a molar mass of 434 Da, which allows diffusion through the porins. No induction of *fec* transport genes was observed [247], and the transport genes encoding cytoplasmic membrane transport activities had to be constitutively overexpressed from a multicopy plasmid to provide sufficient amounts of transport proteins. A direct involvement of FecA in induction was shown with *fecA* missense mutants, which induced *fec* transcription but did not transport ferric citrate and expressed *fec* genes constitutively in the absence of ferric citrate [247].

Compared with other *E. coli* ferric siderophore receptors, the FecA protein contains an N-terminal extension. When this extension is removed by genetic means (residues 47–101, numbering includes the signal peptide) induction is abolished but FecA fully retains transport activity. Synthesis of the N-terminal FecA (1–100) and FecA (1–127) fragments inhibit induction but not transport [246]. This proves that the N-terminus of mature FecA (741 amino acids, signal peptide comprises 33 residues) extending from residue 1 to approximately 120 [246] participates in signal transduction. The N-terminal portion of FecA is localized in the periplasm [246] and most likely interacts with the FecR regulatory protein of which residues 101–317 are contained in the periplasm, residues 86–100 span the cytoplasmic membrane, and the N-terminal portion is in the cytoplasm [248]. The transmembrane topology of FecR suggests that it transmits the signal, elicited by ferric citrate on FecA, across the cytoplasmic membrane into the cytoplasm. Cytoplasmic N-terminal fragments of FecR, the smallest of which consists of 59 residues, induce *fec* transport gene transcription constitutively [248], showing that the cytoplasmic segment participates in transcription regulation and the periplasmic region responds to the signal. Signal transduction is inhibited by carbonylcyanide-*m*-chlorophenylhydrazone (CCCP), which dissipates the electrochemical potential of the cytoplasmic membrane [246]. The energized cytoplasmic membrane is required for signal transduction across the outer membrane and not for signal transduction across the cytoplasmic membrane, since the *fecA4* mutant that transcribes the *fec* transport genes in the absence of ferric citrate and TonB is unaffected by CCCP [246]. The TonB-ExbB-ExbD complex transmits the energy from the cytoplasmic membrane to the outer membrane for signal transduction, as for ferric citrate transport across the outer membrane. The molecular mechanisms underlying both metabolic processes must not be the same, since the *fecA4* mutant signals but does not transport.

FecR does not directly act on the promoter upstream of the *fecA* gene that regulates transcription of *fecA* and of the *fecBCDE* genes downstream of *fecA* (see Fig. 7). The *fecI* gene upstream of *fecR* encodes a sigma factor with a distant sequence similarity to sigma 70 factors. The FecI type of sigma factors are widespread in different genera and are designated as ECF factors derived from extracytoplasmic functions, because they all seem to be involved in extracytoplasmic metabolic activities [249]. Purified FecI mediates specific binding of the RNA polymerase core enzyme to the promoter region upstream of *fecA*, as revealed by DNA mobility band shift experiments, and promotes *fecA* transcription in vitro [250]. *fecA* promoter DNA band shift caused by cell lysates requires synthesis of FecA, FecI, and FecR and growth of cells with ferric citrate [250].

fecIR and *fecABCDE* form separate transcripts [251]. Transcription of *fecIR* is repressed by iron and the Fur protein but is uninfluenced by ferric citrate, whereas *fecABCDE* transcription is regulated by iron and Fur and by ferric citrate via FecI and FecR. FecI FecR regulate *fec* transport genes transcription, but they display no autoregulation [252].

The iron transport genes are regulated by the internal iron concentration and by external ferric citrate. This is a very economic way of using ferric citrate as an iron source. When it is not needed or not present, the transport system can be almost totally shut off by cytoplasmic iron. If the iron concentration in the cytoplasm falls below a certain limit, the FecIR genes are transcribed. To turn on the ferric citrate transport system, the carrier has to be in the culture medium. Trace amounts of FecA, by far the most highly synthesized transport protein, are sufficient for initiation of *fec* transport gene transcription, whereas the FecBCDE transport proteins are completely dispensable. The FecA and FecR proteins exhibit not only regulatory functions, they display vectorial activities in that they transmit information through three cell compartments (Fig. 8). Binding of ferric citrate to FecA induces a signal that is transferred from the cell surface into the periplasm and across the cytoplasmic membrane into the cytoplasm. It is likely that the information flux involves coupled conformational changes of FecA and FecR. The conceptual question that remains to be answered is how FecI is activated by FecR. No phosphorylation or proteolytic processing of FecI has been found.

Pseudomonas putida WCS358 transports ferric pseudobactin BN8. An outer membrane protein, PupB, was identified as part of the transport system. Synthesis of PupB is induced by ferric pseudobactins BN7 and BN8. Synthesis of a chimeric receptor protein composed of an N-terminal fragment of PupB and a C-terminal fragment of PupA, which is involved in pseudobactin 358 transport, was induced by pseudobactin 358 but not by pseudobactin BN8. Substrate specificity is conferred by the PupA fragment and induction by PupB, since PupA exhibits no inducing activity. This result indicates a regulatory function of the N-terminus of PupB in PupB and PupBA synthesis [253]. Two additional regulatory genes are involved in regulation of *pupB* transcription: PupI, which is homologous to FecI, and PupR, homologous to FecR (see Table 2). The following mechanism of *pupB* transcription initiation has been suggested. When ferric pseudobactin BN8 is in the medium, it is transported in a TonB-dependent manner by the PupB protein across the outer membrane. During transport, a signal, probably the conformational change of the receptor, is transduced to the regulatory protein, presumably PupR, which no longer represses PupI activity, which in turn activates *pupB* gene transcription.

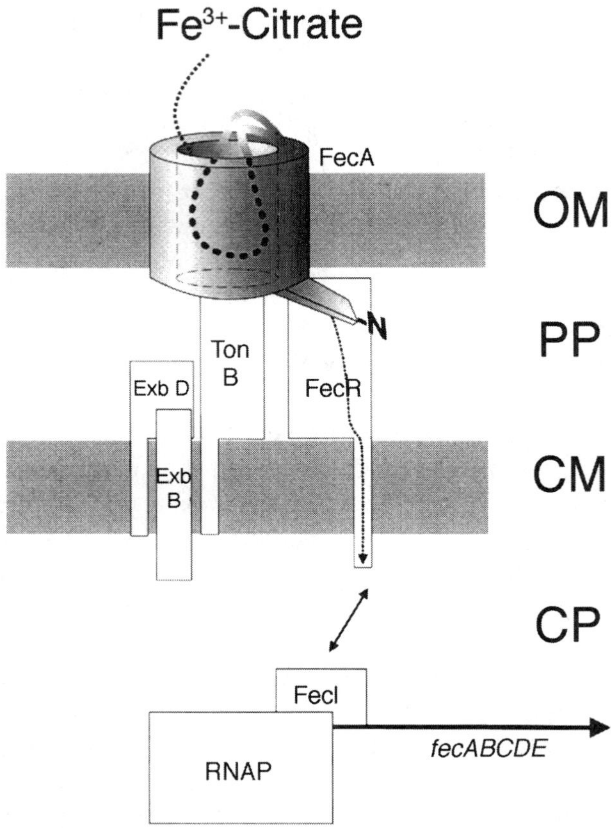

FIG. 8. Description of the transmembrane signaling during transcription initiation of the *fecABCDE* genes catalyzing ferric citrate-mediated iron transport in *E. coli*. OM, outer membrane; PP, periplasm; CM, cytoplasmic membrane; CP, cytoplasm; RNAP, RNA polymerase core enzyme. The solid arrow indicates the direction of transcription; the dashed arrow the signal pathway from FecA to FecR and FecI. FecA is assumed to form a channel that is regulated by a gating loop and interacts with its N-terminal end with FecR. FecI binds RNAP to the *fecA* promoter.

6.4. Indirect Regulation of Gene Transcription by Iron and Fur

There are very few examples of iron regulation of gene transcription independent of Fur. In most cases, these examples have not been studied in depth really to exclude participation of Fur. However, there are examples that Fur controls synthesis of a regulator, and some of them will be discussed in this section. An overview of the different regulators is given in Table 2. The FecIR system and the related systems in *Pseudomonas* have been addressed in Sec. 6.3.

In many bacteria, it is possible to isolate Fur mutants in the presence of high manganese concentrations [254]. In vitro, it was shown that Mn^{2+} can substitute for Fe^{2+} as a corepressor of Fur [232]. At low Mg^{2+} concentrations, high levels of Mn^{2+} are accumulated in the cell. Mn^{2+} replaces Fe^{2+} on Fur and represses iron uptake irrespective of the iron demands of the cell. A Fur mutant which constitutively expresses the iron transport systems is able to satisfy its iron needs even in the presence of Mn^{2+} and seems to have a growth advantage under these conditions [254]. In *E. coli*, the *fur* gene could be inactivated by a transposon insertion and by deletion. In several other organisms like *Pseudomonas aeruginosa* [255], *Neisseria gonorrhoeae* [256], and *Synechococcus* spp. [257], it was not possible to isolate *fur* deletion mutants by gene replacement. However, it was possible with Mn^{2+} to select point mutants with a changed regulatory phenotype [254,255]. The mutants in *P. aeruginosa* were defective in ferric siderophore uptake and oxic growth and had a lowered level of superoxide dismutase and catalase [255]. In *Synechococcus* spp., the attempt to inactivate the *fur* gene insertionally led to heteroallelic mutants which still retained intact copies of *fur*. These mutants constitutively produced flavodoxin and hydroxamates [257]. In some organisms, flavodoxin synthesis is stimulated under low-iron growth conditions; possibly as a substitute for iron-containing ferredoxins. These observations indicate that Fur not only regulates iron uptake but positively regulates vital metabolic functions. Also in *E. coli*, Fur is necessary under certain conditions for growth. Fur mutants are unable to grow on succinate as the sole carbon source [254]. This coincides with the stimulation of *fur* expression by the catabolite activator protein (CAP) [233]. In addition, it has been

TABLE 2

Regulators of Bacterial Iron Metabolism

Regulator (type)	Action	Fur dependence	Targets [reference]	Organisms [reference]
Fe^{2+}-Fur (repressor)	−	Yes	Siderophore uptake and biosynthesis	Many gram-negative bacteria
	+	Yes	*sodB* [258], acid-tolerance response [229], oxic stress defense [128]	*E. coli*, *S. typhimurium*
Fe^{2+}-DtxR (repressor) similar to	−		Siderophore biosynthesis, toxin production	*C. diphtheriae* [239]
Fe^{2+}-IdeR	−		Siderophore biosynthesis	*M. smegmatis* [245]
	+		Oxic stress defense	
FecI/FecR (sigma factor)	+	Yes	Fe^{3+}-dicitrate transport	*E. coli*
PupI/PupR (sigma factor)	+	Yes	Pseudobactin BN7 transport	*P. putida* [12,253]
	+	Yes	Pyoverdin biosynthesis, uptake, and exotoxin A	*P. aeruginosa* [12,259]
PvdS (sigma factor) similar to PfrI	+	Yes	Pseudobactin 358 biosynthesis and uptake	*P. putida* [12]
and PbrA	+	Yes		*P. fluorescens* [12]

PfrA (similar to AlgQ)	+	No	Pseudobactin 358 biosynthesis	*P. putida* [260]
PchR (AraC-type) similar to YbtA	±	Yes	Pyochelin biosynthesis and uptake	*P. aeruginosa* [261,262]
	±	Yes	Yersiniabactin biosynthesis and uptake	*Y. pestis* [263]
IrgB (LysR-type)	+	Yes	Unknown	*V. cholerae* [270]
AngR	+	Yes	Anguibactin biosynthesis, uptake	*V. anguillarum* [271–273]
TAF	+			
RNAα	−		Anguibactin uptake	
PfeR/PfeS (two-component sensor kinase)	+	Yes	Uptake of enterobactin	*P. aeruginosa* [163,266]
RepA/RepB (two-component sensor	+	Unknown	Synthesis of a fluorescent siderophore	*P. viridiflava* [267]
kinase) similar to BarA/UvrY	+	Unknown	Synthesis of enterobactin	Uropathogenic *E. coli* [268]
BvgA/BvgS (two-component sensor kinase)	−	Unknown	Repression of alcaligin biosynthesis	Certain strains of *B. bronchiseptica* [269]

+, Regulator with activating activity; −, regulator with inhibiting activity.

shown that Fur is a positive regulator for the expression of the acid-tolerance response which helps *Salmonella* to survive in the acidic milieu of the stomach on its way to the gut [229]. Especially interesting was the characterization of the mutant Fur(H90R) which was unable to repress enterobactin production but still showed a normal acid-tolerance response [229].

Although the positive regulation of SodB (the iron-containing superoxide dismutase) by Fur was observed in *E. coli* some time ago [258], the mechanism has not been elucidated. In addition, it has been shown that a *fur* mutation in a strain defective in its DNA repair through a *recA* mutation is lethal under oxic conditions [128]. This shows that Fur has a strong influence on the antioxidant defense systems. These are some examples which stress the importance of Fur as a general regulator in cellular metabolism; however, the regulatory mechanisms are unknown.

Fur control of regulators, besides the *E. coli* FecIR system, has been mainly observed in pseudomonads. In contrast to *E. coli* where only the Fec system can be induced by the presence of its siderophore, in pseudomonads, all of the known transport systems are only induced in the presence of the appropriate siderophore. Pseudomonads seem to be able to use such a great diversity of different siderophores that specific regulatory systems have been evolved to induce only those transport systems where the related siderophore is present. Since only iron-loaded siderophores seem to act as inducer, the hierarchy of induction is directed by the affinity of the siderophore to iron.

A group of very similar positive regulators of fluorescent siderophore biosynthesis have been found in *P. putida* (PfrI), *P. fluorescens* (PbrA), and *P. aeruginosa* (PvdS) [12]. They have a certain similarity to FecI and seem to act as sigma factors. For positive regulation, the appropriate siderophore is necessary; however, no FecR homologue has been identified to date.

Virulence factors and exotoxin A production in *P. aeruginosa* are regulated by iron and many attempts in recent years aimed at elucidating the regulatory mechanism have been made. Neither the *toxA* gene nor the regulator gene *regA* has any Fur box in its promoter regions. The iron-Fur-regulated PvdS protein was shown to be necessary for the production of exotoxin by allowing transcription of the *regAB* genes. Transcription of the *toxA* promoter is not stimulated by PvdS [259].

In *P. putida*, in addition to the activator PfrI, a second positive regulator, PfrA, of pseudobactin 358 biosynthesis was identified [260]. This protein had 58% identity to AlgQ, a positive regulator of alginate production in *P. aeruginosa*. PfrI could complement an *algQ* mutant in *P. aeruginosa* for alginate production. Siderophore production in this *algQ* mutant was only 60% reduced [260]. Alginate production is a virulence factor in cystic fibrosis. *algQ* is influenced by a set of different regulators, but not by Fur, and it is not known how it activates alginate biosynthesis.

In addition to FecI homologues, other types of Fur-controlled regulatory proteins are found in pseudomonads. PchR is a regulator of pyochelin synthesis and transport and belongs to the AraC type regulators [261]. PchR is required for production of FptA, the outer membrane receptor for pyochelin. FptA and PchR expression is pyochelin dependent. Since siderophores such as enterobactin and pyoverdin with superior iron-binding capacities suppress pyochelin-dependent upregulation of FptA, it is assumed that ferric pyochelin is the regulatory active component. It is not known whether ferric pyochelin is interacting directly with PchR or if another sensor recognizes the presence of ferric pyochelin. Activating and repressing PchR activities were observed, as is to be expected for an AraC-type regulator [262]. A homologue of PchR, the regulator YbtA for synthesis of the yersiniabactin receptor Psn, was characterized recently from *Yersinia pestis* [263]. Also in this case the presence of YbtA and yersiniabactin was necessary for full expression of Psn (pesticide receptor). A model is proposed where yersiniabactin directly interacts with YbtA to modulate its activity [263]. The similarity between the two systems includes a chemical similarity between the structurally unusual siderophores pyochelin and yersiniabactin [264,265].

P. aeruginosa is able to use ferric enterobactin as an iron source and the transport system is induced in the presence of the siderophore. Upstream of the receptor gene *pfeA*, the *pfeRS* operon was identified to code for a two-component histidine kinase-type regulation [163]. A *pfeR* mutant failed to induce the PfeA receptor in the presence of enterobactin. No autoregulation by PfeR was observed. A *pfeA* knockout mutant remained inducible with enterobactin, indicating that enough enterobactin reaches the sensor protein PfeS [266].

The histidine kinase two-component regulator RepAB was shown to activate fluorescent siderophore production in *P. viridiflava*. The genes *repA* and *repB* were highly similar to *lemA*/*gacA* described for *P. syringae* and *P. fluorescens* where these genes control synthesis of extracellular enzymes and secondary metabolites like antibiotics and antifungal activities [267]. The sensor LemA has a complicated domain structure with an additional response regulator domain. In an uropathogenic strain of *E. coli*, the homologue of LemA, BarA (= AirS), was shown to be necessary for siderophore biosynthesis [268]. The same mutation in *E. coli* K-12 had no influence on siderophore production.

BvgS has a similar complex domain structure as LemA and is a global regulator of virulence determinants and toxins in *Bordetella*. A repression of siderophore biosynthesis has been observed by BvgS in certain *Bordetella bronchiseptica* strains [269]. In all these cases, it is not known how these two-component regulators influence siderophore synthesis and uptake and whether Fur is integrated into the regulatory network.

In the pathogen *Vibrio cholerae*, the outer membrane protein IrgA with similarities to the enterobactin receptor FepA was identified as a virulence determinant. It is assumed to be a TonB-dependent siderophore receptor, although no siderophore could be identified. Upstream of *irgA*, a regulatory gene *irgB* was identified [270]. IrgB is homologous to the LysR family of regulators. IrgB activates the expression of *irgA*. *irgA* and *irgB* are negatively regulated by Fur and iron which bind to the divergent overlapping promoter region [270].

Another interesting regulatory principal has been demonstrated in the fish pathogen *Vibrio anguillarum*, where the plasmid-encoded siderophore anguibactin is a virulence factor [2]. The expression of the biosynthesis and transport genes is determined by a number of regulators: an antisense RNAα, the positive regulator AngR, ill-defined trans-acting factors (TAF), anguibactin, and the chromosome-encoded Fur protein [271]. *fatA*, the gene for the anguibactin receptor, and *fatB*, the gene for the binding protein, are repressed under iron-rich conditions. The antisense RNAα is found mainly under these conditions, and it was observed that iron did not stimulate transcription initiation of RNAα, but it enhanced the half-life of the antisense RNA [272]. Expression of *fatA* and *fatB* were repressed by RNAα [272,273]. This type of regulation has not been observed in any other iron transport system; however, possibly nobody was looking for this type of regulation.

This enumeration shows that different types of known regulatory principles as LysR-type regulators, AraC-type regulators, two-component histidine kinase regulators, antisense RNA, and specialized sigma factors have been found under Fur control.

7. OUTLOOK

Bacteria acquire iron from all kinds of sources in their environment—ferrous iron, ferric iron, ferric iron complexed by siderophores of bacteria and fungi, heme, hemoglobin, hemopexin, ferric iron bound to transferrin, and lactoferrin of the bacterial hosts.

Gram-negative bacteria select the iron source by receptor proteins in the outer membrane that display a high specificity for the compounds, most notably to be seen with the transferrin and lactoferrin receptors of *Neisseria* and *Haemophilus* which recognize only the transferrins and lactoferrins of their hosts. It also seems that the transport mechanism across the outer membrane is very similar for all substrates, but, in fact, the mechanism is not well enough understood in molecular terms to be sure that this is true. Since gram-positive bacteria do not possess an outer membrane but a permeable cell wall, selection for the iron source to be taken up occurs at the cytoplasmic membrane. The specificity of transport across the cytoplasmic membrane of gram-positive and of gram-negative bacteria is lower than across the outer membrane of gram-negative bacteria. For example, *E. coli* K-12 contains only a single transport system for the various ferric hydroxamate compounds and a single transport system for the various ferric catecholate compounds across the cytoplasmic membrane but various substrate-specific receptors in the outer membrane.

The transport mechanism of iron across the cytoplasmic membrane, regardless whether it takes place via ferric siderophores, heme, or as ferric iron (not ferrous iron), always occurs via periplasmic binding proteins, which in gram-positive bacteria, which do not have a periplasm, are anchored to the outside of the cytoplasmic membrane by a lipid of the murein lipoprotein type. In the past, research mainly focused on the ecological aspect of how a certain strain acquires iron in a certain environment. The power of molecular biology allowed us to identify transport proteins and to suggest transport mechanisms from gene sequences by homology search with better understood systems; mainly those of *E. coli*. The field lacks sufficient in-depth studies aimed at understanding transport mechanisms in molecular terms. In the future, such studies should include isolation of iron transport proteins and the entire iron transport complexes and their reconstitution in artificial lipid bilayer membranes, as has been done with the maltose and the histidine transport systems. They should involve systematic mutational analyses, isolation of proteins, and determination of their three-dimensional structure by x-ray analysis and nuclear magnetic resonance.

Siderophore formation involves biosynthesis and secretion, of which the latter is entirely unknown. Evidence for a weak membrane association of biosynthetic enzymes was found [274], suggesting that siderophores are immediately secreted to avoid the potentially cell-damaging complexation of iron in the cytoplasm. Since siderophores are usually charged molecules, it is unlikely that they are released by diffusion. Rather there should be siderophore export systems through the cytoplasmic membrane. To identify genes, their products, and functions in siderophore secretion, conditional mutants in secretory genes may be isolated which, for example, secrete siderophores at 30°C and are viable but not at 40°C where they are killed.

Regarding regulation, repression by the Fur and the DtxR proteins loaded with Fe^{2+} is quite well understood, but induction by Fe^{2+}Fur, regulation by iron without Fur and DtxR, and the various regulatory mechanisms exerted by ferric siderophores have still to be studied on the molecular level. The ferric citrate regulation of *E. coli* and the similar ferric pseudobactin regulation of *P. putida* are examples for mechanistically novel regulatory devices that may be uncovered by such studies.

ACKNOWLEDGMENTS

The authors' work was supported by the Deutsche Forschungsgemeinschaft and the Fonds der Chemischen Industrie. We thank Karen A. Brune for critically reading the manuscript and H. Pilsl and S. Plantör for computer drawing of Figures 2 and 8, respectively.

ABBREVIATIONS

afu	Actinobacillus ferric iron uptake in *Actinobacillus pleuropneumoniae*
alg	alginate biosynthesis and regulation in *Pseudomonas aeruginosa*
angR	anguibactin, biosynthesis and uptake regulation in *Vibrio anguillarum*
ara	arabinose metabolism in *E. coli*
ATP	adenosine 5′-triphosphate
barA	bacterial adaptive response in *E. coli*
bfd	iron-sulfur protein encoded upstream of *bfr* in *E. coli*
bfr	bacterioferritin
btu	vitamin B_{12} uptake system
bvgRS	Bordetella virulence genes regulatory system in *Bordetella*
cbr	chrysobactin regulator, uptake system for achromobactin in *Erwinia chrysanthemi*
CCCP	carbonyl cyanide-3-chlorophenylhydrazone
cir	colicin I and catecholate receptor in *E. coli*
CM	cytoplasmic membrane
cor	cobalt resistance, Mg^{2+} transporter
CR	conserved region
dps	DNA-binding protein, starvation
dtxR	diphtheria toxin regulator in *Corynebacterium diphtheriae*
ECF	extracytoplasmic functions
exbBD	excretion of colicin B inhibitor, components of the Ton-complex
fat	ferric anguibactin transport in *Vibrio anguillarum*
fbp	ferric iron binding protein in *Neisseria*
fec	ferric citrate uptake system in *E. coli*
feo	ferrous iron uptake system in *E. coli*
fep	ferric enterobactin uptake system in *E. coli*
fes	ferric enterobactin esterase in *E. coli*
fhu	ferric hydroxamate uptake system in *E. coli*
fiu	ferric iron uptake, catecholate receptor in *E. coli*

fptA	ferric pyochelin transport, receptor in *Pseudomonas aeruginosa*
ftn	ferritin in *E. coli*
fur	ferric iron uptake regulator
gacA	global antibiotic and cyanide control response regulator in *Pseudomonas*
has	haeme uptake in *Serratia marcescens*
hemPRSTU	heme uptake system in *Yersinia enterocolitica*
hgbA	hemoglobin receptor in *Haemophilus ducreyi*
hhuA	hemoglobin/haptoglobin uptake, receptor in *Haemophilus influenzae*
his	histidine synthesis and uptake in *E. coli*
hit	Haemophilus iron transport in *Haemophilus influenzae*
hmbR	hemoglobin receptor in *Haemophilus influenzae*
hupA	hemoglobin uptake in *Haemophilus ducreyi*
hutA	heme uptake receptor in *Vibrio cholerae*
hxu	heme/hemopexin uptake in *Haemophilus influenzae*
ideR	iron-dependent regulation in *Mycobacterium*
irg	iron-regulated protein in *Vibrio cholerae*
lamB	lambda and maltose receptor in *E. coli*
lbp	lactoferrin binding protein in *Neisseria*
lemA	lesion formation regulator in *Pseudomonas syringae*
lysR	lysine biosynthesis regulator in *E. coli*
mal	maltose uptake and metabolism in *E. coli*
NADP	nicotinamide adenine dinucleotide phosphate
nif	nitrogen fixation
OM	outer membrane
pbrA	pseudobactin regulator in *Pseudomonas fluorescens*
PBT	periplasmic binding protein-dependent transport
pchR	pyochelin regulator in *Pseudomonas aeruginosa*
pfe	Pseudomonas receptor for ferric enterobactin in *Pseudomonas aeruginosa*
pfr	Pseudomonas ferric regulator in *Pseudomonas putida*
PP	periplasm
psn	pesticin and yersinabactin receptor in *Yersinia pestis*
pup	pseudobactin uptake in *Pseudomonas fluorescens*
pvd	pyoverdin uptake in *Pseudomonas aeruginosa*

rep	regulation of extracellular protease and pectate lyase in *Pseudomonas viridiflava*
rum	rhizobactin uptake in *Morganella morganii*
SDS	sodium dodecylsulfate
sfu	Serratia ferric iron uptake system in *Serratia marcescens*
shl	Serratia hemolysin in *Serratia marcescens*
shu	Shigella heme uptake in *Shigella dysenteriae*
sod	superoxide dismutase in *E. coli*
soxR	superoxide, oxidative stress regulator in *E. coli*
TAF	trans-acting factor
tbp	transferrin binding protein in *Neisseria*
tol	tolerance to colicins in *E. coli*
tonB	T-one resistance in *E. coli*
viu	vibriobactin uptake in *Vibrio cholerae*
ybtA	yersiniabactin AraC-like regulator
yfu	yersinia ferric iron uptake in *Yersinia enterocolitica*

REFERENCES

1. V. Braun and K. Hantke, in *Handbook of Microbial Iron Chelates*, (G. Winkelmann, ed.), CRC Press, Boca Raton, FL, 1991, pp. 107–138.
2. J. H. Crosa, *Ann. Rev. Microbiol.*, *38*, 69–89 (1984).
3. D. Expert, C. Enard, and C. Masclaux, *Trends Microbiol.*, *4*, 232–236 (1996).
4. C. A. Genco and P. J. Desai, *Trends Microbiol.*, *4*, 179–184 (1996).
5. C. N. Cornelissen and F. P. Sparling, *Mol. Microbiol.*, *14*, 843–850 (1994).
6. R. Kingsley, W. Rabsch, P. Stephens, M. Roberts, R. Reissbrodt, and P. H. Williams, *FEMS Immunol. Med. Microbiol.*, *11*, 257–264 (1995).
7. S. D. Gray-Owen and A. B. Schryvers, *Trends Microbiol.*, *4*, 185–191 (1996).
8. X. Tao, N. Schiering, H.-Y. Zeng, D. Ringe, and J. R. Murphy, *Mol. Microbiol.*, *14*, 191–197 (1994).

9. S. M. Payne, *Trends Microbiol.*, *1*, 66–69 (1993).
10. K. G. Wooldrigde and P. H. Williams, *FEMS Microbiol. Rev.*, *12*, 325–348 (1993).
11. R. D. Perry, *Trends Microbiol.*, *1*, 142–147 (1993).
12. V. Venturi, P. Weisbeek, and M. Koster, *Mol. Microbiol.*, *17*, 603–610 (1995).
13. V. Braun and G. Winkelmann, *Prog. Clin. Biochem. Med.*, *5*, 67–99 (1987).
14. B. F. Matzanke, I. Berner, E. Bill, A. X. Trautwein, and G. Winkelmann, *Biol. Met.*, *4*, 181–185 (1991).
15. W. Boos and J. M. Lucht, in *Escherichia coli and Salmonella typhimurium*, (F. C. Neidhardt, ed.), ASM Press, Washington, D.C., 1996, pp. 1175–1209.
16. H. Nikaido, in *Escherichia coli and Salmonella typhimurium*, (F. C. Neidhardt, ed.), ASM Press, Washington, D.C., 1996, pp. 29–47.
17. V. Braun and H. C. Wu, in *Bacterial Cell Wall*, (J.-M. Ghuysen and R. Hakenbeck, eds.), Elsevier, Amsterdam, 1994, pp. 319–341.
18. H. Killmann, R. Benz, and V. Braun, *EMBO J.*, *12*, 3007–3016 (1993).
19. R. Koebnik, K. Hantke, and V. Braun, *Mol. Microbiol.*, *7*, 383–393 (1993).
20. G. S. Moeck, B. S. F. Bazzaz, M. F. Gras, T. S. Ravi, M. J. H. Ratcliffe, and J. W. Coulton, *J. Bacteriol.*, *176*, 4250–4259 (1994).
21. G. S. Moeck, M. J. H. Ratcliffe, and J. W. Coulton, *J. Bacteriol.*, *177*, 6118–6125 (1995).
22. C. Bös and V. Braun, *FEMS Microbiol. Lett.* In press.
23. H. Killmann and V. Braun, *J. Bacteriol.*, *174*, 3479–3486 (1992).
24. H. Killmann and V. Braun, *FEMS Microbiol. Lett.*, *119*, 71–76 (1994).
25. H. Killmann, G. Videnov, and G. Jung, *J. Bacteriol.*, *177*, 694–698 (1995).
26. H. Killmann and V. Braun, unpublished observations (1997).
27. P. Boulanger, M. le Maire, M. Bonhivers, S. Dubois, M. Desmadril, and L. Letellier, *Biochemistry*, *35*, 14216–14224 (1996).
28. K. Hantke and V. Braun, *J. Bacteriol.*, *135*, 190–197 (1978).

29. G. Schultz, F. Ullrich, K. J. Heller, and V. Braun, *Mol. Gen. Genet.*, *216*, 230–238 (1989).
30. H. Killmann, K. Hantke, and V. Braun, unpublished observations (1997).
31. M. Bonhivers, A. Ghazi, P. Boulanger, and L. Letellier, *EMBO J.*, *15*, 1850–1856 (1996).
32. K. P. Locher and J. P. Rosenbusch, *J. Biol. Chem.*, *272*, 1448–1451 (1997).
33. L. Letellier, K. P. Locher, L. Plancon, and P. J. Rosenbusch, *J. Biol. Chem.*, *272*, 8836 (1997).
34. S. K. Armstrong, C. L. Francis, and M. A. McIntosh, *J. Biol. Chem.*, *265*, 14536–14543 (1990).
35. S. K. Armstrong and M. A. McIntosh, *J. Biol. Chem.*, *270*, 2483–2488 (1995).
36. C. K. Murphy, V. I. Kalve, and P. E. Klebba, *J. Bacteriol.*, *172*, 2736–2746 (1990).
37. J. Liu, J. M. Rutz, P. E. Klebba, and J. B. Feix, *Biochemistry*, *33*, 13274–13283 (1994).
38. J. M. Rutz, J. Liu, J. A. Lyons, J. Goranson, S. K. Armstrong, M. A. McIntosh, J. B. Feix, and P. E. Klebba, *Science*, *258*, 471–475 (1992).
39. R. E. Hancock and V. Braun, *J. Bacteriol.*, *125*, 409–415 (1976).
40. R. J. Kadner, K. Heller, J. W. Coulton, and V. Braun, *J. Bacteriol.*, *143*, 256–264 (1980).
41. V. Braun and H. Wolff, *FEBS Lett.*, *34*, 77–80 (1973).
42. V. Braun, K. Schaller, and H. Wolff, *Biochim. Biophys. Acta*, *323*, 87–97 (1973).
43. W. Köster and V. Braun, *J. Biol. Chem.*, *265*, 21407–21410 (1990).
44. V. Braun, *J. Bacteriol.*, *171*, 6387–6390 (1989).
45. V. Braun and C. Herrmann, *Mol. Microbiol.*, *8*, 261–268 (1993).
46. C. Bradbeer, *J. Bacteriol.*, *175*, 3146–3150 (1993).
47. J. T. Skare, B. M. M. Ahmer, C. L. Seachord, R. P. Darveau, and K. Postle, *J. Biol. Chem.*, *268*, 16302–16308 (1993).
48. K. J. Heller, R. J. Kadner, and K. Günter, *Gene*, *64*, 147–153 (1988).
49. H. Schöffler and V. Braun, *Mol. Gen. Genet.*, *217*, 378–383 (1989).
50. P. E. Bell, C. D. Nau, J. T. Brown, J. Konisky, and R. J. Kadner, *J. Bacteriol.*, *172*, 3826–3829 (1990).

51. K. Günter and V. Braun, *FEBS Lett.*, *274*, 85–88 (1990).
52. E. Fischer, K. Günter, and V. Braun, *J. Bacteriol.*, *171*, 5127–5134 (1989).
53. I. Traub, S. Gaisser, and V. Braun, *Mol. Microbiol.*, *8*, 409–423 (1993).
54. R. A. Larsen, M. G. Thomas, G. E. Wood, and K. Postle, *Mol. Microbiol.*, *13*, 627–640 (1994).
55. V. Braun, S. Gaisser, C. Herrmann, K. Kampfenkel, H. Killmann, and I. Traub, *J. Bacteriol.*, *178*, 2836–2845 (1996).
56. I. Traub and V. Braun, *FEMS Microbiol. Lett.*, *119*, 65–70 (1994).
57. K. Kampfenkel and V. Braun, *J. Bacteriol.*, *174*, 5485–5487 (1992).
58. K. Kampfenkel and V. Braun, *J. Biol. Chem.*, *268*, 6050–6057 (1993).
59. W. Bitter, J. Tommassen, and P. J. Weisbeek, *Mol. Microbiol.*, *7*, 117–130 (1993).
60. I. Stojiljkovic and N. Srinivasan, *J. Bacteriol.*, *64*, 5008–5014 (1996).
61. H. Hoffmann, E. Fischer, H. Schwarz, and V. Braun, *Arch. Microbiol.*, *145*, 334–341 (1986).
62. G. S. Moeck, P. Tawa, H. Xiang, A. A. Ismail, J. L. Turnbull, and J. W. Coulton, *Mol. Microbiol.*, *22*, 459–471 (1996).
63. F. A. Quiocho, *Philos. Trans. R. Soc. Lond. Ser. B*, *326*, 341–351 (1990).
64. R. Tam and M. H. Saier, Jr., *Microbiol. Rev.*, *57*, 320–346 (1993).
65. G. L. Gilliland and F. A. Quiocho, *J. Mol. Biol.*, *146*, 341–362 (1981).
66. M. M. Flocco and S. L. Mowbray, *J. Biol. Chem.*, *269*, 8931–8936 (1994).
67. J. Y. Zou, M. M. Flocco, and S. L. Mowbray, *J. Mol. Biol.*, *233*, 739–752 (1993).
68. L. A. Actis, M. E. Tolmasky, L. M. Crosa, and J. H. Crosa, *Mol. Microbiol.*, *17*, 197–204 (1995).
69. R. E. Kerppola, V. K. Shyamala, P. Klebba, and G. F.-L. Ames, *J. Biol. Chem.*, *266*, 9857–9865 (1991).
70. A. L. Davidson and H. Nikaido, *J. Biol. Chem.*, *266*, 8946–8951 (1991).

71. A. L. Davidson, S. S. Laghaeian, and D. E. Mannering, *J. Biol. Chem.*, *271*, 4858–4863 (1996).
72. M. Kammler, C. Schön, and K. Hantke, *J. Bacteriol.*, *175*, 6212–6219 (1993).
73. K. Hantke, unpublished results (1997).
74. R. M. Tsolis, A. J. Bäumler, F. Heffron, and I. Stojiljkovic, *Infect. Immun.*, *64*, 4549–4556 (1996).
75. C. J. Bult, O. White, G. J. Olsen, L. Zhou, R. D. Fleischmann, G. G. Sutton, J. A. Blake, L. M. FitzGerald, R. A. Clayton, J. D. Gocayne, A. R. Kerlavage, B. A. Dougherty, J. F. Tomb, M. D. Adams, C. I. Reich, R. Overbeek, E. F. Kirkness, K. G. Weinstock, J. M. Merrick, A. Glodek, J. L. Scott, N. S. M. Geoghagen, J. F. Weidman, J. L. Fuhrmann, D. Nguyen, T. R. Utterback, J. M. Kelley, J. D. Peterson, P. W. Sadow, M. C. Hanna, M. D. Cotton, K. M. Roberts, M. A. Hurst, B. P. Kaine, M. Borodovsky, H.-P. Klenk, C. M. Fraser, H. O. Smith, C. R. Woese, and J. C. Venter, *Science*, *273*, 1058–1073 (1996).
76. M. M. Gibson, D. A. Bagga, C. G. Miller, and M. E. Maguire, *Mol. Microbiol.*, *5*, 2753–2762 (1991).
77. S. Kühn, V. Braun, and W. Köster, *J. Bacteriol.*, *178*, 496–504 (1996).
78. I. Stojiljkovic and K. Hantke, *EMBO J.*, *11*, 4359–4367 (1992).
79. C. M. Fraser, J. D. Gocayne, O. White, M. D. Adams, R. A. Clayton, R. D. Fleischmann, C. J. Bult, A. R. Kerlavage, G. Sutton, J. M. Kelley, J. L. Fritchman, J. F. Weidman, K. V. Small, M. Sandusky, J. Fuhrmann, D. Nguyen, T. R. Utterback, D. M. Saudek, C. A. Phillips, J. M. Merrick, J.-F. Tomb, B. A. Dougherty, K. F. Bott, P.-C. Hu, T. S. Lucier, S. N. Peterson, H. O. Smith, C. A. Hutchison III, and J. C. Venter, *Science*, *270*, 397–403 (1995).
80. R. Himmelreich, H. Hilbert, H. Plagens, E. Pirkl, B.-C. Li, and R. Herrmann, *Nucl. Acids Res.*, *24*, 4420–4449 (1996).
81. L. Fecker and V. Braun, *J. Bacteriol.*, *156*, 1301–1314 (1983).
82. R. Burkhardt and V. Braun, *Mol. Gen. Genet.*, *209*, 49–55 (1987).
83. J. W. Coulton, P. Mason, and D. D. Allat, *J. Bacteriol.*, *169*, 3844–3849 (1987).
84. W. Köster and V. Braun, *Mol. Gen. Genet.*, *217*, 233–239 (1989).
85. P. J. Wookey, S. Hussein, and V. Braun, *J. Bacteriol.*, *146*, 1158–1161 (1981).

86. M. R. Rohrbach, S. Paul, and W. Köster, *Mol. Gen. Genet.*, *248*, 33–42 (1995).
87. M. R. Rohrbach, V. Braun, and W. Köster, *J. Bacteriol.*, *177*, 7186–7193 (1995).
88. A. Mademidis, H. Killmann, W. Kraas, I. Flechsner, G. Jung, and V. Braun, unpublished observations (1997).
89. W. Köster and V. Braun, *Mol. Gen. Genet.*, *223*, 379–384 (1990).
90. W. Groeger and W. Köster, unpublished results (1997).
91. B. Böhm, H. Boschert, and W. Köster, *Mol. Microbiol.*, *20*, 223–232 (1996).
92. W. Köster, *Biol. Met.*, *4*, 23–32 (1991).
93. W. Saurin, W. Köster, and E. Dassa, *Mol. Microbiol.*, *12*, 993–1004 (1994).
94. E. Dassa and M. Hofnung, *EMBO J.*, *4*, 2287–2293 (1985).
95. K. Becker, W. Köster, and V. Braun, *Mol. Gen. Genet.*, *223*, 159–162 (1990).
96. G. Schultz-Hauser, W. Köster, H. Schwartz, and V. Braun, *J. Bacteriol.*, *174*, 2305–2311 (1992).
97. A. Angerer, S. Gaisser, and V. Braun, *J. Bacteriol.*, *172*, 572–578 (1990).
98. L. Zimmermann, A. Angerer, and V. Braun, *J. Bacteriol.*, *171*, 238–243 (1989).
99. P. Adhikari, S. A. Berish, A. J. Nowalk, K. L. Veraldi, S. A. Morse, and T. A. Mietzner, *J. Bacteriol.*, *178*, 2145–2149 (1996).
100. S. A. Berish, D. R. Kapczynski, and S. A. Morse, *Nucleic Acids. Res.*, *18*, 4596 (1990).
101. P. Adhikari, S. D. Kirby, A. J. Nowalk, K. L. Veraldi, A. B. Schryvers, and T. A. Mietzner, *J. Biol. Chem.*, *270*, 25142–25149 (1995).
102. E. M. Saken and J. Heesemann, EMBL Database, Acc. Nr.: Z47200 (1995).
103. N. Chin, J. Frey, C.-F. Chang, and Y.-F. Chang, *FEMS Microbiol. Lett.*, *143*, 1–6 (1996).
104. C. Y. Chen, S. A. Berish, S. A. Morse, and T. A. Mietzner, *Mol. Microbiol.*, *10*, 311–318 (1993).
105. A. J. Nowalk, S. B. Tencza, and T. A. Mietzner, *Biochemistry*, *33*, 12769–12775 (1994).

106. W. Köster, unpublished results (1997).

107. V. Laufberger, *Bull. Soc. Chim. Biol.*, *19*, 1576–1582 (1937).

108. P. M. Harrison and P. Arosio, *Biochim. Biophys. Acta*, *1275*, 161–203 (1996).

109. E. I. Stiefel and G. D. Watt, *Nature*, *279*, 81–83 (1979).

110. J. Yariv, *Biochem. J.*, *211*, 527 (1983).

111. S. C. Andrews, P. M. Harrison, and J. R. Guest, *J. Bacteriol.*, *171*, 3940–3947 (1989).

112. F. Frolow, A. J. Kalb, and J. Yariv, *Nat. Struct. Biol.*, *1*, 453–460 (1994).

113. N. F. Inglis, K. Stevenson, A. H. Hosie, and J. M. Sharp, *Gene*, *150*, 205–206 (1994).

114. M. C. Pessolani, D. R. Smith, B. Rivoire, J. McCormick, S. A. Hefta, S. T. Cole, and P. J. Brennan, *J. Exp. Med.*, *180*, 319–327 (1994).

115. C. N. Penfold, P. L. Ringeling, S. L. Davy, G. R. Moore, A. G. McEwan, and S. Spiro, *FEMS Microbiol. Lett.*, *139*, 143–148 (1996).

116. P. A. Denoel, M. S. Zygmunt, V. Weynants, A. Tibor, B. Lichtfouse, P. Briffeuil, J. N. Limet, and J. J. Letesson, *FEBS Lett.*, *361*, 238–242 (1995).

117. M. R. Cheesman, A. J. Thomson, C. Greenwood, G. R. Moore, and F. Kadir, *Nature*, *346*, 771–773 (1990).

118. S. C. Andrews, N. E. Le Brun, V. Barynin, A. J. Thomson, G. R. Moore, J. R. Guest, and P. M. Harrison, *J. Biol. Chem.*, *270*, 23268–23274 (1995).

119. M. Izuhara, K. Takamune, and R. Takata, *Mol. Gen. Genet.*, *225*, 510–513 (1991).

120. P. D. Hempstead, A. J. Hudson, P. J. Artymiuk, S. C. Andrews, M. J. Banfield, J. R. Guest, and P. M. Harrison, *FEBS Lett.*, *350*, 258–262 (1994).

121. N. E. Le Brun, S. C. Andrews, J. R. Guest, P. M. Harrison, G. R. Moore, and A. J. Thomson, *Biochem J.*, *312*, 385–392 (1995).

122. F. H. Kadir and G. R. Moore, *FEBS Lett.*, *276*, 81–84 (1990).

123. M. A. Michaux, A. Dautant, B. Gallois, T. Granier, B. L. d'Estaintot, and G. Precigoux, *Proteins*, *24*, 314–321 (1996).

124. C. J. Carrano, R. Böhnke, and B. F. Matzanke, *FEBS Lett.*, *390*, 261–264 (1996).

125. A. J. Hudson, S. C. Andrews, C. Hawkins, J. M. Williams, M. Izuhara, F. C. Meldrum, S. Mann, P. M. Harrison, and J. R. Guest, *Eur. J. Biochem.*, *218*, 985–995 (1993).

126. I. Stojiljkovic, A. J. Bäumler, and K. Hantke, *J. Mol. Microbiol.*, *236*, 531–545 (1994).

127. R. P. Garg, C. J. Vargo, X. Cui, and D. M. Kurtz, Jr., *Biochemistry*, *35*, 6297–6301 (1996).

128. D. Touati, M. Jacques, B. Tardat, L. Bouchard, and S. Despied, *J. Bacteriol.*, *177*, 2305–2314 (1995).

129. S. Wai, K. Nakayama, K. Umene, T. Moriya, and K. Amako, *Mol. Microbiol.*, *20*, 1127–1134 (1996).

130. M. Almiron, A. J. Link, D. Furlong, and R. Kolter, *Genes Dev.*, *6*, 2646–2654 (1992).

131. M. Pena and G. Bullerjahn, *J. Biol. Chem.*, *270*, 22478–22482 (1995).

132. M. Bozzi, G. Mignogna, S. Stefanini, D. Barra, C. Longhi, P. Valenti, and E. Chiancone, *J. Biol. Chem.*, *272*, 3259–3265 (1997).

133. R. Böhnke and B. F. Matzanke, *BioMetals*, *8*, 223–230 (1995).

134. A. Hartmann and V. Braun, *Europ. J. Biochem.*, *99*, 517–524 (1979).

135. E. Fischer, B. Strelow, D. Hartz, and V. Braun, *Arch. Microbiol.*, *153*, 329–336 (1990).

136. M. Fontecave, J. Coves, and J. L. Pierre, *Biometals*, *7*, 3–8 (1994).

137. T. A. Rouault, C. D. Stout, S. Kaptain, J. B. Harford, and R. D. Klausner, *Cell*, *64*, 881–883 (1991).

138. E. Paraskeva and M. Hentze, *FEBS Lett.*, *389*, 40–43 (1996).

139. J. Green, B. Bennett, P. Jordan, E. Ralph, A. Thomson, and J. Guest, *Biochem. J.*, *316*, 887–892 (1996).

140. N. Khoroshilova, H. Beinert, and P. J. Kiley, *Proc. Natl. Acad. Sci. USA*, *92*, 2499–2503 (1995).

141. J. Wu, W. R. Dunham, and B. Weiss, *J. Biol. Chem.*, *270*, 10323–10327 (1995).

142. E. Hidalgo and B. Demple, *EMBO J.*, *13*, 138–146 (1994).

143. P. Gaudu, N. Moon, and B. Weiss, *J. Biol. Chem.*, *272*, 5082–5086 (1997).

144. L. Zheng, R. H. White, V. L. Cash, R. F. Jack, and D. R. Dean, *Proc. Natl. Acad. Sci. USA*, *90*, 2754–2758 (1993).

145. L. Zheng and D. R. Dean, *J. Biol. Chem.*, *269*, 18723–18726 (1994).

146. D. H. Flint, J. F. Tuminello, and T. J. Miller, *J. Biol. Chem.*, *271*, 16053–16067 (1996).

147. D. H. Flint, *J. Biol. Chem.*, *271*, 16068–16074 (1996).

148. V. Braun, R. Gross, W. Köster, and L. Zimmermann, *Mol. Gen. Genet.*, *192*, 131–139 (1983).

149. W. Köster and V. Braun, *Mol. Gen. Genet.*, *204*, 435–442 (1986).

150. M. J. Friedrich, L. C. DeVeaux, and R. J. Kadner, *J. Bacteriol.*, *167*, 928–934 (1986).

151. B. P. Surin, H. Rosenberg, and G. B. Cox, *J. Bacteriol.*, *161*, 189–198 (1985).

152. C. F. Higgins, P. D. Haag, K. Nikaido, F. Ardeshir, G. Garcia, and G. F. Ames, *Nature*, *298*, 723–727 (1982).

153. H. Staudenmaier, B. Van Hove, Z. Yaraghi, and V. Braun, *J. Bacteriol.*, *171*, 2626–2633 (1989).

154. W. Köster, L. A. Actis, L. S. Waldbeser, M. E. Tolmasky, and J. H. Crosa, *J. Biol. Chem.*, *266*, 23829–23833 (1991).

155. R. Schneider and K. Hantke, unpublished results (1997).

156. M. A. Jalal, M. B. Hossain, D. van der Helm, J. Sanders-Loehr, L. A. Actis, and J. H. Crosa, *J. Am. Chem. Soc.*, *111*, 292–296 (1989).

157. J. H. Crosa, *Microbiol. Rev.*, *53*, 517–530 (1989).

158. L. A. Actis, M. E. Tolmasky, D. H. Farrell, and J. H. Crosa, *J. Biol. Chem.*, *263*, 2853–2860 (1988).

159. C. F. Earhart, in *Escherichia coli and Salmonella typhimurium*, (F. C. Neidhart, ed.), ASM Press, Washington, D.C., 1996, pp. 1075–1090.

160. G. Stevenson, A. Kessler, and P. R. Reeves, *FEMS Microbiol. Lett.*, *125*, 23–30 (1995).

161. D. A. Bastin, G. Stevenson, P. K. Brown, A. Haase, and P. R. Reeves, *Mol. Microbiol.*, *7*, 725–734 (1993).

162. B. A. Ozenberger, M. S. Nahlik, and M. A. McIntosh, *J. Bacteriol.*, *169*, 3638–3646 (1987).

163. C. R. Dean and K. Poole, *Mol. Microbiol.*, *8*, 1095–1103 (1993).

164. G. Massad, J. E. Arceneaux, and B. R. Byers, *Biometals*, *7*, 227–236 (1994).
165. D. T. Richardson and S. F. Park, *Microbiol.*, *141*, 3181–3191 (1995).
166. H. P. Fiedler, Tübingen, personal communication (1997).
167. J. Konisky, S. Soucek, K. Frick, J. K. Davies, and C. Hammond, *J. Bacteriol.*, *127*, 249–257 (1976).
168. K. Hantke, *Mol. Gen. Genet.*, *191*, 301–306 (1983).
169. K. Hantke, *FEMS Microbiol. Lett.*, *67*, 5–8 (1990).
170. N. A. Curtis, R. L. Eisenstadt, S. J. East, R. J. Cornford, L. A. Walker, and A. J. White, *Antimicrob. Agents Chemother.*, *32*, 1879–1886 (1988).
171. T. Maejima, M. Inoue, and S. Mitsuhashi, *Antimicrob. Agents Chemother.*, *35*, 104–110 (1991).
172. J. R. Butterton, J. A. Stoebner, S. M. Payne, and S. B. Calderwood, *J. Bacteriol.*, *174*, 3729–3738 (1992).
173. J. R. Butterton and S. B. Calderwood, *J. Bacteriol.*, *176*, 5631–5638 (1994).
174. C. Sauvage, T. Franza, and D. Expert, *J. Bacteriol.*, *178*, 1227–1231 (1996).
175. T. Franza and D. Expert, *J. Bacteriol.*, *173*, 6874–6881 (1991).
176. B. Mahe, C. Masclaux, L. Rauscher, C. Enard, and D. Expert, *Mol. Microbiol.*, *18*, 33–43 (1995).
177. T. G. Spiro, G. Bates, and P. Saltman, *J. Am. Chem. Soc.*, *89*, 5559–5562 (1967).
178. S. Hussein, K. Hantke, and V. Braun, *Eur. J. Biochem.*, *117*, 431–437 (1981).
179. V. Braun and C. Herrmann, unpublished observation (1997).
180. G. Schultz-Hauser, B. Van Hove, and V. Braun, *FEMS Microbiol. Lett.*, *95*, 231–234 (1992).
181. B. R. Otto, A. M. Verweij van Vught, and D. M. MacLaren, *Crit. Rev. Microbiol.*, *18*, 217–233 (1992).
182. I. Stojiljkovic and K. Hantke, *Mol. Microbiol.*, *13*, 719–732 (1994).
183. M. Mills and S. Payne, *J. Bacteriol.*, *177*, 3004–3009 (1995).
184. A. G. Torres and S. M. Payne, *Mol. Microbiol.*, *23*, 825–833 (1997).

185. D. P. Henderson and S. M. Payne, *Mol. Microbiol.*, *7*, 461–469 (1993).

186. D. P. Henderson and S. M. Payne, *J. Bacteriol.*, *176*, 3269–3277 (1994).

187. I. Stojiljkovic, V. Hwa, L. de Saint Martin, P. O'Gaora, X. Nassif, F. Heffron, and M. So, *Mol. Microbiol.*, *15*, 531–541 (1995).

188. I. Stojiljkovic, J. Larson, V. Hwa, S. Anic, and M. So, *J. Bacteriol.*, *178*, 4670–4678 (1996).

189. C. Elkins, C. J. Chen, and C. E. Thomas, *Infect. Immun.*, *63*, 2194–2200 (1995).

190. M. K. Stevens, S. Porcella, J. Klesney Tait, S. Lumbley, S. E. Thomas, M. V. Norgard, J. D. Radolf, and E. J. Hansen, *Infect. Immun.*, *64*, 1724–1735 (1996).

191. I. Maciver, J. L. Latimer, H. H. Liem, U. Muller Eberhard, Z. Hrkal, and E. J. Hansen, *Infect. Immun.*, *64*, 3703–3712 (1996).

192. L. A. Lewis, E. Gray, Y.-P. Wang, B. A. Roe, and D. W. Dyer, *Mol. Microbiol.*, *23*, 737–749 (1997).

193. C.-J. Chen, P. F. Sparling, L. A. Lewis, D. W. Dyer, and C. Elkins, *Infect. Immun.*, *64*, 5008–5014 (1996).

194. M. S. Hanson, S. E. Pelzel, J. Latimer, U. Muller Eberhard, and E. J. Hansen, *Proc. Natl. Acad. Sci. USA*, *89*, 1973–1977 (1992).

195. L. D. Cope, S. E. Thomas, J. L. Latimer, C. A. Slaughter, U. Muller Eberhard, and E. J. Hansen, *Mol. Microbiol.*, *13*, 863–873 (1994).

196. R. Schönherr, R. Tsolis, T. Focareta, and V. Braun, *Mol. Microbiol.*, *9*, 1229–1237 (1993).

197. L. D. Cope, R. Yogev, U. Muller Eberhard, and E. J. Hansen, *J. Bacteriol.*, *177*, 2644–2653 (1995).

198. G. P. Jarosik, I. Maciver, and E. J. Hansen, *Infect. Immun.*, *63*, 710–713 (1995).

199. S. Letoffe, J. M. Ghigo, and C. Wandersman, *Proc. Natl. Acad. Sci. USA*, *91*, 9876–9880 (1994).

200. S. Letoffe, J. M. Ghigo, and C. Wandersman, *J. Bacteriol.*, *176*, 5372–5377 (1994).

201. R. Binet and C. Wandersman, *Mol. Microbiol.*, *22*, 265–273 (1996).

202. J.-M. Ghigo, S. Letoffe, and C. Wandersman, *J. Bacteriol.*, *179*, 3572–3579 (1997).

203. B. R. Otto, J. G. Kusters, J. Luirink, F. K. Degraaf, and B. Oudega, *Infect. Immun.*, *64*, 4345–4350 (1996).

204. E. D. Weinberg, *Physiol. Rev.*, *64*, 65–102 (1984).

205. P. A. Mickelsen and P. F. Sparling, *Infect. Immun.*, *33*, 555–564 (1981).

206. P. A. Mickelsen, E. Blackman, and P. F. Sparling, *Infect. Immun.*, *35*, 915–920 (1982).

207. A. B. Schryvers and B. C. Lee, *Can. J. Microbiol.*, *35*, 409–415 (1989).

208. A. B. Schryvers and L. J. Morris, *Mol. Microbiol.*, *2*, 281–288 (1988).

209. J. E. Anderson, P. F. Sparling, and C. N. Cornelissen, *J. Bacteriol.*, *176*, 3162–3170 (1994).

210. S. W. Irwin, N. Averil, C. Y. Cheng, and A. B. Schryvers, *Mol. Microbiol.*, *8*, 1125–1133 (1993).

211. A. Pettersson, A. Maas, and J. Tommassen, *J. Bacteriol.*, *176*, 1764–1766 (1994).

212. A. Pettersson, P. van der Ley, J. T. Poolman, and J. Tommassen, *Infect. Immun.*, *61*, 4724–4733 (1993).

213. A. Petterson, J. T. Poolman, P. van der Ley, and J. Tommassen, *Antonie van Leeuwenhoek*, *71*, 129–136 (1997).

214. A. B. Schryvers and S. Gray Owen, *J. Infect. Dis.*, *165(Suppl.1)*, 5103–5104 (1992).

215. J. A. Ogunnariwo and A. B. Schryvers, *Avian. Dis.*, *36*, 655–663 (1992).

216. S. D. Gray Owen, S. Loosmore, and A. B. Schryvers, *Infect. Immun.*, *63*, 1201–1210 (1995).

217. P. J. Desai, A. Angerer, and C. A. Genco, *J. Bacteriol.*, *178*, 5020–5023 (1996).

218. A. Locke and E. R. Main, *J. Infect. Dis.*, *48*, 419–435 (1931).

219. C. M. Litwin and S. B. Calderwood, *Clin. Microbiol. Rev.*, *6*, 137–149 (1993).

220. C. E. Lankford, *Crit. Rev. Microbiol.*, *2*, 273–331 (1973).

221. J. F. Ernst, R. L. Bennett, and L. R. Rothfield, *J. Bacteriol.*, *135*, 928–934 (1997).

222. K. Hantke, *Mol. Gen. Genet.*, *182*, 288–292 (1981).

223. S. Schäffer, K. Hantke, and V. Braun, *Mol. Gen. Genet.*, *200*, 110–113 (1985).

224. M. J. Casadaban and S. N. Cohen, *Proc. Natl. Acad. Sci. USA*, *76*, 4530–4533 (1979).

225. S. Wee, J. B. Neilands, M. L. Bittner, B. C. Hemming, B. L. Haymore, and R. Seetharam, *Biol. Met.*, *1*, 62–68 (1988).

226. V. Braun, S. Schäffer, K. Hantke, and W. Tröger, *Coll. Mosbach*, *41*, 164–179 (1990).

227. M. Y. Hamed and J. B. Neilands, *J. Inorg. Biochem.*, *50*, 193–210 (1993).

228. T. Saito, D. Duly, and R. J. Williams, *Eur. J. Biochem.*, *197*, 39–42 (1991).

229. H. K. Hall and J. W. Foster, *J. Bacteriol.*, *178*, 5683–5691 (1996).

230. T. Saito, M. R. Wormald, and R. J. Williams, *Eur. J. Biochem.*, *197*, 29–38 (1991).

231. M. Y. Hamed and J. B. Neilands, *J. Inorg. Biochem.*, *53*, 235–248 (1994).

232. V. De Lorenzo, S. Wee, M. Herrero, and J. B. Neilands, *J. Bacteriol.*, *169*, 2624–2630 (1987).

233. V. De Lorenzo, M. Herrero, F. Giovannini, and J. B. Neilands, *Eur. J. Biochem.*, *173*, 537–546 (1988).

234. E. Le Cam, D. Frechon, M. Barray, A. Fourcade, and E. Delain, *Proc. Natl. Acad. Sci. USA*, *91*, 11816–11820 (1994).

235. I. Stojiljkovic and K. Hantke, *Mol. Gen. Genet.*, *247*, 199–205 (1995).

236. L. Holm, C. Sander, H. Ruterjans, M. Schnarr, R. Fogh, R. Boelens, and R. Kaptein, *Protein Eng.*, *7*, 1449–1453 (1994).

237. R. Schneider and K. Hantke, *Mol. Microbiol.*, *8*, 111–121 (1993).

238. J. Boyd, M. N. Oza, and J. R. Murphy, *Proc. Natl. Acad. Sci. USA*, *87*, 5968–5972 (1990).

239. M. P. Schmitt and R. K. Holmes, *Infect. Immun.*, *59*, 3903–3908 (1991).

240. X. Tao and J. R. Murphy, *J. Biol. Chem.*, *267*, 21761–21764 (1992).

241. X. Tao and J. R. Murphy, *Proc. Natl. Acad. Sci. USA*, *91*, 9646–9650 (1994).

242. X. Qiu, C. L. Verlinde, S. Zhang, M. P. Schmitt, R. K. Holmes, and W. G. Hol, *Structure*, *3*, 87–100 (1995).

243. X. Ding, H. Zeng, N. Schiering, D. Ringe, and J. R. Murphy, *Nat. Struct. Biol.*, *3*, 382–387 (1996).

244. K. Günter-Seeboth and T. Schupp, *Gene*, *166*, 117–119 (1995).

245. O. Dussurget, M. Rodriguez, and I. Smith, *Mol. Microbiol.*, *22*, 535–544 (1996).

246. I. Kim, A. Stiefel, S. Plantör, A. Angerer, and V. Braun, *Mol. Microbiol.*, *23*, 333–344 (1997).

247. C. Härle, I. Kim, A. Angerer, and V. Braun, *EMBO J.*, *14*, 1430–1438 (1995).

248. M. Ochs, S. Veitinger, I. Kim, D. Welz, A. Angerer, and V. Braun, *Mol. Microbiol.*, *15*, 119–132 (1995).

249. M. A. Lonetto, K. L. Brown, K. E. Rudd, and M. J. Buttner, *Proc. Natl. Acad. Sci. USA*, *91*, 7573–7577 (1994).

250. A. Angerer, S. Enz, M. Ochs, and V. Braun, *Mol. Microbiol.*, *19*, 163–174 (1995).

251. S. Enz, V. Braun, and J. Crosa, *Gene*, *163*, 13–18 (1995).

252. M. Ochs, A. Angerer, S. Enz, and V. Braun, *Mol. Gen. Genet.*, *250*, 455–465 (1996).

253. M. Koster, W. van Klompenburg, W. Bitter, J. Leong, and P. Weisbeek, *EMBO J.*, *13*, 2805–2813 (1994).

254. K. Hantke, *Mol. Gen. Genet.*, *210*, 135–139 (1987).

255. D. J. Hasset, P. A. Sokol, M. L. Howell, J. F. Ma, H. T. Schweizer, U. Ochsner, and M. L. Vasil, *J. Bacteriol.*, *178*, 3996–4003 (1996).

256. C. E. Thomas and P. F. Sparling, *J. Bacteriol.*, *178*, 4224–4232 (1996).

257. M. Ghassemian and N. A. Straus, *Microbiol.*, *142*, 1469–1476 (1996).

258. J. A. Fee, *Mol. Microbiol.*, *5*, 2599–2610 (1991).

259. U. A. Ochsner, Z. Johnson, I. L. Lamont, H. E. Cunliffe, and M. L. Vasil, *Mol. Microbiol.*, *21*, 1019–1028 (1996).

260. V. Venturi, C. Ottevanger, J. Leong, and P. J. Weisbeek, *Mol. Microbiol.*, *10*, 63–73 (1993).

261. D. E. Heinrichs and K. Poole, *J. Bacteriol.*, *175*, 5882–5889 (1993).

262. D. E. Heinrichs and K. Poole, *J. Bacteriol.*, *178*, 2586–2592 (1996).

263. J. D. Fetherston, S. W. Bearden, and R. D. Perry, *Mol. Microbiol.*, *22*, 315–325 (1996).

264. H. Drechsel, H. Stephan, R. Lotz, H. Haag, H. Zähner, K. Hantke, and G. Jung, *Liebigs Ann.*, *1995*, 1727–1733 (1995).

265. C. E. Chambers, D. D. McIntyre, M. Mouck, and P. A. Sokol, *BioMetals*, *9*, 157–167 (1996).

266. C. R. Dean, S. Neshat, and K. Poole, *J. Bacteriol.*, *178*, 5361–5369 (1996).

267. C. H. Liao, D. E. McCallus, J. M. Wells, S. S. Tzean, and G. Y. Kang, *Can. J. Microbiol.*, *42*, 177–182 (1996).

268. J. P. Zhang and S. Normark, *Nature*, *273*, 1234–1236 (1996).

269. P. C. Giardina, L.-A. Foster, J. M. Musser, B. J. Akerley, J. F. Miller, and D. W. Dyer, *J. Bacteriol.*, *177*, 6058–6063 (1995).

270. M. B. Goldberg, S. A. Boyko, and S. B. Calderwood, *Proc. Natl. Acad. Sci. USA*, *88*, 1125–1129 (1991).

271. P. C. Salinas and J. H. Crosa, *Gene*, *160*, 17–23 (1995).

272. Q. Chen and J. H. Crosa, *J. Biol. Chem.*, *271*, 18885–18891 (1996).

273. L. S. Waldbeser, Q. Chen, and J. H. Crosa, *Mol. Microbiol.*, *17*, 747–756 (1995).

274. F. M. Hamtash, M. Immeraan, and C. Earhart, *Microbiol.*, *143*, 147–156 (1997).

275. J. M. Hornung, H. A. Jones, and R. D. Perry, *Mol. Microbiol.*, *20*, 725–739 (1996).

4

Molecular Biology of Iron Transport in Fungi

Sally A. Leong[1] *and Günther Winkelmann*[2]

[1]USDA/ARS and Department of Plant Pathology, University of Wisconsin—Madison, Madison, WI 53706, USA

[2]Microbiology & Biotechnology, University of Tübingen, D-72076 Tübingen, Germany

1. INTRODUCTION 148
2. FUNGAL SIDEROPHORES 149
3. BIOCHEMISTRY OF SIDEROPHORE TRANSPORT 153
4. MOLECULAR BIOLOGY OF SIDEROPHORE BIOSYNTHESIS AND TRANSPORT 156
    - 4.1. Genetics of Hydroxamate Siderophore Biosynthesis 156
        - 4.1.1. Rhodotorulic Acid 156
        - 4.1.2. Ferrichrome 158
    - 4.2. Molecular Biology of Siderophore Biosynthesis 160
        - 4.2.1. *sid1* 160
        - 4.2.2. *sid2* 161
    - 4.3. Genetics of Siderophore Regulation in *Ustilago* 163
        - 4.3.1. *urbs1* 164
        - 4.3.2. Model for Regulation of *sid1* 167
        - 4.3.3. Iron-Mediated Regulation of *sid1* 169

4.4. Iron Transport in *Ustilago* 170

5. MOLECULAR BIOLOGY OF IRON TRANSPORT IN YEAST 171
5.1. *FRE1* 171
5.2. *FRE2* 173
5.3. A New Genetic Strategy 174
5.4. An Integrated Picture 174
5.5. Analogies with Iron Transport Systems in Other Organisms 175
5.6. Regulation of Iron Transport in Yeast 175
5.7. Low-Affinity Iron Transport in Yeast 176

6. CONCLUSIONS 176

ACKNOWLEDGMENT 178

ABBREVIATIONS 178

REFERENCES 179

1. INTRODUCTION

Siderophores are common not only in the prokaryotic bacteria but also among the eukaryotic fungi and some higher eukaryotic organisms; that is, monocotyledonous plants. They can be defined as low-molecular virtually ferric-specific iron chelators that are excreted under iron-limiting conditions and regulated by the internal iron concentration. Several recent books have focused on the various siderophore structures and their function and regulation [1–4]. The nomenclature of siderophores is still not uniform and in most cases the prefix *deferri-* or *desferri-* is used to denote the iron-free ligand. It has recently been suggested the prefix *pro-* be used for the iron-free forms of the bacterial ferrioxamines, that is, proferrioxamine B, instead of deferri-ferrioxamine B [5]. Because many fungal siderophores have been first isolated and elucidated as the ferric complex, the name consequently refers to the intact iron complex. However, owing to new isolation techniques for most ligands, the tendency to name the iron-free siderophores is increasing.

2. FUNGAL SIDEROPHORES

Historically, ferrichrome, isolated from culture filtrates of the fungus *Ustilago sphaerogena*, was the first siderophore [6]. Siderophore screening within the Basidiomycetes and Ascomycetes revealed a variety of structural variations of the prototype ferrichrome molecule (Fig. 1a). Thus, the cyclic -Orn-Orn-Orn-Gly-Gly-Gly- motif may be altered to give other cyclic hexapeptide siderophores. One or two Gly residues may be replaced by Ser or Ala resulting in a number of structurally related cyclic hexapeptide ferrichromes (reviewed in Ref. 7). Thus the most common siderophores of the ferrichrome class, collectively called ferrichromes, are ferrichrome, ferricrocin, ferrichrome C, ferrichrysin, ferrirubin, ferrirhodin, and ferrichrome A. Whereas in the former four ferrichromes, the three δ-*N*-acyl residues consist of acetyl groups, the latter three possess instead a *cis*-anhydromevalonyl (ferrirubin), *trans*-anhydromevalonyl (ferrihodin), or the acidic *trans*-β-methylglutaconyl residue (ferrichrome A), respectively.

Most fungi of the *Aspergillus* and *Penicillium* group produce rather a collection of different ferrichromes than a single compound which corresponds to the general observation of structural variations within secondary metabolites. Thus, the eukaryotic fungi have similar characteristic and important traits of secondary metabolism as found within the prokaryotic bacteria.

In addition to the well-known ferrichromes, the group of van der Helm has described a new subclass called asperchromes [8], containing a combination of δ-*N*-acetyl and δ-*N*-anhydromevalonyl residues in one molecule, suggesting that in this organism the biosynthetic enzymes are not as stringent with regard to hydroxamate acyl at the δ-*N*-ornithine termini. Also, the size of the cyclic peptide ring may vary. Thus, from a culture of *Neovossia indica*, a heptapeptide, named tetraglycylferrichrome, was isolated containing four Gly residues in addition to the triornithine sequence [9]. However, as has been pointed out earlier, the fact that residue 6 in all ferrichromes is always glycine, suggests a structural role for the cyclic peptide conformation [10]. An unusual derivative of ferrirhodin is des(diserylglycyl)ferrirhodin (DDF) representing an open-ring peptide siderophore which has been isolated from low-iron cultures of *Agaricus ochraceus* [11] and *A. bisporus* (white mushroom) [12].

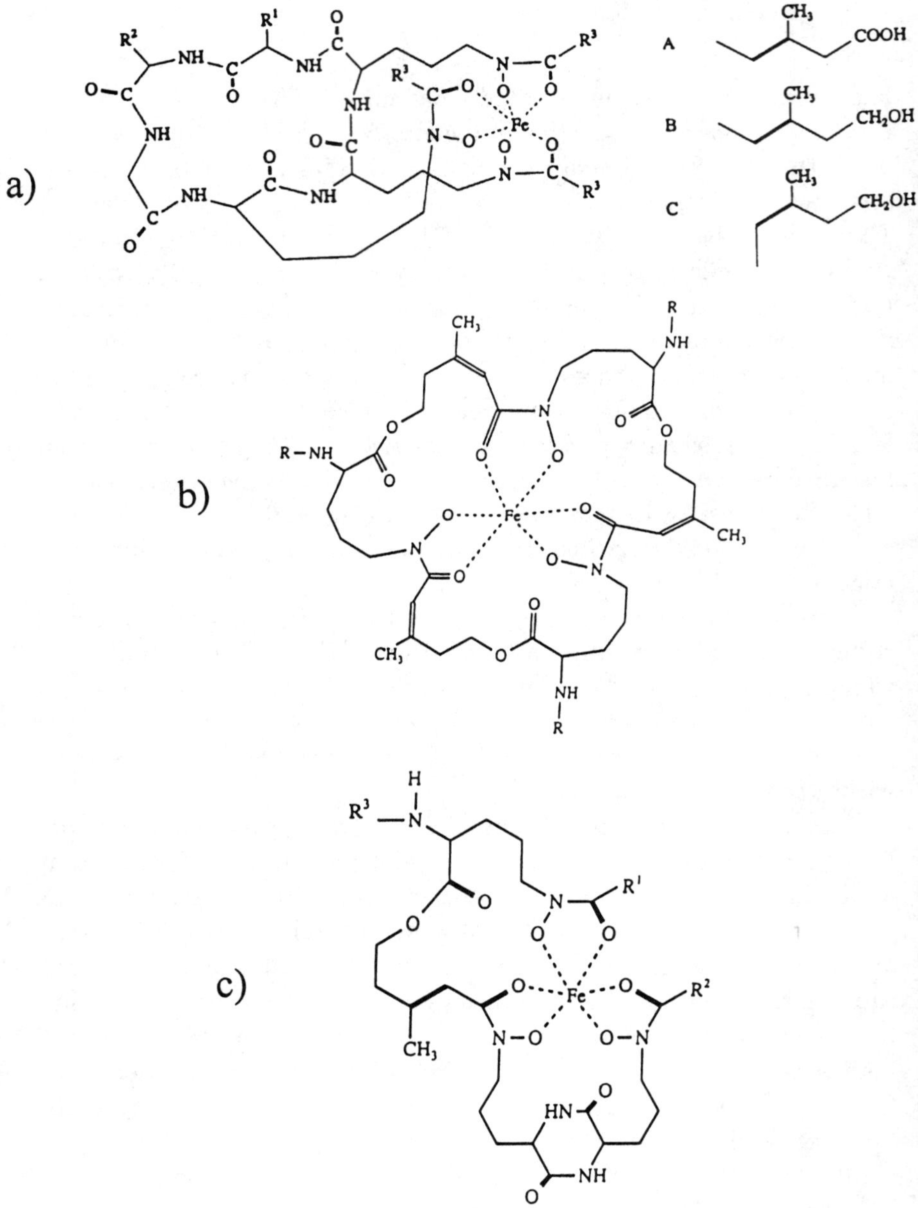

FIG. 1. (a) Ferrichrome ($R^1 = R^2 = H$, $R^3 = CH_3$), ferrichrome A ($R^1 = R^2 = H$, $R^3 =$ A), ferricrocin ($R^1 = H$, $R^2 = CH_2OH$, $R^3 = CH_3$), ferrichrome C ($R^1 = H$, $R^2 = CH_3$,

d) 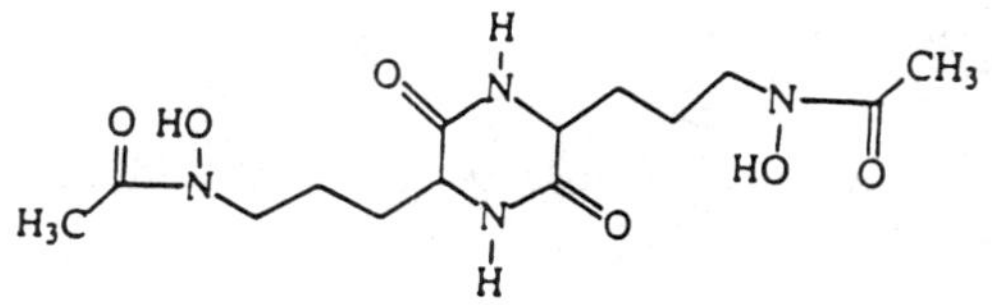

e) 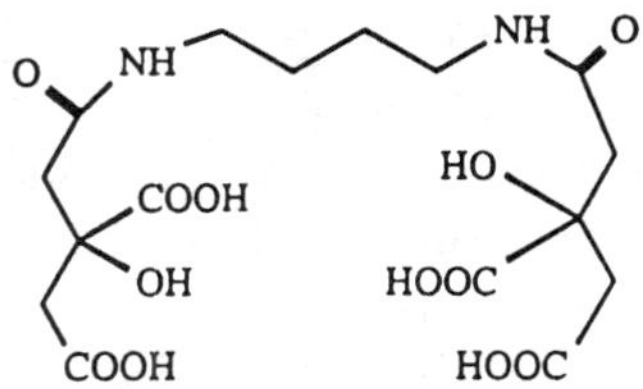

$R^3 = CH_3$), ferrichrysin ($R^1 = R^2 = CH_2OH$, $R^3 = CH_3$), ferrirubin ($R^1 = R^2 = CH_2OH$, $R^3 = B$), ferrirhodin ($R^1 = R^2 = CH_2OH$, $R^3 = C$), asperchrome B1 ($R^1 = R^2 = CH_2OH$, $R^{3*} = 2B, 1CH_3$), asperchrome D1 ($R^1 = R^2 = CH_2OH$, $R^{3*} = 1B, 2\,CH_3$) (* = position not confirmed); (b) fusarinine C (R = H), triacetylfusarinine C (R = acetyl); (c) coprogen ($R^1 = R^2 = B$, R^3 = acetyl), coprogen B ($R^1 = R^2 = B$, $R^3 = H$), neocoprogen I ($R^1 = B$, $R^2 = CH_3$, R^3 = acetyl), neocoprogen II ($R^1 = R^2 = CH_3$, R^3 = acetyl); (d) rhodotorulic acid; (e) rhizoferrin.

In addition to the above-mentioned cyclic peptides of the ferrichromes and asperchromes, a group of cyclic esters have been isolated from strains of *Fusarium* and related genera, which are called fusarinines or fusigens [13,14]. Triacetylfusarinine C (triacetylfusigen) is the prototype siderophore of the fusarinine class [15]. Both, the nonacetylated form (fusarinine C = fusigen) and the acetylated form (*N,N′,N″*-triacetyl-fusarinine C = triacetylfusigen) are commonly found in low-iron culture filtrates of fungi (see Fig. 1b). As the chemical stability of these ester-containing siderophores is generally low, a variety of hydrolysis products can be expected in culture filtrates; that is, linear trimer, dimer, and monomer. However, it is still unresolved whether the linear forms represent active siderophores, precursors, or breakdown products. Owing to the high-stability constant of triacetylfusarinine C, iron exchange reactions have been observed which seem to be responsible for iron transfer from other siderophores or even from the monomer to the cyclic trimer which finally completes the transport of iron into the cell [16]. The essential building block for all fusarinines is the monomer *cis*-

fusarinine, although the *trans*-fusarinine monomer has also been found in fungal cultures [17–19].

Trans-configurated fusarinines represent essential constituents of the coprogen class (see Fig. 1c) which so far comprise 11 structurally different and naturally occurring derivatives [20]. Coprogen contains two *trans*-fusarinine moieties connected head to head to form a diketopiperazine unit (dimerum acid) and a third *trans*-fusarinine molecule esterified to the terminal OH group of dimerum acid [21]. Rhodotorulic acid (see Fig. 1d), produced by the basidiomycetous yeast *Rhodotorula*, is structurally related to coprogen, consisting of a diketopiperazine unit possessing acetyl instead of anhydromevalonyl residues.

A common feature of all these fungal siderophores is the hydroxamate group known to be a strong iron binding bidentate. Three bidentates are required to form an octahedral ferric trishydroxamate complex. The chelate effect increases the formation constant so that hexadentate ligands predominate in nature. Owing to the presence of only two bidentates in rhodotorulic acid, three ligand molecules (dihydroxamates) are required to bind two iron atoms giving a Fe_2L_3 complex. Thermodynamic aspects and formation constants have been discussed in detail by Crumbliss [22].

Most interestingly, not all fungi make use of the hydroxamate group as an iron-binding ligand. As shown in our laboratory, fungi of the Zygomycetes group favor carboxylate-type siderophores. The prototype of the fungal carboxylate-type siderophores, rhizoferrin (see Fig. 1e), was first isolated from *Rhizopus microsporus* [23–25], a member of the Mucorales (Zygomycetes). Subsequent screening of related families within the order Mucorales revealed that rhizoferrin is widespread among these fungi. Even in the related order Enteromophthorales, rhizoferrin production has been detected [26]. The coordination chemistry of this carboxylate-type siderophore, the iron(III) complex and its metal analogues has been studied recently [27]. It is interesting to note that rhizoferrin-producing fungi of the Mucorales have been shown to contain ferritin as an iron storage protein [28], which so far has not been observed in hydroxamate-producing fungi. Mössbauer studies confirmed that most Ascomycetes and Basidiomycetes can use siderophores as internal iron reservoirs which are also incorporated into conidiospores during spore formation [29,30]. The Zygomycetes seem to form different ferritin types, mycelia-specific (zygoferritin) and spore-specific ferritins (mycoferritin), of which zygoferritin exhibits iron cores similar to bacterioferritin [31]. Thus, transport and storage of iron are two closely linked strategies of fungal metabolism, which deserve further attention when iron homeostasis is considered.

3. BIOCHEMISTRY OF SIDEROPHORE TRANSPORT

The transport of fungal siderophores has been extensively discussed in previous reviews [2,20,32]. Most of the research in this area is based on kinetic studies. Different siderophore-mediated iron transport mechanisms are possible in fungi, depending on the siderophores used and the organisms studied. A characteristic feature of all fungal transport systems is the pronounced specificity of transport. Most fungi recognize only certain siderophores which are preferably excreted by the producing fungus [33,34]. Detailed structure-function relationships have been elaborated with radiolabeled ferrichromes and its derivatives [35,36]. The fact that structurally different ferrichromes are taken up in *Neurospora* and related *Aspergillus* and *Penicillium* strains with different efficiencies has led to the view that the overall structure of the molecules and the number and kind of *N*-acyl residues play an important role in recognition and transport. How the transport system of *Neurospora crassa* responds to structural changes of the δ-*N*-acyl residues of ferrichromes and coprogens is shown in Fig. 2. Although exchange of the anhydromevalonyl versus acetyl groups in coprogen decreased transport rates (Fig. 2a), the gradual replacement of the δ-*N*-acetyl groups by anhydromevalonyl groups in the ferrichrome series inhibited transport rates (Fig. 2b), suggesting that the structural determinant for transport of each siderophore type is different [37]. From this we may conclude that different membrane-associated recognition sites are operative in fungi. However, subsequent to this step it was found that both siderophore types, although structurally unrelated behaved competitively during ferrichrome transport. This unexpected behavior was explained on the basis of different receptors (binding sites) but common transport systems in fungi.

When coprogen was labeled with a photoaffinity label (*p*-azidobenzoyl), transport and competition in the dark was observed, and on

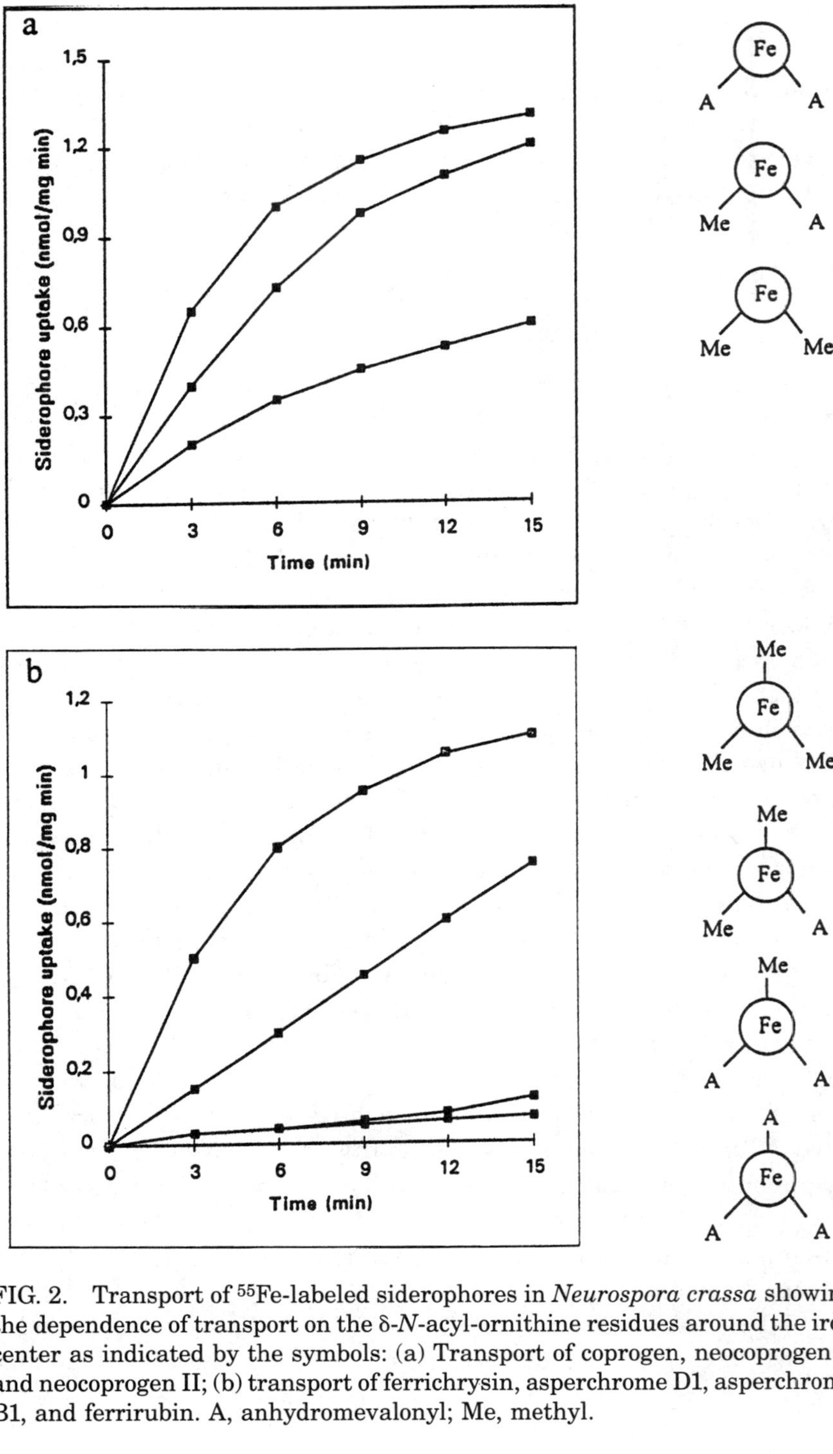

FIG. 2. Transport of ^{55}Fe-labeled siderophores in *Neurospora crassa* showing the dependence of transport on the δ-*N*-acyl-ornithine residues around the iron center as indicated by the symbols: (a) Transport of coprogen, neocoprogen I, and neocoprogen II; (b) transport of ferrichrysin, asperchrome D1, asperchrome B1, and ferrirubin. A, anhydromevalonyl; Me, methyl.

illumination, transport of ferrichrysin, a ferrichrome-type siderophore, was also inhibited, indicating that the transport of both siderophores was affected [38]. Although no siderophore transport protein could be identified in fungi so far, the kinetic data are in favor of one or more membrane-located transport system(s).

A comparison of membrane proteins in *Neurospora crassa* grown under iron-sufficient and iron-deficient conditions revealed no difference in the protein bands when analyzed by sodium dodecylsulfate (SDS)/polyacrylamide gel electrophoresis [39], which precludes overproduction of transport proteins as found in the outer membranes of gram-negative bacteria. However, transport rates of low-iron-grown cells was enhanced about fivefold, suggesting that either the number of transport proteins per cell or the activity has been increased during iron-deficient growth. A reductive transport mechanism was suggested for ferrichrome A uptake in *U. sphaerogena* [40]. According to this suggestion, binding to a specific recognition site may still occur, but the metal is removed from the strong chelate prior to its entry into the cell.

A specific aspect of siderophore transport is the stereospecificity of transport and recognition. The availability of enantioferrichrome enabled us to study the stereospecific recognition of ferrichrome in *Neurospora* and some related fungi [41]. Although a clear stereospecificity of ferrichrome uptake could not be observed in *Escherichia coli*, fungi behaved in a strictly enantioselective manner [42]. A similar conclusion has been drawn from studies with chromium-substituted ferrichromes in *Ustilago* [43].

It is well known that fungi possess an effective proton translocating plasma membrane adenosine triphosphatase (ATPase) which contributes to a membrane potential (inside negative) and which in turn is responsible for proton symport mechanisms. There is evidence that the membrane potential is also the driving force for siderophore transport in fungi [44]. Significant transport inhibition occurred after transient membrane depolarization which could be made permanent by simultaneously blocking the membrane ATPase by ATPase inhibitors. Owing to the strong activity of proton pumps in *Neurospora crassa*, a proton symport is the most likely mechanism for siderophore transport [45] which would also correspond to the view of a common transporter and different binding sites. As has been mentioned earlier, the fate of siderophores after entering the fungal cell is dependent on the type of siderophores and the organism studied. However, for all hydroxamate siderophores, a reductive removal of iron by biological reductants (NADH,

NADPH) seems possible, as the redox potential of these siderophores is around −400 to −450 mV versus nernst hydrogen electrode (NHE) [46]. Thus, enzymatic release of iron from siderophores has been studied in *N. crassa* and *A. fumigatus* [47].

Recent reports in the literature show that siderophores can be isolated from various fungi of different origin, that is, mycorrhiza [48], soil-borne plant pathogens [49], and dermatophytes [50], so that their function in the environment is obvious.

4. MOLECULAR BIOLOGY OF SIDEROPHORE BIOSYNTHESIS AND TRANSPORT

Despite a long history of studies on the biochemistry of biosynthesis and uptake of siderophores in fungi [51–54], progress on the molecular biology of these processes is very recent and of limited scope by comparison to the extensive work that has been done on siderophore systems of many bacteria (Chap. 3). This delay can be attributed in part to slow progress in the development of the tools needed for molecular genetic studies in fungi [55,56]. Although a gene transfer system was already in place for *Saccharomyces cerevisiae* in the late 1970s, these tools became routine in other fungi only a decade later. In many cases, these tools are still not available. In addition, Mendelian genetic analysis is not possible in many fungi. Finally, most researchers focusing on fungal siderophore systems have simply not embraced a molecular biology approach.

4.1. Genetics of Hydroxamate Siderophore Biosynthesis

Mutational studies of siderophore biosynthesis have been pursued in two fungi, *Rhodotorula pilimanae* [57] and *Ustilago maydis* [52].

4.1.1. Rhodotorulic Acid

R. pilimanae produces the dihydroxamate rhodotorulic acid (see Fig. 1d), a compound made by many basidiomycete yeasts. Biochemical studies conducted by Akers and Neilands [58] demonstrated that mo-

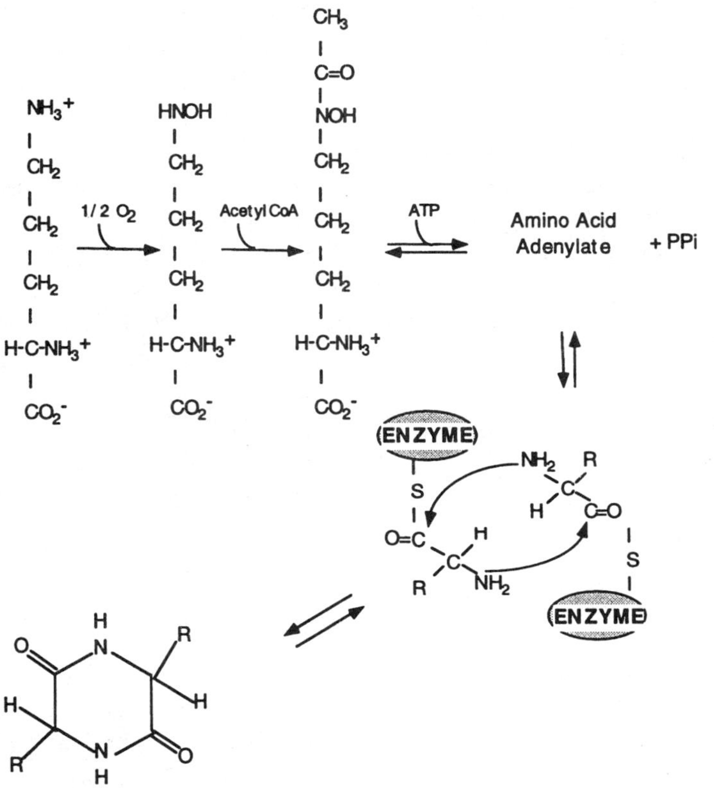

FIG. 3. Biosynthetic pathway of rhodotorulic acid.

lecular oxygen is used to hydroxylate ornithine to give δ-*N*-hydroxyornithine (Fig. 3). In the second step, δ-*N*-hydroxyornithine is acetylated by δ-*N*-hydroxyornithine:acetyl-CoA-*N*-transacetylase. The enzyme was isolated from *R. pilimanae* [59]. Two molecules of δ-*N*-acetyl-*N*-hydroxyornithine are then condensed to form the diketopiperazine ring characteristic of rhodotorulic acid. A δ-*N*-acetyl-N-hydroxyornithine-dependent ATP-pyrophosphate (PPi) exchange activity has been observed in *Rodotorula* extracts [60] as well as extracts from other fungi [61,62] suggesting that cyclization may occur via covalent bonding of the substrate as a thioester to an enzyme carrier such as occurs in nonribosomal peptide synthesis.

Ethyl methane sulfonate mutagenesis of the *R. pilimanae* led to the isolation of 12 mutants that were defective in biosynthesis of rhodotorulic acid [57]. These were identified by their inability to feed iron to the *Escherichia coli* mutant RW193 which is unable to produce its own siderophore enterobactin [63]. This bacterial strain grows poorly in nutrient agar medium containing 0.1 mmol deferriferrichrome A but demonstrates normal growth when rhodotorulic acid or a colony of *R. pilmaneae* is placed on the medium. Of the 12 mutants obtained, 9 continued to produce low levels of rhodotorulic acid, whereas three mutants produced undetectable amounts of the siderophore. One mutant was studied further and shown to produce a novel hydroxamate lacking a free α amino group and containing δ-*N*-hydroxyornithine. Based on this result, it was postulated that this mutant is blocked in the cyclization step of biosynthesis. Unfortunately, further studies of this pathway were not pursued owing to the lack of a genetic and transformation system for *R. pilimaneae*.

4.1.2. Ferrichrome

In a second study, a series of mutants blocked in ferrichrome siderophore production in *U. maydis* were obtained using the mutagen *N*-methyl-*N′*-nitro-*N*-nitrosoguanidine [64]. Mutants were identified by their inability to support the growth of a *Salmonella typhimurium* LT-2 enb-7 indicator strain defective in production of its endogenous siderophore enterobactin [65]. This assay takes advantage of the ability of *S. typhimurium* to take up the fungal siderophore ferrichrome. Like *E. coli*, *S. typhimurium* is not able to take up ferrichrome A; therefore, this siderophore is “invisible” in the assay. Subsequent genetic and biochemical characterization of these mutants identified two complementation groups. *sid1* mutants were unable to produce ferrichrome and ferrichrome A and were found to be defective in ornithine-N^5-oxygenase [64,66], whereas *sid2* mutants were unable to produce ferrichrome but still produced ferrichrome A [64]. Thus, *sid1* mutants were blocked in a common step in ferrichrome and ferrichrome A biosynthesis, whereas *sid2* mutants were blocked in a branch pathway step required for ferrichrome biosynthesis (Fig. 4). Crosses involving *sid1* and *sid2* mutants revealed tight linkage of the two genes (7% recombination) suggesting these siderophore genes are clustered in *U. maydis*.

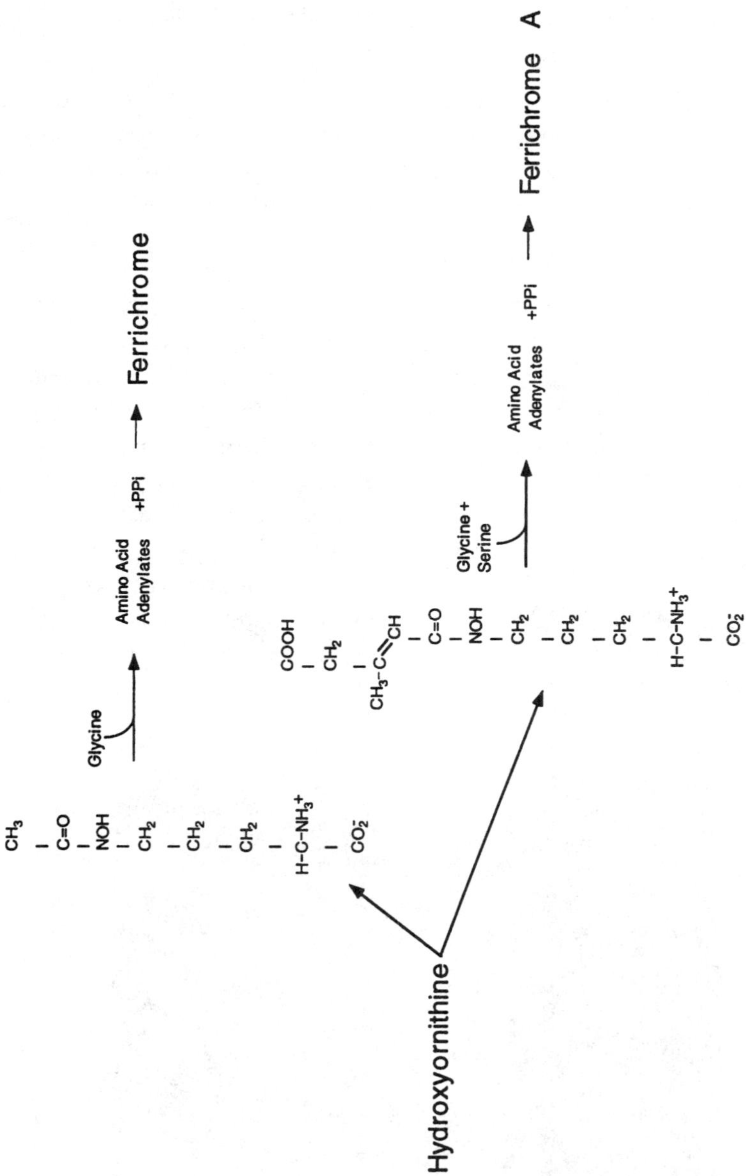

FIG. 4. Biosynthetic pathway of ferrichromes. Steps leading to the biosynthesis of δ-*N*-acetyl-δ-*N*-hydroxyornithine are as in Fig. 3.

4.2. Molecular Biology of Siderophore Biosynthesis

Molecular genetic tools including a stable gene transfer system [67], electrophoretic karyotype [68], and the ability to conduct single-step gene replacement [69] were developed for *U. maydis* making possible the cloning and characterization of *sid1* and *sid2*.

4.2.1. sid1

The *sid1* gene was isolated by introducing a library of wild-type *U. maydis* genomic DNA into a *sid1* mutant and asking for restoration of the ability to produce a growth halo on the *Salmonella* indicator [64] (Fig. 5). Ferrichrome-producing colonies were isolated and the transforming cosmid DNA recovered by transformation of *E. coli*. Subcloning led to a minimum-sized complementing *Ssp*I fragment of 7.1 kb [66]. DNA sequence analysis of the fragment revealed an open reading frame encoding a protein of 570 amino acids with sequence identity at the N-terminus with ornithine-N^5-oxygenase of *Pseudomonas aeruginosa* [70] (Fig. 6). Disruption of the gene led to a loss in siderophore produc-

FIG. 5. Growth halo (arrow) produced by *U. maydis sid1* mutant transformed with wild-type *sid1* gene.

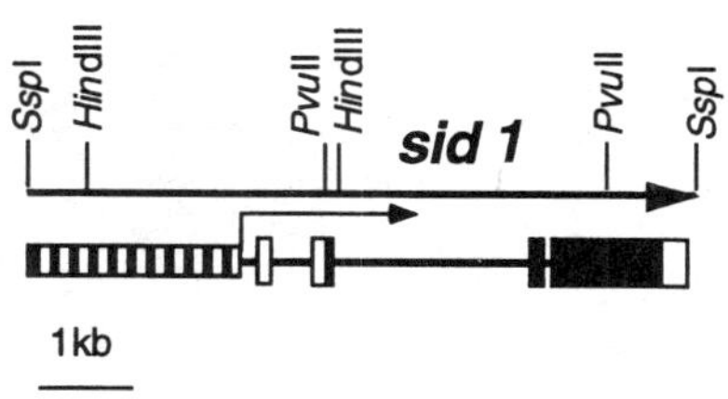

FIG. 6. *sid1* gene of *U. maydis*. Arrow above gene marks the start of transcription. Stripped box indicates the promoter region and boxes downstream of arrow indicate exons: open box, untranslated exon; black box, translated exon.

tion and ornithine-N^5-oxygenase activity [66]. Overexpression of the cDNA from the *U. maydis* glyceraldehyde-3-*P*-dehydrogenase (GAPDH) promoter gave rise to 5X elevated levels of enzyme (B. Mei and S. A. Leong, unpublished finding). Taken together, these data confirm that the structural gene for ornithine-N^5-oxygenase had been cloned from *U. maydis*.

4.2.2. sid2

Despite numerous attempts to clone *sid2* by direct complementation with a *Ustilago* genomic DNA library, no candidate clones were obtained. These results suggested that the gene might be very large and/or the genomic DNA library employed did not contain the intact gene. As mentioned above, genetic analysis showed that *sid1* and *sid2* are tightly linked. As the ratio of physical to genetic distance in *Ustilago* is similar to that in yeast (2–3 kb/cM), attempts were made to locate *sid2* by transforming the *sid2* mutant with clones overlapping and extending down the genome from *sid1* (Fig. 7). Clones obtained from a new cosmid library of *Ustilago* genomic DNA prepared in a genome walking vector [71] were analyzed in this manner, and clones containing *sid2* were obtained. The organization of the *sid1/sid2* locus is shown in Fig. 7.

DNA sequence analysis of the *sid2* minimal complementing region has been completed and revealed an open reading frame of 3947 amino acids (G. Gentil, W. Yuan, A. Budde, H. von Dohren, and S. A. Leong, unpublished findings). Significant sequence identity was found throughout the gene to numerous, conserved motifs found in nonribosomal

A

B

```
Domain 1          VGLDS............LGDSVS
Domain 2          AGLDS............LGLVLS
Domain 3          LGIDS..............L(220 amino acids)GLDS

Consensus  XGLDS
```

FIG. 7. (A) Ferrichrome siderophore gene cluster of *U. maydis*. (B) Sequence identities of the *U. maydis sid2* gene and acyl carrier domains of nonribosomal peptide synthetases.

peptide synthetases [72]. *sid2* is organized into three amino acid-activating domains of about 1000 amino acids in length. Presumably Sid2p participates in production of the ferrichrome cyclic peptide either by synthesizing the triglycyl peptide or triornithyl peptide or perhaps the entire hexapeptide (see Fig. 4). The latter possibility is made plausible by the finding of two putative acyl carrier motifs in each of the three amino acid-activating domains of Sid2p. Thus, each domain might be used more than one time. Disruption of *sid2* gave rise, as expected, to a non-ferrichrome-producing phenotype.

The mechanism of peptide bond formation in ferrichrome has not been investigated in *U. maydis*. In *Aspergillus quadricinctus* [73], iron-regulated δ-*N*-acetyl-*N*-hydroxyornithine and glycine-dependent ATP-PP exchange activities were identified, implicating enzymatic activation to aminoadenylates to form the peptide (see Fig. 4). The purified enzyme had a molecular weight (MW) of 9.6×10^5 Da as determined by gel filtration. The enzyme complex could be dissociated into subunits of 9×10^4 Da having acetylhydroxyornithine activating function and 4×10^5 Da having glycine-activating function. Future studies will focus on determining the enzymatic activity encoded by *sid2*. Based on the size of the protein encoded by *sid2*, it may encode the glycine-activating enzyme. The function of the amino acid-activating domains and the multiple acyl carrier motifs is being studied through direct assay of the wild-type enzyme and mutant enzymes created by deletion of domains and site-directed mutagenesis of the acyl carrier motifs.

Based on recent findings from other siderophore biosynthetic systems in bacteria, we anticipate finding other needed functions to produce ferrichrome, including δ-*N*-hydroxyornithine:acetyl-CoA-*N*-transacetylase [74], phosphopantethiene transferase [75], and a multidrug (MDR) transporter for export of siderophores [76]. Factors required for activation of transcription of siderophore genes is also anticipated. Evidence for transcriptional activation of siderophore genes has been found in several prokaryotes [77–79]. A two-component positive regulatory system is required for expression of the ferric enterobactin receptor of *P. aeruginosa* [79,80] and the iron dicitrate transport system of *E. coli* [81]. In the case of the *P. aeruginosa* ornithine-N^5-oxygenase gene, *pvdA*, a novel sigma factor *pvdS* is required for transcription of the oxygenase gene under iron deficiency conditions of growth [82]. Similar regulatory pathways may also be operative in *U. maydis*.

4.3. Genetics of Siderophore Regulation in *Ustilago*

Like in many prokaryotes [83] (Chapt. 3), a repressive mechanism of siderophore gene regulation is clearly operative in *U. maydis*. Siderophore production is repressed at 10 μM added iron in a low-iron minimal medium [84]. Likewise, the abundance of the *sid1* transcript was greatly reduced under iron-repressive growth conditions [66] and, when the *sid1* promoter was fused to a reporter gene such as β-glucuronidase (GUS) of *E. coli* [85], the reporter activity was elevated 16-fold under low-iron growth conditions when compared with high-iron growth conditions [86] (Fig. 8).

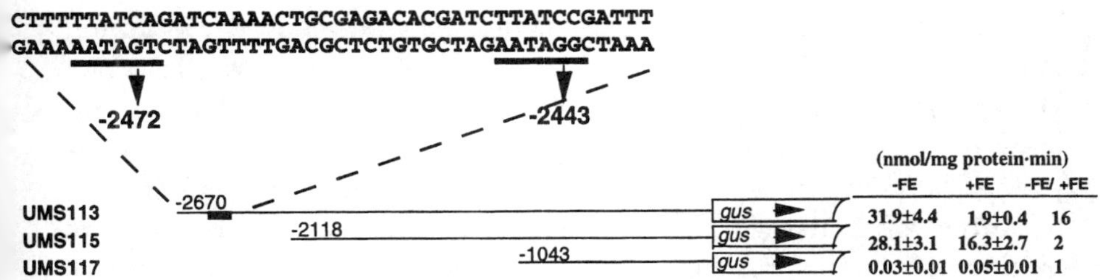

FIG. 8. GUS activity in *U. maydis* strains carrying a single copy of the *sid1* promoter linked to the *E. coli* GUS gene at the *sid1* locus. Strains 115 and 117 contain deletions of the promoter created by Bal31 nuclease.

Mutants constitutive for siderophore biosynthesis were isolated after *N*-methyl-*N'*-nitro-*N*-nitrosoguanidine mutagenesis using *S. typhimurium* LT-2 TA2701 as a biological indicator [87]. In contrast to *S. typhimurium* enb-7, this strain is unable to utilize ferrichrome owing to a mutation in the outer membrane receptor *fhuA* for ferrichrome uptake; thus its growth is inhibited by siderophores secreted by *Ustilago*. By conducting the assay under high-iron growth conditions, when siderophores are normally not secreted by *Ustilago*, it was possible to isolate mutants that constitutively expressed siderophore by their ability to inhibit growth of the *Salmonella* indicator. These mutants expressed siderophores in both low- and high-iron growth media [87]. Three mutants, *urbs1-1*, *urbs1-2*, and *urbs1-3*, were studied further and the mutations were shown to be recessive and segregate as single Mendelian traits. Complementation analysis demonstrated that the three mutations were allelic.

4.3.1. urbs1

Introduction of a cosmid library of genomic DNA into mutant *urbs1-2* led to the eventual identification of a 4.5-kb *Kpn*I/*Bgl*II fragment that was able to complement the mutant in trans [87]. DNA sequence analysis revealed a single open reading frame of 950 amino acids containing two putative zinc finger motifs belonging to the GATA family of transcription factors and an unusal tract of 21 histidines within 28 amino acids at the C-terminus (Fig. 9). Disruption of *urbs1* led to constitutive expression of siderophores and accumulation of the *sid1* mRNA [66] indicating that Urbs1 either directly or indirectly regulates transcription of *sid1*.

To determine whether Urbs1 interacts directly or indirectly with the *sid1* promoter, a series of constructs were developed containing portions of the *sid1* promoter fused to GUS [86]. This analysis revealed a sequence contained in a 0.7-kb *Ssp*I/*Hin*dIII fragment that was required for iron regulation of *sid1* (see Fig. 6). Further deletions studies using Bal31 nuclease showed that only 300 bp of this fragment are required for iron regulation of *sid1*. DNA sequence analysis of this region showed the presence of two GATA sequences with 28 nucleotides of one another (see Fig. 8).

To investigate whether Urbs1 interacts directly with the *sid1* pro-

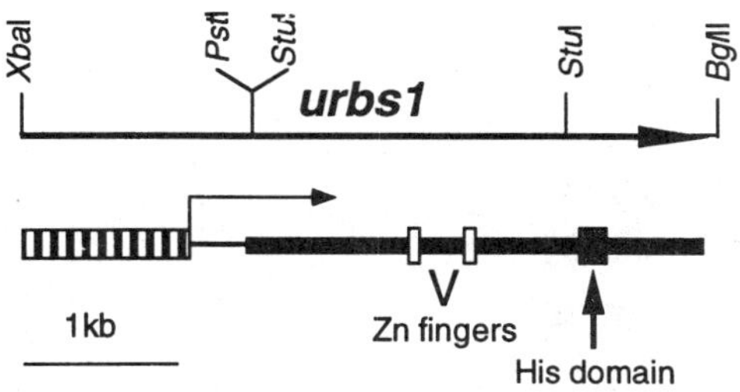

FIG. 9. *urbs1* gene of *U. maydis*. Thick line indicates the extent of the open reading frame. Arrow above gene marks start of transcription. Stripped box indicates promoter region. Open boxes indicated zinc finger motifs belonging to GATA family of transcription factors. Shaded box indicates histidine tract.

moter, the 0.7-kb *Ssp*I/*Hin*dIII fragment containing the two GATA sites was cloned both upstream and downstream of a yeast upstream activating sequence (UAS) for *GAL4* and the yeast *CYC1* basic promoter linked to *lacZ* of *E. coli* (Fig. 10) (Z. An and S. A. Leong, unpublished findings). Urbs1 was supplied in trans by expression from the yeast *GAL1* promoter. Analysis of the levels of *lacZ* activity under glucose repression and after galactose induction, suggested that Urbs1 directly affects expression of *sid1* (Z. An and S. A. Leong, unpublished findings).

Direct interaction of Urbs1 with sequences within the 0.7-kb fragment was obtained by electrophoretic mobility shift analysis using whole cell extracts of *U. maydis* [86]. Specific binding was observed to a 63-bp oligonucleotide containing the two GATA sequences but not to an oligonucleotide in which the two GATA sequences were mutated to CTGA. Furthermore, mutation of these sequences in the genome of *U. maydis* led to deregulated expression of *sid1* as measured using a GUS reporter. These data strongly suggest that Urbs1 binds directly to these GATA sequences in vivo to mediate repression of *sid1*.

Reciprocal studies were conducted with Urbs1 and mutant forms of Urbs1. GAP repair [88] was used to locate the mutations in Urbs1 alleles [89]. All three alleles contained mutations in the C-terminal finger domain leading to single amino acid changes (Fig. 11) [89]. *urbs1-2* also contained a frameshift mutation at the C-terminus leading to the addition of 114 amino acids not normally present in Urbs1. Urbs1-1 and Urbs1-3 showed greatly reduced DNA-binding activity

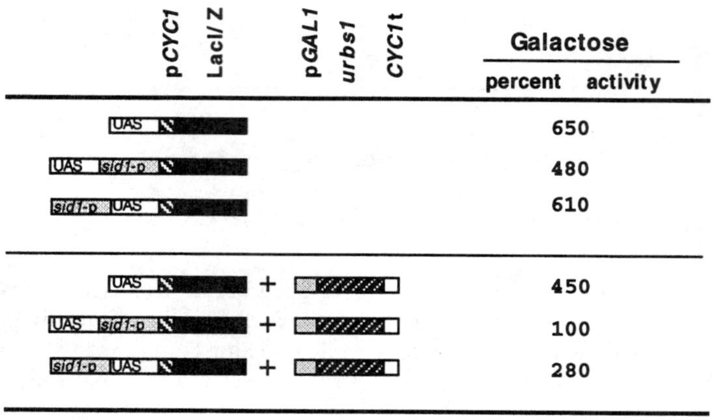

FIG. 10. Expression of *urbs1* in yeast demonstrates repression of a yeast promoter containing the 0.7-kb iron-responsive region from *U. maydis sid1* gene. Yeast strain YPH500 (*ura-3*, *lys-2*, *ade-2*, *trp-1*, *his-3*, and *leu-2*) was cotransformed with one plasmid containing the *sid1* iron-responsive region fused to *lacIZ* and another containing Urbs1 under *GAL1* control using the *GAL1* promoter and *CYC1* terminator. Binding of Urbs1 to 0.7-*Hin*dIII-*Ssp*I fragment of *sid1* promoter blocks expression of *lacZ* from a promoter containing the *GAL1* UAS and *CYC1* basic promoter. β-Galactosidase activity is presented as a percentage of the activity observed when the promoter is repressed by *urbs1*.

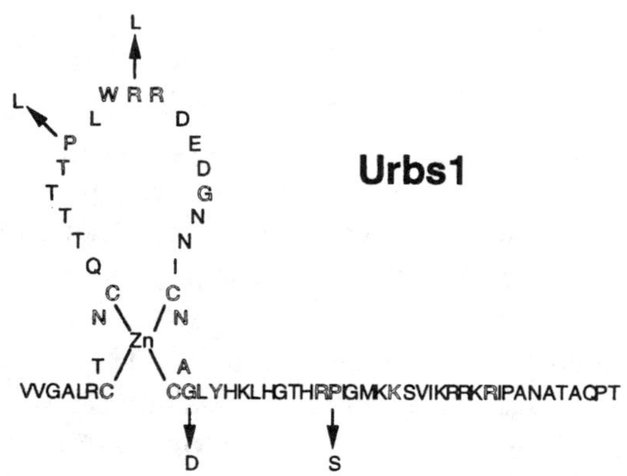

FIG. 11. C-terminal finger domain of Urbs1. Open characters show consensus amino acid residues among GATA family transcription factors. Arrows indicate mutations of the three mutant alleles created by *N*-methyl-*N*′nitro-*N*-nitrosoguanidine mutagenesis and an arginine residue that was subjected to site-directed mutagenesis.

when compared with wild-type Urbsl. The activity of Urbsl-2 was not investigated. Site-directed mutations were created in the highly conserved arginine residues 394 and 492 at the tip of each finger domain (Fig. 11). In GATA-1, these have been shown to participate in DNA binding as well as playing a role in maintaining the integrity of the finger structure [90]. Although the mutation in the C-terminal finger severely affected DNA binding, the parallel mutation in the N-terminal finger had only a modest effect on DNA binding. Regulation of siderophore production was constitutive in Arg492Leu and normal in Arg394Leu mutants. These data confirm the importance of the second finger domain in DNA recognition and siderophore gene regulation. Thus Urbsl, apparently requires only a single functional C-terminal finger for its function, and the N-terminal finger may play a more subtle role in DNA binding by affecting the stability or specificity of binding.

4.3.2. *Model for Regulation of sid1*

A major unresolved question is how Urbsl mediates regulation of *sid1* from a site so remote from the start of transcription. The two GATA sites ("distal GATA") are 1.6 kb upstream of the transcription start site. Interestingly, when a 1.1 kb of DNA is deleted between the distal GATA and the basic promoter, constitutive expression of the GUS reporter was observed [86] suggesting that either the spacing of these two regions and/or a sequence(s) critical to repression of *sid1* was eliminated in this deletion.

Two mechanisms of regulation have been suggested to explain the available data. In one case, Urbsl may cause the DNA to loop through binding to other GATA sites found in the region between the distal GATA and the start of transcription (Fig. 12A). The resultant loop could then preclude the binding of general transcription factors or derail RNA polymerase II. Loop formation may not require binding of Urbsl to itself but could occur through heterotypic interactions with other proteins that interact with the downstream promoter region. In the second model, Urbsl is predicted to serve as an organizer of inactive chromatin, thus masking regulatory sites required for activation of *sid1* (Fig. 12B). Neither mechanism is mutually exclusive and elements of both may apply to the regulation of *sid1*.

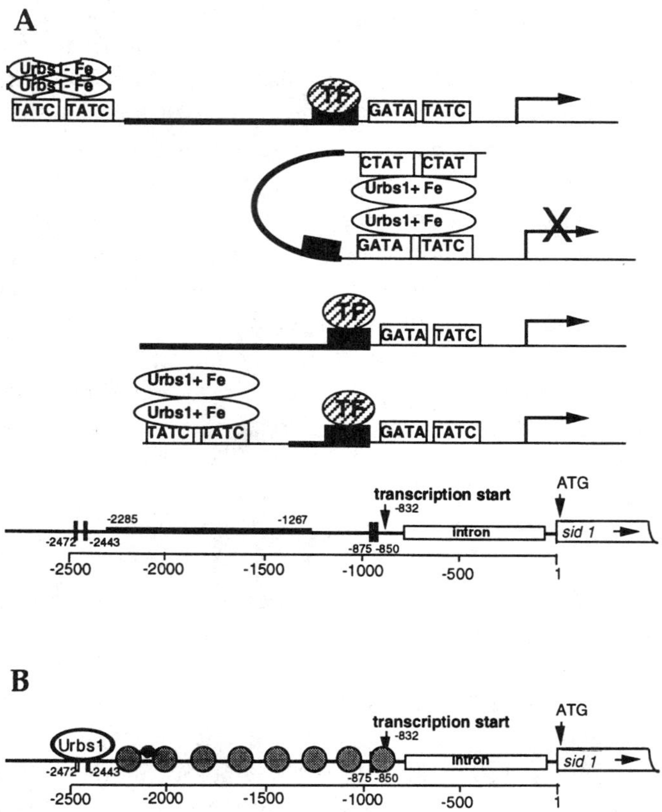

FIG. 12. Models for regulation of *sid1* by Urbs1. (A) Loop model for the action of regulatory action of Urbs1 on *sid1* expression. Urbs1, in the presence of iron, interacts with a high-affinity site (−2443) and a low-affinity site (−850/−832) to form a loop in the *sid1* promoter region. Loop formation could also involve GATA sequences in the first intron. Deletion of the high-affinity site or alteration of the spacing of the binding sites and/or removal of a specific sequence by deletion of the −2285 to −1267 region prevents loop formation leading to constitutive expression of *sid1*. (B) Chromatin remodeling model. Binding of Urbs1 to the iron-responsive element and possibly other sites in the *sid1* promoter organizes chromatin (shaded circles) across *sid1* promoter thus precluding the binding of activators and other factors required for transcription of *sid1*.

4.3.3. *Iron-Mediated Regulation of sid1*

The role of iron in the regulation of siderophore genes in *Ustilago* is also unknown. Iron may modulate siderophore gene expression at one or more levels. Iron may serve as a corepressor of genes regulated by Urbs1 and/or as a physiological effector that indirectly affects expression of siderophore genes by altering the structure and/or cellular location of Urbs1 or other unknown factors involved in siderophore gene regulation. As with Fur, the prokaryotic homologue of Urbs1, iron may directly activate Urbs1 in DNA binding [83]. In this context, it is interesting to note that Urbs1 contains several motifs that could participate in chelation of iron: (1) the finger domains which contain four cysteines for coordination of a metal ion; (2) the histidine track at the C-terminus; and (3) five cysteines in the region intervening the two finger domains.

To investigate direct interaction of metals with Urbs1, the effect of chelating agents on the ability of Urbs1 to shift DNA was examined. Inclusion of deferriferrichrome (1 μg/μL, 1.4 mM) or ethylenediaminetetraacetic acid (EDTA) up to 5 mM did not affect the gel mobility shift assay (Z. An, Q. Zhao, and S. A. Leong, unpublished findings). However, at high levels, EDTA began to reduce the intensity of the gel-shifted band. The biological implication of these results is not yet clear, as EDTA can chelate many cations in addition to metals suspected to associate with Urbs1. We know that the ferrichrome siderophores are not corepressors of siderophore genes, as inclusion of ferrichrome in the gel mobility shift analysis at 2 μg/μL (3 mM) did not affect the gel shift assay. Furthermore, the GUS reporter analysis was done in a strain in which the wild-type *sid1* locus was replaced with the GUS fusion construct, thus inactivating the *sid1* gene. Nevertheless, normal regulation by iron of GUS activity expressed from the *sid1* promoter was observed (Fig. 8) [86].

Systematic mutagenesis of the key amino acid residues implicated in metal chelation is underway to assess their role in metal binding and function of Urbs1. In parallel, Urbs1 is being purified from *Ustilago* in order to determine what metals are normally associated with the protein. To date, none of the members of the GATA family of transcription factors has been isolated in its native form and the metal associated with the finger domains has yet to be determined. Omichinski et al. [91] prepared a synthetic C-terminal finger peptide of chicken GATA-1 and then substituted different divalent metal ions including Zn^{2+}, Fe^{2+}, Co^{2+} and Cd^{2+}. In all cases, the metal-complexed peptide was able to shift the target oligonucleotide; however, the iron finger showed a two-fold increase in the binding constant relative to the zinc-finger peptide. This finding leads one to question whether some GATA family members contain iron fingers rather than zinc fingers in vivo. In the case of Urbs1, the binding of iron to its fingers would be coordinated with its ability to bind to target genes under its repressive control. In the case of GATA-1, an iron-containing finger might be associated with a need further to coordinate globin synthesis with the availability of iron to the cell; globin contains 70% of the iron in the human body [92].

As noted above, bacterial systems of regulation of siderophore biosynthesis and transport are elaborate and Fur is only one part of this complex system. The possibility of parallel regulatory systems in fungi is very likely. Fungi also have the added complexity of organelles. Often transcription factors are differentially localized in the cell to regulate their action. For example, during phosphate sufficiency in yeast, the transcription activator Pho4p is retained in the cytoplasm as a phosphorylated protein [93]. When cells are starving for phosphate, Pho4p is translocated to the nucleus in a dephosphorylated state where it activates transription of genes required for phosphate uptake and utilization [93]. As *urbs1* is constitutively expressed both into RNA and protein, a mechanism to regulate the activity of Urbs1 must exist. Whether this is simply the binding of iron by Urbs1 or a more elaborate and fine-tuned system as described for PHO4 in yeast remains to be determined.

4.4. Iron Transport in *Ustilago*

Transport of ^{55}Fe mediated by ferrichrome, ferrichrome A, and citrate has been studied in wild-type *sid1* and *urbs1* mutants (S. A. Leong and G. Winkelmann, unpublished findings). In all cases, the *urbs1* mutant showed increased accumulation of the radiolabeled metal consistent with deregulation of iron uptake. This mutant also shows an orange colony phenotype on iron-containing (1 mM) agar medium [87]. Extraction of cells for siderophores confirmed that the color is due to the presence of greatly elevated levels of both ferrichrome and ferrichrome A (A. D. Budde and S. A. Leong, unpublished findings). The orange color was clearly intracellular as protoplasts formed from the *urbs1* mutant grown on 1 mM $FeSO_4$ contained an orange-colored cytoplasm. Ferrichrome A is not usually associated with wild-type cells. Whether the cell-associated siderophores represent a siderophore component that has not yet been secreted and/or siderophores that have been transported back into cells is unclear. As noted above, the mechanism of ferrichrome A-iron uptake in *U. sphaerogena* involves reduction of iron at the membrane and recycling of the deferrisiderophore into the medium [40]. Thus, the presence of ferrichrome A in cells is unexpected.

Attempts to isolate siderophore transport mutants in *Ustilago* have to date been unsuccessful. However, a new and very powerful selection strategy for isolation of high-affinity iron transport mutants in yeast was recently described (see below) and should be applicable with slight modification to the isolation of similar classes of mutants in *Ustilago* as well as other fungi. We turn now to a description of recent progress in the understanding of the molecular biology of iron transport in yeast.

5. MOLECULAR BIOLOGY OF IRON TRANSPORT IN YEAST

In contrast to filamentous fungi and most yeasts, *Saccharomyces cerevisiae* does not produce siderophores to take up iron from the environment [94]. Instead, high-affinity iron transport is mediated by a complex system of reductases and oxidases coupled to a putative ferric ion transporter (Fig. 13) [95]. A number of gene homologues have been identified in the fission yeast *Schizosaccharomyces pombe* and ferric reductase activity has been reported in *Cryptococcus neoformans* indicating conservation of the high-affinity transport system in these two yeasts [96–98]. Moreover, this iron transport system shows many parallels to that in higher organisms such as humans [94].

5.1. *FRE1*

Early studies determined that extracellular iron presented in the ferrous state could be taken up faster than ferric ion [99]. Ferric ion was

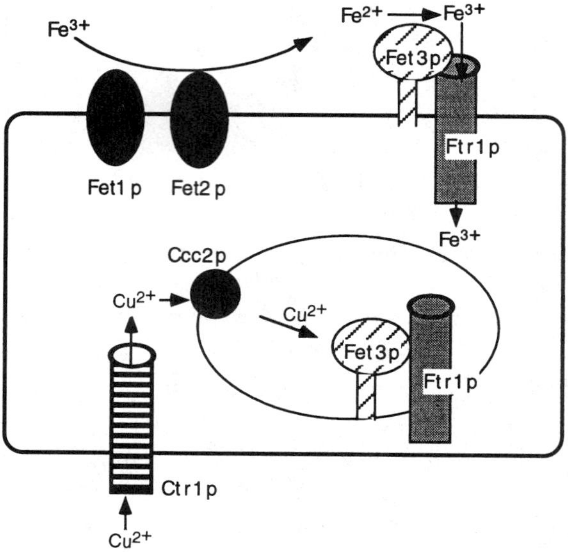

FIG. 13. Yeast model of high-affinity iron transport. Ferric ion is reduced by the redundant reductases Fetlp and Fet2p. The resulting ferrous ion is reoxidized by Fet3p and transported into the cytoplasm via Ftrlp. Copper transported into cells via Ctrlp is required for high-affinity iron uptake as Fet3p is a copper-dependent oxidase. Copper is inserted into Fet3p in a post-Golgi compartment. Ccc2p is required to deliver copper to this compartment. Secretion of Fet3p and Ftrlp is coupled.

shown to be transported by a reductive process when supplied in the siderophores rhodotorulic acid and ferrioxamine B or citrate and rates of uptake did not depend on the stability constants of the chelators for ferric ion [99,100]. An externally directed membrane-associated redox system induced during iron deficiency was implicated in the reductive process [99]. Genetic evidence for such a ferric reductase was obtained by isolation of a mutant that was sensitive to iron deprivation [101]. This mutant lacked ferric reductase activity and the ability to take up ferric ion. Both phenotypes cosegregated suggesting a single gene defect and the mutation was named *fre1-1*. Complementation of the mutant with a gene library led to the isolation of *FRE1*. *FRE1* encodes a 686-amino acid protein with sequence similarity to human gp91-*phox*, a component of the human NADPH phagocyte oxidoreductase, and contains two transmembrane domains [102]. Disruption of the gene gave rise to a mutant with a greatly reduced ability to reduce ferric ion and take up ferric ion but with no obvious change in the ability to take up ferrous ion. *FRE1* was surmized to participate in high-affinity, reductive transport of ferric ion, as *fre1* mutants were able to grow at a normal rate in the presence of high concentrations of ferric ion. The promoter region of *FRE1* was fused to *lacZ* and shown to be negatively regulated by increasing iron concentration consistent with a role in high-affinity iron uptake [102]. Promoter deletion studies identified a directly repeated sequence TTTTCGTCA(T/C)C which may be involved in iron regulation of the gene.

To determine whether the reductive system of transport is more widely conserved in lower eukaryotes, *frp1*, a homologue of *FRE1*, was isolated from fission yeast by complementation of a strain deficient in ferric iron uptake. *frp1* was shown to have several putative transmembrane domains and significant homology to *FRE1* within those regions homologous to gp91-*phox* [96,102]. Disruption of *frp1* created a strain unable to take up ferric ion but unaffected in its ability to take up ferrous ion. Like *FRE1*, *frp1* is induced by iron starvation [96]. Promoter deletion studies identified sequences critical to iron-dependent regulation between −90 and −43 with respect to the transcription initiation site [96]. However, no sequence similarity was found between this region and the putative iron regulatory sequence of *FRE1*.

5.2. *FRE2*

The yeast genome sequencing project led to the identification of *FRE2*, a homologue of *FRE1* and *frp1* on chromosome 11 [103]. Based on its sequence identity, *FRE2* was predicted to encode a second ferric reductase responsible for the residual activity seen in *FRE1* mutants. This prediction was fulfilled by examining the iron-dependent growth phenotype and ferric reductase activity of single and double mutants defective in *FRE1* and *FRE2*. As expected, the double, but not the single mutants, were unable to grow in iron-deficient medium and cell pellets were unable to reduce ferric ion. *FRE2* showed iron-dependent regulation similar to *FRE1*. Through fusion to a reporter gene, only 156 bp of DNA upstream of the AUG initiator was determined to mediate iron regulation of *FRE2*.

5.3. A New Genetic Strategy

A second major advance in the molecular biology of yeast iron transport came with the development of a novel selection strategy to isolate mutants defective in this process [104]. *FRE1* mutants were isolated using a colony color assay to detect strains unable to reduce iron [101]. Although this method was effective, it is a screen for mutants, thus limiting the number of candidate mutants that can be surveyed. The innovation came by fusing the iron-regulated *FRE1* promoter to *HIS3* in a *his3* null mutant to create an iron-dependent auxotrophy. By selecting for spontaneous mutants able to grow prototrophically in the presence of iron, several classes of mutants altered in high-affinity iron transport were obtained. These include *CTR1*, a membrane-associated protein required for high-affinity copper uptake [105,106]; *FET3*, a membrane-associated protein with homology to copper-dependent oxidases such as ceruloplasmin [107]; *FTR1*, a membrane-associated ferric ion transporter [95]; and *AFT1*, a transcription activator of genes involved in high-affinity iron uptake [108,109].

5.4. An Integrated Picture

Elegant studies of these various genes and their products have given rise to an integrated picture of their functional interrelationship (see Fig. 13). Frelp and Fre2p are redundant reductases required to reduce extracellular iron to ferrous ion which is a substrate of the oxidase Fet3p. Fet3p is physically associated with Ftrlp in the membrane and delivers ferric ion to this partner for transport into the cytoplasm [95]. The Fet3p/Ftrlp complex is sufficient to account for all high-affinity ferrous ion transport in yeast [95,97]. Ftrlp is also required for the normal trafficking of Fet3p through the cytoplasm to the plasma membrane and the concommitent loading of the Fet3p with copper in the post-Golgi compartment by Ccc2p, a P-type ATPase [95,110]. Ctrlp is needed to transport copper into cells for Fet3p function [105], thus providing a link between copper and iron uptake.

5.5. Analogies with Iron Transport Systems in Other Organisms

Homologues of Ftrl and Fet3 have been isolated in fission yeast [96,98]. However, the fission yeast genes cannot complement the respective yeast mutants indicating that functional interaction of the heterologous proteins is not productive. CCC2 is a homologue of the Menkes-Wilson disease gene, a P-type ATPase required for delivery of copper to ceruloplasmin [110]. Ceruloplasmin is a ferrooxidase that may be required to deliver iron to the circulation. Aceruloplasminemia, the lack of a functional ceruloplasmin in humans, leads to iron loading in various organs and severe tissue damage [95].

5.6. Regulation of Iron Transport in Yeast

Like siderophore-uptake systems, high-affinity iron transport is repressed when cells are sufficient in iron [99,101,108,109,111]. One class of mutants deregulated for expression of the *FRE1::HIS3* fusion construct was shown to be dominant and defective in *AFT1*. DNA sequence analysis of *AFT1* revealed a 690-amino acid polypeptide with a glutamine-rich C-terminus consistent with a role as a transcription activator [108]. Disruption of *AFT1* led to a lack of expression of putative target genes *FRE1*, *FRE2*, and *FET3* consistent with a role in gene activation. By contrast, these genes were expressed constitutively in the *AFT1*^{up} dominant mutant used to clone the gene. Electrophoretic mobility shift analysis was used to identify the consensus binding site T/C/G/ACACCCAG [109]. This site is present in the putative target genes *FET3, FRE1*, *FRE2*, *FTR1*, *CCC2*, and *FTH1* (a homologue of *FTR1*) [109]. In vivo footprinting demonstrated that Aft1p is associated with target-binding site of *FET3* in cells grown in the absence, but not the presence, of repressing quantities of iron [109]. Since *AFT1* was expressed constitutively in these different iron concentrations, the data suggest that the ability of Aft1p to bind DNA is somehow affected by iron.

5.7. Low-Affinity Iron Transport in Yeast

The mechanism of low-affinity iron transport has received less attention to date. Excretion of anthranilic and 3-hydroxyanthranilic acids into the extracellular medium has been correlated with an extracellular ferric ion-reducing activity [112]. Secretion is constitutive and was somewhat greater in iron-containing than in iron-deficient media. After reduction to ferrous ion, low-affinity iron uptake occurs via Fet4p [113]. *FET4* was isolated by expressing a yeast cDNA library under control of a *GAL1* promoter in a *fet3* mutant. Transformants able to grow on an iron-deficient, galactose-containing medium were identified. Overexpression of *FET4* lead to elevated levels of ferrous ion uptake in both the wild-type and the *fet3* mutant. Sequence analysis of *FET4* revealed a polypeptide of 552 amino acids, of which 51.5% are hydrophobic suggesting an integral membrane localization. Disruption of *FET4* resulted in mutants able to conduct high-affinity iron uptake but unable to transport iron via the low-affinity uptake system. These data demonstrated that these two systems operate independently and additively. Additional genes for iron transport are suggested as the *fet3*/*fet4* mutant was still able to grow [113].

6. CONCLUSIONS

Basic studies on the molecular biology of high-affinity iron transport in microorganisms are providing an essential framework for understanding how living cells maintain iron homeostasis. Studies in microorganisms are providing a detailed understanding of this process at the molecular and cellular level. As noted above, many parallels have been found between iron uptake in yeast and humans, and these similarities are providing molecular clues to better understand genetic diseases in humans.

Maintenance of iron homeostasis through limiting iron uptake at the membrane level is a universal theme in biology. Excess iron within cells is deleterious [114], as iron can act catalytically to produce DNA-damaging and lipid-peroxidizing oxygen radicals. Mutations in *fur*, a gene regulating high-affinity iron uptake in *E. coli*, led to eightfold elevated levels of intracellular iron [115] and increased rates of cell death and mutagenesis during aerobic growth [115,116]. Likewise, the *AFT1*^{up} mutation in yeast leads to deregulated iron uptake and reduced cell growth in iron-replete medium [108]. In both systems, the effects of iron overload are aggravated by mutations in DNA repair systems demonstrating that DNA damage is prevalent under these conditions [116,117]. To date, it has not been possible to isolate a *Ustilago* strain containing mutations in both *sid1* and *urbs1* suggesting that deregulation of iron uptake without a means of intracellular iron chelation by siderophores may be lethal. In humans, hemachromatosis leads to iron loading in the liver, heart, and pancreas resulting in cirrhosis of the liver, pancreatic fibrosis, and cardiac dysfunction, as well as increased risk of heptocellular carcinoma [118].

From a practical standpoint, fundamental studies on the molecular biology of iron transport in fungi are providing genes and enzymes that can be engineered to create novel chelators for clinical use. Desferral continues to be the best treatment for iron overload in humans, yet its use is limited by a lack of oral efficacy [119]. Can siderophore genes from fungi be used in gene therapy to engineer circulating human cells that produce a chelator in vivo? Studies of novel siderophore drug conjugates developed by the laboratory of Marvin Miller [120] have shown the potential of drug conjugates for delivery of antimicrobials to specific target populations of pathogenic microorganisms. Many of these conjugates are based on the structure of ferrichrome, a product of *sid1* and *sid2*. Thus, a biotransformation strategy may be possible for large scale production of these conjugates in the future.

The ability of fungal pathogens to acquire iron from the host or its environment may influence spore viability, spore germination, colonization of a habitat, and virulence. Studies on the molecular biology of iron transport in genetically tractable fungi such as *Ustilago* and yeast are providing a framework for related studies in less tractable fungal systems of medical and agricultural importance. For example, all fungi with the exception of the Zygomycetes produce δ-*N*-hydroxyornithine-containing siderophores [121]. Thus, the *sid1* gene should be rather ubiquitous among fungi and may be a useful target for drug development if siderophores are found to be virulence factors of fungi pathogenic to humans. A basic understanding of the enzymology and regulation of expression of this gene is requisite to the success of such a strategy. In this context, it is interesting to note that *Pseudomonas aeruginosa* has been shown to require siderophores as virulence factors [122] and that attachment of this bacterium to host cells in the lung is enhanced during iron starvation [123]. One of the most difficult fungal organisms to control today is *Aspergillus fumigatus*, an opportunistic pathogen which infects the lung. *A. fumigatus* produces the siderophore ferricrocin [124]. It is tempting to speculate that siderophores are required for infection of the lung by this fungus and that other virulence genes will show iron-regulated expression as in *P. aeruginosa*.

We hope that this brief overview of the molecular biology of fungal iron transport will stimulate new investigations in fungi of medical and agricultural importance. This is a young and exciting field with many new avenues of research to explore.

ACKNOWLEDGMENT

Some of the research described in this chapter was supported by the U.S. Department of Agriculture and Public Health service Grant GM33716 to S.A.L. and by the Deutsche Forschungsgemeinschaft to G.W.

ABBREVIATIONS

AFT	activator for ferrous ion transport
Bgl	restriction enzyme from *Bacillus*
CCC2	copper-transporting P-type ATPase with similarity to human Menkes-Wilson genes
CoA	coenzyme A
CTR	high-affinity copper transporter into the cell
DDF	des(diserylglycyl)ferrirhodin
EDTA	ethylenediaminetetraacetic acid
FET	ferrous ion transport
fhu	ferrihydroxamate uptake
FRE	ferric reductase
frp	ferric reductase
FTH	homologue of FTR1

FTR	ferrous transport suppressor
Fur	ferric uptake regulator
GAL	galactose utilization
GAPDH	glyceraldehyde-3-*P*-dehydrogenase
GUS	β-glucuronidase
Hind	restriction enzyme from *Haemophilus*
HIS	histidine utilization
Kpn	restriction enzyme from *Klebsiella*
lac	β-galactosidase
MDR	multidrug transporter
NADH	nicotinamide adenine dinucleotide (reduced)
NADPH	nicotinamide adenine dinucleotide phosphate (reduced)
NHE	nernst hydrogen electrode
Pho4P	transcription factor that activates phosphate pathway
PPi	pyrophosphate
pvd	pyoverdine
SDS	sodium dodecylsulfate
sid	siderophore
Ssp	restriction enzyme from *Sphaerotilus*
UAS	upstream activating sequence
urbs	*Ustilago* regulator of biosynthesis of siderophores

REFERENCES

1. G. Winkelmann, D. van der Helm, and J. B. Neilands (eds.), *Iron Transport in Microbes, Plants and Animals*, VCH Verlagsgesellschaft, Weinheim, Germany, 1978.
2. G. Winkelmann (ed.), *Handbook of Microbial Iron Chelates*, CRC Press, Boca Raton, FL, 1991.
3. G. Winkelmann and D. R. Winge (eds.), *Metal Ions in Fungi*, Marcel Dekker, New York, 1994.
4. G. Winkelmann and C. J. Carrano (eds.), *Transition Metals in Microbial Metabolism*, Harwood Academic Publishers, Reading, UK, 1997.
5. G. Feistner, *BioMetals*, *8*, 193–196 (1995).
6. J. B. Neilands, *J. Am. Chem. Soc.*, *74*, 4846–4847 (1952).

7. M. A. F. Jalal and D. van der Helm, in *Handbook of Microbial Iron Chelates*, (G. Winkelmann, ed.) CRC Press, Boca Raton, FL, 1991, pp. 235–269.
8. M. A. F. Jalal, R. Mocharla, C. L. Barnes, M. B. Hossain, D. R. Powell, D. L. Eng-Wilmot, S. L. Grayson, B. A. Benson, and D. van der Helm, *J. Bacteriol.*, *158*, 683–688 (1984).
9. G. Deml, K. Voges, G. Jung, and G. Winkelmann, *FEBS Lett.*, *173*, 53–57 (1984).
10. D. van der Helm, M. A. F. Jalal, and M. B. Hossain, in *Iron Transport in Microbes, Plants and Animals*, (G. Winkelmann, D. van der Helm and J. B. Neilands, eds.), VCH, Weinheim, Germany, 1987, pp. 135–165.
11. M. A. F. Jalal, J. L. Galles, and D. van der Helm, *J. Org. Chem.*, *50*, 5642–5645 (1985).
12. D. L. Eng-Wilmot, J. P. Adjimani, and D. van der Helm, *J. Inorg. Biochem.*, *48*, 183–195 (1992).
13. J. M. Sayer and T. Emery, *Biochemistry*, *7*, 184–190 (1968).
14. H. Diekmann, *Arch. Microbiol.*, *58*, 1–5 (1967).
15. M. B. Hossain, D. L. Eng-Wilmot, R. A. Loghry, and D. van der Helm, *J. Am. Chem. Soc.*, *102*, 5766–5773 (1980).
16. J. P. Adjimani and T. Emery, *J. Bacteriol.*, *169*, 3664–3668 (1987).
17. H. Diekmann, *Arch. Microbiol.*, *73*, 65–76 (1967).
18. M. A. F. Jalal, S. K. Love, and D. van der Helm, *J. Inorg. Biochem.*, *28*, 417–430 (1986).
19. M. A. F. Jalal, S. K. Love, and D. van der Helm, *J. Inorg. Biochem.*, *29*, 259–267 (1987).
20. D. van der Helm and G. Winkelmann, in *Metal Ions in Fungi* (G. Winkelmann and D. R. Winge, eds.), Marcel Dekker, New York, 1994, pp. 39–98.
21. W. Keller-Schierlein and H. Diekmann, *Helv. Chim. Acta*, *53*, 2035–2044 (1970).
22. A. Crumbliss, in *Handbook of Microbial Iron Chelates* (G. Winkelmann, ed.) CRC Press, Boca Raton, FL, 1991, pp. 177–233.
23. H. Drechsel, J. Metzger, S. Freund, G. Jung, J. R. Boelaert, and G. Winkelmann, *BioMetals*, *4*, 238–243 (1991).
24. H. Drechsel, G. Jung, and G. Winkelmann, *BioMetals*, *5*, 141–148 (1992).

25. C. J. Carrano, A. Thieken, and G. Winkelmann, *BioMetals*, *9*, 185–189 (1996).
26. A. Thieken and G. Winkelmann, *FEMS Microbiol. Lett.*, *94*, 37–42 (1992).
27. C. J. Carrano, H. Drechsel, D. Kaiser, G. Jung, B. F. Matzanke, G. Winkelmann, N. Rochel, and A. Albrecht-Gary, *Inorg. Chem.*, *35*, 6429–6436 (1996).
28. N. David and K. Easterbrook, *J. Cell. Biol.*, *48*, 15–28 (1971).
29. B. F. Matzanke, E. Bill, A. X. Trautwein and G. Winkelmann, *J. Bacteriol.*, *169*, 5873–5876 (1987).
30. B. F. Matzanke, E. Bill, A. X. Trautwein and G. Winkelmann, *BioMetals*, *1*, 18–25 (1988).
31. B. F. Matzanke, in *Metal Ions in Fungi* (G. Winkelmann and D. R. Winge, eds.), Marcel Dekker, New York, 1994, pp. 179–214.
32. G. Winkelmann, in *Iron Chelation in Plants and Soil Microorganisms* (L. L. Barton and B. C. Hemming, eds.), Academic Press, San Diego, 1993, pp. 219–239.
33. T. Emery, *Biochemistry*, *10*, 1483–1488 (1971).
34. C. Wiebe and G. Winkelmann, *J. Bacteriol.*, *123*, 837–842 (1975).
35. H. Huschka, H. U. Naegeli, H. Leuenberger-Ryf, W. Keller-Schierlein, and G. Winkelmann, *J. Bacteriol.*, *162*, 715–721 (1985).
36. H. Huschka, M. A. F. Jalal, D. van der Helm, and G. Winkelmann, *J. Bacteriol.*, *167*, 1020–1024 (1986).
37. G. Winkelmann, I. Berner, and H. Huschka, *J. Plant Nutr.*, *11*, 883–892 (1988).
38. C. T. Bailey, E. M. Kime-Hunt, C. J. Carrano, H. Huschka, and G. Winkelmann, *Biochim. Biophys. Acta*, *883*, 299–305 (1986).
39. H. Huschka and G. Winkelmann, *BioMetals*, *2*, 108–113 (1989).
40. D. J. Ecker and T. Emery, *J. Bacteriol.*, *155*, 616–622 (1983).
41. G. Winkelmann, *FEBS Lett.*, *97*, 43–46 (1979).
42. G. Winkelmann and V. Braun, *FEMS Microbiol. Lett.*, *11*, 237–241 (1981).
43. J. Leong, J. B. Neilands, and K. N. Raymond, *Biochem. Biophys. Res. Commun.*, *60*, 1066–1071 (1974).
44. H. Huschka, G. Müller, and G. Winkelmann, *FEMS Microbiol. Lett.*, *20*, 125–129 (1983).

45. G. Winkelmann and H. Huschka, *J. Plant Nutr.*, *7*, 479–487 (1984).
46. G. B. Wong, M. J. Kappel, K. N. Raymond, B. F. Matzanke, and G. Winkelmann, *J. Am. Chem. Soc.*, *105*, 810–815 (1983).
47. J. F. Ernst and G. Winkelmann, *Biochim. Biophys. Acta*, *500*, 27–41 (1977).
48. K. Haselwandter, B. Dobernigg, W. Beck, G. Jung, A. Cansier, and G. Winkelmann, *BioMetals*, *5*, 51–56 (1992).
49. S. Konetschny-Rapp, G. Jung, H. Huschka, and G. Winkelmann, *BioMetals*, *1*, 90–98 (1988).
50. H. Mor, Y. Kashman, G. Winkelmann, and I. Barash, *BioMetals*, *5*, 213–216 (1992).
51. T. Emery, *Adv. Enzymol.*, *35*, 135–185 (1971).
52. B. Mei and S. A. Leong, in *Metal Ions in Fungi* (G. Winkelmann and D. R. Winge, eds.), Marcel Dekker, New York, 1994, pp. 117–147.
53. H. J. Plattner and H. Diekmann, in *Metal Ions in Fungi* (G. Winkelmann and D. R. Winge, eds.), Marcel Dekker, New York, 1994, pp. 99–116.
54. G. Winkelmann, in *Handbook of Microbial Chelates* (G. Winkelmann, ed.), Academic Press, Boca Raton, FL, 1991, pp. 275–306.
55. C. Guthrie and G. R. Fink (eds.), *A Guide to Yeast Genetics and Molecular Biology*, Academic Press, New York, 1991.
56. J. R. Fincham, *Microbiol. Rev.*, *53*, 148–170 (1989).
57. A. Liu and J. B. Neilands, *Structure and Bonding*, *58*, 97–106 (1984).
58. H. A. Akers and J. B. Neilands, in *Biological Oxidation of Nitrogen* (J. W. Garrod, ed.), Elsevier/North-Holland, Amsterdam, 1978, pp. 429–436.
59. J. B. Neilands, K. Knopka, B. Schwyn, M. Coy, R. T. Francis, B. H. Paw, and A. Bagg, in *Iron Transport in Microbes, Plants and Animals* (G. Winkelmann, D. van der Helm, and J. B. Neilands, eds.), VHC, Weinheim, Germany, 1987, pp. 1–33.
60. T. Anke and H. Diekmann, *FEBS Lett.*, *27*, 259–262 (1972).
61. K.-D. Siegmund, H. J. Plattner, and H. Diekmann, *Biochim. Biophys. Acta*, *1076*, 123–129 (1991).
62. T. Anke and H. Diekmann, *Arch. Microbiol.*, *95*, 227–236 (1974).

63. J. B. Neilands and K. Nakamura, in *Handbook of Microbial Chelates* (G. Winkelmann, ed.), Academic Press, Boca Raton, FL, 1991, pp. 1–14.
64. J. Wang, A. Budde, and S. A. Leong, *J. Bacteriol.*, *171*, 2811–2818 (1989).
65. M. Lucky, J. R. Pollack, R. Wayne, B. N. Ames, and J. B. Neilands, *J. Bacteriol.*, *111*, 731–738 (1972).
66. B. Mei, A. D. Budde, and S. A. Leong, *Proc. Natl. Acad. Sci. USA*, *90*, 903–907 (1993).
67. J. Wang, D. W. Holden, and S. A. Leong, *Proc. Natl. Acad. Sci. USA*, *85*, 865–869 (1988).
68. T. Kinscherf and S. A. Leong, *Chromosoma*, *96*, 427–433 (1988).
69. J. Kronstad, J. Wang, S. Covert, D. Holden, S. McKnight, and S. A. Leong, *Gene*, *79*, 97–106 (1989).
70. P. Visca, A. Ciervo, and N. Orsi, *J. Bacteriol.*, *176*, 1128–1140 (1994).
71. Z. An, M. L. Farman, S. Taura, and S. A. Leong, *Gene*, *176*, 93–96 (1996).
72. H. Kleinkauf and H. von Döhren, *Eur. J. Biochem.*, *216*, 335–351 (1996).
73. W. Hummel and H. Diekmann, *Biochim. Biophys. Acta*, *657*, 313–320 (1981).
74. D. E. Ong and T. F. Emery, *Arch. Biochem. Biophys.*, *148*, 77–83 (1972).
75. R. H. Lambalot, A. M. Gehring, R. S. Flugel, P. Zuber, M. Lacelle, M. A. Marahiel, R. Reid, C. Khosla, and C. T. Walsh, *Chem. Biol.*, *3*, 923–936 (1996).
76. B. J. McMorran, M. E. Merriman, I. T. Rombel, I. L. Lamont, *Gene*, *176*, 55–59 (1996).
77. Q. Chen, A. M. Wertheimer, M. E. Tolmasky, and J. H. Crosa, *Mol. Microbiol.*, *22*, 127–134 (1996).
78. V. Venturi, C. Ottevanger, J. Leong, and P. J. Weisbeek, *Mol. Microbiol.*, *10*, 63–73 (1993).
79. C. R. Dean and K. Poole, *Mol. Microbiol.*, *8*, 1095–1103 (1993).
80. C. R. Dean, S. Neshat, and K. Poole, *J. Bacteriol.*, *178*, 5361–5369 (1996).

81. B. Van Hove, H. Staudenmaier, and V. Braun, *J. Bacteriol.*, *172*, 6749–6758 (1990).
82. L. Leoni, A. Ciervo, N. Orsi, and P. Visca, *J. Bacteriol.*, *178*, 2299–2313 (1996).
83. A. Bagg and J. B. Neilands, *Microbiol. Rev.*, *51*, 509–518 (1987).
84. A. D. Budde and S. A. Leong, *Mycopathologia*, *108*, 125–133 (1989).
85. S. R. Gallagher (ed.), *GUS Protocols*, Academic Press, London, 1992.
86. Z. An, B. Mei, W. M. Yuan, and S. A. Leong, *EMBO J.*, *16*, 1742–1750 (1997).
87. C. Voisard, J. Wang, J. McEvoy, P. Xu, and S. A. Leong, *Mol. Cell. Biol.*, *13*, 7091–7100 (1993).
88. R. Rothstein, in *A Guide to Yeast Genetics and Molecular Biology* (C. Guthrie and G. R. Fink, eds.), Academic Press, New York, 1991, pp. 281–301.
89. Z. An, Q. Zhao, J. McEvoy, W. M. Yuan, J. Markley, and S. A. Leong, *Proc. Natl. Acad. Sci. USA.*, *94*, 5882–5887 (1997).
90. J. G. Omichinski, M. Clore, O. Schaad, G. Felsenfeld, C. Trainor, E. Appella, S. J. Stahl, and A. M. Gronenborn, *Science*, *261*, 438–446 (1993).
91. J. G. Omichinski, C. Trainor, T. Evans, A. M. Groenborn, G. M. Clore, and G. Felsenfeld, *Proc. Natl. Acad. Sci. USA*, *90*, 1676–1680 (1993).
92. T. Emery, *Iron and Your Health*, CRC Press, Boca Raton, FL, 1991.
93. E. M. O'Neill, A. Kaffman, E. R. Jolly, and E. K. O'Shea, *Science*, *271*, 209–212 (1996).
94. D. M. De Silva, C. C. Askwith, and J. Kaplan, *Physiol. Rev.*, *76*, 31–47 (1996).
95. R. Stearman, D. S. Yuan, Y. Yamaguchi-Iwai, R. D. Klausner, and A. Dancis, *Science*, *271*, 1552–1557 (1996).
96. D. G. Roman, A. Dancis, G. J. Anderson, and R. D. Klausner, *Mol. Cell. Biol.*, *13*, 4342–4350 (1993).
97. C. Askwith and J. Kaplan, *J. Biol. Chem.*, *272*, 401–405 (1997).
98. K. J. Nyhus, A. T. Wilborn, and E. S. Jacobson, *Infect. Immun.*, *65*, 434–438 (1997).

99. E. Lesuisse, F. Raguzzi, and R. R. Crichton, *J. Gen. Microbiol.*, *133*, 3229–3236 (1987).
100. E. Lesuisse and P. Labbe, *J. Gen. Microbiol.*, *135*, 257–263 (1989).
101. A. Dancis, R. D. Klausner, A. G. Hinnebusch, and J. G. Barriocanal, *Mol. Cell. Biol.*, *10*, 2294–2301 (1990).
102. A. Dancis, D. G. Roman, G. J. Anderson, A. G. Hinnebusch, and R. D. Klausner, *Proc. Natl. Acad. Sci. USA*, *89*, 3869–3873 (1992).
103. E. Georgatsou and D. Alexandraki, *Mol. Cell. Biol.*, *14*, 3065–3073 (1994).
104. R. D. Klausner and A. Dancis, *FEBS Lett.*, *355*, 109–113 (1994).
105. A. Dancis, D. S. Yuan, D. Haile, C. Askwith, D. Eide, C. Moehle, J. Kaplan, and R. D. Klausner, *Cell*, *76*, 393–402 (1994).
106. A. Dancis, D. Haile, D. S. Yuan, and R. D. Klausner, *J. Biol. Chem.*, *269*, 25660–25667 (1994).
107. C. Askwith, D. Eide, A. Van Ho, P. S. Bernard, L. Li, S. Davis-Kaplan, D. M. Sipe, and J. Kaplan, *Cell*, *76*, 403–410 (1994).
108. Y. Yamaguichi-Iwai, A. Dancis, and R. D. Klausner, *EMBO J.*, *14*, 1231–1239 (1995).
109. Y. Yamaguichi-Iwai, R. Stearman, A. Dancis, and R. Klausner, *EMBO J.*, *15*, 3377–3384 (1996).
110. D. S. Yuan, R. Stearman, A. Dancis, T. Dunn, T. Beeler, and R. D. Klausner, *Proc. Natl. Acad. Sci. USA*, *92*, 2632–2636 (1995).
111. D. Eide, S. Davis-Kaplan, I. Jordan, D. Sipe, and J. Kaplan, *J. Biol. Chem.*, *267*, 20774–20781 (1992).
112. E. Lesuisse, M. Simon, R. Klein, and P. Labbe, *J. Gen. Microbiol.*, *138*, 85–89 (1992).
113. D. R. Dix, J. T. Bridgham, M. A. Broderius, C. R. Byersdorfer, and D. J. Eide, *J. Biol. Chem.*, *269*, 26092–26099 (1994).
114. B. Halliwell and J. Gutteridge, *Biochem. J.*, *219*, 1–14 (1984).
115. K. Keyer and J. A. Imlay, *Proc. Natl. Acad. Sci. USA*, *93*, 13635–13640 (1996).
116. D. Touati, M. Jacques, B. Tardat, L. Bouchard, and S. Despied, *J. Bacteriol.*, *177*, 2305–2314 (1995).
117. C. Philpott, Y. Iwai, R. D. Klausner, and A. Dancis, *Abstracts, 1996 Yeast Genetics and Molecular Biology Meeting*, Madison, WI, 1996, p. 43.

118. D. H. G. Crawford, L. W. Powell, J. W. Halliday, and B. A. Leggett, *Annu. Rev. Nutr.*, *16*, 139–160 (1996).
119. M. J. Pippard, in *The Development of Iron Chelators for Clinical Use* (R. J. Bergeron and G. M. Brittenham, eds.), Academic Press, Boca Raton, FL, 1994, pp. 57–74.
120. M. J. Miller and F. Malouin, in *The Development of Iron Chelators for Clinical Use* (R. J. Bergeron and G. M. Brittenham, eds.), Academic Press, Boca Raton, FL, 1994, pp. 275–306.
121. G. Winkelmann, *Mycol. Res.*, *7*, 529–534 (1992).
122. A. Scharfman, H. Kroczynski, C. Carnoy, E. Van Brussel, G. Lamblin, R. Ramphal, and P. Roussel, *Infect. Immun.*, *64*, 5417–5420 (1996).
123. J.-M. Meyer, A. Neely, A. Stintzi, C. Georges, and I. A. Holder, *Infect. Immun.*, *64*, 518–523 (1996).
124. H. Diekmann and E. Krezdorn, *Arch. Microbiol.*, *106*, 191–194 (1975).

5

Soil Microorganisms and Iron Uptake by Higher Plants

Janette Palma Fett,* Kristin LeVier,* and Mary Lou Guerinot

Department of Biological Sciences, Dartmouth College, Hanover, NH 03755, USA

1. INTRODUCTION 188
2. IRON UPTAKE SYSTEMS OF SELECTED PLANT-ASSOCIATED MICROORGANISMS 189
    2.1. Plant Beneficial Microorganisms 189
        2.1.1. Rhizobia 189
        2.1.2. Mycorrhizal Fungi 195
        2.1.3. Pseudomonads 196
    2.2. Plant Pathogens 198
        2.2.1. *Agrobacterium* 198
        2.2.2. *Erwinia* 198
3. IRON UPTAKE SYSTEMS OF PLANTS 200

**Current addresses*: Janette Palma Fett, Departamento de Botanica/Centro de Biotecnologia, UFRGS, Porto Alegre 90046-900, Brazil
Kristin LeVier, Department of Biology, Building 68-633, Massachusetts Institute of Technology, Cambridge, MA 02139, USA

3.1. Strategy I Plants 200
3.2. Strategy II Plants 201

4. UTILIZATION OF MICROBIAL SIDEROPHORES BY PLANTS 203

5. UTILIZATION OF PHYTOSIDEROPHORES BY MICROBES 205

6. CONCLUSIONS 206

ACKNOWLEDGMENTS 207

ABBREVIATIONS 207

REFERENCES 208

1. INTRODUCTION

Iron, an essential nutrient, is not readily available in the soil. Although abundant, iron is found mainly as stable, insoluble oxyhydroxide polymers that effectively limit free Fe(III) to an equilibrium concentration of approximately 10^{-17} M at neutral pH, a value far below that required for the optimal growth of microbes or plants (estimated to be 10^{-8} to 10^{-6} M) (as reviewed in Ref. [1]). Thus, many soil inhabitants have developed various strategies for acquiring iron. Basically, organisms have three means at their disposal to dissolve Fe(III) oxides: protonation, reduction, and chelation. Protonation (defined here to mean release of protons) drives more Fe(III) into solution by shifting the equilibrium in favor of dissociation of the $Fe(OH)_3$ complexes; the solubility of Fe(III) increases 1000-fold with each one-unit decrease in pH. Chelation largely involves the release of siderophores, small molecular weight, high-affinity Fe(III) chelators synthesized and secreted by microorganisms and some plants under conditions of iron limitation. Once chelated, the Fe(III) remains in solution. Reduction of Fe(III) to Fe(II) relies on membrane-bound Fe(III) reductases and offers the benefit that Fe(II) is much more soluble in the rhizosphere than Fe(III) at a given pH. For comparison, Fe(II) is approximately 10^{16}-fold more soluble than Fe(III) at neutral pH.

In this chapter, we will briefly review the strategies employed by various soil microbes and plants to acquire iron and then explore how the various iron-uptake strategies of a particular group either enhance or inhibit the iron-uptake abilities of other soil inhabitants (see Fig. 1). For example, what is the ecological significance of microbial siderophores in soils and on plant surfaces? Microbial siderophores have been postulated to serve as iron sources for plants, as biocontrol agents, and as carbon sources for various microbes in addition to their obvious role as an iron source for the microbes that produced them. Grasses also release their own siderophores, called phytosiderophores, and these may serve functions similar to those suggested for microbial siderophores. In addition, plants release organic acids that can also aid in dissolution of $Fe(OH)_3$ [2]. Although a variety of siderophores are produced by soil microbes and plants under laboratory conditions, there have been few reports of siderophore isolation from soil samples (e.g., see Ref. [3]); questions of siderophore abundance in the soil remain largely unanswered. The residence time of a siderophore in the soil is another unknown. Are siderophores which are most abundant in the soil made in the greatest amounts, are they the most resistant to degradation, or are they the least utilized of the siderophores? Certainly, the soil is a complex environment and siderophores are likely to play an important role in the life cycles of both microbes and plants. We would like to know more about the role of siderophore production by various biocontrol microorganisms in protecting plants from deleterious pathogens. We would also like to know whether plants themselves have mechanisms for withholding iron from potential pathogens, as has been well described for animal-pathogen interactions.

2. IRON UPTAKE SYSTEMS OF SELECTED PLANT-ASSOCIATED MICROORGANISMS

2.1. Plant Beneficial Microorganisms

2.1.1. Rhizobia

Rhizobia (used here to refer collectively to members of the genera *Azorhizobium*, *Bradyrhizobium*, *Rhizobium*, and *Sinorhizobium*) live in the soil or engage in a nitrogen-fixing symbiosis with a suitable

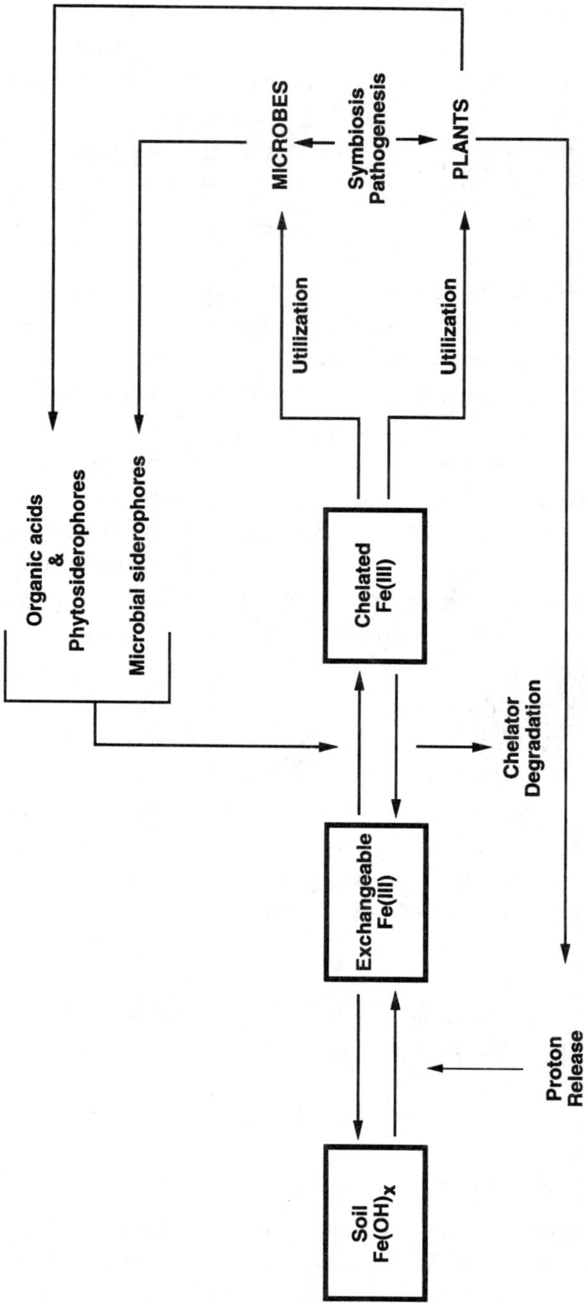

FIG. 1. Schematic drawing depicting processes influencing iron utilization by plants and microbes in terrestrial habitats.

legume host plant. Each environment presents unique challenges with respect to iron acquisition. As free-living soil microorganisms, rhizobia must have a way to solubilize iron as well as a way to compete for this nutrient with other organisms present in the rhizosphere. *In planta*, iron is likely to be tightly chelated in various storage forms, because free iron catalyzes the production of potent oxidants that damage biomolecules (e.g., see Ref. [4]). In addition, the ability of plant pathogens to acquire iron has been shown to be an important virulence factor (see Sec. 2.2 and [5]), so rhizobia must have mechanisms for accessing iron which one assumes is generally unavailable to invading pathogens.

As a starting point in the study of rhizobial iron acquisition, investigators have established that some rhizobial strains produce siderophores (reviewed in Ref. [6]; see also Refs. [7–17]). When present, siderophore production is strain specific; that is, there does not appear to be any particular siderophore made by all rhizobia. For example, *Rhizobium meliloti* DM4 produces a carboxylate-type siderophore [18], *R. meliloti* 1021 produces a dihydroxamate-type siderophore [19], and *Bradyrhizobium japonicum* 61A152 releases citric acid as a siderophore under iron-deficient growth conditions [20]. Citrate production appears to be unusual among bradyrhizobia, because only a minority of the strains analyzed have this capability; only 5 of 43 strains tested release citrate in response to iron deficiency, and 2 other strains show constitutive release [8,11,20]. A bacterium which is closely related to *Bradyrhizobium japonicum*, *Sinorhizobium tropici*, has two genes encoding citrate synthase, the first, *cssA*, located on the chromosome [21] and the second, *pcsA*, located on the symbiotic plasmid [22]. The plasmid-borne *pcsA* gene has increased expression under low iron conditions, suggesting a role for this gene in iron uptake. Interestingly, when the *pcsA* gene is transferred to wild-type *R. leguminosarum* bv. *phaseoli* cells, the resulting strain forms approximately 40% more nodules on host plants and grows more rapidly than does the parental strain [22]. This implies that the capacity to produce citrate under iron-limiting conditions confers a superior nodulation ability on *S. tropici* and *R. leguminosarum*. We have recently cloned a gene, *gltA*, encoding citrate synthase from *B. japonicum* 61A152; the expression of this gene is not iron regulated [23]. There does not appear to be another closely related citrate synthase gene in the *B. japonicum* genome; experiments are in progress to knock out the *gltA* gene and analyze the mutant for altered iron-uptake ability.

Although there is definitely a competitive advantage to producing siderophores, the ability to utilize siderophores which are produced by other organisms can also provide a clear benefit [24]. For example, *Pseudomonas aeruginosa* synthesizes and uses two of its own siderophores but has also been demonstrated to utilize the heterologous bacterial siderophores enterobactin, aerobactin, ferroxamine B and a number of siderophores produced by other pseudomonads [25]. In order to utilize a siderophore, an organism must have a siderophore-specific iron-uptake system composed of four main components: a high-affinity outer membrane protein (OMP), the inner membrane-anchored TonB protein, a periplasmic binding protein, and several inner membrane-associated proteins (reviewed in Ref. [26]). The genes encoding these proteins are coordinately derepressed under conditions of iron deficiency, and many are negatively regulated by the product of the *fur* (*f*erric *u*ptake *r*egulation) gene (reviewed in Ref. [27]). Although there have been a number of reports of iron-regulated OMPs in rhizobia [7,9,10,13–15,28], most simply state that such proteins were observed and that they may serve as Fe(III)-siderophore receptors. Reigh and O'Connell [14] did correlate the presence of specific iron-repressible OMPs with the production and release of specific siderophores, suggesting that Fe(III)-siderophore transport in rhizobia may be similar to transport in *E. coli* and other gram-negative bacteria. We have now purified the three major iron-regulated OMPs from *B. japonicum* strain 61A152 [23], which releases citric acid as a siderophore under conditions of iron deficiency [20] and is also able to utilize Fe(III) from two fungal siderophores, rhodoturulate and ferrichrome [29]. We have also cloned and characterized a gene, *fegA*, encoding one of these OMPs [30]. The *fegA* gene encodes a protein with similarity to hydroxamate-type siderophore receptors, is iron-regulated, and has a presumptive Fur-binding site in its promoter.

Now that rhizobial siderophore production and utilization have been documented in the laboratory, the next question to be addressed is whether siderophore production is important in the soil, *in planta* or in both situations. The best way to address this question is through an analysis of mutants which no longer produce or transport siderophores. Gill et al. [31] analyzed the ability of *R. meliloti* mutants, defective in various aspects of the rhizobactin siderophore system, to form nodules and fix nitrogen. Thirty days after inoculation, there was no significant difference in either nodule number or plant dry weight for plants inoculated with wild-type bacteria compared with those inoculated with the mutants, although plants inoculated with the wild type had greater nitrogen-fixing activity. It is important to point out that there is no direct evidence that bacterial siderophores are produced or utilized *in planta* in any of the rhizobial/legume symbioses.

Nadler et al. [32] isolated a mutant of *Rhizobium leguminosarum* biovar *viciae*, Pop116, with a complex phenotype that included formation of ineffective nodules and a defect in high-affinity iron uptake. The strain has recently been shown to be unable to synthesize or to export siderophores [33]. Surprisingly, the defect in Pop116 lies in the cytochrome-*c* biogenesis operon *cycHJKL*, and, at this time, the exact cause of the siderophore defect is not clear. One possibility is that a cytochrome is involved in some step in the biosynthesis of the siderophore known to be produced by the parental strain. A similar observation for the requirement of a functional cytochrome biogenesis pathway for pyoverdine production has been made in *Pseudomonas fluorescens* ATCC 17400 [34]. A *Tn5* mutant of *P. fluorescens* that cannot grow under iron-limiting conditions and only excretes low amounts of pyoverdines was shown to have a defect in a cytochrome biogenesis gene with great similarity to those previously described. These investigators suggested that one of the genes, *cytA*, might be involved in transport of both heme and pyoverdine from the cytoplasm to the periplasm.

In the course of sequencing the region identified by the pop116 mutation, two genes, *feuQP*, whose products are similar to two-component sensor and regulator proteins were found to lie upstream of *cycHJKL*. A *feuQ* mutant was then constructed and this strain, like pop116, had a greatly reduced rate of iron uptake. However, *feuQ* does not control expression of a *cycH-lacZ* fusion and *feuQ* mutants fixed nitrogen as well as the wild type. Further investigation is required to determine what genes are under the control of the *feuQP* system.

Although some information is available regarding the types of carbon and nitrogen sources available to bacteria in the nodule (termed bacteroids), little is known about sources of iron within the nodule. Bacteroids have an increased iron demand relative to their free-living state for synthesis of iron-containing proteins necessary for nitrogen fixation. For example, experiments using a soybean split-root system showed that induction of iron-deficiency responses (proton release and

Fe(III) reduction) is much stronger in the roots of plants which have actively fixing nodules than in the roots of uninoculated plants [35].

There are two general strategies by which microsymbionts could acquire iron within host plants. One strategy is the direct utilization of host iron compounds such as Fe(III)-citrate. Alternatively, microorganisms could synthesize reductants or chelators to dissociate iron from host iron complexes such as Fe(III)-organic acid complexes [36] and ferritin (reviewed in Ref. [37]), and thus make iron available for their own use. In the nodule, bacteroids are surrounded by a membrane of plant origin, the peribacteroid membrane (PBM), which has selective permeability to metabolites. The membrane-enclosed packets are called symbiosomes and form a discrete compartment within the plant cell cytoplasm. Iron presented as a number of chelates, including Fe(III)-citrate, has now been shown to be transported at much higher rates across the PBM than across bacteroid membranes, suggesting the presence of an iron storage pool inside the symbiosome in the peribacteroid space (PBS) [38]. Pulse chase experiments confirmed the presence of such a pool. The PBS iron storage pool may be the previously described nonheme iron fraction in the PBS which represents approximately 25% of the total soybean nodule iron [39,40]. Wittenberg et al. [40] state that this iron fraction is bound by low molecular weight siderophore compounds and might be involved in iron storage in the symbiosome. As the Fe(III)-siderophore fractions isolated by Wittenberg et al. [40] differed in size and optical spectra when isolated from nodules of plants inoculated with three different strains of bradyrhizobia, the PBS siderophore compounds are hypothesized to be of bacterial origin. Cultured *B. japonicum* cells have not been demonstrated to produce classic siderophores for iron acquisition but it is possible that bacteroids produce symbiosis-specific siderophore(s).

Because iron could be transported into the PBS from a variety of chelates and because the PBM is derived from the plant plasma membrane, LeVier et al. [38] reasoned that iron may be reduced by a membrane-bound PBM reductase. They further suggested that Fe(II) would then move passively into the symbiosome driven by a concentration gradient. Fe(II) would then be reoxidized and rechelated when it reached the PBS, perhaps being bound by the low molecular weight siderophore fraction in the PBS. As needed, iron would move into the bacteroid. In support of this model, LeVier et al. showed that the PBM does indeed have Fe(III)-chelate reductase activity.

2.1.2. *Mycorrhizal Fungi*

Mycorrhizas are the most widespread association between microorganisms and higher plants. The fungal mycelium acts as a functional extension of a plant's root system, extending the nutrient-acquisition zone far beyond that of the root itself [41]. Thus, mycorrhizal fungi play an important role in the mineral nutrition of plants. They are best known for providing plants with phosphorus, but it is becoming increasingly clear that they can also be important in the iron nutrition of plants. Mycorrhizal plants generally have higher iron contents and greater rates of iron uptake than their nonmycorrhizal counterparts (e.g., see Refs. [42–44]). However, in experiments with the fungus *Pisolithus tinctorius* and slash pine, mycorrhizal roots had lower rates of uptake of ^{55}Fe from ferrioxamine B and from a nonpurified siderophore isolated from the fungus than did nonmycorrhizal roots [45]. This result may be attributed to the fact that mycorrhizal roots had a higher iron content than nonmycorrhizal roots, and as such they might have exerted some type of negative feedback effect on iron uptake. Conversely, it might suggest a competition between the host and the plant for available iron.

The observation that mycorrhizal fungi can improve the iron nutrition of plants raises the question of how mycorrhizal fungi themselves take up iron. The production of siderophores has now been reported for ericoid, ectomycorrhizal, and ectendomycorrhizal fungi using both bioassays and biochemical analyses of culture supernatants from iron-limited cells [46–50]. Isolation of siderophores from three different ericoid species using chromatography and high pressure liquid chromatography (HPLC) has shown that the ascomycete *Hymenoscyphus ericae* and the hyphomycete *Oidiodendron griseum* produce the hydroxamate ferricrocin as their principal siderophore, whereas an endophyte of *Rhodothamnus chamaecistus* produces mainly the hydroxamate fusigen [48]. Ferricrocin has also been purified from three isolates of the ascomycete *Wilcoxina* that commonly infects a variety of conifers and deciduous trees [50]. Other nonmycorrhizal fungi such as *Aspergillus* and *Neurospora* also produce ferricrocin (as discussed in Ref. [50]).

Vesicular arbuscular (VA) mycorrhizal fungi cannot be maintained without host plants, making it difficult to study iron uptake by the fungi themselves. Information on the effects of VA mycorrhizal fungi on plant iron nutrition is limited, but a recent study with maize showed that the VA mycorrhizal fungi did improve iron acquisition in maize grown under severe iron deficiency [44]. Similarly, infection of grapevine roots with the VA mycorrhizal fungus *Glomus mosseae* resulted in higher leaf iron concentrations [51]. The single report on production of siderophores by VA mycorrhizal fungi was based on results obtained from nonaxenic plants growing in nutrient solution and must be interpreted with caution [42].

2.1.3. Pseudomonads

Fluorescent pseudomonads are common inhabitants of the rhizosphere and phyllosphere and have been shown either to cause (e.g., *Pseudomonas syringae*) or to inhibit (e.g., *P. fluorescens* and *P. putida*) plant disease. They are characterized by the production of yellow-green pigments called pyoverdines or pseudobactins that fluoresce under ultraviolet (UV) light and function as siderophores. Some strains of fluorescent pseudomonads also produce pyochelin and salicylic acid that also serve as siderophores. Here we will focus on the plant growth-promoting fluorescent pseudomonads. Numerous examples of their beneficial effects on plant growth have been reported, such as increased seedling emergence and vigor, improved root system development, and increased yields (as reviewed in Refs. [52 and 53]). These beneficial effects on plant growth, of course, cannot be cleanly separated from the biological control activity of these pseudomonads. The ability of fluorescent pseudomonads to inhibit the growth of other bacteria and fungi, including phytopathogens, has been variously attributed to antibiotic production, siderophore production, hydrogen cyanide (HCN) production, and superior colonization (as reviewed in Refs. [52 and 53]). It is likely that all of these contribute to the overall competitiveness of fluorescent pseudomonads. Pyoverdine siderophores produced by various strains of *Pseudomonas* spp. have been shown to suppress certain soilborne diseases, most notably wilts caused by *Fusarium oxysporum* [54] and seedling diseases caused by *Pythium* spp. [55].

One question which arises when thinking about biocontrol agents is how commonly the growth of a particular bacterium or fungus is limited by iron availability. A sensor that is responsive to iron has been constructed by fusing a promoterless ice nucleation activity gene (*inaZ*) to an iron-regulated (induced in response to iron limitation) promoter of a genomic region involved in pyoverdine production [56]. Experiments with a *P. fluorescens* strain carrying this construct showed that cells encountered an iron-limited environment immediately after inoculation onto bean roots in an agricultural environment. However, after 1 or 2 days, iron became more available as evidenced by a decrease in reporter gene activity [57].

Probably the best studied of the fluorescent pseudomonads is plant growth-promoting *P. putida* WCS358. At least 15 genes are involved in the biosynthesis of the siderophore produced by this strain which is referred to as pseudobactin 358. Fe(III)-pseudobactin is taken into the cell via binding to the PupA outer membrane receptor [58]. The importance of being able to utilize pseudobactin 358 was nicely illustrated in a study which examined the competitiveness of strain WCS358 and strain WCS374, which cannot utilize pseudobactin 358. During four successive plant growth cycles on radish, strain WCS358 significantly reduced rhizosphere population densities of the wild type WCS374 by up to 30 times, whereas WCS374 harboring the siderophore receptor gene *pupA* on a plasmid maintained its population density [59]. WCS358, in addition to producing its own siderophore, can also utilize a large number of siderophores produced by other pseudomonads. Not surprisingly, this strain has multiple outer membrane receptors for these siderophores [60]. Although PupA synthesis is derepressed when cells are grown under iron deficiency, expression of the receptor for the heterologous siderophore pseudobactin BN7, PupB, requires both iron limitation and the presence of the siderophore itself [61]. Two ECF (*e*xtra*c*ytoplasmic *f*unction)-type sigma factors, PupI and PfrI, are involved in positive control of siderophore expression in *P. putida* WCS358, with PupI governing the expression of PupB [62] and PfrI regulating the synthesis of pseudobactin 358 [63]. In addition, a *fur* gene has been isolated from *P. putida* WCS358, but all attempts to construct a *fur* null mutant failed [63], thus preventing a definitive test of the role *fur* plays in iron regulation in this strain. The inability to construct a *fur* deletion mutant strongly suggests that *fur* is an essential gene in *P. putida*. This also appears to be true for *P. aeruginosa* [64], *Neisseriae gonorrheae* [65], *N. meningitidis* [66], and *Vibrio anguillarum* [67].

2.2. Plant Pathogens

Iron is a known virulence factor in animal pathogenesis, where extracellular iron is tightly chelated and not readily available to invading pathogens [68,69]. Lack of iron serves as a signal to activate the expression of a number of virulence factors in the pathogen, such as toxins, in addition to signaling the activation of siderophore systems. In order to address whether iron is a virulence factor in plant pathogenesis, we need to know what forms of iron are available to an invading pathogen *in planta* and we would also like to know whether iron indeed is limiting for proliferation of the pathogen in its host.

2.2.1. *Agrobacterium*

Some of the earliest work on the role of iron in plant pathogenesis was carried out with *Agrobacterium tumefaciens*, the organism responsible for tumorous growth (crown gall) on plants. *Agrobacterium* is able to transform plant cells, and it is widely used as a vector for the incorporation of foreign DNA into plants, producing transgenic plants that express genes of interest (for a recent review, see Ref. [70]). Under low-iron culture conditions, *A. tumefaciens* produces the linear catechol siderophore agrobactin, which is able to reverse iron starvation in this organism [71]. Addition of agrobactin to the growth medium of iron-deficient pea and bean plants has been shown to stimulate plant iron uptake significantly [72]. Autoradiograms of the plants were used to show that the iron taken up under the influence of agrobactin was translocated into the shoot rather than being passively adsorbed or precipitated onto the root surface. Synthesis of chlorophylls *a* and *b* was stimulated with the addition of agrobactin, indicating that the iron taken up was physiologically available to the plants. It is likely that agrobactin was serving as a substrate for the root plasma membrane bound Fe(III)-chelate reductase.

2.2.2. *Erwinia*

Certainly the best-studied system with respect to the role of iron in plant-microbe interactions is that of the plant pathogen *Erwinia chrysanthemi*, which causes soft rot in a variety of plants (for a review see Ref.

[5]). During infection, *E. chrysanthemi* colonizes intercellular spaces and dissolves cell walls using pectinolytic enzymes which it releases (as reviewed in Ref. [73]). Under iron deficiency, cultured *E. chrysanthemi* 3947 synthesizes two siderophores, chrysobactin [*N*-[*N*2-(2,3-dihydroxybenzoyl)]-*D*-[lysyl]-*L*-serine] [74] and achromobactin [75]. The ferrichrysobactin outer membrane receptor gene and the genes required for chrysobactin biosynthesis have been identified and are clustered together in an operon (*fct cbsCEBA*) [76,77] that is negatively regulated by iron [78]. The *cbsA*, *cbsB*, *cbsC*, and *cbsE* genes are the counterparts of *E. coli entA*, *entB*, *entC*, and *entE*, with the functions encoded by these genes being completely interchangeable. The outer membrane receptor protein for ferrichrysobactin, Fct, shows some similarity to receptors for hydroxamate-type siderophores despite the fact that chrysobactin itself is a catechol [79]. There are now a number of examples of this, indicating that one cannot predict the substrate for a transporter merely based on amino acid similarity to other transporters of known specificity.

Iron acquisition mediated by chrysobactin is essential for dissemination of the pathogen throughout its plant host. First of all, mutants of *E. chrysanthemi* defective in chrysobactin transport and biosynthesis produce only localized symptoms; that is, they cannot induce systemic infections on saintpaulia plants [80]. Second, intercellular fluids from diseased plants contain chrysobactin, the first example for direct release of a microbial siderophore *in planta* [81]. The third piece of evidence that chrysobactin is important comes from use of a chromosomal *lacZ* fusion to the promoter of the chrysobactin operon. Induction of LacZ activity is detected 10 h following leaf inoculation, providing the first direct evidence that sites of infection are sensed by the bacteria as iron-restricted environments [82]. So, like the situation in animals, lack of iron in the host organism triggers the synthesis of siderophores by the pathogen which then provides the iron needed to mount a successful invasion. Most interestingly, pectinase expression appears to be regulated by iron, further supporting the central role of iron in plant pathogenesis [83].

So what is the role of the second siderophore, achromobactin? Chrysobactin can compete efficiently with strong chelators while achromobactin cannot. Furthermore, achromobactin is produced under conditions when chrysobactin synthesis is completely repressed [75].

We know very little about whether the plant has mechanisms for withholding iron from invading pathogens. Growth of several mutant strains of *Erwinia chrysanthemi* was inhibited by plant polyphenols (tannins), and the growth inhibition was explained by the inability of the mutants to assimilate iron in the presence of polyphenols [84]. The mutant bacterial strains used in the experiments were altered in their siderophore-mediated iron transport pathway. Each polyphenol molecule has several chelating groups and can bind several Fe(III) ions, and each Fe(III) can itself coordinate with up to three chelating groups belonging to different polyphenol molecules. A lattice is consequently formed, which eventually results in the coprecipitation of iron and polyphenols and the removal of Fe(III) from solution. It was suggested that polyphenols can withhold iron (and other metal ions) from pathogens in a manner similar to that of iron-binding proteins in animals, such as transferrin, conalbumin, and lactoferrin. Iron withholding by polyphenols could be involved in the protection of leaves and fruits and in the protection against rots of many durable heartwoods. One interesting hypothesis is that polyphenols in the bark could constitute a chemical barrier preventing microorganisms from reaching the underlying living tissues (phloem, cambium, and xylem) just as conalbumin in egg white prevents microorganisms from reaching the developing embryo in the yolk [84].

3. IRON UPTAKE SYSTEMS OF PLANTS

3.1. Strategy I Plants

All plants except the grasses use a reductive strategy to acquire iron that is similar to the system used by the yeast *Saccharomyces cerevisiae* (reviewed in Ref. [85]). This strategy has been referred to as strategy I. The initial reduction of Fe(III), carried out by a plasma membrane-bound Fe(III)-chelate reductase, is followed by transport of Fe(II) across the root epidermal cell membrane. What is the substrate of the Fe(III) chelate reductase? It is likely that various microbial siderophores can serve as substrates. Genes encoding the Fe(III) reductase have not yet been cloned from plants, but *Arabidopsis thaliana* mutants that exhibit defects in Fe(III) reduction have been identified [86]; map-based cloning is now in progress. Both the Fe(III) reductase [87] and the Fe(II) transport activities [88] are enhanced under iron deficiency. Proton release, which lowers rhizosphere pH and thereby increases the solubility of Fe(III), is also enhanced under iron deficiency [89,90].

We have recently cloned a gene, *IRT1*, that encodes an Fe(II) transporter from *Arabidopsis thaliana* by complementing an iron uptake-deficient strain of *Saccharomyces cerevisiae* [91]. *IRT1* mRNA is expressed at a high level in roots of iron-deficient plants; no expression is seen in roots of iron-sufficient plants or in shoots of either iron-deficient or iron-sufficient plants. The deduced protein product of IRT1 is predicted to have eight membrane-spanning domains and to contain a presumptive metal binding motif, HGHGHGHG. Related sequences have been found in the genomes of rice, yeast, nematodes, and humans, and some of these genes have been shown to be zinc transporters [85,92,93].

3.2. Strategy II Plants

Under iron deficiency, grasses synthesize and release nonprotein amino acids called phytosiderophores. Grasses also induce a high-affinity uptake system for Fe(III)-phytosiderophore complexes that transports the Fe(III)-phytosiderophores into the root as intact complexes [94]. This strategy is termed strategy II and is considered to be more efficient than strategy I. For example, grasses can grow on calcareous soils that will not support the growth of dicots. One reason for this may be that strategy II is less pH dependent than strategy I. The phytosiderophores that have been characterized are all structurally related and have been generically termed mugineic acids (MAs), after the first phytosiderophore that was characterized. All MAs share the same pathway from *L*-methionine to 2-deoxymugineic acid (as reviewed in Ref. [94]). Subsequent biosynthetic steps depend on the plant species and cultivar. The amounts of MAs produced by different plants also vary considerably, with sorghum producing 20 times less than barley [95]. The classification of C4 grasses with low rates of phytosiderophore release, such as maize and sorghum, as strategy II plants has been questioned [96]. It has been proposed that only iron-efficient C3 species of grasses, with high rates of phytosiderophore release, should be considered strategy II plants [96]. However, experiments with both chlorosis-resistant and chlorosis-susceptible (high and low siderophore release, respectively) cultivars of oats (C3) and an iron-efficient genotype of maize (C4) have failed to show either Fe(III) reduction or chelate splitting when Fe(III)-siderophores were supplied, indicating that these plants should not be considered strategy I plants [97].

There are a number of points to consider when trying to evaluate the ecological importance of phytosiderophores. First, what are concentrations of phytosiderophores in the rhizosphere? A monoclonal antibody raised against nicotianamine, thought to be a precursor in the biosynthesis of MAs, recognizes a number of MAs [98], and might be used to estimate levels in the soil. However, extracting MAs from the soil for use in an enzyme-linked immunosorbent assay, (ELISA)-assay, is not straightforward. Second, do rhizosphere microorganisms affect the distribution of phytosiderophores? As might be expected, there are a number of bacteria which can utilize phytosiderophores as a sole source of carbon as well as a source of iron [99]. In at least one case, utilization of phytosiderophores as an iron source appears to be indirect, with ligand exchange occurring between the microbial siderophore and the phytosiderophores (see Sec. 5 and Ref. [100]). This raises a third question: How successfully do phytosiderophores compete with microbial siderophores for Fe(III)? Plants which have been appropriately examined release appreciable amounts of phytosiderophores with a distinct diurnal rhythm to the release [101–103]. Release occurs over 4–6 hs, beginning about 2 h after the onset of light. It has been suggested that the release of a single large quantity once a day may increase the probability that phytosiderophores will mobilize Fe(III) before being degraded by rhizosphere microorganisms. Release is also localized; physical separation on the root between zones of active release and uptake of phytosiderophores and zones of heavy colonization by bacteria may also protect phytosiderophores from competition for iron with microbial siderophores. A final question concerns specificity; that is, do phytosiderophores mobilize other elements besides Fe(III)? There is recent evidence that maize can take up Zn-phytosiderophore complexes [104].

Efforts are underway to isolate and characterize the genes involved in the acquisition of iron by Gramineae. In barley, a differential hybridization approach has identified cDNAs specific to iron deficiency; some of these genes may encode products which are involved in the synthesis, secretion, or uptake of mugineic acids (MAs) [105]. Although *ids1*, the first cDNA clone characterized, was found to encode a protein which is similar to metallothionein, two of the other clones identified (*ids2* and *ids3*) encode proteins which show similarity to 2-oxoglutarate-dependent dioxygenase, suggesting that they may function in the hydroxylation process leading to the synthesis of mugineic acid.

4. UTILIZATION OF MICROBIAL SIDEROPHORES BY PLANTS

The relative importance of microbial siderophores in plant iron nutrition and the possible mechanism(s) by which plants make use of siderophore-bound iron are still controversial subjects, in part because most earlier studies were carried out under nonaxenic conditions. We will discuss the use of microbial siderophores by strategy I and strategy II plants, focusing on the most recent studies in the field.

In cotton plants, the use of a fluorescent ferrioxamine B analogue to study iron uptake by nonaxenically and axenically grown plants showed that Fe(III) could be removed from the chelate and taken up by the plants; no uptake of the chelate was observed [106]. Experiments with axenic cucumber, however, have led to the conclusion that both iron and the siderophore ferrioxamine B (FOB) were taken up by the plants in a very efficient manner. In those experiments, a bioassay was used to detect the siderophore in xylem exudates, where it was shown to be present within 2 h of treatment [107]. Experiments conducted with metabolic inhibitors suggested that the FOB uptake involved passive uptake mechanisms. The possibility of Fe(III)-siderophore entry into the plant through breaks in the symplasm, at sites of lateral root emergence, was considered unlikely, because siderophore was detected in exudates only when FOB, but not DFOB (the deferrated form), was included in the nutrient solution. Passive diffusion of either FOB or DFOB was also predicted to be negligible owing to the low solubility in lipids of these compounds. Facilitated diffusion was suggested to be the most likely mechanism for the siderophore-mediated transport observed in those experiments [107]. However, it should also be considered that in long-term experiments, higher concentrations of siderophores in the rhizosphere soil than in the bulk soil could result in increased levels of exchangeable iron, which can be freshly precipitated as amorphous $Fe(OH)_3$ at the rhizoplane [97]. Siderophores of low stability could act as a readily accessible source of Fe(III) for root exudates and enhance iron uptake.

One possible mechanism for iron assimilation from microbial siderophores by plants is the functioning of plant nitrate reductase as a siderophore reductase. A partially purified siderophore reductase activity from soybean seedlings had characteristics of a nitrate reductase and would probably copurify with nitrate reductase [108]. Nitrate reductase purified from squash cotyledons was shown to work as a ferrisiderophore reductase for FOB, ferrichome A, ferrichrome, ferrirhodotorulic acid, and ferrischizokinen [109,110]. Because nitrate reductase is localized in the cytoplasm, this enzyme may function to reduce Fe(III)-siderophore complexes only if they have been taken up by diffusion or by a siderophore transporter system. To determine whether there may be other reductases on root cell membranes which can catalyze Fe(III)-siderophore reduction, experiments were conducted with several cellular subfractions. A variety of microbial Fe(III)-siderophores were reduced by NAD(P)H oxido-reductases located in the plasma membrane as well as in microsomal and cytosolic cell fractions from root cells, indicating that at least one or more reductases are able to utilize siderophores as an electron acceptor [111]. This demonstrates that plants can remove Fe(III) from siderophores either at the cell surface or in the cytosol. However, these activities are possibly nonspecific and may not correspond to the actual siderophore reductase(s) that function in vivo.

Attempts to determine whether strategy II plants that normally release their own siderophores can also utilize microbial siderophores have yielded conflicting data, in large part because of the use of nonaxenic conditions. Experiments which take into account the effects of rhizosphere microorganisms suggest that microbial siderophores do not serve directly as sources of iron for phytosiderophore-releasing plants [112]. However, microbial siderophores may serve indirectly after degradation by microbes and release of Fe(III). Although there are very few documented cases of the utilization of siderophores as carbon sources [113–115], it seems reasonable to suppose that there are probably many microbes capable of degrading siderophores. DeAngelis et al. [114] have raised the intriguing question of whether microbes display nutritional selectivity in deciding whether to utilize a particular siderophore as a carbon or as an iron source. Based on their observations of a single strain, they suggest that organisms might not catabolize those siderophores which they can utilize as iron sources. Catabolism of siderophores is an issue which can be raised not only in soil systems but in aquatic and host systems as well.

The use of iron from microbial siderophores by plants via an indirect mechanism that involves ligand exchange has also been considered. This hypothesis gained support in a recent study with barley and maize [116]. The fungal siderophore rhizoferrin was found to be an efficient source of iron for both plants. Uptake and translocation rates from the fungal siderophore, as well as from other microbial siderophores and iron chelates, were parallel to the diurnal rhythm of phytosiderophore release by the plants. It was suggested that Fe(III)-rhizoferrin is taken up by strategy II plants via an indirect mechanism in which there is ligand exchange between the microbial siderophore and phytosiderophores, which are then taken up by the plant. The hypothesis was verified by in vitro ligand-exchange experiments. It is quite possible that this mechanism can also account for the utilization of iron from microbial siderophores reported previously, and that the active uptake observed was actually uptake of Fe-phytosiderophore complexes after exchange of the metal from the microbial siderophores to phytosiderophores.

Regardless of the mechanism by which plants make use of iron chelated by microbial siderophores, it has long been observed that the addition of bacterial siderophores to calcareous soil can be used as a remedy for lime-induced chlorosis. In peanut, chlorosis was prevented by the application of 19 mg Fe (supplied as *Pseudomonas* siderophores) per kilogram of soil or 45 mg Fe kg^{-1} soil of FOB [117]. However, the same plant chlorophyll levels were obtained by the addition of only 1.2 mg Fe kg^{-1} soil when iron was supplied as FeEDDHA. Although less efficient, microbial siderophores could be important sources of iron to soil grown plants, since synthetic chelators, such as EDDHA and EDTA, are not naturally present in soils.

5. UTILIZATION OF PHYTOSIDEROPHORES BY MICROBES

Although several studies on the use of microbial siderophores as iron sources by plants have been pursued, little attention has been given to the possibility of microbes using phytosiderophores as iron sources. As we have already discussed, many microbes can use siderophores which they themselves do not synthesize using a siderophore-specific iron-uptake system to internalize the ferrisiderophore complexes. Alternately, those microorganisms that use a reductive strategy to acquire iron might be able to use phytosiderophores as a substrate for their Fe(III) chelate reductase. Using radiolabeled Fe(III)-MA as an iron source, ^{55}Fe was shown to be taken up by three strains of *Pseudomonas putida* (WCS358, B1, and St3) under iron-deficient conditions; uptake decreased 70% when the bacteria were grown in the presence of iron [100]. No differences in uptake between iron-deficient and iron-sufficient cells were observed for a siderophore-deficient *Pseudomonas* strain of WCS358, JM218. Taken together, these results suggest that utilization of Fe(III)-MA complex as a source of iron by fluorescent *Pseudomonas* occurred via an indirect mechanism (ligand exchange). When a Fe(III)-MA complex was mixed with free pseudobactin, ligand exchange occurred rapidly, with a half-life for Fe(III)-MA of 2 h. Conversely, no ligand exchange was seen when fully chelated pseudobactin was challenged with an equimolar concentration of MA. Values reported in the literature for the stability constant of Fe(III)-MA complexes vary widely, but in general microbial siderophores have a higher affinity for iron than MAs. Thus, the physical separation along the root of zones of active release and uptake of MAs from zones heavily populated by microorganisms, could protect the phytosiderophores not only from microbial degradation but also from competition for iron with pseudobactin and other microbial siderophores (as discussed earlier).

6. CONCLUSIONS

Better understanding of the interactions between microbes and plants in the rhizosphere remains one of the most challenging problems in terrestrial ecology. Iron is a limiting nutrient for both plant and microbial growth. We need to develop a working model of competition for iron in the rhizosphere. Such knowledge may eventually allow us to manipulate the survival and growth of microorganisms introduced into the soil for various purposes, such as for biocontrol of plant pathogens or in aid of improving nitrogen fixation. Furthermore, recent advances in model systems such as yeast and arabidopsis have already begun to facilitate efforts to engineer plants that compete more effectively for scarce nutrients such as iron. It may well be possible to achieve biased rhizospheres with well-matched sets of plants and associated microflora that are suited to a particular habitat.

ACKNOWLEDGMENTS

Work on this topic in the laboratory of M.L.G. has been supported by grants from the National Science Foundation, the U.S. Department of Energy, and the U.S. Department of Agriculture. We thank Rob McClung for reading several drafts and all members of the Guerinot Laboratory for stimulating discussions.

ABBREVIATIONS

cbs	chrysobactin biosynthesis
cDNA	complementary deoxyribonucleic acid
css	chromosomal citrate synthase
cycHJKL	cytochrome-*c* biogenesis operon
cytA	cytochrome
DFOB	deferrated ferrioxamine B
ECF	extracellular function
EDDHA	ethylenediamine-di(*o*-hydroxy)-phenylacetic acid
EDTA	ethylenediaminetetraacetic acid
ELIZA	enzyme-linked immunosorbent assay
ent	enterobactin
fct	ferrichrysobactin transport
fegA	ferric iron gathering
feu	ferric uptake
FOB	ferrioxamine B
fur	ferric uptake regulation
gltA	glutamate
HCN	hydrogen cyanide

HPLC	high-pressure liquid chromatography
ids	iron deficiency specific
ina	ice nucleation activity
irt	iron-regulated transporter
lacZ	lactose
MA	mugineic acid
NAD	nicotinamide adenine dinucleotide
OMP	outer membrane protein
PBM	peribacteroid membrane
PBS	peribacteroid space
pcs	plasmid citrate synthase
pfr	*Pseudomonas* ferric regulator
pup	pseudobactin uptake protein
tonB	T one
Tn5	transposon 5
VA	vesicular arbuscular

REFERENCES

1. M. L. Guerinot and Y. Yi, *Plant Physiol.*, *104*, 815–820 (1994).
2. D. L. Jones, P. R. Darrah, and L. V. Kochian, *Plant Soil*, *180*, 57–66 (1996).
3. H. A. Akers, *Appl. Environ. Microbiol.*, *45*, 1704–1706 (1983).
4. D. Touati, M. Jacques, B. Tardat, L. Bouchard, and S. Despied, *J. Bacteriol.*, *177*, 2305–2314 (1995).
5. D. Expert, C. Enard, and C. Masclaux, *Trends Microbiol.*, *4*, 232–236 (1996).
6. M. L. Guerinot, *Plant Soil*, *130*, 199–209 (1991).
7. K. C. Carson, A. R. Glenn, and M. J. Dilworth, *Arch. Microbiol.*, *161*, 333–339 (1994).
8. K. C. Carson, S. Holliday, A. R. Glenn, and M. J. Dilworth, *Arch. Microbiol.*, *157*, 264–271 (1992).
9. E. Fabiano, G. Gualtieri, C. Pritsch, G. Polla, and A. Arias, *Plant Soil*, *164*, 177–185 (1994).
10. R. S. Jadhav and A. Desai, *FEMS Microbiol. Lett.*, *115*, 185–190 (1994).

11. D. Lesueur, H. G. Diem, and J. M. Meyer, *J. Appl. Bacteriol.*, *74*, 675–682 (1993).
12. M. G. Manjanatha, T. E. Loynachan, and A. G. Atherly, *Soil Biol. Biochem.*, *24*, 151–155 (1992).
13. H. N. Patel, R. N. Chakraborty, and S. B. Desai, *Curr. Microbiol.*, *28*, 119–121 (1994).
14. G. Reigh and M. O'Connell, *J. Bacteriol.*, *175*, 94–102 (1993).
15. N. Roy, P. Bhattacharyya, and P. K. Chakrabartty, *Microbiology*, *140*, 2811–2820 (1994).
16. S. Suneja, K. S. Yadav, and H. R. Sharma, *Crop Res.*, *8*, 621–626 (1994).
17. D. van Rossum, A. Muyotocha, H. W. van Verseveld, A. H. Stouthamer, and F. C. Boogerd, *Plant Soil*, *163*, 177–184 (1994).
18. M. J. Smith, J. N. Shoolery, B. Schwyn, I. Holden, and J. B. Neilands, *J. Am. Chem. Soc.*, *107*, 1739–1743 (1985).
19. M. Persmark, P. Pittman, J. S. Buyer, B. Schwyn, P. R. Gill, and J. B. Neilands, *J. Am. Chem. Soc.*, *115*, 3950–3956 (1993).
20. M. L. Guerinot, E. J. Meidl, and O. Plessner, *J. Bacteriol.*, *172*, 3298–3303 (1990).
21. I. Hernández-Lucas, M. A. Pardo, L. Segovia, J. Miranda, and E. Martínez-Romero, *Appl. Environ. Microbiol.*, *61*, 3992–3997 (1995).
22. M. A. Pardo, J. Lagunez, J. Miranda, and E. Martínez, *Mol. Microbiol.*, *11*, 315–321 (1994).
23. K. LeVier, *Iron Acquisition in Bradyrhizobium japonicum*, PhD thesis, Dartmouth College, Hanover, NH, 1996.
24. J. S. Buyer and L. J. Sikora, *Plant Soil*, *129*, 101–107 (1990).
25. C. R. Dean and K. Poole, *Mol. Microbiol.*, *8*, 1095–1103 (1993).
26. V. Braun, *FEMS Microbiol. Lett.*, *16*, 295–307 (1995).
27. M. L. Guerinot, *Ann. Rev. Microbiol.*, *48*, 743–772 (1994).
28. P. R. Gill. Jr., and J. B. Neilands, *Mol. Microbiol.*, *3*, 1183–1189 (1989).
29. O. Plessner, T. Klapatch, and M. L. Guerinot, *Appl. Environ. Microbiol.*, *59*, 1688–1690 (1993).
30. K. LeVier and M. L. Guerinot, *J. Bacteriol.*, *178*, 7265–7275 (1996).

31. P. R. Gill, Jr., L. L. Barton, M. D. Scoble, and J. B. Neilands, *Plant Soil*, *130*, 211–217 (1991).
32. K. D. Nadler, A. W. B. Johnston, J. W. Chen, and T. R. John, *J. Bacteriol.*, *172*, 670–677 (1990).
33. K. H. Yeoman, M.-J. Delgado, M. Wexler, J. A. Downie, and A. W. B. Johnston, *Microbiology*, *143*, 127-134 (1997).
34. A. Gaballa, N. Koedam, and P. Cornelis, *Mol. Microbiol.*, *21*, 777–785 (1996).
35. R. E. Terry, K. U. Soerensen, V. D. Jolley, and J. C. Brown, *Plant Soil*, *130*, 225–230 (1991).
36. L. O. Tiffin, *Plant Physiol.*, *41*, 510–514 (1966).
37. S. C. Andrews, P. Arosio, W. Bottke, J.-F. Briat, M. von Darl, P. Harrison, J.-P. Laulhere, S. Levi, S. Lobréaux, and S. J. Yewdall, *J. Inorg. Biochem.*, *47*, 161–174 (1992).
38. K. LeVier and M. L. Guerinot, *Plant Physiol.*, *111*, 893–900 (1996).
39. J. B. Wittenberg and B. A. Wittenberg, *Annu. Rev. Biophys. Biophys. Chem.*, *19*, 217–241 (1990).
40. J. B. Wittenberg, B. A. Wittenberg, D. A. Day, M. K. Udvardi, and C. A. Appleby, *Plant Soil*, *178*, 161–169 (1996).
41. H. Marschner, *Mineral Nutrition of Higher Plants*, 2nd Ed., Academic Press, London, 1995.
42. W. A. Cress, G. V. Johnson, and L. L. Barton, *J. Plant Nutr.*, *9*, 547–556 (1986).
43. J. R. Leake, G. Shaw, and D. J. Read, *New Phytol.*, *114*, 651–657 (1990).
44. R. B. Clark and S. K. Zeto, *J. Plant Nutr.*, *19*, 247–264 (1996).
45. C. Leyval and C. P. P. Reid, *New Phytol.*, *119*, 93–98 (1991).
46. P. J. Szaniszlo, P. E. Powell, C. P. P. Reid, and G. R. Cline, *Mycologia*, *73*, 1158–1174 (1981).
47. R. Schüler and K. Haselwandter, *J. Plant Nutr.*, *11*, 907–913 (1988).
48. K. Haselwandter, B. Dobernigg, W. Beck, G. Jung, A. Cansier, and G. Winkelmann, *BioMetals*, *5*, 51–56 (1992).
49. K. Haselwandter, *Crit. Rev. Biotechnol.*, *15*, 287–291 (1995).
50. V. Prabhu, P. F. Biolchini, and G. L. Boyer, *BioMetals*, *9*, 229–234 (1996).

51. L. Bavaresco, M. Fregoni, and C. Fogher, in *Iron Nutrition in Soils and Plants* (J. Abadía, ed.), Kluwer, Dordrecht, The Netherlands, 1995, pp. 83–89.

52. D. J. O'Sullivan and F. O'Gara, *Microbiol. Rev.*, *56*, 662–676 (1992).

53. M. Höfte, M. Vande Woestyne, and W. Verstraete, in *Biochemistry of Metal Micronutrients in the Rhizosphere*. (J. A. Manthey, D. E. Crowley, and D. G. Luster, eds.), Lewis, Boca Raton, FL, 1994, pp. 81–92.

54. P. Lemanceau, and C. Alabovette, *Biocontrol Sci. Technol.*, *3*, 219–234 (1993).

55. S. Buysens, K. Heungens, J. Poppe, and M. Höfte, *Appl. Environ. Microbiol*, *62*, 865–871 (1996).

56. J. E. Loper and S. E. Lindow, *Appl. Environ. Microbiol.*, *60*, 1934–1941 (1994).

57. J. E. Loper and M. D. Henkels, *Appl. Environ. Microbiol.*, *63*, 99–105 (1997).

58. W. Bitter, J. D. Marugg, L. A. de Weger, J. Tommassen, and P. J. Weisbeek, *Mol. Microbiol.*, *5*, 647–655 (1991).

59. J. M. Raaijmakers, I. van der Sluis, M. Koster, P. A. H. M. Bakker, P. J. Weisbeek, and B. Schippers, *Can. J. Microbiol.*, *41*, 126–135 (1995).

60. M. Koster, W. Ovaa, W. Bitter, and P. Weisbeek, *Mol. Gen. Genet.*, *248*, 735–743 (1995).

61. M. Koster, J. van de Vossenberg, J. Leong, and P. J. Weisbeek, *Mol. Microbiol.*, *8*, 591–601 (1993).

62. M. Koster, W. van Klompenburg, W. Bitter, J. Leong, and P. Weisbeek, *EMBO J.*, *13*, 2805–2813 (1994).

63. V. Venturi, C. Ottevanger, M. Bracke, and P. Weisbeek, *Mol. Microbiol.*, *15*, 1081–1093 (1995).

64. R. W. Prince, C. D. Cox, and M. L. Vasil, *J. Bacteriol.*, *175*, 2589–2598 (1993).

65. S. A. Berish, S. Subbarao, C.-Y. Chen, D. L. Trees, and S. A. Morse, *Infect. Immun.*, *61*, 4599–4606 (1993).

66. C. E. Thomas and P. F. Sparling, *Mol. Microbiol.*, *11*, 725–737 (1994).

67. M. E. Tolmasky, A. M. Wertheimer, L. A. Actis, and J. H. Crosa, *J. Bacteriol.*, *176*, 213–220 (1994).

68. C. M. Litwin and S. B. Calderwood, *Clin. Microbiol. Rev.*, *6*, 137–149 (1993).
69. S. M. Payne, *Trends Microbiol.*, *1*, 66–69 (1993).
70. J. Sheng and V. Citovsky, *Plant Cell*, *8*, 1699–1710 (1996).
71. S. A. Ong, T. Peterson, and J. B. Neilands, *J. Biol. Chem.*, *254*, 1860–1865 (1979).
72. J. O. Becker, E. Messens, and R. W. Hedges, *FEMS Microbiol. Ecol.*, *31*, 171–175 (1985).
73. F. Barras, F. Van Gijsegem, and A. K. Chatterjee, *Annu. Rev. Phytopathol.*, *32*, 201–234 (1994).
74. M. Persmark, D. Expert, and J. B. Neilands, *J. Biol. Chem.*, *264*, 3187–3193 (1989).
75. B. C. Mahé, C. Masclaux, L. Rausher, C. Enard, and D. Expert, *Mol. Microbiol.*, *18*, 33–43 (1995).
76. T. Franza, C. Enard, F. van Gijsegem, and D. Expert, *Mol. Microbiol.*, *5*, 1319–1329 (1991).
77. T. Franza and D. Expert, *J. Bacteriol.*, *173*, 6874–6881 (1991).
78. D. Expert, C. Sauvage, and J. B. Neilands, *Mol. Microbiol.*, *6*, 2009–2017 (1992).
79. C. Sauvage, T. Franza, and D. Expert, *J. Bacteriol.*, *178*, 1227–1231 (1996).
80. C. Enard, A. Diolez, and D. Expert, *J. Bacteriol.*, *170*, 2419–2426 (1988).
81. C. Neema, J.-P. Laulhere, and D. Expert, *Plant Physiol.*, *102*, 967–973 (1993).
82. C. Masclaux and D. Expert, *Plant J.*, *7*, 121–128 (1995).
83. C. Masclaux, N. Hugouvieux-Cotte-Pattat, and D. Expert, *Mol. Plant-Microbe Interact.*, *9*, 198–205 (1996).
84. I. Mila, A. Scalbert, and D. Expert, *Phytochemistry*, *42*, 1551–1555 (1996).
85. D. J. Eide and M. L. Guerinot, *ASM News*, *63*, 199–205 (1997).
86. Y. Yi and M. L. Guerinot, *Plant J.*, *10*, 835–844 (1996).
87. H. F. Bienfait, R. J. Bino, A. M. van der Bliek, J. F. Duivenvoorden, and J. M. Fontaine, *Physiol. Plant.*, *59*, 196–202 (1983).
88. T. C. Fox, J. E. Shaff, M. A. Grusak, W. A. Norvell, Y. Chen, R. L. Chaney, and L. V. Kochian, *Plant Physiol.*, *111*, 93–100 (1996).

89. E. Alcantara, M. D. de la Guardia, and F. J. Romero, *Plant Physiol.*, *96*, 1034–1037 (1991).

90. V. Römheld, C. Müller, and H. Marschner, *Plant Physiol.*, *76*, 603–606 (1984).

91. D. Eide, M. Broderius, J. Fett, and M. L. Guerinot, *Proc. Natl. Acad. Sci. USA*, *93*, 5624–5628 (1996).

92. H. Zhao and D. Eide, *Proc. Natl. Acad. Sci. USA*, *93*, 2454–2458 (1996).

93. H. Zhao and D. Eide, *J. Biol. Chem.*, *271*, 23203–23210 (1996).

94. J. F. Ma and K. Nomoto, *Physiol. Plant.*, *97*, 609–617 (1996).

95. V. Römheld and H. Marschner, *Plant Soil*, *123*, 147–153 (1990).

96. C. M. Lytle and V. D. Jolley, *J. Plant Nutr.*, *14*, 341–362 (1991).

97. H. Marschner and V. Römheld, in *Iron Nutrition in Soils and Plants* (J. Abadía, ed.). Kluwer, Dordrecht, The Netherlands, 1995, pp. 375–388.

98. S. Mihashi, L. J. Sikora, J. S. Buyer, N. Fushiya, and S. Mori, *Plant Cell Physiol.*, *33*, 151–156 (1992).

99. S. Watanabe and H. Wada, *Jpn. J. Soil Sci. Plant Nutr.*, *60*, 413–417 (1989).

100. E. Jurkevitch, Y. Hadar, Y. Chen, M. Chino, and S. Mori, *BioMetals*, *6*, 119–123 (1993).

101. S. Takagi, K. Nomoto, and T. Takemoto, *J. Plant Nutr.*, *7*, 469–477 (1984).

102. F. Zhang, V. Römheld, and H. Marschner, *Soil Sci. Plant Nutr.*, *37*, 671–678 (1991).

103. A. Walter, A. Pich, G. Scholz, H. Marschner, and V. Römheld, *J. Plant Physiol.*, *147*, 191–196 (1995).

104. N. von Wirén, H. Marschner, and V. Römheld, *Plant Physiol.*, *111*, 1119–1125 (1996).

105. N. Nakanishi, N. Okumura, Y. Umehara, N. K. Nishizawa, M. Chino, and S. Mori, *Plant Cell Physiol.*, *34*, 401–410 (1993).

106. E. Bar-Ness, Y. Hadar, Y. Chen, A. Shanzer, and J. Libman, *Plant Physiol.*, *99*, 1329–1335 (1992).

107. Y. Wang, H. N. Brown, D. E. Crowley, and P. J. Szaniszlo, *Plant Cell Environ.*, *16*, 579–585 (1993).

108. C. L. Tipton and J. Thomsen, *Plant Physiol.*, *79*, 432–435 (1985).

109. D. Castignetti and J. J. Smarrelli, *Biochem. Biophys. Res. Com.*, *125*, 52–58 (1984).
110. J. J. Smarrelli and D. Castignetti, *Biochim. Biophys. Acta*, *882*, 337–342 (1986).
111. D. E. Crowley, Y. C. Wang, C. P. P. Reid, and P. J. Szaniszlo, *Plant Soil*, *130*, 179–198 (1991).
112. E. Bar-Ness, Y. Hadar, Y. Chen, V. Römheld, and H. Marschner, *Plant Physiol.*, *100*, 451–456 (1992).
113. D. Castignetti, A. S. Siddiqui, and K. W. Olsen, *BioMetals*, *3*, 197–203 (1990).
114. R. DeAngelis, M. Forsyth, and D. Castignetti, *BioMetals*, *6*, 234–238 (1993).
115. G. Winkelmann, K. Schmidtkunz, and F. A. Rainey, *BioMetals*, *9*, 78–83 (1996).
116. Z. Yehuda, M. Shenker, V. Römheld, H. Marschner, Y. Hadar, and Y. Chen, *Plant Physiol.*, *112*, 1273–1280 (1996).
117. E. Jurkevitch, Y. Hadar, and Y. Chen, *Soil Sci. Soc. Am. J.*, *52*, 1032–1037 (1988).

6

Iron Transport in Graminaceous Plants

Satoshi Mori

Laboratory of Plant Molecular Physiology,
The University of Tokyo, Tokyo 113, Japan

1.	INTRODUCTION	216
	1.1. Iron Transport in the Roots of Strategy I Plants	216
	1.2. Iron Transport in the Roots of Strategy II Plants	219
2.	IRON TRANSPORT FROM ROOT TO SHOOT	219
3.	IRON TRANSPORT FROM SHOOT TO ROOT	222
4.	EFFECT OF IRON STATUS ON IRON TRANSPORT IN PLANTS	223
5.	LIGHT EFFECT ON IRON TRANSPORT IN PLANTS IN COMPARISON WITH PHOSPHATE TRANSPORT	227
6.	LIGHT-DEPENDENT IRON TRANSPORT INTO CHLOROPLASTS	231
7.	CONCLUSIONS	233
	ACKNOWLEDGMENTS	235

ABBREVIATIONS 235
REFERENCES 236

1. INTRODUCTION

Iron deficiency in cultivated plants is a world wide problem in calcareous soil, which accounts for 30% of the cultivated soil around the world. There is sufficient Fe in this soil, but it is sparingly soluble in soil water under aerobic soil conditions at pH above 7. From an evolutionary point of view, however, plants have acquired the ability to absorb the sparingly soluble Fe from the rhizosphere. The pioneering work of Takagi [1], who discovered mugineic acid, paved the way to the establishment of two Fe-acquisition systems in the plant kingdom, strategy I and strategy II [2]. Strategy I (Fig. 1) is typical for dicots and monocots, except grasses, and is characterized by increased root-reducing capacity. On the other hand, strategy II (Fig. 2) is associated with the roots of graminaceous plants. In this system, natural chelators, called mugineic acid families (MAs), are released from the roots. The two strategies of the Fe-acquisition mechanism proposed by Römheld and Marschner [2] are still effective as leading hypotheses. I have modified the mechanism of strategy I and strategy II based on new evidence obtained after their proposal.

1.1. Iron Transport in the Roots of Strategy I Plants

Caffeic acid from soybean [3] and tomato [4], pistidic acid from pigeon pea [5], and other Fe chelators, such as alfafuran from alfalfa [6,7], are secreted from roots and solubilize Fe of soil minerals as Fe(III)-chelate. On the other hand, proton extrusion to the rhizosphere by the function of proton adenosine triphosphatases (ATPase) of the plasma membrane occurs in some plants from the epidermal cells of the root tips. The roots under Fe deficiency show increased cell wall ingrowth into cytoplasm (transfer cells) or an increase in the number of root hairs which are supposed to increase the surface area of the epidermal cell membranes

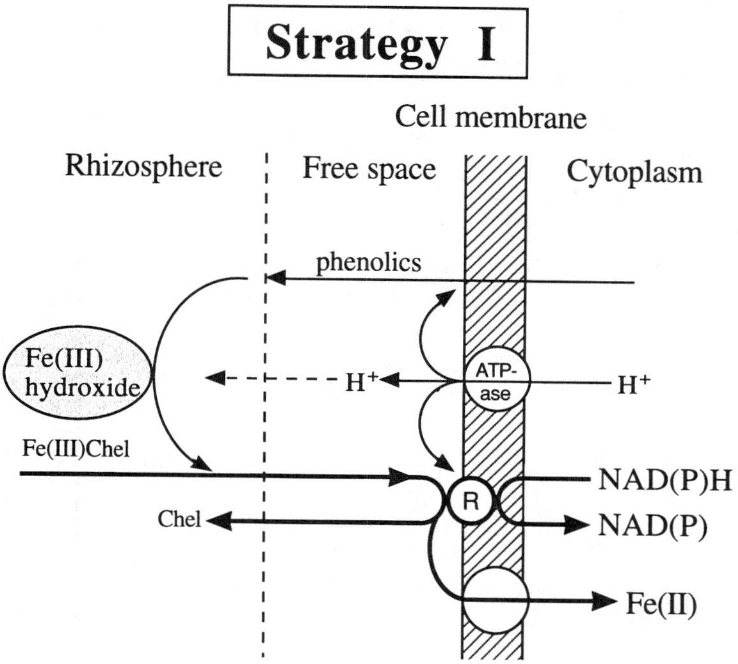

FIG. 1. Fe absorption mechanism of strategy I in the roots of dicots and nongraminaceous monocots.

for proton extrusion [8–10]. This decreased pH of the rhizosphere increases the concentration of soluble Fe(III) according to the stability constant of $Fe(OH)_3$ (e.g., Fe(III) = $10^{-38}/[OH]^3 = 10^{-17}$ M at pH 7). The soluble Fe is moved in the transpiration stream in the soil to the root surface, then Fe(III) is reduced to Fe(II) by reductase in the root cell membranes, and the Fe(II) ion is absorbed by the Fe(II) transporter. Recently, Fe(II)-transporter-like cDNA has been cloned in *Arabidopsis thaliana* [11]. The gene expression was enhanced by Fe-deficiency treatment. On the other hand, the genes of Fe(III)-reductase (frohA and frohB) have been cloned from *Arabidopsis thaliana* (N. Robinson, personal communication), although the complete analysis of these genes is still unsolved. Fe(III)-reductase in *Saccharomyces cerevisiae* has been cloned as the *FRE1* gene [12].

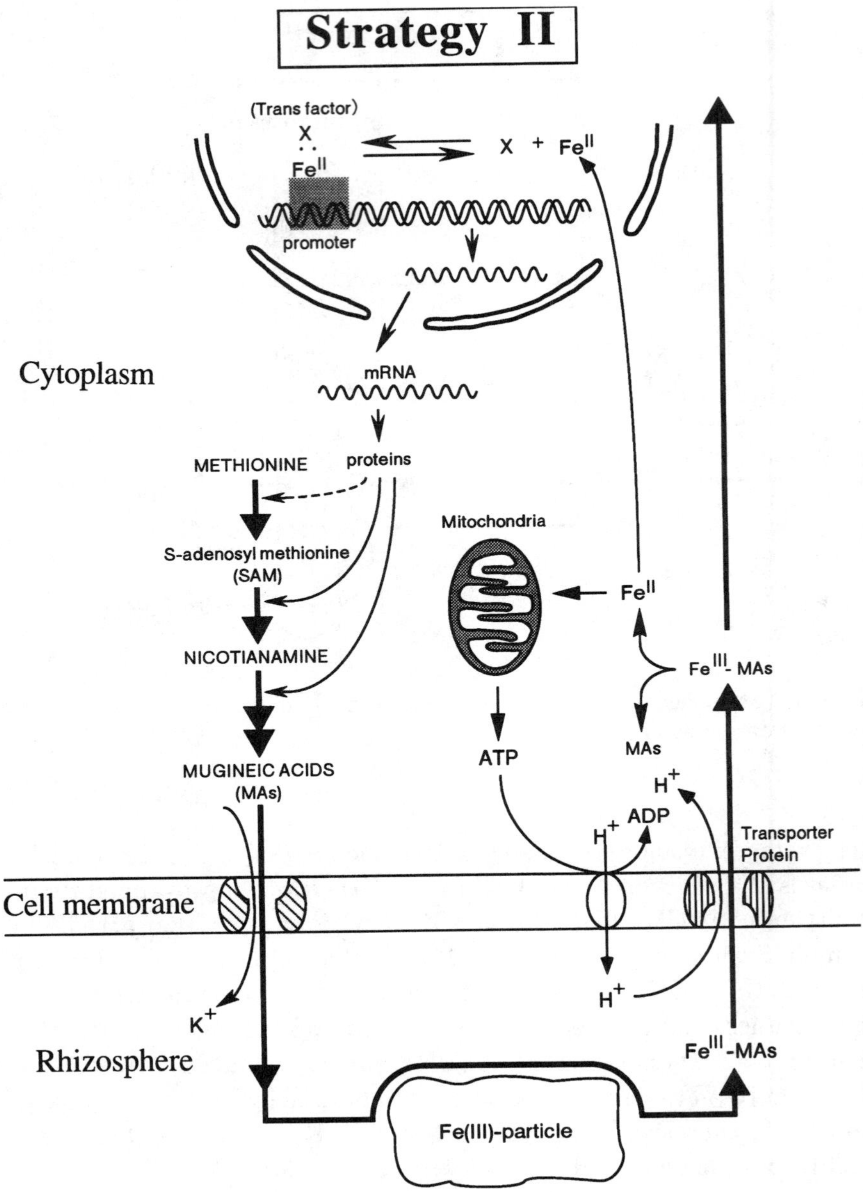

FIG. 2. Fe absorption mechanism of strategy II in the roots of graminaceous plants.

1.2. Iron Transport in the Roots of Strategy II Plants

Fe-deficient graminaceous plants secrete MAs into the rhizosphere, and this natural chelator binds with sparingly soluble Fe(III) in the soil medium converting it to water-soluble Fe(III)-MAs; thus transporting Fe into the root cells through a highly specific transport system for Fe(III)-Mas [13,14]. The specific transporter was proposed based on the indirect evidence that root membrane transport of ^{59}Fe supplied as ^{59}Fe(III)-MAs was not inhibited by the strong synthetic Fe(II)-chelator, basophenanthroline disulfonate (BPDS) [2]. The MAs group comprises six compounds: mugineic acid (MA), avenic acid (AVA), 3-hydroxymugineic acid (HMA), 3-epihydroxymugineic acid (epiHMA), 2′-deoxymugineic acid (DMA), and distichonic acid (DA) [15]. Methionine was identified as a primary precursor [16,17], and the biosynthetic pathway from methionine to DMA [18–22] and from DMA to other MAs has been established [23–25].

2. IRON TRANSPORT FROM ROOT TO SHOOT

The iron-uptake rate from roots to shoots is increased more than 10-fold when Fe is supplied by Fe(III)-MAs as compared with Fe(III)-EDTA or Fe(III)-desferal in rice [13] or barley [2,14].

Until Takagi's discovery of mugineic acid, rapid Fe transport from root to shoot had not been reported. We detected radioactivity in the new leaves of barley only within 10 min after supplying ^{59}Fe(III)-epiHMA in the water-culture solution (Fig. 3). After a 4-h treatment, Fe was strongly transported to the tiller (Fig. 4, no. 1), the newest leaf (Fig. 4, no. 2), and other leaves (Fig. 4, nos. 3, 4, and 5). Thus, tiller(s) and newly developing leaf are recognized as strong sinks for Fe. Iron accumulation depended on the age of the leaves, as reported by Brown et al. [26]. Iron absorbed from one root is also translocated to the other roots. When Fe was absorbed from a single new root, Fe was transported to the other roots, especially to the root tips (Fig. 5a). Root primordia were strongly labeled (Fig. 5b). The sectioning and staining or SEM (scanning electron microscopy) image of the Fe-deficient root tips revealed the emergence of many root primordia [27]. Therefore, root meristems

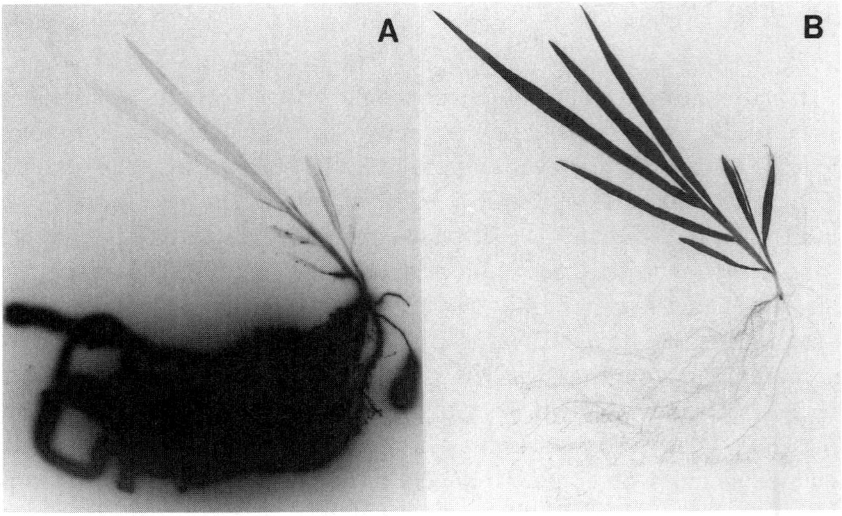

FIG. 3. Transport of Fe from the roots during 10 min supply of ^{59}Fe(III)-epiHMA in barley roots. Basal part of the roots were not dipped in the culture solution of ^{59}Fe. (A) Radio autograph. (B) Gross image of the plant.

are doubtless a strong sink of Fe. The leaf veins connected to the single root are strongly labeled (Fig. 5c). When Fe was absorbed from an old single root, Fe was transported evenly to the other roots (Fig. 6). Absorbed Fe from a single root (Fig. 6, big arrow) was also transferred to the corresponding leaves, but first only to the veins (thin arrows) directly connected to that root. The labeled vein is not always the central vein but sometimes a peripheral vein. An interesting phenomenon was observed; that is, Fe from the roots deposited at once to the end of the basal part of the leaves where meristematic cells exist (arrowhead). This meristematic zone will develop to leaves or roots in the future. This zone was also strongly labeled when ^{59}Fe was supplied from a cut leaf, as can be seen in Fig. 7. Therefore, this part also seems to be the place where the distribution of Fe is controlled; to which leaves or roots should the Fe coming from a leaf or from a root be distributed. The strong labeling of this zone was also observed in the phosphate transport from the cut leaf (Fig. 8, arrowhead). Therefore, I will call this zone the “discrimination center” of nutrient transport.

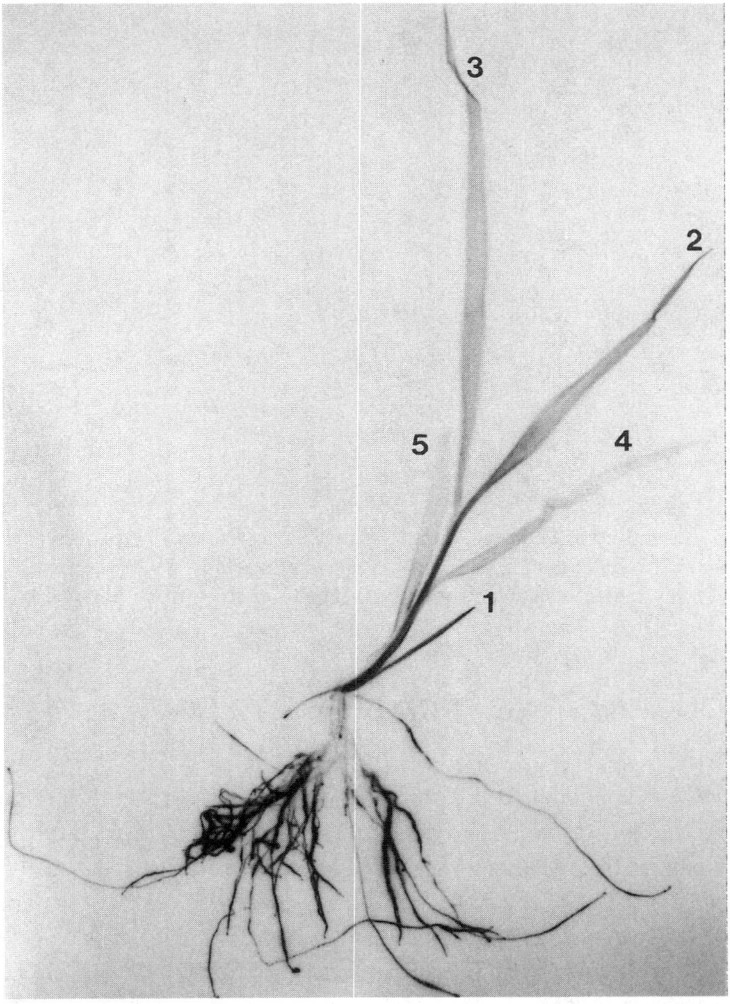

FIG. 4. Transport of Fe from the roots during 4-h supply of $^{59}Fe(III)$-epiHMA in barley roots. The newest leaf (no. 2) and tiller (no. 1) are strongly labeled. Basal parts are not dipped in the culture solution of ^{59}Fe.

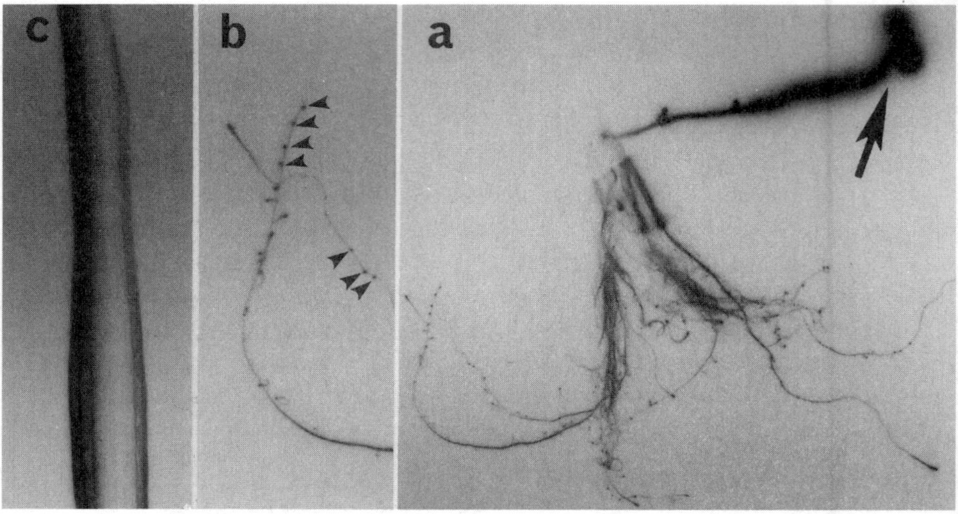

FIG. 5. Transport of Fe from a single new root (arrow in a) to other roots (a,b), and leaves (c). Special veins in the leaves are labeled. Root primordia in the enlarged image of root-tips are also strongly labeled (arrowheads in b).

3. IRON TRANSPORT FROM SHOOT TO ROOT

When Fe was supplied from the cut part of a leaf of the main shoot, Fe was transported through the phloem to the basal part of the leaf and retranslocated to the roots and other leaves, especially the tillers (Fig. 9, arrowheads). Most of the absorbed Fe was possibly mainly transported through an apoplasmic mechanism and partly through the veins. Within 4 h, the front of the transported Fe is still in the middle part of the leaf (Fig. 9, double arrowheads), and after 8 h, it has reached the basal part of the leaf (Fig. 10). Retranslocation of Fe throughout the discrimination center is first to the tillers (Fig. 10, nos. 1 and 2) and then to the newest leaf (Fig. 10, no. 4) of the main shoot. Transport of Fe to the root meristems is very clear in the case of Fe-deficient plants, where many branched root tips are strongly labeled (Fig. 10). This specific Fe transport from shoot to root is very similar to the phosphate transport from shoot to root, as shown in Fig. 8.

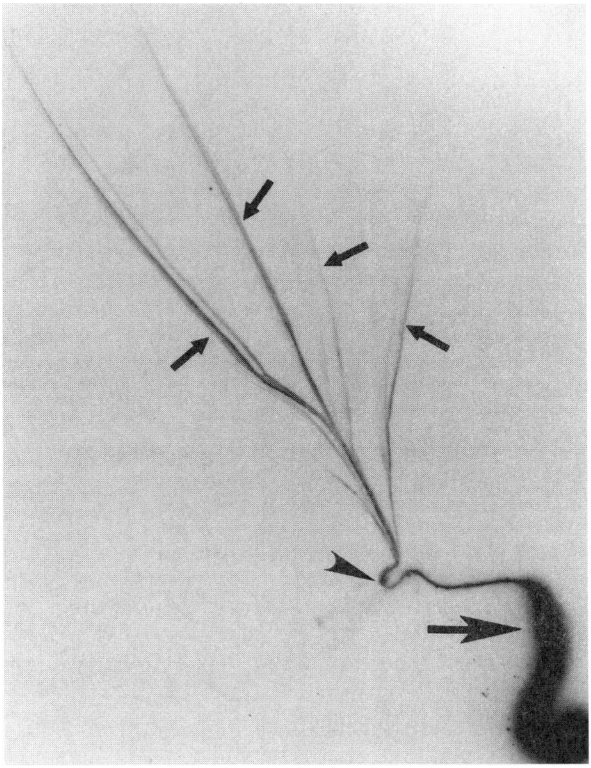

FIG. 6. Transport of Fe from an old single root (big arrow) to other roots and leaves. Special veins of the special leaves (small arrows) are labeled although other roots are evenly labeled. Arrowhead shows a discrimination center.

4. EFFECT OF IRON STATUS ON IRON TRANSPORT IN PLANTS

The role of Fe(III)-MAs in Fe transport into plant roots has been studied in detail. However, no information is available on their role in Fe transport in the foliage of the plants despite the presence of MAs in plant shoots [27]. Moreover, the environmental factors that regulate Fe transport in higher plants have not been studied. We initially chose to investigate the influence of Fe status on Fe transport in intact rice plants.

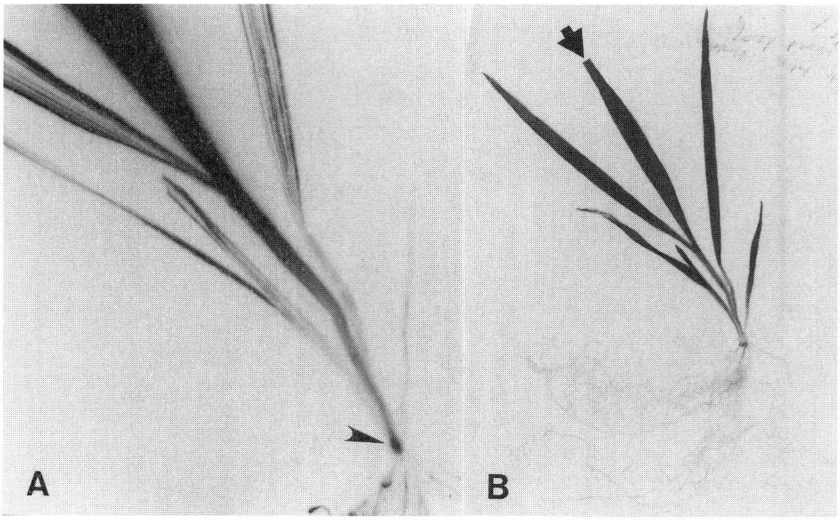

FIG. 7. Transport of Fe of 4-h supply from Fe-sufficient barley leaf. The discrimination center (arrowhead) is strongly labeled. Leaf veins are asymmetrically labeled. (A) Radioautograph. (B) Gross image of the plant.

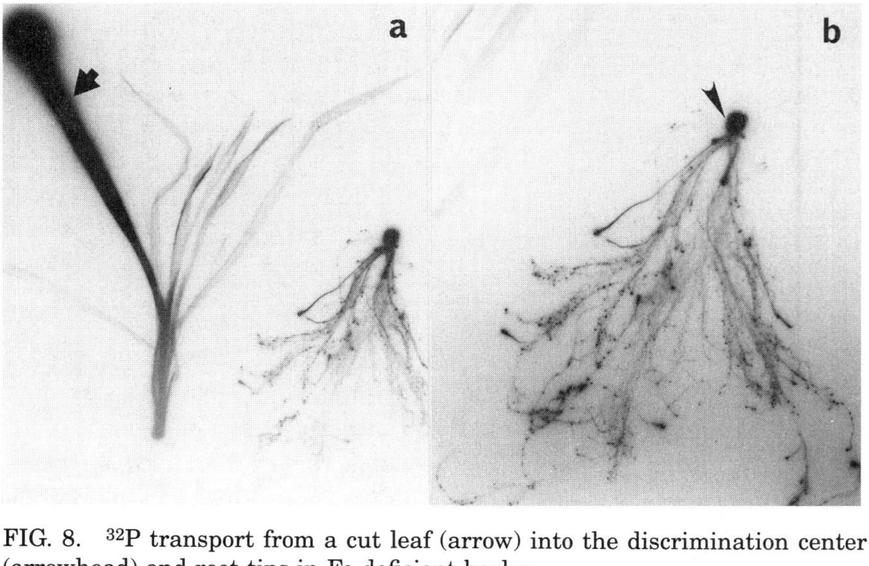

FIG. 8. ^{32}P transport from a cut leaf (arrow) into the discrimination center (arrowhead) and root tips in Fe-deficient barley.

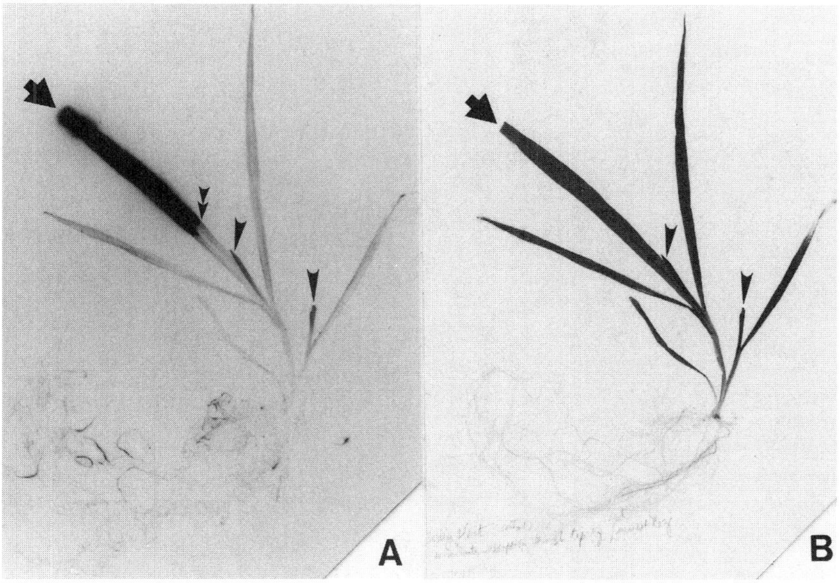

FIG. 9. Transport of Fe from a cut leaf (arrows) during 30-min supply of ^{59}Fe(III)-epiHMA. Fe was transported to the two tillers (arrowheads) other leaves and roots. However, most radioactivity still remained in the middle part of the ^{59}Fe-supplied leaf. (A) Radioautograph. (B) Gross image of the plant.

Distribution of ^{59}Fe in the shoots between two rice plants, one cultured under Fe deficiency and the other under sufficient Fe, was compared by supplying ^{59}Fe(III)-epiHMA in the culture medium. After absorption for 4 h, radioactivity in the Fe-deficient rice shoot was stronger than that in the Fe-sufficient one (Fig. 11a). After 12 h, the difference could still be observed (Fig. 11b). However, after 24 h, the difference was no longer clear (data not shown). These data coincide with the fact that Fe-deficient graminaceous plants can absorb [14] and transport [28] much higher amounts of Fe than Fe-sufficient plants. This is probably caused by the increased activity of the (Fe(III)-epiHMA)-transporter in the Fe-deficient root cells and the increase in shoot demand for Fe as well.

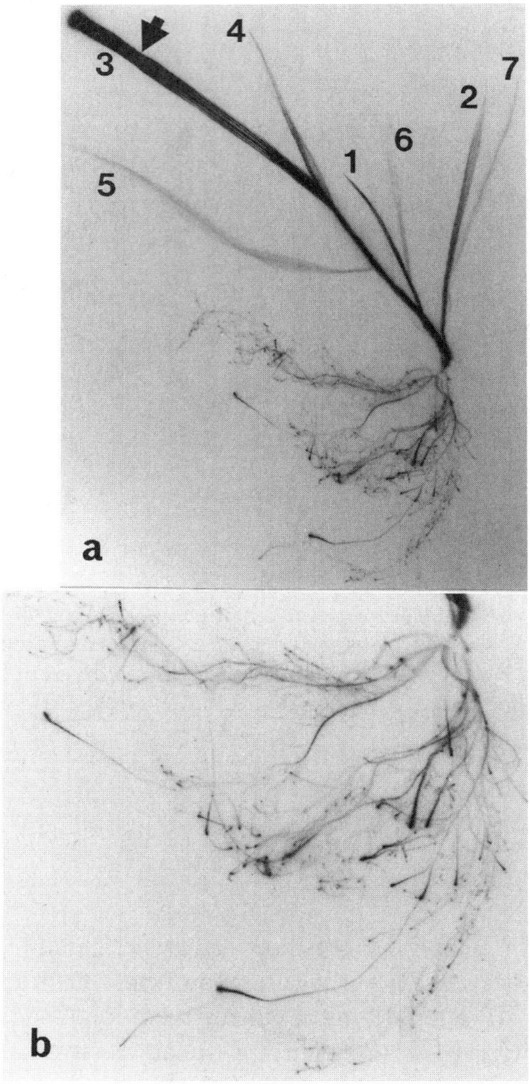

FIG. 10 (a) Fe transport after 4-h supply of $^{59}Fe(III)$-epiHMA from a cut leaf (arrow, no. 3) to tillers (nos. 1 and 2) and other leaves (nos. 4–7). (b) Many branched root tips are strongly labeled.

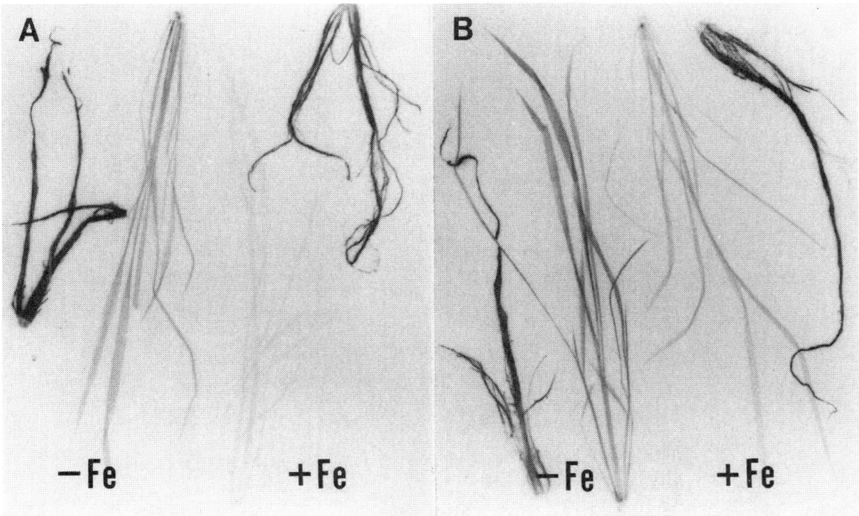

FIG. 11. Comparison of Fe distribution in the shoot between Fe-sufficient and Fe-deficient rice plants. (A) Fe was supplied for 4 h. (B) Fe was supplied for 12 h. +Fe: Fe-sufficient rice. −Fe: Fe-deficient rice. Roots and shoots are separated.

5. LIGHT EFFECT ON IRON TRANSPORT IN PLANTS IN COMPARISON WITH PHOSPHATE TRANSPORT

Next we investigated the influence of light on Fe transport in intact barley plants. ^{59}Fe(III)-epiHMA was supplied to the roots and the newest leaf was covered with aluminum foil or all the leaves except the newest leaf were covered with aluminum foil. Fe was not transported to the covered leaf/leaves (Fig. 12a, c). When the middle part of each leaf was partly covered with aluminum foil, Fe transport in the covered area of leaves, indicated by arrows, was drastically reduced in the absence of light, with only the veins being labeled in the covered area (Fig. 13) [29]. A similar phenomenon was observed in ^{32}P transport under shadowing (Fig. 14). This suggested that the radial Fe transport in the leaf from xylem to apoplasm or from apoplasm to the mesophyll cells is light regulated.

More than 90% of the Fe in plants is located in the chloroplasts [30], and Fe is essential for the chlorophyll biosynthesis [31] and light-

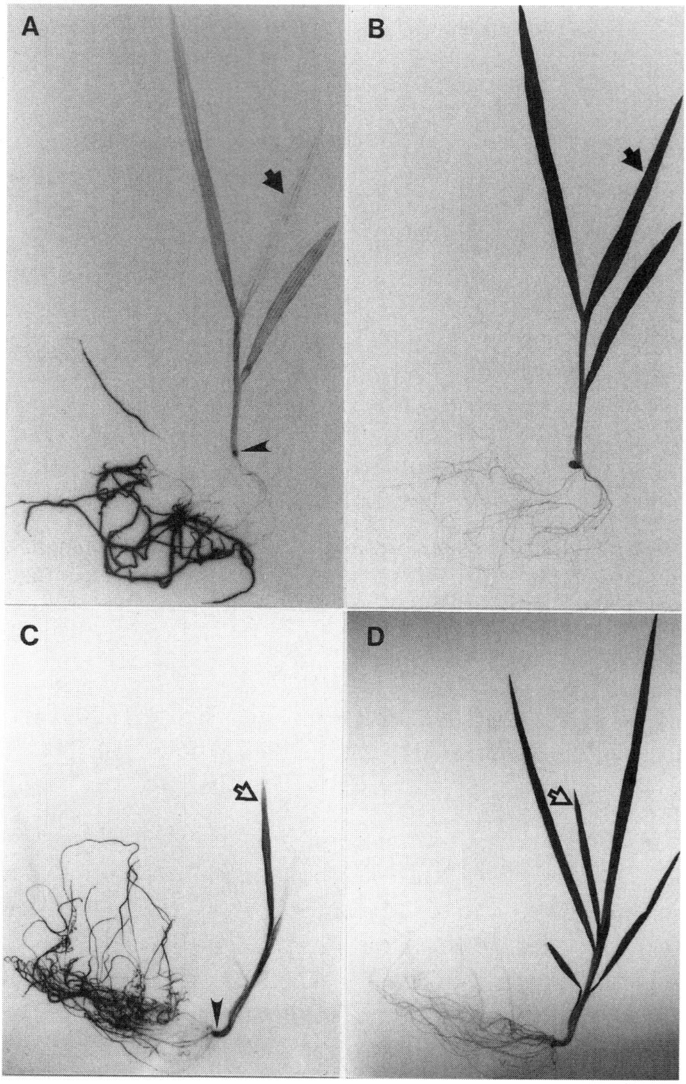

FIG. 12. Effect of shadowing on Fe transport in plants. (C) All leaves except the newest leaf (arrows) were covered. (A) Only the newest leaf (arrow) was covered. (B, D) Gross images of the plants. Arrowhead in (C) is a strongly labeled discrimination center.

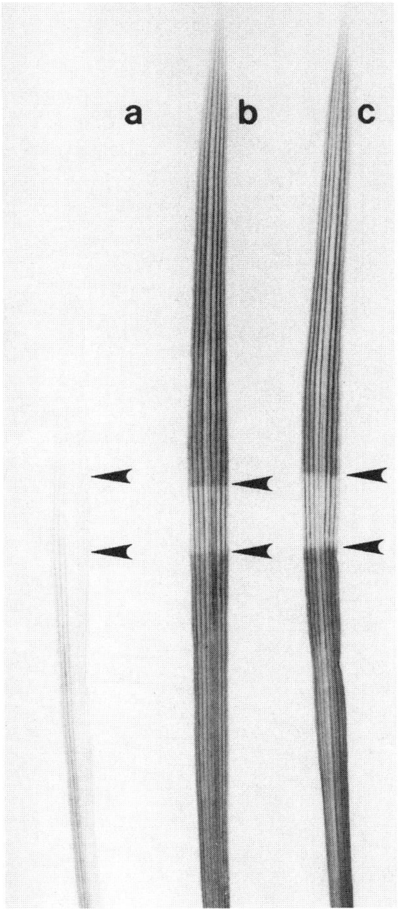

FIG. 13. Fe transport in the partially covered leaf. Only veins of the covered area are labeled. From left to right: (a) 30-min, (b) 1-h, and (c) 4-h treatment with $^{59}Fe(III)$-epiHMA in barley roots, respectively. Leaves between arrowheads are covered with aluminum foil.

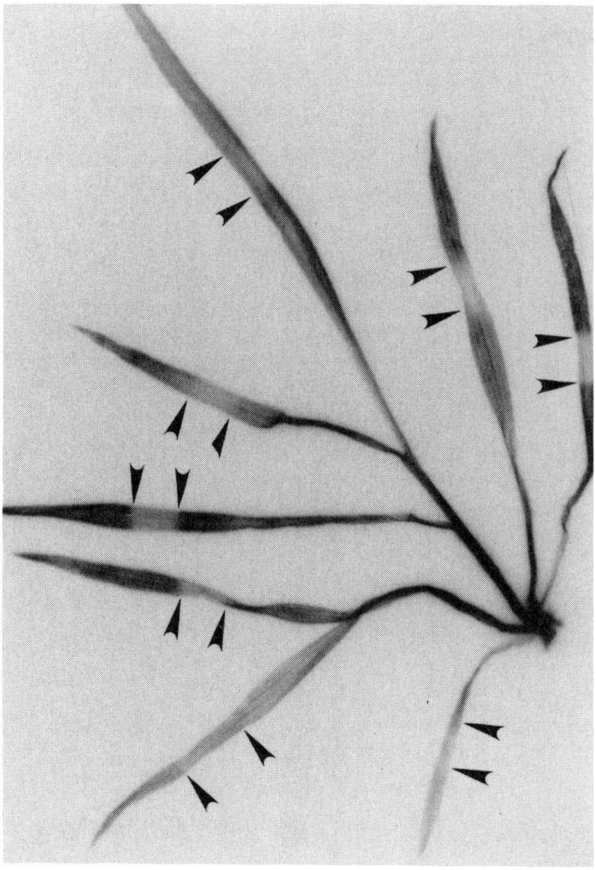

FIG. 14. Effect of shadowing on ^{32}P transport in barley. Middle part of every leaf (arrowheads) was partially covered with aluminum foil. Only veins of all partially covered areas are labeled. ^{32}P-supplied roots are not shown.

harvesting systems. Therefore, we speculated that chloroplasts in the mesophyll cell might be the most important organelles as a sink of Fe transport which is regulated by light. Then we prepared chloroplasts from Fe-sufficient barley leaves for in vivo experiments of Fe transport.

6. LIGHT-DEPENDENT IRON TRANSPORT INTO CHLOROPLASTS

Iron absorption by the irradiated chloroplasts was much higher than by those kept in darkness. Iron uptake increased from 0.4 μg Fe mg^{-1} chlorophyll at 0 min to 1.15 μg Fe mg^{-1} chlorophyll at 60 min in the presence of light, whereas in darkness, it remained almost constant (around 0.4 μg Fe mg^{-1} chlorophyll) over the same period (Fig. 15). When samples were kept under the condition of 20 min light-dark cycles, Fe initially absorbed during 20 min of irradiation was seen to leave the chloroplasts (efflux) during the subsequent 20 min of darkness (Fig. 15; see from 20-

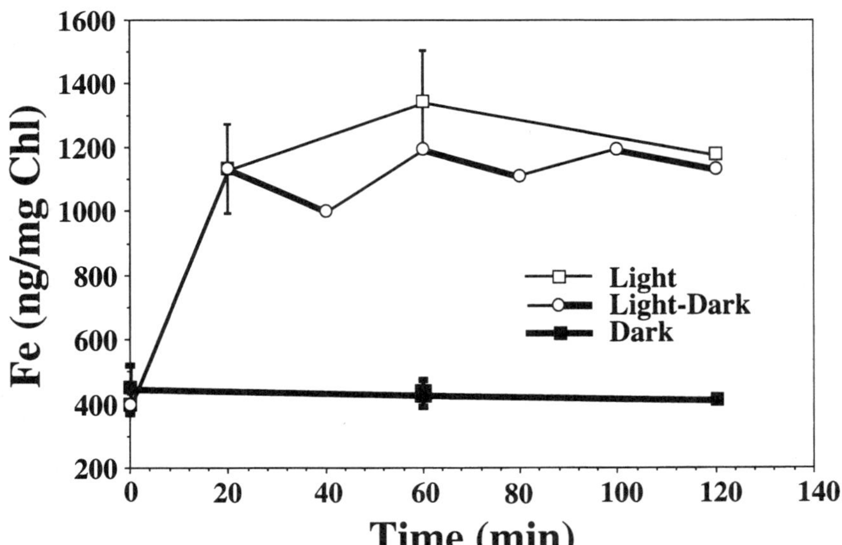

FIG. 15. Iron transport into chloroplasts under the conditions of light, darkness, or a light-dark cycle.

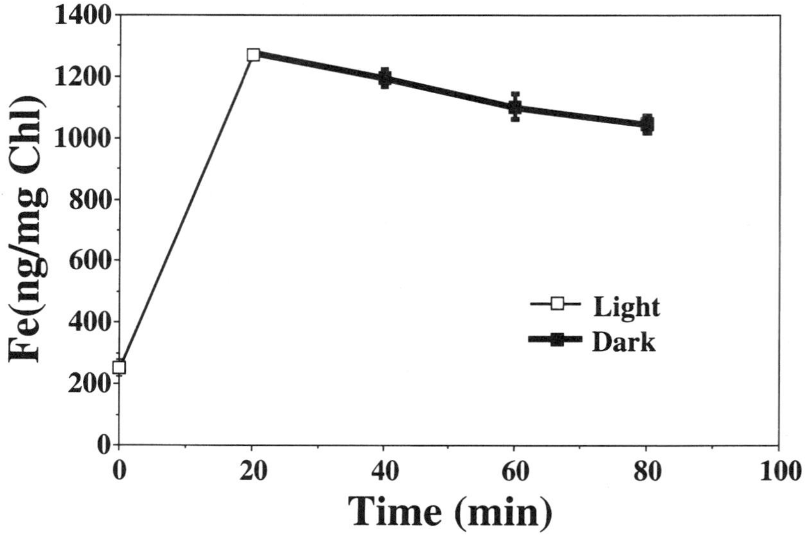

FIG. 16. Release of ^{59}Fe from chloroplasts under dark conditions and after Fe (^{59}Fe(III)-epiHMA) was absorbed into chloroplasts under light.

to 40-min time interval indicated by the bold line). Resupplying light for another 20 min (from 40 to 60 min) restored the initial status. Once light induced and then kept continuously under dark, the release of Fe from chloroplast was almost linear (Fig. 16). These light-stimulated Fe influx and dark-induced-mimicking Fe efflux were neither light- nor dark-triggered (on-off) reactions, but rather they were light- or dark-dependent reactions, since both needed continuous light or continuous darkness. Light-mediated Fe absorption doubled at a higher temperature (19°C) than at 0°C (data not shown). This light-dependent Fe absorption by chloroplasts was inhibited almost completely by dichlorophenyldimethylurea (DCMU) (10^{-5} M) under light [29] indicating that the transport is indirectly light energy dependent, probably not through some phytochrome-mediated signal transduction systems. However, neither carbonylcyanide-*m*-chlorophenylhydrazone (CCCP) nor dicyclohexylcarbodiimide (DCCD) inhibited Fe transport into the chloroplast. No heavy metals inhibited Fe influx. BPDS (10^{-5} M) inhibited Fe influx but a higher amount (10^{-2} M) of BPDS could not completely inhibit it [32]. This suggested that Fe was taken up by chloroplasts not only through a strategy I-like system (see Fig. 1) but also by other Fe transport systems.

7. CONCLUSIONS

From our studies, we conclude that the Fe fluxes in chloroplasts are regulated by light and dark. Chloroplasts have the tendency to efflux Fe if light energy is not supplied. This phenomenon is probably responsible for the results of our autoradiographic studies; that is, dark treatment interrupted the Fe transport from the leaf veins to the mesophyll cells (see Fig. 13). Since the mesophyll chloroplasts are a strong sink for Fe, high Fe demand by illuminated chloroplasts will strongly influence Fe transport into mesophyll cells from the apoplasm in the leaf, whereas Fe efflux from the dark-treated chloroplasts will interrupt the Fe demand of mesophyll cells from the apoplasm. Zang et al. [33] reported that when lower leaves were covered with aluminum foil, the Fe content of the top chlorotic trifoliate leaves of bean plants cultured in Fe-deficient medium was increased and the chlorosis was partly cured. They speculated that Fe in the covered leaves was transferred to other parts which demanded Fe. Our autoradiography data (see Figs. 7, 9, and 10) support their speculation that Fe can be retranslocated from leaf to leaf. This phenomenon may occur more strongly under dark conditions from chloroplast to cytoplasm, cytoplasm to apoplast, and then apoplast to phloem. The overall mechanism of Fe transport in graminaceous plants is schematically represented in Fig. 17.

The effects of various inhibitors were examined to unravel the driving forces underlying the light-regulated transmembrane Fe influx. BPDS is a well-known inhibitor of Fe transport in strategy I plants [2] complexing with Fe(II) and resulting in the failure of Fe transport across the membrane, thus being suggestive of the presence of Fe(III)-reductase in a membrane. BPDS incompletely inhibited Fe transport into chloroplasts, indicating that the light-dependent influx of Fe into barley chloroplasts is controlled by different mechanisms working at the same time. Some is possibly carried by an Fe(II)-transporter after the function of ferric reductase [34] and the rest of it by another unknown mechanism like strategy II (see Fig. 2). First, we examined

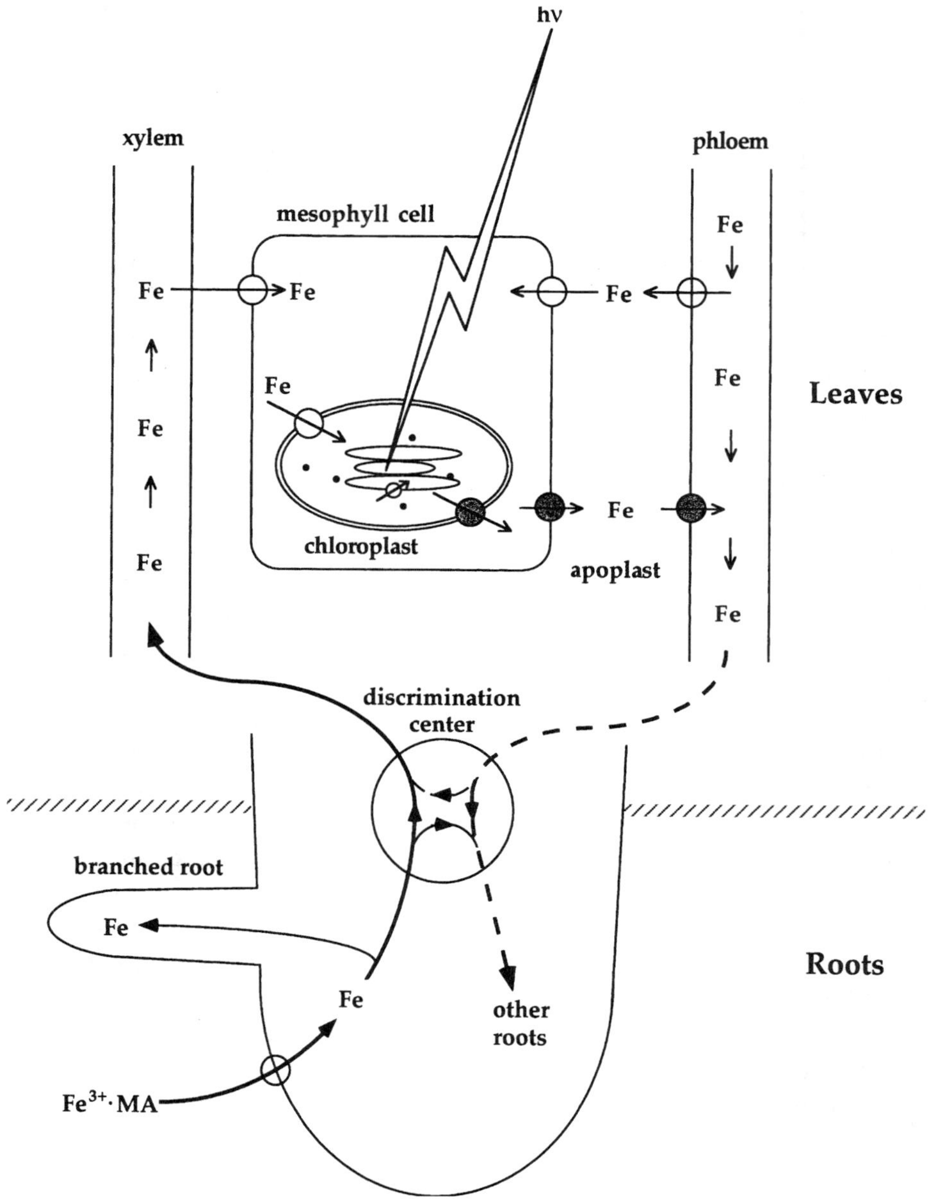

FIG. 17. Schematic representation of Fe transport in barley under dark or light conditions. A white circle means transporters activated under light and a black circle in the membrane means transporters possibly activated under dark.

whether the reduced product, that is, the Fe(II) ion, is transported by the Fe(II)-transporter. Recently, the transporter activity of an Fe(II)-transporter-like gene cloned in *Arabidopsis thaliana* [11] was inhibited by 10 mM of heavy metal ions. However, in our experiments, no heavy metal ions inhibited Fe transport (supplied as ^{59}Fe(III)-epiHMA) into the chloroplast. Thus, the character of the Fe(II)-transporter in the chloroplast envelope may be different from that in the plant plasma membrane. Neither CCCP nor DCCD inhibited Fe transport into chloroplasts.

Recently, the total genomic-DNA of *Synechocystis* sp. PCC6803 has been sequenced (http://www.kazusa.or.jp/cyanobase/). Screening the database by the key word "iron" revealed: ferrous iron transport protein B (feoB) and Fe(III)-dicitrate transport system permease proteins, FecC(fecC), FecD(fecD), FecE(fecE), and FecB(fecB). *Synechocystis* and chloroplast are proposed to be of the same evolutionary origin. Thus, a molecular genetic approach to the chloroplast Fe-transporter using these *Synechocystis* genes as probes would help to clarify the mechanism of light-regulated Fe transport into the chloroplast. The following questions remain to be solved in the future: (1) Does the chloroplast envelope have a ferric reductase? (2) Is the Fe(III)-MAs transporter, like that in the roots of strategy II plants, also acquired in the chloroplast envelope of barley? (3) Does the Fe(III)-dicitrate transport system work in the chloroplast envelope? (4) Is the Fe(II)-transporter directly regulated by light?

ACKNOWLEDGMENTS

The author is deeply indebted to Prof. N. K. Nishizawa and Dr. N. Bughio for arrangement and discussion of the autoradiographs.

ABBREVIATIONS

ADP	adenosine-5′-diphosphate
ATP	adenosine-5′-triphosphate
AVA	avenic acid
BPDS	basophenanthroline disulfonate

CCCP	carbonylcyanide-*m*-chlorophenylhydrazone
cDNA	complementary deoxyribonucleic acid
Chl	chlorophyll (a + b)
DA	distichonic acid
DCCD	dicyclohexylcarbodiimide
DCMU	dichlorophenyldimethylurea
DMA	2′-deoxymugineic acid
EDTA	ethylenediamine-*N,N,N′,N′*-tetraacetate
epiHMA	3-epihydroxymugineic acid
FRE1	ferric reductase gene of *Saccharomyces cerevisiae*
HMA	3-hydroxymugineic acid
MAs	mugineic acid family of phytosiderophores
mRNA	messenger ribonucleic acid
NAD	nicotinamide adenine dinucleotide
NADP	nicotinamide adenine dinucleotide phosphate
NADPH	nicotinamide adenine dinucleotide phosphate (reduced)
SAM	*S*-adenosyl methionine
SEM	scanning electron microscopy

REFERENCES

1. S. Takagi, *Soil Sci. Plant Nutr.*, *22*, 423–433 (1976).
2. V. Römheld and H. Marschner, *Plant Physiol.*, *80*, 175–180 (1986).
3. J. C. Brown and J. E. Ambler, *Agron. J.*, *65*, 311–314 (1973).
4. R. A. Olsen, J. H. Bennet, D. Blume, and J. C. Brown, *J. Plant Nutr.*, *3*, 905–921 (1981).
5. K. Ae, J. Arihara, K. Okada, T. Yoshihara, and C. Johansen, *Science*, *248*, 477–480 (1990).
6. Y. Masaoka, M. Kojima, S. Sugihara, T. Yoshihara, M. Koshino, and A. Ichihara, *Plant Soil*, *155/156*, 75–78 (1993).
7. A. Noguchi, T. Yoshihara, A. Ichihara, S. Sugihara, M. Koshino, M. Kojima, and Y. Masaoka, *Biosci. Biotech. Biochem.*, *58*, 2312–2313 (1994).
8. D. Kramer, *Naturwissenschaften*, *65*, 339 (1978).
9. D. Kramer, V. Römheld, E. Lansberg, and H. Marschner, *Planta*, *147*, 335–339 (1980).

10. W. Schmidt and M. Bartels, *Plant Physiol.*, *110*, 217–225 (1996).

11. D. Eide, M. Broderium, J. Fett, and M. L. Guerinot, *Proc. Natl. Acad. Sci. USA*, *93*, 5624–5628 (1996).

12. A. Dancis, D. G. Roman, G. J. Anderson, A. G. Hinnebush, and R. D. Klausner, *Proc. Natl. Acad. Sci. USA*, *89*, 3869–3873 (1992).

13. S. Takagi, Production of phytosiderophores, in *Iron Chelation in Plants and Soil Microorganisms* (L. Barton and H. Hemming, eds.), Academic Press, Tokyo, 1993, pp. 111–131.

14. S. Mihashi and S. Mori, *Biol. Metals*, *2*, 146–154 (1989).

15. K. Nomoto, Y. Sugiura and S. Takagi, in *Iron Transport in Microbes, Plants and Animals* (G. Winkelman, D. van der Helm, and J. B. Neilands, eds.), VCH, Weinheim, Germany, 1990, pp. 401–425.

16. S. Mori and N. K. Nishizawa, *Plant Cell Physiol.*, *28*, 1081–1092 (1987).

17. S. Kawai, S. Takagi, and Y. Sato, *J. Plant Nutr.*, *11*, 633–642 (1988).

18. S. Shojima, N. K. Nishizawa, S. Fushiya, T. Irifune, and S. Mori, *Plant Physiol.*, *93*, 1497–1503 (1990).

19. K. Kanazawa, K. Higuchi, S. Fushiya, N. K. Nishizawa, M. Chino, and S. Mori, in *Iron Nutrition in Soils and Plants* (J. Abadía, ed.), Kluwer, Dordrecht, The Netherlands, 1995, pp. 37–41.

20. K. Kanazawa, K. Higuchi, N. K. Nishizawa, S. Fushiya, and S. Mori, *J. Exp. Botany*, *46*, 1241–1244 (1995).

21. K. Higuchi, K. Kanazawa, N. K. Nishizawa, M. Chino, and S. Mori, *Plant Soil*, *165*, 173–179 (1994).

22. K. Higuchi, K. Kanazawa, N. K. Nishizawa, and S. Mori, *Plant Soil*, *178*, 171–177 (1996).

23. S. Mori, N. K. Nishizawa, and J. Fujigaki, *Jpn. J. Genet.*, *65*, 343–352 (1990).

24. S. Mori and N. Nishizawa, *Plant Cell Physiol.*, *30*, 1057–1061 (1989).

25. J. M. Ma and K. Nomoto, *Plant Physiol.*, *102*, 373–378 (1993).

26. A. L. Brown, S. Yamaguchi, and J. Leal-Diaz, *Plant Physiol.*, *40* (Suppl.), 35–38 (1965).

27. S. Mori, N. K. Nishizawa, H. Hayashi, M. Chino, E. Yoshimura, and J. Ishihara, *Plant Soil*, *130*, 143–156 (1991).

28. V. Römheld, in *Iron Transport in Microbes, Plants and Animals* (G.

Winkelman, D. van der Helm, and J. B. Neilands, eds.), VCH, Weinheim, Germany, 1990, pp. 353–374.

29. N. Bughio, M. Takahashi, E. Yoshimura, N. K. Nishizawa, and S. Mori, *Plant Cell Physiology*, *38*, 101–105 (1997).
30. N. Terry and J. Abadia, *J. Plant Nutr.*, *9*, 609–646 (1986).
31. D. V. Wettstein, S. Gough, and G. C. Kannangara, *Plant Cell*, *7*, 1039–1057 (1995).
32. N. Bughio, N. Takahashi N, E. Yoshimura, N. K. Nishizawa, and S. Mori, *Plant Soil*, In press.
33. C. Zang, V. Römheld, and H. Marschner, *J. Plant Physiol.*, *146*, 268–272 (1995).
34. Q. J. Groom, S. K. Procter, and N. J. Robinson, *A. thaliana* frohA and FrohB gene in EMBL/GenBank/DDBJ databases Y09581.

7

Coordination Chemistry of Siderophores: Thermodynamics and Kinetics of Iron Chelation and Release

Anne-Marie Albrecht-Gary[1]
and Alvin L. Crumbliss[2]

[1]Laboratoire de Physico-Chimie Bioinorganique, URA405 du CNRS, Institut de Chimie, Strasbourg, France

[2]Department of Chemistry, Duke University, Durham, NC, 27708 USA

1. INTRODUCTION 240

2. SIDEROPHORE LIGANDS: STRUCTURAL FEATURES 241
   2.1. Metal Selectivity 241
   2.2. Siderophore Binding Groups 242
   2.3. Siderophore Denticity 243
   2.4. Siderophore Architecture 246

3. Fe(III) AND Fe(II) COMPLEXES: SOLUTION STRUCTURES 248
   3.1. Ultraviolet-Visible Absorption Spectrophotometry 248
   3.2. Fluorescence Emission Spectroscopy 253
   3.3. Circular Dichroism Measurements 259
   3.4. Mass Spectrometry Methods 267
   3.5. Nuclear Magnetic Resonance Spectroscopy 268

4. Fe(III) AND Fe(II) COMPLEXES: STABILITY 268

4.1. Methods of Expressing Iron-Chelate Stability Constants 268
4.2. Summary of Fe(III)-Siderophore Stability Constants 270
4.3. Fe-Siderophore Redox Potentials 273
4.4. Synthetic Models for the Siderophores 277
4.4.1. Synthetic Hydroxamic Acid Siderophore Models 278
4.4.2. Synthetic Catechol Siderophore Models 285
4.4.3. Other Synthetic Siderophore Models 287
5. IRON(III) UPTAKE AND RELEASE MECHANISM: KINETICS IN SOLUTION 296
5.1. Reactivity at Aquated Fe^{3+} 296
5.1.1. Initial Chelation Step 296
5.1.2. Second and Third Chelate Formation Steps 301
5.2. Siderophore Complex Dissociation Reactions 306
5.3. Kinetics and Mechanism of Selected Model Systems 309
5.4. Ligand Exchange Reactions at Fe(III) 310
6. PERSPECTIVES 313
ACKNOWLEDGMENTS 314
ABBREVIATIONS 314
REFERENCES 316

1. INTRODUCTION

Microbial iron bioavailability is largely determined by the coordination chemistry associated with the siderophores, a class of organic compounds which have a high and specific chelating affinity for Fe(III). The processes involved in siderophore-mediated microbial Fe transport are solubilization, transport to the cell membrane, transport across the cell membrane, and deposition at an appropriate site within the cell. The important chemical events are selective chelation of Fe(III) in the presence of other environmentally prevalent metal ions, molecular recognition of the Fe(III)-siderophore complex, transport of the Fe in complexed form across the membrane, and deposition of the Fe at an appropriate site within the cell. This latter process ultimately involves ligand exchange, which may (or may not) be preceded by Fe(III) reduction and/or hydrolysis of the siderophore ligand. The important chemical characteristics of the siderophores which are related to these processes are affinity and selectivity for Fe(III) and lability of the Fe(III) complex at a specific site (e.g., the cell surface or interior). This chapter will review the literature, emphasizing reports published since 1990, which is relevant to the coordination chemistry, kinetics, and thermodynamics of these chemical processes involved in siderophore-mediated Fe(III) transport.

For a complementary approach to the kinetics and thermodynamics of Fe(III) chelation related to siderophore mediated Fe(III) transport, and reference to the earlier literature, the reader is referred to related reviews [1–4]. Three more recent related reviews have also been published [5–7].

2. SIDEROPHORE LIGANDS: STRUCTURAL FEATURES

2.1. Metal Selectivity

The task of the siderophore is specifically to solubilize Fe in the presence of other environmentally prevalent metal ions. This is accomplished by selective chelation of Fe(III). In physicochemical terms, this may be expressed as complex stability constants that are higher with Fe^{3+} than with Al^{3+}, Ca^{2+}, Cu^{2+}, Zn^{2+}, and other environmental metal ions. A mechanism for selective chelation of Fe(III) includes charge, but in the case of Al(III), other factors must be involved, particularly since the Al concentration on the earth's surface exceeds that of Fe. Consequently, selectivity is an important issue. Important characteristics of Fe(III) are hardness, size, d-electron configuration, charge, and electronegativity. Fe(III) is a harder cation than Al(III), as illustrated by its greater electronegativity and stronger affinity for OH^-, the archetypical hard ligand [8]. Consequently, ligands most likely to form stable complexes with Fe(III) usually contain hard O donor atoms. Fe(III) is a d^5 ion which forms high-spin complexes with siderophore ligands [9].

The lack of crystal field stabilization energy means that most of the first-sphere bond stability comes from an electrostatic interaction between the +3 metal ion and the negative charge (or dipole) of the O donor atom. Fe(III) has a normal coordination number of 6, with no strong directional preference for the donor atoms. For the siderophores and model siderophore structures that have been determined, the octahedral first coordination shell is irregular and the Fe(III) ion has the appropriate size to fit the bidentate bite size of the siderophore donor moieties [5,10]. The smaller size of Al(III) relative to Fe(III) may cause an increase in electronic repulsion between ligating atoms, thus decreasing the stability of the Al(III) siderophore complex and effectively increasing the thermodynamic selectivity of the siderophore for Fe(III).

Another consequence of the different charge/radius ratios for Fe(III) and Al(III) is the fact that the free energy of formation of the hydrated metal ion ($\Delta G^0{}_f[M^{3+}_{aq}]$) is less negative for Fe^{3+}_{aq} (-10.6 kJ mol^{-1}) than for Al^{3+}_{aq} (-486 kJ mol^{-1}) [11]. Since a stability constant for complex formation in aqueous solution represents the difference in free energy of formation between the aquated metal ion and the metal complex, the more strongly aquated Al^{3+} ion will not be as likely to have as great a ΔG^0 between the hexaaquo ion and the complex as will Fe^{3+} [12].

2.2. Siderophore Binding Groups

There are three essential features associated with chelator structure. These include the type of metal-binding group, the number of binding groups in a particular molecule (ligand denticity; Sec. 2.3), and the stereochemical arrangement of the binding groups (molecular architecture; Sec. 2.4).

Almost all of the chelators which have been identified to serve a siderophore function contain a hydroxamic acid (**1**), catechol (**2**), and/or α-hydroxycarboxylic acid (**3**) metal-binding group. For example, fungi, *Streptomyces*, and some enteric bacteria produce hydroxamic acid siderophores. Enteric and other related bacteria produce catechol siderophores, and *Rhizobium* bacteria and *Rhizopus* and related fungi produce siderophores with α-hydroxycarboxylic acid Fe(III)-binding groups [13]. A common feature of each of these moieties is their ability to form stable five-membered chelate ring complexes with Fe(III), as shown in Eqs. (1) to (3).

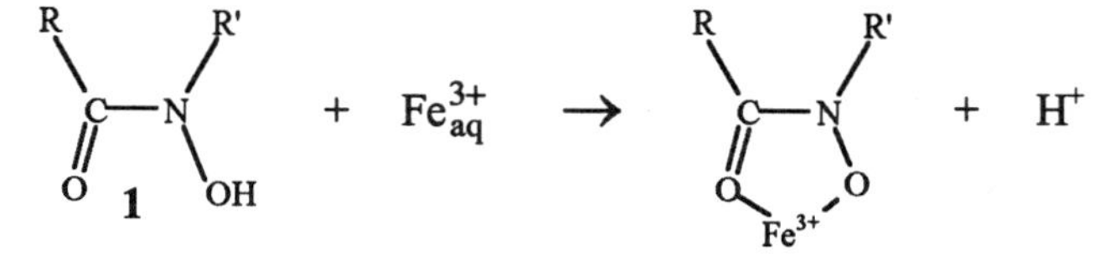

(1)

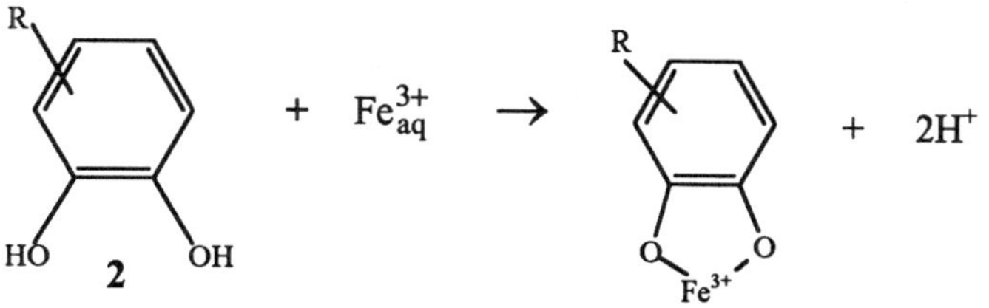

(2)

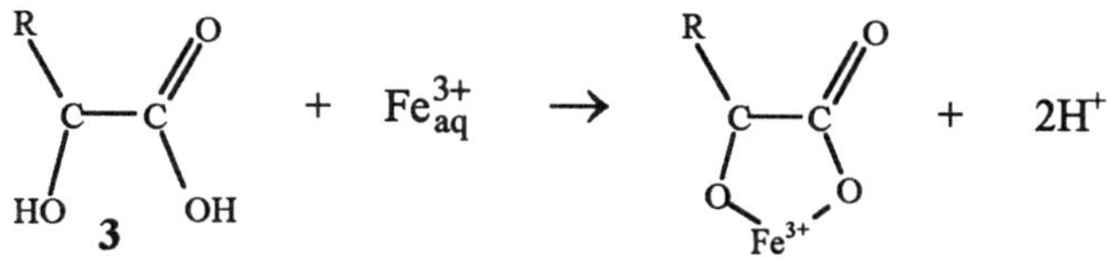

(3)

Several hexadentate siderophore ligands have structures which incorporate mixed Fe(III)-binding groups; for example, aerobactin (**4**; 2 hydroxamic acids and 1 α-hydroxycarboxylic acid in a linear structure), pyoverdin PaA (**5**; 2 hydroxamic acids and 1 catechol in an exocyclic structure), and azotobactin (**6**; 1 hydroxamic acid, 1 catechol and 1 α-hydroxycarboxylic acid in a linear structure). Several phytosiderophores are examples of siderophores which do not possess either a hydroxamate or catecholate binding group [14]. Rhizoferrin (**7**) contains two α-hydroxycarboxylic acids and two carboxylic acid-binding groups [15–17]. Mugineic acid (**8**) is a hexadentate ligand with carboxylamine-, hydroxyl-, and amine-binding groups [18].

2.3. Siderophore Denticity

Ligand denticity is an important feature of siderophore design, and most siderophores are hexadentate. Normally, in coordination chemistry, a chelate effect is operative whereby increased ligand denticity results in an increase in complex stability. However, for catecholate-

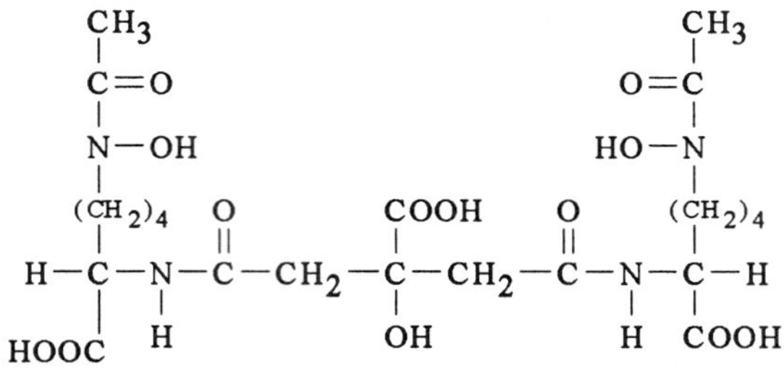

aerobactin
4

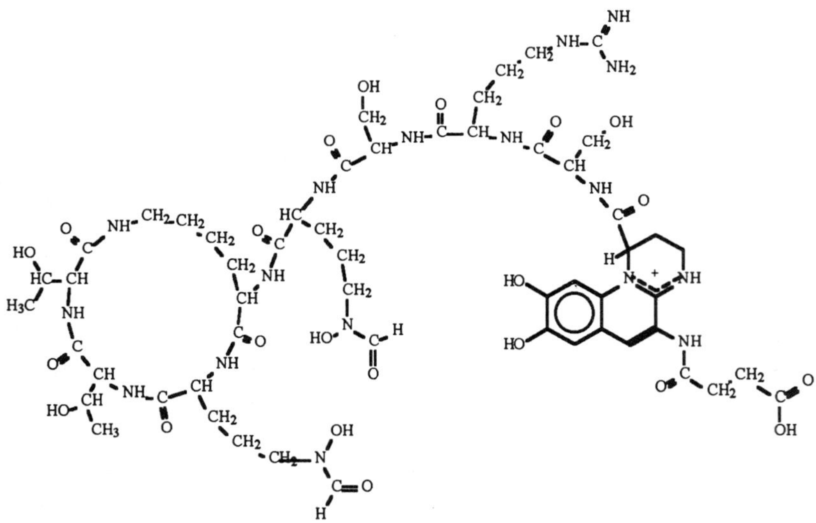

pyoverdin PaA
5

and hydroxamate-based siderophores, there is no evidence for a significant chelate effect being operative [1,19,20]. This may be due to the relatively long chain lengths between hydroxamate- and catecholate-binding groups, which results in their acting independently.

Although a stability enhancement does not seem to be the reason for the evolution of the siderophores as polydentate (usually hexadentate) Fe(III) ligands, there is a concentration effect which favors ligands of higher denticity. This can be illustrated from the following two rearranged equilibrium quotient expressions showing the concentration of free acetohydroxamate [A^-] and deferriferrioxamine B (**9**) [L^{3-}] anion

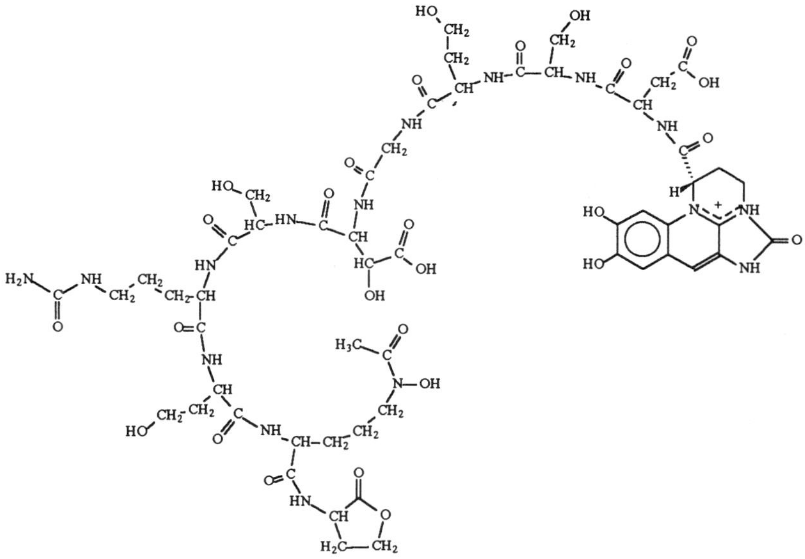

azotobactin δ
6

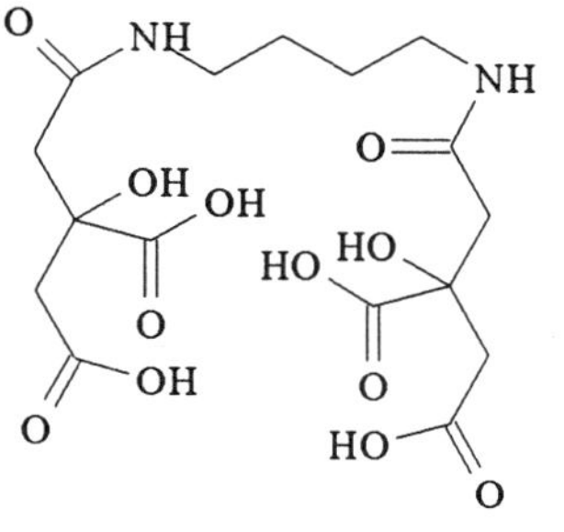

rhizoferrin
7

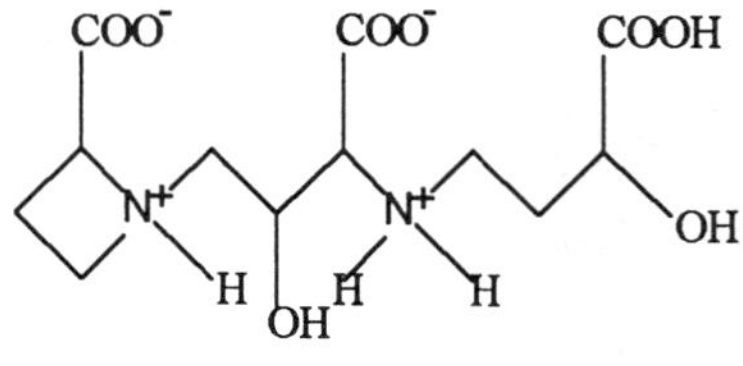

mugineic acid
8

required to establish a given equilibrium ratio of complexed to free Fe^{3+}_{aq}. For example, since log β_3 is 28.29 for acetohydroxamic acid and log β is 30.6 for deferriferrioxamine B [19], an arbitrarily selected ratio of

$$[A^-] = \left\{\frac{[FeA_3]}{[Fe^{3+}_{aq}]\beta_3}\right\}^{1/3} \tag{4}$$

$$[L^{3-}] = \left\{\frac{[FeL]}{[Fe^{3+}_{aq}]\beta}\right\} \tag{5}$$

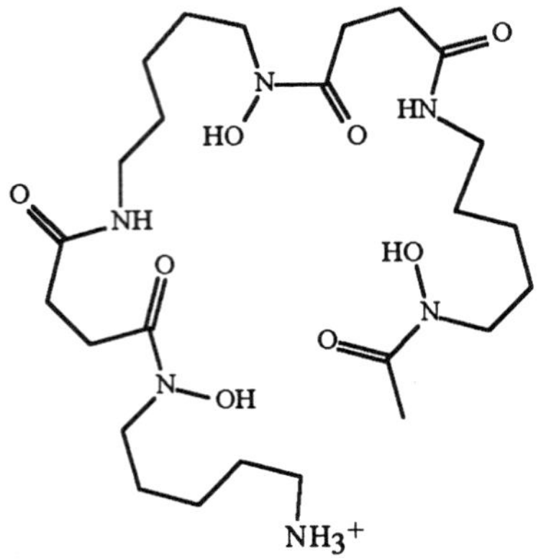

deferriferrioxamine B
9

complexed (Fe(III)) to free (Fe(III)) of 10^3 requires a 10^{18} higher concentration (in moles/liter) of bidentate acetohydroxamic acid relative to hexadentate deferriferrioxamine B. A denticity-related concentration effect is also reflected in pM values for acetohydroxamic acid (18.3) and ferrioxamine B (27.5) calculated at pH 7, $[Fe^{3+}_{aq}]$ = 1 μM and [ligand] = 1 mM. This concentration effect means that fewer moles of polydentate siderophore need to be excreted by an organism than a monocatecholate or monohydroxamate. This analysis is also true on a gram basis, since when β and β_3 are large, the concentration effect is greater than the molecular weight differences between siderophore structures of various denticity.

2.4. Siderophore Architecture

The combination of Fe(III)-binding groups to produce a polydentate ligand results in a range of different siderophore architectures, which include linear acyclic (e.g., deferriferrioxamine B (**9**)), exocyclic (e.g., deferriferrichrome (**10**)), and endocyclic (e.g., deferriferrioxamine E (**11**)). Although there are trihydroxamic acid examples of each of these architectures, no endocyclic structure has been identified for the catecholate siderophores. The synthetic siderophore mimics include other architectures (acyclic, tripodal, exocyclic, endocyclic, and macrobi-

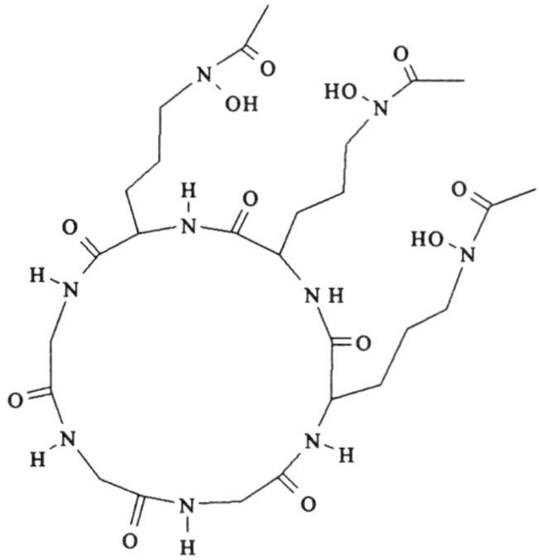

deferriferrichrome
10

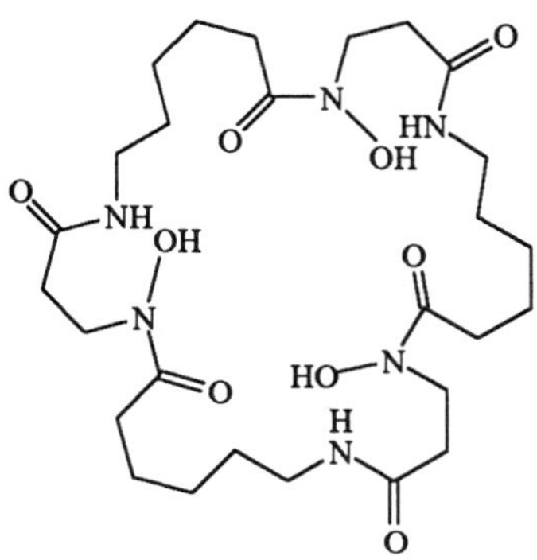

deferriferrioxamine E
11

cyclic). Ligand architecture will influence the stability and the lability of a Fe^{3+}-siderophore complex through preorganization of the chelation cavity [21], variations in ligand flexibility, and the relative ease of assuming a pseudo-octahedral coordination geometry.

Not all siderophore structures are hexadentate in spite of the increased Fe(III) scavenging efficiency that this arrangement provides. Ferrioxamine H is a pentadentate ligand with two hydroxamic acids and one carboxylic acid-binding group [22]. Tetradentate dihydroxamic acids include linear structures, such as rhodotorulic acid (**12**) [20], and

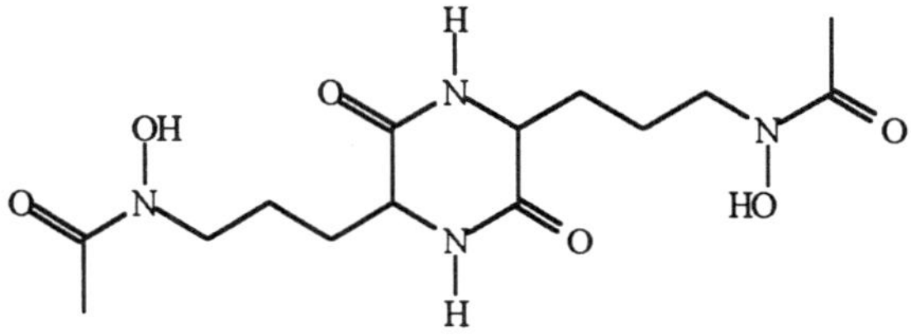

rhodotorulic acid
12

cyclic structures, such as the 20-membered ring alcaligin (**13**) [23–27] and putrebactin [28], and 22-membered ring bisucaberin [23]. These tetradentate ligands achieve Fe(III) coordination saturation by forming

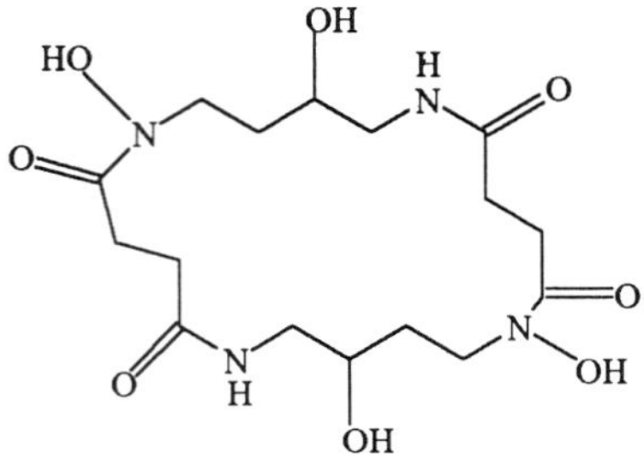

alcaligin
13

complexes with the stoichiometry Fe_2L_3. Bidentate siderophores which form FeL_3 complexes include monohydroxamic acids and thiohydroxamic acids: *N*-methylphenylacetohydroxamic acid [29]; 1-α-aspartyl-1-α-*N*-hydroxyaspartyl-*D*-cycloserine [30]; and *N*-methyl-thioformohydroxamate [31].

3. Fe(III) AND Fe(II) COMPLEXES: SOLUTION STRUCTURES

3.1. Ultraviolet-Visible Absorption Spectrophotometry

The ligand to metal charge transfer band with catecholates, hydroxamates, and dihydroxyquinolines is characteristic of the octahedral coordination of Fe(III). The wavelength maxima and the corresponding absorbtivities are given in Table 1 for various $Fe(III)L_3$ complexes. The chemical structures (**14–21**) of these bidentate ligands [32–34] are presented in Fig. 1.

In Table 2, we have gathered the ultraviolet (UV)-visible absorption data for some natural or synthetic tricatecholate (**22–34**, **42**), tri-(dihydroxyquinoline) (**37**), and trihydroxamate compounds (**35–36**, **38–41**).

The chemical structures of five catecholamide siderophores [35] are presented in Fig. 2. The structures of the tricatecholate [36–39] and trihydroxamate natural or synthetic ligands listed in Table 2 are described in Fig. 3. Tripodal [40,41] and cryptate trihydroxamate [42]

TABLE 1

Absorption Properties of $Fe(III)L_3$ Complexes with Bidentate Ligands

Ligand	No.	λ_{max} (nm)	ε_{max} (M^{-1} cm^{-1})	Reference
Hydroxamates				
β-methyl-*D*-glucofururono-hydroxamate	**14**	465	2.388×10^3	32
Catecholates				
MTA	**15**	446	5.7×10^3	33
		510	5.4×10^3	33
ETA	**16**	448	5.79×10^3	33
		509	5.1×10^3	33
PTA	**17**	448	5.8×10^3	33
		511	5.0×10^3	33
2,2′-Dihydroxybiphenyl Derivatives				
BPS	**18**	510	9.64×10^2	34
BPAS	**19**	480	7.00×10^2	34
BPDAS	**20**	470	1.810×10^3	34
BPANS	**21**	460	8.25×10^2	34

synthetic siderophores are given in Fig. 4, as well as deferriferrioxamine B derivatives [43].

Using UV-visible absorption spectrophotometry and taking advantage of the ligand to Fe(III) charge transfer bands (LMCT) with either catecholate-type or hydroxamate-binding sites, it has been possible to characterize the siderophore ferric complexes and to determine their corresponding stability constants. With catecholate-type natural or synthetic ligands, the determination of the acido-basic properties of the metal free compounds is possible using a combination of potentiometric and spectrophotometric measurements. As an example, we

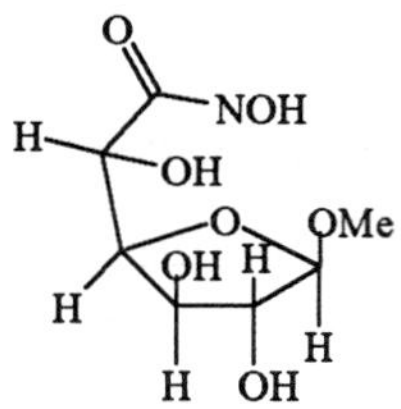

β-methyl-*D*-glucofururonohydroxamic acid
14

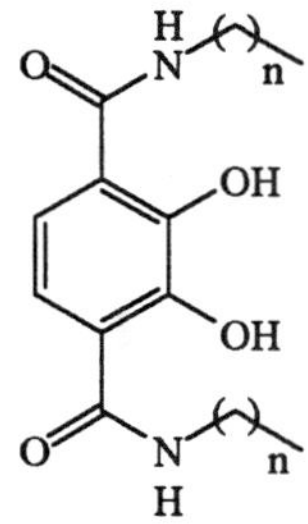

MTA	n = 0	**15**
ETA	n = 1	**16**
PTA	n = 2	**17**

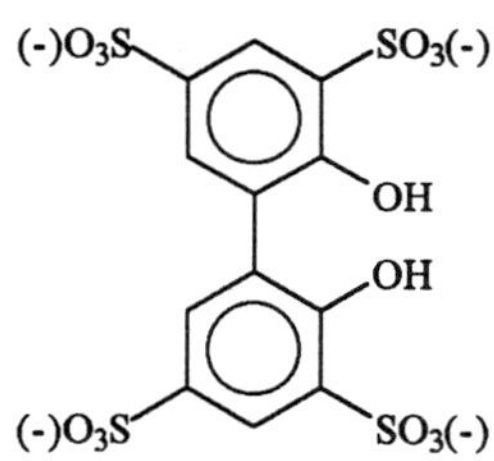

BPS
18

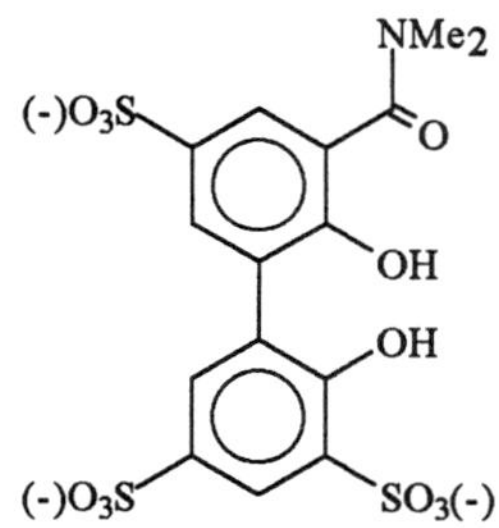

BPAS
19

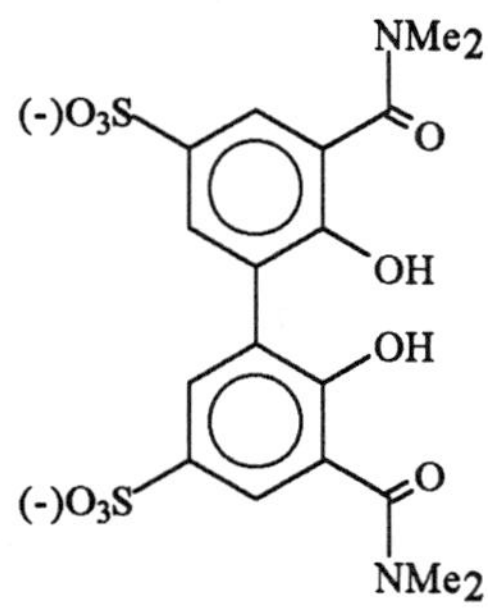

BPDAS
20

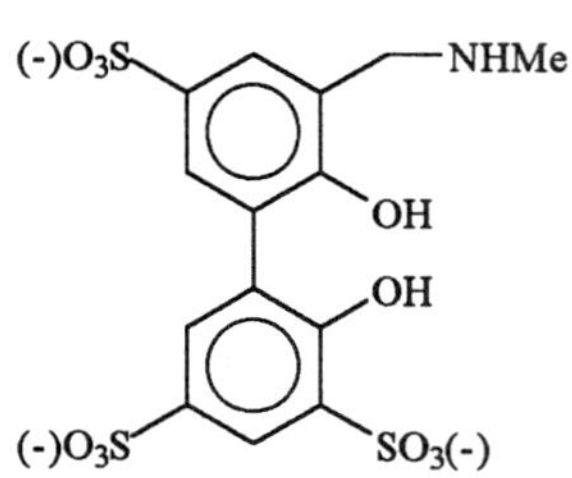

BPANS
21

FIG. 1 Chemical structures of hydroxamate, catecholate, and 2,2′-dihydroxybiphenyl bidentate ligands.

TABLE 2

Absorption Properties of Various Fe(III)L Complexes with Hexadentate Natural or Synthetic Ligands[a]

Ligand	No.	λ_{max} (nm)	ε_{max} (M^{-1} cm^{-1})	Reference
Tricatecholates				
MECAM	**29**	344	12.7×10^3	38
		496	4.7×10^3	38
enterobactin	**27**	338	15.1×10^3	38
		498	5.7×10^3	38
DHBS linear trimer	**28**	494	5.4×10^3	36
lysine-based linear ligand	**42**	495	5.2×10^3	39
Ar(L-leu)[b]	**30**	502	4.65×10^3	37
Ar(L-ala)[b]	**31**	502	4.37×10^3	37
Ar(N-Me-L-leu)[b]	**32**	496	4.79×10^3	37
Tr(L-leu)[b]	**33**	500	5.03×10^3	37
Tr(L-ala)[b]	**34**	496	5.04×10^3	37
agrobactin	**23**	507	3.6×10^3	35
agrobactin A	**25**	495	3.65×10^3	35
parabactin	**22**	516	3.1×10^3	35
parabactin A	**24**	522	3.32×10^3	35
vibriobactin	**26**	496	3.4×10^3	35
Tri-8-Hydroxyquinoline Ligands				
N-TRENSOX	**37**	595	1.2×10^2	41
Trihydroxamates				
trihydroxamate cryptate	**36**	430	2.75×10^3	42
TRENDROX	**35**	442	2.934×10^3	40
ferrioxamine B analogues[c]				
ala derivative	**38**	425	2.460×10^3	43
leu derivative	**39**	423	2.120×10^3	43
asp derivative	**40**	430	2.450×10^3	43
glu derivative	**41**	420	2.320×10^3	43
Mixed Ligands				
pyoverdin PaA	**5**	403	1.9×10^4	44
		460	6.5×10^3	44
		540	3.5×10^3	44

[a]Solvent: water.
[b]Solvent: methanol/water (15/85).
[c]Solvent: methanol/water (80/20).

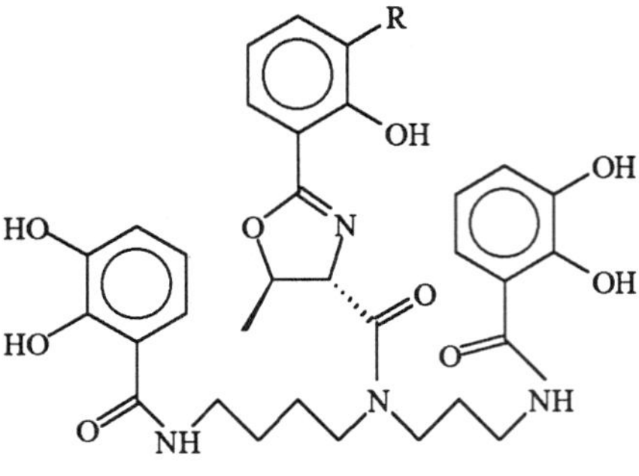

L-parabactin (R = -H)
22
L-agrobactin (R = -OH)
23

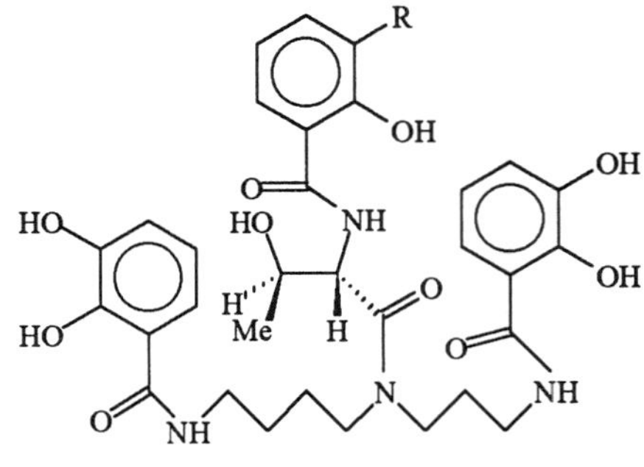

L-parabactin A (R = -H)
24
L-agrobactin A (R = -OH)
25

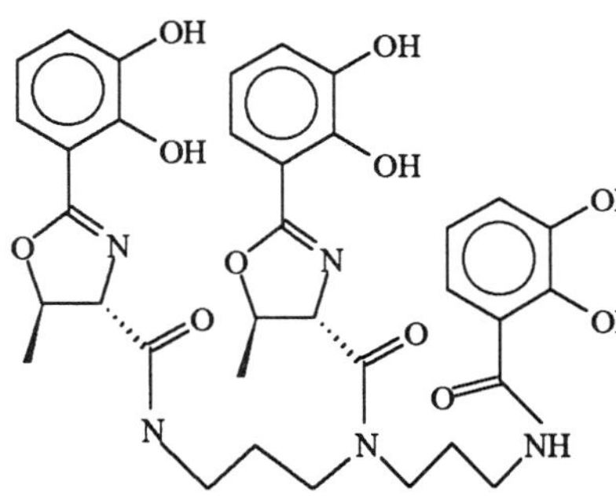

L-vibriobactin
26

FIG. 2 Catecholamide siderophores.

present the spectrophotometric study [44] of pyoverdin PaA (**5**) and of its Fe(III) complexes over a wide range of p[H]. Pyoverdin PaA is a siderophore excreted by *Pseudomonas aeruginosa* [45,46] which consists of a peptide bound via its N-terminus to a chromophore derived from 2,3-diamino-6,7-hydroxyquinoline. It forms Fe(III) complexes by means of three bidentate chelating sites which are the catechol group on the chromophore, a hydroxamate function in the middle of the peptide chain, and a second terminal hydroxamic acid [45]. The spectrophotometric titration at various p[H] of pyoverdin PaA (**5**) as a free ligand is given in Fig. 5a. With the potentiometric and the spectrophotometric data, it was possible to determine the two protonation constants of the catechol-type chromophore. The spectrophotometric titration of the Fe(III) pyoverdin PaA complex at various proton concentrations led to the identification of three thermodynamic Fe(III) species $LFe(III)^-$, LHFe and LH_2Fe^+, and their respective electronic spectra (Fig. 5b). The LMCT band centered at about 450 nm is related to the binding of the hydroxamate groups of the siderophore. At longer wavelength (about 550 nm), the absorption band corresponds to the binding of the dihydroxyquinoline function of pyoverdin PaA.

3.2. Fluorescence Emission Spectroscopy

The bacteria *Pseudomonas* and *Azotobacter* excrete fluorescent siderophores [46–50] (pseudobactins (**43**), pyoverdins (**5**), and azotobactins (**6**)) constituted by a peptidic chain bearing a dihydroxyquinoline-type

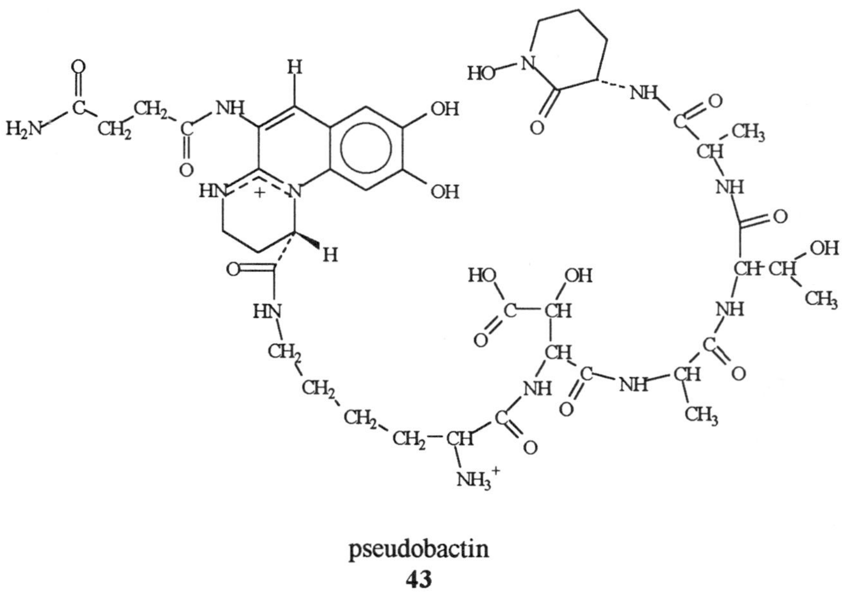

pseudobactin
43

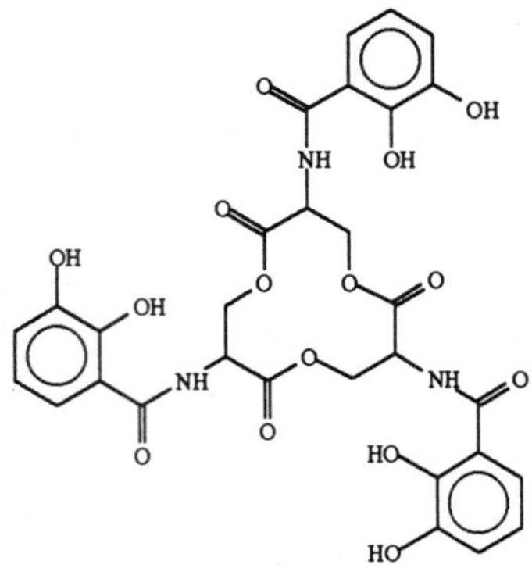

enterobactin
27

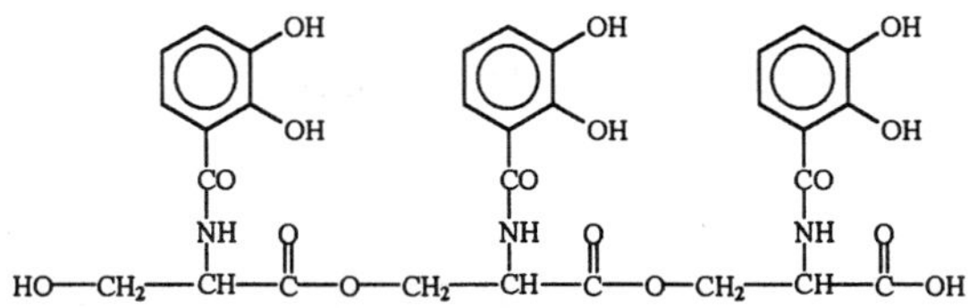

DHBS linear trimer
28

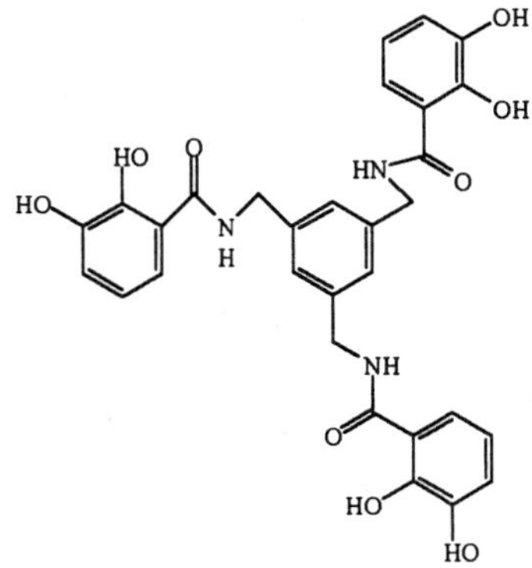

MECAM
29

FIG. 3 Enterobactin, and tripodal and linear synthetic tricatecholate analogues (**27–34**).

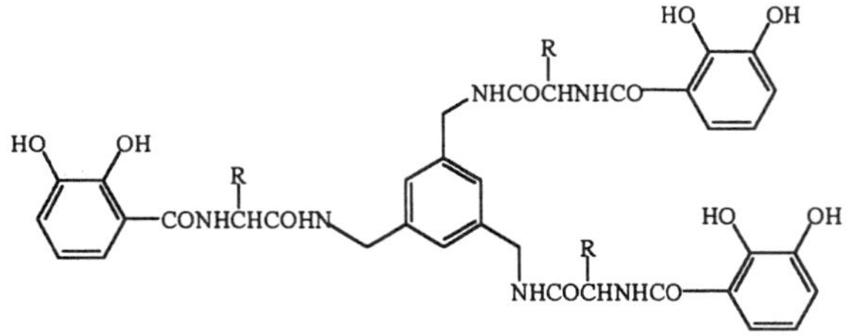

Ar(L-leu): R = i-Bu Ar(L-ala): R = Me

30 **31**

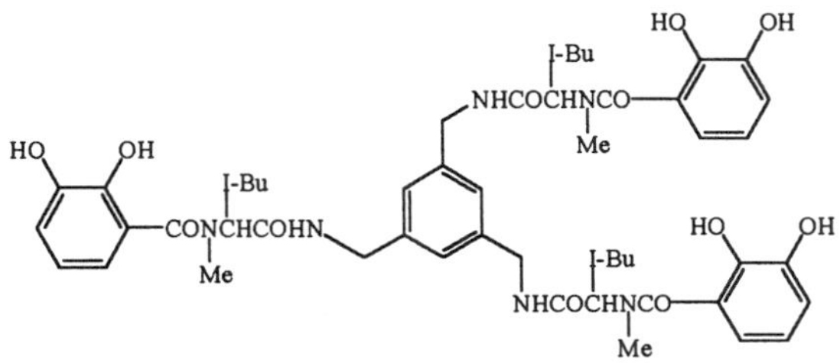

Ar(N-Me-L-leu)

32

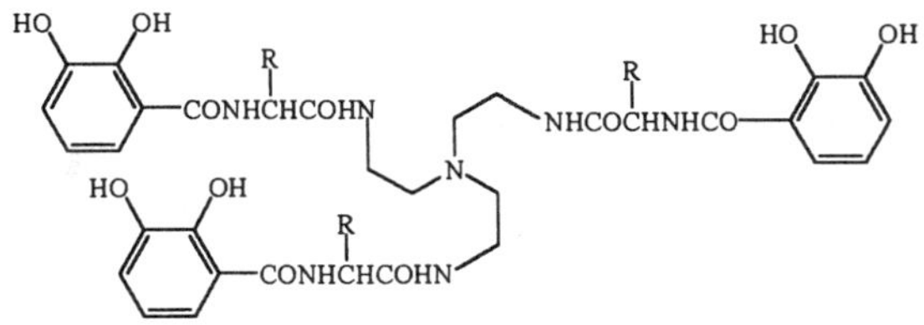

Tr(L-leu): R = i-Bu Tr(L-ala): R = Me

33 **34**

FIG. 3 Continued

fluorophore and two other bidentate-binding groups (hydroxamate or α-hydroxycarboxylates). The fluorescence of the free siderophores is quenched by the coordination of the Fe(III), as shown in Fig. 6 for azotobactin δ (**6**), which is a very promising Fe(III) biosensor [51].

The simplest fluorescent hydroxamate Fe(III) ligand [52] proposed is a monohydroxamate function labeled with an anthracene fluorophore (**44**) (Fig. 7). Biomimetic fluorescent analogues of deferriferrioxamine B (**9**) were examined for the determination of Fe(III) under physiological conditions [53]. Three different fluorophores were bound to the terminal amine group of deferriferrioxamine B: 7-nitrobenz-2-oxa-1,3-diazol (NBD, **45**), [2,3-*c*]-naphto-2-cyano-2*H*-pyrol (NCP, **46**), and *N*-methyl-

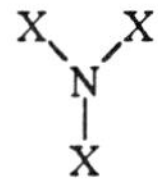

X = $CH_2CH_2NHC(O)CH_2CH_2C(O)N(O)R$

TRENDROX

35

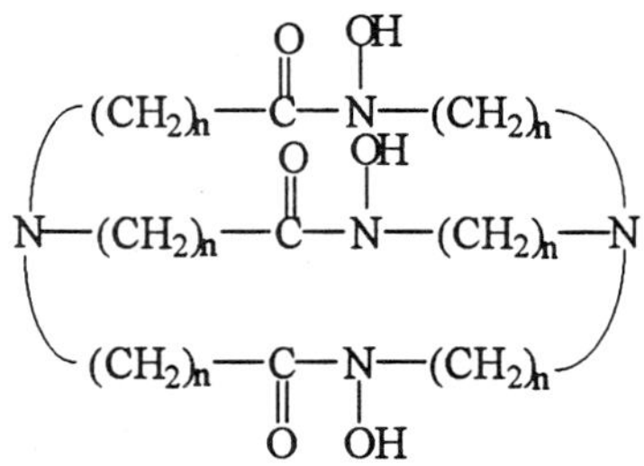

36

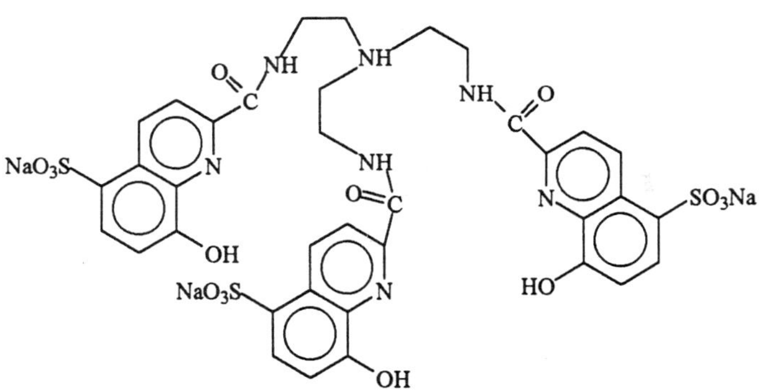

N-TRENSOX

37

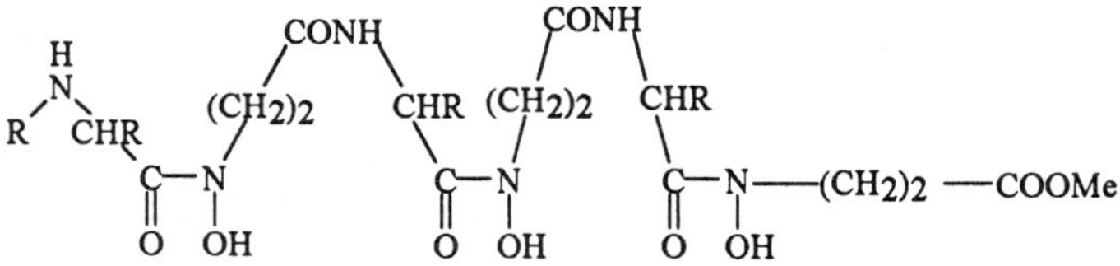

R = Me R = Bu

38 **39**

FIG. 4 Linear, tripodal, and cryptate trihydroxamate and tripodal quinoline ligands (**35–41**).

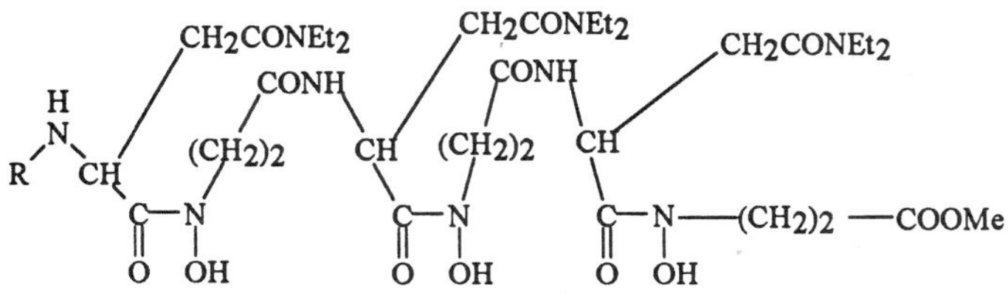

40

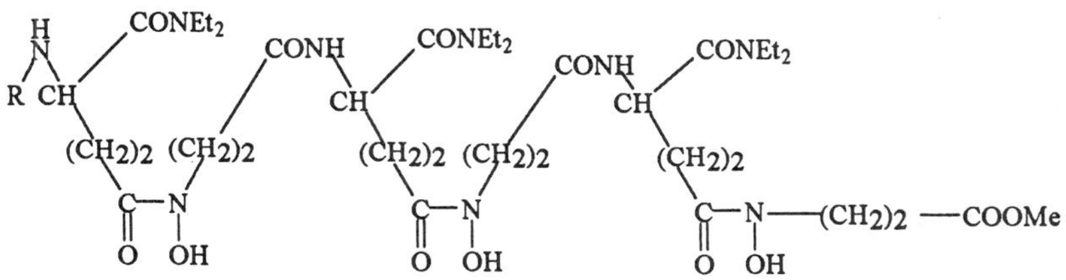

41

FIG. 4 Continued

anthranilic (MA, **47**) (Fig. 7). The authors demonstrated the application of NBD-DFO (**45**) for the specific detection and monitoring of Fe(III) in solutions. Interactions of this fluorescent analogue with ferritin and transferrin were also described [53].

Tripodal biomimetic analogues of the microbial Fe carrier deferriferrichrome (**10**) were labeled with a fluorescent marker at a site which does not interfere with Fe binding or membrane receptor recognition [54]. Variation in the nature of the amino acids (glycine, alanine, and leucine) linking the tetrahedral carbon used as an anchor and the hydroxamate coordination groups was examined as well as changes in the spacer between the fluorophore and the anchor. The alanine derivative with a long spacer (**48**) is presented in Fig. 7. Incubation of the bacterium *Pseudomonas putida* with the nonfluorescent Fe(III) complex of the leucine derivative showed a release of the fluorescent unloaded carrier. These results demonstrate the involvement of the ferrichrome receptor and Fe(II) transport systems of the bacterium and provided evidence for a shuttle mechanism [54].

A tripyrenyl trihydroxamate fluorescent ligand (see Fig. 7) with a tripodal structure has also been synthesized (**49**) [55]. The binding of Fe(III) was shown to induce a quasi-total quenching of the pyrene fluorescence. This compound belongs to a new class of molecular systems which can act as sensitive and selective probes for Fe(III) detection.

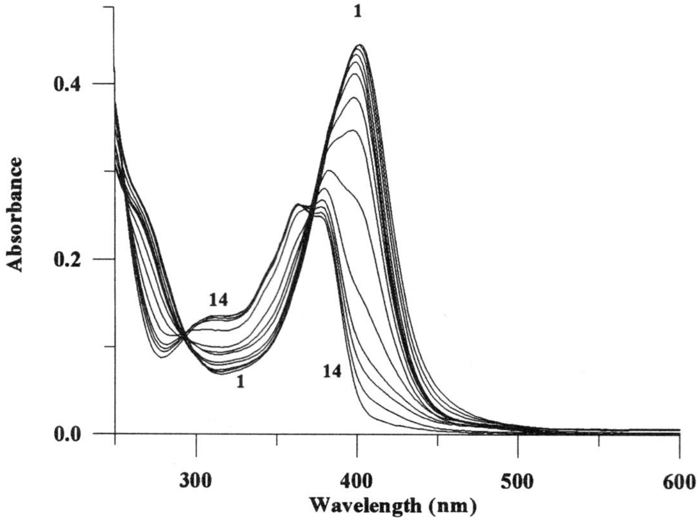

(a)

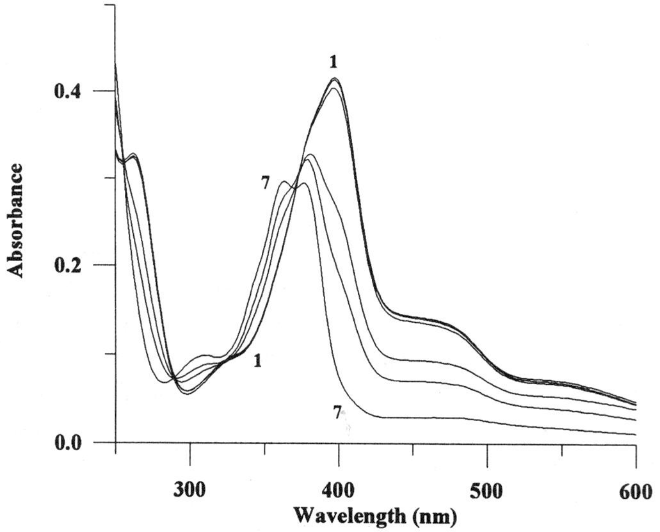

(b)

FIG. 5 Evolution of the absorption spectra of pyoverdin PaA (**5**) and of its Fe(III) species at different proton concentrations. Solvent: water; T = 25.0 ± 0.1°C; I = 0.1, l = 0.2 cm (a) $[PaA]_0 = 7.7 \times 10^{-5}$ M. spectrum 1: p[H] = 9.8; spectrum 14: p[H] = 1.7; (b) $[PaAFe(III)]_0 = 1.3 \times 10^{-4}$ M. spectrum 1: p[H] = 5.1; spectrum 7: p[H] = 1.2. (Reproduced with permission from [44].)

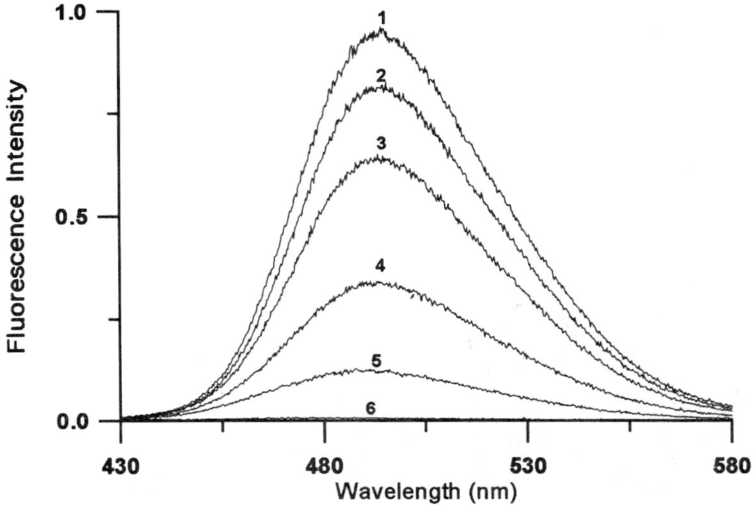

FIG. 6 Fluorescence quenching of azotobactin δ (**6**) by addition of Fe(III) [51]. Solvent: water; p[H] = 4.4 (acetate buffer); T = (25 ± 1)°C; l = 1 cm. [azotobactin δ] = 2×10^{-6} M; λ_{exc} = 380 nm; [Fe(III)]/[azotobactin δ] = spectrum 1: 0; 2: 0.12; 3: 0.25; 4: 0.37; 5: 0.49; 6: 0.61; 7: 0.73; 8: 0.86; 9: 0.98.

3.3. Circular Dichroism Measurements

Bergeron and Weimar [35] have examined the stereospecificity of Fe(III) uptake by *L*-parabactin (**22**), excreted by the soil bacterium *Paracoccus denitrificans*. The chirality of the Fe(III) chelates of the natural *L*-parabactin A (**24**) and *L*-parabactin (**22**), and of the synthetic *D*-parabactin A, was determined by circular dichroism (Table 3) and related to the Fe transport efficiency.

Karpishin et al. [56] synthesized symmetric bidentate ligands (*N,N′*-bis (S)-α methylbenzyl-2,3-dihydroxyterephtalamide] (**50**) which form tricatecholate ferric complexes with a pure Λ conformation in aqueous solution (Fig. 8). Using CD spectrometry, the authors [56] confirmed the Δ conformation of ferric enterobactin. S-**50** Fe(III) shows a positive band at 541 nm which is opposite in the Fe(III) enterobactin complex. These bands arise from ligand to metal charge transfer transitions. The chirality of S-**50** Fe(III) is unchanged between the solid state and solution. In the same publication [56], the chirality of various ferric

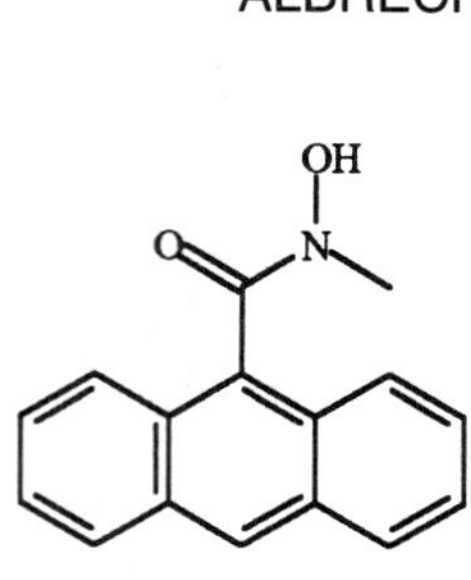

44

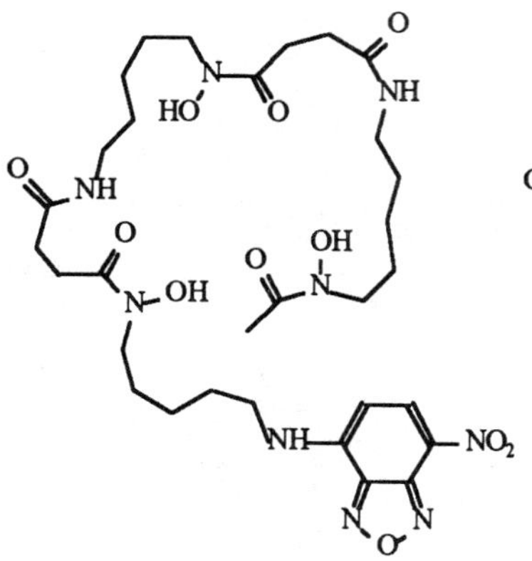

NBD-DFO
45

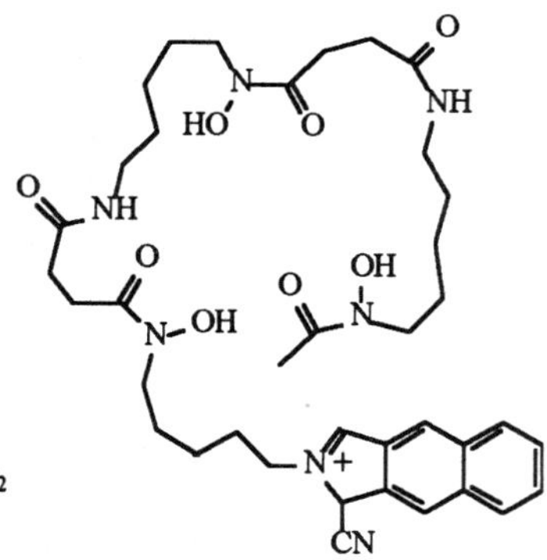

NCP-DFO
46

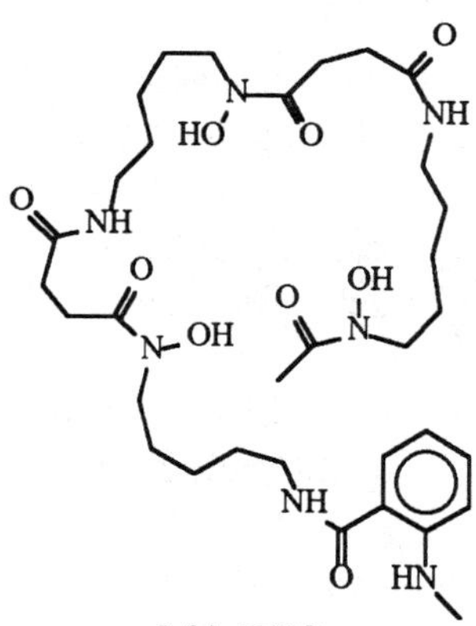

MA-DFO
47

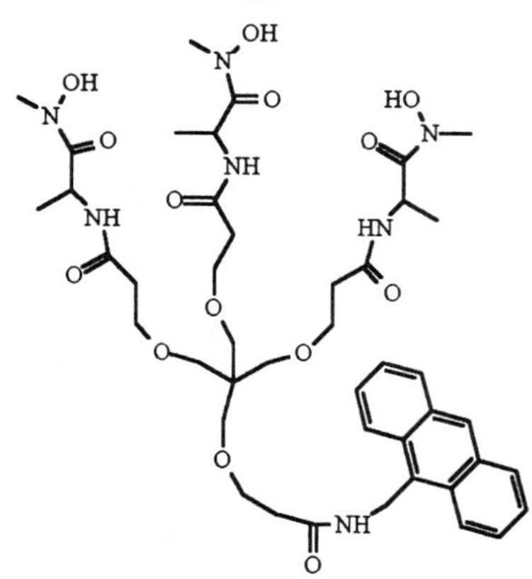

anthryl-DFC
48

FIG. 7 Fluorescent biomimetic analogs of deferriferrioxamine B (**45–47**) and of deferriferrichrome (**48–49**).

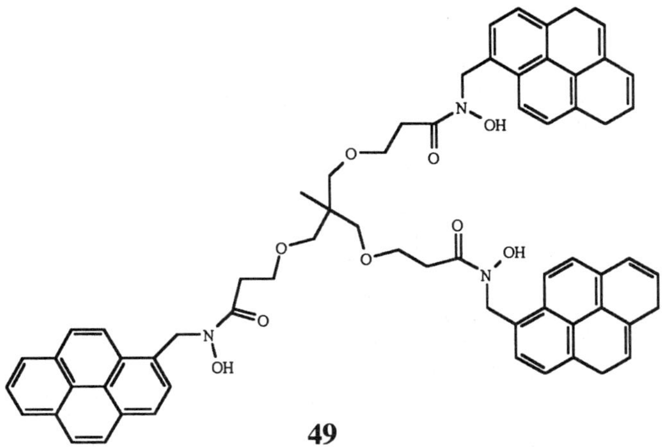

49

FIG. 7 Continued

species with related ligands (**52–54**) was examined using CD spectrometry. The corresponding CD data are presented in Table 4.

Chiral models of enterobactin (**27**) have also been synthesized by Akiyama and Ikeda [39]. They are linear triscatecholate ligands based on *L*- and *D*-lysines (**42**). The schematic representation of the corresponding *L*-lysine ferric complex is given in Fig. 9 as well as the CD

TABLE 3

Circular Dichroism Properties of Two Natural Catecholamide Siderophores and of a Related Chiral Analog

Ferric complex	No.	λ_{max} (nm)	$\Delta\varepsilon$ (M^{-1} cm^{-1})
L-Parabactin Λ enantiomer	**22**	550	+2.2
L-Parabactin A Λ enantiomer	**24**	550	+1.2
D-Parabactin A Δ enantiomer	**24**	550	−1.2

Source: Reproduced with permission from [35].

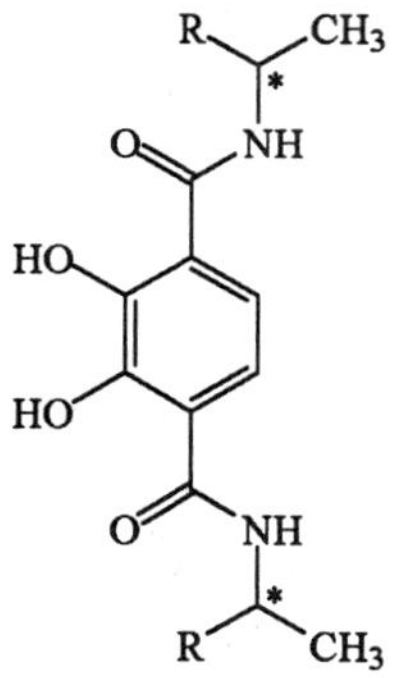

R = Ph: $H_2(PhMe)_2TAM$
50
R = Et: $H_2(EtMe)_2TAM$
51

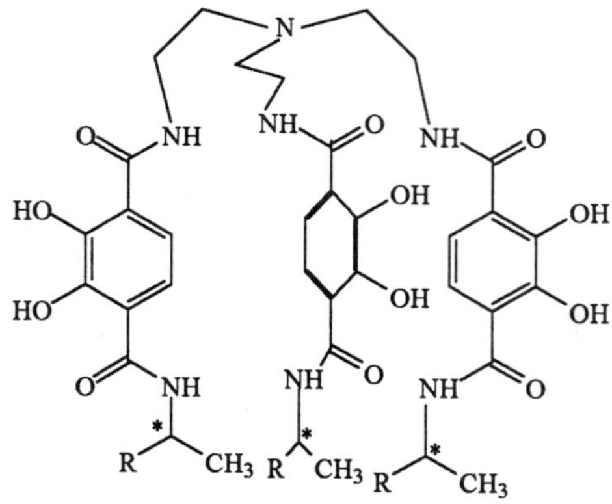

R = Ph: $H_6TREN(PhMe)TAM$
52
R = Et: $H_6TREN(EtMe)TAM$
53

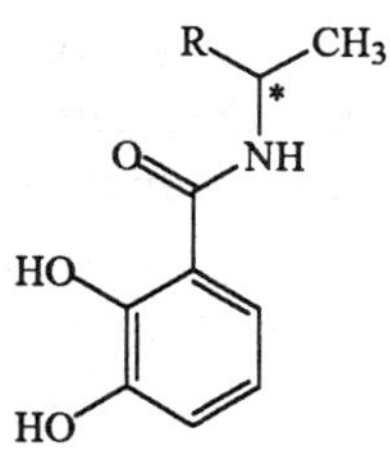

R = Ph: $H_2(PhMe)CAM$
54

FIG. 8 Chiral catecholate derivatives [56].

spectra of the enantiomeric pair *L*- and *D*-**42** Fe(III) complexes ($\Delta\varepsilon$ = +3.0 at 535 nm).

Tor et al. [37] have studied a tripodal tricatecholate ligand based on an aromatic ring with peptidic *L*-leucine arms (**30**) (Fig. 10). The visible spectrum of the corresponding Ar(*L*-leucine) ferric complex showed a LMCT band at 502 nm which is characteristic of Fe^{3+} octahedrally coordinated to three catecholates. The CD spectrum (Fig. 10) shows in this region two Cotton effects ($\Delta\varepsilon$ = +4.2 at 432 nm and −2.3 at 550 nm), which are close to those observed for the Fe(III) enterobactin

TABLE 4

Circular Dichroism Results of Various Terephthalamide and Catecholamide Ligands Compared to the Natural Siderophore, Enterobactin

Ferric complex	Major diastereomer	λ_{max} (nm)	$\Delta\varepsilon$ (M^{-1} cm^{-1})
R-$\mathbf{54}_{Fe}$	Δ	535	−1.6
S-$\mathbf{54}_{Fe}$	Λ	535	+1.6
R-$\mathbf{53}_{Fe}$	Δ	535	−0.55
S-$\mathbf{52}_{Fe}$	Λ	536	+1.5
R-$\mathbf{51}_{Fe}$	Δ	542	−1.1
S-$\mathbf{50}_{Fe}$	Λ	541	+3.8
Enterobactin **27**	Δ	533	−4.0

Source: Reproduced with permission from [56].

complex [57] ($\Delta\varepsilon$ = +4.0 at 420 nm and −4.0 at 535 nm). This demonstrates that the predominant absolute configuration of Ar(*L*-leu) Fe^{3+} is identical with that of the enterobactin ferric complex; namely, Δ-*cis* [3].

Yakirevitch et al. [43] have described a new family of chiral ferrioxamine analogues which were synthesized to establish whether the ferrioxamine B receptor was capable of chiral discrimination. The investigators aimed at synthetic ferrioxamine analogues (see Fig. 4) that would preferentially adopt a *cis* configuration and exhibit significant chiral preference. Quite pronounced differences were observed in the chiral purity of the Fe(III) complexes. In contrast with derivatives **38** and **39** with poor chiral preference, the analogue **41** showed a strong chiral preference for the Δ configuration when using the natural *L*-glutamic acid (Fig. 11).

Chiral artificial siderophores have been proposed by Akiyama et al. [58] as models for ferrichrome (**55–57**). These tripodal *L*-alanyl-*L*-alanyl-β-(*N*-hydroxy)alanine-based trihydroxamic acids form stable chiral Fe(III) complexes. The CD characteristics are given in Table 5.

55 $N[CH_2CO$-Ala-Ala-β(HO)Ala-OMe$]_3$

56 $N[CH_2CH_2NH$-Ala-Ala-β(HO)Ala-Ala-Ac$]_3$

57 $N[CH_2CH_2NH$-Ala-β(HO)Ala-Ala-Boc$]_3$

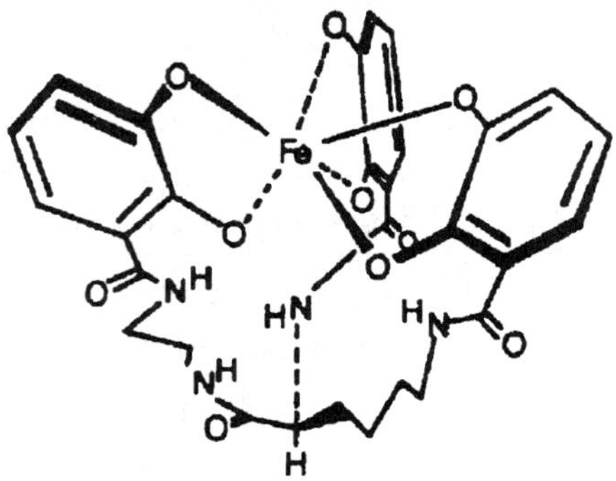

42-Fe(III)

(a)

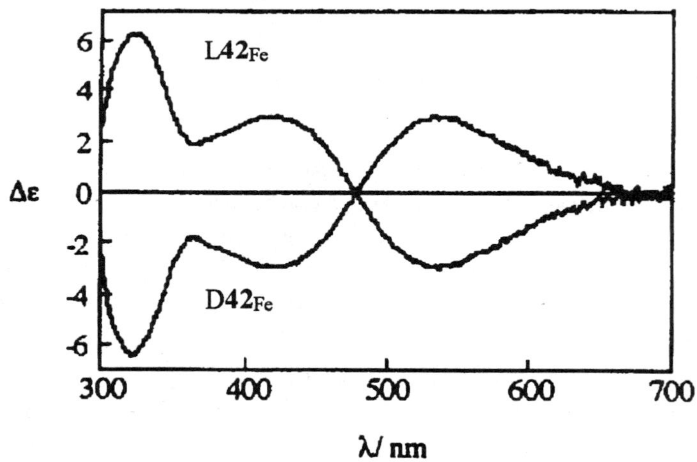

(b)

FIG. 9 (a) Triscatecholate *L*-lysine iron(III) complex (**42**) in the Δ-cis configuration and (b) CD spectra of *L*42- and *D*42-Fe(III) complexes. (Reproduced with permission from [39].)

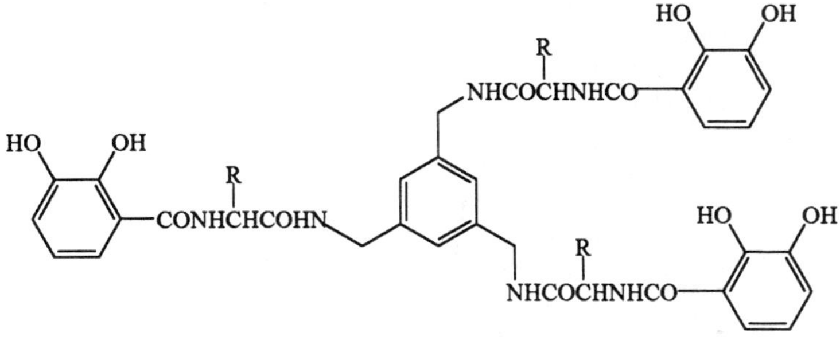

Ar(L-leu) ; R = i-Bu
30

(a)

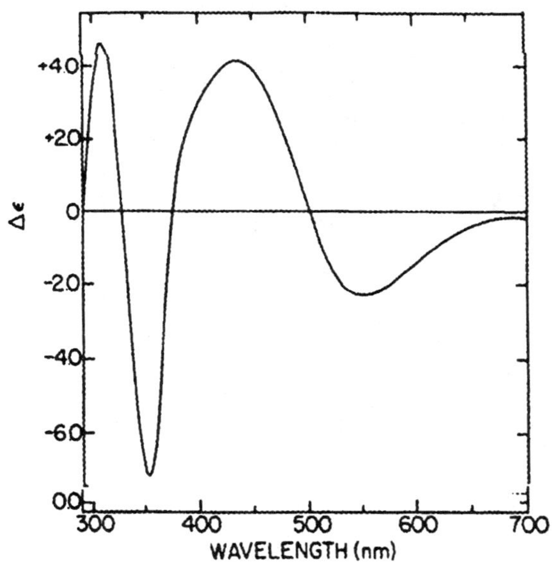

(b)

FIG. 10 (a) A chiral enterobactin analogue (**30**) and (b) the CD spectrum of its ferric complex. (Reproduced with permission from [37].)

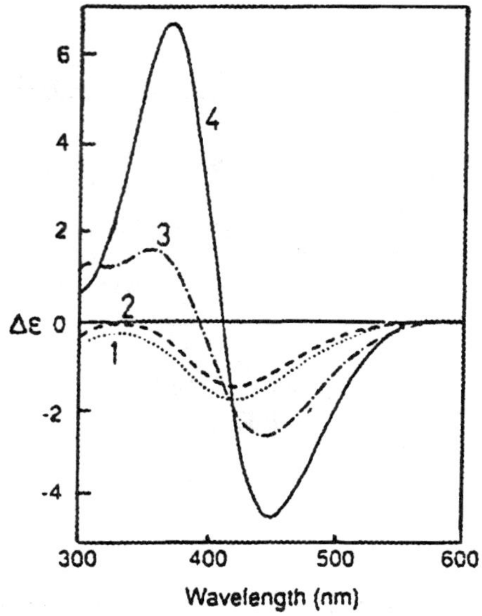

FIG. 11 CD spectra of the ferric complexes with deferriferrioxamine B analogues. Spectrum 1: **38**-Fe(III); Spectrum 2: **39**-Fe(III); Spectrum 3: **40**-Fe(III); Spectrum 4: **41**-Fe(III). (Reproduced with permission from [43].)

TABLE 5

Circular Dichroism Data for Ferrichrome and Chiral Synthetic Trihydroxamate Ligands

Complex	Configuration	λ (nm)	Δε (M^{-1} cm^{-1})
Ferrichrome (Fe(III)-10)	Λ	465	+2.4
		360	−3.7
Fe(III)-**55**	Λ	465	+2.1
		380	−6.2
Fe(III)-**56**	Δ	445	−4.0
		365	+2.3
Fe(III)-**57**	Δ	455	−5.3
		375	+3.0

Source: Reproduced with permission from [58].

3.4. Mass Spectrometry Methods

Mass spectrometric measurements, particularly fast atom bombardment (FAB) mass spectrometry, have been widely used for the structure elucidation of new siderophores. As an example, the structure of pyoverdins C, D, E, and, PaA (**5**), excreted by *Pseudomonas aeruginosa*, have been elucidated by mass spectromety combined with nuclear magnetic resonance (NMR) and degradation studies [45,50]. The same combination of methods has been used to establish the structure of chromopeptide siderophores that are excreted by the bacterium *Azotobacter vinelandii* [46,59]. A new citrate-based dihydroxamate siderophore, acinetoferrin (**58**), excreted by *Acinetobacter haemolyticus*, has been identified by gas chromatography-mass spectrometry (GC-MS) combined with ^{1}H and ^{13}C NMR [60].

New developments of the mass spectrometric method will provide a very efficient tool to probe the solution structure of ferric complexes

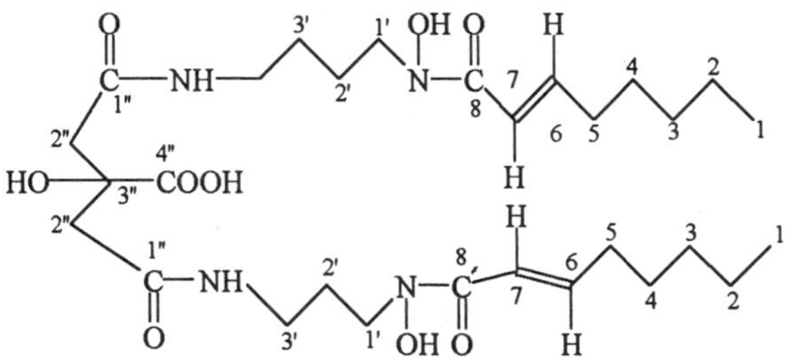

acinetoferrin
58

[61]. Electrospray mass spectrometry (ESMS), which preserves the integrity of the species during the ionization process, is able to provide information about the nature of the complexes in solution at equilibrium. Caudle et al. [62,63] demonstrated by ESMS that for a series of dihydroxamic acids (**59**), when n $\leqslant$ 6, the 1:1 Fe(III) complex is a dimer

$$\textbf{59} \quad CH_3N(OH)C(O)(CH_2)_nC(O)N(OH)CH_3 \quad n = 2\text{–}8.$$

($Fe_2L_2(H_2O)_4^{2+}$) in solution and when n $\geqslant$ 6 it is a monomer ($FeL(H_2O)_2^+$). The 1:1 Fe(III) rhodotorulic acid (**12**) complex was identified as a monomer ($FeL(H_2O)_2^+$) [62].

A FAB mass spectrometric study carried out on hydroxamate siderophores (ferrioxamine B, ferrichrome, ferricrocin, ferrirhodin, coprogen, and triacetyfusarinine C) demonstrates that the complex parent ion loses a ^{53}Fe in the matrix [64]. In addition, the results show that the complex and the free ligand lose an ^{16}O by a reduction process. The loss of an O in the free ligand and in the Fe(III) species has also been observed by ESMS on diferric trihydroxamate helicates [65].

3.5. Nuclear Magnetic Resonance Spectroscopy

^{1}H and ^{13}C NMR spectroscopy has been widely used, particularly in combination with FAB mass spectrophotometry, in order to establish the chemical structures of microbial siderophores and their synthetic analogues [46,50,59].

The salicylate mode of Fe(III) bonding by the ligand *N,N′,N″*-tris(2,3-dihydroxybenzoyl)-1,3,5-tris(aminomethyl)benzene (MECAM) (**29**) has been demonstrated using the shift of a CD_3 substituent on the catecholate moiety. Lower pH values induce a change from the catecholate mode to the salicylate mode of bonding which corresponds to a large paramagnetic shift of the ligand [66]. Dynamic ^{1}H and ^{13}C NMR experiments were carried out in order to observe the facile inversion of Ga(III) enantiomer complexes with siderophore-type ligands. A catecholamide bidentate ligand (**15**) with *N*-tert-butyl and a *N′*-benzyl substituents were considered to elucidate the mechanism of inversion. The rate constants and the activation parameters of the inversion of the enantiomers were calculated from NMR data. A trigonal twist mechanism via a trigonal prismatic transition state was confirmed [67].

4. Fe(III) AND Fe(II) COMPLEXES: STABILITY

4.1. Methods of Expressing Iron-Chelate Stability Constants

Formation constants define the stability of an Fe-siderophore complex in solution. In addition, these constants are a measure of the siderophore ligand's selectivity for Fe^{3+}_{aq} in the presence of competing metal ions and its ability to solubilize Fe and prevent precipitation by hydrolysis. The Fe^{3+}_{aq} stability constant may also be a deciding factor in determining the mechanism for Fe(III) release from the siderophore by influencing the dissociation rate constants and the Fe(III/II) reduction potential.

In aqueous solution, chelation of Fe^{3+}_{aq} (in the absence of other metal ions) by a siderophore involves a competition between Fe^{3+}_{aq} and H^+ for the deprotonated ligand. Although Eq. (6) represents the experimentally observed competitive equilibrium, literature compilations of equilibrium constants are expressed as β_{FeLH} values according to the hypothetical equilibrium reaction shown below for a hexadentate triprotic siderophore. β_{FeLH} values are informative for analyzing the influence of different Fe(III)-binding groups and ligand architecture on Fe(III) complex stability. However, some degree of caution must be exercised in their use as β values defined above are concentration quotients, and the units will change with changes in ligand denticity.

$$LH_n + Fe^{3+}_{aq} \longrightarrow FeL + nH^+ \tag{6}$$

$$K_n = \frac{[FeL][H^+]^n}{[Fe^{3+}_{aq}][LH_n]} \tag{7}$$

$$L^{3-} + Fe^{3+}_{aq} \longrightarrow FeL \tag{8}$$

$$\beta_{110} = \frac{[FeL]}{[Fe^{3+}_{aq}][L^{3-}]} \tag{9}$$

Consequently, β values should only be directly compared for ligands of the same denticity. Furthermore, the equilibrium expression in Eqs. (8,9), which defines β, does not take into account the competition with H^+ which is present in a true environmental situation (Eq. (6)). Both of these problems are addressed in an empirical approach which does not require knowledge of the ligand K_a values [3]. A pM value is defined from equilibrium reaction (6) as the negative log of the free or uncomplexed Fe^{3+}_{aq} concentration (pM = $-\log[Fe^{3+}_{aq}]$) calculated from the observed H^+ dependent formation constant (Eq. (7)) for a fixed set of experimental conditions: pH 7.4; $[H_nL]_{tot} = 10\ \mu M$ and $[Fe^{3+}_{aq}]_{tot} = 1\ \mu M$. The larger the pM value for a particular ligand, the more stable is the metal complex at these conditions. Precipitation as Fe(III) hydroxide is indicated by a pM value below the limit set by the K_{sp} of $Fe(OH)_3$.

The most common methods for determining Fe-siderophore complex stability constants include potentiometric titration, electrochemistry, and spectrophotometry [68,69]. In addition, competition experiments with well-characterized chelators (e.g., ethylenediamine-*N,N,N′,N′*-tetraacetate [EDTA] [70]) are often used so that the large β values may be determined at conditions where measurable quantities of both reactants and products are present [71,72].

4.2. Summary of Fe(III)-Siderophore Stability Constants

Table 6 is a compilation of stability constants and pM values for a number of siderophore structures arranged by Fe(III)-chelating groups. Inspection of Table 6 reveals high β and pM values for these natural chelators spanning 26 and 16 orders of magnitude, respectively. The hexadentate tricatecholate enterobactin (**27**) forms the most stable complex with Fe(III). The hexadentate trihydroxamates all fall within a narrow range of β values (29–32.5), independent of structure: linear, exo-cyclic and endo-cyclic. The cyclic deferriferrioxamine E (**11**) forms the most stable trihydroxamate complex, although coprogen (**60**) has a

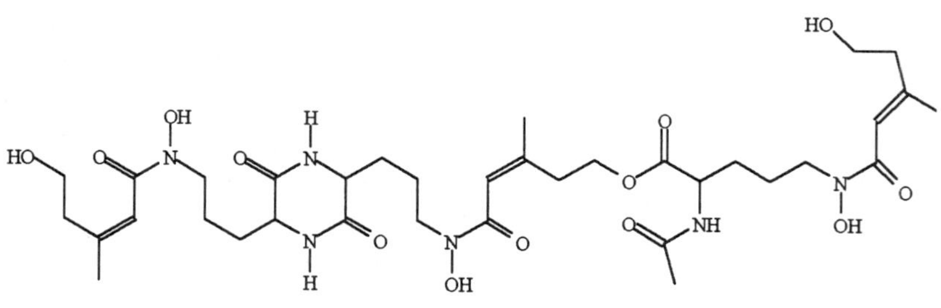

coprogen
60

comparable pM value. The mixed donor siderophores pyoverdin PaA (**5**) and azotobactin (**6**) are efficient Fe(III) sequestering agents, with azotobactin having the highest pM value in Table 6 other than enterobactin (**27**). The large β value estimated for alterobactin A suggests this is an avid Fe(III) scavenger synthesized by marine organisms, but a definitive determination of the β value must await direct determination of the ligand pK_a values [73].

The tetradentate dihydroxamates, rhodotorulic acid (**12**) and alcaligin (**13**), must form di (or higher nuclearity) Fe complexes in order to satisfy the six coordination sites on Fe. Both siderophores form Fe_2L_3 complexes with β_{230} constants given in Table 6. The stability per Fe center (log $\beta_{230}/2$) is comparable with the linear and cyclic trihydroxamic acid complexes. The cyclic alcaligin ligand is highly preorganized and results in a more stable complex than the linear rhodotorulic acid.

The crystal structure of Fe_2(alcaligin)$_3$ has a number of interesting features which are related to the stability of the complex [27]. This is the first x-ray crystal structure of a dihydroxamate siderophore complex of Fe(III) and the first example of an Fe_2L_3 complex which has a monobridged rather than triple-bridged helicate structure. The terminally bound alcaligin ligand demonstrates a high degree of preorganization in that it has exactly the same conformation as the free ligand. This high degree of preorganization is reflected in the solution stability constants for the complex. Consider the stepwise equilibria for alcaligin and rhodotorulic acid in Eqs. (10) and (11). K_1 is 100-fold greater for alcaligin than for rhodotorulic acid (log K_1 = 23.5 [27] vs 21.6 [20]), which

$$Fe^{3+}_{aq} + L^{2-} \rightleftharpoons FeL^{+} \qquad K_1 \qquad (10)$$

$$2\,FeL^{+} + L^{2-} \rightleftharpoons Fe_2L_3 \qquad K_2 \qquad (11)$$

is reflective of preorganization in the former case and not the latter. The effect of preorganization is lost in the second step (Eq. (11)), since the alcaligin must undergo significant reorganization in order to form the ligand bridge (log K_2 = 17.74 [27] vs 19.8 [20]). However, preorganization positively affects the overall stability of the alcaligin complex, since $\beta_{230} = (K_1)^2(K_2)$.

Equilibria involving the second coordination sphere of a siderophore complex are also of interest. Ferrioxamine B has a pendant amine group not involved in first-sphere coordination. Protonation of this amine group (pK_a = 10.40 [74,75]) significantly influences the hydrophilicity of the Fe(III) complex. This protonated amine side chain can be used as a recognition element in host-guest complexation in the second coordination shell of ferrioxamine B involving synthetic and natural ionophores [76–82]. These stable supramolecular assemblies have enhanced lipophilicity and the ionophore hosts can be utilized as selective carriers in liquid membrane transport [80].

TABLE 6

Fe^{3+}-Siderophore Stability Constants in Aqueous Solution[a]

Siderophore	No.	log β_{110}[b]	pM[c]	Reference
Tricatecholate				
enterobactin	**27**	≅49	35.5	38
Monocatecholate				
chrysobactin		35.4[d]		195
Trihydroxamates				
coprogen	**60**	30.2	27.5	179
ferrichrysin		29.96	25.8	74
ferricrocin		30.4	26.5	179
ferrichrome	**10**	29.07	25.2	74,179
ferrichrome A		32.0		180
ferrioxamine B	**9**	30.60[e]	26.6	19
		30.99[f]		75
ferrioxamine D_1		30.76		74
ferrioxamine D_2		31.40	26.69	112
ferrioxamine E	**11**	32.49[g]	27.7	181
		32.21	27.27	112
Dihydroxamates				
rhodotorulic acid	**12**	62.2[h]	21.8	20
		21.55[b]		
alcaligin	**13**	64.66[h]	23.0	27
		23.5[b]		
α-Hydroxycarboxylates				
mugineic acid	**8**	32.5–33.3		182
rhizoferrin	**7**	25.3	19.7	183
Mixed				
aerobactin[i]	**4**	22.5	23.3	184
alterobactin A[j]		⩾49 (est)[k]		73
azotobactin[j]	**6**	28.1	27.8	185
ferrithiocin[l]	**62**	31.04[m]	22.14	196
parabactin[n]	**22**	≅48		197
pyoverdin PaA[o]	**5**	30.8	27	44

See page 273 for an explanation of footnote symbols.

4.3. Fe-Siderophore Redox Potentials

Siderophores have a much lower affinity for Fe^{2+}_{aq} than for Fe^{3+}_{aq}, and the kinetics for ligand exchange at high-spin Fe^{2+} are much faster than for high-spin Fe^{3+} [83]. Consequently, reduction of the Fe(III)-siderophore complex followed by ligand exchange is a chemically reasonable mechanism for Fe release at the cell membrane or interior.

Redox potentials are important in assessing whether Fe(III/II) reduction is thermodynamically feasible at biological conditions. Electrochemical methods may be used to determine ligation and environmental (e.g., pH, solvent) effects on the redox potential. Stability constants may also be determined by electrochemical methods [69] and the pH dependence of the redox potential can be used to determine the H^+ stoichiometry of the electrode reaction and equilibrium protonation constants for the complex [2].

Table 7 is a compilation of available redox potentials for Fe^{3+}-siderophore complexes. These data illustrate the strong influence of

---

[a]Data collected in aqueous solution at 25°C and $I = 0.1$ M except where noted.
[b]See Eq. (9).
[c]pM = $-\log\,[Fe^{3+}_{aq}]$ when $[Fe^{3+}]_T = 10^{-6}$ M, [siderophore] = 10^{-5} M, pH = 7.4.
[d]$\log \beta_3$ for $Fe^{3+}_{aq} + 3\ L^{2-} \rightleftharpoons FeL_3^{3-}$.
[e]$\log \beta$ for $Fe^{3+}_{aq} + HDFO^{2-} \rightleftharpoons Fe(HDFO)^{+}$.
[f]$\log \beta$ for $Fe^{3+}_{aq} + DFO^{3-} \rightleftharpoons Fe(DFO)$.
[g]20 °C, $I = 0.1$ M.
[h]$\log \beta_{230}$ for $2\ Fe^{3+}_{aq} + 3\ L^{2-} \rightleftharpoons Fe_2L_3$.
[i]Fe(III)-binding groups: two hydroxamic acids and one α-hydroxycarboxylic acid in a linear structure.
[j]Fe(III)-binding groups: one catechol, one hydroxamic acid, and one α-hydroxycarboxylic acid.
[k]$\log \beta_{110}$ indirectly estimated based on EDTA competition experiments at pH 6 and an estimation of the unknown alterobactin pK_a values.
[l]Fe(III)-binding groups: one phenol, one N atom in thiazoline ring; one carboxylic acid.
[m]$\log \beta_{220}$ for $Fe^{3+}_{aq} + 2\ L^{2-} \rightleftharpoons FeL_2^{-}$. Corresponding values for the synthetic analogue norferrithiocin: $\log \beta_{220} = 29.09$; pM = 20.51 [196].
[n]Fe(III)-binding groups: two catechols and one phenol.
[o]Fe(III)-binding groups: two hydroxamic acids and one catechol.

TABLE 7

Fe^{3+}-Siderophore Redox Potentials in Aqueous Solution

Siderophore	No.	$E_{1/2}$ (V vs. NHE)[a]	Reference
Tricatecholate			
enterobactin	**27**	−0.750	187,189,202
Monocatecholate			
2,3-dihydroxybenzoylserine	**61**	−0.350	93
Trihydroxamates			
coprogen	**60**	−0.447	179
ferricrocin		−0.412	179
ferrichrome	**10**	−0.400	188
ferrichrome A		−0.440	188
ferrioxamine B	**9**	−0.468	88,186,187
N,N′,N″-triacetylfusarinine C		−0.468	198
Dihydroxamates			
rhodotorulic acid	**12**	−0.359	2,20
α-Hydroxycarboxylates			
mugineic acid	**8**	−0.102	14
rhizoferrin[b]	**7**	−0.082	183
Mixed			
aerobactin[c]	**4**	−0.336	2,184
ferrithiocin[d]	**62**	−0.166	196
parabactin[e]	**22**	−0.673	98
parabactin A[e]	**24**	−0.400	98
pseudobactin[f]	**43**	−0.482	190
pyoverdin PaA[g]	**5**	−0.510	44

[a]Aqueous solution at pH 7.
[b]Calculuated at pH 7 from differences in K_{app} for Fe^{2+} and Fe^{3+}.
[c]Fe(III)-binding groups: two hydroxamic acids and one α-hydroxycarboxylic acid in a linear structure.
[d]Fe(III)-binding groups: one phenol, one N atom in thiazoline ring, and one carboxylic acid.
[e]Fe(III)-binding groups: two catechols and one phenol. Hydrolysis of the oxazoline ring of ferric *L*-parabactin leads to the open chain threonyl structure of *L*-parabactin A.
[f]Fe(III)-binding groups: one catechol, one hydroxamic acid, and one α-hydroxycarboxylic acid.
[g]Fe(III)-binding groups: two hydroxamic acids and one catechol.

the binding group on the redox potential, with relatively little influence from ligand architecture (although the latter statement is based on limited data). The catechol donor groups produce complexes with the most negative potentials. The potentials for enterobactin (**27**) and parabactin (**22**) at pH 7 are beyond the range for physiological reductants. The hexadentate trihydroxamic acids have more positive potentials centered around −0.430 mV (vs normal hydrogen electrode [NHE]). The α-hydroxycarboxylates exhibit the most positive reduction potentials listed in Table 7. These are phytosiderophore structures, and it has been suggested that reduction of Fe(III)-chelate complexes by reducing agents excreted at the root is involved in Fe uptake by plants [84,85]. The siderophore structures with mixed metal-binding groups listed in Table 7 exhibit redox potentials which are reflective of a ligand field strength which is a weighted average of the individual binding groups involved.

Siderophores with denticity less than six tend to have more positive Fe(III/II) redox potentials than their hexadentate counterparts. The FeL_3 complex with the bidentate monomer 2,3-dihydroxybenzoylserine (**61**) has a redox potential shifted 400 mV positive from its tri-

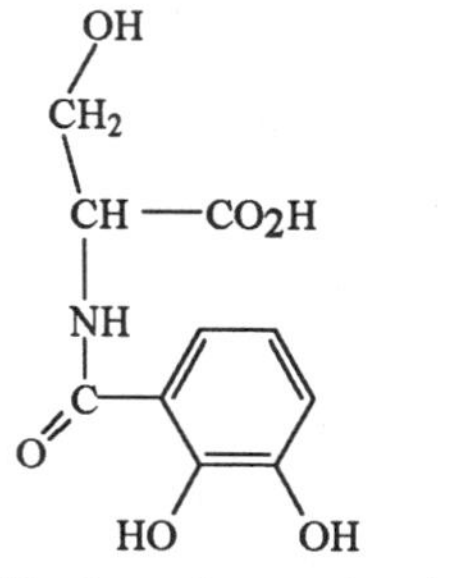

2-3-dihydroxybenzoylserine
61

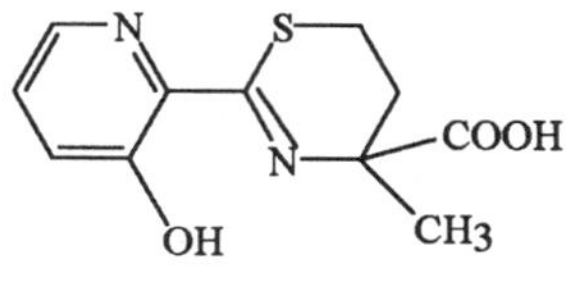

ferrithiocin
62

meric analogue, enterobactin (**27**). Tridentate ferrithiocin (**62**) has a more positive potential than the other hexadentate mixed-donor siderophores, although direct comparison with a hexadentate analogue is not possible. The Fe^{3+} complex of the tetradentate dihydroxamate rhodotorulic acid (Fe_2L_3: L = **12**) has a redox potential shifted more positive than the trihydroxamic acids listed in Table 7. This suggests a redox mechanism for this siderophore complex. Rhodotorulic acid has been shown to donate Fe to the *Rhodotorula pilimanae* cell membrane external surface [86]. This organism cannot utilize Fe from a trihydroxamic acid siderophore, which suggests that the more positive redox potential for the rhodotorulic acid complex may point to Fe release at the cell surface by a reductive mechanism [87]. This conclusion should be made cautiously, however, as differences in siderophore carrier-cell receptor interactions and differences in the dynamics of siderophore ligand dissociation influenced by the different Fe complex structures (Fe_2L_3 for dihydroxamates and FeL for the trihydroxamates) may also be involved.

pH can also influence the redox potentials of the siderophores. The redox potential for ferrioxamine B (**9**) increases by more than 300 mV with a decrease in pH to 3.5 [88]. The midpoint potential for reduction at pH 4.5 is about −180 mV, which is in the range of biological reductants. Ferrioxamine B electrochemistry becomes irreversible in this range, probably because of a more facile proton-assisted dissociation of Fe from the less stable Fe(II) complex.

Reports have been made of Fe(III) reduction in hydroxamate siderophores by reductants from cell-free extracts [89–92]. This supports the hypothesis that, at biologically relevant conditions, reductive release of Fe(II) from its hydroxamate siderophore complex may occur, particularly in the presence of an appropriate ligand to trap the Fe(II) released. A model has been proposed for ferrichrome (**10**) and ferrichrome A-mediated Fe(III) uptake in *Ustilago sphaerogena* involving Fe(III/II) reduction both in the cell interior Fe(III)-ferrichrome) and the membrane surface Fe(III)-ferrichrome A) [87].

Although the redox potentials for the Fe^{3+} tricatecholates are clearly too negative for reaction with physiological reductants, this does not eliminate the possibility that redox is involved in the Fe release mechanism. There is the possibility that a transformation of the siderophore ligand occurs while it is bound to the Fe(III) which results in a positive shift in redox potential, or that there is a significant reduction in pH which would also shift the redox potential positive. Either or both of these processes may be operative. The presence of a soluble enterobactin esterase in *Escherichia coli* [3], and the fact that the redox potential of the tris Fe(III) complex of 2,3-dihydroxybenzoylserine (**61**) is shifted 400 mV positive relative to enterobactin (**27**) [93], suggest that the first of these two processes may be operative for enterobactin-mediated Fe(III) delivery to *E. coli*. However, studies with synthetic analogues of enterobactin with no hydrolyzable ester bonds and pH 7 redox potentials that are too negative for physiological reduction [2,3] have also been reported to deliver Fe(III) to *E. coli* and *Bacillus subtilis* [94–97]. This complicates accepting ligand hydrolysis followed by reduction as the in vivo mechanism for enterobactin delivery of Fe(III) to *E. coli*. However, the redox potentials for enterobactin and several of the biomimetic models are very pH sensitive. The redox potential of enterobactin may be as high as + 170 mV at pH 4 [2,3]. Consequently, redox may play an important role in enterobactin-mediated microbial Fe(III) delivery. The coordination chemistry of enterobactin and its relationship to the mechanism for Fe(III) uptake in *E. coli* has been reviewed [1,2,66].

Parabactin (**22**) also has a redox potential which is too negative for physiological reduction (see Table 7). However, hydrolysis of the oxazoline group to produce parabactin A (**24**) shifts the redox potential positive to within the range for physiological reduction, which is consistent with the possibility for redox delivery of Fe in *Paracoccus denitrificans* [98,99].

Although consideration of Fe-siderophore redox potentials at pH 7 (see Table 7) is certainly important, it may not tell the whole story. As discussed here, the environment (e.g., pH, hydrophobicity) can appreciably influence the redox potential. Changes in the ligation of Fe brought about by partial ligand unwrapping, ternary complex formation, and ligand hydrolysis will also influence the redox potential. It has also been stated that it is the redox potential of the {ferrisiderophore oxidoreductase enzyme-ferric siderophore} complex rather than the potential of the free ferric siderophore complex that is important when considering redox as a process involved in Fe release [100].

4.4. Synthetic Models for the Siderophores

Synthetic siderophore mimics are investigated for numerous reasons. These include the development of more effective agents to deliver Fe to plants, physicochemical studies of structure-reactivity relationships, and therapeutic applications. Deferriferrioxamine B (**9**) as the methylsulfonate salt (Desferal) is currently used to treat patients with Fe overload [101,102]. There is an intense interest in improving the efficacy of ferrioxamine B in this application. Also, in the treatment of Fe(III)-overload diseases there is a critical need for effective chelating agents that do not promote microbial growth. Consequently, there is an interest in developing synthetic siderophore analogues that can act as therapeutic agents, that can exhibit bacteriostatic effects, and that can serve as microbial growth promoters.

4.4.1. *Synthetic Hydroxamic Acid Siderophore Models*

Van der Helm and coworkers have defined a set of absolute structure-stability criteria for the hydroxamate siderophores based on x-ray crystal structures of synthetic monohydroxamates and their Fe(III) and Ga(III) complexes [10,103–107]. In an absolute sense, the most stable hydroxamate complex is the one with (1) the shortest C−N bond; (2) the smallest difference in C=O and N−O distances; (3) the smallest difference in $Fe-O_C$ and $Fe-O_N$ distance; and (4) the smallest difference in the distances between the Fe(III) ion and the planes of the oxime and carbonyl O atoms [103]. These criteria are consistent with the following resonance structure view of the Fe(III)-hydroxamate chelate ring. Sta-

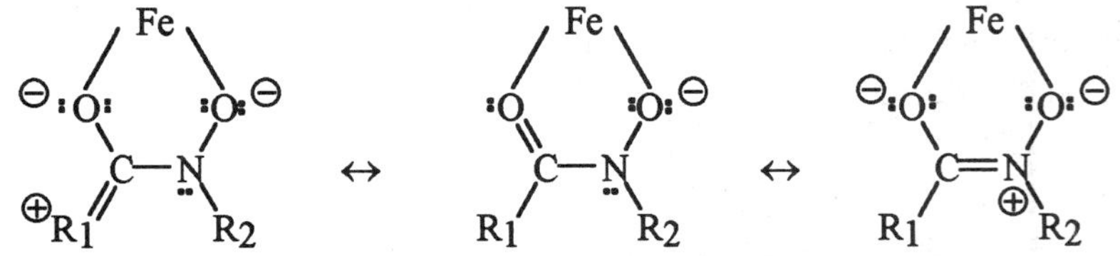

63 **64** **65**

bility is enhanced when the carbonyl C (R_1) and N (R_2) substitutents will stabilize a positive charge build-up on N and a negative charge build-up on the carboxyl O atom. This translates into an inductive electron-releasing group on the N atom (e.g., an alkyl group) and an inductive or resonance electron-releasing group on the C atom. The predictions of absolute Fe(III)-hydroxamate stability based on these structural criteria are consistent with observed stability constants (β) for synthetic monohydroxamic acid complexes of Fe(III) in aqueous solution [108–111].

Application of the absolute structure-stability criteria to a comparison of Fe(III), Al(III), and Ga(III) complexes of synthetic and natural hydroxamate siderophores suggests an absolute stability order Al(III) > Ga(III) > Fe(III). This is not consistent with equilibrium measurements in aqueous solution [8]. However, when one takes into account the differences in free energies of formation of the aqueous ions ($-\Delta G_f^0$; Al(III) > Ga(III) > Fe(III) [11]) the absolute structure-stability criteria are consistent with the observed decrease in affinity of the hydroxamates for the +3 metal ions in aqueous solution in the order Fe(III) > Ga(III) > Al(III).

4.4.1.1. Ferrioxamine models. A series of new cyclic un-"natural" siderophores has been prepared by the directed fermentation of a *Streptomyces olivaceus* strain by supplementing the production medium with a variety of unnatural diamines and diamino acid precursors [112]. Ten new cyclic deferriferrioxamine structures were produced and characterized; eight trihydroxamic acids and two dihydroxamates. In all cases, the stability of the Fe(III) complexes is lower than that observed for deferriferrioxamine E (**11**), the natural siderophore produced by this bacterial strain.

Ferrioxamine B has been derivatized by coupling with an additional catechol (**66**–**68**) or hydroxypyridinone (**69**) moiety with a high Fe(III)-binding affinity [113]. Addition of a catechol group increases the

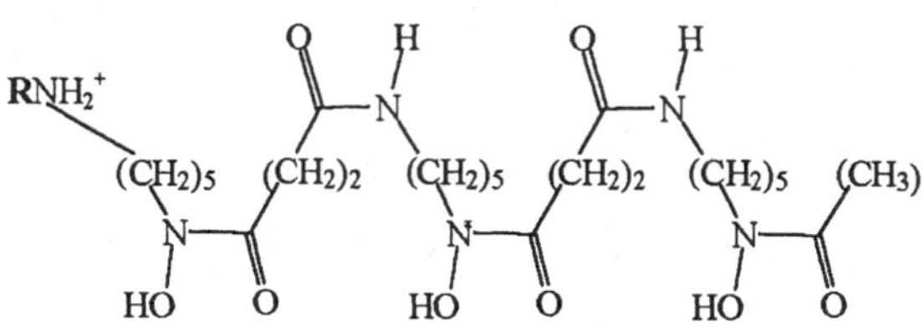

deferriferrioxamine B derivatives

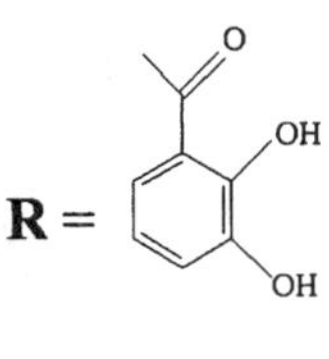

DFOCAM
66

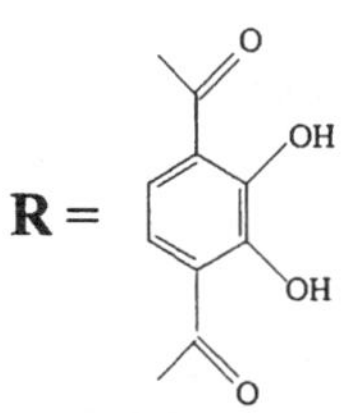

DFOCAMC
67

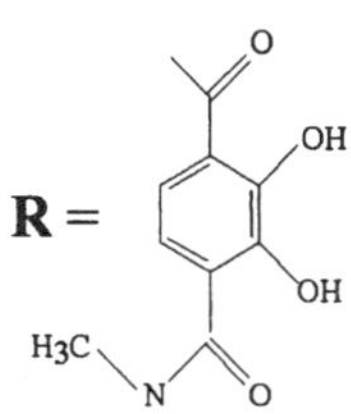

DFOMTA
68

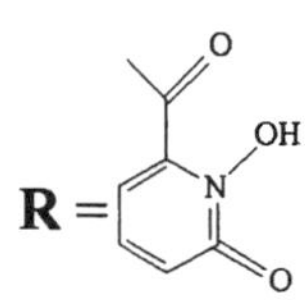

DFOHOPO
69

TABLE 8

Fe^{3+} Stability Constants with Ferrioxamine B Derivatives

Ferrioxamine B derivative	No.	log β_{110}	pM
Deferriferrioxamine B	**9**	30.60	26.6
DFOCAM	**66**	35.4	28.2
DFOCAMC	**67**	34.9	26.7
DFOMTA	**68**	34.8	28.5
DFOHOPO	**69**	30.7	26.7

Source: Compiled from [113].

Fe(III) affinity of ferrioxamine B (Table 8). Owing to differences in acidities of the functional groups and the inability of Fe(III) to expand its coordination number to 8, coordination for **66**, **67**, and **68** changes from trishydroxamate coordination at low pH to monocatechol bishydroxamate coordination at high pH. For **69**, the change is from monohydroxypyridine bishydroxamate coordination at low pH to trishydroxamate coordination at high pH. The real benefit from use of these derivatives may be in their increased kinetic efficiency in removing Fe from other carriers [114]. The catecholate derivatives, as mixed hydroxamate/catecholate ligands, are useful as pyoverdin (**5**) models and their log β and pM values are comparable to this siderophore (compare Tables 6 and 8).

Akiyama and coworkers have synthesized two linear and one cyclic peptide-based models for the deferriferrioxamines through fragment condensation of Ala-Ala-β(OH)Ala and Ala-β(OH)Gly-Ala units [115]. The resulting trihydroxamic acids have a cyclic (**70**) structure with eight-atom spacers between hydroxamate groups, and linear structures with eight (**71**) and seven (**72**) atom spacers (Table 9) [115,116]. Tris chelate complexes with Δ-C-*cis,cis* configuration were formed in solution in all cases. Addition of a single methylene group to **72** greatly increases stability for **71**. However, an Fe(III) complex with **71** is still ~10^3 less stable than ferrioxamine B (**9**), which has nine atoms between hydroxamate groups. A factor of 10 is gained in stability on changing to

TABLE 9

Fe^{3+} Stability Constants
with Peptide Based Trihydroxamic Acids

Ligand	No.	log β_{110}[a]	$K_{Fe(HL)}$[b]
-[Ala-Ala-β(OH)Ala-]$_3$- (cyclic)	**70**	28.3	4.14×10^4
Ac-[Ala-Ala-β(OH)Ala-]$_3$OH	**71**	27.4	1.44×10^3
Ac-[Ala-β(OH)Gly-Ala]$_3$OH	**72**	17.0	2.24×10^6

[a]Determined by competition experiments with EDTA. T = 25°C; I = 0.1 M (KCl).
[b]$K_{Fe(HL)}$ for the equilibrium reaction (22); M^{-1}.
Source: Data from [115,116].

cyclic structure **70**. This may reflect some small degree of preorganization, or it may be a result of the slightly higher basicity of the hydroxamate groups in **70**. In any event, the Fe(III) complex of **70** is significantly less stable than that of ferrioxamine E (**11**; by ~ 10^4) and the un-"natural" cyclic ferrioxamines prepared by directed fermentation [112].

Two retrohydroxamate ferrioxamine models were coupled to mono-6-amino-6-deoxy-β-cyclodextrin by an amide linkage (**73**,**74**) to produce water-soluble artificial siderophores [117]. Fe(III) stability constants in aqueous solution (log β_{110} Fe-**73** = 32, and log β_{110} Fe-**74** = 30) were

73 Me_3C-O-C(O)[-NH-$(CH_2)_5$-C(O)-N(OH)-$(CH_2)_2$-C(O)-]$_3$-NH-β-CD

74 MeC(O)[-NH-$(CH_2)_5$-C(O)-N(OH)-$(CH_2)_2$-C(O)-]$_3$-NH-β-CD

estimated from EDTA exchange reactions. The relative higher stability of Fe-**73** is proposed to be a result of a structure with the Me_3C-O-moiety inserted into the β-cyclodextrin cavity. EDTA ligand exchange kinetics for Fe-**73** and Fe-**74** are essentially equivalent. Both Fe-**73** and Fe-**74** showed growth promotion for *Aureobacterium flavescens*, a mutant auxotroph for hydroxamate siderophores which does not synthesize such siderophores [117].

4.4.1.2. Ferrichrome models. A series of three ferrichrome models (**55–57**; see Sec. 3.3) based on *L*-alanyl-*L*-alanyl-β-(*N*-hydroxy)alanine tripodal hydroxamic acids have been prepared (Table 10) [58]. Stability constants with Fe(III) are estimated to be comparable with ferrichrome

TABLE 10

Fe^{3+} Stability Constants with Synthetic Ferrichrome Models

Ligand	No.	log β_{110}[a]	pM	Relative exchange rate[b]	Reference
Ferrichrome	**10**	29.07	25.2	1	74,179
N[CH_2C(O)-Ala-Ala-β(OH)Ala-OMe]$_3$	**55**	29[c]		4	58
N[CH_2CH_2NH-Ala-Ala-β(HO)Ala-Ala-Ac]$_3$	**56**	28[c]		44	58
N[CH_2CH_2NH-Ala-β(HO)Ala-Ala-Boc]$_3$	**57**			18	58
	75	28.11	23.3[d]		118
	76	26.42	21.5[d]		118
	77	29.12	22.1[d]		118
TRENDROX	**35**	32.9	27.8		40

[a]T = 25°C in aqueous solution.
[b]Relative pseudo-first-order rate constants obtained at a single [EDTA] for the exchange of FeL with EDTA at pH 5.4, I = 0.1 and 25°C in water.
[c]Estimated value from EDTA exchange reactions at pH 5.4, I = 0.1 in water at 25°C.
[d]100% excess ligand; pH = 7.4; [Fe(III)] = 10^{-6} M.

but with somewhat higher kinetic lability at pH 5.4 relative to exchange with EDTA.

Martell and coworkers have investigated a series of trihydroxamic acid ligands with a carbon-anchored tripodal structure (**75**–**77**) [118].

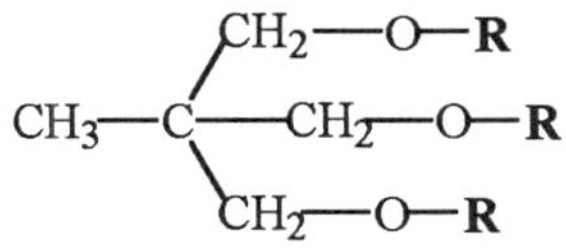

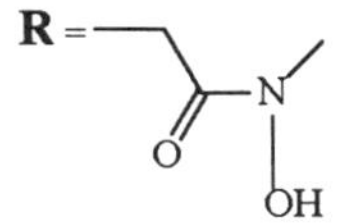

75

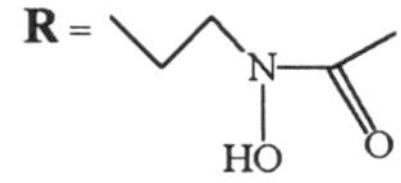

76

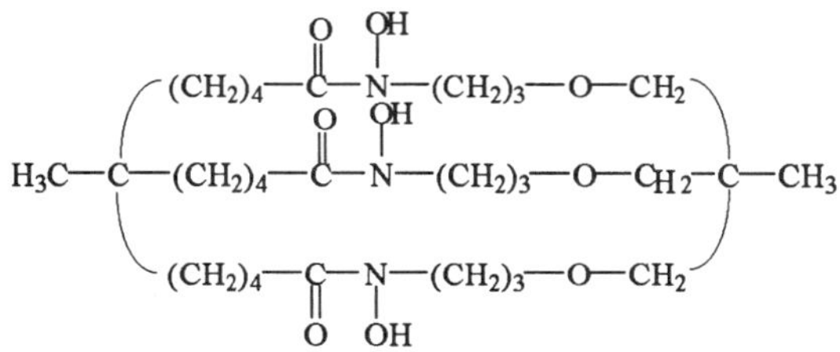

77

Fe(III) complex stabilities approach that of ferrichrome (Table 10). Fages and coworkers have modified the synthesis of **75** to include a pyrene chromophore bound to the terminal N atom for fluorescent detection of Fe(III) (see **49**; Sec. 3.2) [55]. When the carbon-anchored tripodal structure is closed at both ends to produce a cryptand (**77**), stability increases, presumably due to ligand preorganization. However, although **77** has a log β_{110} equivalent to ferrichrome, the pM value is 3 log units lower (Table 10) [118]. A more effective tripodal backbone for trihydroxamic acid structures appears to be that based on TREN. The hexadentate TRENDROX ligand (**35**) has a significantly higher log β_{110} and pM value than ferrichrome (Table 10) [40].

Santos and coworkers have proposed **78** as a ferrichrome mimic

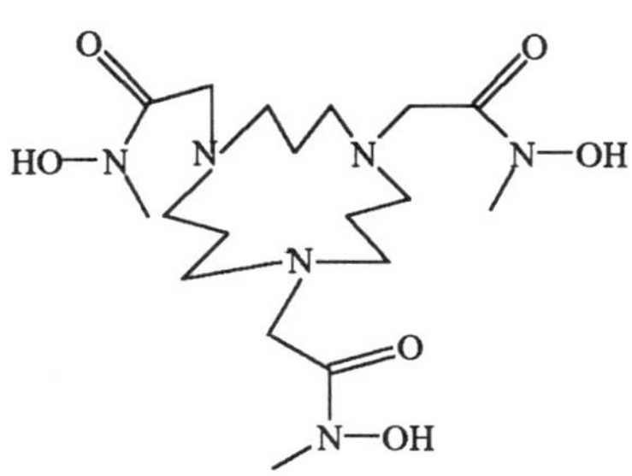

78

[119]. A hexacoordinate complex $FeHL^+$ with one ring N atom protonated is the major species over the pH range 5–9 and has a log β_{111} of 24.2. Since neither a β_{110} nor a pM value are available, direct comparison with ferrichrome is not possible. The $E_{1/2}$ for $FeLH^+$ is −290 mV (pH 7 and 20/1 ligand/Fe), which is more positive than ferrichrome and the other trihydroxamic acids listed in Table 7. Log β_{112} for $Fe(II)LH_2^{2+}$ was determined from cyclic voltammetry as 19.9. Ligand **78** is biologically active and will deliver Fe to *Anthrobacter flavescens* and *E. coli*.

4.4.1.3. Other synthetic dihydroxamic acid siderophore models. Chelation of Fe(III) in the presence of excess dihydroxamic acid ligand will produce Fe_2L_3. Such species can exist in a singly bridged structure (**79**),

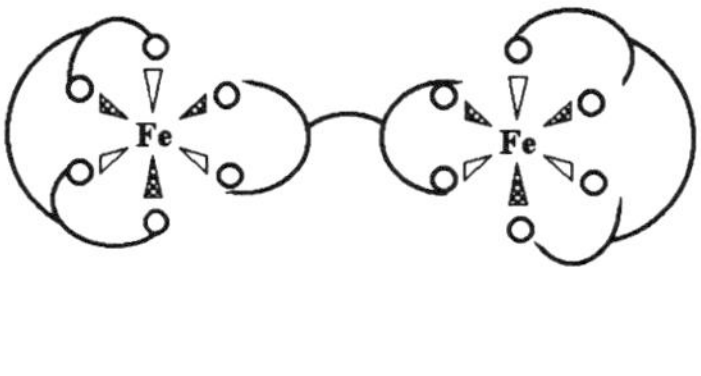

79

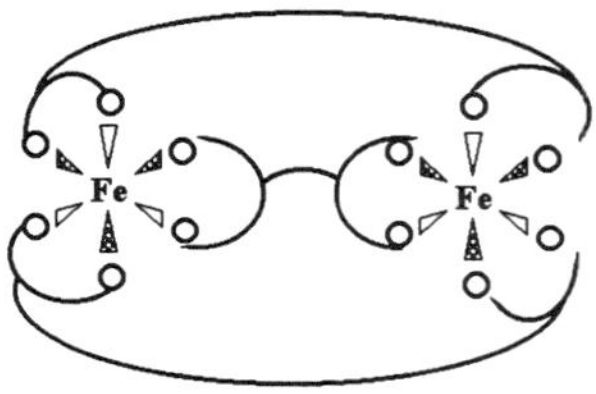

80

as established for $Fe_2(alcaligin)_3$ [27], or as a triply-bridged helicate structure (**80**), as established for an open-chain synthetic dihydroxamic acid [120]. Crystallographic data are not available for Fe_2(rhodotorulic acid)$_3$, although by analogy it is often assumed that a helicate structure applies. This still leaves the structure of a linear dihydroxamate complex in the condition of limited ligand availability (or when the Fe is in the process of dissociating from its ligand) undetermined as $FeL(H_2O)_2^+$ or $Fe_2L_2(H_2O)_4^{2+}$. The 1:1 Fe:rhodotorulic acid complex structure was established to be monomeric (Fe(rhodotorulic acid)$(H_2O)_2^+$) in aqueous solution at low pH by electrospray mass spectroscopy (ESMS) [62].

For the series of dihydroxamic acids $RN(OH)C(O)\text{-}(CH_2)_n\text{-}C(O)N(OH)R$, ($R = CH_3$, H (**59**); see Sec. 3.4) log β_{110} for $FeL(H_2O)_2^+$ is 22 when n = 8 and $R = CH_3$, and 20 when n = 8 and R = H [121,122]. This is comparable to that found for rhodotorulic acid (**12**) (see Table 6), and the difference in stability between the two synthetic ligands is consistent with *N*-alkyl group stabilization of resonance form **65**. Log β_{220} for the di-Fe species $Fe_2L_2(H_2O)_4^{2+}$ (L = **59**, n = 2 or 4, $R = CH_3$) is **44**, which is the same stability per Fe center as the monomeric structure. Hydrolysis measurements of coordinated water in $FeL(H_2O)_2^+$ and $Fe_2L_2(H_2O)_4^{2+}$

show an enhanced second pK_a value for the dimer, consistent with the formation of a hydroxo or oxo bridge [122].

Santos and coworkers report the synthesis of a new hydroxamic acid piperazine-1,1′-bis(*N*-methylacetohydroxamic acid) (**81**), which forms a 3/2 complex with Fe(III) (log β_{230} = 61.7 at I = 1.0, 25°C) [123]. Stability is comparable to that observed for rhodotorulic acid (**12**) (see Table 6). Electrochemical measurements suggest that the two Fe centers have the same redox potential, are noninteracting, that reduction coincides with monoprotonation, and that dissociation of the Fe(II) species $Fe_2L_3H^-$ occurs rapidly. The stability of the Fe(II) complex was estimated from voltammetric measurements (log β_{231} (Fe(II)) = 32.6 at I = 1.0, 25°C). Expansion of the six-membered ring platform of **81** by a

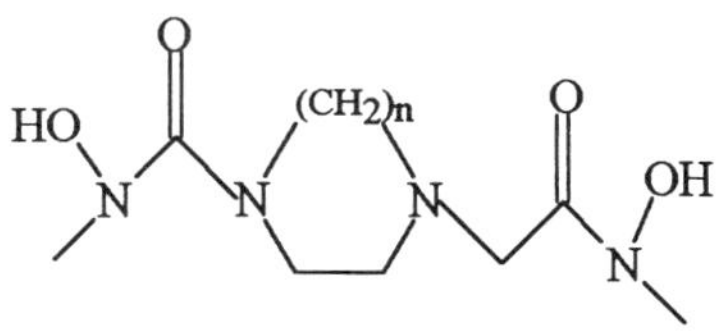

n = 2 n = 3
81 **82**

-CH_2- group to form **82** relieves the steric strain associated with the boat form of the piperazine ring and forms a more stable Fe(III) complex (log β_{230} = 65.4 at I = 1.0, 25°C) and Fe(II) complex (log β_{231} Fe(II) = 36.2 at I = 1.0, 25°C) [124].

4.4.2. *Synthetic Catechol Siderophore Models*

Raymond's laboratory has synthesized a number of hexadentate catechol models for enterobactin based on an aryl or tripodal triamine (TREN) platform [125–127]. Stability constants and pM values are listed in Table 11 for selected examples. The $(Et)_3$MECAM structure (**84**) has the highest log β_{110} and approaches that for enterobactin (see Table 6). The most efficient Fe(III) scavenger of this synthetic series, however, is the closed cryptand derivative of TRENCAM (**85**), bicapped-TRENCAM (**86**), which exhibits a pM value that approaches enterobactin. Consideration of this series of enterobactin mimics provides some insights into the role of ligand predisposition on Fe(III)-siderophore complex stability. Comparison of enterobactin (**27**) (see Table 6) with linear entero-

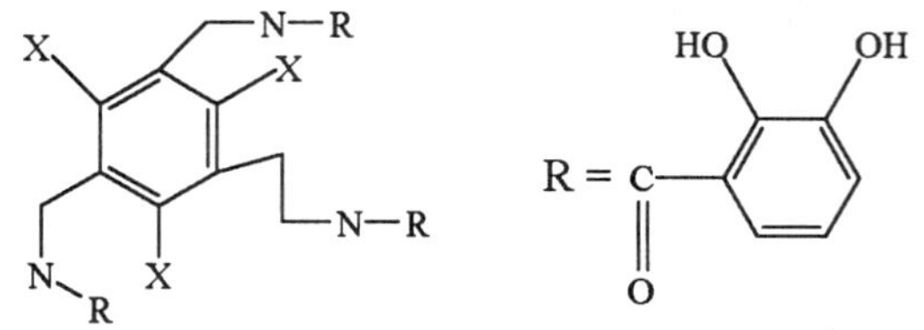

X = H	MECAM **29**
$X = CH_3$	Me_3MECAM **83**
$X = C_2H_5$	Et_3MECAM **84**

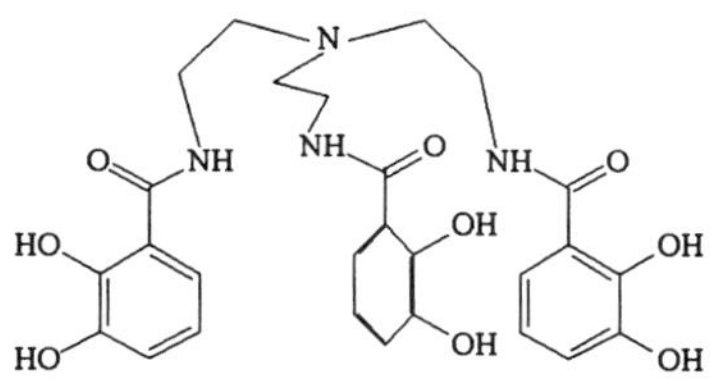

TRENCAM
85

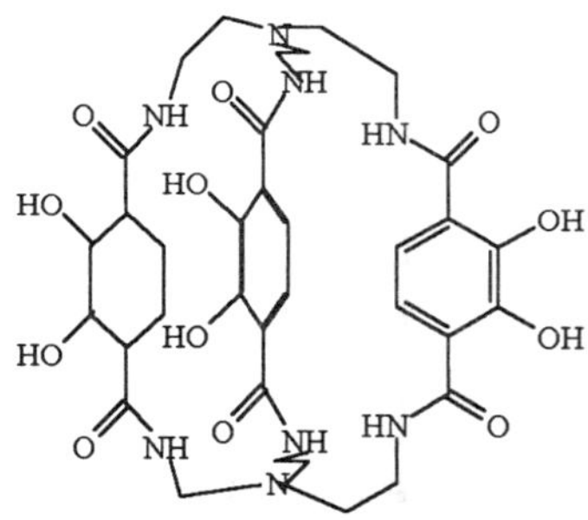

bicapped TRENCAM
86

TABLE 11

Fe^{3+} Stability Constants for Synthetic Enterobactin Models[a]

Ligand	No.	$\log \beta_{110}$	pM	Reference
MECAM	**29**	43	29.1	125,126
Me_3MECAM	**83**	43		127
Et_3MECAM	**84**	47		126
Linear enterobactin	**28**	43		126
TRENCAM	**85**	43.6	27.8	125
Bicapped TRENCAM	**86**	43.1	30.7	125

[a]Aqueous solution at 25°C.

bactin (**28**) (Table 11) shows a 10^6 increase in stability for the more conformationally predisposed enterobactin [126]. The importance of limiting the conformational space of a chelator is illustrated in the 10^4 increase in log β_{110} on going from MECAM (**29**) to the more constrained $(Et)_3$MECAM (**84**) (Table 11) [126].

Akiyama and coworkers have synthesized a lysine-based tricatecholate (**42**; Table 2) model for enterobactin with a log β_{110} = 46, estimated from EDTA exchange equlibria at pH 5.4 and I = 0.1 [39]. Tse and Kishi synthesized enterobactin models based on a platform derived from the mono-orthoformate of *Scyllo*-inositol (**87, 88**) [128]. The R and

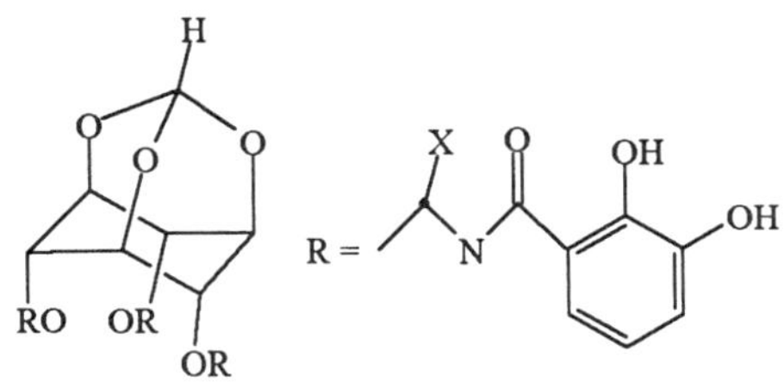

X = H
87

X = C_6H_5
88

S chiral structures (**88**) show a preference for Λ and Δ Fe(III) coordination with an estimated log β ~ 48, close to that of enterobactin. This estimation is two orders of magnitude higher than for the more flexible achiral analogue (**87**). Apparently the steric bulk of the phenyl groups predispose the catechol groups for chelation in **88** relative to **87**. Ethylenediamine-bis(3-aminocatechol) (ethylcam) has been synthesized and its $Fe_2(ethylcam)_3^{6-}$ and $Fe_2(ethylcam)_2(OH)_2^{4-}$ complexes proposed as models for bis(catecholamide) siderophores [129]. The coordinatively saturated $Fe_2(ethylcam)_3^{6-}$ has a triply bridged helicate structure.

4.4.3. *Other Synthetic Siderophore Models*

It has been suggested that the hydroxypyridinones, which are produced by several microorganisms, are also siderophores [3,130]. There is considerable interest in using these ligands as oral therapeutic agents in the treatment of Fe(III) overload associated with thalassemia and sickle cell anemia [131]. Table 12 compares log β_n values for a number of hydroxypyridinone derivatives (Fig. 12) along with acetohydroxamic

TABLE 12

Fe^{3+} Stability Constants for Bidentate Chelators[a]

Ligand	No.	log β_1	log β_2	log β_3	pM[b]	Reference
3H4P	**89**	14.26	25.73	34.91		133
		14.2	25.8	35.1		199
MHP	**90**	15.4[c]	27.43[c]	37.2[c]		135
DMHP	**91**	15.1	26.61	35.88	19.4	130
		14.92[c]	27.15[c]	37.2[c]		135
DEHP	**92**	15.2	26.96	36.8		132
EMHP	**93**	15.33[c]	27.83[c]	37.7[c]		135
PMHP	**94**	15.34[c]	27.91[c]	37.7[c]		135
3H2P	**95**	11.48	21.25	29.26		133
MH2P	**96**	11.8	21.63	29.99		133
EH2P	**97**			32.3		136
DH2P	**98**	10.22	17.70	21.66		133
AHA[d]		11.42	21.10	28.33	12.5	134
Catechol		20.4	35.5	44.9	15.1	134

[a]Aqueous solution at T = 25°C; I = 0.1 M.
[b]100% excess ligand; pH = 7.4; [Fe(III)] = 10^{-6} M.
[c]Aqueous solution at T = 25°C; I = 0.2 M.
[d]Acetohydroxamic acid.

acid and catechol [130,132–136]. The 3-hydroxy-4-pyridinones form more stable Fe(III) complexes than do the 3-hydroxy-2-pyridinones owing to an increased negative charge density on the carbonyl O atom via resonance. This is supported by an analysis of the crystal structures of the free ligands and their tris Fe(III) complexes [12]. Alkyl substitution at the 1 and/or 2 position greatly influences the lipophilicity of the ligand and the Fe(III) complexes but does not significantly influence the stability of the complexes [135]. The 3-hydroxy-4-pyridinone derivatives have β_n values significantly higher than the hydroxamic acid-binding group but are still well below that of catechol. However, based on the single pM value reported, it may be that the 3-hydroxy-4-pyridinone are better Fe(III) scavengers than catechol at environmental conditions [130]. (However, the pM value reported in Ref. 130 was not

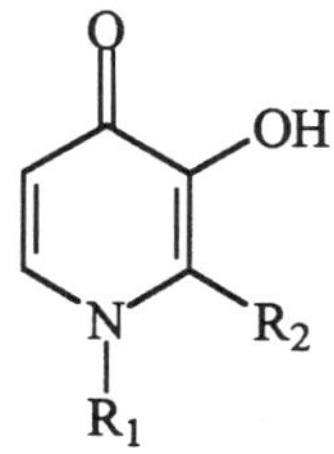

3-hydroxy-4-pyridinones

R_1	R_2		R_1	R_2	
H	H	3H4P **89**	C_2H_5	C_2H_5	DEHP **92**
H	CH_3	MHP **90**	C_2H_5	CH_3	EMHP **93**
CH_3	CH_3	DMHP **91**	C_3H_7	CH_3	PMHP **94**

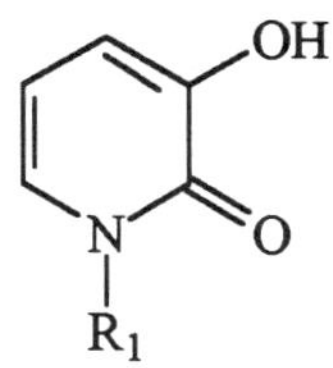

3-hydroxy-2-pyridinones

R_1 = H	3H2P **95**
R_1 = CH_3	MH2P **96**
R_1 = C_2H_5	EH2P **97**

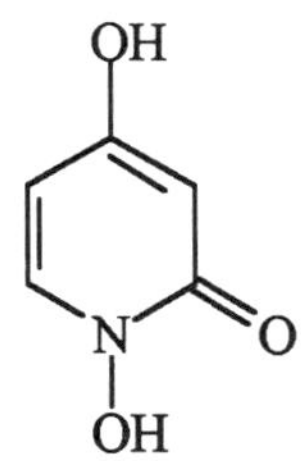

1,2-dihydroxy-2-pyridinone
DH2P
98

FIG. 12 Hydroxypyridinone structures.

calculated at the conditions normally used to define pM.) Certainly the fact that the hydroxypyridinones have a higher acidity than the catechols (and therefore less competition from proton binding) and high Fe(III) affinity makes them attractive environmental Fe(III) chelators.

The hydroxypyridinones provide an unusual example whereby a

hexadentate chelator has a lower Fe(III) affinity than the corresponding bidentate unit. The Fe(III) complex of **97** has a log β_3 = 32.3, whereas the hexadentate analogue built on the TREN backbone (**99**) (Fig. 14, see below) has a log β_{110} = 28.8 [136]. (Although this comparison involves β_3 and β_{110} constants with different units, the analysis is substantiated by pM values.) The crystal structure of Fe-**99** shows that the tertiary amine platform has the N atom lone pair pointed toward the Fe [137]. Molecular mechanics calculations suggest that the lower β value for the hexadentate ligand is due to a lack of ligand preorganization [138].

For many years, numerous laboratories have studied the chelating properties of EDTA and its derivatives. The incorporation of carboxylic acid and phenol moieties into the ethylenediamine backbone structure provides the most efficient Fe(III) binding for this class of chelators, although selectivity for Fe(III) is not always optimum. Fig. 13 is a selective compilation of log β_{110} values for this group of EDTA derivatives [139,140].

Table 13 is a summary of log β_{110} and pM values for a group of chelators based on a cyclic aliphatic amine platform with pendant carboxylic acid, phenol, or hydroxypyridine groups [141,191]. The cyclic platform represents an increase in Fe(III) binding relative to EDTA. The most effective Fe(III) chelation comes from the combination of a cyclic amine with pendant phenol (**105**) or hydroxypyridine donors (**106**) [141].

The TREN (tris(2-aminoethyl)amine) backbone (Fig. 14) has been used as a platform to connect the 2,2′-dihydroxyphenyl chelate moiety with an amine or amide linkage to form hexadentate chelators TBPAS (**113**) and TBPANS (**114**) [142]. The amide linkage allows either salicylate or dihydroxyphenyl chelation to occur. The sulfonyl group increases the acidity of the phenolic groups as well as the water solubility of the ligand. Fe(III) stability constants are presented in Table 14. Data are also included for comparison with the corresponding monomer unit (as ΔpM), as well as tricatecholate (TRENCAM, **85**) and trihydroxamate (TRENDROX, **35**) ligands which are also based on the TREN (trisamide) backbone.

An investigation of the Fe(III) complexes of TBPAS (**113**) and TBPANS (**114**) [142] leads to a consideration of several structural effects: complexation involving a seven-membered ring, differences in conformational flexibility, and differences in total electrostatic charges

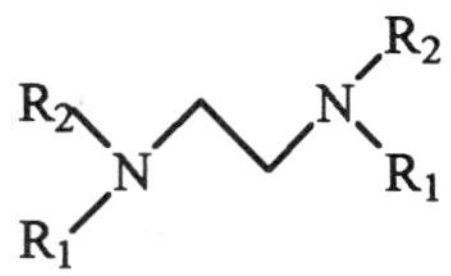

R_1	R_2	Ligand	Log β_{110}	Reference
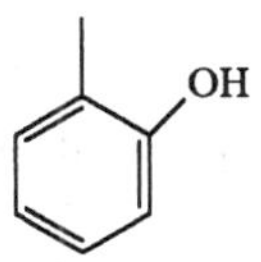	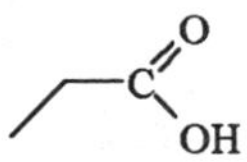	HPED, **100**	31.8	137
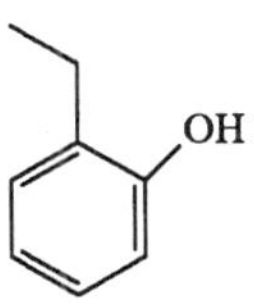	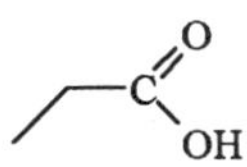	HBED, **101**	39.01	137
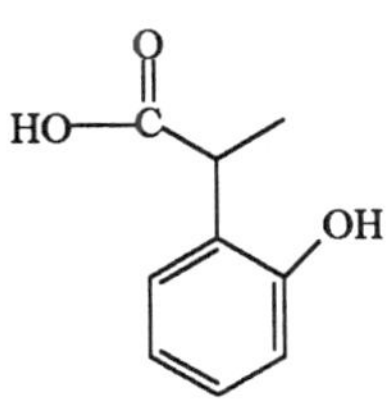	H	EHPG, **102**	35.54 (racemic)	137
			33.28 (meso)	137
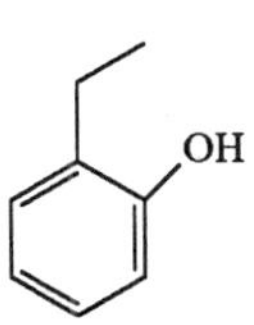	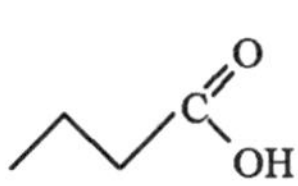	HBEP, **103**	31.16	138

FIG. 13 EDTA derivatives and their Fe(III) complex stability constants.

from the TREN anchor (depending on whether bonding to the biphenol monomer occurs through a CO or CH_2 group). Judging from log β_{110} values, the additional flexibility of an amine connecting link rather than an amide link provides for a more stable complex (log β_{110} (TBPANS) > log β_{110} (TBPAS)) (Table 14). However, owing to differences in pK_a values for the two ligands, their pM values are nearly equal. Consideration of ΔpM illustrates that complexation by a hexadentate ligand synthesized from bidentate units enhances Fe(III) binding. The largest effect is seen for TBPAS over TBPANS. At pH 7.4 where the pM values were calculated, the secondary amine links are likely to be protonated, which would make the hexadentate ligand more conformationally rigid and provide electronic repulsion with the Fe^{3+} ion. Data in Table 14 illustrate that tribiphenolates (**113, 114**) are less effective than tricate-

TABLE 13

Fe^{3+} Stability Constants with Cyclic Aliphatic Amine Ligands Containing Pendant Donor Groups[a]

Ligand	No.	Log β_{110}	pM[b]
NOTA[c]	**104**	28.3	23.7
TACN-Me_2HB[c]	**105**	51.3[d]	33.3[d]
TACN-MeHP	**106**	50.0	39.3
DOTA	**107**	29.4	23.3
TRITA	**108**	27.5	21.3
TETA	**109**	26.5	20.9
N_3O-Ac	**110**	26.8	22.4
N_3O_2-Ac	**111**	19.82	16.1
N_3O_3-Ac	**112**	22.5	19.0

[a]Solvent: water; T = 25°C; I = 0.1 M.
[b]100% excess ligand; pH = 7.4; [Fe(III)] = 10^{-6} M.
[c]Ref. [191].
[d]Solvent: ethanol:water (75%:25%); T = 25°C; I = 0.1 M.
Source: Compiled from [141] and [191].

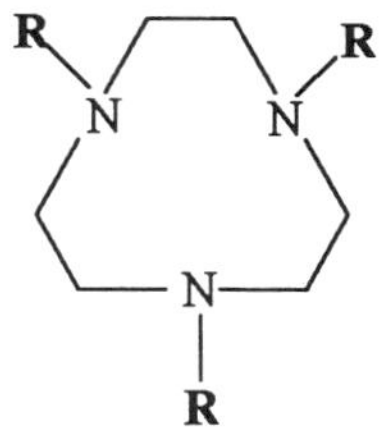

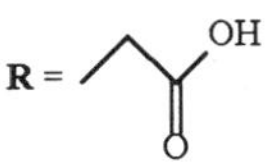

NOTA
104

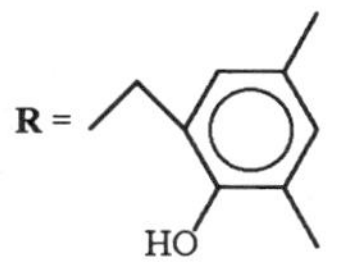

TACN-Me_2HB
105

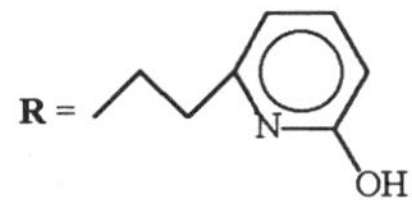

TACN-MeHP
106

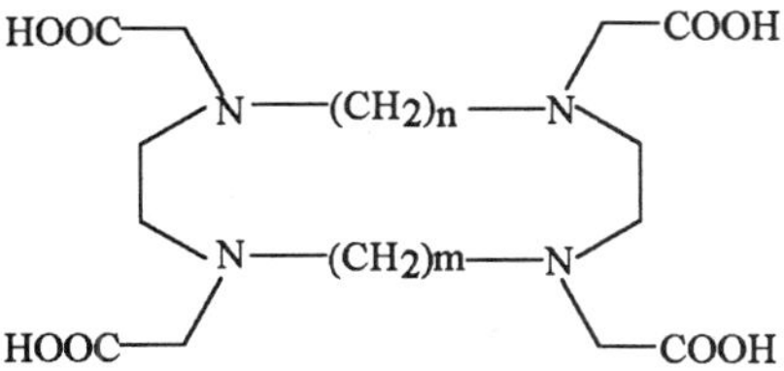

	n	m
DOTA **107**	2	2
TRITA **108**	2	3
TETA **109**	3	3

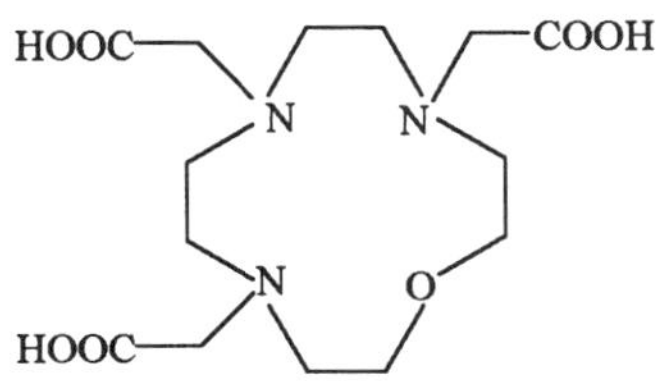

N_3O-Ac
110

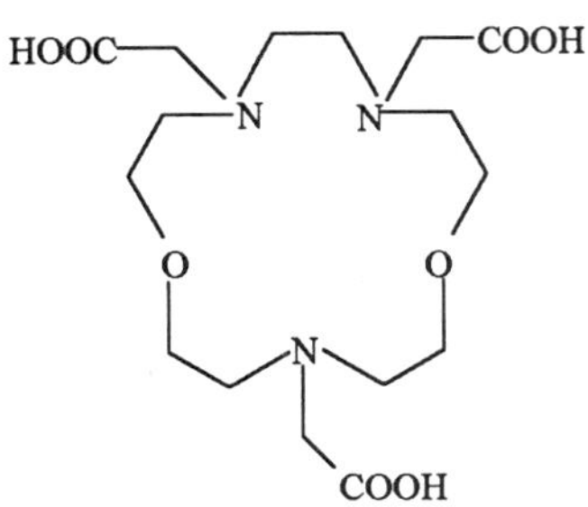

N_3O_2-Ac
111

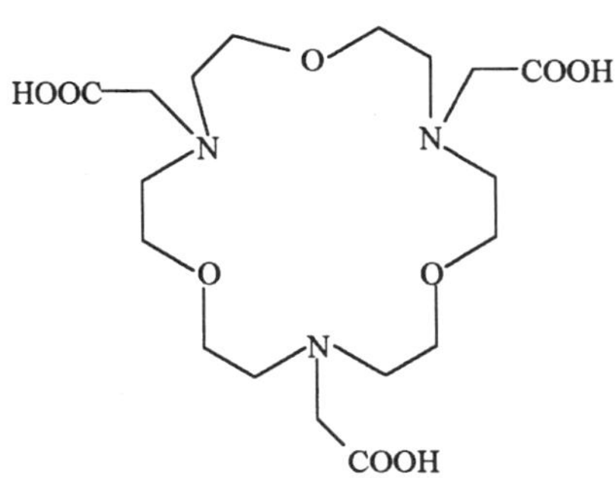

N_3O_3-Ac
112

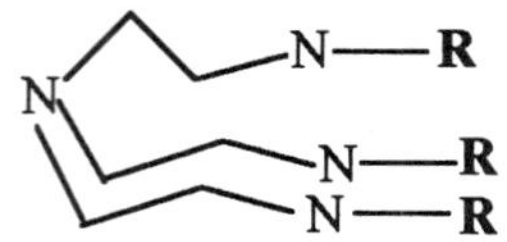

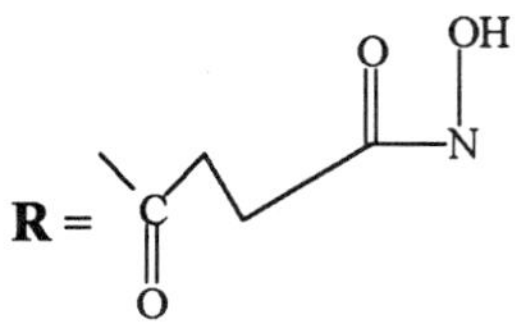

TRENDROX
35

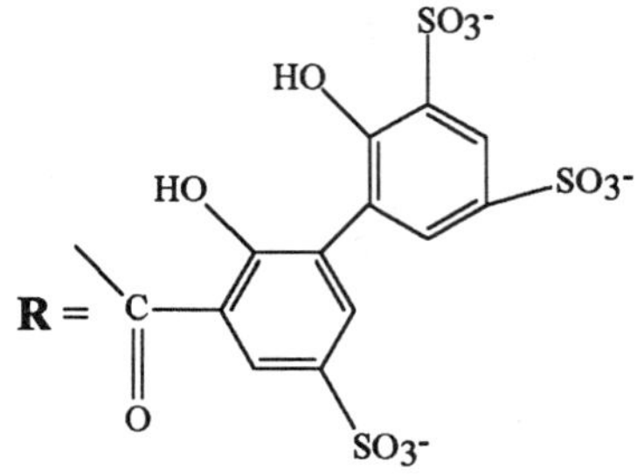

TBPAS
113

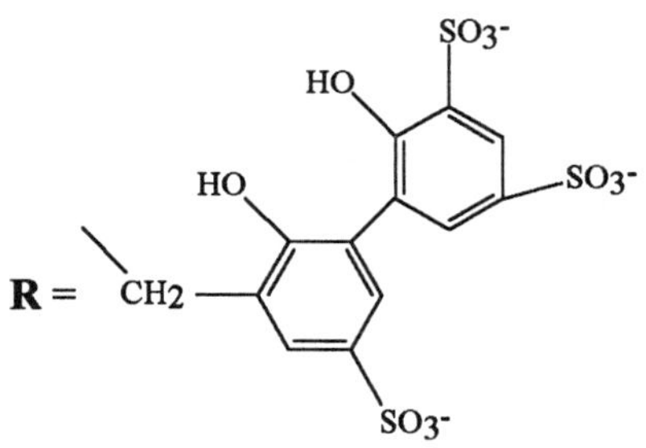

TBPANS
114

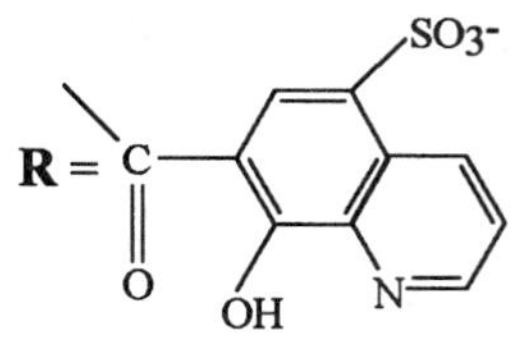

O-TRENSOX
115

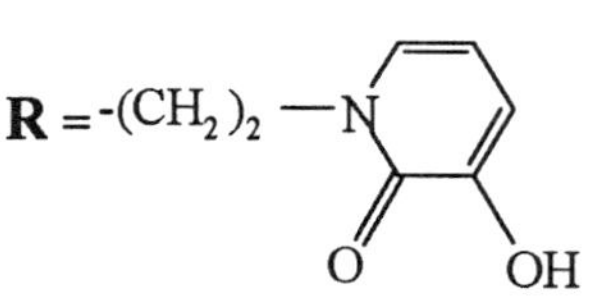

CP130
99

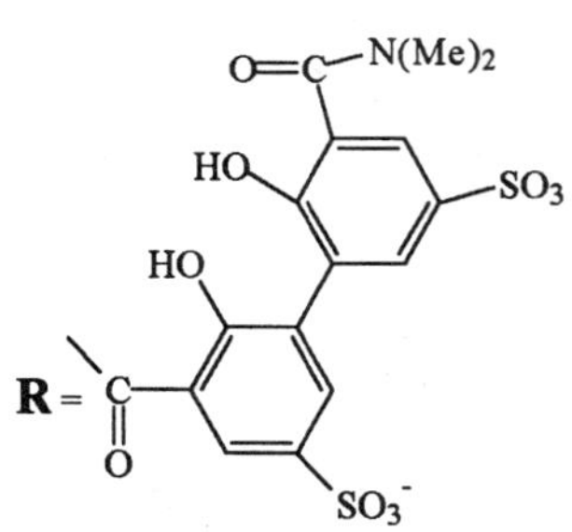

TBPDAS
116

FIG. 14 Synthetic Fe(III) chelators based on the TREN backbone.

TABLE 14

Fe^{3+} Stability Constants with Ligands Based on the TREN Backbone

Ligand	No.	log β_{110}	pM	ΔpM[a]	Reference
TBPANS	**114**	31.8	22.1	6.8	142
TBPAS	**113**	28.7	22.8	10.1	142
O-TRENSOX	**115**	30.9	29.5		143
N-TRENSOX	**37**		21.6		143
TRENCAM	**85**	43.6	27.8	11.8	200
TRENDROX	**35**	32.9	27.8	14.5	40

[a]Difference in pM values between the hexadentate trimer ligand and the corresponding bidentate monomer unit; ΔpM = pM(tripod) − pM(corres. monomer) [34].

cholate (**85**) and trihydoxamate (**35**) ligands built on the TREN backbone. This may be due to the less efficient chelation provided by the seven-membered ring of the 2,2′-dihydroxybiphenyl moiety in contrast with the five-membered ring formed from the catecholate and hydroxamate moieties. Recent data show that TBPAS and TBPANS act as tetradentate ligands and that one pendant arm is free [41].

The unsymmetrical 8-hydroxyquinoline bidentate chelate moiety can be connected to the TREN backbone in either of two ways to give the potentially hexadentate ligands *O*-TRENSOX (**115**) (Fig. 14; Table 14) and *N*-TRENSOX (**37**) (Fig. 4; Table 14) [143]. The ligands are sulfonated in order to increase acidity and water solubility. The Fe(III) complex of *O*-TRENSOX appears to exhibit highly promising properties for plant nutrition [144] and the free ligand may be suitable for Fe chelation therapy [145]. As may be expected, connection of the 8-hydroxyquinoline moiety to the TREN backbone provides a more stable complex than the 2,2′-dihydoxyphenyl moiety. Based on pM values, *O*-TRENSOX (**115**) is a better Fe(III) ligand than is TBPAS (**113**). *O*-TRENSOX also exhibits a higher affinity for Fe(III) than *N*-TRENSOX. This may be due to preorganization of the *O*-TRENSOX ligand for Fe(III) complexation due to H-bonding of the *ortho* amide proton and the phenolate group; such preorganization is not possible for *N*-TRENSOX. The Fe(III) *O*-TRENSOX complex exhibits a relatively positive quasireversible redox potential (+0.087 V; pH 4.2–10.5) [143].

The Fe(III) *O*-TRENSOX chelate shows promise in applications as an artificial siderophore [144]. Since Fe uptake by plants may involve extracellular reduction of Fe(III) [146,147], the affinity of a synthetic carrier for Fe(II) may also be important in this application. *O*-TRENSOX has a relatively high Fe(II)-binding constant (log β_{110} (Fe(II)) = 19.3; pM = 17.9) [143], which is higher than the log β_{110} (Fe(II)) for EDTA (14.3), citrate (4.2), and mugineic acid (**8**) (10.1) [148]. The efficiency of *O*-TRENSOX in delivering Fe to plants may be illustrated by the relative rates of Fe uptake by ferritin in maize: Fe(III)citrate > Fe(III)*O*-TRENSOX > Fe(III)EDTA [144].

5. IRON(III) UPTAKE AND RELEASE MECHANISM: KINETICS IN SOLUTION

Siderophore selectivity and sensitivity to low Fe concentrations is achieved by a high and specific stability constant (β) for Fe^{3+} chelation. However, selective transport to the cell is only part of the process of microbial Fe bioavailability. Deposition at the cell surface or cell interior involves a ligand exchange reaction which may or may not be preceded by Fe^{3+} reduction. In this section, we address aspects of the kinetics of ligand exchange reactions at a Fe^{3+}_{aq} center which are relevant to siderophore-mediated microbial Fe transport. The larger topic of the kinetics and mechanisms of Fe(III) reactions in aqueous solution has been reviewed recently [149].

5.1. Reactivity at Aquated Fe^{3+}

5.1.1. *Initial Chelation Step*

In controlled laboratory conditions in aqueous solution at low pH, there are three major Fe(III) species present whose relative abundance depends on pH and total Fe(III) concentration: $Fe(H_2O)_6^{3+}$, $Fe(H_2O)_5OH^{2+}$, and $Fe_2(H_2O)_8(OH)_2^{4+}$. These species can provide parallel paths to Fe(III) complex formation via substitution of coordinated H_2O. The intimate mechanism [83,150,151] for ligand exchange at Fe^{3+}_{aq} will change, depending on the composition of the inner coordination shell. It is instructive to first consider water exchange kinetics at $Fe(H_2O)_6^{3+}$ (Eq. (12), $k = 1.7 \times 10^2\ s^{-1}$ at 25°C, $\Delta H^* = 64.0$ kJ/mol, $\Delta S^* = 12.1$ J/°Kmol, $\Delta V^* = -5.4$ cm³/mol) and $Fe(H_2O)_5OH^{2+}$ (Eq. (13), $k = 1.4 \times 10^5\ s^{-1}$ at 25°C, $\Delta H^* = 42.4$ kJ/mol, $\Delta S^* = 5.3$ J/°Kmol, $\Delta V^* = 7.0$ cm³/mol) [152–

$$Fe(H_2O)_6^{3+} + H_2O^* \rightleftharpoons Fe(H_2O^*)(H_2O)_5^{3+} + H_2O \tag{12}$$

$$Fe(OH)(H_2O)_5^{2+} + H_2O^* \rightleftharpoons Fe(OH)(H_2O^*)(H_2O)_4^{2+} + H_2O \tag{13}$$

154]. The mechanism for water exchange at $Fe(H_2O)_6^{3+}$ proceeds via an associative interchange (I_a) mechanism, based largely on a consideration of the activation volume (ΔV^*) for the process [152,155]. Incorporation of OH^- into the inner coordination shell labilizes water exchange at Fe^{3+}_{aq} by a factor of 750, and based on the ΔV^*, the mechanism for the process changes to dissociative interchange (I_d) [152,155]. Other strong electron donors in the inner coordination shell labilize coordinated water and increase the dissociative character of the water exchange process. For example, water exchange at $FeL(OH_2)^-$ (L = *o*-phenylenediamine-*N,N,N',N'*-tetraacetate) proceeds at a rate which is about 10^5 times faster than $Fe(H_2O)_6^{3+}$ via a dissociative process ($k = 1.2 \times 10^7\ s^{-1}$; $\Delta V^* = 4.6$ cm³/mol) [156]. Other relevant examples are cited below.

The dimer $Fe_2(H_2O)_8(OH)_2^{4+}$ can provide a significant pathway to complex formation, particularly under conditions of higher Fe^{3+}_{aq} concentrations. Investigations of the reactivity of this species are of interest as a model for reactivity at polynuclear hydrated Fe(III) centers. Undoubtedly, at the environmental pH values encountered by the siderophores, polynuclear Fe(III) centers of high molecular weight are dominant. Such species are not well defined and therefore less amenable to quantitative mechanistic investigation. Consideration of the reasonably well-characterized $Fe_2(H_2O)_8(OH)_2^{4+}$ dimer, in relation to reactivity at $Fe(H_2O)_6^{3+}$ and $Fe(H_2O)_5OH^{2+}$, provides an opportunity to investigate the influence of multinuclear centers and coordinated OH^- on reactivity at a Fe(III) center. Unfortunately, rate and activation parameters for $Fe(OH)_2(H_2O)_8^{4+}$ water exchange are not available for interpretation. However, some relevant solution equilibrium data are available.

The structure of $Fe_2(H_2O)_8(OH)_2^{4+}$ is uncertain. Literature kinetic data have been interpreted in terms of a di-μ-hydroxy structure (**117**)

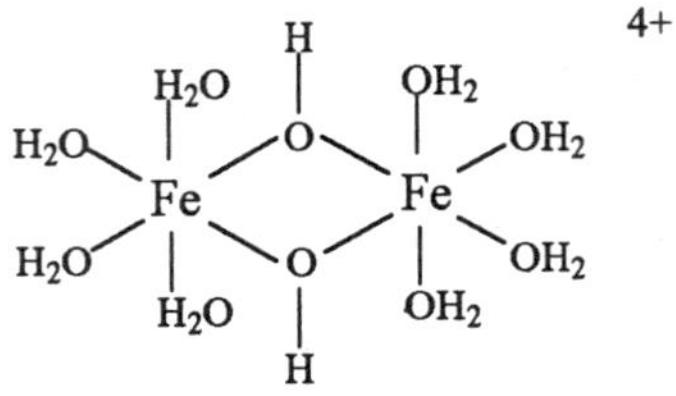

117

[157–160], although a μ-oxo structure has also been proposed. Relevant dimer equilibria and their equilibrium constants are as follows, where $K_{2Fe} = 6 \times 10^{-4}$ M at 25°C and $I = 2.67$ M($NaClO_4$) [149,161], $K_a = 1.4 \times 10^{-3}$ M [162] and and $K_b = 8.7 \times 10^{-1}$ [162].

$$2\,Fe(H_2O)_6^{3+} \rightleftharpoons Fe(H_2O)_4(OH)_2Fe(H_2O)_4^{4+} + 2H^+ \quad K_{2Fe} \tag{14}$$

$$Fe(H_2O)_4(OH)_2Fe(H_2O)_4^{4+} \rightleftharpoons 2Fe(H_2O)_5OH^{2+} \quad K_a \tag{15}$$

$$Fe(H_2O)_4(OH)_2Fe(H_2O)_4^{4+} + H^+ \rightleftharpoons Fe(H_2O)_6^{3+} + Fe(H_2O)_5OH^{2+} + H_2O \quad K_b \tag{16}$$

Ligand substitution reactions at aquated Fe^{3+} and $FeOH^{2+}$ are dominated by the dynamics of water exchange from the inner coordination shell. Therefore, to a first approximation, the second-order rate constant for ligand substitution for coordinated water at either $Fe(H_2O)_6^{3+}$ or $Fe(H_2O)_5OH^{2+}$ may be described as in Eq. (17), where k is the experi-

$$k = K_{os}Sk_{H_2O} \tag{17}$$

mentally observed second-order rate constant for initial chelate bond formation, K_{os} is the outer-sphere association constant for ligand entry into the Fe(III) primary solvation shell, S is a statistical coefficient that accounts for solvent shell composition, and k_{H_2O} is the first-order rate constant for water exchange at $Fe(H_2O)_6^{3+}$ or $Fe(H_2O)_5OH^{2+}$. In most environmentally relevant cases, complexation involves the attack of an uncharged siderophore ligand on the Fe(III) center. Using uncharged *N*-methylacetohydroxamic acid and $Fe(H_2O)_6^{3+}$ or $Fe(H_2O)_5OH^{2+}$ as an example, K_{os} has been calculated as 0.1 M^{-1} [110]. K_{os} has been estimated at 3×10^{-2} M^{-1} for $Fe(H_2O)_5OH^{2+}$ and H_4DFO^+, where electrostatic repulsion between like charges diminishes outer sphere association [163]. A reasonable solvation number for the primary solvation shell of either $Fe(H_2O)_6^{3+}$ or $Fe(H_2O)_5OH^{2+}$ is 10–15. The statistical parameter, S, can be assigned an approximate value of 1/6. Since (K_{os}S) is expected to be <1, complex formation is predicted to proceed with diminished velocity relative to water exchange if water exchange dynamics control the ligand exchange process.

Table 15 is a compilation of formation rate constants for model siderophore-binding sites (hydroxamic acid and catechol), a di- and trihydroxamate, and a mixed catechol-hydroxamate siderophore. Assuming the validity of Eq. (17), and therefore water exchange controlled dynamics, and the rate constants for Eqs. (12) and (13), reasonable values for K_{os} for the association of a neutral ligand with the solvation shell of a +3 or +2 metal ion are obtained from these data. There is some variation in k_1, suggestive of an associative interchange (I_a) intimate mechanism, as is the case for water exchange at $Fe(H_2O)_6^{3+}$

TABLE 15

First Chelate Ring Formation Rate Constants for $Fe(H_2O)_6^{3+}$ and $Fe(H_2O)_5OH^{2+}$

Ligand	No.	k_1[a] $M^{-1}\ s^{-1}$	k_{1h}[b] $M^{-1}\ s^{-1}$	Reference
TIRON[c]		2.3	1.7×10^3	168
$CH_3C(O)N(OH)H$		1.2	2.0×10^3	108
$CH_3C(O)N(OH)CH_3$		1.8	2.7×10^3	110
$\{CH_3N(OH)C(O)\}_2$-$(CH_2)_8$		4.0×10^1	1.8×10^4	121
$\{HN(OH)C(O)\}_2$-$(CH_2)_8$		1.7	1.0×10^3	121
Deferriferrioxamine B	**9**	2.0×10^{-1}	2.0×10^2	166
			3.6×10^3	167
FeH_2DFO^{2+}[d]		1.3	1.6×10^3	149
Pyoverdin PaA	**5**	6	7.7×10^3	44

[a]Second-order rate constant at 25°C for the reaction: $Fe(H_2O)_6^{3+} + LH_x \longrightarrow FeL(H_2O)_n^{(3-x)+} + xH^+$.
[b]Second-order rate constant at 25°C for the reaction: $Fe(H_2O)_5OH^{2+} + LH_x \longrightarrow FeL(H_2O)_n^{(3-x)+} + (x-1)H^+$.
[c]1,2-dihydroxy-3,5-benzenedisulfonate.
[d]Second-order rate constants for the reaction of tetracoordinate monoferric ferrioxamine B with $Fe(H_2O)_6^{3+}$ and $Fe(H_2O)_5OH^{2+}$ to form diferric ferrioxamine B.

[152,155]. There is also some variation in k_{1h}, indicating a certain level of entering group participation in the transition state. However, since it appears that water exchange at $Fe(H_2O)_5OH^{2+}$ is dissociative (I_d), this small entering group effect may occur as a result of H-bonding in the solvation shell with coordinated OH^-, which would influence K_{os}. Ligand substitution at $Fe(H_2O)_5OH^{2+}$ is 10^3 faster than at $Fe(H_2O)_6^{3+}$ owing to the labilizing effect of the OH^- ligand in the inner coordination shell. This emphasizes the strong influence of pH on the reactivity of Fe^{3+}_{aq}. The fact that the ratio of water exchange rate constants ($k_{H_2O}{}^{FeOH^{2+}}/k_{H_2O}{}^{Fe^{3+}} = 750$; [152–154]) is comparable to the ratio k_{1h}/k_1 for the examples given in Table 15 is further support for the dominance of water exchange dynamics in Fe(III) siderophore complex formation.

Table 16 lists rate constants for complex formation reactions via the hydroxy species $Fe(H_2O)_5OH^{2+}$ and $Fe(H_2O)_4(OH)_2Fe(H_2O)_4^{4+}$. Some caution should be exercised in interpreting these data from different laboratories, as in some cases evidence for a pathway due to $Fe(H_2O)_4(OH)_2Fe(H_2O)_4^{4+}$ comes entirely from nonlinearity in the pseudo-first-order constant dependence on [Fe(III)]. This is consistent with a $Fe(H_2O)_4(OH)_2Fe(H_2O)_4^{4+}$ path but can also indicate other experimental problems. However, by comparing the ratio of rate constants for reaction via the monomeric and hydroxy bridged path ($\mathbf{R} = k_{1h}'/k_{1h}$; column 6 in Table 16) for data from the same laboratory, a reasonable assessment of the influence of dimer formation on reactivity at an Fe(III) center can be made. Data are arranged in Table 16 according to the structure of the chelating moiety assumed to form the first Fe(III) chelate ring.

Both $Fe(H_2O)_5OH^{2+}$ and $Fe(H_2O)_4(OH)_2Fe(H_2O)_4^{4+}$ are significantly more reactive than $Fe(H_2O)_6^{3+}$ (cf. Tables 15 and 16). Reactivity at $Fe(H_2O)_4(OH)_2Fe(H_2O)_4^{4+}$ and $Fe(H_2O)_5OH^{2+}$ varies with entering ligand by a factor of 34 000 and 3100, respectively. This suggests either some associative (I_a) character in these substitution reactions (in contrast to the I_d water exchange mechanism at $Fe(H_2O)_5OH^{2+}$), or influence from a preequilibrium step involving K_{oS}. Chelating ligand discrimination between reaction at $Fe(H_2O)_5OH^{2+}$ and $Fe(H_2O)_4(OH)_2Fe(H_2O)_4^{4+}$ varies ($\mathbf{R} < 1$ to 30). Structural and/or charge effects may be operative. This can be seen by comparing the following entry pairs in Table 16: (2/3) (12/13), 14/15), (12/14), and (13/15). This permits us to compare reactivities for the same entering group ligand structure but with a different charge. Comparison of ratios of rate constants k_{1h} and k_{1h}' for each of these entry pairs illustrates that a more negative (or less positive) charge on the entering ligand results in an increase in reactivity at both $Fe(H_2O)_5OH^{2+}$ and $Fe(H_2O)_4(OH)_2Fe(H_2O)_4^{4+}$.

5.1.2. Second and Third Chelate Formation Steps

Subsequent ring closure steps are more rapid owing to the labilizing effect afforded by a hydroxamate or catecholate group in the inner coordination shell in much the same way that OH^- enhances coordinated water lability in $Fe(H_2O)_5OH^{2+}$ and $(H_2O)_4Fe\text{-}\mu\text{-}(OH)_2\text{-}Fe(H_2O)_4^{4+}$. These subsequent steps may vary in rate depending on the identity of the chelate group and ligand architecture. Since subsequent ring closure steps are rapid relative to initial chelate ring formation, they are often not kinetically detectable and rate constants for these steps are determined from kinetic investigations of the dissociation reaction. In this section, we will consider mechanistic changes which may occur for each ring closure step of the complexation reaction. Data will be presented which illustrate an increase in dissociative character for the second and third chelate ring formation step. We begin by considering the stepwise complexation of Fe^{3+}_{aq} by synthetic monohydroxamic acids and catechol (Table 17), and then we will consider polydentate complex formation.

For all three ligands in Table 17, k_{1h}, k_2, and k_3 are comparable and about 10^3 greater than k_1. This reflects the labilizing effect of an electron donor in the inner coordination shell on the dissociating water ligand and suggests that the hydroxide, hydroxamate, and catecholate anions are equivalent in this regard. This labilization effect further suggests an increase in dissociative character with increasing substitution at the Fe(III) center, which is reflected in the ΔV^*_n data presented in Table 17 for the acetohydroxamic acid reaction. The ΔV^*_n becomes less negative as the stepwise formation of the tris complex proceeds. It is interesting to note that although reactivity of $Fe(CH_3C(O)N(O)H)(H_2O)_4^{2+}$ and $Fe(H_2O)_5OH^{2+}$ toward H_2O substitution is the same ($k_{1h} = k_2$), the ΔV^*_n data suggest more dissociative character for the reaction at $Fe(H_2O)_5OH^{2+}$. A small degree of water leaving group participation in the transition state is suggested by a linear free energy (LFER) analysis of these reactions [110]. A consequence of the fact that stepwise complex formation rate constants for the different ligands in Table 17 are nearly

TABLE 16

Formation Kinetics via Hydroxy Complex Paths Involving $Fe(H_2O)_5OH^{2+}$ and $Fe_2(H_2O)_8(OH)_2^{4+}$

Entry No.	Ligand: Type	Ligand: Name	k_{1h}[a] $M^{-1}\,s^{-1}$	k_{1h}'[b] $M^{-1}\,s^{-1}$	R = [c] k_{1h}'/k_{1h}	I(M)[d]	Reference
	Hydroxamic Acid						
1		Acetohydroxamic acid	5.8×10^3	6.5×10^3	1.1	0.5	192
			5.7×10^3	8.2×10^3	1.4	1.0	192
		Azotobactin δ (**6**)[e]					
2		H_4L	2.1×10^4	6.1×10^5	29	2.0	51
3		H_5L^+	5.9×10^3	—	—	2.0	51
		Pyoverdin PaA (**5**)[e]					
4		H_6L^{2+}	7.7×10^3	5.0×10^4	6.8	2.0	44
	Catechol						
5		TIRON[f]	3.1×10^3	1.1×10^4	3.5	1.0	201
	Biphenol						
6		BPA (H_2L)[g]	3.7×10^2	4.1×10^2	1.1	0.5	34
7		BPAS (**19**, H_2L^{3-})	8.5×10^2	4.9×10^2	0.6	0.5	34
8		BPDA (H_2L)[h]	2.6×10^2	1.8×10^2	6.9	0.5	34
9		BPDAS (**20**, H_2L^{2-})	5.4×10^2	1.3×10^2	2.4	0.5	34
10		TBPDAS (**116**, H_7,L^+)[i]	2.8×10^2	8.9×10^2	3.2	0.5	142

	Di-hydroxy						
11		Squaric acid (HL^-)[j]	5×10^4	4.5×10^5	9.0	1.0	193
	Quinoline						
		Daunomycin					
12		H_2L	1.7×10^2	3.9×10^1	0.2	0.5	194
13		$Fe(HL)^{2+}$[k]	1.6×10^1	1.8×10^1	1.1	0.5	194
		Quinizarin-2-sulfonate					
14		H_2L^-	3.2×10^3	1.2×10^3	0.4	0.5	194
15		$Fe(HL)^{+}$[k]	4.8×10^2	6.0×10^2	1.3	0.5	194

[a]Second-order rate constant at 25°C for reaction at hydroxy monomer $Fe(H_2O)_5OH^{2+}$.
[b]Second-order constant at 25°C for reaction at the dihydroxy bridged dimer $Fe_2(H_2O)_8(OH)_2^{4+}$.
[c]Ratio of rate constants for reaction at hydroxy dimer relative to monomer.
[d]Ionic strength, M.
[e]Initial chelate bond formation occurs at the hydroxamate group. H_4L, H_5L^+, and H_6L^{2+} represent different protonated forms of the ligand.
[f]1,2-Dihydroxy-3,5-benzenedisulfonate.
[g]Structure 19 without sulfonate groups.
[h]Structure 20 without sulfonate groups.
[i]Trimer of BPDA.
[j]Ligand pK_a requires that there is a proton ambiguity in the kinetic results; rate constants represent a sum of reactions: $FeOH^{2+}_{aq} + HL^- \longrightarrow$ and $Fe^{3+}_{aq} + L^{2-} \longrightarrow$. The magnitude of the rate constants suggest that reactivity occurs exclusively at $FeOH^{2+}$.
[k]Rate data for addition of a second Fe(III) center to daunomycin or quinizarin-2-sulfonate via reaction with $Fe(H_2O)_5OH^{2+}$ and $Fe(H_2O)_4(OH)_2Fe(H_2O)_4^{4+}$.

TABLE 17

Stepwise Formation Rate Constants for Monocatechol and Hydroxamic Acid Complexes of Fe(III)

Reaction[a]		TIRON[b] k_n/ $M^{-1}\,s^{-1}$	$CH_3C(O)N(OH)CH_3$[c] k_n/ $M^{-1}\,s^{-1}$	$CH_3C(O)N(OH)H$[d] k_n/ $M^{-1}\,s^{-1}$	ΔV_n^*/cc mol^{-1}[e]
$Fe(H_2O)_6^{3+} + LH_x \longrightarrow FeL(H_2O)_4^{+3-x} + xH^+$	k_1	2.3	1.80	1.2	−6.3
$Fe(H_2O)_5OH^{2+} + LH_x \longrightarrow FeL(H_2O)_4^{+3-x} + (x-1)H^+$	k_{1h}	1.7×10^3	2.65×10^3	2.0×10^3	+5.2
$FeL(H_2O)_4^{+3-x} + LH_x \longrightarrow FeL_2(H_2O)_2^{+3-2x} + xH^+$	k_2	3.2×10^3	8.1×10^2	1.6×10^3	−2.2
$FeL_2(H_2O)_2^{+3-2x} + LH_x \longrightarrow FeL_3^{+3-3x} + xH^+$	k_3	7.7×10^3	7.5×10^2	1.7×10^3	−2.2

[a]When LH_x = TIRON, x = 2 and the charges on the sulfonate groups are ignored for simplicity; when $LH_x = CH_3C(O)N(OH)CH_3$ or $CH_3C(O)N(OH)H$ x = 1.

[b]TIRON = 1,2-dihydroxy-3,5-benzenedisulfonate [168].

[c]Ref. 110.

[d]Refs. 108 and 111.

[e]Volume of activation [165].

identical is that differences in the stepwise formation equilibrium constants are entirely due to differences in dissociation rate constants.

Formation rates for full complexation of Fe^{3+}_{aq} by the bidentate ligands cannot be compared with the polydentate siderophores, since the molecularity of the second and third chelation steps are second order in the former case and first order in the later. In each case, subsequent chelation steps are successively more rapid so that initial chelate bond formation at the Fe(III) center is rate limiting. However, certain steric features associated with polydentate siderophores may come into play in some instances. In the case of the hydroxamic acid chelating group, rotation about the C$-$N bond from a **E** configuration, which is unfavorable for chelation, to a favorable **Z** configuration may influence the kinetics of five-membered chelate ring formation (Eq. (18)). The Z/E ratio, and rotational rate constant (k_{rot}) for *N*-methyl-

(18)

acetohydroxamic acid have been measured at 4 and 3 s^{-1}, respectively, at 298K [110]. Biruš and coworkers [164] report a Z/E ratio of 0.25 for deferriferrioxamine B and k_{rot} constants of 9, 15, and 12 s^{-1} for the three hydroxamate groups starting at the protonated amine end of the molecule. For acetohydroxamic acid and *N*-methylacetohydroxamic acid, the Z/E ratio is favorable for five-membered ring chelate formation after initial bond formation at the carbonyl O atom [108]. For these two monohydroxamate ligands, ring closure is rapid relative to C$-$N bond rotation [108,110,111,165]. However, the stepwise ferrioxamine B chelation reaction involving monohydroxamate coordination to bis-hydroxamate coordination was found to occur in two kinetically detectable

$$\begin{aligned} Fe(H_2O)_4H_3DFO^{3+} &\rightleftharpoons \{Fe(H_2O)_3H_3DFO^{3+}\}^* \\ &\rightleftharpoons Fe(H_2O)_2H_2DFO^{2+} + H^+ \end{aligned} \tag{19}$$

steps (Eq. (19)) [166,167]. The synthetic dihydroxamic acid ligand $CH_3N(OH)\text{-}C(O)\text{-}[CH_2]_8\text{-}C(O)\text{-}N(OH)CH_3$ (HA—AH) was found to behave similarly to deferriferrioxamine B (Eq. (20)) [121]. The rate constant for conversion of the mono complex to the intermediate { }* in Eq. (20) is 12

$$\begin{aligned} Fe(H_2O)_4(A\text{—}AH)^{2+} &\rightleftharpoons \{Fe(H_2O)_3(A\text{—}AH)^{2+}\}^* \\ &\rightleftharpoons Fe(H_2O)_2(A\text{—}A)^{+} + H^+ \end{aligned} \tag{20}$$

s^{-1} [121]. This is consistent with the intermediate species { }* being a half-formed five-membered chelate ring, where ring closure proceeds at a rate similar to C−N bond rotation. In the case of ferrioxamine B, an upper limit of 3 s^{-1} was obtained for the rate constant for the conversion of the mono complex to the intermediate { }* (Eq. (19)) [166]. This is consistent with ring closure being slow relative to C−N bond rotation and with an unfavorable Z/E ratio for five-membered chelate ring formation. Why five-membered ring closure is slow for deferriferrioxamine B and $CH_3N(OH)$-C(O)-$[CH_2]_8$-C(O)-$N(OH)CH_3$ (both of which have the same number of atom spacers between hydroxamate binding groups) and rapid for the monohydroxamates is open to speculation. Steric bulk and hydration of the long chain substituent on the hydroxamate N atom are likely to play a role.

5.2. Siderophore Complex Dissociation Reactions

Since formation rate constants are dominated by the dynamics of water loss from the Fe(III) inner coordination shell, relatively little difference in formation rate constants are seen for different siderophore structures. Consequently, differences in stability constants (β) are largely reflected in differences in dissociation rate constants, which can be significant for different siderophore structures. A comparison of the dissociation kinetics for a monocatecholate, and mono-, di-, and trihydroxamic acids will serve to illustrate the influence of donor group, electronic substituent effects, solvation effects, denticity, and other ligand architectural effects.

The stepwise $[H^+]$-dependent dissociation rate constants for each monocatechol and monohydroxamate ligand in Table 18 exhibit a range $>10^6$ M^{-1} s^{-1}. Each stepwise constant is 10^1–10^2 times greater for catechol relative to acetohydroxamic acid, which is up to 35 times greater than for *N*-methylacetohydroxamic acid [108,110,111,168]. This reflects a different chelator donor strength (catechol vs hydroxamate), and the electronic effect of an electron-donating methyl group on the N atom of the hydroxamate moiety, which increases the electron density on the carbonyl O atom (see resonance forms (**63–65**)). The latter influences dissociation kinetics, because cleavage of the Fe(III)-carbonyl O bond is the second and rate limiting step in hydroxamate moiety dissociation [108,109]. Solvation effects may also play a role in determining this difference in dissociation rate constants, since the dissociating N-CH_3

TABLE 18

Stepwise Dissociation Rate Constants for Monocatechol and Mono-, Di-, and Trihydroxamic Acid Complexes of Fe(III)[a]

Reaction[b]	k_{-n}	Dissociating ligand[c]: TIRON[d]	AHA[e]	NMAHA[f]	di-HA[g]	ferrioxamine B[h]
$FeL_3^{+3-3x} + xH^+ \longrightarrow FeL_2(H_2O)_2^{+3-2x} + LH_x$	$k_{-3}/M^{-1}\ s^{-1}$	1.7×10^7	1×10^5	8.6×10^3	na	2.9×10^2
$FeL_2(H_2O)_2^{+3-2x} + xH^+ \longrightarrow FeL(H_2O)_4^{+3-x} + LH_x$	$k_{-2}/M^{-1}\ s^{-1}$	$>2.8 \times 10^5$	1.4×10^3	1.0×10^2	2.1×10^{-1}	3.4×10^{-1}
$FeL(H_2O)_4^{+3-x} + xH^+ \longrightarrow Fe(H_2O)_6^{3+} + LH_x$	$k_{-1}/M^{-1}\ s^{-1}$	1.2	1.1×10^{-1}	3.2×10^{-3}	0.8×10^{-3}	1.9×10^{-3}
$FeL(H_2O)_4^{+3-x} \longrightarrow Fe(H_2O)_5OH^{2+} + LH_x$	k_{-1h}/s^{-1}	1.3	0.8×10^{-1}	7.1×10^{-3}	0.6×10^{-3}	2.1×10^{-3}

[a]25°C.

[b]When LH_x = TIRON, x = 2 and the charges on the sulfonate groups are ignored for simplicity; when $LH_x = CH_3C(O)N(OH)CH_3$ or $CH_3C(O)N(OH)H$ x = 1; when LH_x = the dihydroxamic acid $CH_3N(OH)C(O)[CH_2]_8C(O)N(OH)CH_3$ x = 1 and the reaction in row 2 corresponds to the dissociation of one of the two hydroxamate groups and the reactions in rows 3 and 4 correspond to the dissociation of the second hydroxamate group; when LH_x = deferriferrioxamine B, FeL_3 represents ferrioxamine B, $FeHDFO^+$, which undergoes a stepwise dissociation of each hydroxamate group, charges on the complexes in rows 1–4 are $FeHDFO^+$, FeH_2DFO^{2+}, FeH_3DFO^{3+}, and FeH_3DFO^{3+}, respectively.

[c]TIRON = 1,2-dihydroxy-3,5-benzenedisulfonate; AHA = acetohydroxamic acid, $CH_3C(O)N(OH)H$; NMAHA = N-methylacetohydroxamic acid, $CH_3C(O)N(OH)CH_3$; di-HA = $CH_3N(OH)C(O)[CH_2]_8C(O)N(OH)CH_3$; ferrioxamine B = H_4DFO^+.

[d]Ref. 168.

[e]Refs. 108 and 111.

[f]Ref. 110.

[g]Ref. 121.

[h]Refs. 166 and 167.

substitutent is more likely to perturb bulk or solvation shell water structure than is the N-H substituent.

Data in Table 18 demonstrate an effect of ligand denticity on dissociation rates, which may also be a manifestation of ligand stereochemistry and solvation effects. Penetration of a single catechol or hydroxamate group into the solvation shell or bulk solvent may be more facile than penetration of a bulky long chain polydentate ligand. Dissociation of *N*-methylacetohydroxamic acid from FeL_3 has a H^+ dependent second-order rate constant 30 times greater than the corresponding dissociation rate constant for the hexacoordinate ferrioxamine B. This may be a denticity effect, solvation effect, and/or a hydrophobic effect of the connecting hydrocarbon chain on the approach of H^+_{aq}. This effect is also seen in the dissociation rate constants for a tetra-coordinate Fe(III). Bis(*N*-methylacetohydroxamato)Fe(III) dissociates a hydroxamate moiety with a second-order rate constant 500 times greater than found for the dihydroxamic acid complex and 300 times greater than for tetradentate ferrioxamine B (Table 18). This denticity effect is not evident in the dissociation of the last hydroxamate group to give the free Fe^{3+}_{aq} (Table 18), which may be due to the fact that the noncoordinated hydroxamate moieties are well removed from the Fe(III) reaction center.

A linear correlation exists between ΔG^* and ΔG^0 for each dissociation step of the hydroxamate complexes presented in Table 18 [110,121]. This indicates a high degree of correlation of the stepwise rate and equilibrium constants and certainly supports a mechanistic similarity between each step of the H^+-dependent dissociation of tris-, bis-, and mono(hydroxamato)iron(III) complexes. Trends in the data for the monocatechol system suggest that it also follows this correlation [168]. The slope of the $\Delta G^*/\Delta G^0$ correlation defined by 12 data points, including ferrioxamine B, is 1.5, which suggests a late transition state in each step of the dissociation process with minimal participation from the departing ligand [121].

Fe^{3+}-siderophore dissociation kinetics are undoubtedly influenced by medium or environmental effects. The discussion in the preceding paragraphs is based on data obtained at low pH to drive the ligand dissociation reaction by protonation. Ionic strength and the presence of electrolyte anions capable of ternary complex formation will influence the dynamics of siderophore dissociation [1]. SDS micelles enhance the ferrioxamine B dissociation rate [169]. The ferrioxamine B complex was found to reside in the micelle Stern layer where an increase in localized H^+ concentration accelerates ligand dissociation. This may have implications as a pathway for siderophore dissociation at a cell surface.

5.3. Kinetics and Mechanism of Selected Model Systems

Einaga and coworkers investigated the thermodynamics and kinetics of Fe(III) chelation by a phenol substituted amino hydroxamic acid, N^{α}-salicyl-2-amino-N'-hydroxypropanamide (H_2slalh, **118**) [170]. Spec-

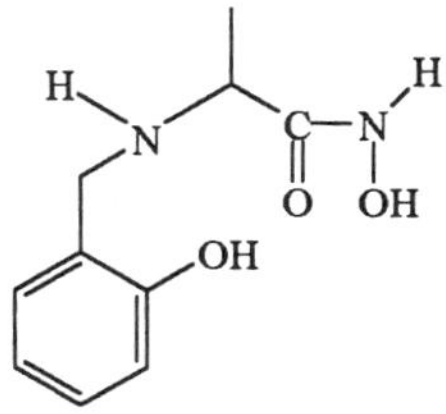

H_2slalh
118

tral changes as a function of pH (25°C; 0.1 M $NaClO_4$) were interpreted in terms of bidentate complex formation ($Fe(H_2slalh)(OH_2)_4^{3+}$) involving the hydroxamate moiety below pH 1.8 (log β = 10.29) and tetradentate complex formation ($Fe(slalh)(OH_2)_2^{+}$) involving hydroxamate, amine, and phenolate coordination above pH 1.8 (log β = 23.17). The parent α-aminohydroxamic acid (*L*-alaninehydroxamic acid; Halh) has a Fe(III) formation constant log β = 10.70, consistent with bidentate chelation of a single hydroxamate group. (This is in contrast with a previously reported value of log β = 17.15 [171]. The higher value suggests tridentate complex formation at their conditions.) Einaga and coworkers [170] report second-order rate constants $1.44 \times 10^2\ M^{-1}\ s^{-1}$ for reaction of H_2slalh with Fe^{3+}_{aq} below pH 1.8 and $3.88 \times 10^2\ M^{-1}\ s^{-1}$ above pH 1.8. These are large rate constants for ligand substitution reactions at $Fe(H_2O)_6^{3+}$ (see Table 15). They report a second order rate constant of $1.05 \times 10^3\ M^{-1}\ s^{-1}$ for H_2slalh reaction with $FeOH^{2+}_{aq}$, consistent with established trends for reaction at the hydroxy species. These investigators also report a large rate constant for *L*-alaninehydroxamic acid reaction with $Fe(H_2O)_6^{3+}$ ($4.27 \times 10^2\ M^{-1}\ s^{-1}$), suggesting participation from some hydroxy containing species. Einaga and coworkers observe a nonlinear dependence of the pseudo-first-order rate constant on pH over the range 1.3–2.5, which they interpret as due to initial rate limiting $Fe(H_2O)_6^{3+}$-carbonyl O atom bond formation below pH 1.8 and initial rate limiting phenolate O atom bond formation with $Fe(H_2O)_6^{3+}$ and $Fe(H_2O)_5OH^{2+}$ above pH 1.8. This interpretation is highly speculative, as the pH dependence observed for H_2slalh is the same as that observed for other Fe(III) reactions with ligands containing a single hydroxamate group [108,109]. A change in the site of Fe(III) attack on the H_2slalh ligand from the carbonyl O atom to the phenolate O atom at pH 1.8 seems unlikely, given the investigators' report that the pK_a of the phenolic proton is 10.49.

The issue of initial bond formation in siderophore chelation reactions is important. Albrecht-Gary and coworkers have shown that initial chelation for pyoverdin PaA (**5**) occurs at the hydroxamate rather than catecholate group [44]. Derivatization of ferrioxamine B by the addition of a catecholate group influences its ability to scavenge Fe from other chelate complexes, presumably by increasing the efficiency of the initial bond formation [113,114].

5.4. Ligand Exchange Reactions at Fe(III)

Fe removal from its siderophore complex must be a relatively facile process in vivo at the appropriate time and place. Consequently, there is an interest in processes which will enhance the Fe^{3+} dissociation rate, such as H^+ catalysis, the labilizing effect of anions or low molecular weight chelators through ternary complex formation, and/or oxidation-reduction [1]. In order to study siderophore dissociation at more biologically relevant pH values near 7, Fe exchange between competing ligands must be investigated in order to drive the dissociation reaction and prevent hydrolysis of uncomplexed Fe^{3+}_{aq}. In addition, in the context of therapeutic uses of siderophores or siderophore mimics, and in considering the mechanism of the role of siderophores in bacterial infection, it is of interest to determine the dynamics of Fe removal from various biological carriers and storage complexes (e.g., transferrin) by siderophores.

At physiological pH, Fe^{3+} and citrate combine to form a polymer of

MW 2.1×10^5 [172]. Fe(III) removal from its citrate complex is useful in assessing the ability of a siderophore to mobilize Fe(III) in humans. Fe(III) citrate is present in blood plasma, and at 100 μM is believed to be one of the major components of a pool of Fe(III) that is not bound to transferrin [173,174]. Faller and Nick [175] have investigated the kinetics and mechanism of direct Fe removal from its citrate polymer complex by deferriferrioxamine B (**9**) and 3-hydroxy-1,2-dimethyl-4-pyridinone (DMHP, **91**) at 37°C in the presence of 100 mM NaCl. These two chelators are of particular interest, as the methylsulfonate salt of deferriferrioxamine B is marketed by Ciba-Geigy under the name Desferal as a therapeutic agent for the removal of Fe(III) from Fe-overload patients, and DMHP (also referred to as L1 in the clinical literature) has been investigated for its ability to mobilize Fe(III) in humans [176]. Kinetic data for both chelators are consistent with the following scheme (L = **9**, **91**) involving ternary complex formation, followed by rate limit-

$$\begin{aligned} \{Fe_x(\text{citrate})_y\}_n + L &\overset{K}{\rightleftharpoons} L\text{-}Fe(\text{citrate})\text{-}\{Fe_{x-1}(\text{citrate})_{y-1}\}_n \\ &\xrightarrow{k} FeL + \text{citrate} + \{Fe_{x-1}(\text{citrate})_{y-1}\}_n \end{aligned} \tag{21}$$

ing dissociation to form $FeHDFO^+$ or $Fe(DMHP)_3$. The data demonstrate that Fe(citrate) depolymerization is not rate limiting and that the key step is rapid formation of the ternary complex. DMHP is more efficient in this regard (K_{DMHP} = 2 mM^{-1}; K_{DFO} = 0.1 mM^{-1}) and as a result is kinetically more effective in removing Fe from citrate polymer at lower, more physiologically relevant concentrations. The rate-limiting dissociation of the ternary complex to form FeL slightly favors deferriferrioxamine B (k_{DMHP} = 0.019 s^{-1}; k_{DFO} = 0.040 s^{-1}). Ternary complex formation is known to be an important step in chelate ligand exchange reactions [83], including those involving siderophores [1]. Care should be taken in comparing the numerical values of the constants reported by Faller and Nick [175] with other systems as the presence of Cl^- is known to influence Fe^{3+} ligand exchange kinetics [1]. However, these results clearly demonstrate the kinetic efficiency of a bidentate chelator in removing Fe from a polymer even though its thermodynamic affinity for Fe may not be optimal (log β and pM values favor Fe(III) complexation by deferriferrioxamine B over DMHP; see Tables 6 and 12).

Tripodal trihydroxamic acid ligands (**119–122**) with different chain lengths (13–19 spacer atoms between hydroxamate groups) were pre-

$$N[(CH_2)_2\text{-}NHC(O)(CH_2)_n\text{-}N(OH)\text{-}C(O)\text{-}(CH_2)_2NH\text{-}C(O)OCMe_3]_3$$

119: n = 2; **120**: n = 3; **121**: n = 4; **122**: n = 5

pared via condensation of *N*-hydroxysuccinimide esters of *N*-acyl-*N*-benzyloxyaminoalkanoic acids with tris(2-aminoethyl)amine (TREN) [177]. Relative kinetic lability for the Fe(III) complexes was determined from initial rates for exchange with EDTA in 50% dimethylformamide (DMF) at pH 6.9. Exchange kinetics with EDTA are dependent on spacing between hydroxamate groups over a factor of 7 in the order **120** = **121** > **122** > **119**. These tripodal ligands show significant growth activity toward *Aureobacterium flavescens* but far less than that exhibited by ferrioxamine B. This may be related to the observation that ferrioxamine B exchanges Fe(III) with EDTA at a much slower rate than do ligands **119–122**.

Akiyama and coworkers investigated the kinetic lability of Fe(III) complexes of cyclic (**70**) and linear (**71**) peptide-based trihydroxamic acids (see Table 9) [115,116]. Exchange kinetics with EDTA are proposed to proceed according to the following mechanism involving ternary complex formation (Eqs. (22) and (23); x = 3 at pH 2.1 and 2 at pH 5.4).

$$Fe(L) + H^+ \rightleftharpoons Fe(HL)^+ \qquad K_{Fe(HL)} \tag{22}$$

$$Fe(HL)^+ + H_xEDTA^{(4-x)-} \rightleftharpoons (HL)Fe(H_xEDTA)^{n-} \longrightarrow H_3L + FeEDTA^- + yH^+ \tag{23}$$

Although $K_{Fe(HL)}$ is larger for **70** than for **71** (see Table 9), the linear complex Fe-**71** exchanges ligands about three times faster than the cyclic complex Fe-**70**. This suggests that dissociation of the ternary complex is rate limiting. Although the linear vs cyclic structural effect on Fe(III) exchange is small, the investigators report that peptide-based hydroxamate ligand exchange with EDTA is 10–30 times greater than for ferrioxamine B exchange with EDTA. To summarize these kinetic data with the thermodynamic data discussed earlier (Sec. 4.4.1.1), the peptide-based trihydroxamate complexes Fe-**70** and Fe-**71** are less stable and more labile than ferrioxamine B.

The di-Fe complex $Fe_2(\mathbf{82})_3$ (Sec. 4.4.1.2) was found to undergo ligand exchange with EDTA by a consecutive two-step reaction mechanism in which both Fe sites are kinetically equivalent and the release of the first Fe occurs at a slightly faster rate than the second [178]. The investigators propose that the first Fe release involves two conformers in equilibrium which react by different mechanisms. No direct evidence is given for ternary intermediate complex formation, although they are most certainly formed.

6. PERSPECTIVES

Clearly the coordination chemistry of Fe(III) and Fe(II) plays a pivotal role in siderophore mediated Fe transport and bioavailability. Central features include (1) ligand architecture and stereochemistry, which influences the thermodynamic affinity and selectivity for Fe(III) chelation; (2) Fe chelate stereochemistry, which influences molecular recognition by the cell receptor and Fe lability; (3) oxidation/reduction of the Fe center which also influences Fe chelate stability and lability; (4) kinetics and mechanism of ligand exchange involving Fe(III) and Fe(II); and (5) second coordination shell effects and supramolecular assembly formation, which are relevant to molecular recognition and membrane transport. More research needs to be done if we are to gain a fundamental understanding of how the simplest organisms acquire Fe from the environment. For example, we need a better understanding of reaction pathways for Fe release from its extremely stable siderophore complex at the cell surface and the cell interior. More redox potentials of Fe(III) siderophore complexes need to be determined at conditions relevant to the cell surface receptor and cell interior. The role of ligand structure and Fe-chelate structure in molecular recognition of the siderophore complex by the microbial cell needs to be further elucidated. Answers to these and other issues will be obtained by further study of both natural and synthetic siderophores. In addition, there should be further exploitation of natural siderophore ligand architecture and coordination chemistry in the design of chelating agents for therapeutic purposes. More work needs to be done on siderophore Fe chelation and coordination chemistry in relation to human disease caused by bacterial and fungal infection. It is our hope that this review of the coordination chemistry associated with siderophore mediated Fe transport will stimulate further activity in these and related areas.

ACKNOWLEDGMENTS

The authors thank Mrs P. Giunti for assembling the manuscript. A.L.C. thanks the NSF and ACS/PRF for their support of his research in this area and the French Academy of Sciences, CNRS, the European School of Chemistry in Strasbourg, and Elf Aquitaine for a Professeur de l'Académie Chaire Elf-Académie des Sciences award and l'Université Joseph Fourier de Grenoble for a Professeur Associé, during which much of this review was written. A.M.A.-G. thanks the CNRS and the European School of Chemistry in Strasbourg-France for their financial support.

ABBREVIATIONS

3H2P	3-hydroxy-2-pyridinone
3H4P	3-hydroxy-4-pyridinone
AHA	acetohydroxamic acid
BPA	3-*N,N′*-dimethylcarbamoyl-2,2′-dihydroxybiphenyl
BPANS	3-methylaminomethyl-2,2′-dihydroxybiphenyl-5,3′,5′-trisulfonate
BPAS	3-*N,N′*-dimethylcarbamoyl-2,2′-dihydroxybiphenyl-5,3′,5′-trisulfonate
BPDA	3,3′-*N,N′*-bis-dimethylcarbamoyl-2,2′-dihydroxy-biphenyl
BPDAS	3,3′-*N,N′*-bis-dimethylcarbamoyl-2,2′-dihydroxy-biphenyl-5,5′-disulfonate
BPS	2,2′-dihydroxybiphenyl-3,5,3′,5′-tetrasulfonate
CAM	catecholamide
CD	circular dichroism
DEHP	1,2-diethyl-3-hydroxy-4-pyridinone

DFC	deferriferrichrome
DFO	deferriferrioxamine B
DFOCAM	*N*-(2,3-dihydroxybenzoyl)deferrioxamine B
DFOCAMC	*N*-(2,3-dehydroxy-4-carboxybenzoyl)deferrioxamine B
DFOHOPO	*N*-[(1,2-dihydro-1-hydroxy-2-oxopyridine-6-yl)-carbonyl]deferrioxamine B
DFOMTA	*N*-[2,3-dihydroxy-4-(methylamido)benzoyl]deferri-oxamine B
DH2P	1,4-dihydroxy-2-pyridinone
DHBS	*N*-(2,3-dihydroxybenzoyl)serine
DMF	dimethylformamide
DMHP	1,2-dimethyl-3-hydroxy-4-pyridinone
DMPHP	3-hydroxy-1,2-dimethyl-4-pyridinone
DOTA	1,4,7,10-tetraazacyclododecane-*N,N′,N″,N‴*-tetraacetic acid
EDTA	ethylenediamine-*N,N,N′,N′*-tetraacetate
EH2P	1-ethyl-3-hydroxy-2-pyridinone
EHPG	1,2-ethylenebis(hydroxyphenylglycine)
EMHP	1-ethyl,2-methyl-3-hydroxy-4-pyridinone
ESMS	electrospray mass spectrometry
ETA	2,3-dihydroxy-*N,N′*-diethylterephtalamide
FAB	fast atom bombardment
GC-MS	gas chromatography-mass spectrometry
HALH	*L*-alaninehydroxamic acid
HBED	*N,N′*-bis(2-hydroxybenzyl)ethylenediamine-*N,N′*-diacetic acid
HBEP	*N,N′*-bis(hydroxybenzyl)-*N,N′*-ethylenediaminedi-3-propionic acid
HPED	*N,N′*-bis(2-hydroxyphenyl)-*N,N′*-ethylenediaminediacetic acid
LFER	linear free energy
LMCT	ligand to metal charge transfer
MA	*N*-methylanthranilic acid
MECAM	*N,N′,N″*-tris(2,3-dihydroxybenzoyl)-1,3,5-tris(amino-methyl)benzene
MH2P	1-methyl-3-hydroxy-2-pyridinone
MHP	2-methyl-3-hydroxy-4-pyridinone
MTA	2,3-dihydroxy-*N,N′*-dimethylterephtalamide

NBD	7-nitrobenz-2-oxa-1,3-diazol
NCP	[2,3-*c*]-naphto-2-cyano-2*H*-pyrol
NHE	normal hydrogen electrode
NMAHA	*N*-methylacetohydroxamic acid
NMR	nuclear magnetic resonance
NOTA	1,4,7-triazacyclononane-*N*,*N*′,*N*″-triacetic acid
PMHP	1-propyl-2-methyl-3-hydroxy-4-pyridinone
PTA	2,3-dihydroxy-*N*,*N*′-dipropylterephtalamide
SDS	sodium dodecylsulfate
TACN	*N*,*N*′,*N*″-tris-1,4,7-triazacyclononane
TAM	2,3-dihydroxyterephthalamide
TBPANS	*tris*[2-aminoethyl-(2,2′-dihydroxybiphenyl-5,3′,5′-trisulfonato-3-carbonyl)]amine
TBPAS	*tris*[2-aminoethyl-(3-*N*,*N*′-dimethylcarbamoyl-2,2′-dihydroxybiphenyl-5,3′,5′-trisulfonate-3-carbonyl)]amine
TBPDAS	*tris*[2-aminoethyl-(3′-*N*′-dimethylcarbamoyl-2,2′-dihydroxybiphenyl-5,5′-disulfonate-3-carbonyl)]amine
TETA	1,4,8,11-tetraazacyclotetradecane-*N*,*N*′,*N*″,*N*‴-tetraacetic acid
TIRON	1,2-dihydroxy-3,5-benzenedisulfonate
TREN	*tris*(2-aminoethyl)amine
TRENCAM	*tris*[2-aminoethyl-(2,3-dihydroxybenzoyl)]amine
TRENDROX	*tris*[2-aminoethyl-(3-hydroxyaminocarbonyl)-propanoyl]amine
TRENSOX	*tris*[2-aminoethyl-(8-hydroxyquinoline-5-sulfonato-7-carbonyl)]amine
TRITA	1,4,7,10-tetraazacyclotridecane-*N*,*N*′,*N*″,*N*‴-tetraacetic acid

REFERENCES

1. A. L. Crumbliss, in *Handbook of Microbial Iron Chelates* (G. Winkelmann, ed.), CRC Press, Boca Raton, FL, 1991, p. 177.
2. B. F. Matzanke, G. Müller-Matzanke, and K. N. Raymond, in *Physical Bioinorganic Chemistry Series: Iron Carriers and Iron Proteins* (T. M. Loehr, ed.), VCH, New York, 1989, p. 1.

3. K. N. Raymond, G. Müller, and B. F. Matzanke, in *Topics in Current Chemistry*, Vol. 123 (F. L. Boscheke, ed.), Springer-Verlag, New York, 1984, p. 49.
4. R. C. Hider, *Struct. Bond.*, *58*, 25 (1984).
5. D. van der Helm and G. Winkelmann, in *Metal Ions in Fungi* (G. Winkelmann and D. R. Winge, eds.), Marcel Dekker, New York, 1994, p. 39.
6. J. R. Telford and K. N. Raymond, in *Comprehensive Supramolecular Chemistry*, Vol. 1 (J. M. Lehn and G. W. Gokel, eds.) Pergamon Press, London, 1996, p. 245.
7. K. N. Raymond and J. R. Telford, in *Bioinorganic Chemistry: An Inorganic Perspective of Life; NATO ASI Series C: Mathematical and Physical Sciences, 459* (D. P. Kessissoglou, ed.), Kluwer, Dordrecht, The Netherlands, 1995, p. 25.
8. A. L. Crumbliss and J. M. Garrison, *Comments Inorg. Chem.*, *8*, 1 (1988).
9. W. T. Oosterhuis, *Struct. Bonding*, *20*, 59 (1974).
10. D. van der Helm, M. A. F. Jalal, and M. B. Hossain, in *Iron Transport in Microbes, Plants and Animals* (G. Winkelmann, D. van der Helm, and J. B. Neilands, eds.), VCH, Weinheim, Germany, 1987, p. 135.
11. W. M. Latimer, *The Oxidation States of the Elements and their Potentials in Aqueous Solutions*, Prentice-Hall, New York, 1952.
12. G. Xiao, D. van der Helm, R. C. Hider, and P. S. Dobbin, *J. Chem. Soc. Dalton Trans.*, 3266 (1992).
13. A. Thieken and G. Winkelmann, *FEMS Microbiol. Lett.*, *111*, 281 (1993).
14. Y. Sugiura and K. Nomoto, *Struct. Bond.*, *58*, 107 (1984).
15. H. Drechsel, J. Metzger, S. Freund, G. Jung, J. R. Boelart, and G. Winkelmann, *BioMetals*, *4*, 238 (1991).
16. H. Drechsel, G. Jung, and G. Winkelmann, *BioMetals*, *5*, 141 (1992).
17. H. Drechsel, M. Tschierske, A. Thieken, G. Jung, H. Zahner, and G. Winkelmann, *J. Ind. Microbiol.*, *14*, 105 (1995).
18. Y. Sugiura and H. Tanaka, *J. Am. Chem. Soc.*, *103*, 6979 (1981).
19. G. Schwarzenbach and K. Schwarzenbach, *Helv. Chim. Acta*, *46*, 1390 (1963).

20. C. J. Carrano, S. R. Cooper, and K. N. Raymond, *J. Am. Chem. Soc.*, *101*, 599 (1979).
21. D. J. Cram, *Science*, *240*, 760 (1986).
22. S. Adapa, P. Huber, and W. Keller-Schierlein, *Helv. Chim. Acta*, *65*, 1818 (1982).
23. T. Nishio and Y. Ishida, *Agric. Biol. Chem.*, *54*, 1837 (1990).
24. T. Nishio, N. Tanaka, J. Hiratake, Y. Katsube, Y. Ishida, and J. Oda, *J. Am. Chem. Soc.*, *110*, 8733 (1988).
25. T. J. Brickman, J. G. Hansel, M. J. Miller, and S. K. Armstrong, *BioMetals*, *9*, 191 (1996).
26. C. H. Moore, L. A. Foster, D. G. Gerbig, Jr., D. W. Dyer, and B. W. Gibson, *J. Bacteriol.*, *177*, 1116 (1995).
27. Z. Hou, C. J. Sunderland, T. Nishio, and K. N. Raymond, *J. Am. Chem. Soc.*, *118*, 5148 (1996).
28. R. G. Arnold, M. R. Hoffman, T. J. DiChristina, and F. W. Picardal, *Appl. Environ. Microbiol.*, *56*, 2811 (1990).
29. F. H. Hulcher, *Biochemistry*, *21*, 4491 (1982).
30. W. G. McCullough and R. S. Merkal, *Bacteriology*, *137*, 243 (1979).
31. S. J. Bell, S. A. Friedman, and J. Leong, *Antimicrob. Agents Chemother.*, *15*, 384 (1979).
32. S. Ursic, B. Zorc, Y. Pilepic, and D. Vikic-Topic, *Croat. Chem. Acta*, *65*, 851 (1992).
33. T. M. Garrett, P. W. Miller, and K. N. Raymond, *Inorg. Chem.*, *28*, 128 (1989).
34. G. Serratrice, A. Zeghli, C. G. Béguin, P. Baret, and J.-L. Pierre, *New J. Chem.*, *17*, 297 (1993).
35. R. J. Bergeron and W. R. Weimar, *J. Bacteriol.*, *172*, 2650 (1990).
36. R. C. Scarrow, D. J. Ecker, C. Ng, S. Liu, and K. N. Raymond, *Inorg. Chem.*, *30*, 900 (1991).
37. Y. Tor, J. Libman, A. Shanzer, C. E. Felder, and S. Lifson, *J. Am. Chem. Soc.*, *114*, 6661 (1992).
38. L. D. Loomis and K. N. Raymond, *Inorg. Chem.*, *30*, 906 (1991).
39. A. Akiyama and T. Ikeda, *Chem. Lett.*, 849 (1995).
40. C. Y. Ng, S. J. Rodgers, and K. N. Raymond, *Inorg. Chem.*, *28*, 2062 (1989).

41. H. Boukhalfa, "Complexation du fer par des ligands tripodes à bases de sous unités 8-hydroxyquinoléine et 2,2′-dihydroxybiphenyle: Aspects thermodynamiques et cinétiques," PhD Thesis, Université Joseph Fourier, Grenoble (1996).
42. Y. Sun and A. E. Martell, *Tetrahedron*, *46*, 2725 (1990).
43. P. Yakirevitch, N. Rochel, A.-M. Albrecht-Gary, J. Libman, and A. Shanzer, *Inorg. Chem.*, *32*, 1779 (1993).
44. A.-M. Albrecht-Gary, S. Blanc, A. Z. Ocaktan, and M. A. Abdallah, *Inorg. Chem.*, *33*, 6391 (1994).
45. P. Demange, C. Wendenbaum, C. Linget, C. Mertz, M. T. Cung, A. Dell, and M. A. Abdallah, *BioMetals*, *3*, 155 (1990).
46. P. Demange, P. Bateman, A. Dell, and M. A. Abdallah, *Biochemistry*, *27*, 2745 (1988).
47. M. A. Abdallah, in *Handbook of Microbial Iron Chelates* (G. Winkelmann, ed.), CRC Press, Boca Raton, FL, 1991, p. 139.
48. M. Teintze, M. B. Hossain, C. I. Barnes, J. Leong, and D. van der Helm, *Biochemistry*, *20*, 6446 (1981).
49. S. Wendenbaum, P. Demange, A. Dell, J. M. Meyer, and M. A. Abdallah, *Tetrahedron Lett.*, *24*, 4877 (1983).
50. G. Briskot, H. Taraz, and H. Budzikiewicz, *Liebigs Ann. Chem.*, 1989, 375 (1989).
51. T. Palanché, "Mécanismes de coordination du fer: Applications biologiques et analytiques," Ph.D. Thesis, Université Louis Pasteur, Strasbourg, France (1995).
52. M. Y. Chae and A. W. Clarnik, *J. Fluoresc.*, *2*, 225 (1992).
53. S. D. Lytton, B. Mester, J. Libman, A. Shanzer, and Z. I. Cabantichik, *Anal. BioChem.*, *205*, 326 (1992).
54. H. Weizman, O. Ardon, B. Mester, J. Libman, O. Dwir, Y. Hadar, Y. Chen, and A. Shanzer, *J. Am. Chem. Soc.*, *118*, 12368 (1996).
55. F. Fages, B. Bodenant, and T. Weil, *J. Org. Chem.*, *61*, 3956 (1996).
56. T. B. Karpishin, T. D. P. Stack, and K. N. Raymond, *J. Am. Chem. Soc.*, *115*, 6115 (1993).
57. H. J. Rogers, C. Synger, B. Kimber, and P. M. Bayley, *Biochim. Biophys. Acta*, *497*, 548 (1977).
58. M. Akiyama, Y. Hara, and H. Gunji, *Chem. Lett.*, 225 (1995).
59. A. Schaffner, R. Hartmann, K. Taraz, and H. Budzikiewicz, *Z. Naturforsch.*, *51c*, 139 (1996).

60. N. Okujo, Y. Sakakibara, T. Yoshida, and S. Yamamoto, *BioMetals*, *7*, 170 (1994).
61. M. Jaquinod, E. Leize, N. Potier, A.-M. Albrecht-Gary, A. Shanzer, and A. van Dorsselaer, *Tetrahedron Lett.*, *34*, 2771 (1993).
62. M. T. Caudle, R. D. Stevens, and A. L. Crumbliss, *Inorg. Chem.*, *33*, 6111 (1994).
63. M. T. Caudle, R. D. Stevens, and A. L. Crumbliss, *Inorg. Chem.*, *33*, 843 (1994).
64. E. De Hoffmann and V. Stroobant, *Biol. Mass. Spectrom.*, *20*, 142 (1991).
65. S. Blanc, P. Yakirevitch, E. Leize, M. Meyer, J. Libman, A. van Dorsselaer, A.-M. Albrecht-Gary, and A. Shanzer, *J. Am. Chem. Soc.*, *119*, 4934 (1997).
66. M. E. Cass, T. M. Garrett, and K. N. Raymond, *J. Am. Chem. Soc.*, *111*, 1677 (1989).
67. B. Kersting, J. R. Telford, M. Meyer, and K. N. Raymond, *J. Am. Chem. Soc.*, *118*, 5712 (1996).
68. A. E. Martell and R. J. Motekaitis, *Determination and Use of Stability Constants*, 2nd ed., VCH, New York, 1992.
69. M. Meloun, J. Havel, and E. Hogfeldt, *Computation of Solution Equilibria*, Ellis Horwood, Halstead Press, New York, 1988.
70. A. E. Martell and R. M. Smith, *Critical Stability Constants*, Vol. 3, Plenum Press, New York, 1974, p. 301.
71. S. J. Barclay, B. H. Huynh, and K. N. Raymond, *Inorg. Chem.*, *23*, 2011 (1984).
72. D. C. Harris, *Quantitative Chemical Analysis*, 3rd ed., Freeman, New York, 1987.
73. R. T. Reid, D. H. Live, D. J. Faulkner, and A. Butler, *Nature*, *366*, 455 (1993).
74. G. Anderegg, F. L'Eplattenier, and G. Schwarzenbach, *Helv. Chim. Acta*, *46*, 1409 (1963).
75. A. Evers, R. D. Hancock, A. E. Martell, and R. J. Motekaitis, *Inorg. Chem.*, *28*, 2189 (1989).
76. I. Spasojević, I. Batinić-Haberle, P. L. Choo, and A. L. Crumbliss, *J. Am. Chem. Soc.*, *116*, 5714 (1994).
77. I. Batinć-Haberle and A. L. Crumbliss, *Inorg. Chem.*, *34*, 928 (1995).

78. I. Batinć-Haberle, I. Spasojević, R. A. Bartsch, and A. L. Crumbliss, *J. Chem. Soc. Dalton Transactions*, 2503 (1995).

79. I. Batinić-Haberle, I. Spasojević, and A. L. Crumbliss, *Inorg. Chem.*, *35*, 2352 (1996).

80. A. L. Crumbliss, I. Batinić-Haberle, and I. Spasojević, *Pure Appl. Chem.*, *68*, 1225 (1996).

81. I. Batinić-Haberle, I. Spasojević, and A. L. Crumbliss, *Inorg. Chim. Acta*, *260*, 35 (1997).

82. I. Batinić-Haberle, I. Spasojević, Y. Lang, R. A. Bartsch, and A. L. Crumbliss, *Inorg. Chem.* Submitted.

83. R. G. Wilkins, *Kinetics and Mechanism of Reactions of Transition Metal Complexes*, 2nd ed., VCH, Weinheim, Germany, 1991.

84. F. Bienfait, in *Iron Transport in Microbes, Plants and Animals* (G. Winkelmann, D. van der Helm, and J. B. Neilands, eds.), VCH, Weinheim, Germany, 1987, p. 339.

85. V. Römheld, in *Iron Transport in Microbes, Plants and Animals* (G. Winkelmann, D. van der Helm, and J. B. Neilands, eds.), VCH, Weinheim, Germany, 1987, p. 353.

86. C. J. Carrano and K. N. Raymond, *J. Bacteriol.*, *136*, 69 (1978).

87. T. Emery, in *Iron Transport in Microbes, Plants and Animals* (G. Winkelmann, D. van der Helm, and J. B. Neilands, eds.), VCH, Weinheim, Germany, 1987, p. 235.

88. R. Helman and G. D. Lawrence, *J. Electroanal. Chem.*, 276 (*Bioelectrochemistry and Bioenergetics*, *22*, 187) (1989).

89. J. B. Straka and T. Emery, *Biochim. Biophys. Acta*, *569*, 277 (1979).

90. J. F. Ernst and G. Winkelmann, *Biochim. Biophys. Acta*, *500*, 27 (1977).

91. J. E. L. Arceneaux and B. R. Byers, *J. Bacteriol.*, *141*, 715 (1980).

92. K. A. McCreary and C. Ratledge, *J. Gen. Microbiol.*, *113*, 67 (1979).

93. I. G. O'Brien, G. B. Cox, and F. Gibson, *Biochem. Biophys. Acta*, *237*, 537 (1971).

94. S. Heidinger, V. Braun, V. L. Pecoraro, and K. N. Raymond, *J. Bacteriol.*, *153*, 109 (1983).

95. D. J. Collins, C. Lewis, and J. M. Swan, *Aust. J. Chem.*, *27*, 2593 (1974).

96. W. C. Hollifield, Jr. and J. B. Neilands, *Biochemistry*, *17*, 1922 (1978).
97. G. S. Lodge, C. G. Gaines, J. E. L. Arcenaux, and B. R. Byers, *Biochem. Biophys. Res. Commun.*, *97*, 1291 (1980).
98. J. P. Robinson and J. V. McArdle, *J. Inorg. Nucl. Chem.*, *43*, 1951 (1981).
99. R. C. Hider, A. R. Mohd-Nor, J. Silver, I. E. G. Morrison, and L. V. G. Rees, *J. Chem. Soc. Dalton Trans.*, 609 (1981).
100. J. B. Neilands, *J. Biol. Chem.*, *270*, 26723 (1995).
101. R. J. Bergeron and G. M. Brittenham (eds.), *The Development of Iron Chelators for Clinical Use*, CRC Press, Boca Raton, FL, 1994.
102. A. E. Martell, W. F. Anderson, and D. G. Badman (eds.), *Development of Iron Chelators for Clinical Use*, Elsevier/North Holland, New York, 1981.
103. A. Dietrich, K. A. Fidelis, D. R. Powell, D. van der Helm, and D. L. Eng-Wilmot, *J. Chem. Soc. Dalton Trans.*, 231 (1991).
104. A. Dietrich, D. R. Powell, D. L. Eng-Wilmot, M. B. Hossain, and D. van der Helm, *Acta Crystallogr., Sect. C*, *46*, 816 (1990).
105. R. Mocharla, D. R. Powell, C. L. Barnes, and D. van der Helm, *Acta Crystallogr., Sect. C.*, *39*, 868 (1983).
106. D. van der Helm and M. Poling, *J. Am. Chem. Soc.*, *98*, 82 (1976).
107. M. B. Hossein, D. van der Helm, and M. Poling, *Acta Crystallogr., Sect. B*, *39*, 258 (1983).
108. B. Monzyk and A. L. Crumbliss, *J. Am. Chem. Soc.*, *101*, 6203 (1979).
109. C. P. Brink and A. L. Crumbliss, *Inorg. Chem.*, *23*, 4708 (1984).
110. M. T. Caudle and A. L. Crumbliss, *Inorg. Chem.*, *33*, 4077 (1994).
111. M. Biruš, Z. Bradić, N. Kujundžić, M. Pribanić, P. C. Wilkins, and R. G. Wilkins, *Inorg. Chem.*, *24*, 3980 (1985).
112. S. Konetschny-Rapp, G. Jung, K. N. Raymond, J. Meiwes, and H. Zahner, *J. Am. Chem. Soc.*, *114*, 2224 (1992).
113. Z. Hou, D. W. Whisenhunt, Jr., J. Xu, and K. N. Raymond, *J. Am. Chem. Soc.*, *116*, 840 (1994).
114. S. J. Rogers and K. N. Raymond, *J. Med. Chem.*, *26*, 439 (1983).
115. Y. Hara and M. Akiyama, *Inorg. Chem.*, *35*, 5173 (1996).

116. M. Akiyama, A. Katoh, M. Iijima, T. Takagi, K. Natori, and T. Kojima, *Bull. Chem. Soc. Jpn.*, *65*, 1356 (1992).
117. M. Akiyama, A. Katoh, J. Kato, K. Takahashi, and K. Hattori, *Chem. Lett.*, 1189 (1991).
118. R. J. Motekaitis, Y. Sun, and A. E. Martell, *Inorg. Chem.*, *30*, 1554 (1991).
119. M. A. Esteves, M. C. T. Vaz, M. L. S. Simoes Goncalves, E. Farkas, and M. A. Santos, *J. Chem. Soc. Dalton*, 2565 (1995).
120. R. C. Scarrow, D. L. White, and K. N. Raymond, *J. Am. Chem. Soc.*, *107*, 6540 (1985).
121. M. T. Caudle, L. P. Cogswell, III, and A. L. Crumbliss, *Inorg. Chem.*, *33*, 4759 (1994).
122. M. T. Caudle, C. D. Caldwell, and A. L. Crumbliss, *Inorg. Chim. Acta*, *240*, 519 (1995).
123. M. A. Santos, M. A. Esteves, M. C. T. Vaz, and M. L. S. S. Goncalves, *J. Chem. Soc. Dalton*, 927 (1993).
124. M. A. Santos, M. A. Esteves, M. C. T. Vaz, and M. L. S. S. Goncalves, *Inorg. Chim. Acta*, *214*, 47 (1993).
125. T. M. Garrett, T. J. McMurry, M. W. Hosseini, Z. E. Reyes, F. E. Hahn, and K. N. Raymond, *J. Am. Chem. Soc.*, *113*, 2965 (1991).
126. T. D. P. Stack, Z. Hou, and K. N. Raymond, *J. Am. Chem. Soc.*, *115*, 6466 (1993).
127. K. N. Raymond, *Pure Applied Chem.*, *66*, 773 (1994).
128. B. Tse and Y. Kishi, *J. Am. Chem. Soc.*, *115*, 7892 (1993).
129. E. J. Enemark and T. D. P. Stack, *Inorg. Chem.*, *35*, 2719 (1996).
130. R. J. Motekaitis and A. E. Martell, *Inorg. Chim. Acta*, *183*, 71 (1991).
131. R. C. Hider, J. B. Porter, and S. Singh, in *The Development of Iron Chelators for Clinical Use* (R. J. Bergeron and G. M. Brittenham, eds.), CRC Press, Boca Raton, FL, 1994, p. 353.
132. R. Ma, J. J. Reibenspies, and A. E. Martell, *Inorg. Chim. Acta*, *223*, 21 (1994).
133. E. T. Clarke and A. E. Martell, *Inorg. Chim. Acta*, *196*, 185 (1992).
134. R. M. Smith and A. E. Martell, *Critical Stability Constants, Vols. 1–6*, Plenum, New York (1974, 1975, 1976, 1977, 1982, 1989).
135. P. S. Dobbin, R. C. Hider, A. D. Hass, P. D. Taylor, P. Sargon, J. B.

Porter, G. Xiao, and D. van der Helm, *J. Med. Chem.*, *36*, 2448 (1993).

136. M. Streator, P. D. Taylor, R. C. Hider, and J. Porter, *J. Med. Chem.*, *33*, 1749 (1990).
137. G. Xiao, D. van der Helm, R. C. Hider, and P. S. Dobbin, *Inorg. Chem.*, *34*, 1268 (1995).
138. G. Xiao, D. van der Helm, R. C. Hider, and B. L. Rai, *J. Phys. Chem.*, *100*, 2345 (1996).
139. R. Ma and A. E. Martell, *Inorg. Chim. Acta*, *209*, 71 (1993).
140. P. A. Ganeshpure, D. Kar, and S. Satish, *Transition Met. Chem.*, *17*, 212 (1992).
141. A. E. Martell, R. J. Motekaitis, E. T. Clarke, R. Delgado, Y. Sun, and R. Ma, *Supramol. Chem.*, *6*, 353 (1996).
142. G. Serratrice, C. Mourral, A. Zeghli, C. G. Béguin, P. Baret, and J. L. Pierre, *New J. Chem.*, *18*, 749 (1994).
143. P. Baret, C. G. Beguin, H. Boukhalfa, C. Caris, J.-P. Laulhere, J.-L. Pierre, and G. Serratrice, *J. Am. Chem. Soc.*, *117*, 9760 (1995).
144. C. Caris, P. Baret, C. G. Béguin, G. Serratrice, J.-L. Pierre, and J.-P. Laulhere, *Biochem. J.*, *312*, 879 (1995).
145. G. Lescoat, N. Chenoufi, G. Zanninelli, O. Loreal, N. Pasdeloup, P. Brissot, C. Caris, P. Baret, and J.-L. Pierre, *Hepatology* (St. Louis), *20*, 182A (1994).
146. M. Shenker, R. Ghirlando, I. Oliver, M. Helmann, Y. Hadar, and Y. Chen, *Soil Sci. Soc. Am. J.*, *59*, 837 (1995).
147. D. E. Crowley, Y. C. Wang, C. P. Reid, and P. J. Szanizlo, *Iron Nutrition and Interactions in Plants*, Kluwer, London, 1991.
148. C. Nomoto, Y. Sugiura, and S. Takagi, in *Iron Transport in Microbes, Plants and Animals* (Winkelmann, G., Van der Helm, D. and Neilands, J., eds.), VCH, Weinheim, Germany, 1987, p. 401.
149. M. Biruš, N. Kujundžić, and M. Pribanić, *Prog. Reaction Kinetics*, *18(3)*, 171 (1993).
150. C. H. Langford and H. B. Gray, *Ligand Substitution Processes*, W. A. Benjamin, New York, 1965.
151. F. Basolo and R. G. Pearson, *Mechanisms of Inorganic Reactions*, 2nd ed., Wiley, New York, 1967.
152. R. W. Swaddle and A. E. Merbach, *Inorg. Chem.*, *20*, 4212 (1981).

153. M. Grant and R. B. Jordan, *Inorg. Chem.*, *20*, 55 (1981).
154. H. W. Dodgen, G. Liu, and J. P. Hunt, *Inorg. Chem.*, *20*, 1002 (1981).
155. A. Cusanelli, U. Frey, D. T. Richens, and A. E. Merbach, *J. Am. Chem. Soc.*, *118*, 5265 (1996).
156. M. Mizuno, S. Funahashi, N. Nakasuka, and M. Tanaka, *Inorg. Chem.*, *30*, 1550 (1991).
157. J. M. Knudsen, E. Larsen, J. E. Moreira, and O. F. Neilsen, *Acta Chem. Scand., Ser A,*, *A29*, 833 (1975).
158. H. K. Pan, D. J. Yarusso, G. S. Knapp, M. Pineri, A. Meagher, J. M. D. Coey, and S. L. Cooper, *J. Chem. Phys.*, *79*, 4736 (1983).
159. C. Ropars, M. Rougee, M. Momenteau, and D. Lexa, *J. Chim. Phys. Phys-Chim. Biol.*, *65*, 816 (1957).
160. H. Po and N. Sutin, *Inorg. Chem.*, *10*, 428 (1971).
161. R. S. Sapleszko, R. C. Patel, and E. Matijevic, *J. Phys. Chem.*, *81*, 1061 (1977).
162. R. M. Milburn, *J. Am. Chem. Soc.*, *79*, 537 (1957).
163. I. Batinić-Haberle, M. Biruš, and M. Pribanić, *Inorg. Chem.*, *30*, 4882 (1991).
164. M. Biruš, M. Gabričević, O. Kronja, and B. Klaic, *Inorg. Chem.*, *34*, 3110 (1995).
165. M. Biruš and R. van Eldik, *Inorg. Chem.*, *30*, 4559 (1991).
166. B. Monzyk and A. L. Crumbliss, *J. Am. Chem. Soc.*, *104*, 4921 (1982).
167. M. Biruš, Z. Bradić, G. Krznarić, N. Kujundžić, M. Pribanić, P. C. Wilkins, and R. G. Wilkins, *Inorg. Chem.*, *26*, 1000 (1987).
168. Z. Zhang and R. B. Jordan, *Inorg. Chem.*, *35*, 1571 (1996).
169. I. Batinić-Haberle, I. Spasojević, and A. L. Crumbliss, *Inorg. Chem.*, *33*, 3151 (1994).
170. T. Ozawa, K. Iwai, K. Jitsukawa, H. Masuda and H. Einaga, *Polyhedron*, *13*, 1523 (1994).
171. E. Farkas, J. Szoke, T. Kiss, H. Kozlowski, and W. Bal, *J. Chem. Soc., Dalton Trans.*, 2247 (1989).
172. T. G. Spiro, L. Pape, and P. Saltman, *J. Am. Chem. Soc.*, *89*, 5555 (1967).

173. M. Grootveld, J. D. Bell, B. Halliwell, O. I. Aruoma, A. Bomford and P. J. Sadler, *J. Biol. Chem.*, *264*, 4417 (1989).
174. H. G. Parkes, R. E. Allen, A. Furst, D. R. Blake, and M. C. Grootveld, *J. Pharm. Biomed. Anal.*, *9*, 29 (1991).
175. B. Faller and H. Nick, *J. Am. Chem. Soc.*, *116*, 3860 (1994).
176. M. A. Kline and C. Orvig, *Clin. Chem.*, *38*, 562 (1992).
177. A. Katoh and M. Akiyama, *J. Chem. Soc. Perkin Trans.*, 1839 (1991).
178. M. A. Santos, M. A. Esteves, and J. M. G. Marinho, *J. Chem. Soc. Dalton*, 3123 (1993).
179. G. B. Wong, M. J. Kappel, K. N. Raymond, B. Matzanke, and G. Winkelmann, *J. Am. Chem. Soc.*, *105*, 810 (1983).
180. B. F. Anderson, D. A. Buckingham, G. B. Robertson, J. Webb, K. S. Murray, and D. E. Clark, *Nature*, *262*, 772 (1976).
181. G. Anderegg, F. L'Eplattenier, and G. Schwarzenbach, *Helv. Chim. Acta*, *46*, 1400 (1963).
182. T. Murakami, K. Ise, M. Hayakawa, S. Kamei, and S. Takag, *Chem. Lett.*, 2139 (1989).
183. C. J. Carrano, H. Drechsel, D. Kaiser, G. Jung, B. Matzanke, G. Winkelmann, N. Rochel, and A. M. Albrecht-Gary, *Inorg. Chem.*, *35*, 6429 (1996).
184. W. R. Harris, C. J. Carrano, and K. N. Raymond, *J. Am. Chem. Soc.*, *101*, 2722 (1979).
185. S. Blanc, T. Palanché, and A.-M. Albrecht-Gary, unpublished results, 1996.
186. H. Bickel, G. E. Hall, W. Keller-Schierlein, V. Prelog, E. Vischer, and A. Wettstein, *Helv. Chim. Acta*, *43*, 2129 (1960).
187. S. R. Cooper, J. V. McArdle, and K. N. Raymond, *Proc. Natl. Acad. Sci. USA*, *75*, 3551 (1978).
188. E. F. Wawrousek and J. V. McArdle, *J. Inorg. Biochem.*, *17*, 169 (1982).
189. C. W. Lee, D. J. Ecker, and K. N. Raymond, *J. Am. Chem. Soc.*, *107*, 6920 (1985).
190. E. Bar-Ness, Y. Chen, Y. Hadar, H. Marschner, and V. Römheld, in *Iron Nutrional Interest in Plants* (Y. Chen and Y. Hadar, eds.), Kluwer, Dordrecht, The Netherlands, 1991, p. 271.
191. E. T. Clarke and A. E. Martell, *Inorg. Chim. Acta*, *181*, 273 (1991).

192. M. Biruš, N. Kujundžić, and M. Pribanić, *Inorg. Chim. Acta*, *55*, 65 (1980).

193. M. J. Sisley and R. B. Jordan, *Inorg. Chem.*, *30*, 2190 (1991).

194. S. S. Massoud and R. B. Jordan, *Inorg. Chem.*, *30*, 4851 (1991).

195. M. Persmark and J. B. Neilands, *BioMetals*, *5*, 29 (1992).

196. K. Langemann, D. Heineke, S. Rupprecht, and K. N. Raymond, *Inorg. Chem.*, *35*, 5663 (1996).

197. J. B. Neilands, T. Peterson, and S. A. Leong, *ACS Symp. Ser.*, *140*, 263 (1980).

198. J. P. Adjimani and T. Emery, *J. Bacteriol.*, *169*, 3664 (1987).

199. R. C. Scarrow, P. E. Riley, K. Abu-Dari, D. L. White, and K. N. Raymond, *Inorg. Chem.*, *24*, 952 (1985).

200. S. J. Rogers, C. W. Lee, C. Y. Ng, and K. N. Raymond, *Inorg. Chem.*, *26*, 1622 (1987).

201. J. Xu and R. B. Jordan, *Inorg. Chem.*, *27*, 1502 (1988).

202. V. L. Pecoraro, W. P. Harris, G. B. Wong, C. J. Carrano, and K. N. Raymond, *J. Am. Chem. Soc.*, *105*, 4623 (1983).

8

Biomimetic Siderophores: From Structural Probes to Diagnostic Tools[a]

Abraham Shanzer and Jacqueline Libman*[†]

Department of Organic Chemistry, The Weizmann Institute of Science, Rehovot 76100, Israel

1. INTRODUCTION 330
2. METHODOLOGY 333
3. ENTEROBACTIN: THE MINIMALISTIC APPROACH TOWARD PERFECTION 335
4. HYDROPHILIC FERRICHROME ANALOGUES AS SPECIES-SPECIFIC MICROBIAL GROWTH PROMOTERS 338
5. HYDROPHILIC FERRIOXAMINE ANALOGUES AS VARIABLE RANGE MICROBIAL GROWTH PROMOTERS 340

---

[a]This chapter is dedicated to the memory of Dr. Jacqueline Libman who untimely passed away on March 30, 1997.

*To whom correspondence should be addressed.

6. FLUORESCENT LABELED SIDEROPHORE ANALOGUES: PROBES FOR IRON TRAFFICKING IN REAL TIME 343

7. LIPOPHILIC FERRICHROME ANALOGUES OR REVERSED SIDEROPHORES AS IRON SCAVENGERS 346

8. SUMMARY AND CONCLUSION 349

ACKNOWLEDGMENTS 350

ABBREVIATIONS 351

REFERENCES 351

1. INTRODUCTION

Most living systems require iron for their growth and development. Iron is the second most abundant metal on the earth crust. Yet, it occurs mostly as insoluble oxide and is poorly soluble in water at ambient conditions. Therefore, biological systems had to devise a variety of strategies to solubilize this essential metal and incorporate it into their functional constituents. Even more so, ingenious strategies had to be developed to maintain intracellular iron levels within very narrow, physiologically permissible windows. Siderophores are low molecular weight iron(III) binders of microbial origin that are excreted into the environment where they effectively scavenge iron(III) [1–6]. Whereas the free siderophores fail to penetrate biological membranes, the structurally well-defined siderophore-iron(III) complexes are taken up by microbial cells via highly specific membrane receptors. They transverse the biological membranes in energy-dependent processes and release the bound iron(III) to the cytoplasm (Fig. 1). Release of iron(III) may involve reduction, ligand exchange, or decomposition of the siderophore to its constituents. Whatever the detailed mechanism, the dependence of most living systems on iron for growth and development, the large variety and wide distribution of siderophores, and of their matching membrane receptors offer a superb playground for the study of recognition phenomena [1–8].

The increasing interest in microbial iron uptake derives from

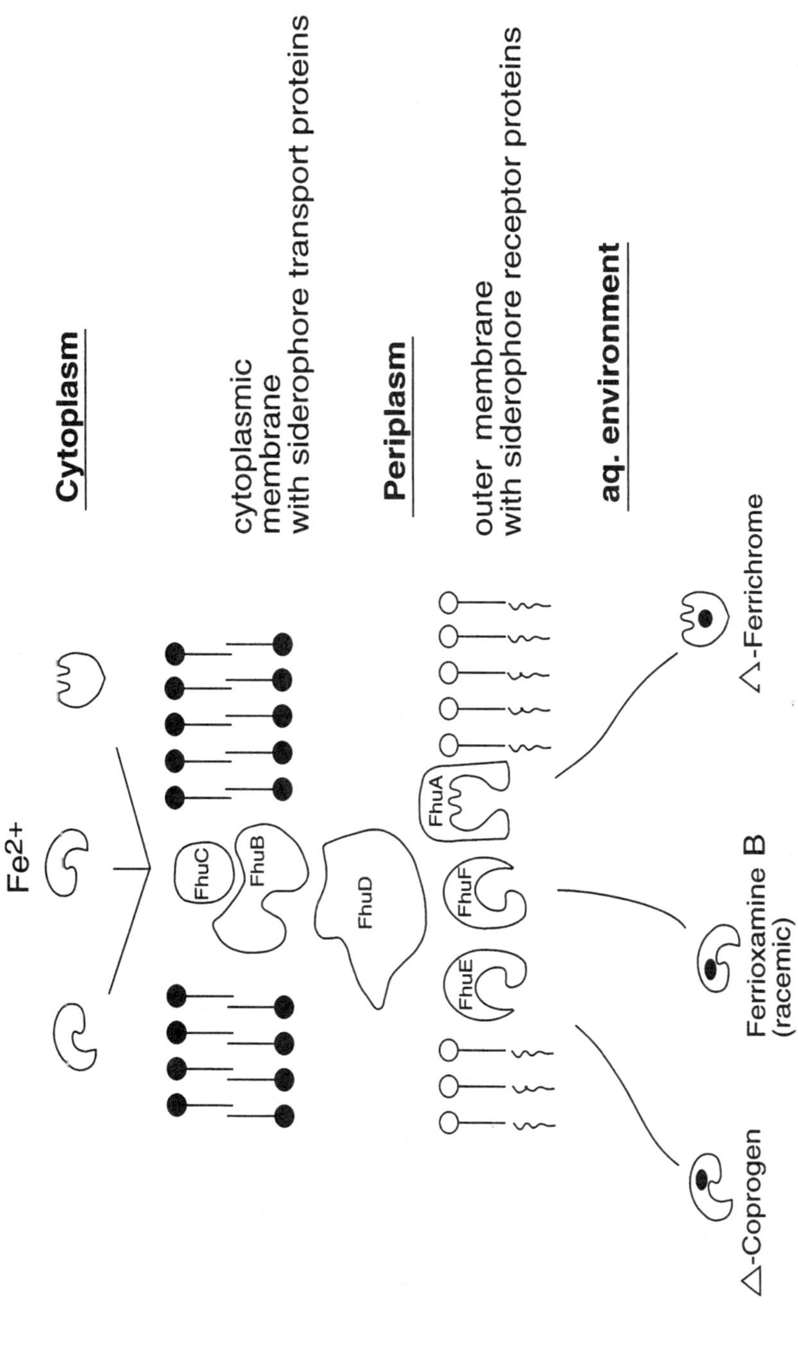

FIG. 1. Schematic representation of siderophore-mediated iron(III) uptake across microbial membranes.

several considerations. First, microorganisms represent superb model systems because of their large variety, and because some of the principles operating in these systems might also be of relevance to higher organisms. Second, microorganisms are directly relevant to humans. Microorganisms have been proven to be invaluable in food technologies since ancient times, and they are gaining increasing importance as precision workers for the production of hormones and enzymes alike. Finally, pathogenic microorganisms are increasingly threatening human health as result of their augmenting resistance to currently available drugs and the increasing numbers of immunocompromised patients. Phytoplankton is increasingly being linked to the composition of the atmosphere, and iron-stimulated growth of phytoplankton is believed to reduce atmospheric CO_2, minimize the green house effect, and reduce global warming.

Our interest in microbial iron uptake, and particularly siderophore-mediated microbial iron uptake, was propelled by all these considerations and by our interest in biomimetic chemistry. Thus our goal was to reproduce the properties of natural molecules with the simplest possible synthetic compounds, and we specifically aimed at synthesizing biomimetic iron carriers that would reproduce the performance of the natural compounds both in vitro and in vivo. Biomimetic siderophores appeared to us to be superb targets. By pursuing these targets, we hoped to provide an instructive exercise in molecular recognition, including chiral recognition, to teach us about the structural requirements for specific iron recognition to occur, and to provide insight into the structural requirements for siderophore-iron recognition by specific microbial receptors and iron-uptake systems. It was further hoped that modifying the biomimetic iron carriers' envelopes could provide either growth inhibitors or growth promoters. Iron-free, lipophilic carriers were envisioned to inhibit growth by effectively penetrating cellular membranes by diffusion, scavenging iron from intracellular stores, and egress the cells with their iron loads. Iron-loaded, hydrophilic carriers were anticipated to act as growth promoters when fit to introduce iron via microbial iron-uptake systems. The latter family of compounds could provide versatile structural probes to differentiate between analogous receptors of different organisms or act as Trojan horses to smuggle visualizing labels or cytotoxic agents into microorganisms, thereby providing new generations of diagnostic tools and possibly antimicrobials. It was also hoped that the structural principles operating in natural siderophores would assist in the synthesis of binders for metal ions other than iron, including metal ions of biological relevance such as copper, vanadium, or zinc. In this chapter we will highlight our contributions in each of the above-mentioned directions by concentrating on a few examples from our laboratories rather than screening the literature.

2. METHODOLOGY

In our search for biomimetic siderophores that would mimic some of the properties of the natural siderophores both in vitro and in vivo, we were confronted with a formidable task: namely, the requirement to match a "key" to a "lock" of still unknown structure. In order to succeed in this endeavor, it is necessary that the first generation of compounds show at least some biological activity. Achieving any biological response, even if very modest, is essential to provide a point of departure for future optimizations. In an attempt to meet these stringent demands, we developed a strategy that combines three elements in an iterative fashion [9–13]:

1. A modular, "lego-type" design and synthesis of the target molecules which enables systematic modifications till optimal performance is achieved
2. The use of theoretical calculations for the detailed design of the molecules
3. Fast and reliable assays to probe the performance of the compounds prepared

The modular assembly facilitated systematic modifications such that each generation of compounds profited from progress made with earlier ones. Theoretical modeling helped to select from the many possible structures the most promising for synthesis, to interpret the experimental findings, and to suggest improved structures for further reiteration processes.

We employed whole microorganisms (actually mutants that do not produce siderophores) as test systems, and compared the activity of the synthetic siderophores with that of the natural counterparts. In order to identify the nature of the receptors involved in the carrier-mediated iron(III) uptake, we applied a combination of methods that relied on either of two principles [14,15]: (1) growth promotion and cellular $^{55}Fe^{3+}$ uptake in competition with the natural siderophores and/or (2) comparison of the response of wild-type organisms with that of mutants that lack specific receptors [5,16].

According to these principles, four families of biomimetic siderophore analogues, analogues of enterobactin [17], ferrichrome [18], ferrioxamine B [19], and coprogen [19] (Fig. 2), were prepared and their properties examined.

We selected the above four siderophores as target structures because they may be regarded as prototypes of the more than 200 siderophores known so far; many of which had been characterized by x-ray diffraction studies [1–6]. They encompass representatives of the two

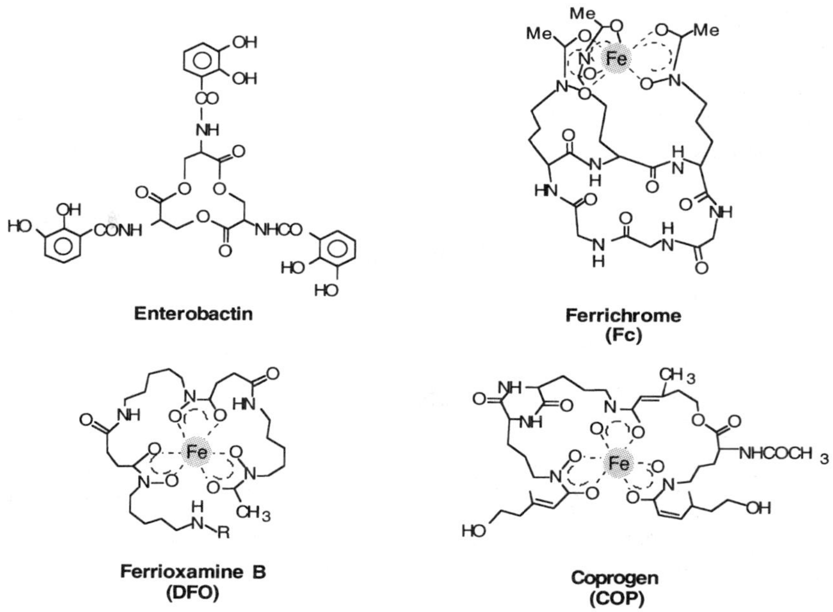

FIG. 2. Representative siderophores: enterobactin [3], ferrichrome [20], ferrioxamine [1], and coprogen. The iron(III) complexes of enterobactin and coprogen adopt the Δ-configuration, and the ferrichrome complex the Λ-configuration. Ferrioxamine B lacks chiral centers and forms five isomeric complexes, each as a racemic mixture [21].

major classes; those that are based on catecholates as ion-binding sites and those that make use of hydroxamates. Moreover, they include the two major topologies: the tripodal one and the linear one, both of which embed iron(III) into an octahedral ion-binding cavity, and they include derivatives that possess chiral centers which determine the absolute configuration of the iron(III) complexes. The absolute configuration of siderophore-iron(III) complexes may often be crucial in determining microbial activity, since many siderophore receptors exhibit high enantioselectivity [8,22,23].

3. ENTEROBACTIN: THE MINIMALISTIC APPROACH TOWARD PERFECTION

Our first target molecules were enterobactin analogues [17,24]. We chose enterobactin as the first system because of its high symmetry and exceptionally high iron(III)-binding efficiency [3]. Enterobactin is a tripode-like molecule consisting of a 12-membered trilactone ring which serves as an anchor and three symmetrically projecting side chains that bear the ion-binding catecholates [3]. When binding iron(III), the catecholates generate an octahedral cavity of Δ-cis configuration. Pioneering work of Raymond and coworkers demonstrated that the proximity of the amide NH group to the catecholate group is essential for efficient binding to occur, and they synthesized the first triscatecholates with enterobactin-like activity [3,25–27]. Yet, the still large difference between the binding efficiency of the natural and the synthetic compounds led us to speculate that there must be additional structural elements that are responsible for enterobactin's superiority.

In an attempt to identify the stereochemical peculiarities of enterobactin, we proceeded in a stepwise fashion. We first established the very nature of enterobactin's trilactone anchor by synthesizing the parent trilactone and establishing its structure by x-ray diffraction [28]. The trilactone ring was found to adopt a C_3-symmetrical conformation and to have its lactone-carbonyl and lactone-ether groups pointing to opposite faces of the molecule's ring plane. We then extended the parent trilactone symmetrically with three benzamide groups, thereby obtaining a structural analogue of enterobactin, trisbenzamide (TBA), with the trilactone functioning as anchor [24]. TBA, in contrast to enterobactin, did lend itself to full spectroscopic examination and structural characterization. It thus served us as a sounding board to test the reliability of EFF (empirical force field) calculations and to link experiment and theory. Once the validity of such calculations had been established for TBA, they were expanded to the natural enterobactin in order to provide insight into its structural characteristics and their effect on the molecule's binding properties (Fig. 3).

The above studies [24] revealed that TBA as well as free enterobactin adopt a propeller-type conformation with the side chains axially positioned and the amide-NH groups H-bonded to the lactone oxygen of the ring. On ion binding, a rearrangement of the H-bonding network occurs: the H-bonds between the amide-NH groups and the ring oxygens become weakened and replaced by H-bonds between the amide-NH groups and the catecholate anions. The H-bonding network of enterobactin thus shapes the molecule in the free state to adopt a conformation prone to ion binding and rearranges to stabilize the complex once

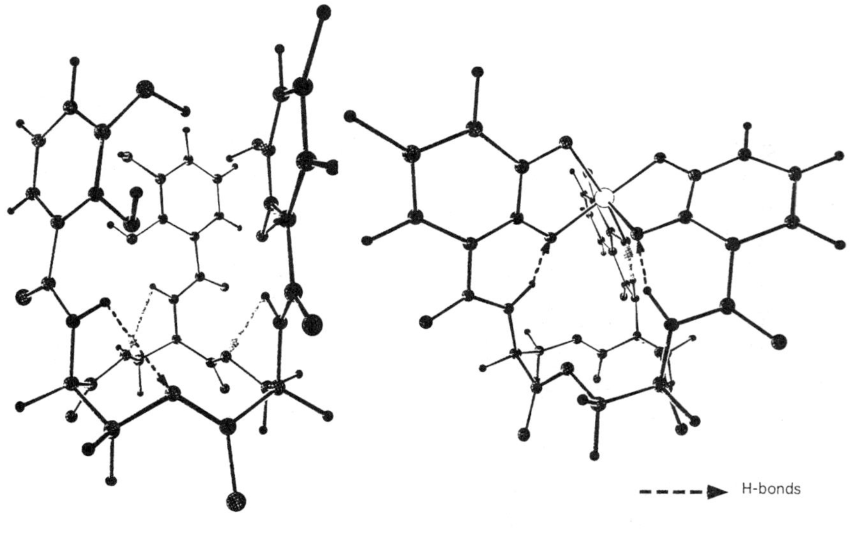

FIG. 3. Calculated lowest energy conformation of enterobactin (left) and of its iron(III) complex (right) [24]. The H-bonds between the amide NH protons and the lactone oxygens in the free ligand (left) shift on complexation to form H-bonds between the amide NH protons and the catecholate oxygens.

formed. (We like to term this H-bond-shift "wiper-shift", as it is reminiscent of the movement of windshield wipers.) Much to our satisfaction, recent x-ray diffraction studies of the crystalline V(IV)-enterobactin complex [29] confirmed the details of the structure of the enterobactin-iron(III) complex that had been derived from our calculations [24].

In our search for biomimetic enterobactin analogues, we aimed to reproduce the conformational characteristics of the natural compound with synthetically versatile and easily accessible molecules. The idea was to reproduce the H-bond network of enterobactin with simple, tripodal structures by replacing the H-bonds between the side chains and the ring with belts of H-bonds between the chains (Fig. 4). Guided by this rational, we synthesized chiral molecules that make use of trisamines as "anchors" and extend amino acid residues as structural units (H-bond formers) [30]. Indeed, these trispeptides proved to form interstrand H-bonds between different kinds of amide linkages, and to thereby adopt propeller-like arrangements of preferred chiral sense, as established by a combination of experimental and theoretical tools

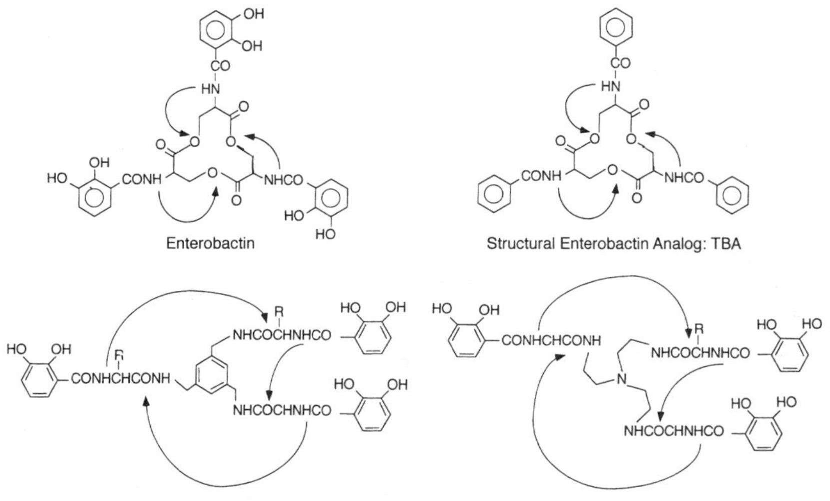

FIG. 4. H-bond networks that impose propeller-like conformations in enterobactin and in its structural analogue TBA [24] (top) and in biomimetic enterobactin analogues [17] (bottom).

[30,31]. Extension of these structures with catechol groups provided binders that form iron(III) complexes of preferential Δ-cis configuration, as does the enterobactin-iron(III) complex [32]. Moreover, the iron(III)-binding efficiency of these compounds exceeded that of the nonchiral precursor, and proved to be related to the strength of their H-bond networks, where some of the derivatives approached the binding efficiency of the natural siderophore within three orders of magnitude [17]. It is interesting to note that an alternative approach toward predisposition and enhanced binding efficiency has recently been put forward. This approach relies exclusively on van der Waals forces rather than H-bonds [33].

The thorough analysis of enterobactin demonstrated the advantage of preorganization via noncovalent, intramolecular interactions for the generation of highly efficient ion binders.

4. HYDROPHILIC FERRICHROME ANALOGUES AS SPECIES-SPECIFIC MICROBIAL GROWTH PROMOTERS

In an attempt to establish whether the principles developed for enterobactin analogues could be of general value, we extended them for the design of biomimetic ferrichrome analogues [20]. Ferrichrome, in contrast to enterobactin, is asymmetrical and makes use of hydroxamates instead of catecholates as ion-binding sites, and it is biologically active as its Λ-cis rather than Δ-cis iron complex [20]. In order to obtain ferrichrome analogues, we (1) replaced the nonsymmetrical hexapeptide ring of the natural compound by a C_3-symmetrical triscarboxylate as anchor, (2) compensated for the concurrent loss of chirality by extending the anchor with amino acids as variable chiral elements, and (3) terminated the chains with hydroxamate groups [18]. In spite of their greatly simplified structures, some of these binders (type 2, m = 2; Fig. 5) proved to mimic ferrichrome fully in its iron-binding properties and microbial activity [14,34,35].

Taking advantage of the favorable solubility properties of the ferrichrome analogues and of their metal complexes, we examined their structures in further detail [18]. We demonstrated that networks of

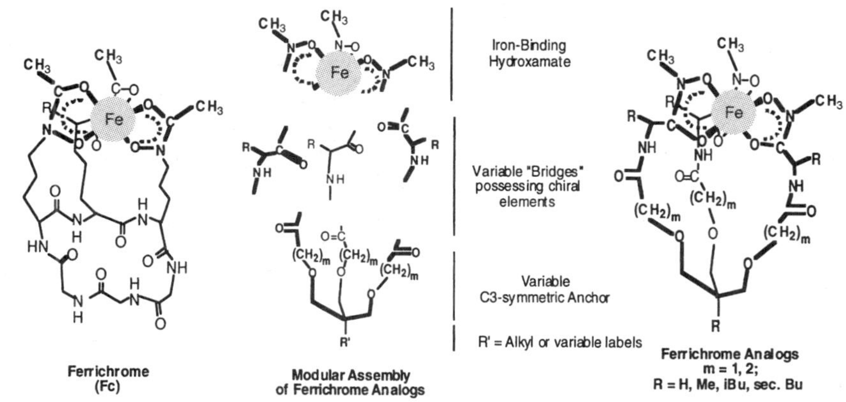

FIG. 5. Design of ferrichrome analogues.

inter- and intrastrand H-bonds shape the conformations of the free ligands toward propeller-like arrangements in both families of analogues prepared (type 1, m = 1 and type 2, m = 2; Fig. 5). When loaded with metal ions, van der Waals forces, H-bonds, or a combination of both stabilize the complexes formed, and thereby promote the formation of well-defined 1:1 complexes of high isomeric and optical purity [13,18]. In the absence of such intramolecular forces, mixtures of polynuclear and polymeric complexes are obtained.

Some of the hydrophilic type 2 analogues, but none of the type 1 analogues, exerted quite remarkable microbial activity [13,14]. Within the bioactive family of compounds, some fully mimicked ferrichrome as iron carrier, whereas others acted as inhibitors [14]. Moreover, the binders' activity differed from organism to organism. Thus the L-ala derivative (m = 2, R = Me; Fig. 5) acted as iron(III) carrier and growth promoter in *Arthrobacter flavescens* [34] but as inhibitor in *Pseudomonas putida* [35]. The gly derivative (m = 2, R = H; Fig. 5), on the other hand, acted as iron(III) carrier even in *Pseudomonas putida* [35,36]. In close analogy to the natural siderophore [37], the enantiomeric (D-ala) derivative (type 2, m = 2; Fig. 5) did not show ferrichrome-like activity in any of the organisms examined.

The scale of activities of the different ferrichrome analogues enabled us to identify those domains that are critical for attaining biological activity [14]. They showed that the domain around the iron(III)

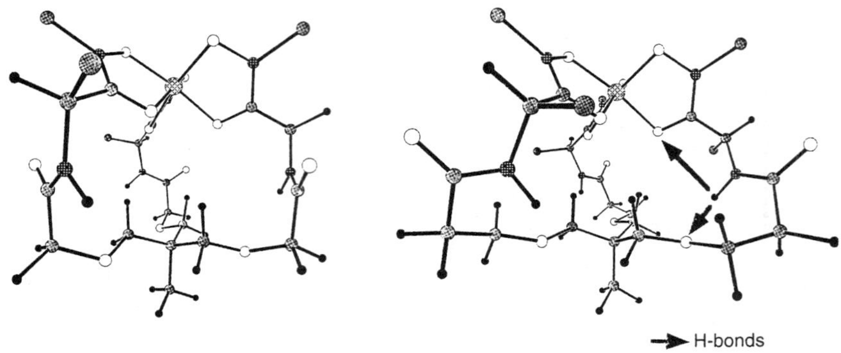

FIG. 6. Calculated lowest energy conformation of ferrichrome analogues (left, biologically inactive type 1 analogues; right, bioactive, type 2 analogues, with intramolecular H-bonds) [13].

center and its chiral sense have to be preserved, whereas the anchor may be modified. The pronouncedly different behavior of type 1 and type 2 analogues are believed to derive from long-range conformational effects: in type 2 analogues the iron(III)-binding domain is well exposed, whereas in type 1 analogues, the side chains shield the iron(III)-binding domain prohibiting recognition (Fig. 6) [13].

5. HYDROPHILIC FERRIOXAMINE ANALOGUES AS VARIABLE RANGE MICROBIAL GROWTH PROMOTERS

The success with the ferrichrome analogues encouraged us to approach a more complex problem: the synthesis of ferrioxamine analogues (Fig. 7) [38]. The ferrioxamines, similarly to ferrichrome, make use of hydroxamate groups as ion-binding sites [1,38,39]. Yet, at variance with the ferrichromes, the ferrioxamines are linear structures that often lack chiral centers. They all possess three hydroxamate groups on a string which are bridged by amide or ester containing methylene chains. The most studied representative is the linear ferrioxamine B, which became the most used drug for the treatment of iron overload. Ferrioxamine B forms a total of five isomers when binding three-valent metal ions,

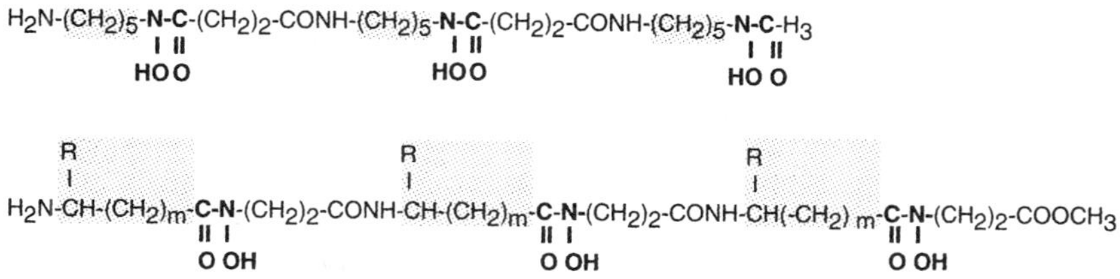

m = 0; R = Me (ala), iBu (leu), CH_2CONEt_2 (asp)
m = 2; R = $CONEt_2$ (glu)

FIG. 7. Natural ferrioxamine B (top) [38] and biomimetic ferrioxamine analogues (bottom) [19].

as established by examination of its kinetically inert Cr(III) complexes [21].

In an attempt to examine the possibility of geometrical discrimination by the ferrioxamine receptor, we aimed at ferrioxamine B analogues that would form a smaller number of configurational isomers when binding iron(III) [19]. Considering the exclusive formation of cisoid complexes from both enterobactin and ferrichrome, we settled for ferrioxamine analogues of cisoid configuration. Toward this goal we shortened the bridges between the binding groups and reversed the directionality of the hydroxamate linkages relative to that in the natural compound (Fig. 7). This reversal was conceived not to affect the compounds' biological activity (as earlier shown with ferrichrome analogues) [34,35,37], but greatly to facilitate synthesis by relying on natural amino acids as building blocks. The use of natural amino acids as synthons also enabled us to introduce chiral centers that direct the configuration of the complexes to either the left- or right-handed chiral sense (Fig. 7) [19] and to examine whether the ferrioxamine B receptor could differentiate between enantiomers. Considering the preponderance of chiral siderophores, we speculated that "archaic" ferrioxamine B might have been chiral and consisted of chiral lysine residues, which merely lost their carboxyl groups in the course of evolution.

We thus assembled chiral ferrioxamine analogues from three identical monomers, each composed of a chiral amino acid linked to 3-*O*-benzyl-hydroxy-amino-propionic acid via a hydroxamate group (Fig. 7) [19]. All analogues prepared were found to bind iron(III) in a 1:1 stoichiometry and to adopt preferentially the Δ-cis configuration when composed of *L*-amino acids [19].

The ferrioxamine analogues possessing a single methylene bridge (m = 0, R = Me, iBu, CH_2CONEt_2; Fig. 7) facilitated iron(III) uptake in several organisms such as *Erwinia herbicola* [15] and *Pseudomonas putida* [40] by making use of both the ferrioxamine B and coprogen B receptors. The glutamic acid derivatives (m = 2; Fig. 7) acted as iron(III) carriers in *Erwinia herbicola* but as selective inhibitor of the ferrioxamine receptor in *Pseudomonas putida*. The inhibition of the D-glu derivative of Λ-cis configuration was found te be two- to threefold more effective than that of its enantiomer. The latter observation demonstrates that the ferrioxamine receptor indeed exerts some chiral discrimination, although far from that reported for the receptors of enterobactin [22] or ferrichrome [14,23]. This result might indicate conformational tolerance by the receptor, poor match between the synthetic analogues and the receptor, or a combination of both.

The activity of the lower homologues (m = 0; Fig. 7) as ferrioxamine and coprogen analogues, independent of the nature of the amino acid component, is indicative of their fit to both siderophore receptors. This may be attributed to the small molecular diameters of these iron(III) complexes that can easily adopt themselves to the respective recognition sites. The differential action of the higher homologues (m = 2; Fig. 7) is attributed to the more stringent stereochemical requirements of these compounds. The latter findings indicate that related receptors of different organisms may differ in their stereochemical demands [14].

The above results illustrate the advantage of the modular strategy of design and synthesis and of theoretical calculations in the search for bioactive molecules. This approach provided molecular probes that differentiate between the structural requirements for receptor recognition and for transport and led to both (1) artificial substrates that act as "master keys" and recognize more than one type of receptor, and (2) substrates that act as "custom-made keys" and differentiate between the same kind of receptor in different organisms. The latter observations indicate that there are structural differences between analogous receptors of different organisms. Yet, the very nature of receptor-siderophore interactions is still a matter of speculation, and structural information on representative receptors will be needed to help unravel this problem.

6. FLUORESCENT LABELED SIDEROPHORE ANALOGUES: PROBES FOR IRON TRAFFICKING IN REAL TIME

The pronounced species-specificity of some of the synthetic siderophore analogues led us to contemplate whether it would be possible to use such compounds as diagnostic tools for the identification of specific organisms. Realization of this idea would require the development of derivatives that would signal the microbial iron uptake processes. The presence of a tetrahedral carbon as anchor in the ferrichrome analogues proved here to be of advantage, as it provided us with a site for labeling, which would neither interfere with iron binding nor with receptor recognition. The bioactive ferrichrome analogues (glycyl derivative, m = 2, R = H, and alanyl derivative, m = 2, R = Me; Fig. 5), were thus labeled with a fluorescent label at the carbon anchor (Figure 8) [41,42]. To our satisfaction, iron(III) binding by the labeled derivatives expressed itself by quenching of the label's fluorescence. This quenching process was reversed on addition of a competing chelator, indicating that the quenching process occurs intramolecularly.

Incubation of *Pseudomonas putida* JM218 with the iron(III)-loaded, labeled glycyl derivative (R = H; Fig. 9) resulted in cellular iron uptake and the appearance of anthracene fluorescence in the cultures' growth media. Incubation with the alanyl derivative (R = Me; Fig. 9) failed to bring about any significant iron uptake or any significant fluorescence (Fig. 9) [42]. This behavior is parallel with that of the nonlabeled analogues. The combination of the fluorescence and iron-uptake results indicate that the labeled glycyl derivative acts as ferrichrome analogue, delivers iron(III) to the microorganism, and is then released to the medium, whereas the alanyl derivative fails to deliver iron(III), acts as inhibitor of the ferrichrome receptor, and remains bound to the membrane receptor as an intact complex. Quite remarkably, other *Pseudomonas* strains such as *Pseudomonas fluorescens S680* or *Pseudomonas fluorescens WCS3742* failed to respond to the labeled ferrichrome analogues, highlighting their pronounced specificity [42].

The above data demonstrate the potential of labeled iron carriers as species-specific diagnostic tools. The energy dependence of siderophore-mediated iron uptake renders these agents specific to live organisms.

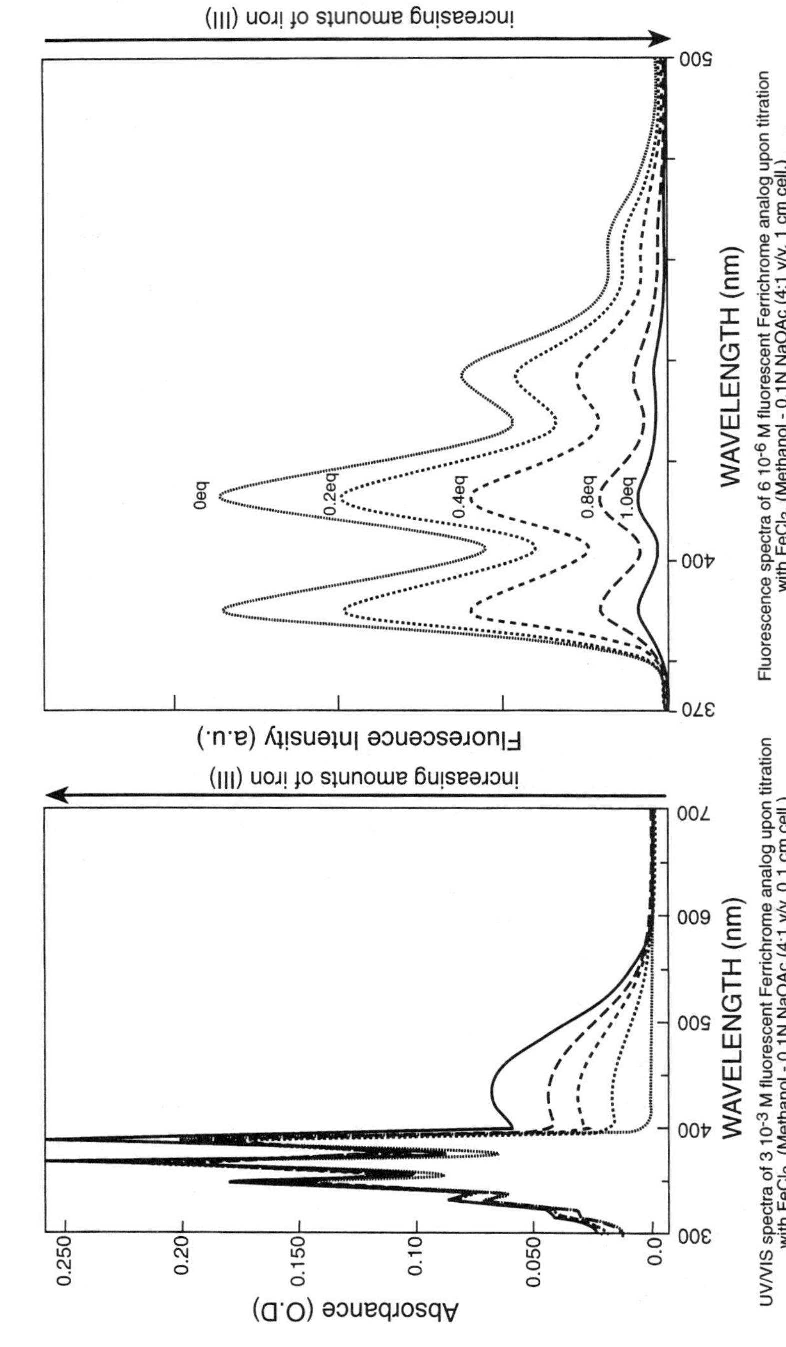

FIG. 8. UV/VIS (left) and fluorescence titration (right) of fluorescent ferrichrome analogues with iron(III) [42].

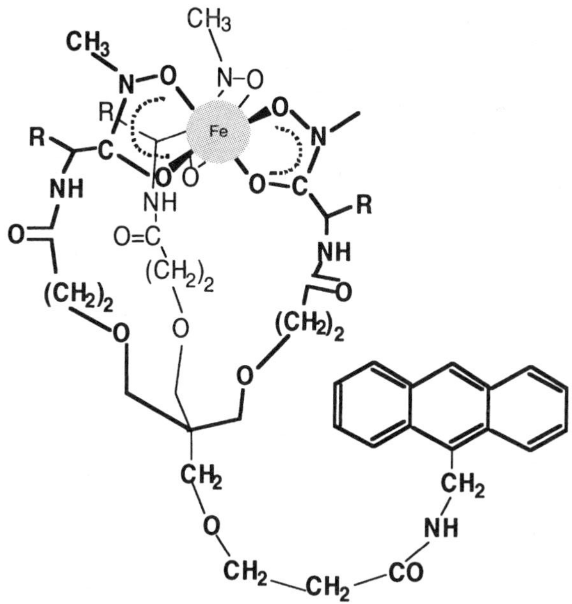

R = H; R = Me

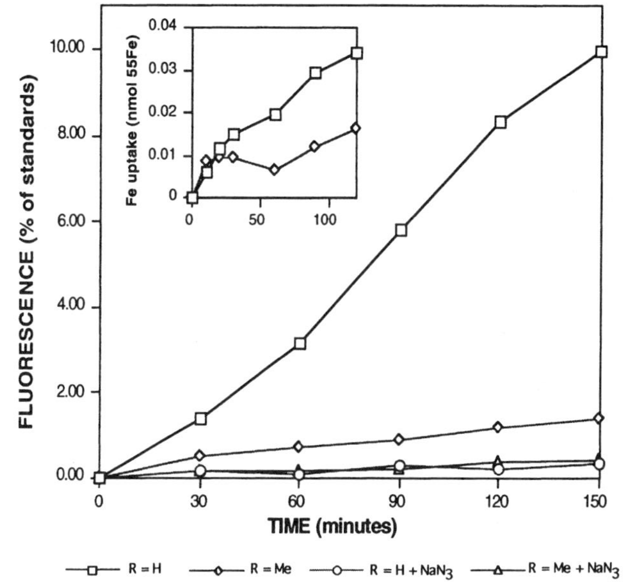

FIG. 9. Appearance of anthracene fluorescence in the culture media of *Pseudomonas putida* JM218 on treatment with labeled ferrichrome analogues. Insert: ^{55}Fe uptake with the ferrichrome analogues [42].

The latter point is believed to be of particular relevance for future applications.

These new fluorescent siderophores open new potential in confocal microscopy for tracing iron(III) trafficking and mapping intracellular compartmentation [43]. Recent work has been devoted to expansion of the variability of fluorescent probes by linking probes of different sizes, optical properties, and polarities (such as derivatives of 7-nitrobenzofurazane (NBD), fluorescein, and rhodamine) [44]. It has been shown that activity toward *Pseudomonas putida* has not been altered by these manipulations. This opens the possibility of building an assay or library of labeled siderophores that can give a real-time diagnosis by distinguishing between different species of microorganisms.

The modular design and synthesis of the biomimetic analogues greatly facilitates labeling with spectroscopic markers, thereby providing derivatives where recognition and transport are coupled with optical signaling.

7. LIPOPHILIC FERRICHROME ANALOGUES OR REVERSED SIDEROPHORES AS IRON SCAVENGERS

Since iron is an essential element for virtually all living species, it occurred to us that modifying the siderophore analogues with lipophilic substituents could have favorable permeation properties across cellular membranes and could exert sequestration of essential iron(III). This could prove useful for arresting parasite growth [45,46], including stains of *Plasmodium falciparum*. The desired chemical properties were introduced by reproducing the iron-binding cavity of natural ferrichrome by systematic substitution of their hydrophilic envelopes for more hydrophobic ones. In contrast to natural carriers, which participate in receptor-mediated iron uptake in cells and act as growth promoters, our new synthetic carriers were designed to penetrate cellular membranes by diffusion, scavenge intracellular iron, and thereby act as growth inhibitors.

Because the functions of such binders are opposite, or reversed, to those of natural siderophores, we designated the term *reversed sidero-*

phores (RSs) for them. The RS design was based on tripodal topology that mimics the natural ferrichrome model (see Fig. 5). These synthetic ferrichromes (SFs) are assembled in a modular fashion with amino acid residues of variable hydrophobicity. Such an approach facilitates systematic chemical modifications for obtaining optimal antimalarial performance [47,48]. Moreover, by tagging the molecules with fluorescent or radioactive markers, the drug can be traced in various cellular compartments [49].

To assess the siderophoric hypothesis, we chose to apply the agents to *P. falciparum* as potential antimalarial drugs. The malarial parasites complete their erythrocyte cycle of growth in 48 h and are highly dependent on an immediate and substantial supply of iron. Therefore, scavenging iron promises to be an effective tool for arresting parasite growth and opens new avenues for chemotherapy. This is of particular importance, since currently used drugs are steadily losing their efficacy owing to increasing drug resistance displayed by parasites [50].

In vitro and in vivo preliminary experiments [49,51–55] confirm that reversed siderophores arrest growth and proliferation of the malarial parasite. Moreover, the ostensible difference in the permeability properties of the host cell membrane in infected cells as compared with noninfected cells opens the possibility of drug design with selective access to infected cells [56,57]. The extent of growth inhibition was found to depend on the relative iron(III)-binding or iron(III)-extraction efficiency of the siderophores but mainly on their hydrophobicity, as determined by their partition coefficients between saline and octanol (Table 1). This conclusion was supported by the fact that in open systems where the erythrocytic membrane does not represent a hydrophobic barrier, the less hydrophobic chelators proved to be as effective as the hydrophobic ones.

The observed growth inhibition was demonstrated to derive from iron(III) deprivation. Application of the reversed siderophores as their iron(III) complexes instead of as their free ligands completely reversed the inhibitory effect on parasitic growth. Since the activity of these chelators derives from iron(III) deprivation, they were expected to have similar activity on quinine-sensitive as well as quinine-resistant strains of *P. falciparum*, and indeed the most resistant as well as the most sensitive strains proved to be equally effected by the reversed siderophores.

TABLE 1

Chemical and Antimalarial Properties of Reversed Siderophores (SF series)

$Et\text{-}C\text{-}\{CH_2O(CH_2)_n\text{-}CO\text{-}(NRCHR'CO)_m\text{-}NOHCH_3\}_3$

Compound	n	m	R	R′	Relative binding[a]	Extraction efficiency[b]	Partition coefficients[c]	Hydrophobicity[d]	IC_{50}[e] (μg/mL)
	2	0	—	—	47	83	1.7		4
SF1-L-ala	2	1	H	L-Me	110	88	0.53	1.0	40
SF1-D-ala	2	1	H	D-Me					45
SF1-pro	2	1	L-$(CH_2)_3$-		14	38	0.65	1.5	>50
SF1-leu	2	1	H	L-iBu	10	35	12.5	3.5	17
SF1-ileu	2	1	H	L-sc.iBu	25	52	14.0	5.0	<2
SF-leu	1	1	H	L-iBu	20	62	28.6	3.5	4
SF1-phe	2	1	H	Bn	72		40.0	5.0	0.6
Dfo					100				43

[a]Relative binding efficiencies were determined by competition with EDTA in aqueous MeOH (MeOH-0.1 *N* NaOAc) using UV as an analytical tool.

[b]Extractions were performed by equilibrated solution of ligand in $CHCl_3$ with aqueous solutions of $FeCl_3$, citric acid, 40 nM tris at pH 6.9. Percentages are given relative to total ligand present.

[c]Partition experiments were performed by equilibrating the free ligands between equal volumes of *n*-octanol and saline.

[d]The hydrophobicity scale given is that of the amino acid residues (according to Tanford/Segret scale) as given by Eisenberg [58].

[e]Ligand concentrations at which the growth of *Plasmodium falciparum* is reduced to 50%.

Although these hydroxamate binders were found to exhibit their antiplasmodial action by iron(III) deprivation, they did not deplete serum from iron(III). Pre-incubation of whole serum with the most effective of the reversed siderophores did not lower its iron(III) contents.

The reversed siderophores were effective against all stages of parasitic growth and against a variety of multidrug-resistant strains of *P. falciparum* [59]. The most potent agent of this synthetic ferrichrome series, SFI-phe, was not toxic to mammalian cells in culture and was 70-fold more potent and 20-fold faster acting than desferrioxamine [48].

Although the results appear promising in relation to possible iron scavengers as antimalarials, several very basic key questions are still unanswered. These include the source of erythrocytic iron(III) which is scavenged by the reversed siderophores and the nature of the metabolic process which is curtailed by iron(III) deprivation. Elucidation of these problems is hoped to shed light on the iron(III) metabolism of the malaria-causing parasite and allow the development of even superior chelators of potential antimalarial activity.

8. SUMMARY AND CONCLUSION

Concentrating on few natural siderophores, we have outlined the biomimetic approach toward the preparation of synthetic analogues that simulate the performance of natural compounds in vitro and in vivo. We followed the biomimetic methodology that relies on identifying the essential features of the natural compounds and on reproducing these features with the simplest possible synthetic structure. In the course of this undertaking, the origin of the outstanding iron-binding selectivity and efficiency of enterobactin could be traced to the nature of the dihydroxybenzamido-binding site and to the presence of an H-bonding network. The latter shapes the molecule to adopt an ion-binding configuration prior to binding and stabilizes the complex once formed. The generation of the H-bonding network proved also to be favorable to the iron(III)-binding effect in ferrichrome analogues. The latter point illustrates the difference and the advantage of biomimetics relative to classical trial and error approaches in the design of biologically active molecules. Following the rational of biomimetics, identifying the functional principles of one system allows to successful mimicry of another related system. When applied to biological systems, biomimetic iron(III) carriers serve as probes to trace the structural requirements for receptor recognition. Biomimetic iron(III) carriers, which may systematically be modified, can provide "biological kits" to trace differences between various organisms. These could lead to several possibilities: (1) changing the ion-binding envelopes, creating iron scavengers (growth inhibitors) or growth promoters; and (2) fine tuning of the ion-binding envelope provides either highly selective growth promoters or broad-range active microbial growth promoters. Owing to their modular design, they also allow the attachment of fluorescent labels, which further facilitate mapping of cellular iron(III)-uptake processes. Libraries of fluorescently bound siderophores may provide a diagnostic technique for real-time identification of pathogenic microorganisms. Following microbial identification, substitution of the fluorescent probe by a cytotoxic agent can serve as a species-specific drug-targeting agent. The great advantage of such a prodrug is that it is recognized solely by the microbial receptors, thereby lowering drug dosage requirements. It may further be speculated that siderophore-drug conjugate penetrating via specific receptors will reduce the problem of drug resistance caused by membranal mutations.

ACKNOWLEDGMENTS

The authors thank O. Ardon, Z. I. Cabantchik, C. Caris, Y. Chen, C. E. Felder, Y. Hadar, E. Jurkevitch, S. Lifson, B. Mester, I. Mor, R. Nudelman, Y. Tor, and H. Weizman for the performance of the work, and Mrs. R. Lazar for her outstanding technical assistance.

This work was supported in part by the U.S.-Israel Binational Science Foundation, The Minerva Foundation, The Israel Academy of Sciences, and the Gesellschaft für Biotechnologische Forschung mbH, Braunschweig, Germany, which are all gratefully acknowledged.

A. S. is the holder of the Siegfried and Irma Ullman Professorial Chair in Chemistry.

ABBREVIATIONS

Bn	benzyl
Bu	*n*-butyl
COP	coprogen
Dfo	desferrioxamine
EDTA	ethylenediamine-*N*,*N*,*N′*,*N′*-tetraacetate
EFF	empirical force field
Fc	ferrichrome
Fhu	membranal protein
iBu	isobutyl
Me	methyl
NBD	7-nitrobenzofurazane
RS	reversed siderophores
SFI-phe	synthetic ferrichrome (type I) phenyl alanine
SFs	synthetic ferrichromes
TBA	trisbenzamide
Tris	2-amino-2(hydroxymethyl)-1,3-propanediol
UV	ultraviolet

REFERENCES

1. V. Prelog, *Pure and Appl. Chem.*, *6*, 327 (1963).
2. J. B. Neilands, *Structure and Bonding*, *58*, 1–24 (1984).
3. K. N. Raymond, G. Müller, and B. F. Matzanke, *Top. Cur. Chem.*, *123*, 49–102 (1984).
4. R. C. Hider, *Structure and Bonding*, *58*, 25–87 (1984).
5. G. Winkelmann, D. van der Helm, and J. B. Neilands, eds., *Iron Transport in Microbes, Plants and Animals*, VCH Verlagsgesellschaft mbH, Weinheim, Germany, 1987.
6. B. F. Matzanke, G. Müller-Matzanke, and K. N. Raymond, *Iron Carriers and Iron Proteins*, Vol. 5, VCH Publishers: New York, 1989; pp. 1–121.
7. T. Emery, *Metal Ions in Biological Systems*, *7*, 77–126 (1978).

8. A. Bagg and J. B. Neilands, *Microbiol. Rev.*, *51*, 509–518 (1987).
9. S. Lifson, C. E. Felder, A. Shanzer, and J. Libman, in *Synthesis of Macrocycles: The Design of Selective Complexing Agents*, Vol. 3 (R. M. Izatt and J. J. Christensen, eds.), Wiley, New York, 1987, pp. 241–307.
10. A. Shanzer, and J. Libman, in *Handbook of Microbial Iron Chelates* (G. Winkelmann, ed.), CRC Press, Boca Raton, FL, 1991, pp. 309–338.
11. J. Libman, Y. Tor, I. Dayan, A. Shanzer, and S. Lifson, *Isr. J. Chem.*, *32*, 31–40 (1992).
12. A. Shanzer, J. Libman, and S. Lifson, *Pure and Appl. Chem.*, *64*, 1421–1435 (1992).
13. S. Lifson, C. E. Felder, J. Libman, and A. Shanzer, in *NATO ASI Series*, Vol. 426 (G. Wipff, ed.), Kluwer, The Netherlands, 1994, pp. 349–376.
14. A. Shanzer, J. Libman, P. Yakirevitch, Y. Hadar, Y. Chen, and E. Jurkevitch, *Chirality*, *5*, 359–365 (1993).
15. I. Berner, P. Yakirevitch, J. Libman, A. Shanzer, and G. Winkelmann, *Biol. Metals*, *4*, 186–191 (1991).
16. V. Braun, in *The Enzymes of Biological Membranes*, Vol. 3 (A. N. Martonosi, ed.), Plenum Press, New York, 1985, pp. 617–652.
17. Y. Tor, J. Libman, A. Shanzer, C. E. Felder, and S. Lifson, *J. Am. Chem. Soc.*, *114*, 6661–6671 (1992).
18. I. Dayan, J. Libman, Y. Agi, and A. Shanzer, *Inorg. Chem.*, *32*, 1467–1475 (1993).
19. P. Yakirevitch, N. Rochel, A.-M. Albrecht-Gary, J. Libman, and A. Shanzer, *Inorg. Chem.*, *32*, 1779–1787 (1993).
20. T. Emery, *Biochemistry*, *10*, 1483–1488 (1971).
21. J. Leong, and K. N. Raymond, *J. Am. Chem. Soc.*, *97*, 293–296 (1975).
22. J. B. Neilands, T. J. Erickson, and W. H. Rastetter, *J. Biol. Chem.*, *256*, 3831–3832 (1981).
23. G. Winkelmann and V. Braun, *FEMS Microbiol. Lett.*, *11*, 237–241 (1981).
24. A. Shanzer, J. Libman, S. Lifson, and C. E. Felder, *J. Am. Chem. Soc.*, *108*, 7609–7619 (1986).
25. W. R. Harris and K. N. Raymond, *J. Am. Chem. Soc.*, *101*, 6534 (1979).

26. W. R. Harris, K. N. Raymond, and F. L. Weitl, *J. Am. Chem. Soc.*, *103*, 2667 (1981).

27. V. L. Pecoraro, F. L. Weitl, and K. N. Raymond, *J. Am. Chem. Soc.*, *103*, 5133 (1981).

28. A. Shanzer, J. Libman, and F. Frolow, *J. Am. Chem. Soc.*, *103*, 7339–7340 (1981).

29. T. B. Karpishin, T. M. Dewey, and K. N. Raymond, *J. Am. Chem. Soc.*, *115*, 1842–1851 (1993).

30. Y. Tor, J. Libman, A. Shanzer, C. E. Felder, and S. Lifson, *J. Am. Chem. Soc.*, *114*, 6653–6661 (1992).

31. M. G. Paterlini, T. B. Freedman, L. A. Nafie, Y. Tor, and A. Shanzer, *Biopolymers*, *32*, 765–782 (1992).

32. Y. Tor, J. Libman, A. Shanzer, and S. Lifson, *J. Am. Chem. Soc.*, *109*, 6517 (1987).

33. T. D. P. Stack, Z. Hou, and K. N. Raymond, *J. Am. Chem. Soc.*, *115*, 6466–6467 (1993).

34. A. Shanzer, J. Libman, R. Lazar, Y. Tor, and T. Emery, *Biochem. Biophys. Res. Commun.*, *157*, 389–394 (1988).

35. E. Jurkevitch, Y. Hadar, Y. Chen, J. Libman, and A. Shanzer, *J. Bacteriol.*, *174*, 78–83 (1992).

36. The achiral glycine forms racemic mixtures of complexes, where the Δ-cis isomer is believed to be taken up and constantly replenished by fast equilibration in the medium.

37. T. Emery, L. Emery, and R. K. Olsen, *Biochem. Biophys. Res. Commun.*, *119*, 1191–1197 (1984).

38. W. Keller-Schierlein, and V. Prelog, *Helv. Chim. Acta*, *44*, 709–713 (1961).

39. W. Keller-Schierlein, and V. Prelog, *Helv. Chim. Acta*, *44*, 1981–1985 (1961).

40. E. Jurkevitch, Y. Hadar, Y. Chen, P. Yakirevitch, J. Libman, and A. Shanzer, *Microbiology-UK*, *140*, 1697–1703 (1994).

41. H. Weizman, M.Sc. Thesis, Feinberg Graduate School, The Weizmann Institute of Science, 1993.

42. H. Weizman, O. Ardon, B. Mester, J. Libman, O. Dwir, Y. Hadar, Y. Chen, and A. Shanzer, *J. Am. Chem. Soc.*, *118*, 12368 (1996).

43. E. Bar-ness, Y. Hadar, Y. Chen, A. Shanzer, and J. Libman, *Plant Physiol.*, *99*, 1329–1335 (1992).

44. R. Nudelman, O. Ardon, Y. Hadar, Y. Chen, J. Libman, and A. Shanzer, *J. Med. Chem.*, Submitted (1997).

45. Z. I. Cabantchik, *Parasitol. Today*, *11*, 74–78 (1995).

46. Z. I. Cabantchik, H. Glickstein, J. Golenser, M. Loyevsky, and A. Tsafack, *Acta Haematol.*, *95*, 70–77 (1996).

47. A. Shanzer, J. Libman, S. Lytton, H. Glickstein, and Z. I. Cabantchik, *Proc. Natl. Acad. Sci.*, *88*, 6585–6589 (1991).

48. A. Tsafack, J. Libman, A. Shanzer, and Z. I. Cabantchik, *Antimicrob. Agents Chemother.*, *40*, 2160–2166 (1996).

49. M. Loyevsky, S. D. Lytton, B. Mester, J. Libman, A. Shanzer, and Z. I. Cabantchik, *J. Clin. Invest.*, *91*, 218–224 (1993).

50. D. F. Clyde, *Epidem. Rev.*, *9*, 250 (1986).

51. C. Raventos-Suarez, S. Pollack, and R. L. Nagel, *Am. J. Trop. Hyg.*, *31*, 919 (1982).

52. M. D. Scott, A. Ranz, F. A. Kuypers, B. H. Lupin, and S. R. Meshnick, *Br. J. Haematol.*, *75*, 598–602 (1990).

53. S. D. Lytton, B. Mester, I. Landau, J. Libman, A. Shanzer, and Z. I. Cabantchik, *Am. J. Hematol.*, *43*, 217–220 (1993).

54. S. Pollack, R. N. Rassan, D. E. Davidson, and A. Escajadillo, *Proc. Soc. Exp. Biol. Med.*, *184*, 162 (1987).

55. S. D. Lytton, J. Libman, A. Shanzer, and Z. I. Cabantchik, *Blood*, *84*, 910–915 (1994).

56. Z. I. Cabantchik, *Blood (Editorial Rev.)*, *74*, 1464–1471 (1989).

57. H. M. Rodriguez, and M. Jungery, *Nature*, *324*, 388 (1986).

58. D. Eisenberg, *Annu. Rev. Biochem.*, *53*, 595–623 (1984).

59. S. D. Lytton, B. Mester, Y. Dayan, H. Glickstein, J. Libman, A. Shanzer, and Z. I. Cabantchik, *Blood*, *81*, 214–222 (1993).

9

The Physical Chemistry of Bacterial Outer-Membrane Siderophore Receptor Proteins

Dick van der Helm

Department of Chemistry & Biochemistry, University of Oklahoma, 620 Parrington Oval, Norman, OK 73019

1. INTRODUCTION 356

2. RECEPTOR PROTEINS WITH KNOWN SEQUENCES 357
   - 2.1. General Characteristics 357
   - 2.2. *Escherichia coli* Receptor Proteins 360
   - 2.3. Receptor Proteins from Pseudomonads 361
   - 2.4. Receptor Proteins from *Yersinia enterocolitica* 362
   - 2.5. Other Receptor Proteins 362
   - 2.6. Other Uptake Systems 363

3. HOMOLOGY, MODELS, AND CHANNELS 364
   - 3.1. N- and C-Terminal Residues and Homology Boxes I, II, III, and IV 364
   - 3.2. Homology and Phylogenetics 364
   - 3.3. FepA Model 365
   - 3.4. FhuA and Other Models 369

3.5. Channel Formation 371

4. INTERACTIONS AND SPECIFICITY 372
4.1. Binding Proteins in Periplasm 372
4.2. TonB Interactions 374
4.3. Purification of Receptor Proteins 376
4.4. Physical Studies 378
4.5. Binding Studies 378
4.6. The Substitution of Iron with Gallium 380
4.7. Effect of Ca on Binding 381
4.8. Specificity, Stereospecificity, and Binding Pocket 382

5. MECHANISM 386
5.1. Structural Information 386
5.2. Bayer Zones 387
5.3. Energy Requirement and pH Dependence 388
5.4. Possible Mechanisms 389

6. CONCLUSIONS 391

ACKNOWLEDGMENTS 391

ABBREVIATIONS 391

REFERENCES 392

1. INTRODUCTION

A remarkable amount of information has been obtained for the high-affinity iron transport in gram-negative bacteria. This is primarily obtained from genetic and in vivo experiments. Originally, the emphasis was on the outer membrane receptors, but recently all other aspects of the transport have been addressed. In vitro results for the proteins involved in the transport are still scarce. This chapter addresses only the chemistry of the outer-membrane receptors. The review therefore describes the properties and interactions of the receptors as molecules. However, this material will take into account, where appropriate, genetic information and in vivo results. The subject was last reviewed in 1982 [1].

2. RECEPTOR PROTEINS WITH KNOWN SEQUENCES

2.1. General Characteristics

By the end of 1996, the sequences of 22 outer membrane ferric siderophore receptor proteins (FSRPs) from gram-negative bacteria were published (Table 1). Many other FSRPs have been or will be identified in the future. All of them are expressed at high levels only under iron-deficient conditions. Also, all proteins have been proven to be TonB dependent or can be assumed to function in that manner, giving rise to energy-dependent transport of the ferric siderophores. The proteins are similar in size ranging from 651 (FyaA) to 803 (AleB) amino acids and have similar molecular weights.

Gram-negative bacteria are not limited to a single receptor system for ferric siderophores. Initially it was shown that *Escherichia coli* had FSRPs for both homologous (for definition see legend Table 1) siderophores, that is, FepA for Fe(ent) and IutA for ferric aerobactin as well as FSRPs for heterologous siderophores, that is, FhuA for ferrichrome and FecA for Fe citrate and FhuE for several hydroxamates. More recently it was found that other organisms have the same ability. *Pseudomonas aeruginosa* has FptA and FpvA for the homologous pyochelin and pyoverdin but also PfeE for Fe(ent). Similarly, *Yersinia enterocolitica* has FyuA for the homologous yersiniabactin but also receptors for the heterologous ferrioxamine B (FoxA) and ferrichrome (FcuA). *Pseudomonas putida* WCS358 has PupA for the homologous pseudobactin but also PupB for the pseudobactins from *P. putida* BN7 and 8 and by polymerase chain reaction (PCR) amplification at least five additional sequences for other related FSRPs have been indicated [2]. One can expect, therefore, that many organisms will have multiple FSRPs for different ferric siderophores.

Alternatively, a particular ferric siderophore may be taken up by different organisms. Closely and distantly related gram-negative bacteria were tested under iron restriction for reactivity with more than 20 monoclonal antibodies (MAbs) raised for *E. coli* FepA (see Section 3.3) indicating the existence of FepA homologues in a large fraction of these organisms [3]. Limiting the description to those which have been sequenced, it is known that ferrichrome (Fig. 1) has the receptor FhuA in

TABLE 1

Ferric Siderophore Receptor Proteins (FSRP) with Known Amino Acid Sequences

Name	FepA	FhuA	FhuE	IutA	FecA	BfeA	PfeA
Hom/Het.	Hom	Het	Het	Hom	Het	Het	Het
#AA	723	714	693	700	741	676	721
MW	79908	78992	77453	77345	81718		78385
First/Last	Q/F	A/F	A/F	Q/F	A/F	Q/F	A/F
TonB box	Y/12	Y/7	Y/6	Y/6	Y/23	Y?/20	Y/14
Induction	N	N	N	N	Y	N	Y
Ref.	53	54	55	56	57	40	58

Name	FptA	FpvA	PupA	PupB	PupX	PbuA	FyuA
Hom/Het.	Hom	Hom	Hom	Het	Hom	Hom	Hom
#AA	682	770	772	764	783	782	651
MW	78993	86245	86010	83127	86257	85759	71677
First/Last	D/F	Q/F	Q/F	A/F	Q/F	Q/F	Q/F
TonB box	N?/7	Y?/68	Y/68	N?	Y/68	Y	Y/8
Induction	Y	Y	N	Y	N	N	?
Ref.	59	60	61	10	62	6	38

Name	FcuA	FoxA	ViuA	FatA	Fct	RumA	AleB	Cir
Hom/Het.	Het	Het	Hom	Hom	Hom	Het	Hom	Hom
#AA	722	684	650	691	697	683	803	599
MW	77966	75768	72417	71164		~75300	87980	67169
First/Last	A/F	D/F	A/F	L/F	A/W	A/F	V/A	V/F
TonB box	Y/30	Y/1	N	Y/14	Y/7	Y/4	N	Y
Induction	N	N	N	Y	N	N	N	N
Ref.	4	5	63	64,65	66	67	47	68

The distinction between homologous and heterologous uptake systems is that, in the former, the siderophore being transported is produced by the organism, whereas for a heterologous system, the siderophore is not produced by the organism. The boxes I (TonB box), II, III, and IV were originally identified for FepA [53] as residues 12–16, 672–684, 53–94, 114–153. All the proteins are induced by iron deficiency. Some are, in addition, induced by substrate, and this is indicated by Y.

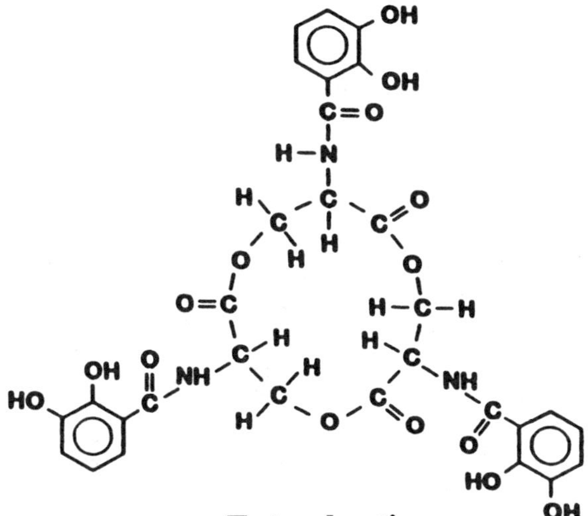

Enterobactin

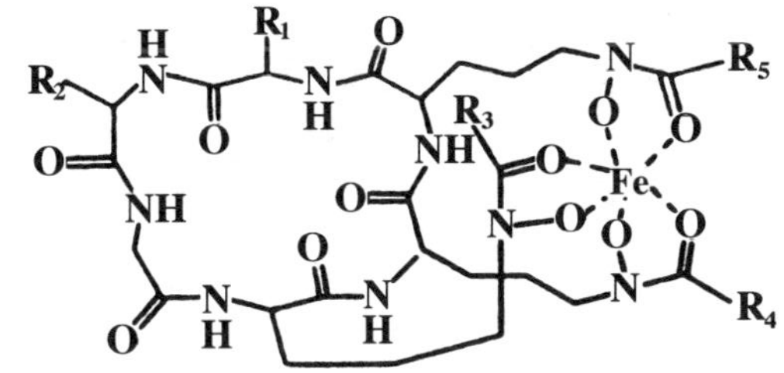

	R_1	R_2	R_3	R_4	R_5
1 Ferrichrome	H	H	CH_3	CH_3	CH_3
2 Ferrichrome A	CH_2OH	CH_2OH	A	A	A
3 Ferricrocin	H	CH_2OH	CH_3	CH_3	CH_3
4 Ferrichrysin	CH_2OH	CH_2OH	CH_3	CH_3	CH_3
5 Ferrirubin	CH_2OH	CH_2OH	B	B	B
6 Ferrirhodin	CH_2OH	CH_2OH	C	C	C

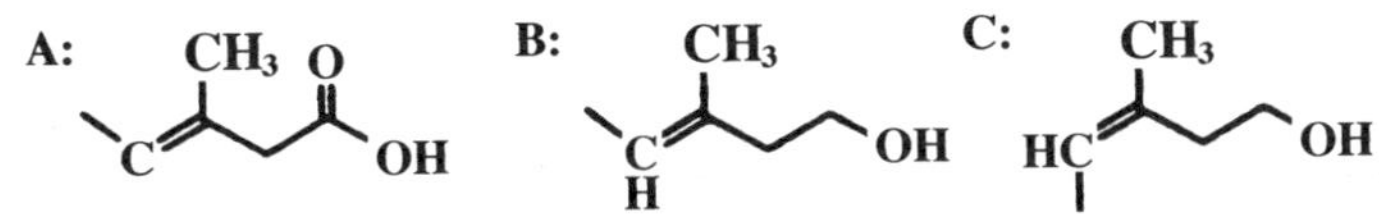

FIG. 1. Chemical structures of enterobactin and ferrichrome. The chemical structures of other siderophores can be found in [177].

E. coli and FcuA in *Y. enterocolitica*. Fe(ent) uses FepA in *E. coli*, PfeA in *P. aeruginosa*, and BfeA in *Bordetella pertussis*. These proteins are homologous but not identical. The homology among the Fe(ent) FSRPs (FepA, PfeA, and BfeA) is rather high (~70%), and even the putative binding regions for the substrate show homology. On the other hand, FhuA and FcuA have only 21% homology. This difference is interesting when one realizes that ferrichrome is a fungal siderophore and thus always uses a heterologous system. Also, the substrate specificity for the two proteins, FhuA and FcuA, is different [4]. Ferrioxamine B is taken up by FoxA in *Y. enterocolitica*, and by FhuE in *E. coli*, but also these two proteins have different substrate specificity [5]. The siderophore produced by *Pseudomonas* M114 may be the same as the one produced by *Pseudomonas* B10 [6], pseudobactin B10. The FSRPs involved PbuA and PupX are almost the same (92.5% identity). Strains M114 and B10 are, however, different.

The FSRPs are produced at high levels only under iron-deficient conditions. This negative control involves in all cases the Fur protein [7] which binds with iron as cofactor to the Fur box, preventing transcription and production of the FSRPs and the enzymes producing the siderophores. A number of the FSRPs require another simultaneous induction system for their production. This second system requires the ligand of the substrate; that is, the deferri-compound. Genetically these are novel and interesting signaling systems. FecA in *E. coli* is induced by citrate and this involves two proteins, FecI and R [8]. The PupB expression in *Pseudomonas* WCS358 uses a similar system [9], but it must be different, because this organism produces FSRPs for other pseudobactins as well [2,10]. In *P. aeruginosa*, FptA production is dependent on pyochelin [11] and FpvA on pyoverdin [12], both homologous siderophores, whereas the heterologous enterobactin (Fig. 1) uses PfeR and S for the production of PfeA [13]. In *Vibrio anguillarum*, the substrate anguibactin enhances the production of FatA [14].

2.2. *Escherichia coli* Receptor Proteins

FepA is the receptor for the homologous Fe(ent) [15]. The protein is also the site of attachment for colicins B and D [16–18].

The first receptor which was identified was the one for ferrichrome. This receptor, TonA, is now more commonly known as FhuA. The protein is also the receptor for phages T1, T5, and ϕ80, colicin M, and the antibiotic analogue of ferrichrome, albomycin [19–22].

FhuE is used by the hydroxamates coprogen, ferrioxamine B, and ferric rhodotorulate [23,24].

Aerobactin, a dihydroxamate built onto citric acid, is produced by pColV-bearing *E. coli* cells. The receptor protein, IutA, is encoded by the plasmid [25]. The function of Fur in the control of the production of the proteins and enzymes was first identified in this system [26].

FecA is the receptor protein for ferric citrate. Citric acid is not a siderophore, although the uptake system has all the properties of a high-affinity system.

Fiu and Cir are two other TonB-dependent outer-membrane proteins in *E. coli*. They are used as receptors for the ferric chelates of monomeric catechols [27,28]. Cir has been sequenced. They are similar to all the other ferric siderophore receptor proteins.

Belonging to the group of ferric siderophore receptor proteins in *E. coli* is BtuB, which is involved in the uptake of vitamin B_{12}. It belongs to this group because the transport is TonB and energy dependent. The protein is homologous with the ferric siderophore receptor proteins. It was, in fact, the first of these proteins which was sequenced [29]. The mature protein BtuB has 594 amino acids and a MW of 66400 Da. The hydropathy plot is similar to those for the FSRPs, indicating a multistrand β-sheet structure, and two different topological models have been proposed [30,31]. Both models contain 16 β-strands. A number of physical chemical studies have been carried out with BtuB, and these are included in this chapter because the results apply to FSRP proteins.

2.3. Receptor Proteins from Pseudomonads

Pseudomonas aeruginosa produces the siderophores pyoverdine PAO and pyochelin [32,33]. The organism also has receptors for heterologous siderophores, and one for Fe(ent), PfeA, has been sequenced. Pyochelin is an unusual siderophore with a thiazoline and thiazolidine ring most likely forming a 2:1 pyochelin Fe(III) complex. The receptor protein for the complex is FptA, The receptor protein for pyoverdine PAO is FpvA.

The pyoverdines, also named pseudobactins, form the largest family of siderophores. They all contain a dihydroxy fluorescent catechol, one or two hydroxamate groups, and one or no hydroxy acid group. The backbone is formed by a linear or cyclic (depsi)peptide formed from both *D*- and *L*-amino acids [34,35]. *P. putida* WCS 358 is a plant growth promoter and the strain is capable of assimilating a large number of ferric pseudobactins. The receptor protein for the homologous siderophore, ferric pseudobactin 358, is PupA. A second receptor protein in *P. putida* WCS358, PupB, was also sequenced. The protein only recognizes ferric pseudobactin BN7 and 8. PupX is the receptor protein in *Pseudomonas* B10 for the homologous ferric pseudobactin B10 [36]. *Pseudomonas* strain M114 is another organism which has outer-membrane receptor proteins for both homologous and heterologous ferric pseudobactins. The receptor for the homologous ferric pseudobactin M114 is PbuA.

2.4. Receptor Proteins from *Yersinia enterocolitica*

The virulence of *Y. enterocolitica* is associated with the production of yersiniabactin [37]. The receptor protein is FyuA [38]. FyuA is also the pesticin receptor. The organism is also able to transport ferrioxamine B. The receptor protein is FoxA. The organism has also a receptor protein, FcuA, for the heterologous ferrichrome siderophore. Phages T5, T1, and ϕ80 and colicin M bind to FhuA but not to FcuA [4]. *Y. enterocolitica* also has a heme-uptake system, and the sequence of this protein HemR is known. It has all the properties of a ferric siderophore receptor. A survey of all known receptor systems in *Y. enterocolitica* has been given [39].

2.5. Other Receptor Proteins

Bordetella pertussis has a receptor protein, BfeA, for the heterologous Fe(ent). Similar proteins to BfeA are indicated in several *Haemophilus* species [40]. *Vibrio cholerae* produces the siderophore vibriobactin [41], a phenolate siderophore with similarities to enterobactin and agrobactin. The outer-membrane receptor protein is ViuA. The major iron-regulated outer-membrane protein, IrgA, which is the virulence factor of the organism, has considerable homology with the outer-membrane iron siderophore receptor proteins but is not involved in iron transport [42]. The virulence of *V. anguillarum* is associated with a plasmid-mediated iron-uptake system. The siderophore which is produced is anguibactin [43], an unusual siderophore with some similarity to pyochelin. It may form a 2:2 complex with Fe^{3+} [44]. The receptor protein is FatA (previously RfeA). *Erwinia chrysanthemi* 3937, a plant pathogen, produces chrysobactin [45], a small catechol siderophore. The outer-membrane receptor protein is Fct. Iron uptake in *Morganelli morganii* is mediated by the fungal siderophore rhizoferrin. It is a siderophore in which two citrates are linked to diaminobutane [46]. The receptor protein for this heterologous siderophore is RumA. *Alcaligenes eutrophus* produces a siderophore alcaligen E. It has a MW of about 1500 Da, but the structure is not known [47]. Alcaligen E is neither a hydroxamate nor a catecholate. The receptor protein is AleB.

2.6. Other Uptake Systems

The iron-hydroxamate-uptake system in *B. subtilis*, a gram-positive bacterium, was investigated [48]. It uses a protein which shows low but significant homology with the periplasmic *E. coli* proteins FepB, FecB, and FhuD. Under anaerobic conditions, iron is reduced to Fe^{2+}, the salts of which are soluble (in contrast to Fe^{3+}) and can therefore be transported to the periplasm by porins. Iron(II) transport in *E. coli* involves a small protein, FeoA, with unknown function and a 70-kDa CM ATPase FeoB [49] (see Chap. 3 of this book). The identification or isolation of fungal transport proteins has been difficult, and a number of negative results have been obtained but not published. The recent advances with the molecular biology of iron acquisition in *S. cerevisiae* [50] may mean a breakthrough for this problem, which is discussed in Chapter 4. Some pathogenic bacteria make use of iron complexes in the host. For instance, iron can be acquired from transferrin or lactoferrin, whereas other host iron sources can be ferritin, hemoglobin, and heme (see Chap. 2). Hemolysin is sometimes required to access these iron pools. The iron-uptake systems show similarity with the ferric siderophore systems. This subject and the sfu system in *Serratia marcescens* [51] is discussed in Chapter 3.

3. HOMOLOGY, MODELS, AND CHANNELS

3.1. N- and C-Terminal Residues and Homology Boxes I, II, III, and IV

Table 1 shows that there is a distinct preference for Ala and Gln as the N-terminal for the FSRPs. The latter is cumbersome when one wants to determine the N-terminal owing to cyclization of the residue. The vast majority of the FSRPs end with Phe at the C-terminal, which is in common with porins [52]. There are few exceptions, porins LamB and *P.* 69 and Fct have Trp as the C-terminal residue, and AleB (Ala). In the porin PhoE, it was shown that the terminal Phe was important for assembly. It was also noted that in porins and FSRPs, the residues 3, 5, 7, and 9 removed from the C-terminal are generally hydrophobic, suggesting that the last 10 residues form a β-strand through the OM [52], whereas the residue at 11 removed is Arg in all FSRPs.

The homology boxes were originally identified when only four sequences were determined [53] (see legend of Table 1), and therefore their identification in the later literature may be biased. However, with many more sequences now known, they still appear significant. Box I, or the TonB box, has received most attention in the literature. It is a short segment of five residues: D Ⓣ I V Ⓥ, with most identity on the circled residues. Some FSRPs do not have an identifiable TonB box, whereas for others, it does not start until residue 66 rather than close to the N-terminal (see Sec. 4.2). Boxes II, III, and IV are properly aligned when all proteins are compared with CLUSTAL, and they stand out from other homology regions, especially box II. Box III should probably be divided into three smaller areas: the beginning and the end while the Arg Gly motive present in all proteins is in the middle. Box IV (sometimes identified as box III in subsequent literature) contains two Glys (i, i + 7), maintained in all proteins, and a considerable number of conserved residues.

3.2. Homology and Phylogenetics

Homology for these proteins is affected by several evolutionary and sometimes opposing forces. (1) If one assumes that all FSRPs have the same number of β-strands, as in the case for the porins, it is to be expected that the β-strands will be more conserved than the periplasmic and extracellular loops; (2) binding regions in FSRPs for the same substrate in different organisms might be conserved, as seems to be the case for FepA, PfeA, and BfeA, thus increasing homology, or, instead, adapt themselves to the substrate. The latter possibly leading to less specific recognition, for example, FhuE and FoxA or FhuA and FcuA, leading to lower homology; and (3) if FSRPs are compared from different organisms rather than for one organism, the homology may be lower owing to differences in the TonB proteins for these organisms, assuming that TonB is involved in the transport process (see below).

Phylogenetic trees for eight to nine FSRPs have been published [4,5,38]. A new one involving 21 FSRPs from Table 1 is given in Fig. 2. (PupX is not included because of its high homology with PbuA and Fct is not found.) ViuA shows only background similarity to the other proteins. It is homologous, however, with the other proteins. For instance, features in homology boxes II, III, and IV and in the C-terminal region, discussed above, are retained in ViuA. AleB and IutA have homology with one another but only background similarity with all the other proteins. Also, FyuA shows little homology. For the other proteins, six groups can be recognized. It is obvious that the effects listed in (1)–(3) above result in certain relationships which are difficult to explain; for instance, the homology of the unspecific FhuE with the specific pseudobactin receptors or the greater homology of PbuA with FptA than with the pseudobactin receptors, whereas others seem to make sense, that is, the group FepA, PfeA, BfeA, which includes Cir, or the homology of FecA and RumA. Only in the case of pseudomonads does the species seem to have a positive effect on homology. A more complete study of homology among the proteins should yield new features.

3.3. FepA Model

Models were proposed for FepA [69], FhuA [70], PupA [62], FoxA [5], and Fct [66]. The hydropathy values for FSRPs indicate transmembrane β-sheets for the secondary structure as observed in porins. The different models show between 26 and 30 of these β-strands, although one may expect the same number of strands for all FSRPs.

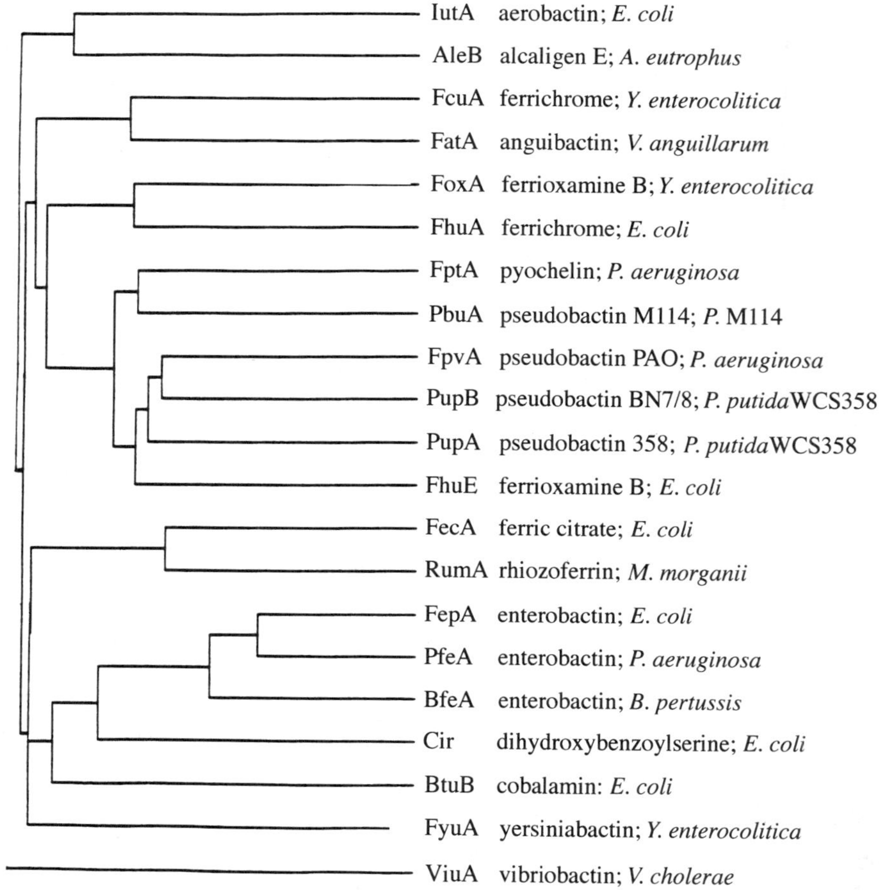

FIG. 2. Phylogenetic tree for 21 ferric siderophore receptor proteins listed in Table 1.

The first model proposed was for FepA. It was based on a thorough study to identify surface epitopes using different MAbs blocking the interaction with ferric enterobactin or inhibiting the killing by colicins B and D. The model is reproduced in Fig. 3. The study concluded that the region from 290 to 339 was the ligand-binding domain (diagonally striped), whereas regions 258–290 and especially 382–400 are close to or interact with the binding site. Subsequently, in mutants in which the ligand-binding site is removed, epitopes in the region 100–142 that are not surface accessible in the wild type are recognized in the mutant [71],

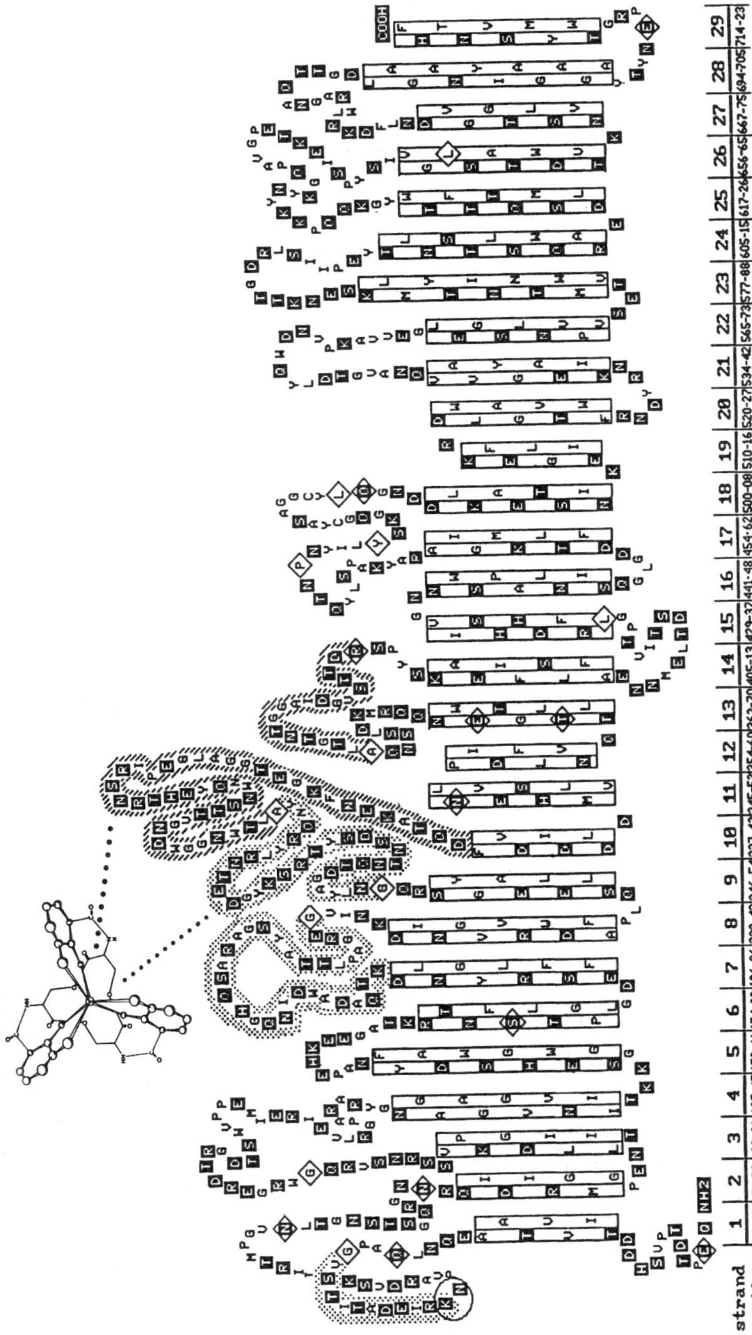

FIG. 3. FepA model. (Reproduced with permission from [69].)

whereas deletion of residues 58–142 eliminates substrate binding. This indicates that the ligand-binding domain, in fact, is a complex assembly of polypeptides [72].

The model contains 29 transmembrane β-strands, although one could expect an even number as is found in the porins (16). The β-sheets found in the crystal structures of porins are somewhat longer than was expected (average β-sheet is 11 residues). This is due to the inclination of the β-sheets with respect to the perpendicular to the membrane as observed in the crystal structures of the porins. The average length of the β-sheets in the FepA model is 9. The TonB box occupies the first β-sheet. It is debatable if this is a contradiction with the ability to interact with TonB. It is possible that the FepA structure changes once the ligand is bound. The black-blocked residues are hydrophilic, and the preponderance of these residues in the loops is obvious. The β-strands can be expected to be amphiphilic (see below), and this model shows this to be generally the case. Also, most periplasmic loops are small, with the largest being between strands 14 and 15. The homology box II starts in strand 27 and extends eight residues downstream into the surface loop. The first 18 residues in the surface loop between strand 1 and 2 and the succeeding strands 2 and 3 are homology box III, whereas 22 residues in the loop between strands 3 and 4 and the 4th strand form homology box IV. The possible homology in boxes II, III, and IV may therefore be biased by the involvement of β-strands.

A study using dipeptide insertions and large deletions generally agrees with the model [73]. A study from the same laboratory using M2 epitope insertions suggested some changes [74] and indicated that residue 55 is on the periplasmic side rather than in the first extracellular loop and that residue 142 is on the periplasmic side rather than in a β-strand, although this might contradict an earlier study for this region [71]. The changes, therefore, involve homology regions III and IV. FepA was point mutated (E 280 C and E 310 C), and the Cys residues in the two proteins were attached to a spin label [75]. An electron spin resonance (ESR) study showed both to be surface localized, but residue 280 to be less accessible than 310. The authors consider this to be compatible with the model, although both residues appear to be equally accessible in Fig. 3. However, the tertiary structure of the binding region will likely be quite different than the one indicated in Fig. 3. Binding of ferric enterobactin resulted in a conformational change in the binding loop. Possible changes on the periplasmic side would not have been noticed in this experiment.

The substrate Fe(ent) has a triple-negative charge, and it is therefore of interest to explore positively charged residues in the binding loop, in particular four arginines (283, 286, 313, and 316), which are conserved in Fe(ent)-binding regions or directly adjacent to it. Any single mutation (in each case to Ala) did not effect binding of the ligand. However, the combined effect of the double mutant Arg 286, Arg 316 had a very obvious effect on binding as well as transport, identifying these residues to be involved in substrate binding [76] (see Sec. 4.8).

3.4. FhuA and Other Models

The FhuA model [70] (Fig. 4) is based on a study of both insertion (2–16 residues) and deletion mutants in which ferrichrome growth activity and colicin M inhibition were explored [77]. In addition, protease activity on whole cells and spheroplasts was determined. This allowed the assignment of nine sites on the cell surface and four sites on the periplasmic side, and this information is incorporated in the model for FhuA. It was also obvious that colicin M binding required more domains than ferrichrome binding, similar to the observation made for FepA [69]. For substrate binding, the region 322–355 and especially Asp 348 is essential [78], and region 196–223 is also important. The model has more β-strands (32) than FepA despite being smaller, and the number of residues in extracellular loops is considerably smaller than in FepA. It is interesting to note that insertions, especially between residues 69 and 135, had significant negative effects on the occurrence of the protein in the membrane. This is the domain which is known as boxes III and IV. The sensitivity of this region was confirmed in a study using epitope insertions [79]. This same study showed that residues 321, 405, and 417 are surface accessible.

A topological analysis of FhuA using MAbs identified regions between residues 1 and 20 and 21 and 59 as being on the periplasmic site and regions between residues 321 and 381 (probably the loop between 316 and 356) and between 417 and 550 (probably the loop between 404 and 433) to be on the cell surface. None of the MAbs inhibited ferrichrome binding [80,81]. The same laboratory also investigated purified

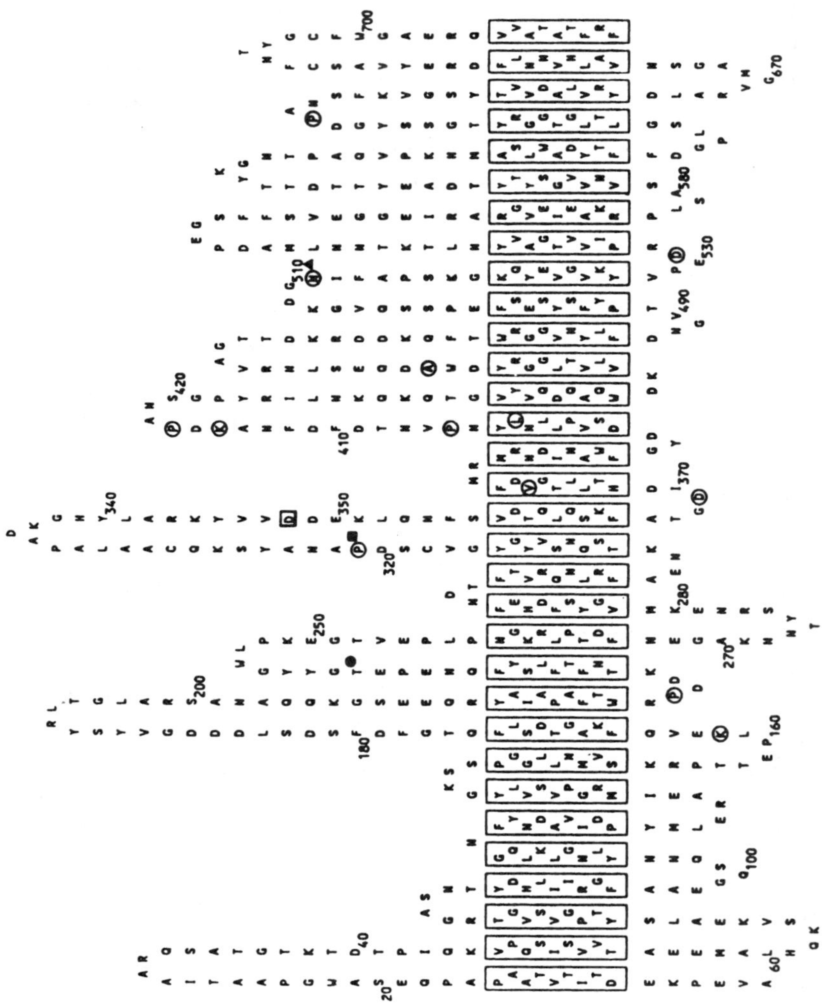

FIG. 4. FhuA model. (Reproduced with permission from [70].)

FhuA and found that with trypsin digestion Lys 67 was protected after binding with ferrichrome but not Lys 5. In the model, Lys 67 is in a periplasmic loop. It indicates, therefore, that ferrichrome binding to FhuA leads to a conformational change on the periplasmic side of FhuA involving Lys 67. The authors postulate that this is a prerequisite for the physical association between TonB and FhuA (see Sec. 5.4). However, it contradicts the observation, from the same laboratory, that deletion mutants Δ060-069 and Δ060-135 of FhuA demonstrate growth promotion by ferrichrome [82].

The FoxA model [5] is built from 30 strands, each nine amino acids in length, and is based on homologies with FhuA. An interesting aspect of the model is that many Phe and Trp residues are in the beginnings and ends of the strands as was observed in the crystal structure of the porins (see below). The PupA model [62] is based on insertion mutants sensitive to a specific endoprotease and it has 26 strands. The long N-terminal is in the periplasm, and the TonB box which starts at 66 (see Table 1) occupies the first strand. Most of the periplasmic loops are short except for the first two with 20 and 16 residues between strands 2 and 3 and 4 and 5. These loops contain parts of homology regions III and IV.

The models show between 26–30 β-strands, which is logical when one considers the MWs of the FSRPs in comparison with those of porins, which have 16 or 18 β-strands. On the other hand, the specific binding behavior of FSRPs may require larger loops, and it is therefore possible that the actual structures of FSRPs can not be extrapolated from those of porins.

3.5. Channel Formation

When residues 200–340 are deleted from FepA, the high-affinity uptake is changed into a diffusion system, which is not dependent on TonB and which is unspecific [71] (both ferric enterobactin and ferrichrome show the same diffusion behavior). The deletion which includes the substrate-binding site does not affect the basic structure of FepA. The deletion mutant forms channels which are considerably larger than OmpF-type pores [83]. A similar experiment was carried out on FhuA. Deleting residues 322–355, implicated in ferrichrome binding (see Sec. 3.4), converted FhuA into an open channel and the uptake no longer required TonB [84,85] and was unspecific. Smaller deletions in the same region gave similar results. It was shown that phage T5 unmasks the same or a very similar channel in FhuA [86].

These results on FSRPs have led to descriptions of the wild-type proteins as gated channels and gated porins [71,83]. These terms may be confusing. Although both porins and the FSRPs use β-strands to transverse the outer membrane, there are important differences between the two classes of proteins. The FSRPs are involved in specific and saturable transport, whereas it is unspecific and diffusion transport in porins. For instance, wild-type FhuA and other FSRPs have a very low transport rate (10/s), whereas normal gated channel transport is much faster (10^7/s). Also, the FSRPs require an energy transduction system involving TonB, whereas the porins do not use such a system. The formation of a channel in deletion mutants gives rise to questions about the mechanism. Does the channel always exist? Is it the result of an interaction by TonB? Is the channel only formed after substrate binding? How and by what cause does the substrate move from the extracellular binding site to the channel? Is the process reversible?

4. INTERACTIONS AND SPECIFICITY

4.1. Binding Proteins in Periplasm

The emphasis of this chapter is on the molecular properties and molecular interactions of the ferric siderophore outer-membrane (OM) receptor proteins and those of their complexes with siderophores. In this section, the interactions on the periplasmic side are discussed first. This involves first the specific periplasmic binding proteins and second the transducing TonB protein.

After the ferric siderophore binds the FSRP, it is transported through the OM and accepted by a periplasmic protein. Further transport involves a dimeric ATPase (FhuC, FepC, FecE in *E. coli*) and hydrophobic cytoplasmic membrane proteins. This information comes primarily from genetic experiments and has been reviewed [87,88] (Fig. 5). It is interesting to note that FhuD is used for three structurally dissimilar ferric siderophores, and also the periplasmic FepB protein is

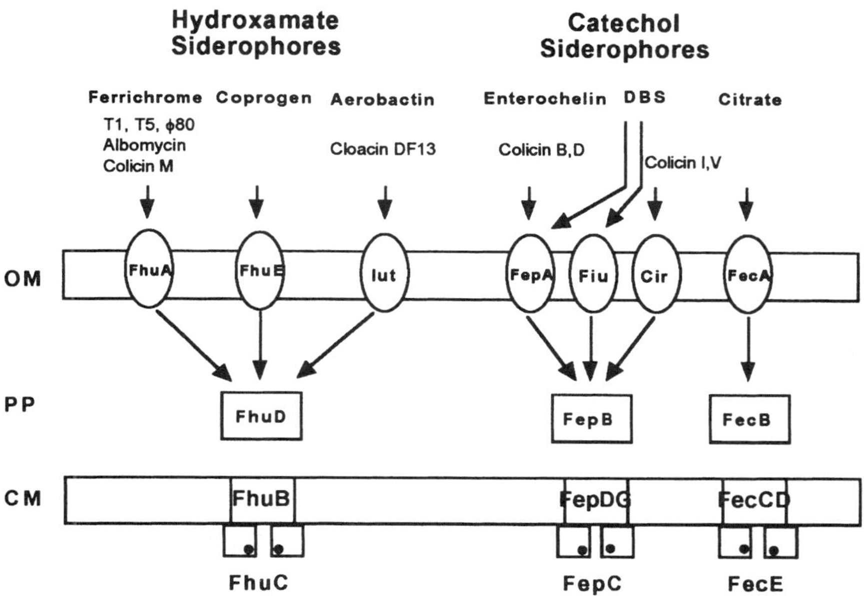

FIG. 5. The organization of transport proteins in *E. coli* for ferric siderophores. (Reproduced with permission from [88].)

used for FepA, Cir, and Fiu. It is thus possible to conclude that the specificity for ferric siderophores resides in the recognition of the outer membrane proteins.

Periplasmic proteins have also been identified for transport systems in other organisms besides *E. coli*. Physical studies on these periplasmic proteins are scarce. Osmotic shock should release those proteins, but binding experiments with radiolabeled (Fe) ferric siderophores yields high unspecific binding to unrelated proteins in the periplasm. Most success has been obtained with FhuD, a 29-kDa protein [89,90]. The protein is protected from proteinase K degradation when it is bound to ferrichrome, aerobactin, or coprogen [91]. This gives direct biochemical evidence that these three substrates which use different FSRPs, FhuA, IutA, and FhuE; all three use FhuD therefore confirming the genetic evidence. Neither ferrichrome A, which is not transported into *E. coli*, nor ferric dicitrate, which uses FecB in the periplasm, gives this protection to FhuD. FhuD was also purified [92] and binding experiments were carried out with the different siderophores showing strong interactions with ferric aerobactin, ferrichrome, and ferric coprogen. The K_d values are between 300–1000 nM. In the case of FepB, a 32-kDa protein [93], a membrane localization vehicle was used to place the protein in the OM [94] of a *fepA* mutant. The cells bound Fe(ent) showing the direct interaction of FepB with the substrate.

4.2. TonB Interactions

All ferric siderophore transport systems and the one for vitamin B_{12} involve energy-coupled transport and signal transduction via the TonB-ExbB-ExbD system of proteins. This is described in Chapter 3 of this book, as well as in several reviews [72,87,88]. Of these three proteins, only TonB extends across the periplasm. Therefore, in view of the intent of this chapter, only the interactions of TonB with the FSRPs for ferric siderophore transport will be discussed. Most of the information is obtained from genetic evidence and physical results are scarce.

TonB mutants were observed at the very beginning of molecular biology [95] as a class of *E. coli* resistant to bacteriophage T1. These mutants are defective in all high-affinity iron transport. TonB from *E. coli* has been sequenced [96]. It has 239 amino acids and a MW of 26 kDa. TonB-PhoA fusions proved it to be anchored in the cytoplasmic membrane with its N-terminal portion while the remainder occupies the periplasmic space [97]. The carboxy-terminus is hydrophobic and essential for activity but seems not to be anchored in the cytoplasmic membrane (CM) [98]. An alternative view comes from modeling, which suggests an amphiphilic transmembrane structure for this segment [72]. The central region of the protein is hydrophilic. There is between residues 72–102 a proline-rich region (see below). Plasmid-encoded TonB is unstable, but in the presence of ExbB, the protein is stable [99,100].

Other TonB proteins have been sequenced. The homology with the *E. coli* protein is for some quite low (28% in *P. putida*) [101,102]. The protein from *P. aeruginosa* [103] has more homology with the *E. coli* protein than the one from *P. putida*. The general topology for all TonB proteins, however, is the same. There is in vivo genetic evidence for the interaction of Q-160 (165) (*E. coli*) with the TonB box in the receptor protein [104,105]. There may be a functional interaction of this residue with the conserved Gly at 239 [106]. However, this residue (160) is not conserved in the TonB proteins. There is also in vivo biochemical evidence for the interaction of TonB with the receptor protein FepA from cross linking studies [107]. Substrate was not required in this case. Another piece of evidence is that FhuA prevents TonB degradation [100,108]. In the region of possible interactions with OM receptor proteins, two glycines are conserved (174 and 186) in all TonB proteins. There are, however, also other residues in the putative interaction region which are identical in all TonB proteins—V163, P164, A167, R204—but the significance of this is not known.

The XAA-Pro dipeptide repeats are interesting. Structurally it would extend or elongate this part of the TonB protein. Nuclear magnetic resonance (NMR) evidence seemed to indicate an interaction of FhuA with a 33-residue peptide containing these repeats [109]. However, when this section was deleted from TonB (Δ66-100), the mutant was fully capable of ϕ80 adsorption, indicating that this section has no role in energy transduction [110]. Still the activity of the deletion mutant TonB (Δ66-100) was less than of wild-type TonB. The (Aax-Pro) repeats in TonB are unusual and their true function still needs to be established.

The transport function of the TonB box in the receptor protein is also controversial. When strains of *E. coli* were treated in vivo with a synthetic TonB box pentapeptide, cell growth in low-iron medium and other TonB-dependent processes were inhibited presumably by saturating the interaction site on TonB [111]. On the other hand, some FSRPs have no recognizable TonB box (see Table 1), and for some others, the homology for this region is low. One could argue that this lack of homology is compensated for by changes in the TonB protein (note the low homology between TonB from *E. coli* and *P. putida*). Also, mutants in the TonB box can be made viable with mutants in TonB [104,112], which is circumstantial evidence for an interaction between TonB and the TonB box. Finally, besides the TonB box, there are other homology boxes in the FSRPs, especially regions III and IV, which could interact with TonB (see Sec. 3.1).

In the literature one can recognize two lines of thought. One is that TonB is the energy transducer from the cytoplasm to the outer membrane [85]. The other [81] is that once the outer-membrane protein is bound to its cognant ferric siderophore, the protein changes its conformation and only in this new conformation does the FSRP interact with TonB resulting in a signal to the cytoplasm. This latter idea removes a paradox which is a result of the former idea. This is the fact that there are just not enough TonB molecules to interact with all FSRPs [72]. The second idea would remove this paradox, because only those FSRPs which are complexed with substrate would interact with TonB.

4.3. Purification of Receptor Proteins

Most of the information on FSRPs has come from genetic experiments and in vivo studies. Complementary information needs to be obtained from physical and structural studies of pure FSRPs. The fact that they are membrane proteins and require detergents to stay in solution makes this a difficult task, and it is the reason for the limited information which has thus far been obtained.

Originally, FepA was produced using a *ent* mutant *E. coli*, BN 3040, which overproduces the protein to some extent. Production requires an iron-deficient medium, because the production is chromosomal [113–115]. Another early method was to use gene fusion with lac and to select *fur* mutants [116]. Both methods have an inherent danger in that other FSRPs are produced as well. The advances in molecular biology in the 1980s allowed construction of overproducing plasmids for just one FSRP [53,117,118] introduced in a selected host. These can be grown in rich media and the production is many times larger. Recently, newer constructs have been used which produce very large amounts of protein as inclusion bodies [119].

Solubilization normally follows the procedure of Schnaitman [120] which consists of sonication, membrane isolation, solubilization of the inner membrane with the detergent Triton X-100, which leaves the outer membrane intact and solubilization of the outer membrane with detergent and EDTA (ethylenediamine-*N*,*N*,*N*′,*N*′-tetraacetate). The EDTA removes Ca^{2+} making it possible to dissolve the OM. The best detergent for this last step is not necessarily Triton but can be *n*-octyl-β-D-glycopyranoside (OG) or another detergent; this has to be experimentally determined, but an uncharged detergent should be used [121–

123]. The purification steps are diethylaminoethyl (DEAE) ion exchange (HPLC) and chromatofocusing (FPLC) [124]. The latter step is necessary to remove all lipopolysaccharide (LPS) [125], and at the same time, is an effective purification step. If necessary, a sizing column can be used as the last step. It is possible to obtain a protein which is better than 99% pure and LPS free. Initially a Mo-catecholate affinity column was used for FepA [115], and recently a colicin B-agarose affinity column was used effectively [126]. Affinity columns using a (Pro-Xaa)n peptide and one using an MAb were not successful [127]. For FhuA, a Ni-nitrolotriacetate (Ni-NTA) agarose column was used after attaching a His tag [81]. Further purification is necessary after the use of an affinity column.

The FSRPs are prone to proteolysis. The first time this was recognized was in the case of FepA where protein "a", later recognized as OmpT, degrades the protein into a 74-kDa product [113,128]. When the protein is produced by a plasmid, it therefore pays to use an *ompT* host [53,121]. FhuA can be hydrolyzed into a product which misses the first nine N-terminal residues [122]. One can prevent problems by using the inhibitor benzamidine during the extraction and purification. The purified proteins are stable for many weeks at 4°C. The identity of the FSRPs or degradation products is determined most securely by N-terminal sequencing [129], and the amount of protein by the bicinchoninic acid (BCA) method in which detergents do not interfere [130].

The first crystallization for an FSRP, FepA, was reported in 1989 [121]. This was done in OG, but since that time better results have been obtained with protein in dodecyldimethylamine oxide (LDAO) [131]. In general, the same techniques are used as with porins and the photosynthetic reaction center. The important strategy appears to be to dissolve the protein in an as low as possible concentration of the detergent in setting up the crystallization experiments. Crystals are obtained relatively easily. For instance, in this laboratory and the laboratory of Deisenhofer, University of Texas, S. W. Medical Center at Dallas, crystals have been obtained for FepA, SeMetFepA, four point mutants of FepA, FepA Fe(ent), FhuA, FhuA ferrichrome, FecA, and PupA. Also, refolded FepA has been crystallized [119]. Large and well-diffracting crystals, however, require many variations in crystallization parameters, such as polyethylene glycol (PEG) and protein concentration, pH, salts, and additives, and a description falls beyond the scope of this chapter. The crystals of derivatives of FSRPs are needed for the crystallographic structure solutions of the proteins.

4.4. Physical Studies

The FSRPs so far purified have a p*I* point between 5.1–5.4 as determined by chromatofocusing. The CD spectrum of purified FhuA [132] indicates that the protein is organized primarily as a β-sheet structure (51%) in agreement with the hydropathy plots. Also, in vitro, ferrichrome inhibits binding with phage T5. Sedimentation studies and size exclusion chromatography indicate monomeric FhuA molecules, each associated with approximately 185 molecules of the detergent dodecyl maltoside [132]. The monomeric form is in agreement with results on partially inactivated FhuA [133]. For FepA, on the other hand, nondenaturing gel chromatography [83] and cross linking [107] might indicate a dimeric or trimeric state. Crystallographic space group information for FepA [131] eliminates the possibility of a trimer, although a dimer is possible but not obligatory. However, it does not seem reasonable that FepA and FhuA would aggregate differently. In another study with purified FhuA [81], CD results show a smaller amount of β-sheet (38%), a number which increases slightly on binding with ferrichrome (42%). The conformational change observed for FhuA on binding with ferrichrome has been described earlier (see Sec. 3.4).

The purified spin-labeled FepAE280C was used for unfolding experiments using ESR [134]. The unfolding process is highly cooperative and can therefore be described by a two-state equilibrium. The initial rates were considerably slower in the presence of ferric enterobactin. The unfolding process was fully reversible in the presence of Triton X-100 and did not require lipids.

4.5. Binding Studies

The first work with purified and partially purified FSRPs was done in Neilands' laboratory [20,114,115]. For purified FepA, a DEAE microassay column was used to determine the dissociation constant K_d (200–500 nM) with Fe(ent). It was found that ferrichrome did not inhibit the binding; however, the ferric salt of the dodecane analogue of enterobactin inhibited it significantly (about 50%) [114]. This was a significant step forward, because it showed that FepA retained its structural integrity once it was removed from the OM and solubilized in detergent. In a later experiment, a gel column was used to determine K_d [115] and a value of ~10 nM was found. The K_d measures the dissociation constant for the reaction: PS $\rightleftharpoons$ P + S, where P is the receptor protein and S the ferric siderophore. Subsequently, values of K_d of 20–30 nM have been reported for pure protein [126] and whole cells [71,76], but also values of 100–400 nM are found for pure protein [75,124] and whole cells [75]. Recently, in the case of PupA with ferric pseudobactin, the K_ds for whole cells (40 nM), OM preparations (200 nM), and purified protein (340 nM) were determined [125]. It seems to indicate that detergent or membrane support has a small effect (200 $\leftrightarrow$ 340 nM), but that the K_d for whole cells (40 nM) might be affected by a cooperative effect with the periplasm. Still discrepant results exist in the literature, possibly as a result of the technique which was used and they need to be resolved. The concentration of Ca^{2+} may have an effect (see below) and stoichiometry has not always been determined.

The specificity of binding of the purified FepA protein was studied with gel filtration [135]. Self-inhibition of labeled ferric enterobactin by unlabeled substrate shows that the binding is at a definite site. Inhibition by FeMECAM was about 50% of that with unlabeled ferric enterobactin, which is similar to the inhibition with the carbocyclic analog of enterobactin mentioned previously. Both FeMECAM and Fe(ent) are transported at the same rate in whole cells [136,137]. Ferrichrome A did not inhibit binding, although it has three negative charges like ferric enterobactin. Also, ferric parabactin did not inhibit, although it is a tricatechol just as is ferric enterobactin. It has, however, a Λ absolute configuration for the iron coordination, whereas Fe(ent) has the Δ configuration. The uptake specificity by FepA was investigated for a number of derivatives of FeMECAM, an analogue of enterobactin in which the triseryl scaffold is replaced by a phenyl group, indicating domains for the substrate [138] (see Sec. 4.8), but these compounds have not yet been used in binding experiments.

FecA has been purified [135], and it shows binding activity for ferric enterobactin but with lower affinity than FepA [124] (K_d ~4000 nM). This is in agreement with uptake studies which show that ferric enterobactin is taken up by UT5600 (*fepA*) at a low rate [135,139]. Both the binding and uptake is enhanced by citrate. MAbs for the binding region of FepA [69] inhibit binding of FepA with ferric enterobactin but not the binding with FecA, showing that the binding regions of FepA and FecA are different.

Ferric *L*-parabactin [140] is taken up by *Paracoccus denitrificans* by a high-affinity system [141]. The OM protein in the organism was partially purified and the K_d for binding with ferric *L*-parabactin was estimated to be 700 nM [123]

Very early an attempt was made to isolate and study the T5 receptor (TonA, FhuA) [142]. However, the first in vitro study on partially purified FhuA was done in 1975 [20]. Phage T5 binding to the receptor was inhibited by ferrichrome. Lower inhibition of T5 binding was observed for ferricrocin, Al-, and Cr-deferriferrichrome and no inhibition was measured for ferrichrome A, ferric enterobactin, ferrioxamine E, and ferric rhodotorulic acid. This is in general agreement with in vivo observations [19]. Later on purified FhuA, overproduced from a plasmid [118], confirmed the inhibition of T5 binding by ferrichrome and showed the deferri compound not to be active in the inhibition. Purified FhuA was used for direct binding and inhibition studies on a sizing column [122]. It showed that the binding of ferrichrome to FhuA occurred at a single site and was specific. The binding was inhibited by ferricrocin and ferrirubin, although slightly less than by unlabeled ferrichrome itself, but no inhibition was found for ferrichrome A nor ferrirhodin. Also these results are in general agreement with uptake studies [122].

4.6. The Substitution of Iron with Gallium

Fe^{3+} has *5d* electrons which are evenly distributed in the five *3d* orbitals. It makes the ion spherically symmetrical. This reflects itself in the coordination of Fe^{3+} in ferric siderophore complexes. The iron coordination is always an irregular octahedron, because crystal field stabilization is not of any significance for these complexes. Ga^{3+} has 10 *3d* electrons, and it is therefore also a spherically symmetrical ion. The ions differ in size by only a very small amount and the association constants for Fe(III) and Ga(III) siderophores differ by only a factor of 10. Fe(III) and Ga(III) complexes of catecholates and hydroxamates are isostructural and accurate crystal structure determinations show only minor differences in geometry for the structures [143,144]. The only major chemical difference between Fe^{3+} and Ga^{3+} is that Fe^{3+} can easily be reduced. However, the transport process across the OM or the binding with the FSRP does not involve oxidation or reduction. Ga(III) siderophore complexes, because they are diamagnetic in contrast to Fe(III) complexes which are paramagnetic, are often used to investigate siderophore structures by NMR [145–148] with success and with the reasonable assumption that the structure information obtained for Ga(III) complexes is directly applicable to Fe(III) siderophore complexes. It is therefore logical to study the binding or binding inhibition of Ga(III) siderophore complexes with FSRPs. Surprisingly it was found that gallichrome showed low inhibition on the binding of ferrichrome with FhuA [122] in qualitative experiments. This is actually in agreement with transport studies which showed that Ga(ent) was taken up at only 25% the rate of Fe(ent) and that *E. coli* cells bound only one sixth as much Ga(ent) as iron complex [137]. Also Ga(parabactin) is not taken up in *P. denitrificans* [141].

Assuming that K_d values are of the order of 10^{-7}, one needs to use concentrations of 10^{-7} M or less for substrate and protein to obtain sensitivity and quantitative answers. This was done with Ga(III) pseudobactin WCS358 and purified PupA [125]. The K_I for GaPs was more than 10 times higher than K_d for FePs with or without $CaCl_2$. This semiquantitative answer, therefore, confirms the unexpected qualitative observations for gallichrome, but it needs to be followed by direct binding experiments and data with other metal ions and other purified FSRPs.

4.7. Effect of Ca on Binding

The first experiments to investigate the effects of Ca^{2+} on substrate binding were done with cobalamin and BtuB [149]. The protein was not isolated but instead broken cell envelopes were used. Ethylenebis-(oxyethylenenitrile)tetraacetic acid (EGTA) buffers were used to generate known "free" Ca^{2+} concentrations between 1 nM and 1 mM. Large changes (20-fold) in the amount of substrate bound were found. An increase of K_d by a factor of 50 was observed with decreasing Ca^{2+} concentration, whereas at lower pH values, a larger concentration of

Ca^{2+} was required than at pH 6.6. Half-maximal binding was 30 nM Ca^{2+} at pH 6.6 and 630 nM at pH 5.4, and maximal binding required 1 and 10 mM, respectively. The specific activity was not much affected. Mg^{2+} did not have the same effect. At higher concentration (>1 mM) of Ca^{2+}, binding decreased. The latter is probably caused by a general decrease of dipolar interaction due to the increase in ionic strength and was also observed for Fe(ent) binding with FepA and ferric pseudobactin binding with PupA (see below).

The Ca^{2+} effect on BtuB was followed by a study to explore the interaction between substrate and Ca^{2+} [31]. It was shown that there is cobalamin-dependent binding of Ca^{2+} to BtuB. The loss of the TonB box diminishes affinity for Ca^{2+} but retains affinity for cobalamin, whereas loss of the first 49 residues (which includes part of homology box III) only binds calcium with lower affinity. The unusual aspect of these experiments is the low specific activity which indicates a stoichiometry of only 0.0001 for substrate/protein.

One may expect similar results for the FSRPs because of their similarity with BtuB. Purified PupA and ferric pseudobactin WCS 358 were investigated at two different pH values (Fig. 6) [125]. Zero concentration of Ca^{2+} was obtained by adding EDTA and low concentrations (0.02–10.0 μM) by adding Ca^{2+} to the buffers. There is an obvious increase of binding when small amounts of Ca^{2+} are added, but the effect is much smaller than observed for BtuB. On adding Ca^{2+}, the K_d value increased twofold, but the B_{max} (Fe-pseudobactin/PupA) (stoichiometry) changed from 0.55 to 0.91, and the latter is thus the driving force for the increased amount of substrate binding which is observed. The figure also shows the slow decrease of binding at higher Ca^{2+} concentrations explained above.

4.8. Specificity, Stereospecificity, and Binding Pocket

The information regarding the specificity of binding and transport for FSRPs is often contradictory and therefore difficult to analyze. FepA and FhuA are both used in TonB-dependent transport by ferric siderophores as well as the much larger colicins and phages. In both cases, there are clear indications that the binding region for these colicins is larger than the binding pocket used by the ferric siderophores, and that

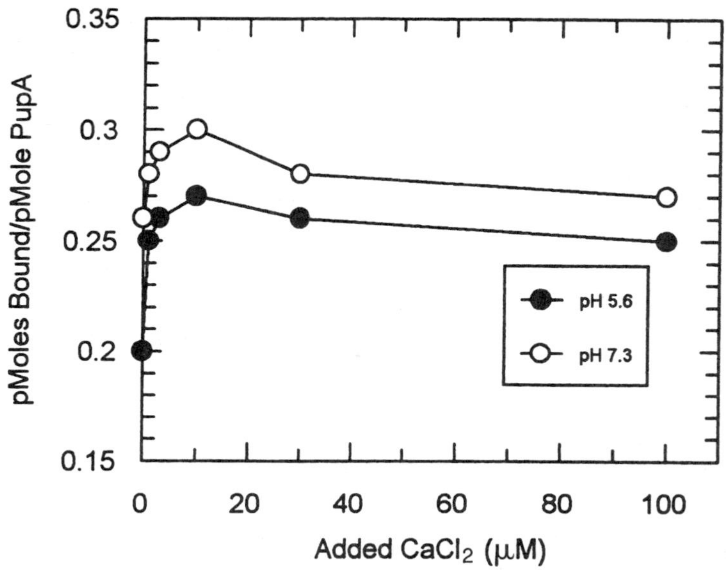

FIG. 6. The effect of low concentrations of Ca^{2+} on the binding of ferric pseudobactin 358 with PupA.

is logical considering the size of these particles. The fact that ferrichrome with FhuA and Fe(ent) with FepA prevents colicin binding and phage infection in vivo indicates that these particles require the ferric siderophore-binding pocket. The ability of analogues of these ferric siderophores to prevent these phenomena could therefore be evidence for a lack of specificity in the binding pocket. It is, however, indirect and qualitative information which needs to be quantitated. Recently, it was shown that the structural requirements for sensitivity to colicin M, phage T1, and phage T5, separately, in the gating loop 322–355 of FhuA are all different [150].

However, other smaller structures quite dissimilar to the natural siderophore of a particular FSRP are sometimes recognized. It is important to prove the identity of the FSRP which is involved considering the multitude of uptake systems which exist or can be induced in an organism. This proof has been given in a number of cases, and those involving FhuA will be listed. Albomycins [151] are natural antibiotics which contain the triornithine hydroxamate part of ferrichrome attached by a peptide bond to lethal moieties. It was shown that ferric albomycin and synthetic analogues use FhuA [22,152]. A semisynthetic analogue of rifamycin (without iron) uses FhuA [153]. Microcin 25 [154], a 2100-Da and 20-amino acid peptide, uses FhuA and TonB [155]. Other examples of this type of recognition by Cir, FiuA, and FhuE are known. Some of the compounds which use FhuA obviously have structures which are very different from ferrichrome, and the fact that these systems require TonB indicates an amazing lack of specificity of recognition between substrate and FSRP. Of course, these observations also require quantitation, in vitro or in vivo, and a study, in each case, of the structural requirements in FhuA. One should, however, also contrast these observations with the specificity which FhuA shows in in vitro binding studies [122] where ferrirubin, but not ferrirhodin, inhibits ferrichrome binding despite the fact that those compounds only differ in the *trans* and *cis* configuration of their *N*-acyl groups, anhydromevalonic acid. Ferrirubin inhibits ferrichrome transport but ferrirhodin does not, and the specificity, or lack of it, is therefore not due to a difference in transport. These contrasting observations on specificity could indicate that there are different ways to trigger the uptake process, but they also reveal our limited knowledge of molecular recognition. The ability in the near future to obtain structures of FSRPs and their complexes with substrates seem to make the FSRPs ideally suited to increase this knowledge.

There is no doubt that certain FSRPs have lower specificity than others. FhuE transports ferrioxamine B, coprogen, and rhodotorulic acid [23]. All are hydroxamates, but they have different structures, although coprogens and ferrioxamines have an overall three-dimensional structural similarity [156]. Both FhuA and FhuE are FSRPs for heterologous siderophores, and it is possible that the evolutionary adaptation for transport was imperfect. On the other hand, there is the example of PupB, transporting only pseudobactins BN7 and 8 among the dozen of pseudobactins tested [10] indicating highly specific recognition. This is also an heterologous system, but possibly it is more evolved by using a signaling system (PupI and R) which is not present for FhuA and E. Also, the low binding and uptake of Ga-siderophores observed for FepA, FhuA, and PupA is difficult to rationalize. There is, therefore, still much work to do to assess specificity. Quite important, however, is that thus far no major differences have been observed between in vivo and in vitro specificities, and that therefore structural and physical chemical studies of purified FSRPs will be significant. Finally, one can ask the enigmatic question: What is the biological advantage for an uptake system which is so specific that it recognizes only one ferric siderophore?

Enantioenterobactin was synthesized using *D*- instead of *L*-serine and the ferric chelate failed to support growth in *E. coli* [157], presumably by not being transported by FepA. Enantioferrichrome, synthesized from *D*-ornithines, was taken up at 50% the rate of ferrichrome via FhuA [158]. Similarly, enantiorhodotorulic acid is taken up at a lower rate (~25%) than the natural compound [24] by FhuE, and the high-affinity uptake of ferric parabactin shows enantiomeric stereospecificity [141,159]. Thus, the configuration around the metal plays a significant role in the specificity, but apparently more so with FepA than with FhuA and FhuE.

The binding pocket of FepA has been investigated by point mutants of FepA and analogues of Fe(ent). Fe(ent) and the analogues are triply negative charged, and one can expect these negative charges to be partially localized on the C=O oxygen atoms of the peptide linkages (see Fig. 1). There is a preponderance of basic residues in the putative substrate-binding region of FepA (see Fig. 3). Indeed, mutations of two acidic Glu residues have no effect on binding and transport [75], but the replacement of two basic (Arg, 286, 316) residues by Ala has a significant effect [76]. This can be taken together with the requirements of the substrate for which it was found that the scaffold is not important, also the NH group can be changed to NCH_3 without affecting uptake [138,160]. On the other hand, reversing the peptide linkage or any additional substitution on the aromatic moieties elminated uptake. The crystal structure of Fe(ent) has not been determined, but the one for the V-chelate is known [161], and it should be a good model for the Fe(ent) structure. Although rotation of the peptide linkages is possible [162], one finds in the V(ent) structure that all carbonyl groups point outward and are available for H bonding. Combining the information about the binding pocket and the substrate would indicate that the NH groups do not interact with the protein but that the three C=O groups form H bonds with Arg-286 and Arg-316 and possibly a third basic residue (R400?). There are two possible problems, however. First, the $N{-}CH_3$ substitutions certainly would rotate all peptide linkages due to steric hinderance and thus directly affect the H-bonding pattern of the C=O groups. Second, if indeed three C=O---H bonds are involved, the loss of one still could place the substrate correctly in the binding pocket, but one would expect some effect on binding and uptake, which is not observed for single mutants. Instead, the double mutant is required to observe a significant difference.

Different FSRPs may well be constructed so that different domains in the substrate are recognized. Obviously, the direct environment of the metal ion is recognized in Fe(ent), but that seems not to be the case for the pseudobactins for which the recognition is quite specific [10,35]. The metal ion surrounding is quite similar for pseudobactins and the differences occur, instead, in the peptide backbone or scaffold for these siderophores, and they appear to form the main domain for recognition in the binding pocket in contrast to Fe(ent) with FepA. It will be interesting to address the issues which were discussed in this section with structural studies, because the results may give us a better understanding about molecular recognition.

5. MECHANISM

5.1. Structural Information

Thus far, no crystal structure has been solved for an FSRP or its complex bound to a ferric siderophore. These structures can be expected in the next few years, and the results will be important pieces in the puzzle to understand the OM transport in gram-negative bacteria and for other membrane transport processes. A discussion of the difficulties and technical aspects of the crystal structure solutions are not appropriate for this chapter, but it can be stated that once one or a few structures have been solved with cumbersome heavy atom techniques, other structure solutions might be less time consuming using the method of molecular replacement.

In the last few years, the crystal structures of several porins have been solved [163–167], and recently these structures were reviewed in relation to their mechanism and function [168]. The common motive is a barrel formed by 16 antiparallel β-sheets (18 for LamB). Such a structure uses peptide H bonds between adjacent β-sheets forming a stable secondary structure. Side chains of sequential residues in a β-sheet point up and down or in the barrel toward the hydrophilic inside of the barrel and out into the lipid environment. The β-sheets are therefore amphiphilic (i and i + 1). On the periplasmic side the β-sheets are connected for the most part by short β-turns forming a smooth surface, whereas on the extracellular sides, large loops exist forming a rough surface. The latter certainly is expected for FSRPs; the former, however, may be different because of the need to interact with TonB. Some unexpected features in the structures of porins are the inclination of the β-sheets with respect to the perpendicular to the membrane which is not 0° but varies from 40° to 60°. Another unexpected feature is the many Phe, Tyr, and Trp's close to the start and the end of the β-sheets. These probably anchor the proteins in the membrane and prevent perpendicular movements of the proteins. When one looks at the outside of the barrel, one goes from the smooth surface to an aromatic girdle then a nonpolar surface and a second aromatic girdle before reaching the rough external surface. The nonpolar surface, including the aromatic girdles, spans about 24–25 Å in agreement with the thickness of the membrane. These same general features can be expected for the FSRPs. It is worth noting that only the model of FoxA takes the aromatic girdles into account. In contrast to porins, one may expect different structures for a particular FSRP between being bound or not bound to its substrate and at different pH and pCa values.

5.2. Bayer Zones

Bayer [169] showed the existence of contacts between the outer and inner membranes by electron microscopy. The number of contacts per cell was estimated to be between 200–400, which is the same magnitude as the number of receptors on the outer membrane. It was also shown that the large majority of phage particles (T5) visualized by electron microscopy associated with these contact zones. When one remembers that ferrichrome and phage T5 use the same receptor, FhuA, (ferrichrome inhibits T5 infection [20,21]) the observation implies that all ferric siderophore transport occurs at contact zones. This poses an apparent contradiction. If transport occurs only at the locations where inner and outer membrane make contact, why are periplasmic proteins such as FhuD, FepB, and others needed, and why is TonB needed? The electron microscopy observations are not without controversy, however. Plasmolysis in 10–20% sucrose is needed to make the observation, although this concentration is not lethal. Does the chemical fixation lead to an artifact? If this is replaced by photocrosslinking, contact zones are still observed but are fewer [170]. Only certain methods of freezing allow the observation of contact zones, but it is possible that some are physically too demanding for the cells. Further work allowed the recognition of two different types of adhesion [170] sites: (1) small singular contact zones and (2) IM/OM appositions where the membranes are separated by a space of 250–500 Å along stretches of several 1000 Å, and these stretches are interspersed with small sites of contact between IM and OM. These are named "bays" and can be considered to be more or less isolated compartments (see Sec. 5.4).

5.3. Energy Requirement and pH Dependence

The outer membrane in gram-negative bacteria has no energy source; however, it operates in an energy-dependent fashion [21,87,171]. The transport across the outer membrane is indeed an active (i.e., energy-dependent) process. This can be concluded from the fact that vitamin B_{12} can be accumulated in the periplasm owing to BtuB transport [172], and that mutations in the periplasmic binding protein FhuD affect neither the transport of ferric aerobactin across the OM nor its accumulation in the periplasmic space [173]. Results with ferric enterobactin uptake showed it to be dependent on an energized cytoplasmic membrane state in conjunction with ATP [15,174]. More recently, working with btuC mutants, it was shown that the proton motive force, generated through the cytoplasmic membrane, is essential for vitamin B_{12} transport across the OM [175]. Still the active transport across the outer membrane is not a unidirectional process, because there is a rapid exchange between periplasmic and extracellular Fe(ent) in *E. coli*-uptake experiments [137]. A mechanism for the transport across the OM should therefore involve either one or several sequential steps or an open gate allowing free exchange. The latter, however, would be inconsistent with the observed accumulation of substrate in the periplasm and the effect of pH.

5.4. Possible Mechanisms

The pH dependence for active transport, the Ca^{2+} effect on binding, the involvement of TonB, and the need for specific periplasmic and CM proteins (all discussed above) resulted in a schematic model for transport [87] shown in Fig. 7.

Any proposed scheme should involve changes in molecular conformation, be it in TonB or the receptor protein or both, but also a pH change should be part of the mechanism. This leads to a paradox, because a conformational change deals with one molecule and is thus a microscopic phenomenon, whereas a pH change is a macroscopic event necessarily involving many molecules. There is no evidence that the transport process is synchronized. Instead it involves one transport protein and one substrate molecule at a time. A possible resolution of this paradox is that the bay regions or compartments observed by electron microscopy involve transport of one molecule and a pH change during transport of just one compartment in which one proton translator molecule in the CM is achieving this pH change.

A second duality exists. Does TonB give a signal from the CM

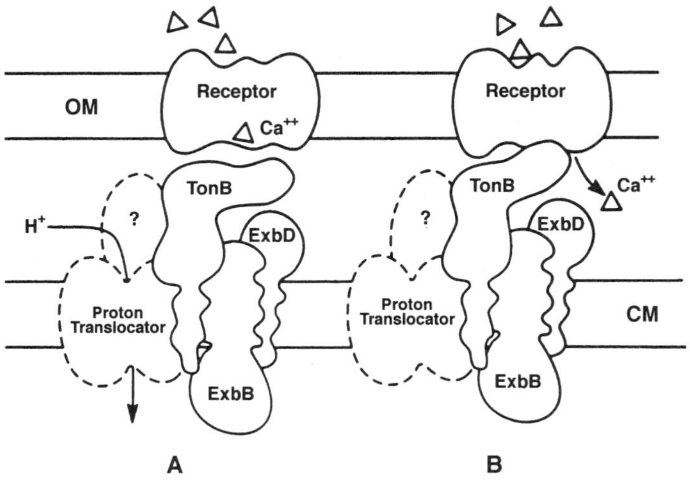

FIG. 7. A model for TonB-dependent transport. (Reproduced with permission from [87].)

toward the OM or does the reverse process occur—a signal from the OM to the CM? In the former, one faces the problem what the initiation of the signal is. The latter seems more logical, which is that binding of the substrate (with/without Ca^{2+}) leads to a conformational change in the receptor protein and only interacts with TonB after binding. This interaction might change the conformation of TonB which is sensed by the proton translocator, changing the pH. This makes effective use of TonB in that only molecules ready for transport require a TonB molecule. The conformational change in the receptor protein may be quite significant and involve not only the TonB box but also homology boxes III and IV.

The substrate has to move, or be moved, at least 25 Å before it is on the periplasmic side. It most likely uses the channel formed by the β-sheets for this process. The pH change may initiate this process by changing the conformation of the binding loop releasing the substrate to the channel and/or widening the channel. But what is driving the transfer through the channel? It is possible that diffusion is all that is necessary. This would allow the process to be reversible, because after release of the substrate by the binding loop, the protein looses its interaction with TonB reversing the pH change. Once the substrate arrives at the periplasmic side of the channel, it can be bound by the specific periplasmic protein (FhuD in the case of FhuA and FepB in the case of FepA). After this the receptor protein would be in its original state and ready to transport another substrate molecule.

This possible molecular mechanism solves a number of paradoxes. It is in agreement with the observation that the ligand-binding loop in FepA exhibited structural changes during uptake [176]. It also agrees with results described earlier. For instance, the ability of internalized Fe(ent) to exchange with extracellular Fe(ent) [137]. The diffusion kinetics once the binding loops were excised in FepA [71] and FhuA [84]. The importance of the TonB box and homology region III and possibly IV and in general the sensitivity of the amino terminal first quarter of the receptor proteins [74,77]. The conformational change on the periplasmic side of FhuA on binding with ferrichrome [81]. It does not explain the TonB-dependent colicin transport or that of antibiotics quite dissimilar in structure to ferric siderophores. Also, some FSRPs do not contain a recognizable TonB box, but this has been discussed before. It does not explain the lower than expected inhibition of Ga siderophores compared with their ferric complexes, and an answer to this question might result in a significant change to the proposed mechanism.

6. CONCLUSIONS

Some information about the molecular characteristics and properties of FSRPs is now known, but much more can and needs to be determined. This chapter invites a large number of well-designed in vitro experiments. For instance, crystallographically, one would like to see the differences in structure between the receptor protein, the protein bound to the receptor with and without Ca^{2+}, and at different pH values. One would like to see the structure of a receptor protein bound to different substrates and to colicin. Also, structures of proteins mutated in the binding region, channel, and homology boxes would give valuable information. Many of these experiments might not be time consuming. In addition, the ability to purify the proteins and the fact that they retain their conformational integrity allows many in vitro chemical experiments which probe the extracellular and periplasmic interactions. Still the complete picture for the specific transport across the outer membrane has to come from the combination of in vitro and in vivo results.

ACKNOWLEDGMENTS

I would like to thank Dr. M.B. Hossain for the construction of the phylogenetic tree and J. Keil for the preparation of the manuscript.

The work in this laboratory is supported by grant GM 21822 from the National Institutes of Health.

ABBREVIATIONS

#AA	number of amino acids
ATP	adenosine 5′-triphosphate
BCA	bicinchoninic acid

BtuB	outer-membrane receptor protein for vitamin B_{12}.
CD	circular dichroism
CM	cytoplasmic membrane
CMC	critical micellar concentration
DEAE	diethylaminoethyl
EDTA	ethylenediamine-*N,N,N′,N′*-tetraacetate
EGTA	ethylenebis(oxyethylenenitrile)tetraacetic acid
ent	enterobactin
ESR	electron spin resonance
FPLC	fast-protein liquid chromatography
FSRP	ferric siderophore outer membrane receptor protein
Fur	ferric uptake regulation
HPLC	high-performance liquid chromatography
IM	inner membrane
LDAO	dodecyldimethylamine oxide
LPS	lipopolysaccharides
MAb	monoclonal antibody
MECAM	synthetic analogue of enterobactin in which the triseryl scaffold is replaced by a phenyl group
MW	molecular weight
NMR	nuclear magnetic resonance
NTA	nitrilotriacetate
OG	*n*-octyl-β-D-glucopyranoside
OM	outer membrane
PCR	polymerase chain reaction
PEG	polyethylene glycol

REFERENCES

1. J. B. Neilands, *Annu. Rev. Microbiol.*, *36*, 285–309 (1982).
2. M. Koster, W. Ovaa, W. Bitter, and P. Weisbeek, *Mol. Gen. Genet.*, *248*, 735–743 (1995).
3. J. M. Rutz, T. Abdullah, S. P. Singh, V. I. Kalve, and P. E. Klebba, *J. Bacteriol.*, *173*, 5964–5974 (1991).
4. R. Koebnik, K. Hantke, and V. Braun, *Mol. Microbiol.*, *7*, 383–393 (1993).

5. A. J. Bäumler and K. Hantke, *Mol. Microbiol.*, *6*, 1309–1321 (1992).
6. J. Morris, D. F. Donnelly, E. O'Neill, F. McConnell, and F. O'Gara, *Mol. Gen. Genet.*, *242*, 9–16 (1994).
7. A. Bagg and J. B. Neilands, *Biochemistry*, *26*, 5471–5477 (1987).
8. B. van Hove, H. Staudenmaier, and V. Braun, *J. Bacteriol.*, *172*, 6749–6758 (1990).
9. M. Koster, W. van Klompenburg, W. Bitter, J. Leong, and P. Weisbeek, *EMBO J.*, *13*, 2805–2813 (1994).
10. M. Koster, J. van de Vossenberg, J. Leong, and P. J. Weisbeek, *Mol. Microbiol.*, *8*, 591–601 (1993).
11. D. E. Heinrichs and K. Poole, *J. Bacteriol.*, *178*, 2586–2592 (1996).
12. K. Gensburg, K. Hughes, and A. W. Smith, *J. Gen. Microbiol.*, *138*, 2381–2387 (1992).
13. C. R. Dean, S. Neshat, and K. Poole, *J. Bacteriol.*, *178*, 5361–5369 (1996).
14. Q. Chen, A. M. Wertheimer, M. E. Tolmasky, and J. H. Crosa, *Mol. Microbiol.*, *22*, 127–134 (1996).
15. A. P. Pugsley and P. Reeves, *J. Bacteriol.*, *130*, 26–36 (1977).
16. S. K. Guterman, *J. Bacteriol.*, *114*, 1217–1224 (1973).
17. A. P. Pugsley and P. Reeves, *Biochem. Biophys. Res. Commun.*, *70*, 846–853 (1976).
18. J. K. Davies and P. Reeves, *J. Bacteriol.*, *123*, 96–101 (1975).
19. R. Wayne and J. B. Neilands, *J. Bacteriol.*, *121*, 497–503 (1975).
20. M. Luckey, R. Wayne, and J. B. Neilands, *Biochem. Biophys. Res. Commun.*, *64*, 687–694 (1975).
21. K. Hantke and V. Braun, *FEBS Lett.*, *49*, 301–305 (1975).
22. A. Hartmann, H.-P. Fiedler, and V. Braun, *Eur J. Biochem.*, *99*, 517–524 (1979).
23. K. Hantke, *Mol. Gen. Genet.*, *191*, 301–306 (1983).
24. B. F. Matzanke, G. I. Müller, and K. N. Raymond, *Biochem. Biophys. Res. Commun.*, *121*, 922–930 (1984).
25. P. H. Williams and P. J. Warner, *Infect. Immun.*, *29*, 411–416 (1980).
26. V. de Lorenzo, F. Giovannini, M. Herrero, and J. B. Neilands, *J. Mol. Biol.*, *203*, 875–884 (1988).

27. K. Hantke, *FEMS Microbiol. Lett.*, *67*, 5–8 (1990).
28. H. Nikaido and E. Y. Rosenberg, *J. Bacteriol.*, *172*, 1361–1367 (1990).
29. K. Heller and R. J. Kadner, *J. Bacteriol.*, *161*, 904–908 (1985).
30. R. J. Ward and R. E. Glass, *Mol. Microbiol.*, *16*, 813–815 (1995).
31. C. Bradbeer and A. Gudmundsdottir, *J. Bacteriol.*, *172*, 4919–4926 (1990).
32. C. D. Cox, *J. Bacteriol.*, *142*, 581–587 (1980).
33. C. D. Cox and P. Adams, *Infect. Immun.*, *48*, 130–138 (1985).
34. H. Budzikiewicz, *FEMS Microbiol. Lett.*, *104*, 209–228 (1993).
35. S. Khalil-Rizvi, S. I. Toth, D. van der Helm, I. Vidavsky, and M. Gross, *Biochemistry*, *36*, 4163–4171 (1997).
36. M. Teintze, M. B. Hossain, C. L. Barnes, J. Leong, and D. van der Helm, *Biochemistry*, *20*, 6446–6457 (1981).
37. H. Haag, K. Hantke, H. Drechsel, I. Stojiljkovic and G. Jung, *J. Gen. Microbiol.*, *139*, 2159–2165 (1993).
38. A. Rakin, E, Saken, D. Harmsen, and J. Heesemann, *Mol. Microbiol.*, *13*, 253–263 (1994).
39. A. Bäumler, R. Koebnik, I. Stojiljkovic, J. Heesemann, V. Braun, and K. Hantke, *Zbl. Bakt.*, *278*, 416–424 (1993).
40. B. Beall and G. N. Sanden, *Microbiology*, *141*, 3193–3205 (1995).
41. G. L. Griffiths, S. P. Sigel, S. M. Payne, and J. B. Neilands, *J. Biol. Chem.*, *259*, 383–385 (1984).
42. M. B. Goldberg, S. A. Boyko, J. R. Butterton, J. A. Stoebner, S. M. Payne, and S. B. Calderwood, *Mol. Microbiol.*, *6*, 2407–2418 (1992).
43. M. A. F. Jalal, M. B. Hossain, D. van der Helm, J. Sanders-Loehr, L. A. Actis, and J. H. Crosa, *J. Am. Chem. Soc.*, *111*, 292–296 (1989).
44. M. B. Hossain, M. A. F. Jalal and D. van der Helm, *J. Chem. Crystallogr.* Submitted.
45. M. Persmark, D. Expert, and J. B. Neilands, *J. Biol. Chem.*, *264*, 3187–3193 (1989).
46. H. Drechsel, J. Metzger, S. Freund, G. Jung, J. R. Boelaert, and G. Winkelmann, *BioMetals*, *4*, 238–243 (1991).
47. A. Gillis, M. A. Kahn, P. Cornelis, J.-M. Meyer, M. Mergeay, and D. van der Lelie, *J. Bacteriol.*, *178*, 5499–5507 (1996).
48. R. Schneider and K. Hantke, *Mol. Microbiol.*, *8*, 111–121 (1993).

49. M. Kammler, C. Schön, and K. Hantke, *J. Bacteriol.*, *175*, 6212–6219 (1993).
50. C. C. Askwith, D. de Silva, and J. Kaplan, *Mol. Microbiol.*, *20*, 27–34 (1996).
51. A. Angerer, S. Gaisser, and V. Braun, *J. Bacteriol.*, *172*, 572–578 (1990).
52. M. Stryuvé, M. Moons, and J. Tommassen, *J. Mol. Biol.*, *218*, 141–148 (1991).
53. M. D. Lundrigan and R. J. Kadner, *J. Biol. Chem.*, *261*, 10797–10801 (1986).
54. J. W. Coulton, P. Mason, D. R. Cameron, G. Carmel, R. Jean, and H. N. Rode, *J. Bacteriol.*, *165*, 181–192 (1986).
55. M. Sauer, K. Hantke, and V. Braun, *Mol. Microbiol.*, *4*, 427–437 (1990).
56. W. J. A. Krone, F. Stegehuis, G. Koningstein, C. van Doorn, B. Roosendaal, F. K. de Graaf, and B. Oudega, *FEMS Microbiol. Lett.*, *26*, 153–161 (1985).
57. U. Pressler, H. Staudenmaier, L. Zimmermann, and V. Braun, *J. Bacteriol.*, *170*, 2716–2724 (1988).
58. C. R. Dean and K. Poole, *J. Bacteriol.*, *175*, 317–324 (1993).
59. R. G. Ankenbauer and H. N. Quan, *J. Bacteriol.*, *176*, 307–319 (1994).
60. K. Poole, S. Neshat, K. Krebes, and D. E. Heinrichs, *J. Bacteriol.*, *175*, 4597–4604 (1993).
61. W. Bitter, J. D. Marugg, L. A. de Weger, J. Tommassen, and P. J. Weisbeek, *Mol. Microbiol.*, *5*, 647–655 (1991).
62. W. Bitter, Ph.D. thesis. Ferric Pseudobactin Transport across the Outer Membrane of Pseudomonas Putida WCS 358. University of Utrecht, The Netherlands, 1992.
63. J. R. Butterton, J. A. Stoebner, S. M. Payne, and S. B. Calderwood, *J. Bacteriol.*, *174*, 3729–3738 (1992).
64. L. A. Actis, M. E. Tolmasky, D. H. Farrell, and J. H. Crosa, *J. Biol. Chem.*, *263*, 2853–2860 (1988).
65. W. L. Köster, L. A. Actis, L. S. Waldbeser, M. E. Tolmasky, and J. H. Crosa, *J. Biol. Chem.*, *266*, 23829–23833 (1991).
66. C. Sauvage, T. Franza, and D. Expert, *J. Bacteriol.*, *178*, 1227–1231 (1996).

67. S. Kühn, V. Braun, and W. Köster, *J. Bacteriol.*, *178*, 496–504 (1996).

68. C. D. Nau and J. Konisky, *J. Bacteriol.*, *171*, 1041–1047 (1989).

69. C. K. Murphy, V. I. Kalve, and P. E. Klebba, *J. Bacteriol.*, *172*, 2736–2746 (1990).

70. R. Koebnik and V. Braun, *J. Bacteriol.*, *175*, 826–839 (1993).

71. J. M. Rutz, J. Liu, J. A. Lyons, J. Goranson, S. K. Armstrong, M. A. McIntosh, J. B. Feix, and P. E. Klebba, *Science*, *258*, 471–475 (1992).

72. P. E. Klebba, J. M. Rutz, J. Liu, and C. K. Murphy, *J. Bioenerg. Biomemb.*, *25*, 603–611 (1993).

73. S. K. Armstrong, C. L. Francis, and M. A. McIntosh, *J. Biol. Chem.*, *265*, 14536–14543 (1990).

74. S. K. Armstrong and M. A. McIntosh, *J. Biochem.*, *270*, 2483–2488 (1995).

75. J. Liu, J. M. Rutz, P. E. Klebba, and J. B. Feix, *Biochemistry*, *33*, 13274–13283 (1994).

76. S. M. C. Newton, J. S. Allen, Z. Cao, Z. Qi, X. Jiang, C. Sprencel, J. D. Igo, S. B. Foster, M. A. Payne, and P. E. Klebba, *Proc. Natl. Acad. Sci.*, *94*, 4560–4565 (1997).

77. G. Carmel, D. Hellstern, D. Henning, and J. W. Coulton, *J. Bacteriol.*, *172*, 1861–1869 (1990).

78. H. Killmann and V. Braun, *J. Bacteriol.*, *174*, 3479–3486 (1992).

79. G. S. Moeck, B. S. F. Bazzaz, M. F. Gras, T. S. Ravi, M. J. H. Ratcliffe, and J. W. Coulton, *J. Bacteriol.*, *176*, 4250–4259 (1994).

80. G. S. Moeck, M. J. H. Ratcliffe, and J. W. Coulton, *J. Bacteriol.*, *177*, 6118–6125 (1995).

81. G. S. Moeck, P. Tawa, H. Xiang, A. A. Ismail, J. L. Turnbull, and J. W. Coulton, *Mol. Microbiol.*, *22*, 459–471 (1996).

82. G. Carmel and J. W. Coulton, *J. Bacteriol.*, *173*, 4394–4403, (1991).

83. J. Liu, J. M. Rutz, J. B. Feix, and P. E. Klebba, *Proc. Natl. Acad. Sci.*, *90*, 10653–10675 (1993).

84. H. Killmann, R. Benz, and V. Braun, *EMBO J.*, *12*, 3007–3016 (1993).

85. V. Braun, H. Killmann, and R. Benz, *FEBS Lett.*, *346*, 59–64 (1994).

86. M. Bonhivers, A. Ghazi, P. Boulanger, and L. Letellier, *EMBO J.*, *15*, 1850–1856 (1996).
87. K. Postle, *J. Bioenerg. Biomemb.*, *25*, 591–601 (1993).
88. V. Braun, *FEMS Microbiol. Rev.*, *16*, 295–307 (1995).
89. R. Burkhardt and V. Braun, *Mol. Gen. Genet.*, *209*, 49–55 (1987).
90. J. W. Coulton, P. Mason, and D. D. Allatt, *J. Bacteriol.*, *169*, 3844–3849 (1987).
91. W. Köster and V. Braun, *J. Biol. Chem.*, *265*, 21407–21410 (1990).
92. M. R. Rohrbach, V. Braun, and W. Köster, *J. Bacteriol.*, *177*, 7186–7193 (1995).
93. M. F. Elkins and C. F. Earhart, *J. Bacteriol.*, *171*, 5443–5451 (1989).
94. D. L. Stephens, M. D. Choe, and C. F. Earhart, *Microbiology*, *141*, 1647–1654 (1995).
95. S. E. Luria and M. Delbrück, *Genetics*, *28*, 491–511 (1943).
96. K. Postle and R. F. Good, *Proc. Natl Acad. Sci. USA*, *80*, 5235–5239 (1983).
97. S. K. Roof, J. D. Allard, K. P. Bertrand, and K. Postle, *J. Bacteriol.*, *173*, 5554–5557 (1991).
98. M. Anton and K. J. Heller, *Gene*, *105*, 23–29 (1991).
99. E. Fischer, K. Günter, and V. Braun, *J. Bacteriol.*, *171*, 5127–5134 (1989).
100. V. Braun, K. Günter, and K. Hantke, *Biol. Metals*, *4*, 14–22 (1991).
101. W. Bitter, J. Tommassen, and P. J. Weisbeek, *Mol. Microbiol.*, *7*, 117–130 (1993).
102. R. A. Larsen, P. S. Myers, J. T. Skare, C. L. Seachord, R. P. Darveau, and K. Postle, *J. Bacteriol.*, *178*, 1363–1373 (1996).
103. K. Poole, Q. Zhao, S. Neshat, D. E. Heinrichs, and C. R. Dean, *Microbiology*, *142*, 1449–1458 (1996).
104. P. B. Bell, C. D. Nau, J. T. Brown, J. Konisky, and R. J. Kadner, *J. Bacteriol.*, *172*, 3826–3829 (1990).
105. H. Schöffler and V. Braun, *Mol. Gen. Genet.*, *217*, 378–383 (1989).
106. A. K. Bruske and K. J. Heller, *J. Bacteriol.*, *175*, 6158–6168 (1993).
107. J. T. Skare, B. M. M. Ahmer, C. L. Seachord, R. P. Darveau, and K. Postle, *J. Biol. Chem.*, *268*, 16302–16308 (1993).

108. K. Günter and V. Braun, *FEBS Lett.*, *274*, 85–88 (1990).

109. S. Brewer, M. Tolley, I. P. Trayer, G. C. Barr, C. J. Dorman, K. Hannavy, C. F. Higgins, J. S. Evans, B. A. Levine, and M. R. Wormald, *J. Mol. Biol.*, *216*, 883–895 (1990).

110. R. A. Larsen, G. E. Wood, and K. Postle, *Mol. Microbiol.*, *10*, 943–953 (1993).

111. M. Tuckman and M. S. Osburne, *J. Bacteriol.*, *174*, 320–323 (1992).

112. K. J. Heller, R. Kadner, and K. Günter, *Gene*, *64*, 147–153 (1988).

113. W. C. Hollifield, Jr., E. H. Fiss and J. B. Neilands, *Biochem. Biophys. Res. Commun.*, *83*, 739–746 (1978).

114. W. C. Hollifield, Jr., and J. B. Neilands, *Biochemistry*, *17*, 1922–1928 (1978).

115. E. H. Fiss, P. Stanley-Samuelson, and J. B. Neilands, *Biochemistry*, *21*, 4517–4522 (1982).

116. K. Hantke, *Mol. Gen. Genet.*, *182*, 288–292 (1981).

117. T. P. Fleming, M. S. Nahlik, J. B. Neilands, and M. A. McIntosh, *Gene*, *34*, 47–54 (1985).

118. H. Hoffmann, E. Fischer, H. Kraut, and V. Braun, *J. Bacteriol.*, *166*, 404–411 (1986).

119. S. Buchanan, B. Smith, L. Venkatramani, D. van der Helm and J. Deisenhofer, *Acta Cryst.*, *A52*, C141 (1996).

120. C. A. Schnaitman, *J. Bacteriol.*, *108*, 553–563 (1971).

121. M. A. F. Jalal and D. van der Helm, *FEBS Lett.*, *243*, 366–370 (1989).

122. R. N. Chakraborty, L. Venkatramani, and D. van der Helm, Unpublished results.

123. R. J. Bergeron, W. R. Weimar, and J. B. Dionis, *J. Bacteriol.*, *170*, 3711–3717 (1988).

124. X. H. Zhou, D. van der Helm, and L. Venkatramani, *BioMetals*, *8*, 129–136 (1995).

125. J. Zhang and D. van der Helm, *Biochemistry*. Submitted.

126. P. E. Klebba, Unpublished results.

127. D. van der Helm, Unpublished results.

128. E. H. Fiss. W. C. Hollifield, Jr., and J. B. Neilands, *Biochem. Biophys. Res. Commun.*, *91*, 29–34 (1979).

129. X. H. Zhou, D. van der Helm, and J. Adjamani, *BioMetals*, *6*, 25–35 (1993).
130. Z. K. Shihabi and R. D. Dyer, *Ann. Clin. Lab. Sci.*, *18*, 235–239 (1988).
131. B. S. Smith, B. Kobel, K. G. Ravichandran, L. Venkatramani, X. H. Zhou, D. van der Helm, and J. Deisenhofer, *Acta Cryst. D*, Submitted.
132. P. Boulanger, M. le Maire, M. Bonhivers, S. Dubois, M. Desmadril, and L. Letellier, *Biochemistry*, *35*, 14216–14224 (1996).
133. G. Schultz, F. Ullrich, K. Heller, and V. Braun, *Mol. Gen. Genet.*, *216*, 230–238 (1989).
134. C. S. Klug, W. Su, J. Liu, P. E. Klebba, and J. B. Feix, *Biochemistry*, *34*, 14230–14236 (1995).
135. X. H. Zhou and D. van der Helm, *BioMetals*, *6*, 37–44 (1993).
136. B. F. Matzanke, D. J. Ecker, T.-S. Yang, B. H. Huynh, G. Müller, and K. N. Raymond, *J. Bacteriol.*, *167*, 674–680 (1986).
137. D. J. Ecker, B. F. Matzanke, and K. N. Raymond, *J. Bacteriol.*, *167*, 666–673 (1986).
138. D. J. Ecker, L. D. Loomis, M. E. Cass, and K. N. Raymond, *J. Am. Chem. Soc.*, *110*, 2457–2464 (1988).
139. M. A. McIntosh, S. S. Chenault, and C. F. Earhart, *J. Bacteriol.*, *137*, 653–657 (1979).
140. T. Peterson and J. B. Neilands, *Tetrahedron Lett.*, *50*, 4805–4808 (1979).
141. R. J. Bergeron, J. B. Dionis, G. T. Elliott, and S. J. Kline, *J. Biol. Chem.*, *260*, 7936–7944 (1985).
142. W. Weidel, G. Koch, and K. Bobosch, *Z. Naturforsch.*, *9*, 573–579 (1954).
143. B. Borgias, S. J. Barclay, and K. N. Raymond, *J. Coord. Chem.*, *15*, 109–123 (1986).
144. A. Dietrich, K. Fidelis, D. R. Powell, D. van der Helm, and D. L. Eng-Wilmot, *J. Chem. Soc., Dalton Trans.*, 231–239 (1991).
145. R. J. Bergeron and S. J. Kline, *J. Am. Chem. Soc.*, *106*, 3089–3098 (1984).
146. M. Llinás, D. M. Wilson, and J. B. Neilands, *Biochemistry*, *12*, 3836–3843 (1973).

147. M. A. F. Jalal, S. K. Love and D. van der Helm, *J. Inorg. Biochem.*, *28*, 417–430 (1986).
148. K. L. Constantine, A. De Marco, M. Madrid, C. L. Brooks III, and M. Llinás, *Biopolymers*, *30*, 239–256 (1990).
149. C. Bradbeer, P. R. Reynolds, G. M. Bauler, and M. T. Fernandez, *J. Biol. Chem.*, *261*, 2520–2523 (1986).
150. H. Killmann, R. Benz, and V. Braun, *J. Bacteriol.*, *178*, 6913–6920 (1996).
151. G. Benz, T. Schröder, J. Kurz, C. Wünsche, W. Karl, G. Steffens, J. Pfitzner, and D. Schmidt, *Angew. Chem. Int. Ed. Engl.*, *21*, 527–528 (1982).
152. E. K. Dolence, A. A. Minnick, C.-E. Lin, and M. J. Miller, *J. Med. Chem.*, *34*, 968–978 (1991).
153. A. P. Pugsley, W. Zimmerman, and W. Wehrli, *J. Gen. Microbiol.*, *133*, 3505–3511 (1987).
154. R. A. Salomón and R. N. Farías, *J. Bacteriol.*, *174*, 7428–7435 (1992).
155. R. A. Salomón and R. N. Farías, *J. Bacteriol.*, *175*, 7741–7742 (1993).
156. M. B. Hossain, M. A. F. Jalal, B. A. Benson, C. L. Barnes, and D. van der Helm, *J. Am. Chem. Soc.*, *109*, 4949–4954 (1987).
157. J. B. Neilands, T. J. Erickson, and W. H. Rastetter, *J. Biol. Chem.*, *256*, 3831–3832 (1981).
158. G. Winkelmann and V. Braun, *FEMS Microbiol. Lett.*, *11*, 237–341 (1981).
159. R. J. Bergeron and W. R. Weimar, *J. Bacteriol.*, *172*, 2650–2657 (1990).
160. S. Heidinger, V. Braun, V. L. Pecoraro, and K. N. Raymond, *J. Bacteriol.*, *153*, 109–115 (1983).
161. T. B. Karpishin and K. N. Raymond, *Angew. Chem. Int. Ed. Engl.*, *31*, 466–468 (1992).
162. G. Xiao, D. van der Helm, R. C. Hider, and B. L. Rai, *J. Phys. Chem.*, *100*, 2345–2352 (1996).
163. M. S. Weiss, T. Wacker, J. Weckesser, W. Welte, and G. E. Schulz, *FEBS Lett.*, *267*, 268–272 (1990).
164. S. W. Cowan, T. Schirmer, G. Rummel, M. Steiert, R. Ghosh, R. A. Pauptit, J. N. Jansonius, and J. P. Rosenbusch, *Nature*, *358*, 727–733 (1992).

165. S. W. Cowan, R. M. Garavito, J. N. Jansonius, J. A. Jenkins, R. Karlsson, N. König, E. F. Pai, R. A. Pauptit, P. J. Rizkallah, J. P. Rosenbusch, G. Rummel, and T. Schirmer, *Structure*, *3*, 1041–1050 (1995).
166. A. Kreusch, A. Neubüser, E. Schlitz, J. Weckesser, and G. E. Schulz, *Prot. Sci.*, *3*, 58–63 (1994).
167. T. Schirmer, T. A. Keller, Y. F. Wang, and J. P. Rosenbusch, *Science*, *267*, 512–514 (1995).
168. B. K. Jap and P. J. Walian, *Physiol. Rev.*, *76*, 1073–1088 (1996).
169. M. E. Bayer, *J. Virol.*, *2*, 346–356 (1968).
170. M. E. Bayer, *J. Struct. Biol.*, *107*, 268–280 (1991).
171. G. E. Frost and H. Rosenberg, *J. Bacteriol.*, *124*, 704–712 (1975).
172. P. R. Reynolds, G. P. Mottur, and C. Bradbeer, *J. Biol. Chem.*, *255*, 4313–4319 (1980).
173. K. G. Wooldridge, J. A. Morrissey, and P. H. Williams, *J. Gen. Microbiol.*, *138*, 597–603 (1992).
174. C. Bradbeer and M. L. Woodrow, *J. Bacteriol.*, *128*, 99–104 (1976).
175. C. Bradbeer, *J. Bacteriol.*, *175*, 3146–3150 (1993).
176. X. Jiang, M. A. Payne, Z. Cao, S. B. Foster, J. B. Feix, S. M. C. Newton, and P. E. Klebba, *Science*, *276*, 1261–1264 (1997).
177. B. Matzanke, in *Handbook of Microbial Iron Chelates* (G. Winkelmann, ed.), CRC Press, Boca Raton, FL, 1990, p. 61.

10

The Iron Responsive Element (IRE) Family of mRNA Regulators

Regulation of Iron Transport and Uptake Compared in Animals, Plants, and Microorganisms

Elizabeth C. Theil

Department of Biochemistry, North Carolina State University, Raleigh, NC 27695-7622, USA

1. INTRODUCTION: mRNA REGULATION AND IREs 404
   1.1. General Features of mRNA Regulation 405
   1.2. Coordinated Regulation of Iron Uptake and Storage: Iso-IREs 406
      1.2.1. mRNA Translation 406
      1.2.2. mRNA Stability/Turnover 407
   1.3. Primary/Secondary Structure of IREs 407
2. SPECIFIC FEATURES OF ISO-IREs: mRNA TRANSLATION REGULATORS 413
   2.1. Sequence of IRE Translation Regulators 413
   2.2. Position of IRE Translation Regulators in mRNA 415
3. SPECIFIC FEATURES OF ISO-IREs: mRNA STABILITY/TURNOVER REGULATORS 416
   3.1. Sequence of IRE Stability/Turnover Regulators 416
   3.2. Position of IRE Translation Regulators of mRNA 417

4. PROTEINS RECOGNIZED BY IREs 417
4.1. Iron-Regulatory Proteins 418
4.2. Initiation Factors 420
4.3. Nucleases? 420
5. REGULATION OF IRON UPTAKE AND STORAGE IN PLANTS AND MICROORGANISMS 421
5.1. Iron Uptake 421
5.1.1. Plants and Yeast 421
5.1.2. Bacteria 422
5.2. Iron Storage 423
5.2.1. Plants and Yeast 423
5.2.2. Bacteria 424
6. PERSPECTIVE 424
ACKNOWLEDGMENTS 426
ABBREVIATIONS 426
REFERENCES 427

1. INTRODUCTION: mRNA REGULATION AND IREs

mRNA is a particularly effective site for regulating gene expression, since proteins are the final gene product and an mRNA target has two advantages: (1) structural and cell specificity and (2) a relatively small regulatory target size. mRNA regulation alters the rate of protein biosynthesis. Compared with DNA, regulation of mRNA exploits the high levels of mRNA cell specificity and the potential for a specific three-dimensional structure intrinsic to a single-stranded molecule. Compared with proteins, mRNA regulation can exploit the relatively smaller target size allowing greater sensitivity to small changes in cell and environmental regulatory signals. In the past, the difficulty of studying single mRNA species and the small amounts available had hampered exploration of mRNA regulation. However, ferritin mRNA regulation is so dramatic that it was discovered before the advantages of biotechnology were readily available [1,2], and it has been a paradigm for mRNA regulation ever since. The conserved mRNA sequence, the IRE (iron-responsive element), which regulates translation of ferritin mRNA and stability of the transferrin receptor (TfR) mRNA, has been found to function in at least four different mRNAs to date (for recent reviews see [3–6]). In addition to basic information about metal-dependent gene expression and mechanisms of mRNA regulation, studies of IREs are providing unique information about the range of coordinated mRNA regulation that is possible.

1.1. General Features of mRNA Regulation

The synthesis of proteins is regulated both by altering rates of DNA transcription (mRNA synthesis) and rates of mRNA translation (mRNA function). Regulation of mRNA function to alter rates of protein biosynthesis is a mechanism of gene regulation that achieves variations in gene expression of enormous flexibility. In cells with nuclei (eukaryotes), the spatial separation of mRNA synthesis (transcription) and function (translation) has led to the use of selective mRNA storage, selective mRNA stabilization (control of turnover rates), and selective binding of ribosomes (control of translation initiation). The results of regulation of mRNA function produce differential rates of protein synthesis that are observed during normal cell development/differentiation, disease, and environmental change. Examples include chloroplast maturation [7,8], the storage and turnover of mRNAs during stress [9], competitive translation of viral mRNAs [10], the turnover of cytokine and oncogene mRNAs [11–14], and iron-induced translation of ferritin mRNA [1–6]. mRNA regulatory sequences are generally, but not exclusively, in the noncoding or untranslated region (UTR) of the mRNA.

In organisms without nuclei (prokaryotes), where mRNA synthesis is physically and temporally coupled to mRNA function, mRNA regulation appears to be less obvious. Nevertheless, regulation of translation also occurs in prokaryotes. For example, alternate mRNA structures that alter mRNA translation rates form during transcription in prokaryotes when environmental amino acid concentrations change [15] and when toxins are produced [16].

mRNAs in animals which contain the IRE sequences are unique, because IRE-dependent mRNA regulation leads to the iron-dependent coordination of protein synthesis of several metabolically related proteins, ferritin, the transferrin receptor, and red cell aminolevulinate synthase (eALAS), the first step in heme biosynthesis [3–6]. Evidence for IRE-dependent regulation of m-aconitase has also been recently obtained [17–19]. In plants, iron regulation of ferritin synthesis is independent of the IRE and appears to bypass mRNA entirely [20–22]; there is no IRE in the gene or mRNA of maize or soybean [23].

1.2. Coordinated Regulation of Iron Uptake and Storage: Iso-IREs

High concentrations of environmental iron used to probe regulation of iron storage and iron uptake demonstrated that cells can coordinate the concentration and safe sequestration of iron while slowing down the incorporation of more iron into the cell: the synthesis of ferritin increases to store the iron and the synthesis of the transferrin receptor decreases to diminish iron uptake. In each case, the information to regulate the synthesis of the two proteins was shown to reside in the mRNA [24–27] in sequences later shown to be related to each other and named the IRE, iron-responsive element [28,29].

The iron effect on the ferritin and TfR-mRNAs is mediated through interactions of the mRNA with cellular protein(s). An IRE-specific binding protein, the IRP, can be modified by phosphorylation, by forming an iron-sulfur cluster, by binding heme, or, in certain cases, by proteosomal degradation, each of which alters the mRNA/protein interaction (see Sec. 4).

The function of the IRE is different in ferritin mRNA and TfR mRNA, since for the ferritin the rate of protein synthesis is regulated through IRE-dependent ribosome binding, but for TfR, the rate of protein synthesis is regulated through IRE-dependent mRNA destabilization or rapid turnover. As a group, IREs share a hairpin loop structure with a conserved hairpin loop sequence, CAGUGX (see Fig. 1). IREs also share recognition for a family of proteins, IRPs, which are aconitase homologues [30] and, in some cases, can have aconitase activity. Phylogenetic conservation of IREs within each type of mRNA and divergence between translation IREs and stability IREs suggests that they are iso-elements (iso-IREs) that may have evolved from progenitor mRNA translation and rapid turnover regulatory elements [3].

1.2.1. mRNA Translation

Translation initiation/ribosome binding is the step which is regulated in ferritin mRNA [1,2,24,25,31,32]. Initiation is a major regulatory step for other mRNAs as well, since initiation is usually the rate-limiting step in protein synthesis.

When iron levels are low, the stabilized ferritin mRNA is found in the supernatant fraction of the cell, but when iron levels are high, the mRNA moves to the polyribosomal fraction [1]. The existence of a *trans* factor which recognizes the ferritin mRNA was demonstrated by comparing the localization of ferritin mRNA and globin mRNA in plant cell extracts (wheat germ) which do not differentially regulate ferritin mRNA to that in animal cell extracts (rabbit reticulocytes) which regulate ferritin mRNA specifically [24,25,31,32]. The *trans* factor has been isolated from animal extracts and called the IRP (iron-regulatory protein) [32,33]. Translation initiation is also the site of iron regulation of erythroid aminolevulinate synthase (eALAS) mRNA [34,35] and m-aconitase mRNA [17,18,19]. Iron regulation of translation and rapid turnover depends on the presence of the IRE [26–29,34,35,37,39].

1.2.2. mRNA Stability / Turnover

IRE-regulated degradation of mRNA (rapid turnover) is the mechanism by which the TfR synthesis is controlled by iron. When cellular iron levels increase, the concentration of TfR-mRNA decreases [26,27], with the consequence that TfR synthesis decreases and iron uptake is slowed. There is only a little change in the transcription rate of TfR DNA, which does not account for the large decrease in TfR concentration when iron levels change [26,27,37–39]. The TfR-IRE element is a member of a superfamily of mRNA rapid-turnover elements which have multiple AU-rich sequences [40,41]. However, the TfR mRNA is currently the only known example of IRE-regulated mRNA turnover and appears to be a composite of the AU-rich elements and an iron-regulatory element [3].

1.3. Primary/Secondary Structure of IREs

The primary and secondary structures are highly conserved among IREs (Table 1; Fig. 1), which illuminated many of the early relationships observed between the different types [26,27,29,30,37–39].

The predicted secondary structure of the IRE family is a set of 9–10 base pairs interrupted by an internal loop which includes a con-

TABLE 1

Phylogenetic Conservation within IRE Types of IREs in Different mRNAs

IRE Type[a]	Phylogenetic sequence conservation (%)
Ferritin	97
Erythroid aminolevulinate synthase (eALAS)	97
m-aconitase	100
Transferrin receptor (TfR)-a	99
TfR-b	96
TfR-c	98
TfR-d	96
TfR-e	99

[a]Ferritin IREs: Human-H [53], human L [54]; rat-H [55], rat L [42]; mouse-H [56]; rabbit L [57]; chick-H [58]; frog-H,M(H^1) [59]. TfR IREs: human [60], rat [61], chick [62]. eALAS IREs: human, mouse [63,64]. Note that succinic dehydrogenase (SDh) [18] was not included in the comparison, because there is only one IRE known [65]; the IRE is absent in human SDh; m-aconitase IREs: human, mouse [17]; porcine [66]; bovine [67].

Data were obtained using the Wisconsin Genetics Computer Group "gap" program, Unix version 8.0. Note the extremely high phylogenetic sequence conservation for each IRE type.

served C residue and ends with a hexaloop which has the sequences CAGUGX (Fig. 1). Confirmation of the secondary structure of the IRE has been difficult, since there are very few reactive sites with either probes of single-stranded RNA (RNAse T1, S1, dimethyl sulfate) or double-stranded RNA structure (RNAse V1) [43]. Tertiary folding of the IRE was indicated by reactivity of the IRE in ferritin mRNA with Fe-EDTA, $Cu(phen)_2$ (1,10-phenanthroline-Cu), $Rh(phen)_2phi^{3+}$, Ru-(tpy)$bpyO^{2+}$, and D_2O exchange during nuclear magnetic resonance (NMR) spectroscopy [43–47], which explains the insensitivity or inaccessibility to large protein nucleases.

The shape selectivity of RNA binding by transition metal complexes has been very useful in probing IRE structure. The effect of mutation on the subdomain structure shown in cleavage patterns from

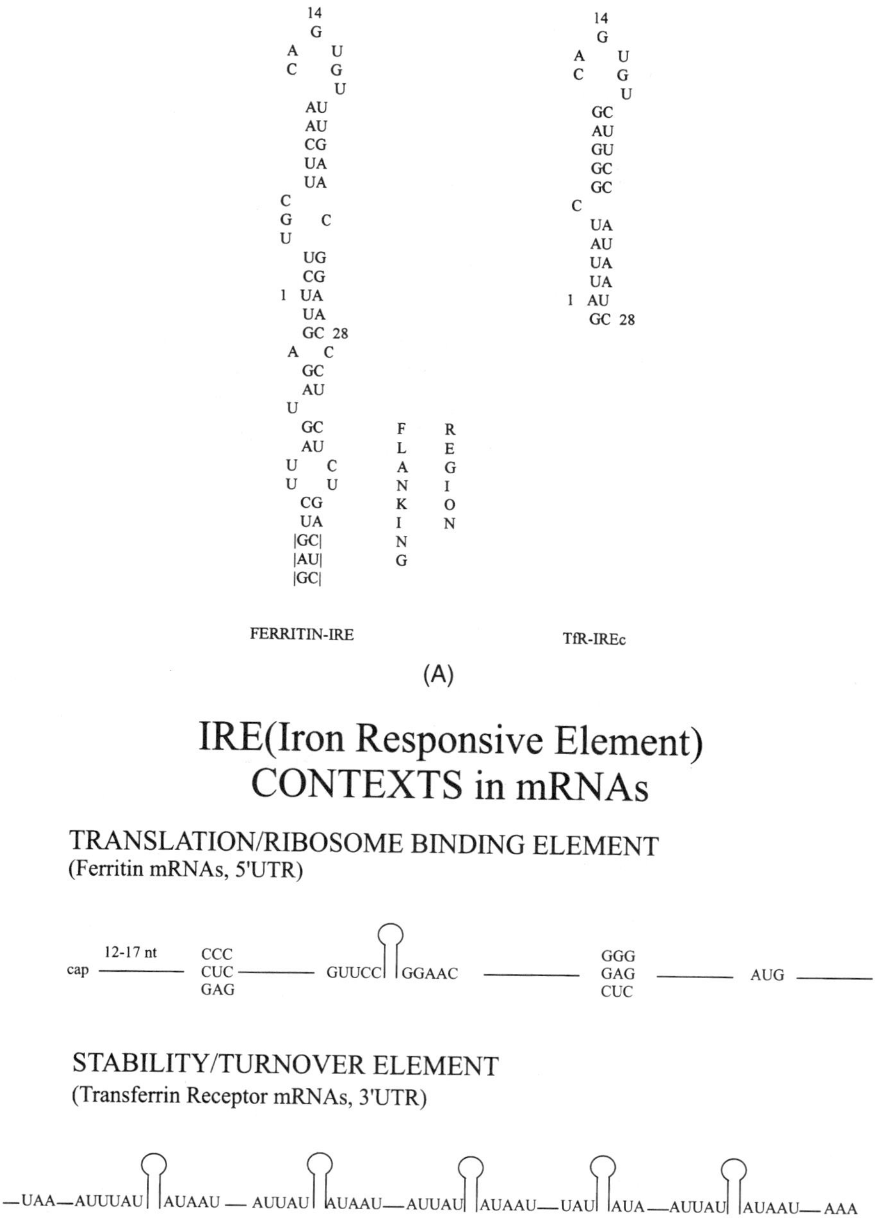

FIG. 1. Secondary structure of ferritin and transferrin receptor IREs. (A) Ferritin-IRE + base-paired flanking region. A conserved triplet of base pairs is shown in brackets. (B) IRE context in mRNA.

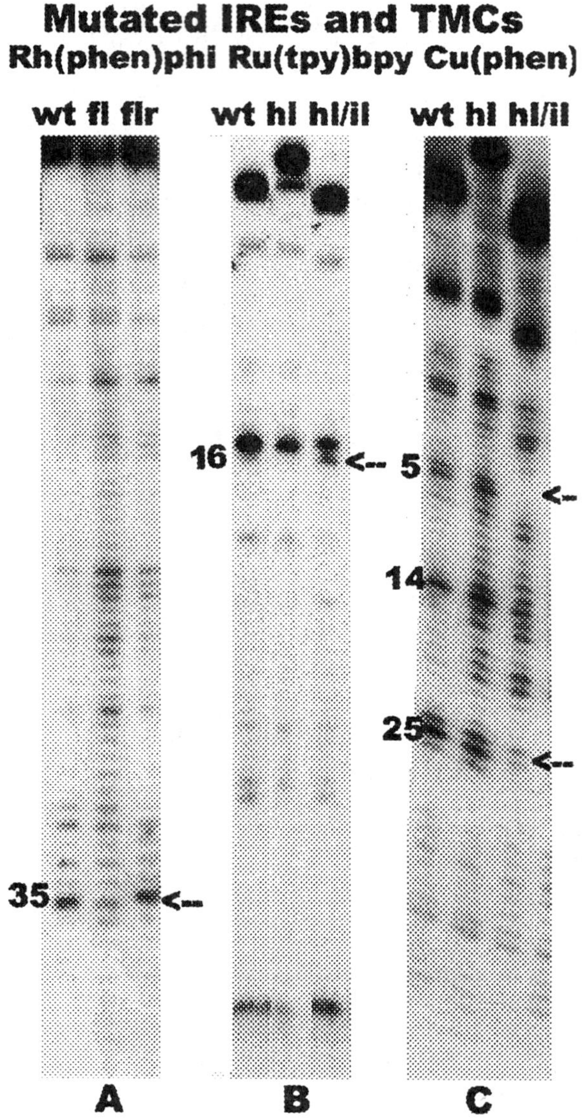

FIG. 2. The use of transition metal complexes to probe the effect of mutation on the folded IRE structure. Full-length (n = 1000 nt), capped ferritin mRNA was analyzed; cleavage sites were detected in ^{32}P-cDNA after reverse transcription of the RNA cleavage products using a prime 40 nucleotides downstream from the end of the IRE (data from [43–45,47]).

three transition metal complexes (TMC) is illustrated in Fig. 2 for four mutant IREs: fl in which the triplet of base pairs GAG/CUC is replaced by CUC/GUC; flr in which CUC/CUC is replaced by CUC/GAG; hl in which CAGUCU is replaced by CAGUAU; and hl/il in which the hairpin loop CAGUGU is enlarged to *CAGUGUU* and the internal loop is changed from UGC••C to C••C [43–47].

High-sequence conservation of IREs suggest a very specific three-dimensional structure which could recognize regulatory proteins such as the IRP protein. "Footprinting" of the IRP showed protection almost exclusively within the IRE itself [43]. The structural specificity of the folded IRE is emphasized by the observations that only two mutations have been observed which function normally in IRP binding (60–90% of wild type) even in a pool of randomly mutated IRE sequences. One mutation is a double mutation which substitutes UA for CG in the hexaloop [48], confirming the importance of base-base interactions at the site also observed by NMR spectroscopy [46]. The second IRE mutation identified which retains high IRP binding is a substitution for the U at the top of the hairpin loop [49], a site previously shown to be outside the folded structure based on nuclease digestion [43]. Single-site base substitution or deletion in the IRE alters IRP binding dramatically (reviewed recently in [3–6]) even in pools of randomly generated oligomers [36,48,49].

Protein nuclease and Fe-EDTA analysis showed that the base-paired regions which flank the ferritin IRE undergo changes in structure when the IRP binds [43]. Mutation of the conserved triplet of base pairs in the ferritin-IRE flanking region (Fig. 3) modulates the IRE structure based on Fe-bleomycin reactivity, as well as rates of protein synthesis [50]. Such data indicate that the ferritin IRE is part of a region of the mRNA with integrated structural features that appear to have a specific three-dimensional structure related to function. The ferritin IRE is also quantitatively the more effective regulator among translation regulators and is the only IRE with base-paired flanking regions (Fig. 3) [34].

The striking structure conservation among IREs (see Table 1) is coupled with mRNA-specific IRE sequence divergence that is also conserved phylogenetically (see Table 2) showing that IREs in the mRNAs encoding different proteins constitute a family of iso-elements. Further comparisons between IREs and other mRNA-regulatory elements sug-

		FL			IRE		FL	
Human-H (17)	CGGCGaGGCU	GAG	uCGUCGGGgu	UCC	ugc UUCAA cagugc UUGGAcGGAAC	CCGGCG-	CUC	gGGCCCGCCG
Mouse-H (16)	GA	GAG	UCGCCGcGG	UUCC	ugc UUCAA cagugc UUGGAcGGAAC	CCGGCG-	CUC	UC
Rat-H (16)	GA	GAG	UCGCCGcGG	UUCC	ugc UUCAA cagugc UUGGAcGGAAC	CCGGCG-	CUC	UC
Chick-H (10)	GGcGGU	GAG	CGgaGCGGG	UUCC	ugc UCAA cagugc UUGGaCGGAAC	GCcgCG	CUC	ACUcUC
Frog-H (10)	GGGAGa	CUC	- uuAGuAGaG	UUCU	ugc UUCAA cagugu UUGAAcGGAAC	cCUCUcu	GAG	gaCUCUC
Frog-H'(M)(8)	GGAGaU	CUC	- uuAGuAGaG	UUCU	ugc UUCAA cagugu UUGAAcGGAAC	cCUCUcu	GAG	aCUCC
Human-L (7)		CCC	gCGGGUCUG	UCUCU	ugc UUCAA cagugu UUGAcGAACAG	AUCCG--	GGG	
Rabbit-L (11)		CCC	GcGGGaCCUGUg	UCU	ugc UUCAA cagugu UUGAAcGGAAC	AGGCCC-	GGG	
Rat-L (11)		CCC	GaGGaUCUGUa	UCU	ugc UCAA cagugu UUGAcGAAC	AGACCCG	GGG	
eALAS (1) (Human & Mouse)			CCGuU	UgGUU	c GUCCUcagugc AGGGCAACAGGAC	uuuGGG		

The number in parenthesis is the distance between the IRE+FL and the cap. Increasing the distance of the IRE+FL from the cap diminishes both translational repression by the IRP protein and translational enhancement (binding of initiation factors). The distance of the ferritin IRE itself varies from 23-33 nucleotides.

FIG. 3. Conserved elements in flanking region (FL) of ferritin IREs. Note that eALAS (shown) and m-aconitase lack the FL and also appear to have a lower range of repression by the IRP binding protein. The conserved triplet of base pairs (see Fig. 1A) in ferritin-IREs is highlighted (see [50]).

gest that translation IREs also belong to a larger class of translation-regulatory elements and the stability IRE element belongs to a more general class of mRNA rapid-turnover/stability elements. Thus, the IREs appear to be composites of both specific, iron-responsive elements combined with more generic features for regulation of mRNA translation or mRNA stability.

2. SPECIFIC FEATURES OF ISO-IREs: mRNA TRANSLATION REGULATORS

Translation IREs have been observed in three different animal mRNAs, ferritin, eALAS, and m-aconitase. (Note that an IRE in succinic dehydrogenase mRNA is found in *Drosophila* but is not found in humans [18,65].) Compared with other mRNA sequences, the IREs display high conservation among different IRE-containing mRNAs (see Table 1). However, within an IRE-mRNA-type group, the mRNA-specific sequence is even more highly conserved (Tables 1 and 2). Specific primary/secondary structure, a single G/C-rich helix near the 5′ end of mRNA, in mRNA translation IREs contrasts with mRNA stability IREs, multiple AU-rich sequences in the 3′-UTR, but the translation IRE structure is common to translation elements of other mRNAs (e.g., [10,51,52]).

2.1. Sequence of IRE Translation Regulators

The sequences of IRE translation regulators are extremely highly conserved within each type of vertebrate mRNAs (96–99% identity); the mRNAs encode ferritin subunits (H and L), m-aconitase, and the aminolevulinate synthase expressed in erythroid cells (see Table 1). For eALAS and m-aconitase IREs sequence information is only available for mammals, but for ferritin IREs information is available for mammals, birds, amphibia, and several invertebrates. When ferritin IREs from the invertebrates *Lymnea* and *Manduca* [68,69] are included in sequence comparisons, the sequence conservation is still very high (82%). In contrast, the mRNA stability IREs in the transferrin receptor mRNAs, although highly conserved phylogenetically among them-

TABLE 2

mRNA-Specific IRE Variation in Vertebrates

	IRE sequence variation[a]			IRE variation in G/C base pair distribution[b]	
	humans	rats	chicks		
	%	%	%	Upper stem	Lower stem
Ferritin	35	45	36	3	2–3
eALAS	43	–[a]	–[a]	3	2
m-aconitase	52	–	–	1	2
TfR-a	74	70	65	2	0
TfR-b	78	83	78	3	0
TfR-c	74	78	87	3	1
TfR-d	100	100	100	3	1
TfR-e	79	74	74	3	0

[a]Ferritin IREs: Human-H [53}, human L [54]; rat-H [55], rat L [42]; mouse-H [56]; rabbit L [57]; chick-H [58]; frog-H,M(H[1]) [59]. TfR IREs: Human [60], rat [61], chick [67]; eALAS IREs: human, mouse [63,64]. For the ALAS comparison to the TfR-IREs only (human IRE sequences are available). m-Aconitase IREs: human, mouse [17]; porcine [66]; bovine [67].

[b]The region of greatest variation between mRNA translation and stability regulatory IRE types is the lower stem (see Fig. 1). Few if any G/C base pairs occur in the lower stem of TfR mRNA stability IREs, and the single G/C base pair in TfR-IREb and TfR-IREd occurs at the base of the IRE stem. The % G+C content of the TfR-IREs is 25.8, 37.9, 44.8, 44.8, and 31.4 for a, b, c, d, and e. Translation regulatory IREs have a G+C content of 46.7% for ferritin IREs including vertebrates (human, mouse, rat, rabbit, chick and frog) and the invertebrate snail, *Lymnaea stagnalis* [69] and 56.6% for eALAS IREs (human and mouse).

Data were obtained using the Wisconsin Genetics Computer Group "gap" program, Unix version 8.0. Note the sequence variation *between* IRE types.

selves, diverge as a group from translation IREs sharing only 35–52% sequence identity (see Table 2).

A relatively high G/C content characterizes the IREs that are translation regulators (see Table 2). Alteration in translation rates of mRNAs in general has been associated with stable double-stranded sequences (high G/C content) [51,52]. Thus, translation IREs are members of a larger class of translation elements.

2.2. Position of IRE Translation Regulators in mRNA

The position of specific sequences in mRNA is associated with function [10,51,52]. At the 5′ end of eukaryotic mRNAs is a set of methylated nucleotides which form a site of binding of initiation factors and formation of the initiation complex (initiator tRNA, ribosome, and, mRNA). Other examples of position-specific sequences in eukaryotic mRNAs are the codons at the beginning and end of the coding sequence which have distinctive behavior with tRNAs and the presence of a polyA$^+$ sequence at the 3′ end of the mRNA.

All translation IREs are present as a single copy in the 5′UTR (see Fig. 1B). Moving the IRE downstream from the cap (>45 nucleotides) diminishes the IRE-regulatory effect on translation [70,71]. For eALAS and m-aconitase mRNAs, the IREs are close to the cap. In ferritin IREs, the appropriate proximity to the cap is achieved by a set of flanking bases which pair to extend the stem of the IRE [44,45] (Figs. 1B and 3). The IRE distance from the cap appears (17–33 nucleotides) in ferritin mRNAs is decreased to 12–17 nucleotides when the base-paired flanking region is included. Evidence for the integration of the FL + IRE includes structural changes in the FL when IRP binds to the IRE [43]; Mg-dependent changes in both the IRE and FL [44] and the effect of FL mutation on both IRE structure and function [50]. The FL stabilizes the IRE structure. For example, the predicted stability of secondary structure is −3.7 kcal/mol for the human H IRE compared with −9.2 kcal/mol for IRE + FL [45]. Mutation in the FL decreases repression by ~ one-half [50].

Quantitative differences occur in regulation by the IREs in ferritin and eALAS mRNAs [34] which may relate to the increased stability of the FL in the ferritin mRNA IRE region. A TfR-IRE inserted near the 5′

cap of a reporter conferred iron-dependent regulation of expression at one fifth the effect of a ferritin translation IRE [38], which emphasizes the quantitative variations in function encoded in each type of sequence.

3. SPECIFIC FEATURES OF ISO-IREs: mRNA STABILITY/TURNOVER REGULATORS

An IRE mRNA stability element has only been found, to date, in the transferrin receptor mRNA. The stability element is highly conserved in terms of sequence identity, copy number (n = 5), and mRNA position (3′UTR) (Fig. 1B) in the mammalian and avian TfR-mRNAs which have been studied (see Table 1). In contrast to the IRE translation elements which are formed by a single loop-stem-bulge-stem structure in the 5′-UTR, with G/C base pairs in upper and lower stems, the IRE mRNA-turnover/stability element is composed of five loop-stem-bulge-stem structures in the 3′UTR which contain AU-rich sequences in the lower stem that are found in rapid-turnover elements of several other mRNAs such as *c-myc*, *c-fos*, interleukin-2 and interleukin-6 (e.g., [11–14,40,41]).

3.1. Sequence of IRE Stability/Turnover Regulators

The set of five IREs in transferrin receptor mRNAs as a group share 72–89% sequence identity (Fig. 1B; Table 2). When each of the five TfR-IREs (a,b,c,d,e) are compared phylogenetically, the sequence conservation for each (a,b,c,d) in human, rat, and chick is even higher (96–99%) (see Table 1).

The structural features shared by all IREs (a folded hairpin loop sequence, a stem with 9–10 base pairs, and a cytosine residue in a bulge/loop) are sufficient to permit recognition of IRPs. However, sequence divergence between the set of five TfR-IREs (35–52% sequence identity) and the translation IREs is significant (see Table 2) and contrasts with the very high phylogenetic sequence identity (96–100%) among the IREs in a particular mRNA (see Table 1). The site of the major sequence differences between the TfR-IREs and translation IREs is in the lower stem. AU base pairs account for 23 of 25 base pairs in the lower stems of the set of five TfR-IREs; 3 of the 5 TfR-IREs (a,b,e) have

100% AU base pairs in the lower stems and two of the TfR-IREs (c,d) (see Fig. 1A) have 80% AU base pairs in the lower stem. In contrast, there are usually 40% AU base pairs in the lower stems of translation IREs (see Table 2).

When the sequences of the five TfR-IREs are considered in a linear array, they form a series of 10 AU-rich sequences (see Fig. 1B). The short hairpin loops common to all IREs are in the center of two runs of AU sequences. In the TfR-regulatory region are non-IRE sequences between IREb and IREc which are also required for the iron-dependent destabilization of the mRNA [27]. The central sequence of the TfR-regulatory element is also AU rich; there are four sets of $(U)_{4\text{-}5}$ sequences between TfR-IREb and TfR-IREc. The sequences which form TfR-IREb and TfR-IREc can also form an alternative secondary structure of similar predicted energy [26,27] (discussed in [3]).

Multiple AU-rich sequences are a feature of a number of mRNA rapid-turnover elements (e.g., [11–14,40,41]). Some of the elements also require stem loop structures [72]. Such observations indicate that in addition to membership in the IRE family, the TfR-IRE-regulatory region belongs to a larger class of mRNA rapid-turnover elements which regulate mRNA stability.

3.2. Position of IRE Translation Regulators in mRNA

The set of TfR-IREs is in the 3′-UTR of transferrin receptor mRNAs (see Fig. 1B). For function as a regulatory element, both the IREs and a non-IRE base-paired sequence in the 3-UTR, between TfR-IREb and TfR-IREc, are required; the sequence in the middle of the IREs also contains a run of U and A residues required for function [26,27,37,38]. The position of the TfR-IREs in the 3′-UTR is shared by the other mRNA stability elements (e.g., *c-fos*, *c-myc*); again emphasizing the membership of stability IREs in a larger, more general class of mRNA stability regulatory elements.

4. PROTEINS RECOGNIZED BY IREs

Nucleic acid-protein complexes with regulatory function usually involve multiple protein-protein interactions as well as protein-DNA or protein-RNA interactions exemplified by DNA transcription and replication complexes, spliceosomes, and ribosomes. Less is known about mRNA-protein-regulatory complexes, but, by analogy, many proteins are likely to be involved. Rather than the protein "on-off" model where mRNA is a passive partner in recognition, the specific primary and tertiary structure of IREs suggests that for IRE-protein recognition, RNA binding of protein is as important as protein binding of RNA.

4.1. Iron-Regulatory Proteins

The best characterized protein/RNA interactions of the IRE is with the IRPs. Cloning has shown the IRPs to be a family of highly conserved proteins [73–83] which are homologues of the aconitases [30]. In fact, one of the IRPs, IRP-1, can acquire aconitase activity by mixing the protein with Fe(II) and sulfide ([80]; see [5,6] for detailed reviews). The ratio of IRP-1 to IRP-2 (IRPa and IRPb) differs in different tissues [36,75,76,78,79]. IRP-2 has a higher specificity for RNA binding than IRP-1 [49,49,83]. The IRE is completely covered by IRP-1 [43].

IRP-1, which can acquire aconitase [80,84] activity, is very closely related [85,86], if not identical, to cytoplasmic aconitase (c-aconitase) and much less similar to mitochondrial aconitase. Analysis of a subset of bovine c-aconitase peptides from each domain of the protein (10% of the total) showed almost complete identity with human or rabbit IRP-1 [86], emphasizing the high probability of equivalence of the proteins, unless IRP-1 and c-aconitase sequences diverge in the c-aconitase regions yet to be sequenced; the complete sequence of c-aconitase is not yet known. The observation that IRP-1 isolated from tissues has different RNA-binding properties than apo-c-aconitase [82] could be attributed to small differences in sequence between c-aconitase and IRP-1 or, more likely, to alterations (oxidation) of residues key for RNA binding during the isolation of c-aconitase.

The mechanism by which cellular iron changes the IRP/IRE interaction is not fully understood. IRP-1 and IRP-2 are phosphorylated by protein kinase C, which appears to change IRE binding under a variety of conditions [87]. The phosphorylation has the additional feature of allowing readily reversible and coordinated regulation of both IRP-1 and IRP-2 [88]. A reversible Fe-IRP interaction, in the form of the FeS cluster-IRP-1 complex, had earlier been proposed as the physiological iron signal, with the apo-protein being the RNA binding form. However, the RNA/IRP-1 interaction is more complex: Only if the FeS form of IRP-1 is treated with 2% β-mercaptoethanol can RNA binding occur in vitro [85,89]. Whether the mercaptoethanol produces the apo-IRP-1 or whether it is a modified FeS complex that binds the RNA is not yet known. In contrast to IRP-1, cytoplasmic iron leads to the proteosomal degradation of IRP-2 protein and also decreases the concentration of IRP-2 mRNA [78,79,81,82]. Thus, the combination of control through phosphorylation and iron-dependent changes in protein conformation (IRP-1) or degradation (IRP-2) provides the potential for reversible as well as irreversible regulatory control of IRE binding.

In whole cells or extracts, a variety of other agents alter the IRE-binding capacity, such as peroxide, NO, and mercaptoethanol (reviewed in [4,5]); heme binding leads to IRP degradation [90]. All the agents can enhance the presence of radicals to which FeS clusters are sensitive. However, oxidative damage to other parts of the proteins, such as cysteine and methionine residues, can also occur under the same conditions, which confounds the interpretation of such observations in the absence of direct information about changes in the protein structure.

The precise location of the IRE-binding site in the IRPs is not known, but a variety of observations suggests that the site in IRP-1 at least is composed of residues at several sites along the polypeptide chain. First a fairly large piece of the N-terminal end of the protein (residues 1–623) is required for RNA binding [91]. Second, ultraviolet-cross-linking studies showed that sites between 121–130 and 480–623 [92] contact the IRE. Finally, mutations which affect RNA binding are at positions distributed at fairly large distances from each other; for example, e.g. Cys 437, 503, 506, Arg 541, Arg 780 [93–95].

The physiological significance of the aconitase activity of IRP-1 remains obscure at this time. Two observations emphasize the complexity of the relationship between aconitase activity and RNA binding. First, when the potential for aconitase activity in IRP-1 was destroyed by mutation S788A, RNA binding was unaffected [96]. Second, IRP-2 binds RNA with even greater specificity than IRP-1 [36,48,49] and has no aconitase activity [78]. Although the interconversion of aconitase to an RNA-binding protein may turn out to have physiological significance, an alternative idea is that IRPs evolved into RNA-binding proteins from aconitase and in contemporary IRPs aconitase activity is simply vestigial. In such a scheme, IRP-2, in which the absence of aconitase activity coincides with higher IRE specificity, would be the more highly evolved.

4.2. Initiation Factors

The binding of initiation factors followed by formation of the initiator tRNA and ribosomes is commonly the rate-limiting step in protein synthesis. Rates of protein synthesis can be varied by differential control at the initiation step, leading to disproportionately high translation of some mRNAs such as those of the polio virus or mRNAs encoding stress proteins. Often the differential rates of translational initiation are only observed in cells with relatively low concentrations of initiation factors and ribosomes such as maturing red cells. When an mRNA is particularly effective at binding ribosomes, translation rates are enhanced and the mRNA is termed "competitive" or under "positive control". The effect is analogous to transcription enhancement or positive control of transcription observed for DNA.

Ferritin translation was shown to be competitive or positively regulated in iron loaded red cells; that is, IRP-independent ferritin synthesis was initiated more efficiently than other mRNAs [2,22]. The effect was also observed in vitro with cell-free extracts from wheat germ which lack the IRP [24,25]. By comparing translation of ferritin mRNAs transcribed in vitro, the positive control element of ferritin mRNA was traced to the IRE region [70]. Experiments with crude initiation factors from wheat germ showed that the ferritin mRNA was differentially reacting with a protein in crude preparations of eIF-4B that contains several other eIF factors. Experiments with isolated eIFs are now in progress.

4.3. Nucleases?

IRE stability elements regulate the mRNA turnover rate by stabilizing the mRNA when iron is low and allowing mRNA degradation when iron is high. The nucleolytic activity which degrades the TfR mRNA when cellular iron concentrations are high has not yet been identified. Nor is much known about the mechanisms of regulated degradation of mRNAs such as *c-myc*, *c-fos*, IL-2, and IL-6 [11–14,40,41]. Degradation of TfR RNA could be caused by specific catalysts with IRE recognition or by a nonspecific nuclease which recognizes an alternate structure of the IRE-regulatory element region. When RNA nucleases have been characterized, in the past, the nucleases have been proteins. However, the association of specific mRNA turnover with ribosomes in some cases [11,97] indicates that the nuclease involved in rapid mRNA turnover could also be RNA.

5. REGULATION OF IRON UPTAKE AND STORAGE IN PLANTS AND MICROORGANISMS

Iron regulates the expression of ferritin and iron-uptake proteins in plants as well as in animals. In spite of the similarity of environmental signals and types of gene products regulating iron metabolism in plants and animals, the genetic mechanisms for response to iron are very different [23]. In microorganisms, iron regulation of iron chelators and membrane-uptake proteins is well known, but very little is known about iron-dependent regulation of ferritin expression in microorganisms. Thus, at this time, whether genetic mechanisms which regulate ferritin expression in microorganisms will match those in plants or those in animals (or neither) is not known.

5.1. Iron Uptake

Changes in iron uptake occur during iron deficiency, for example, which enhance iron uptake in several different ways: increased secretion of protons around the root to increase the local solubility of iron (e.g., [98]), reduction of Fe(III) to Fe(II), which also increases the local solubility of iron, and release of siderophores to increase iron binding (e.g., [99,100]).

5.1.1. Plants and Yeast

Recent studies at the molecular level have identified iron-regulated membrane proteins responsible for reduction of iron and incorporation of Fe(II) in the yeast, *Saccharomyces cerevesiae* (e.g., [101,102]). In the higher plant, *Arabidopsis*, a membrane protein has also been recently identified which is regulated by iron and which is involved in the reduction of Fe(III) [102]. Based on sequence, the membrane proteins involved in iron uptake appear to belong to a large family of proteins in plants and animals which control the transport of ions across cell membranes. In yeast, a membrane protein involved in the transport of Cu is regulated by both iron and copper tying together the metabolism of the two ions [103,104].

5.1.2. Bacteria

Iron uptake has been extensively studied in bacteria in part because a strategy of virulence in infecting bacteria is the production of siderophore chelators to compete with transferrin for iron making iron-uptake/siderophores a potential target for therapy. Host animals respond to bacterial infection by decreasing serum iron through retention of iron in macrophage ferritin. In the process of keeping iron away from bacteria, the host actually makes itself temporarily anemic. Competition for iron between the bacterium and the human emphasizes the crucial role of iron in biology.

The major iron-uptake systems studied in bacteria are regulated either by iron repletion or deficiency and involve different gene products depending an the oxidation state of the iron complex [105]. Ferric citrate induces the synthesis of several membrane proteins which transport ferric citrate into the cell [106]. Transcription of the genes encoding proteins for ferric citrate transport is triggered by a signal transduced from conformational changes in the membrane proteins FecR [107]. Fe(II) induces a distinct set of proteins in prokaryotic cells for uptake and transport of Fe(II) complexes [108].

Deficiency of iron induces the synthesis of a group of proteins that produce siderophores. Ferric iron, for example, is bound $\sim 10^{20}$ times more tightly by siderophores than by transferrin. During iron deficiency, a number of receptors and membrane transporters are produced coordinately with the siderophore to transport the Fe-siderophore complexes into the cell [109]. Some siderophores, for example, deferrioxamine B (desferal) from *Streptomyces*, are not only used by the microorganisms but are used therapeutically to remove excess iron from people with diseases such as hemochromatosis and transfusional iron overload.

A number of questions have assumed increased importance (and increased possibilities for investigation) with the recent expansion of our knowledge about iron uptake in plants and microorganisms. For example, will homologues of the yeast/plant membrane iron-uptake proteins be found in animals? What is the role of nontransferrin iron uptake in animals? Does the metabolic connection between Cu and Fe uptake observed in yeast explain the interconnections previously observed between Fe and Cu metabolism in animals?

5.2. Iron Storage

The concentration of iron required in plant cells and bacterial cells is far above the free solubility of the ferric ions, which is a major oxidation state of the iron in plant cells as in animal cells. Particularly high iron concentrations are present in chloroplasts, nodules, and seeds where developmental changes in ferritin are clear [110–113].

5.2.1. Plants and Yeast

Plants can accumulate excess environmental iron in both vacuoles and in ferritin. Iron induces ferritin synthesis in plants as in animals. The gene product, ferritin protein, is also conserved in plants [114,115]. However, iron acts at the level of transcription in plants in contrast to animal ferritin [20]. Moreover, there is no IRE in plant ferritin genes and the gene organization, although conserved among plants, is distinct from ferritin gene organization in animals [23]. In fact, a chimera of the IRE and soybean ferritin mRNA is poorly regulated under conditions where the natural construct (IRE + animal ferritin mRNA) is highly regulated [21]. The differences between the iron-dependent genetic regulatory mechanisms in plants and animals may relate to the fact that in plants ferritin is synthesized in the cytoplasm and transported to the chloroplast and other plastids (see [23]).

Abscisic acid and oxidative stress both induce ferritin mRNA and protein accumulation in plants in addition to environmental iron [10,11]. The relationship of the induction to iron metabolism is complex, although hormone response can clearly be iron independent. Plant ferritin genes and animal ferritin genes may share an iron-dependent, common genetic regulatory pathway, since both plant and animal genes can be regulated at the level of transcription and by hormones/growth factors. Transcription of animal ferritin genes is regulated by phorbol esters, thyroid-stimulating hormone (TSH), and tumor necrosis factor α (TNF-α) [116–118].

Yeast ferritin regulation has been little studied. Examination of the complete genome of *S. cerevesiae* reveals only a sequence which is similar to prokaryotic ferritin. Possibly the ability of yeast to accumulate iron in vacuoles [119], and the absence of iron-rich organelles such as chloroplasts, may make ferritin less important in yeast than in other eukaryotic cells.

5.2.2. Bacteria

Less is know about the regulation of ferritin genes in bacteria than in plants. In *Escherichia coli*, there are two ferritin genes, one which has a slightly higher sequence similarity to eukaryotic ferritin (FTN) [120] compared with the other (BFr) [121]. The primary structure of each bacterial ferritin is very different from the eukaryotic ferritin sequences, although the clusters of conserved carboxylate groups facing the inner cavity of the assembled protein are conserved [122].

There is no IRE encoded in bacterial ferritin genes. The signals which regulate the expression of bacterial ferritin genes have not been identified. However, an indication of the type of signaling that influences ferritin expression is the observation that ferritin is easily isolated from *Azotobacter vinelandii* [123] that are induced to fix nitrogen in contrast to bacteria grown in a nitrogen-replete environment.

The function of ferritin in bacteria has been studied in somewhat more detail than regulation. An example is the observation that bacterial ferritin (FTN) from *E. coli* appears to protect cells from the effects of dioxygen and excess iron, since the synthesis of a large amount of FTN encoded in a plasmid rescued *E. coli fur* mutants that assimilated large amounts of iron and which died in air without excess FTN [124].

6. PERSPECTIVE

The ferrocentric view of the world predicts that regulation of bioiron would be sophisticated and would provide insight about general cell biology. Facts which amply fulfill the prediction are: the first observation of receptor mediated endocytosis (the transferrin/transferrin receptor complex), the proof of mRNA-specific translational repression (ferritin mRNA storage), and the coordinated regulation of mRNAs encoding metabolically related proteins (ferritin, transferrin receptor).

Studies of events at the cell surface mediated by receptors begun with studies of transferrin now are a huge portion of current research. Studies of mRNA and the proteins which regulate it represent a smaller but rapidly growing area of biology and include not only iron-regulated mRNAs but also organelle-related mRNAs, stress mRNAs, and viral mRNAs. The IRE (iron-responsive) family of mRNA-regulatory isoelements remains unique so far; however, in being bifunctional, IREs regulate both translation (ribosome binding) and rapid-turnover/stability. Other mRNA-regulatory elements appear to be monofunctional.

RNA/protein interactions amplify the regulatory information contained in the IREs. The primary sequences of IREs include a central sequence CAGUG and a C residue four to five nucleotides from the central sequence. The secondary structure is a hairpin loop with the CAGUG forming a terminal loop with the conserved C interrupting the helix stack. The IRE secondary structure is folded into a specific three-dimensional structure. A specific class of proteins that IREs recognize are the IRPs, aconitase homologues. Some IRP proteins can bind iron (an FeS cluster) and acquire aconitase activity (IRP-1); IRP activity is also regulated by iron through degradation of the mRNA and the protein (IRP-2). Other proteins which are recognized by IREs are translation initiation factors (eIFs) and possibly nucleases. Key questions under active investigation are: What is the three-dimensional structure of IREs? What is the difference in the mechanisms for regulating translation and mRNA turnover? Do the different IREs (translation/rapid turnover) bind different IRPs (e.g., IRP-1, IRP-2) in vivo? What is the physiological iron signal for IRPs that do not bind iron? How do other signal transduction pathways (hormones? growth factors? NO?) intersect with the path leading to the IRE/IRP interaction? The answers are being obtained at a very rapid rate and will not only illuminate the molecular basis of iron metabolism but will also lead to new insights in the posttranscriptional regulation of gene expression.

Conservation of IRE sequences appears to be kingdom specific. IREs occur only in animal ferritin genes and not in the plant ferritin genes examined to date. The animal IRE regulates plant ferritin mRNA poorly in chimeras. Coregulating the ferritin gene in the plant cell nucleus with plastid development must occur in plants, since ferritin is in chloroplasts and other plastids, a feature which may make IRE-dependent regulation of mRNA less important in plants. The absence of ferritin mRNA regulation in plants, the differences in ferritin gene organization in plants and animals, and the use of ferritin DNA as the iron target in plants but mRNA as the iron target in animals makes iron regulation of ferritin a fertile ground for yet another new subject: the evolution of genetic regulatory mechanisms.

ACKNOWLEDGMENTS

This work has been supported by the Hematology Extramural Program of the NIH(DK-20251) and the North Carolina Agricultural Research Service. Quantitative data of IRE sequence identity was contributed by Jianzhong Wei.

My many enthusiastic coworkers, students, and postdoctoral associates have made the work not only possible but enriching and enjoyable.

ABBREVIATIONS

BFr	bacterial ferritin
bpy	2,2′-bipyridyl
eALAS	erythroid aminolevulinate synthase
EDTA	ethylenediamine-*N*,*N*,*N*′,*N*′-tetraacetate
eIF	eukaryotic initiation factor
FecR	ferric citrate receptor
FL	base-paired flanking region of ferritin
FTN	eukaryotic-like ferritin
IL	internal loop of IREs
IRE	iron-responsive element
IRP	iron-regulatory protein
mRNA	messenger ribonucleic acid
NMR	nuclear magnetic resonance

phen	1,10-phenanthroline
phi	9,10-phenanthrenequinone diimine
SDh	succinic dehydrogenase
TfR	transferrin receptor
TNF-α	tumor necrosis factor α
tpy	2,2′,2″-terpyridine
tRNA	transfer ribonucleic acid
TSH	thyroid-stimulating hormone
UTR	untranslated region

REFERENCES

1. J. Zahringer, B. S. Baliga, and H. N. Munro, *Proc. Natl. Acad. Sci. USA*, *73*, 857–861 (1976).
2. F. V. Schaefer and E. C. Theil, *J. Biol. Chem.*, *256*, 1711–1715 (1981).
3. E. C. Theil, *Biochem. J.*, *304*, 1–11 (1994).
4. D. P. Mascotti, D. Rup, and R. E. Thach, *Annu. Rev. Nutr.*, *15*, 239–261 (1995).
5. M. W. Hentze and L. C. Kuhn, *Proc. Natl. Acad. Sci. USA*, *93*, 8175–8182 (1996).
6. T. A. Roault and R. D. Klausner, *J. Biol. Inorg. Chem.*, *1*, 494–499 (1996).
7. D. R. Gallie, *Annu. Rev. Plant Physiol. Plant Mol. Biol.*, *44*, 77–105 (1993).
8. P. J. Green, *Plant Physiol.*, *102*, 1065–1070 (1993).
9. S. Lindquist, *Annu. Rev. Gen.*, *22*, 631–677 (1988).
10. S.-Y. Le, J.-H. Chen, N. Sonenberg, and J. V. Maizel, *Virology*, *191*, 858–866 (1992).
11. R. J. Jackson, *Cell*, *74*, 9–14 (1993).
12. D. J. Herrick and J. Ross, *Mol. Cell. Biol.*, *14*, 2119–2128 (1994).
13. C. Y. Brown, C. A. Lagnado, M. Vadas, and G. J. Goodall, *J. Biol. Chem.*, *271*, 20108–20112 (1996).
14. C. Y. Brown, C. A. Lagnado, and G. J. Goodall, *Proc. Natl. Acad. Sci. USA*, *93*, 13721–13725 (1996).

15. C. Yanofsky, *J. Biol. Chem.*, *265*, 751–758 (1988).
16. E. Morfeldt, D. Taylor, A. von Gabain, and S. Arvidson, *EMBO J.*, *14*, 4569–4577 (1995).
17. H.-Y. Kim, T. LaVaute, K. Iwai, R. D. Klausner, and T. A. Rouault, *J. Biol. Chem.*, *271*, 24226–24230 (1996).
18. N. K. Gray, K. Pantopoulos, T. Dandekar, B. A. C. Ackrell, and M. W. Hentze, *Proc. Natl. Acad. Sci. USA*, *93*, 4925–4930 (1996).
19. G. S. Chen, K. L. Schalinske, and R. S. Eisenstein, *J. Nutr.*, *127*, 238–248 (1997).
20. A. M. Lescure, D. Proudhon, H. Pesey, M. Ragland, E. C. Theil, and J. F. Briat, *Proc. Natl. Acad. Sci. USA*, *88*, 8222–8226 (1991).
21. Y. Kimata and E. C. Theil, *Plant Physiol.*, *104*, 263–270 (1994).
22. E. C. Theil, in *Metal Ions in Gene Regulation* (S. Silver and W. Walden, eds.), Chapman and Hall. In press.
23. D. Proudhon, J. Wei, J.-F. Briat, and E. C. Theil, *J. Mol. Evol.*, *42*, 325–336 (1996).
24. G. E. Shull and E. C. Theil, *J. Biol. Chem.*, *257*, 14187–14191 (1982).
25. G. E. Shull and E. C. Theil, *J. Biol. Chem.*, *258*, 7921–7923 (1983).
26. D. Owen and L. C. Kuhn, *EMBO J.*, *6*, 1287–1293 (1987).
27. E. W. Mullner and L. C. Kuhn, *Cell*, *53*, 815–825 (1988).
28. N. Aziz and H. N. Munro, *Proc. Natl. Acad. Sci. USA*, *84*, 8478–8482 (1987).
29. M. W. Hentze, S. W. Caughman, T. A. Rouault, J. G. Barriocanal, A. Dancis, J. B. Harford, and R. D. Klausner, *Science*, *238*, 1570–1572 (1987).
30. T. A. Rouault, C. D. Stout, S. Kaptain, J. B. Harford, and R. D. Klausner, *Cell*, *64*, 881–883 (1991).
31. L. F. Dickey, Y.-H. Wang, G. E. Shull, I. A. Wortman III, and E. C. Theil, *J. Biol. Chem.*, *263*, 3071–3074 (1988).
32. W. E. Walden, S. Daniels-McQueen, P. H. Brown, L. Gaffield, D. A. Russell, D. Bielser, L. C. Bailey, and R. E. Thach, *Proc. Natl. Acad. Sci. USA*, *85*, 9503–9507 (1988).
33. R. A. Rouault, M. W. Hentze, D. J. Haile, J. B. Harford, and R. D. Klausner, *Proc. Natl. Acad. Sci. USA*, *86*, 5768–5772 (1989).

34. C. R. Bhasker, G. Burgiel, B. Neupert, A. Emery-Goodman, L. C. Kuhn, and B. K. May, *J. Biol. Chem.*, *268*, 12699–12705 (1993).

35. O. Melefors, B. Goossen, H. E. Johansson, R. Stripecke, N. K. Gray, and M. W. Hentze, *J. Biol. Chem.*, *268*, 5974–5978 (1993).

36. B. R. Henderson, E. Menotte, and L. C. Kuhn, *J. Biol. Chem.*, *271*, 4900–4908 (1996).

37. J. L. Casey, B. D. Jeso, K. Rao, R. D. Klausner, and J. B. Harford, *Proc. Natl. Acad. Sci. USA*, *85*, 1787–1791 (1988).

38. J. L. Casey, M. W. Hentze, D. N. Koeller, S. W. Caughman, T. A. Rouault, R. D. Klausner, and J. B. Harford, *Science*, *240*, 924–928 (1988).

39. J. L. Casey, D. M. Koeller, V. C. Ramin, R. D. Klausner, and J. B. Harford, *EMBO J.*, *8*, 3693–3699 (1989).

40. G. Shaw and R. Kamen, *Cell*, *46*, 659–667 (1986).

41. D. Caput, B. Beutler, K. Hartog, R. Thayer, S. Brown-Shimer, and A. Cerami, *Proc. Natl. Acad. Sci. USA*, *83*, 1670–1674 (1986).

42. E. A. Leibold and H. N. Munro, *J. Biol. Chem.*, *262*, 7335–7341 (1987).

43. C. M. Harrell, A. R. McKenzie, M. M. Patino, W. E. Walden, and E. C. Theil, *Proc. Natl. Acad. Sci. USA*, *88*, 4166–4170 (1991).

44. Y.-H. Wang, S. R. Sczekan, and E. C. Theil, *Nucleic Acids Res.*, *18*, 4463–4468 (1990).

45. Y.-H. Wang, P.-N. Lin, S. R. Sczekan, R. A. McKenzie, and E. C. Theil, *Biol. Metals*, *4*, 56–61 (1991).

46. H. Sierzputowska-Gracz, R. A. McKenzie, and E. C. Theil, *Nucleic Acids Res.*, *23*, 145–152 (1995).

47. H. H. Thorp, R. A. McKenzie, P.-N. Lin, W. E. Walden, and E. C. Theil, *Inorg. Chem.*, *35*, 2773–2779 (1996).

48. B. R. Henderson, E. Menotti, C. Bonnard, and L. C. Kuhn, *J. Biol. Chem.*, *269*, 17481–17489 (1994).

49. J. Butt, H. Kim, J. P. Basilion, S. Cohen, F. Iwai, C. Philpott, S. Schrul, R. D. Klausner, and T. A. Rouault, *Proc. Natl. Acad. Sci. USA*, *93*, 4345–4349 (1996).

50. D. J. Dix, P.-N. Lin, A. R. McKenzie, W. E. Walden, and E. C. Theil, *J. Mol. Biol.*, *231*, 230–240 (1993).

51. K. R. Blake, A. Murakami, and P. S. Miller, *Biochemistry*, *24*, 6132–6138 (1995).

52. M. Kozak, *Mol. Cell. Biol.*, *9*, 5134–5142 (1989).
53. F. Constanzo, M. Colombo, S. Staempfli, C. Santoro, M. Marone, R. Frank, H. Delius, and R. Cortese, *Nucleic Acids Res.*, *14*, 721–736 (1986).
54. C. Santoro, M. Marone, M. Ferrone, F. Constanzo, M. Colombo, C. Mingati, R. Cortese, and L. Silwengo, *Nucleic Acids Res.*, *14*, 2863–2874 (1986).
55. M. T. Murray, K. White, and H. N. Munro, *Proc. Natl. Acad. Sci. USA*, *84*, 7438–7442 (1987).
56. C. Beaumont, L. Dugast, F. Renaudie, M. Souroujon, and B. Grandchamp, *J. Biol. Chem.*, *264*, 7498–7504 (1989).
57. P. H. Brown, S. Daniels-McQueen, and R. E. Thach, Genbank Accession #14598 (1989).
58. P. W. Stevens, J. B. Dodgson, and J. D. Engel, *Mol. Cell. Biol.*, *7*, 1751-1758 (1987).
59. L. F. Dickey, Y.-H. Wang, G. E. Shull, I. A. Wortman III, and E. C. Theil, *J. Biol. Chem.*, *263*, 3071–3074 (1988).
60. C. Schneider, M. J. Owen, D. Bannville, and J. G. Williams, *Nature*, *211*, 675–678 (1984).
61. K. P. Roberts and M. D. Griswold, *Mol. Endocrinol.*, *4*, 531–542 (1990).
62. L. Chan, N. Grammatikakis, and J. M. Banks, *Nucleic Acids Res.*, *17*, 3763–3771 (1989).
63. T. C. Cox, M. J. Bawden, A. Martin, and B. K. May, *EMBO J.*, *10*, 1891–1902 (1991).
64. T. Dandekar, R. Stripecke, N. K. Gray, B. Goossen, A. Constable, H. E. Johansson, and M. W. Hentze, *EMBO J.*, *10*, 1903–1909 (1991).
65. S. A. Kohler, B. H. Henderson, and L. Kuhn, *J. Biol. Chem.*, *270*, 30781-30786 (1995).
66. L. Zheng, P. C. Andrews, M. A. Hermodson, J. E. Dixon, and H. Zalkin, *J. Biol. Chem.*, *265*, 2814–2821 (1990).
67. N. Xu, W. Liu, G. Woldegiogis, and X. Sun, Genbank Accession number Z49931 (1995).
68. Q.-D. Pham, D. Zhang, D. H. Hufnagel, and J. J. Winzerling, *Gene*, *172*, 255–259 (1996).
69. M. Von Darl, P. M. Harrison, and W. Bottke, *Eur. J. Biochem.*, *222*, 353–366 (1994).

70. D. J. Dix, P.-N. Lin, Y. Kimata, and E. C. Theil, *Biochemistry*, *31*, 2818–2822 (1992).

71. B. Goessen and M. W. Hentze, *Mol. Cell. Biol.*, *12*, 1959–1966 (1992).

72. E. Ikonen, I. Ulmanen, and L. Peltonen, *J. Biol. Chem.*, *267*, 8715–8718 (1992).

73. R. A. Rouault, C. K. Tang, S. Kaptain, W. H. Burgess, D. J. Haile, F. Samaniego, O. W. McBride, J. B. Harford, and R. D. Klausner, *Proc. Natl. Acad. Sci. USA*, *87*, 7958–7962 (1990).

74. H. Hirling, A. Emery-Goodman, N. Thompson, B. Neupert, C. Seiser, and L. C. Kuhn, *Nucleic Acids Res.*, *20*, 33–39 (1992).

75. M. M. Patino, and W. E. Walden, *J. Biol. Chem.*, *267*, 19011–19016 (1992).

76. Y. Yu, E. Radisky, and E. A. Leibold, *J. Biol. Chem.*, *267*, 19005–19010 (1992).

77. F. Samaniego, J. Chin, K. Iwai, T. A. Rouault, and R. D. Klausner, *J. Biol. Chem.*, *269*, 30904–30910 (1994).

78. B. Guo, Y. Yu, and E. A. Leibold, *J. Biol. Chem.*, *269*, 24252–24260 (1994).

79. B. Guo, F. M. Brown, J. D. Phillips, Y. Yu, and E. A. Leibold, *J. Biol. Chem.*, *270*, 16529–16535 (1995).

80. D. J. Haile, T. A. Rouault, C. K. Tang, J. Chin, J. B. Harford, and R. D. Klausner, *Proc. Natl. Acad. Sci. USA*, *89*, 7536–7540 (1992).

81. K. Iwai, R. D. Klausner, and T. A. Rouault, *EMBO J.*, *14*, 5350–5357 (1995).

82. K. Pantopoulos, N. K. Gray, and M. W. Hentze, *RNA*, *1*, 155–163 (1995).

83. B. R. Henderson, C. Seiser, and L. C. Kuhn, *J. Biol. Chem.*, *268*, 27327–27334 (1993).

84. S. Kaptain, W. E. Downey, C. Tang, C. Philpott, D. Haile, D. G. Orloff, J. B. Harford, T. A. Rouault, and R. D. Klausner, *Proc. Natl. Acad. Sci. USA*, *88*, 10109–10113 (1991).

85. D. J. Haile, R. A. Rouault, J. B. Harford, M. C. Kennedy, G. A. Blondin, H. Beinert, and R. D. Klausner, *Proc. Natl. Acad. Sci. USA*, *89*, 11735–11739 (1992).

86. M. C. Kennedy, L. Mende-Mueller, G. A. Blondin, and H. Beinert, *Proc. Natl. Acad. Sci. USA*, *89*, 11730–11734 (1992).

87. R. L. Eisenstein, P. T. Tuazon, K. L. Schalinske, S. A. Anderson, and J. A. Traugh, *J. Biol. Chem.*, *268*, 27363–27370 (1993).

88. K. L. Schalinske and R. S. Eisenstein, *J. Biol. Chem.*, *271*, 7168–7175 (1996).

89. L. Zheng, M. C. Kennedy, G. A. Blondin, H. Beinert, and H. Zalkin, *Biochem. Biophys.*, *299*, 356–360 (1992).

90. L. S. Goessling, D. P. Mascotti, M. Bhattacharyya-Pakrasi, H. Gang, and R. E. Thach, *J. Biol. Chem.*, *269*, 4343–4348 (1994).

91. G. R. Swenson and W. E. Walden, *Nucleic Acids Res.*, *22*, 2627–2633 (1994).

92. J. P. Basilion, T. A. Rouault, C. M. Massinople, and R. D. Klausner, *Proc. Natl. Acad. Sci. USA*, *91*, 574–578 (1994).

93. H. Hirling, B. R. Henderson, and L. C. Kuhn, *EMBO J.*, *13*, 453–461 (1994).

94. C. C. Philpott, D. Haile, T. A. Rouault, and R. D. Klausner, *J. Biol. Chem.*, *268*, 17655–17658 (1993).

95. P. A. DeRusso, C. C. Philpott, K. Iwai, H. S. Mostowski, R. D. Klausner, and T. A. Rouault, *J. Biol. Chem.*, *270*, 15451–15454 (1995).

96. C. C. Philpott, R. D. Klausner, and T. A. Rouault, *Proc. Natl. Acad. Sci. USA*, *91*, 7321–7325 (1994).

97. G. Brawerman, *Cell*, *57*, 9–10 (1989).

98. V. Romheld, in *Iron Transport in Microbes and Animals* (G. Winkelman, D. van der Helm, and J. B. Neilands, eds.) VCH, Weinheim, Germany, 1986, p. 353.

99. K. Kanazawa, K. Higuchi, S. Fushiya, N.-K. Nishizawa, M. Chino, and S. Mori, in *Iron Nutrition in Soils and Plants* (J. Abadia, ed.), Kluwer, Dordrecht, The Netherlands, 1995, p. 37.

100. N. S. Von Wiren, H. Mori, H. Marschner, and V. Romheld, *Plant Physiol.*, *106*, 71–77 (1994).

101. A. Dancis, D. G. Roman, G. J. Anderson, A. Hinnebusch, and R. D. Klausner, *Proc. Natl. Acad. Sci. USA*, *89*, 3869–3873 (1992).

102. D. Eide, M. Broderius, J. Fett, and M. L. Guerinot, *Proc. Natl. Acad. Sci. USA*, *93*, 5624–5628 (1996).

103. C. Askwith, D. Eide, A. Van Ho, P. S. Bernard, L. Li, S. Daais-Kaplan, D. Sioe, and J. Kaplan, *Cell*, *76*, 403–410 (1994).

104. A. Dancis, D. G. Yuan, D. Haile, C. Askwith, D. Eide, C. Moehle, J. Kaplan, and R. D. Klausner, *Cell*, *76*, 393–402 (1994).

105. V. Braun, K. Hantke, K. Erick-Helmerich, W. Koster, U. Pressler, M. Sauer, S. Schaffer, H. Schoffler, H. Staudenmeier, and L. Zimmerman, in *Iron Transport in Microbes and Animals* (G. Winkelman, D. van der Helm, and J. B. Neilands, eds.), VCH, Weinheim, Germany, 1986, p. 36.

106. B. Van Hove, H. Staudenmeier, and V. Braun, *J. Bacteriol.*, *172*, 6749–6758 (1990).

107. K. Wreidt, A. Angerer, and V. Braun, *J. Bacteriol.*, *177*, 3320–3322 (1995).

108. M. Kammler, C. Schon, and K. Hantke, *J. Bacteriol.*, *175*, 6216–6219 (1993).

109. J. B. Neilands, K. Konopka, B. Schwyn, M. Coy, R. T. Francis, B. H. Paw, and A. Bagg in Nomoto, K., Sugiura, Y., and S.-I. Takagi, in *Iron Transport in Microbes, Plants and Animals* (G. Winkelman, D. van der Helm, and J. B. Neilands, eds.), VCH, Weinheim, Germany, 1986, p. 3.

110. S. Lobréaux, O. Massenet, and J.-F. Briat, *Plant Mol. Biol.*, *19*, 563–575 (1992).

111. S. Lobréaux, T. Hardy, and J.-F. Briat, *EMBO J.*, *12*, 651–657 (1993).

112. M. Ragland and E. C. Theil, *Plant Mol. Biol.*, *21*, 555–560 (1993).

113. E. C. Theil and T. Hase, in *Iron Chelation in Plants and Soil Microorganisms*, (L. Barton, ed.), Academic Press, San Diego, 1993, p. 133.

114. S. Lobréaux, S. J. Yewdall, J.-F. Briat, and P. M. Harrison, *Biochem. J.*, *288*, 931–939 (1992).

115. J.-F. Briat, A. M. Laboure, J. P. Laulhere, A. M. Lescure, S. Lobréaux, H. Pesey, D. Proudhon, and O. Wuytswinkel, in *Iron Nutrition in Soils and Plants*, (J. Abadia, ed.), Kluwer, Dordrecht, The Netherlands, 1995, p. 265.

116. C.-C. Chou, R. A. Gatti, M. L. Fuller, P. Concannon, A. Wong, S. Chada, R. C. Davis, and W. A. Salser, *Mol. Cell. Biol.*, *6*, 566–573 (1986).

117. F. Cox, S. J. Gestautas, and B. Rapoport, *J. Biol. Chem.*, *263*, 7060–7067 (1988).

118. L. L. Miller, S. C. Miller, S. V. Torti, Y. Tsuji, and F. M. Torti, *Proc. Natl. Acad. Sci. USA*, *88*, 4946–4950 (1991).

119. F. Raguzzi, E. Lesuisse, and R. R. Crichton, *FEBS Lett.*, *231*, 253–258 (1988).

120. M. Izuhara, K. Takamune, and R. Takata, *Mol. Gen. Genet.*, *225*, 510–513 (1991).

121. S. C. Andrews, P. M. Harrison, and J. R. Guest, *J. Bacteriol*, *171*, 3940–3947 (1989).

122. M. J. Grossman, S. M. Hinton, V. Minak-Bernero, C. Slaughter, and E. I. Stiefel, *Proc. Natl. Acad. Sci. USA*, *89*, 2419–2423 (1992).

123. E. I. Stiefel and G. D. Watt, *Nature*, *279*, 81–83 (1979).

124. D. Touati, M. Jacques, B. Tardat, L. Bouchard, and S. Despied, *J. Bacteriol.*, *177*, 2305–2314 (1995).

11

Structure-Function Relationships in the Ferritins

Pauline M. Harrison, Paul D. Hempstead, Peter J. Artymiuk, and Simon C. Andrews

The Krebs Institute and Department of Molecular Biology and Biotechnology, University of Sheffield, Sheffield, S10 2TN, UK

1. INTRODUCTION 436
2. DIVERSITY AND EVOLUTION OF FERRITINS 438
3. THREE-DIMENSIONAL STRUCTURES OF FERRITINS 444
   3.1. Vertebrate "Isoferritins": Heavy, Medium, and Light Subunits 444
   3.2. L Ferritins 445
      3.2.1. Horse Spleen Ferritin (HoSF) and Recombinant L Homopolymer (rHoLF) 445
      3.2.2. Recombinant Rat L Ferritin (rRaLF) 451
      3.2.3. Recombinant Bullfrog L Ferritin (rBfLF) 451
   3.3. H Ferritins, Heteropolymers, and the Folding and Assembly of Ferritin Subunits 452
      3.3.1. Recombinant Human H Chain Ferritin (rHuHF) 452
      3.3.2. Water Molecules in High-Resolution Structures of H and L Ferritins 453

3.3.3. HuH/LF Heteropolymers 454
3.3.4. Conserved Structural Features in Eukaryotic Ferritins 455
3.3.5. The Structure of Recombinant *Schistosoma mansoni* "Type 1" Ferritin (S.ma-1) 456
3.3.6. The Ferritin of *Escherichia coli* (EcFTN) 456
3.3.7. The Bacterioferritin of *Escherichia coli* (EcBFR) 456
3.3.8. Subunit Folding and Assembly in Mammalian Ferritins 457
4. STRUCTURE-FUNCTION RELATIONSHIPS IN FERRITINS 459
4.1. Metal-Binding Sites and the Effects of Amino Acid Substitutions 459
4.1.1. Metals at Crystal Contacts 459
4.1.2. Intrasubunit Metal Sites 459
4.1.3. Intersubunit Metal Sites in the Three-Fold Channels 462
4.2. The Four-Fold Channels 464
4.3. Iron-Storage Cavities 466
4.4. Ferritin Structure and Iron-Storage Processes 467
4.5. Structural Relationships Between Ferritins and other Four-Helix Bundle Di-iron Carboxylate Proteins 468
5. CONCLUDING REMARKS 471
ACKNOWLEDGMENTS 471
ABBREVIATIONS 471
REFERENCES 472

1. INTRODUCTION

Iron is essential for the growth and development of most organisms but is also potentially toxic and exhibits poor bioavailability. Hence, a means of storing iron is needed. The protein ferritin fulfills this need. Its molecular design, in which a multimeric, self-assembling protein shell concentrates a large amount of iron within a small, confined space, seems to be extraordinarily efficient, and this is indicated by the remarkable conservation of its protein structure. Ferritin has been identified in a wide range of species (see Sec. 2) [1–5]. Its biosynthesis is responsive to iron (see Chapters 10 and 14). All ferritins assemble from 24 protein subunits to form a hollow tetracosamer, with a 30-Å thick protein coat (Fig. 1), which can accommodate over 4000 Fe atoms as an inorganic complex.

Iron is stored within the ferritin shell in a variety of forms, the precise nature of which depends on the biological source or the experi-

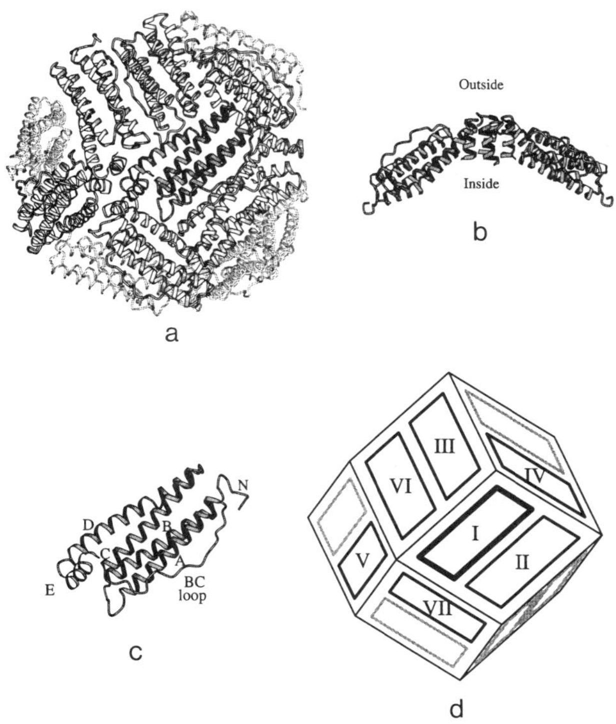

FIG. 1. Structure of rHoLF: (a) half shell showing the relative positions and interfaces between symmetry-related subunits; (b) section through protein shell at a four-fold axis; (c) detail of a single subunit; (d) schematic diagram indicating labeling used for symmetry-related subunits. Drawings were made with the MOLSCRIPT program [77].

mental conditions. In some molecules, the “iron-core” consists of a single well-ordered ferrihydrite particle (5 Fe_2O_3•9 H_2O) occupying virtually the entire available space (80 Å); in others, particularly those with a high phosphate content, it is amorphous [6,7]. The protein shell has also been used in vitro to deposit non-iron containing minerals [8]. Because the atomic structure of the mineral is not orientated with respect to that of the protein, diffraction patterns of ferritin single crystals are a combination of ordered protein diffraction and “powder” diffraction from the randomly orientated cores [9]. Hence, it has not been possible to give a complete structural description of a ferritin molecule, only of iron-free apoferritin.

Ferritins typically have molecular weights of 450 000–500 000 with 154–185 amino acids per subunit. Vertebrate ferritins contain two (H and L) or sometimes three (H, M, and L) subunits of distinct amino acid sequence. Some ferritin subunits have N-terminal extensions which lie on the outside of the assembled shell and target the ferritin to a specific destination (e.g., plastids in plants, yolk sac in snails) [3,10]. A ferritin-like protein, artemin, has C-terminal extensions (Fig. 2 [11]), which fill the cavity and prevent iron storage. Structural studies (see Sec. 3) have shown that each ferritin subunit folds as a bundle of four long helices (A and D) with a fifth short helix (E) lying across one end of the bundle (see Fig. 1) and the subunits pack tightly to form a compact shell. In vitro experiments have indicated that, to store iron, ferritin molecules must provide both a large storage cavity and a means of oxidizing Fe(II). Some or all of their subunits contain di-iron sites at the centers of the four-helix bundles which are responsible for the ferroxidase activity of the protein. The di-iron-binding site comprises a group of seven or eight amino acids, five of which are always present at these centers.

2. DIVERSITY AND EVOLUTION OF FERRITINS

Iron-storage proteins can be divided into two distantly related groups, the heme-free ferritins, found in eukaryotes and prokaryotes, and the heme-containing “bacterioferritins” (BFRs) found in eubacteria and possibly fungi [12]. The ferritins and BFRs have similar molecular architectures and both can serve as iron-sequestering proteins. Their low-sequence similarity (average identity 18.2%) indicates that they are distantly related in evolution (Fig. 2) [13,14].

The recent structural determination of the rubrerythrin (Rr) of *Desulphovibrio vulgaris* has revealed that Rrs are structurally and evolutionarily related to the ferritins and BFRs [15]. Rrs have been

---

FIG. 2. Multiple amino acid sequence alignment of ferritins, BFRs, and Rrs. The "gcg/egcg" programs (with default settings) "Pileup", "Elineup", and "Prettybox" [78] were used to produce and display the optimal alignment of 5 rubrerythrin, 12 bacterioferritin, and 39 ferritin amino acid sequences. The regions corresponding to the α-helices and metal ligands of human H-chain ferritin, *E. coli* BFR, and *D. vulgaris* rubrerythrin are shown below the alignment (1, ligand for iron site A; 2, ligand for iron site B; B, bridging ligand between sites A and B; M, coaxial heme-iron ligand; C, FeS_4 ligand; Y, conserved tyrosine implicated in the ferroxidase activity of ferritins). Residue numbering for each sequence is shown on the right and human H-chain numbering is shown above. The database accession numbers of the sequences used are below. Eukaryotic ferritins: artemin (*Artemia*) P17720; mosquito (*Aedes aegytia*), P41822; lymst_yo (*Lymnaea stagnalis*), P42578; sch_man2 (*Schistosoma mansoni*), P25320; crayfish (*Pacifastacus leniusculus*), X90566; rancal (*Rana catesbeiana*), P07797; lymst_so (*L. stagnalis*), P42577; rancam (*R. catesbeiana*), P07798; rancah (*R. catesbeiana*), P07229; xenla (*Xenopus laevis*), P17663; xen_laev (*X. laevis*), S64727; troutl (*Oncophynchus myleiss*), D86625; salmonl/2 (*Salmo salar*), P49946/P49947; chickh (*Gallus gallus*), P08267; humanh (*Homo sapiens*), P02794; crigr (*Cricetulus griseus*), P29389; rath (*Rattus norvegicus*), P19132; mouseh (*Mus musculus*), P08528; humanl (*H. sapiens*), P02792; horsel (*Equus caballus*), P02791; rabitl (*Oryctolagus cuniculus*), P09451; ratl (*R. norvegicus*), P02793; mousel (*M. musculus*), P49945; sch_manl (*S. mansoni*), P25319; tsaginat (*Taenia saginata*), X95840; pisum (*Pisum sativum*), P19975; soybean (*Glycine max*), P19976; phavu (*Phaseolus vulgaris*), P25699; brassica (*Brassica napus*), U31648; arabadop (*Arabadopsis thaliana*), X94248; maizel/2 (*Zea mays*), P29036/P29390. Prokaryotic ferritins: eco_ftnb/ecoliftn (*E. coli*), D90832/P23887; hiftnl/2 (*Haemophilus influenzae*), P43707/P43708; cjejuni (*Campylobacter jejuni*), D64082; hpylorif (*Helicobacter pylori*), P52093. Bacterioferritins: alceubfr (*Alcaligenes eutrophus*), M69036; syn_bfr2 (*Synechocystis* PCC6803), D90905; ngo_bfrb (*Neisseria gonorrhoeae*), U76634; ecobfr (*E. coli*), P11056; avibfr (*Azotobacter vinelandii*), P22759; brucella (*Brucella melitensis*), P49944; rcap_bfr (*Rhodobacter capsulatus*), 254247; mavbfr (*Mycobacterium avium*), P43314; mlebfr (*Mycobacterium leprae*), P43315; syn_bfrl (*Synechocystis* PCC6803), D90916; ngo_bfra (*N. gonorrhoeae*), U76633; putida_bfr (*Pseudomonas putida*), U66717. Rubrerythrins: cyanophora (*Cyanophora paradoxa* cyanelle), P48379; desvh_nige (*Desulphovibrio vulgaris* strain Hildenborough), U71215; ruby_met (*Methanococcus jannaschii*), U67520; ruby_des (nigerythrin from *D. vulgaris*), P24931; ruby_cpr (*Clostridium perfringens*), P51591.

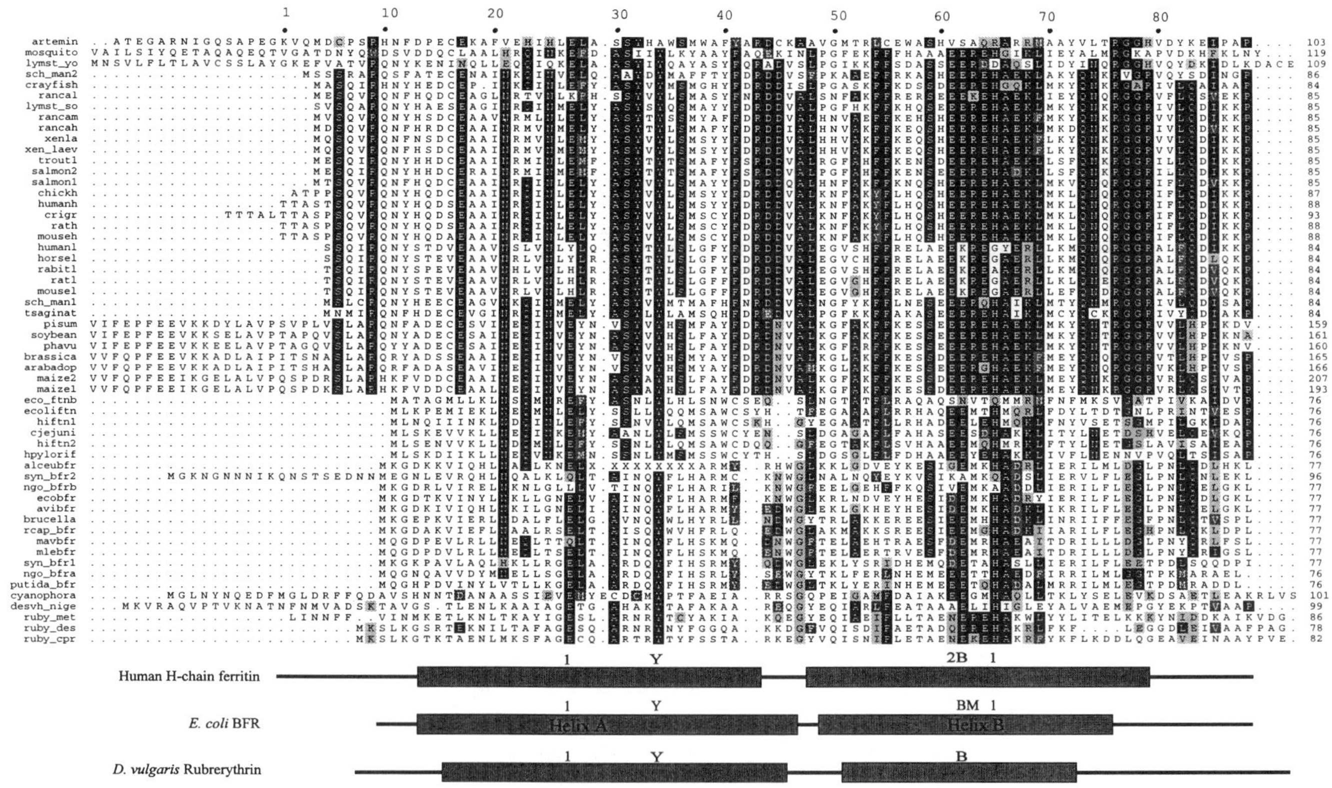
Human H-chain ferritin
E. coli BFR
D. vulgaris Rubrerythrin
Helix A
Helix B
2B
BM

(A)

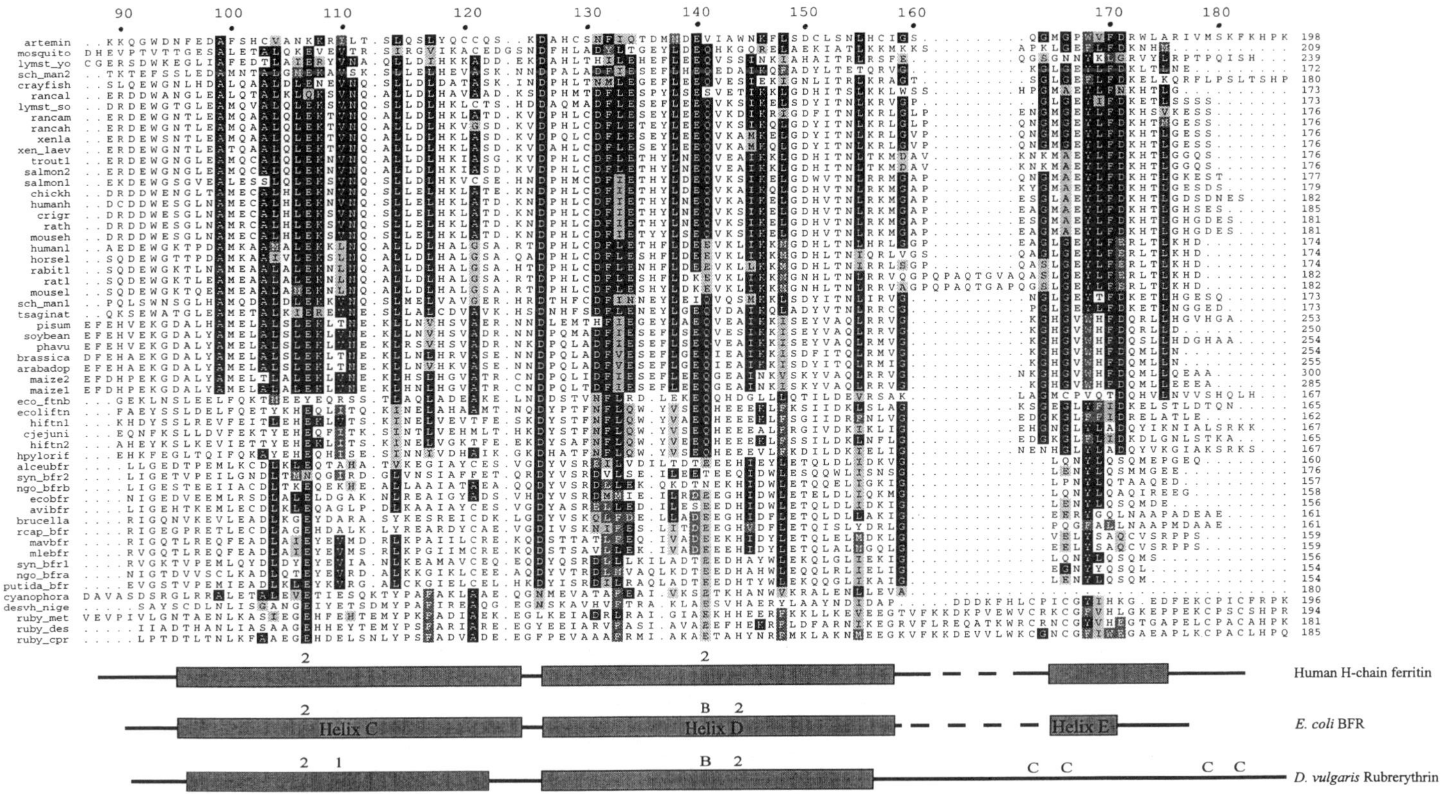
Human H-chain ferritin
E. coli BFR
D. vulgaris Rubrerythrin
Helix C
Helix D
Helix E

(B)

found only in anaerobic bacteria, and recent reports [16] indicate that they act as superoxide dismutases, scavenging oxygen radicals, and that, like ferritin, they possess high ferroxidase activity [17]. Rrs are homodimers of 22-kDa subunits and lack the iron-storage cavity of the ferritins. Each subunit comprises an N-terminal four-helix bundle closely resembling that of BFR with a similar di-iron cluster (see Figs. 9 and 12 in Sec. 4) and a C-terminal rubredoxin-like domain containing an Fe-Cys_4 cluster [15]. A multiple alignment of iron-storage proteins and rubrerythrins highlights the low degree of conservation within the group as a whole (see Fig. 2). The average amino acid sequence identities of *D. vulgaris* Rr with eukaryotic or prokaryotic ferritins or BFRs are just 13.6, 11.6, and 14.0%, respectively, whereas *D. vulgaris* Rr has an average identity of 38% with the other Rrs in Fig. 2.

Rr possesses internal homology within its four-helix bundle domain: the region corresponding to the A and B helices is similar to that of helices C and D [18], indicating that the Rrs (and ferritins and BFRs) have arisen from a primitive, two-helix protein by duplication, fusion, and divergence of their ancestral gene. The Rr-like predicted protein (here called "erythrin") from the cyanelle organelle of *Cyanophora paradoxa* also shows high similarity between helices A and B and C and D (39% identity) but lacks the C-terminal rubredoxin domain and is therefore not a true Rr. Its function is unknown, although it possesses all of the di-iron cluster ligands found in Rrs (see Figs. 2, 9, and 12) and is therefore likely to form a di-iron center. Erythrin is the most distantly related member of the Rr group and is more closely related to the ferritins and BFRs than are the other Rr family members (Fig. 3). The phylogenetic analysis (Fig. 3) provides an indication of the evolutionary relatedness of the various members of the ferritin-bacterioferritin-rubrerythrin (F-B-R) "superfamily". Most of the ferritins can be organized into three major subgroups corresponding to eubacteria, plants, and vertebrates. Given that the common ancestor of the F-B-R super-

FIG. 3. Phylogenetic tree showing the evolutionary relationship between members of the ferritin-bacterioferritin-rubrerythrin superfamily. The programs "Protdist", "Fitch", and "Drawtree" from the "Phylip" package were used to create the phylogenetic tree [79,80]. The input file for Protdist consisted of an abbreviated form of the alignment shown in Fig. 2 (the N-terminal extensions of the plants, *Artemia*, *A. aegytia*, *L. stagnalis* yolk, *Synechocystis* 2, *C. paradoxa* and nigerythrin *D. vulgaris* proteins were omitted, as were the C-terminals of the *Artemia* and rubrerythrin proteins).

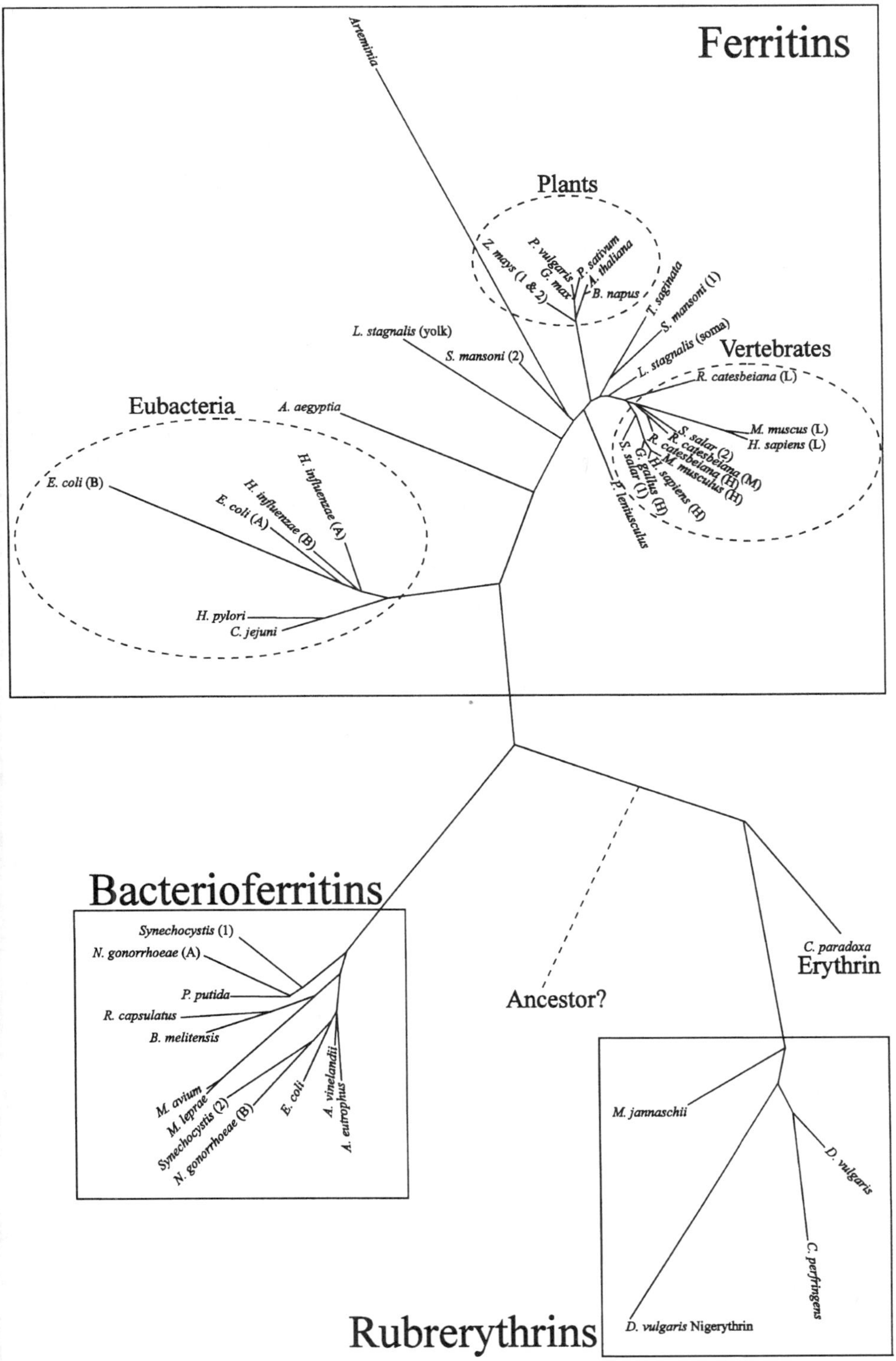

Ferritins
Artemia
Plants
Z. mays (1 & 2)
P. vulgaris
G. max
P. sativum
A. thaliana
B. napus
T. saginata
S. mansoni (1)
L. stagnalis (soma)
Vertebrates
R. catesbeiana (L)
M. muscus (L)
H. sapiens (L)
S. salar (2)
R. catesbeiana (M)
R. catesbeiana (H)
M. musculus (H)
H. sapiens (H)
G. gallus (H)
S. salar (1)
P. leniusculus
L. stagnalis (yolk)
S. mansoni (2)
A. aegyptia
Eubacteria
E. coli (B)
H. influenzae (A)
H. influenzae (B)
E. coli (A)
H. pylori
C. jejuni
Bacterioferritins
Synechocystis (1)
N. gonorrhoeae (A)
P. putida
R. capsulatus
B. melitensis
M. avium
M. leprae
Synechocystis (2)
N. gonorrhoeae (B)
E. coli
A. vinelandii
A. eutrophus
Ancestor?
C. paradoxa
Erythrin
M. jannaschii
D. vulgaris
C. perfringens
D. vulgaris Nigerythrin
Rubrerythrins

family is likely to have been a simple four-helix bundle protein displaying strong sequence identity between its N- and C-terminals, it is possible that erythrin closely resembles the original F-B-R ancestor. Hence, in the phylogenetic tree, the F-B-R ancestor branch has been placed close to the erythrin branch. Low-sequence similarity between helices A and B and C and D of the ferritins and BFRs indicates that, in contrast to the Rrs, these domains have diverged markedly.

Except for *Escherichia coli*, which has two ferritins and one BFR, members of more than one of the three groups of the F-B-R family have not been found within a single organism. Thus, these proteins could have similar, mutually exclusive functions. However, it seems to be common for organisms to have more than one ferritin, BFR or Rr. For instance, *D. vulgaris* has two Rrs; *E. coli*, *Haemophilus influenzae*, vertebrates, and many other eukaryotes possess two (or sometimes three) ferritin subunits [1,4,19]; and *Pseudomonas aeruginosa* [20], *Neisseria gonorrhoeae*, *Synechocystis*, and *Absidia spinosa* [12] contain two BFR subunits (see Figs. 2 and 3). In such cases, the subunits normally coassemble to form heteropolymers, with the only known exception being the *Lymnia stagnalis* "yolk" and "soma"-ferritin subunits, which form discrete homoploymers [21]. The *Saccharomyces cerevisiae* (http://genome-www.stanford.edu) and *Mycoplasma genitalium* (http://www.tigr.org/tdb/mdb/mgdb/mgdb.html) genomes and the nearly complete *Streptococcus pyogenes* (http://dna1.chem.uoknos.edu/strep.html) genome lack any gene encoding an apparent F-B-R family member. Either these organisms possess ferritins (or Rrs) which are not recognized because of their very low sequence similarity to other family members or they do not possess F-B-Rs.

3. THREE-DIMENSIONAL STRUCTURES OF FERRITINS

3.1. Vertebrate "Isoferritins": Heavy, Medium, and Light Subunits

All mammalian tissue ferritins are heteropolymers of various proportions of H and L subunits. Subunit compositions depend on the identity and iron status of the organism; for example, horse spleen and human liver ferritins have an average composition of about $L_{21}H_3$ with a range $L_{24}H_0$–$L_{18}H_6$ [19,22], whereas human brain or heart ferritins average about L_8H_{16} [4]. Iron loading stimulates the synthesis of both H and L subunits with a bias toward L, whereas ferritins from malignant cells tend to be H rich [4].

Unfortunately, no structure of a heteropolymer is available in which both L and H subunits can be distinguished. The ferritins of horse spleen (HoSF) [23–25] and rat liver (RaLiF) [26] both crystallize in the cubic space group F432 in which all subunits are statistically equivalent and the electron density contours can be interpreted solely in terms of the predominant L subunit. To compare H and L subunits, it has been necessary to study recombinant homopolymers. However, so far, no pair of structures of H and L ferritins from the same species has been analyzed and comparisons can be made only on the basis of known sequence and inferred structural homology. From their primary structural alignments, L chains are truncated by four residues at their N- and C-terminals with respect to H chains and to facilitate comparison H chain numbering is used here.

Amphibian red cells contain three types of chain H, L, and M, which may form heteropolymers. In the bullfrog, *Rana catesbeiana*, M and H chains show 84% and M and L chains 62% identity. The structure of the bullfrog L ferritin (BfLF) is described below (see also [27]).

3.2. L Ferritins

3.2.1. *Horse Spleen Ferritin (HoSF) and Recombinant L Homopolymer (rHoLF)*

The first three-dimensional structure of a ferritin to be described was that of horse spleen, which has also formed the prototype for many of the experiments on functionality. Crystallization from 40 mM $CdSO_4$ gave cubic crystals (space-group F432, unit cell side a =184 Å). The structure analysis was made at 6.0-, 2.8-, and 2.6-Å resolution [23,24, 28,29] and a protoporphyrin derivative at 2.6-Å resolution [25,30]. Recently, the isomorphous recombinant horse L ferritin (rHoLF) has been analyzed at 1.9-Å resolution, and the discussion below is based on this high-resolution structure. A schematic diagram of a folded L subunit and the packing of the 24 equivalent subunits in a hollow shell with 432 symmetry are shown in Fig. 1. The subunit comprises a four-helix bundle with a left-handed twist and a short fifth helix lying at about 60° to the bundle axis (helical residues: A, 13–42; B, 48–77; C, 95–125; D, 126–161; E, 162–174 and lengths A, B, and C each about 43 Å, D 52 Å, and E, 16 Å). The N-terminal (residues 5–12 on H chain numbering) lies on the outside surface of the assembled molecule and makes a number of intra- and intersubunit interactions. The subunits are arranged such that their long axes (and A, B, C, and D helices) lie within the shell, whereas the E helices lie across it with their C-terminals on the inside surface of the shell (see Fig. 1b). Helices B and C are joined by a long loop (BC) running along the length of the bundle and subunits are aligned such that the BC loops and the sides of helices A and C form the outside of the molecule. Helices B and D face the inside surface but a break in hydrogen bonding at His136 causes a kink in the long D helix enabling it to extend to the outside surface.

The four-helix bundle is stabilized by a large number of hydrophobic contacts between inward-facing side chains which form two "cores" at either end of the bundle between which lies a hydrophilic region (Figs. 4 and 5a). This region contains a buried salt-bridge, with the other eight intrasubunit bridges all lying on or near the subunit surface (Fig. 6a). The subunit conformation is further stabilized by a large number of side chain-main chain and side chain-side chain hydrogen bonds mainly linking adjacent helix pairs A and B and C and D within the bundle. The only interaction crossing the bundle diagonally is the salt-bridge joining B and D (Fig. 5a). Additional hydrogen bonds join nonhelical regions, with a notable example being between the charged side-chains of Arg9 and Glu17 which link the N-terminal to the A helix (Fig. 6a). Side chains of Glu167 and Glu171 of helix E interact with the bundle through hydrogen bonds to helices C and B, respectively.

HoLF subunits assemble into a compact symmetrical shell by association with six other subunits (see Fig. 1a) reducing their solvent-accessible surface area by 46% [31]. Each subunit forms one long contact (I:II) with an antiparallel neighbor to give a rhomb-shaped symmetrical dimer; two shorter contacts at the other side of the bundle (I:IV near the three-fold axis and I:VII near the four-fold axis) and a tenuous contact across the four-fold axis (I:V) (Fig. 1a). Contacts I:IV and I:VII (respectively involving N- and C-terminal ends of subunit I) are equivalent to

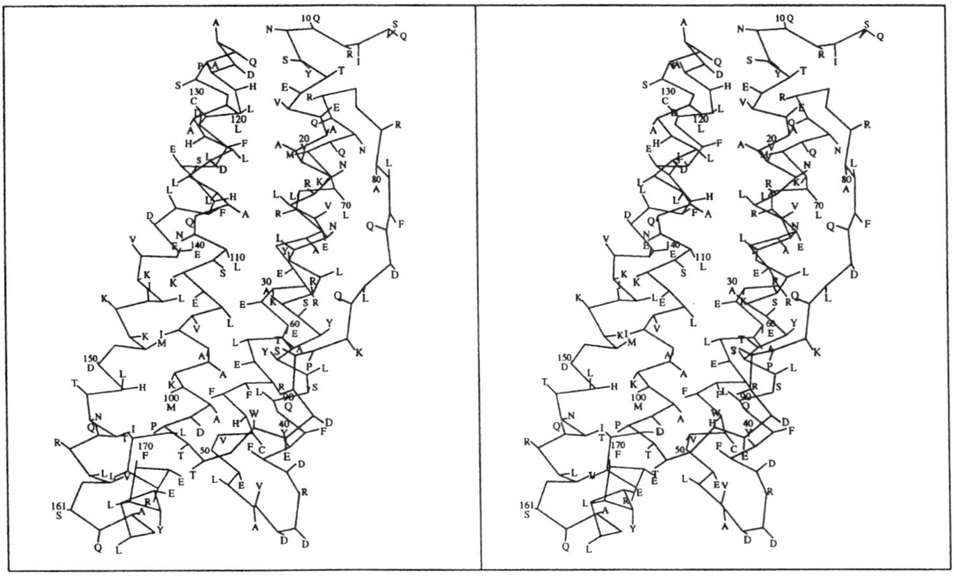

FIG. 4. Stereodiagram of side chain packing in an rHoLF subunit.

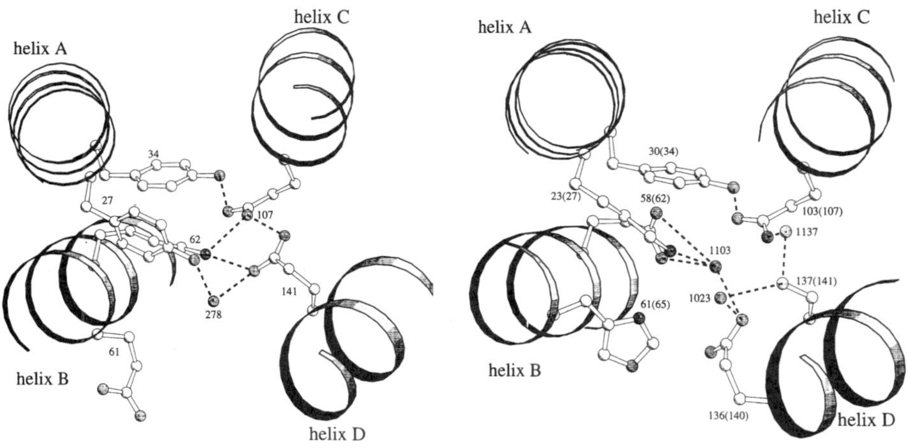

a b

FIG. 5. Central hydrophilic region within a subunit of (a) rHoLF, (b) rBfLF. Note the salt-bridge in (a) and hydrogen-bonded water molecules within both ferritins.

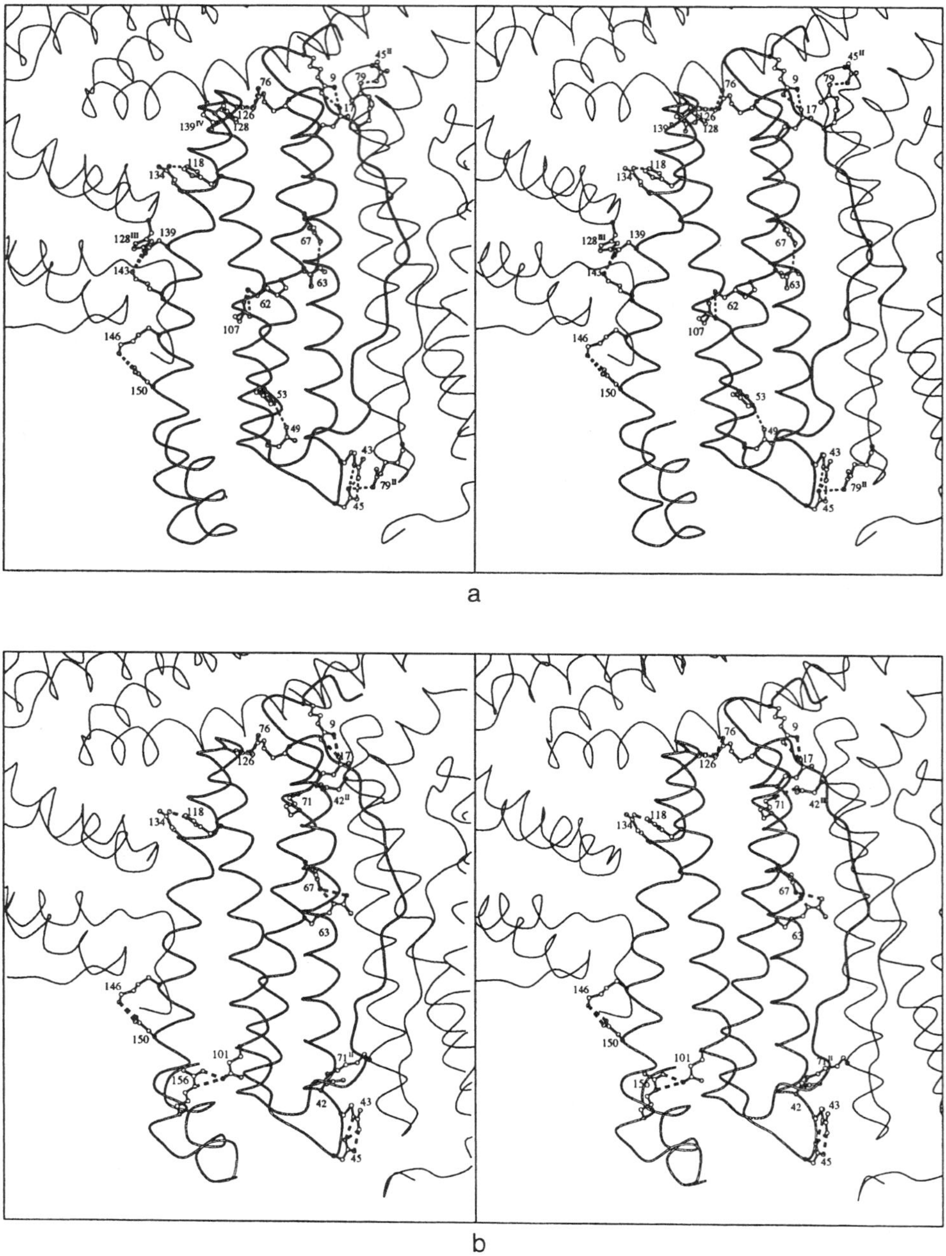

FIG. 6. Stereodiagrams showing ion pairs in (a) rHoLF and (b) rHuHF.

those between III:I and VI:I, respectively, so that interfaces I:II and I:III plus I:VI encompass virtually a complete set of contacts. The I:II association is the more intimate as judged by the 24% greater loss in solvent-accessibility of I:II compared with I:III + VI.

The long I:II interface involves numerous hydrophobic contacts produced by interdigitation of alternating side chains of BC loops and those from A helices of opposing subunits [24]. Helix pairs A(I) and A(II) and B(I) and B(II) are held further apart (about 14 Å) than are A and B or C and D pairs within the bundle (about 10 Å) yielding a partial groove on the inside surface [23,24,31]. At their extreme ends there are ion-pair interactions between Asp45 on the AB turn and Arg79 near the start of the BC loop (Fig. 6a) and in this region there are several water molecules. Toward the center of the interface, B helix side chains form symmetry-related charged networks. The side chains of glutamates 57, 60, 61, and 64 extend into the cavity in a diamond array (Fig. 7a and Sec. 4.3). Glu67 forms an ion-pair with Arg63 of the same subunit, whereas the two guanidino groups of the Arg63 side chains face each other across the two-fold axis and point into the cavity (Fig. 7a). Thus, this central part of the interface is relatively uncrowded and it is here that protoporphyrin IX is found to bind in vitro (nestling in a hydrophobic pocket) with minor side chain rearrangement [25,30] (see Fig. 7c).

The binding of porphyrin to HoSF was first observed experimentally in solution but its physiological significance is unknown [32]. Crystal structure analyses have shown that this binding is accompanied by demetallation of any Sn, Pt, or Fe present within the porphyrin ring system [25,30]. In EcBFR, heme binds in an equivalent position with specific ligation of its iron (see Fig. 7d; Sec. 3.3.7). The second long interface stretching between the three-fold and four-fold axes involves both hydrophilic and hydrophobic interactions. Not far from the three-fold axis is an intersubunit salt-bridge (Asp139[I]–His128[III]) linking the N-end of D(III) to a position near the kink in helix D(I) (see Fig. 6a). Further along the interface N-terminal residues 6–12 of subunit III make several interactions with the C helix of subunit I, including several through water molecules. Still further along the interface the C-end of A and the AB turn of subunit VI are linked to helix D of subunit I and again there is a chain of water molecules between subunits. Contacts between E helices around the four-fold axis are mainly hydrophobic (see Sec. 4.2).

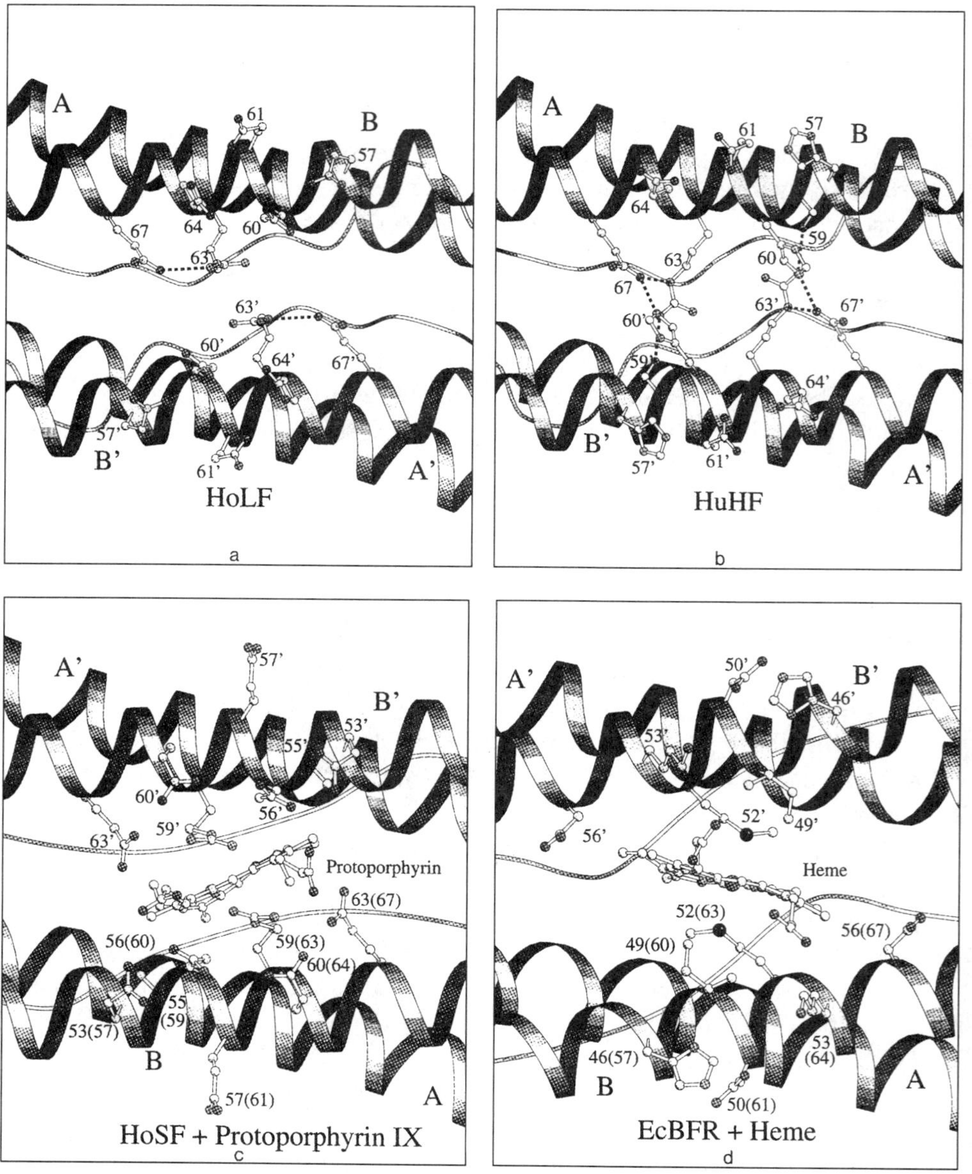

FIG. 7. Central part of I:II intersubunit interfaces of (a) rHoLF; (b) rHuHF; (c) heme derivative of HoSF; (d) EcBFR viewed from the inner surface of the shell along the two-fold axis.

3.2.2. *Recombinant Rat L Ferritin (rRaLF)*

RaLF shares 87% amino acid sequence identity with HoLF and rRalf has a highly similar three-dimensional structure (determined at 2.3-Å resolution [26,33], see Table 1 and Fig. 11a in Sec. 4.2). The hydrophilic region in the center of the subunit has His27 in place of Tyr27 (in HoLF), but it includes the salt-bridge shown in Figs. 5a and 6a for rHoLF. The "diamond" of glutamates 57, 60, 61, and 64 is found as in rHoLF. The DE turn contains an insertion of eight residues, but these must adopt alternative conformations, since no density for them is seen and there is a small shift in the position of E helices with respect to those in rHoLF and rHuHF (Fig. 11a; Sec. 4.2).

3.2.3. *Recombinant Bullfrog L Ferritin (rBfLF)*

BfLF exhibits 52% amino acid sequence identity with HoLF (61% with HuHF). Its 2.2-Å resolution three-dimensional structure [27,34] is extremely similar to that of rHoLF (see Table 1). The D helix is kinked as

TABLE 1

Root-Mean-Square Deviations (Å) Between Cα Following Superposition of Four-Helix Bundles in FBR Proteins (and the Number of Atoms in the Comparison)

	BfLF	HuHF	S.ma-1	EcFTN	EcBFR	Rr	HoSF
HoLF	0.4(148)	0.4(145)	0.6(145)	1.0(139)	1.7(133)	1.6(109)	0.2(148)
BfLF		0.4(145)	0.6(145)	1.0(139)	1.6(134)	1.6(112)	0.4(146)
HuHF			0.5(145)	1.0(139)	1.6(130)	1.8(117)	0.4(145)
S.ma-1				1.0(139)	1.8(135)	1.7(105)	0.6(145)
EcFTN					1.6(128)	1.4(93)	0.9(138)
EcBFR						1.4(123)	1.6(130)
Rr							1.6(112)

Coordinates used from the following PDB files:
EcBFR: 1bcf; BfLF: 1rcd; HoSF: 1ier; HoLF: 1aew; HuHF: 2fha.
Rr coordinates by courtesy of F. deMaré.
EcFTN and S.ma-1 coordinates by courtesy of P. D. Hempstead and S. J. Yewdall.

in rHoLF, but hydrogen bonding is interrupted by Pro136 in place of histidine. Of its eight intrasubunit ion-pairs stabilizing the structure, five are common to rHoLF as is one of its two intersubunit salt-bridges. However, the central hydrophilic region of the subunit is very different with only Tyr34 and Glu61 in common (see Fig. 5b) and in the I:II interface the Arg63 of HoLF and HuHF (see Fig. 7a and b) is substituted by Lys63 and there is no salt-bridge linking Lys63(I) and Glu67. It is unknown whether protoporphyrin binds but there are two putative betaine molecules on the inside of the shell beside the two-fold axis. On the cavity side of the B helices there is a cluster of glutamates resembling those found in rHoLF (see Fig. 7a; Sec. 4.3).

3.3. H Ferritins, Heteropolymers, and the Folding and Assembly of Ferritin Subunits

3.3.1. Recombinant Human H Chain Ferritin (rHuHF)

It is clear that L chain ferritins of diverse origins have very similar secondary, tertiary, and quaternary structures. The structure of HuHF must resemble that of human L chain ferritin (HuLF), since they form copolymers (sequence identity is 56%). Crystals of rHuLF were not of sufficient quality for x-ray analysis but structural similarities among rHuHF, rHoLF, rRaLF, and rBfLF are shown in Table 1 and Fig. 11a (in Sec. 4.2). After engineering crystal contacts in rHuHF (and crystallization from $CaCl_2$) structure determination using HoSF as a starting model was straightforward [33]. Helices A (residues 13–42), B (48–77), and C (95–125) are of the same length in rHuHF and rHoLF but D and E are slightly shorter in rHuHF (D126–159 and E163–174 instead of D126–161 and E162–174), and there is a DE turn of three residues (160–162) in place of a sharp bend. Residue 161 is a *cis*-proline in rHuHF in place of *trans*-serine in rHoLF and rBfLF.

Side chain distributions follow similar patterns in rHuHF as in rHoLF. Within the subunit, apolar residues are concentrated at the two ends of the subunit around a central hydrophilic region, but conservation of inward facing residues in the two structures is not high (40%). Residues at the center of the bundle are shown in Fig. 9a in Sec. 3.3.6. Only three of the seven HoLF residues shown in Fig. 5a are present in HuHF (Tyr34, Glu61, Glu107). However, the conserved Tyr34 and

Glu107 are in virtually identical positions in HuHF and HoLF as are Gln141 and Glu141, respectively. The seven residues of HuHF (Glu27, Tyr34, Glu61, Glu62, His65, Glu107, and Gln141) shown in Fig. 9a are associated with the binding and rapid oxidation of two Fe(II) atoms ("ferroxidase" activity) as discussed in Sec. 4. Because of their conservation in all animal H chains, in most plant ferritins and in the nonheme ferritins of bacteria (see Fig. 2), these seven residues are considered as a signature of non-heme ferritin H chains. Five of them (Glu27, Tyr34, Glu62, His65, Glu107) are present in the heme-containing BFRs (Secs. 3.3.7, 4.1.2, and 4.5). Four of the seven residues (Tyr34, Glu61, Glu62, and His65) are present in BfLF, but these alone do not confer ferroxidase activity [27,34]. The central intrasubunit salt-bridge found in HoLF (see Fig. 5a) and RaLF is lacking in HuHF, but six other ion pairs are conserved (see Fig. 6), three of which are involved in linking helices at or near the ends of the bundle (Arg76–Asp126; Arg156–Glu101 and His118–Glu134) and one links the N-terminal to the start of helix A (Arg9–Glu17). The salt-bridge in the I:III interface of HoLF, Asp139(I)–His128(III) is replaced by hydrogen binding between Asn139(I) and His128(III).

Hydrogen-bond interactions in the intersubunit interfaces of rHuHF and rHoLF are relatively well conserved (71%), especially those in interface I:III (88%). However, there are significant differences in interface I:II (see Fig. 7a and b). Arg63(I) faces across the interface and forms hydrogen bonds to Ser59(II) (Ala in HoLF) and Glu67(I) bonds to His60(II). These residues block the groove between B(I) and B(II) and may prevent binding of protoporphyrin IX to HuHF [30]. The ion pair Arg79(I)–Asp45(II) at the end of the I:II interface of rHoLF is replaced by Lys71(I)–Asp42(II) (Fig. 6a and b). Again, comparing HuHF and BfLF, hydrogen bonding interactions across the I:III plus I:VI interface are better conserved (75%) than either those across I:II (40%) or than the primary structures as a whole (61%).

3.3.2. *Water Molecules in High-Resolution Structures of H and L Ferritins*

Several nonprotein electron-density peaks have been refined as water molecules in the three high resolution structures: (76) in HoLF, (39) in BfLF, and (59) in HuHF. In addition to single waters attached to hydrophilic groups on the outside or inside surfaces, clusters of water molecules are found at the ends of the I:II interface and along the I:III + VI interface (especially I:III near the three-fold axis). Such clusters are present in all three ferritins where they fill gaps between subunits. These waters could provide lubrication and allow the possibility of "breathing" motions in a shell structure which would otherwise be rather rigid. It is notable that several of the waters are found within ±1.5 Å of their positions in the other ferritins: thus, per subunit, 41 waters of HoLF [31] are common to HuHF and of 39 waters located in BfLF 19 are at equivalent positions in all three structures (Fig. 8). Most of the common waters have conserved side chain ligands and/or are attached to equivalent backbone NH or CO. Five of the waters common to all three ferritins lie near the three-fold axes. Trikha et al. [27] drew attention to a chain of waters in BfLF between the three-fold channel and the cluster of glutamates (57, 60, 61, 62, 64) on the B helix. In HoLF, 30 waters have been located on the inside surface on helices B and D.

3.3.3. HuH/LF Heteropolymers

HuLF and HuHF must have a common structural framework, and the amino acid sequence indicates that the central hydrophilic region of the HuLF subunit is like that of HoLF (see Fig. 5a), including the

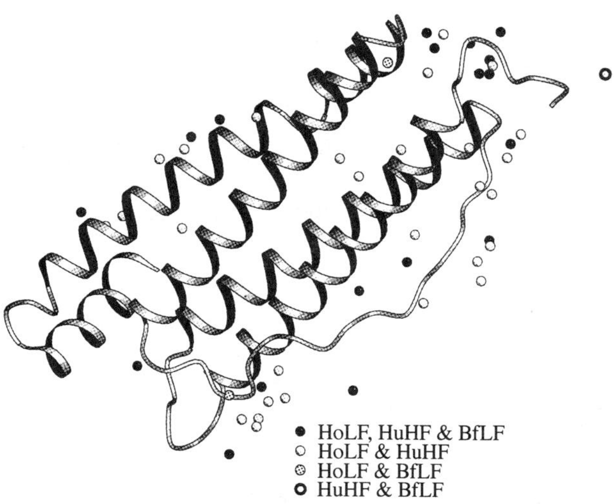

FIG. 8. Matched water molecules in high-resolution structures of rHoLF, rHuHF, and rBfLF lying within 3.6 Å of the protein.

presence of a salt-bridge between Lys62 and Glu107. In contrast, the DE turn may be like that of HuHF. Residues 160–162 are Gly-Pro-Glu in HuLF and Ala-Pro-Glu in HuHF, but in HoLF, the sequence is Gly-Ser-Gln. A high degree of conservation of residues making hydrogen bonds at interfaces (86%) should enable most of the intersubunit interactions found in HuHF to be made in the copolymer. However, within the central part of interface I:II, residues 59 and 60 are Ala and Glu (as in HoLF), respectively, not Ser and His, so the Ser59–Arg63 and His60–Glu67 interactions of HuHF (see Fig. 7b) cannot occur.

It is not clear whether differences in the I:II interface residues of unlike subunits would lead to segregation of like subunits during assembly in vitro, although this is not indicated experimentally [35]. It is unknown whether heme can bind at mixed I:II interfaces.

3.3.4. *Conserved Structural Features in Eukaryotic Ferritins*

There are certain sequence positions (see Fig. 2) which are always or almost always occupied by hydrophobic residues and others by hydrophilic residues. The hydrophobic positions usually correspond to residues that are buried within the subunit (as in Fig. 4). Exceptions include two positions (82 and 85) on the BC loop in HuHF and HoLF where the side chains are buried by making important hydrophobic interactions on the I:II interface and positions 165, 168, and 169 where the residues are involved in interactions around the four-fold axis. Pro127, which is present in most eukaryotic ferritins, plays a part in the tight turn between C and D helices by preventing the extension of helices C and D in their C- and N-terminal directions, respectively. Sequence positions usually occupied by hydrophilic residues nearly all lie on inside or outside surfaces of the shell or in intersubunit interfaces where they frequently make hydrogen bonds across the interface directly or through water molecules.

Owing to conservation of participating residues several of the ion pairs found in HuHF and/or HoLF are potentially conserved in many other eukaryote ferritins. These include Arg9–Glu17, Asp42(I)–Lys71(II), Arg43–Asp45, Asp45(I)–Arg79(II), Arg76–Asp126, and His118–Glu134. However, the two interactions involving Asp45 must be absent from plant ferritins, which have Asn45. None of the ion pairs are found in a prokaryote ferritin, but virtually all ferritins have the potential to form the intrasubunit hydrogen bond between Tyr34 and Glu107 that is shown in Fig. 5a.

3.3.5. *The Structure of Recombinant* Schistosoma mansoni *"Type 1" Ferritin (S.ma-1)*

This trematode ferritin is the only example of an invertebrate ferritin structure. Type 1 ferritin is prevalent in females, type 2 in males [36]. The primary structure of type 1 is 56% identical to that of HuHF and that of type 2 is 51% identical (see Fig. 2). Both have "ferroxidase center" residues and ferroxidase activity has been shown in S.ma-1. The engineering of crystal contacts (by the substitution of Ser86 → Gln) enabled crystallization of S.ma-1 ferritin from $CaCl_2$ in a cubic unit cell isomorphous with that of HuHF. The three-dimensional structure of S.ma-1 ferritin closely resembles that of HuHF (see Table 1), although differences are found particularly near the three-fold axis (Fig. 10), the four-fold axis and in the DE turn.

3.3.6. *The Ferritin of* Escherichia coli *(EcFTN)*

The sequence of EcFTN (see Fig. 2) shows similar patterns of hydrophobic residues to those in Fig. 4. Its structure has been determined [37] and the similarity of its subunit fold to those of eukaryote ferritins confirmed (see Table 1). Like rHuHF, the EcFTN subunit contains a central hydrophilic region with the same seven residues as found in HuHF plus an additional residue, Glu144 (Ala in HuHF). This region binds metals and is associated with ferroxidase activity (Secs. 4.1.2 and 4.5 and Fig. 9b). The subunit arrangement in EcFTN closely resembles that in rHoLF (see Fig. 1), although few of the intersubunit interactions are conserved. A recently reported second *E. coli* ferritin (see legend of Fig. 2), EcFTNb has only three of the seven residues conserved (Glu27, Tyr34, and Gln141) (see Fig. 2) and would not be expected to catalyze Fe(II) oxidation, although a similar subunit fold would be expected from its sequence alignment. Neither its function nor whether it can form copolymers with EcFTN is known. It lacks the "nucleation center" glutamates of HoLF (see Fig. 7a and Sec. 4.3).

3.3.7. *The Bacterioferritin of* Escherichia coli *(EcBFR)*

Unlike the previously described ferritins, EcBFR contains heme as isolated and belongs to the group of BFRs shown in Figs. 2 and 3. X-ray analysis [38] shows that its subunit conformation and shell packing resemble those of other ferritins (see Table 1), although there are interesting differences (see Figs. 7 and Figs. 9–12 in Sec. 4). Again the EcBFR subunit has a hydrophilic region at the center of its four-helix bundle encompassed by two hydrophobic regions. In common with HuHF and EcFTN, it contains Glu27, Tyr34, Glu62, His65, and Glu107, which along with Glu141 and His144, form the ligands of a dinuclear metal site associated with ferroxidase activity (see Figs. 9c and 12). Asp61, which is replaced by Glu, Ala, Gly, Lys, or Val in five of the BFR sequences of Fig. 2, seems not to be of functional importance. The remaining BFR sequences listed in Fig. 2 (except for *Synechocystis* BFR1 and *Neisseria gonorrhoeae* BFRa) have replacements likely to cause loss of ferroxidase activity.

As in the mammalian ferritins, a long mainly hydrophobic I:II intersubunit contact is formed by interdigitation of the A helix and BC loop side chains around molecular dyads and B helices are kept apart to provide space for heme binding. Heme iron atoms have the unique coordination by two symmetry-related Met63 residues (see Fig. 7d), and the protoporphyrin ring atoms are in contact with mainly apolar side chains from A and B helices and the BC loop [30,38]. Met63 is present in only 9 of the 12 BFR sequences shown in Fig. 2, being replaced by Thr in *Synechocystis* BFR1, *N. gonorrhoeae* BFRa and *Pseudomonas putida* BFR. Several of the heme contacts are also replaced, so these proteins may not be hemo-ferritins. However, if heteropolymers can form from the two BFR subunits in these organisms, both ferroxidase and heme functions could be combined in single BFR molecules.

3.3.8. *Subunit Folding and Assembly in Mammalian Ferritins*

Two approaches have been used to investigate subunit folding and assembly. The first has involved the finding of conditions causing dissociation and unfolding of native ferritins (usually HoSF) and the characterization of intermediates of renaturation. In the second approach, the effects on folding and assembly of introducing changes in specific regions of structure by site-directed mutagenesis have been studied in vitro and in vivo. These studies have been carried out on the iron-free apoferritin, since the iron-core is not a required nucleus for assembly [39].

Ferritin is a remarkably stable molecule. Iron-free HoSF dissociates only at pH 2 or at a guanidine HCl concentration of 7 M (pH 3.5 or less) [39–43]. Dissociation is accompanied by loss of secondary structure as judged from spectroscopic changes (CD, ultraviolet (UV)-difference, fluorescence). Removal of denaturants leads to refolding and rapid oligomerization. Association of two subunits at the long interface around two-fold axes was predicted to be the first assembly step [24], and evidence for early dimerization has been found by sedimentation-velocity measurements [43] and by gel-electrophoretic analysis of chemical cross-linked intermediates [42]. By use of this latter approach and by investigating reassociation of chromatographically separated cross-linked oligomers, an assembly pathway was predicted:

$$24\ M_1 \longrightarrow 24\ \mathrm{M}_1 = 8\ \mathrm{M}_2 + 8\ \mathrm{M}_1 = 8\ \mathrm{M}_3 = (4\ \mathrm{M}_6) = 2\ \mathrm{M}_{12} = \mathrm{M}_{24}$$

from unfolded monomers M_1 through structured monomers M_1, dimers, trimers, transient hexamers, and dodecamers. Subunit folding and dimerization were fast relative to formation of higher oligomers. An alternative pathway through tetramers and octamers was preferred by Stefanini et al. [43], who also concluded that the assembly process was highly cooperative.

Cesareni and colleagues have investigated the influence on subunit folding and self-assembly of site-directed changes in the E helix and DE turn of HuHF [44,45]. Introduction of an amber mutation at Gly159 or Pro161 did not prevent assembly to 24-mers in the overexpressing *E. coli* cells, but the isolated ferritin had a melting temperature of ~70°C compared with 88°C for the wild type. Therefore, interactions of the E helix with the four-helix bundle or E helices of other subunits are not essential to regain tertiary and quaternary structures but do contribute to thermostability. Several amino acid substitutions in the E helix could also be tolerated, although substitution of Leu169 at the center of the four-fold interface (see Fig. 11 in Sec. 4.2) by arginine caused a decreased stability of the 24-mer and substitution of positively charged residues at positions 159, 160, or 161 led to folding defects and aggregation. Fusion of the α-peptide of β-galactosidase at the C-terminal did not prevent assembly to complete shells provided there were only two to three hybrid subunits per 24-mer, but some E helix substitutions favored a conformation in which the α-peptide and E helix were "flopped" outside the shell. In other experiments [35], the effects of deletion of residues 1–13 and substitutions Ile85 → Cys (near the two-fold axis), Asp131 → Ile + Glu134 → Phe (near the three-fold axis), and Leu169 → Arg (near the four-fold axis) were examined. None of these substitutions prevented reassembly after denaturation in vitro provided the protein solution was sufficiently concentrated and sufficient time was allowed. In contrast, introduction of a negative charge near the two-fold axis by reaction of rHuHF-I85C with iodoacetic acid led to unstable dimers and failure to assemble into shells. Examination of assembling proteins by gel electrophoresis showed ladders of increasing oligomer size and suggested a stepwise assembly process in which dimers coalesce to give an oligomer nucleus onto which further dimers are added. This could be consistent with assembly dimers around four-fold axes, since coalescence of three symmetrical octamers is topologically impossible [24]. However, it was not certain whether the sub-assembly oligomers were productive intermediates. It is also possible that different pathways are preferred under different experimental conditions. What is clear is that ferritin can accommodate a wide variety of changes at the interfaces and still assemble successfully.

4. STRUCTURE-FUNCTION RELATIONSHIPS IN FERRITINS

4.1. Metal-Binding Sites and the Effects of Amino Acid Substitutions

4.1.1. Metals at Crystal Contacts

Cd^{2+} bridges form intermolecular contacts in both cubic [23,46] and tetragonal [47] HoSF crystals. Although Cd^{2+} is replaceable by other metals (e.g., Ca^{2+} in HuHF [33]), these surface metal sites may be of no physiological significance. Metal bridges feature in the crystals neither of EcBFR [25,38] nor of EcFTN [37].

4.1.2. Intrasubunit Metal Sites

The region in the middle of the four-helix bundle has been found to bind metals in three ferritins (Fig. 9): (1) a single peak of high electron density attributed to Ca^{2+} of crystallization was observed in rHuHF

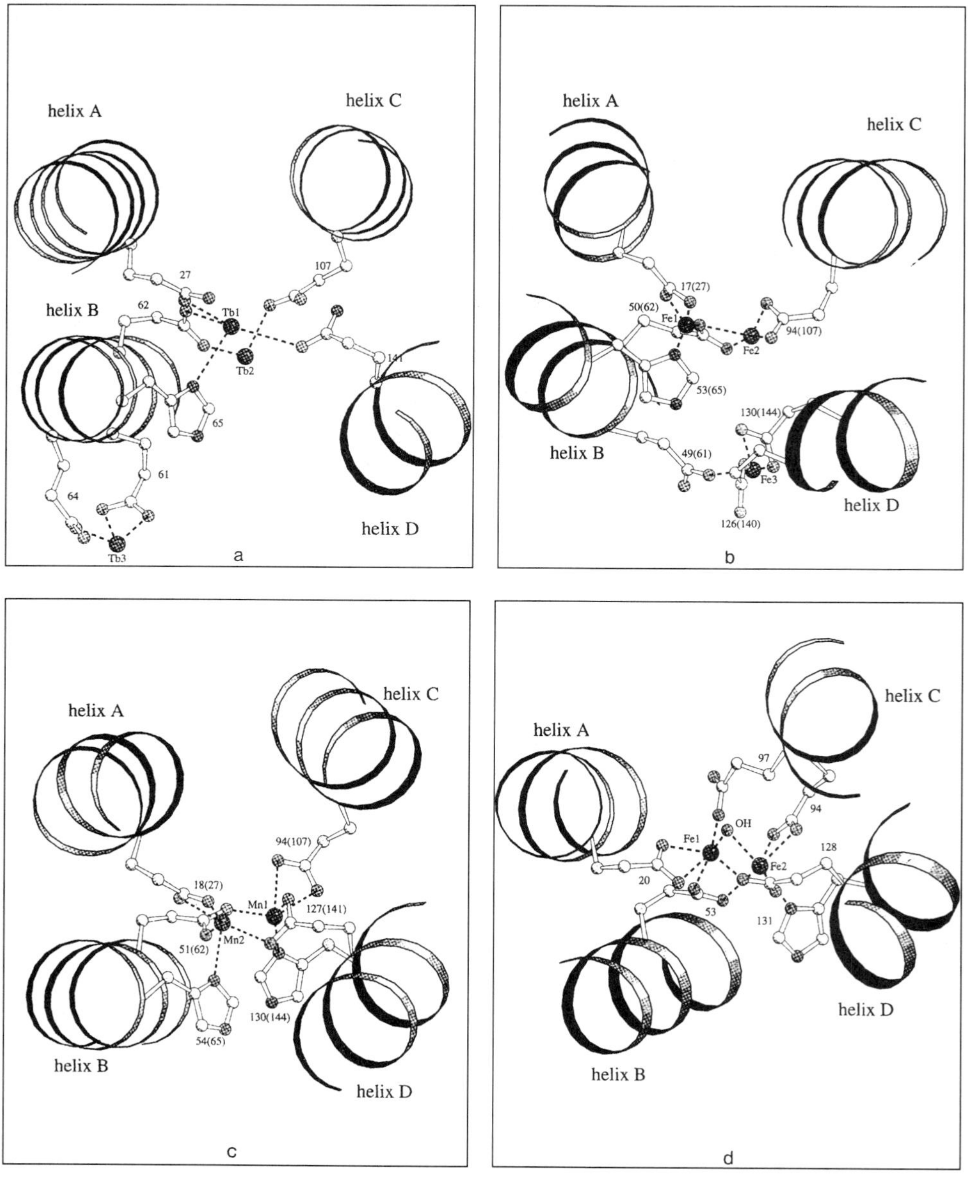

FIG. 9. Intrasubunit metal sites in H-type ferritins and Rr. (a) Tb^{3+} in HuHF; (b) Fe^{n+} in EcFTN; (c) Mn^{2+} in EcBFR; (d) Fe^{3+} in Rr (coordinates kindly provided by F. deMaré).

[33], whereas crystals soaked in $TbCl_3$ showed the presence within the bundle of a pair of Tb^{3+} ions about 3.1 Å apart (probably alternative occupancy) and a third metal site on the inside surface (Fig. 9a); (2) structural analysis of EcBFR at 2.9-Å resolution using crystals grown from solutions containing $MnCl_2$ [38] showed two peaks approximately 4 Å apart and presumably Mn^{2+} ions (Fig. 9c); (3) finally, metal-soaking experiments with crystals of EcFTN grown in the absence of metal ions gave a pair of Fe atoms (3.8-Å separation; Fig. 9b) or of Tb^{3+}, Mn^{2+} or Zn^{2+} in the proposed ferroxidase center [37].

The use of site-directed mutagenesis has indicated that the metal center is responsible for the observed catalysis of Fe(II) oxidation (ferroxidase activity) in HuHF, EcBFR, and EcFTN and that for this catalysis a pair of Fe^{2+} atoms must be present [48–53]. Comparisons of the metal-binding sites in the three ferritins (see Fig. 9) elicit several points of interest. First, the dinuclear sites are structurally similar (α-carbon atoms of Glu27, Tyr34, Glu/Asp61, Glu62, His65, Glu107, Gln/Glu141, Ala/His/Glu144 overlap with the following rms deviations: EcBFR/HuHF, ± 1.0 Å; EcFTN/HuHF, ± 0.6 Å; EcBFR/EcFTN, ± 1.3 Å). Second, site A ligands (Glu27, His65, Glu62) and two of site B ligands (Glu62 and Glu107) are the same in the three ferritins (see Figs. 9 and 12). However, the changes in the third ligand of site B (Glu61 in HuHF, His144 in EcBFR or Glu144 in EcFTN) and residue 141 (Gln in HuHF and EcFTN but Glu in EcBFR) may mean that Fe takes up slightly different positions in the three ferritins, as indeed was found for the three metal derivatives shown in Fig. 9a, b and c.

A third metal site, C, is also seen in HuHF [33] and EcFTN [37] at 7–8 Å distance from the dinuclear center (see Fig. 9). These third sites have different coordination and are at different positions in rHuHF and EcFTN, although they both lie on the inner surface. One of the ligands of Fe at site C (Glu61 in HuHF, Glu144 in EcFTN) can occupy two positions, acting as a ligand of site B in its alternative orientation. It has been suggested that movement of Glu61 or Glu144 may assist the removal of Fe^{3+} from B to C [33,37]. Nevertheless, substitution of Glu61 or Glu144 by Ala is found to have a relatively small effect on the rate of Fe(II) oxidation compared with substitution of E107 or E27 by the same residue [48–53]. So far, no third metal site has been identified in EcBFR, although EPR evidence for such a site has been obtained [54]. A purple color produced during Fe(II) oxidation in BfLF and HuHF was first thought to be due to a tyrosine-Fe(III) charge-transfer complex [52,55]. It has now been suggested [49] that the transient species in HuHF is a diferric-peroxo complex as found in the methane monooxygenase hydroxylase component MMOH [56]. The Fe-tyrosinate may therefore be specific for BfLF.

It is important for the interpretation of the effects of genetically engineered amino acid substitutions that the structure remains unaltered except at the position of substitution. This was shown, for example, when the two substitutions, Glu62 → Gly, His65 → Lys converted the dinuclear center of rHuHF into a HoLF/HuLF-type center complete with engineered Lys62–Glu107 salt-bridge [57]. The low ferroxidase activity must have been due to lack of metal ligands at the center not to a general structural perturbation. The importance of the salt-bridge for stability was shown by increased resistance to denaturation by acid or guanidine HCl [35]. Since the ferroxidase center of H chains is a relatively open region, accessible to water molecules [31], amino acid substitutions within this region may have less structural impact than those in more tightly packed parts of the structure.

4.1.3. Intersubunit Metal Sites in the Three-Fold Channels

At the three-fold axis, the symmetry-related subunits of rHoLF, rBfLF, or rHuHF make a funnel leading from the outside surface (wide part) to the inside surface (narrow part), with the latter containing three Asp131 and three Glu134 to which metal ions can bind (Fig. 10a). In rHuHF crystals, Ca^{2+} is bound by all six carboxy groups in the channel [31,33]. In HoSF, Cd^{2+} and Tb^{3+} have also been found in varous positions on or near the three-fold axis [23,24,31,46] as have Pt and probably Sn, resulting from demetallation of bound protoporphyrin IX complexes [30]. Coordination of metals by His118, Cys130, and Ser135 has also been found [31]. Although BfLF was crystallized from $MnCl_2$, no binding of Mn^{2+} in the three-fold channels was reported even though Asp131 and Glu134 are present [27].

Conservation of Asp131 and Glu134 and also His118 in mammalian and amphibian H and L ferritins ensures that these metal ligands are preserved in the three-fold channels of heteropolymers. All three are conserved in most plant ferritins (except pea which has His131) but not in some animal ferritins; for example, crayfish has Asp118 and Asn131,

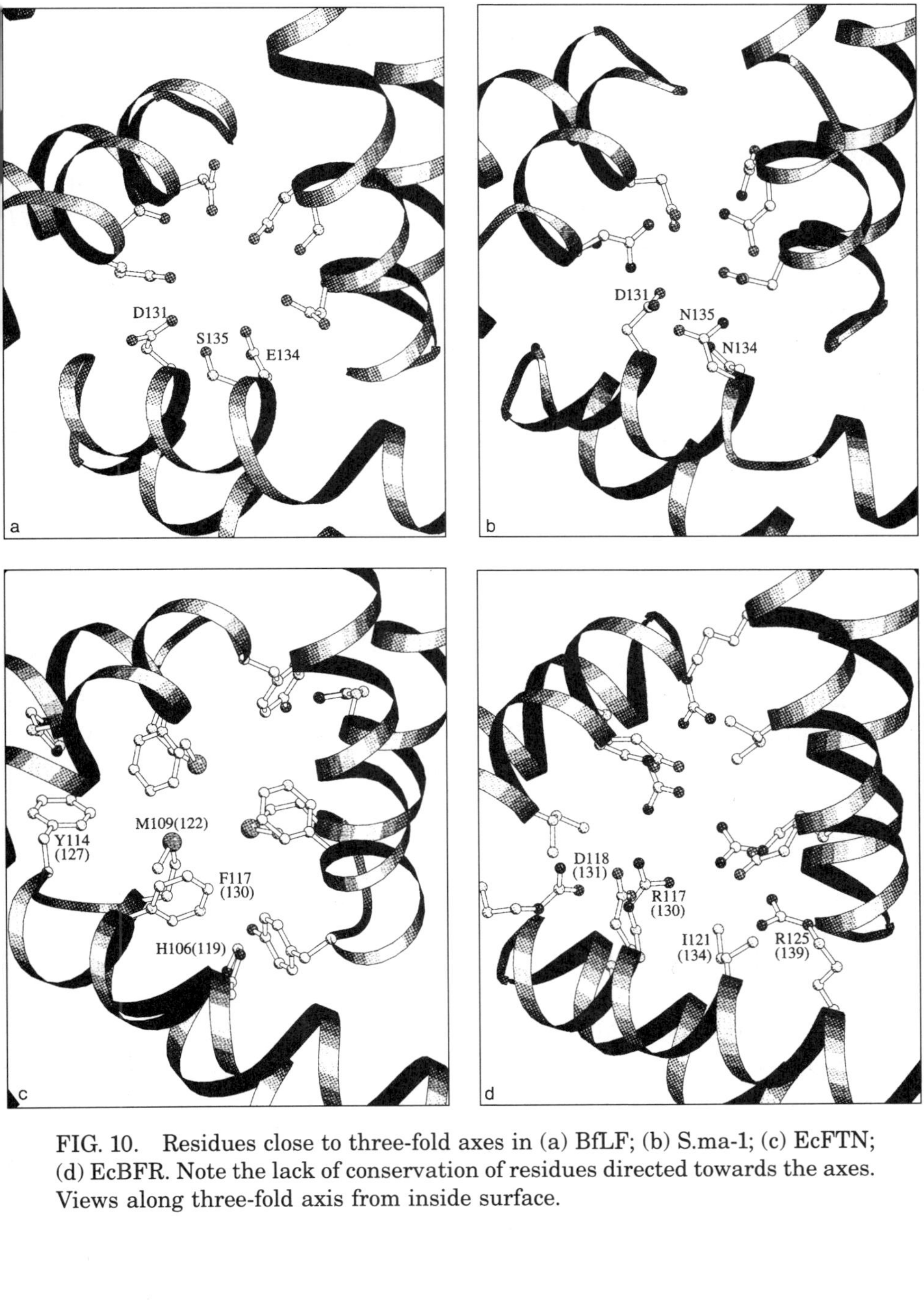

FIG. 10. Residues close to three-fold axes in (a) BfLF; (b) S.ma-1; (c) EcFTN; (d) EcBFR. Note the lack of conservation of residues directed towards the axes. Views along three-fold axis from inside surface.

fresh water snail (*L. stagnalis*) yolk ferritin has Thr131, mosquito has Ile118, and Thr134 and S.ma-1 ferritin has Val118, Asn134, and Asn135 (see Fig. 10b) with no metal binding in these channels. Both heme and nonheme bacterial ferritins have very different three-fold channels. Thus, in EcFTN, the hydrophobic residues Met122 and Phe130 protrude into the channel and no metals are found (see Figs. 2 and 10c). BFRs have a variety of polar or nonpolar residues at positions 131 and 134 (e.g., EcBFR has Asp131 and Ile134; see Figs. 2 and 10d), although one of these two residues is always either Asp or Glu (see Fig. 2). A striking feature of the three-fold channels of EcBFR (see Fig. 10d) is the presence of three Arg130 side chains near the three-fold axis, and these arginines are conserved in several other BFRs (see Fig. 2). Unsurprisingly, no Mn^{2+} of crystallization was reported in the three-fold channels of EcBFR [38]. The function of the three-fold channels in the iron-storage process is discussed in Sec. 4.4.

4.2. The Four-Fold Channels

The confluence of four subunits around the four-fold axes in rHoLF, rHuHF, and rBfLF produces a left-handed bundle of four parallel E helices with a small channel accessible to a 2-Å diameter probe [24,27, 31] (Fig. 11b). The channels are lined by three side chains from each of the four subunits, Leu165, Leu169, and Leu173, in rHoLF; Leu165, Leu169, and His173 in rHuHF; and Met165, Leu169, and His173 in rBfLF (Fig. 11b). The predominantly hydrophobic nature of these channels suggest they may be the entry route for dioxygen. However, this hydrophobic character is not universally conserved (see Fig. 2). For example, S.ma-1 ferritin has Thr169 and Glu173 (see Fig. 2), and both pea and soybean have His165, His169, and Arg173 [10]. Some flexibility in this region of the structure is indicated by the displacements of the E helices along four-fold axes with respect to the rest of the subunit in rHoLF, rRaLF, and rHuHF (Fig. 11a). The DE turn also varies in length and sequence (see Fig. 2). In EcFTN, it is curtailed by two residues with respect to HuHF, and its four-fold channels are lined by Glu165, Phe169, and Glu173.

The four-fold channel of EcBFR is again very different (see Fig. 11c). Here the DE turn and E helix are much shorter (only 6 residues compared with 9–13 residues in the ferritins). The four E helices are twisted in a right-handed sense, and the channels are lined by only two

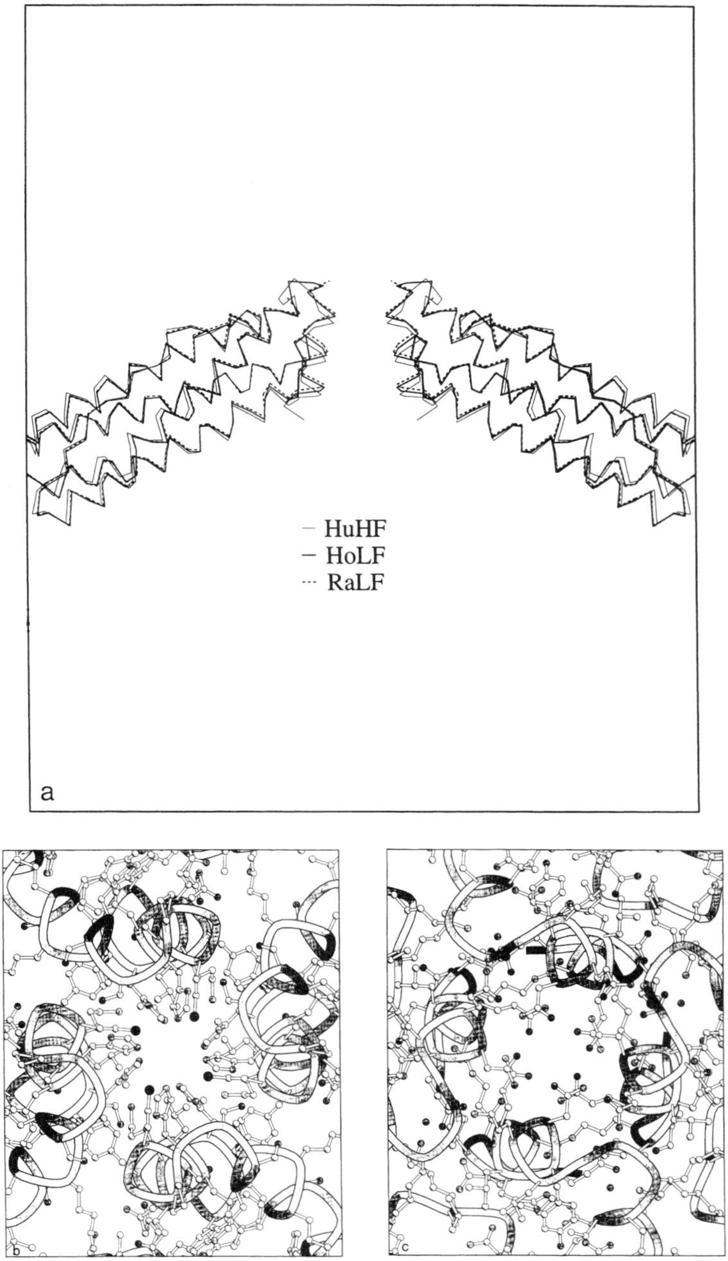

FIG. 11. E helices around four-fold axes in four ferritins and EcBFR. (a) superposition of α-carbon traces of three mammalian ferritins; (b) BfLF; (c) EcBFR. (a) viewed perpendicular to and (b) and (c) along four-fold axis.

residues from each helix, namely, Asn167 and Gln170 (H numbering and alignment of Fig. 2), which are conserved in all but four of the BFR sequences of Fig. 2; namely, those from *Brucella melitensis*, *Rhodobacter capsulatus*, *Mycobacterium avium*, and *Mycobacterium leprae*.

4.3. Iron-Storage Cavities

Since all ferritins contain an iron-sequestration cavity of about 80 Å diameter, a high degree of conservation might be expected of residues pointing into the cavity and potentially in contact with the iron core. However, this is far from the case. Analysis of solvent-accessible surface areas and average side chain radii together with an examination of the structure has provided a list of 14 residues (sequence numbers 49, 50, 53, 57, 59–61, 64, 68, 135, 136, 140, 143, 147) which together with the C-terminal seven residues can be considered to be mainly on the inside surface in HuHF and HoLF [31]. Of these, only Glu61 is conserved in all known eukaryotic sequences, and in prokaryotes, this residue is variable, being usually, but not always, either Glu or Asp. Although in rHuHF these residues are all hydrophilic, this is not true of all the ferritins shown in Fig. 2. However, there are common patterns of certain cavity-facing residues in some groups of ferritins. The "diamond" of glutamates 57, 60, 61, and 64 described above (see Fig. 7) is found in all known L ferritins and also in some plant ferritins. It has been proposed as a nucleation center [24,30,46,58,62]. The efficiency of iron incorporation into site-directed variants of HuLF and HuHF was decreased (E57H, E60H) or increased (E57E, H60E), respectively [58]. The C site ligands E61, E140, E143, and E144 (also E145) are common to all known FTNs. In the BFRs, the heme groups are both close to the ferroxidase centers and protrude into the cavities. However, their presence seems to have little effect on the rate of iron deposition, although it may facilitate iron release [59,60]. Two metal (Cd^{2+}) positions have been found on the inside surface of rHoLF [31] and HoSF [46], but it is unknown whether they have functional significance. One of the sites, near the three-fold axis, has both ligands (His136 and Asp139) conserved in L chains but not H chains (see Fig. 2). The second site has three ligands, Asp42, Glu49, and Cys52, but only the first two of these are conserved in L chains and neither Glu49 nor Lys52 is found in H chains.

4.4. Ferritin Structure and Iron-Storage Processes

For storage by ferritin, Fe^{2+} entering the molecule must be oxidized and deposited within the cavity. Initially, Fe^{2+} oxidation takes place at the intrasubunit di-iron centers. After iron-core nucleation, these centers may be bypassed with Fe^{2+} oxidation now predominantly on the iron core surface [4]. Two alternative pathways to the catalytic centers have been considered: (1) into the cavity via the three-fold channels and then through a short (7–8 Å) intrasubunit channel [24]; and (2) through a narrow intrasubunit channel from the outside surface [33]. This channel is blocked in L chains by Leu110. However, rHoLF and rHuLF can take up Fe^{2+} or Fe^{3+} under certain conditions [49,61,62]. In mammalian heteropolymers, Fe^{2+} oxidation and ferrihydrite nucleation are thought to occur, respectively, on H and L chains [3,4,58,63].

Substitution of Asp131 and Glu134 and chemical modification of residues bordering the three-fold channels (e.g., Cys130) reduce the rate of iron deposition, although not to the same extent as amino acid substitutions at the ferroxidase centers [63–65]. Thus, Fe^{2+} ions probably enter vertebrate ferritins via the three-fold channels prior to oxidation and storage of Fe^{3+}. Their passage into the cavity may be assisted by the negative charge and metal-binding properties (which are found in most animal and plant ferritins) of Asp131 and Glu134. Natural substitutions are always hydrophilic (see Fig. 2), but genetically engineered substitution by hydrophobic residues does not abolish iron incorporation [63]. No appropriate amino acid substitution studies have been performed in EcFTN and the Fe^{2+} entry route is unknown. In EcBFR, changing Asp131 to Ala, making the three-fold channels more hydrophobic, had little effect on iron-uptake rates. Because of the presence of Arg130 side chains, another entry route may be used, perhaps the four-fold channels. Alternatively, access through the channels may be as much a consequence of molecular flexibility as of the precise nature of channel residues.

Iron oxidized at the ferroxidase center is thought to move to the cavity through the short intrasubunit channel with the assistance of the mobile side chains of Glu61 in HuHF or Glu144 in EcFTN [33,37]. Analysis of Mössbauer spectra has shown that Fe^{3+} dimers formed at the oxidation centers of HuHF, EcFTN, or EcBFR are gradually replaced by ferrihydrite clusters of increasing size [52,66]. Furthermore, Fe^{3+} can move from the oxidation center of one molecule into the cavity of another [63,69], perhaps by docking of molecules at their channels. Based on Mössbauer parameters, UV spectra, and considerations of charge neutrality, it has been proposed that the Fe^{3+} dimers formed at the dinuclear centers in HuHF, EcFTN, and EcBFR are first linked by μ-oxo-bridges [48–52,66–68] but that these bridges are broken prior to migration of Fe^{3+} into the cavity [69]. An alternative proposal is that in EcBFR the Fe^{3+} pair is unbridged [54,70].

4.5. Structural Relationships Between Ferritins and other Four-Helix Bundle Di-iron Carboxylate Proteins

The recent structural and metal-binding studies of HuHF [31,33], EcBFR [38], and EcFTN [37] have placed these H-type ferritins in a class of di-iron-carboxylate proteins that includes the R2 subunit of ribonucleotide reductase (RNR-R2) [71–73], MMOH [74,75], and Rr [15]. In all these class 2 di-iron proteins, the dinuclear sites are situated at the centers of four-helix bundles in which the four helices have up-down-down-up connectivities with a long connection between the first and second helix pairs (A and B and C and D in ferritin; see Fig. 1). Instead of the long extended loop of the F-B-Rs, the RNR-R2 and MMOH connection includes a fifth long helix [71,74]. Rr is closely related structurally to the ferritins (see Table 1), although its B helix is shorter (only 22 residues).

The known ligands of iron in the dinuclear centers of Rr, RNR-R2, and MMOH also closely resemble those found for EcFTN and deduced for EcBFR and HuHF. High-resolution x-ray analysis has shown that the coordination and separation of Fe atoms in RNR-R2 and MMOH are dependent on whether the irons are diferrous or diferric [72,73] (Fig. 12). In RNR-R2, the shortening of the Fe-Fe distance on oxidation of RNR-R2 is associated with the formation of an oxo bridge between the iron pair which leads to the preservation of charge neutrality. The 4-Å separation of the Mn^{2+} pair in EcBFR [38] suggests that this complex provides a good model for the diferrous form of the center. The 3.8-Å separation found for Fe atoms in EcFTN [37] and the apparent absence of density for an oxo bridge suggests that the iron may be in the difer-

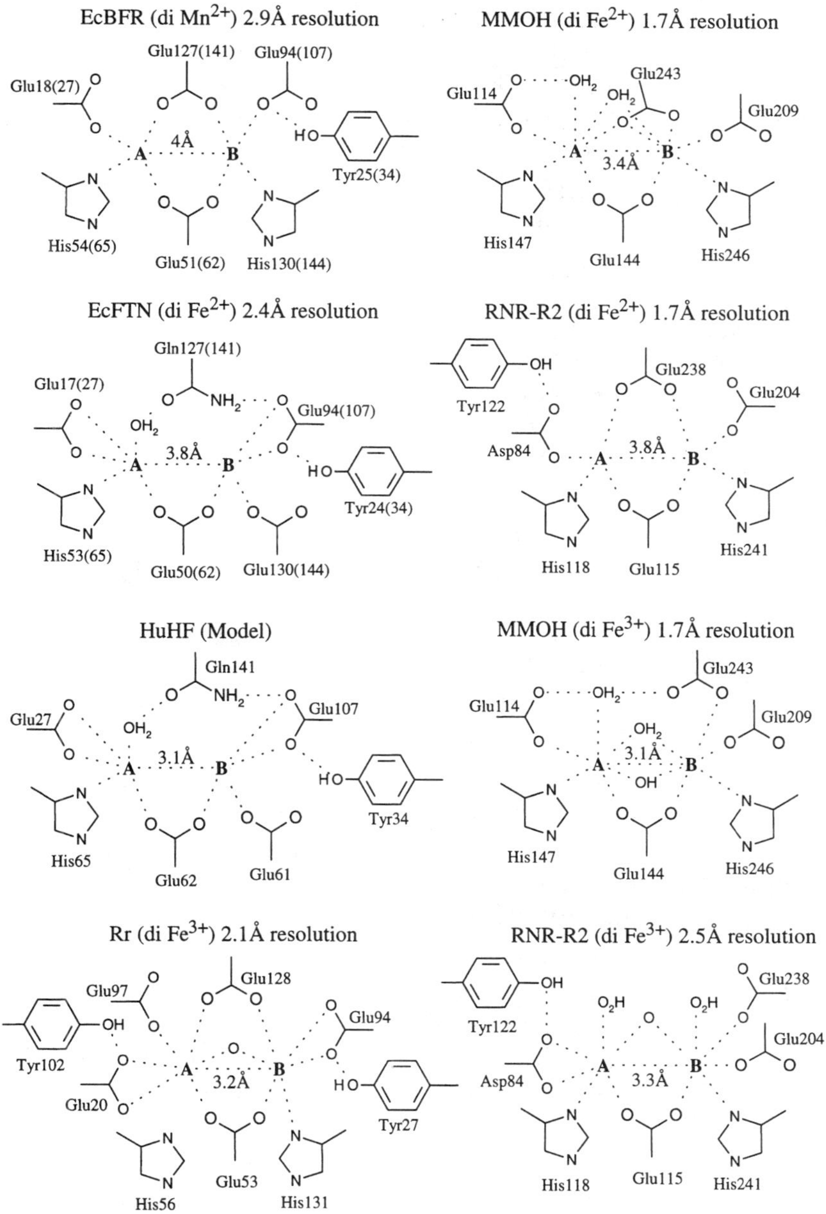

FIG. 12. Schematic diagrams of di-iron centers in rHuHF, EcFTN, EcBFR, Rr, RNR-R2, and MMOH.

rous form in this derivative. Fe^{2+} was diffused into the crystals aerobically, and it was assumed to have been oxidized [37], but the synchrotron radiation used for data collection has been found to reduce diferric RNR-R2 (in crystals frozen to 100°K in glycerol) [73]. Attempts to observe Fe at the dinuclear center of HuHF by x-ray analysis have so far proved to be unsuccessful, probably because, within the time scale of the experiments, Fe had moved into iron-core particles. In Rr, the diferric oxo bridge center is stable and the Fe^{3+}-Fe^{3+} distance is 3.2-Å [15].

Although all di-iron centers in Fig. 12 are similar, no two are identical. For example, EcBFR and RNR-R2 (reduced) differ in one site A ligand (Glu127 in EcBFR, Asp84 in RNR-R2). EcBFR, RNR-R2, and MMOH all have two Glu-X-X-His ligand motifs but EcFTN, HuHF, and Rr have only one. Glu and His ligands of site A follow this sequence in EcFTN and HuHF, as do those of site B in Rr. However, the ligands of site A in Rr and of site B in the two ferritins are all glutamate. Furthermore, EcFTN and HuHF lack the second glutamate bridge found in EcBFR, the oxidized form of Rr, and the reduced forms of RNR-R2 and MMOH. Also in EcFTN and HuHF, one of the glutamate ligands of site B has a different sequence position (Glu144 in EcFTN, Glu61 in HuHF).

Individual variation in the di-iron centers indicates fine tuning to different functions. Oxygen binding and activation at the center leads to tyrosyl radical production or to the hydroxylation of substrate, respectively, in RNR-R2 or MMOH [72,75]. In vitro, Rr can act as a ferroxidase in the presence of Fe^{2+} and O_2, with its di-iron center alternating between the diferric and diferrous forms as exogenous Fe^{2+} is oxidized [17]. The rubredoxin domain is required for this redox cycling. Ferritin's ferroxidase function appears to be different from that of Rr. In in vitro experiments [76], iron newly oxidized by ferritin can be trapped by transferrin (like that oxidized by Rr [17]. However, its iron storage function is associated with the production of stable polynuclear Fe^{3+} clusters, and once formed, these clusters exert a "pull" on the dinuclear iron. Although crystallographic evidence for the formation of oxo-bridged Fe(III) dimers in the ferritins is lacking and may be difficult to obtain because of the comparatively short occupancy of the dinuclear center by Fe^{3+}, the presence of such dimers in wild-type HuHF, EcFTN, and EcBFR is indicated by Mössbauer parameters [51,52,66,68]. In all three ferritins, the percentage of Fe in these species is greatest at 30–60 s after aerobic Fe^{2+} addition to the apo-protein. In the case of HuHF and EcFTN, the provision by HuHF and EcFTN of only three ligands for each Fe^{3+} may facilitate the subsequent dissociation of the oxo-bridges to enable migration of the Fe^{3+}. An alternative suggestion for EcBFR is that Fe^{3+} migration results from charge imbalance (no oxo-bridges) and that the center acts as a redox-driven iron pump [54,70]. However, since Mössbauer species with parameters similar to those of other ferritins and to other di-iron oxo-bridged complexes are also observed in EcBFR [66], the proposed distinctive character of iron storage mechanisms in the hemoferritins requires further investigation.

5. CONCLUDING REMARKS

The comparisons of the three-dimensional and primary structures of the ferritins elicit their remarkable conservation in nature. The importance of an iron-storage function is thus endorsed. The di-iron-carboxylate centers have been associated experimentally with ferroxidase activity in relatively few members of the F-B-R group. Nevertheless, the conservation of the iron ligands in so many diverse species implies that these centers are of key importance for the function of these proteins.

ACKNOWLEDGMENTS

We thank the Wellcome Trust for financial support (to P.J.A., P.M.H. and P.D.H.) and the Biochemical and Biophysical Science Research Council for an Advanced Fellowship to S.C.A. We are very grateful to F. de Maré for provision of rubrerythrin coordinates.

ABBREVIATIONS

BfLF	bullfrog L chain ferritin
BFRs	bacterioferritins
CD	circular dichroism

EcBFR	*E. coli* bacterioferritin
EcFTN	*E. coli* ferritin
F-B-R	ferritin-bacterioferritin-rubrerythrin
H subunit	heavy subunit
HoLF	horse L chain ferritin
HoSF	horse spleen ferritin
HuHF	human H chain ferritin
HuLF	human L chain ferritin
L subunit	light subunit
M subunit	medium subunit
MMOH	methane monooxygenase hydroxylase
RaLiF	rat liver ferritin
RaLF	rat L chain ferritin
rBfLF	recombinant bullfrog L chain ferritin
rHoLF	recombinant horse L chain ferritin
rHuHF	recombinant human H chain ferritin
rHuLF	recombinant human L chain ferritin
RNR	ribonucleotide reductase
Rr	rubrerythrin
rRaLF	recombinant rat L chain ferritin
S.ma-1	recombinant *Schistosoma mansoni* "type 1" ferritin

REFERENCES

1. E. C. Theil, *Annu. Rev. Biochem.*, *56*, 501–527 (1987).
2. P. M. Harrison, S. C. Andrews, P. J. Artymiuk, G. C. Ford, J. R. Guest, J. Hirzmann, D. M. Lawson, J. C. Livingstone, J. M. A. Smith, A. Treffry, and S. J. Yewdall, *Adv. Inorg. Chem.*, *36*, 449–486 (1991).
3. S. C. Andrews, P. Arosio, W. Bottke, J.-F. Briat, M. von Darl, P. M. Harrison, J. P. Laulhere, S. Levi, S. Lobréaux, and S. J. Yewdall, *J. Inorg. Biochem.*, *47*, 161–174 (1992).
4. P. M. Harrison and P. Arosio, *Biochim. Biophys. Acta*, *1275*, 161–203 (1996).
5. R. R. Crichton, in *Inorganic Biochemistry of Iron Metabolism*, Ellis Horwood, Chichester, 1991.

6. S. Mann, J. V. Bannister, and R. J. P. Williams, *J. Mol. Biol.*, *188*, 225–232 (1986).
7. V. J. Wade, A. Treffry, J. P. Laulhere, E. R. Bauminger, M. I. Cleton, S. Mann, J. F. Briat, and P. M. Harrison, *Biochem. Acta*, *1161*, 91–96 (1993).
8. S. Gider, D. D. Awschalom, T. Douglas, S. Mann, and M. Chaparala, *Science*, *268*, 77–80 (1995).
9. F. A. Fischbach, P. M. Harrison, and T. G. Hoy, *J. Mol. Biol.*, *39*, 235–238 (1969).
10. S. Lobréaux, S. J. Yewdall, J. F. Briat, and P. M. Harrison, *Biochem. J.*, *288*, 931–939 (1992).
11. J. DeGraaf, R. Amons, and W. Moeller, *Eur. J. Biochem.*, *193*, 737–750 (1990).
12. C. J. Carrano, R. Böhnke, and B. F. Matzanke, *FEBS Lett.*, *390*, 261–264 (1996).
13. M. J. Grossman, S. M. Hinton, V. Minat-Bernero, C. Slaughter, and E. I. Stiefel, *Proc. Natl. Acad. Sci. USA*, *89*, 2419–2423 (1992).
14. S. C. Andrews, J. M. A. Smith, S. J. Yewdall, J. R. Guest, and P. M. Harrison, *FEBS Lett.*, *293*, 164–168 (1991).
15. F. DeMaré, D. M. Kurtz, and P. Nordlund, *Nature Struct. Biol.*, *3*, 539–546 (1996).
16. Y. Lehmann, L. Meile, and M. Teuber, *J. Bacteriol.*, *178*, 7152–7158 (1996).
17. F. Bonomi, D. M. Kurtz, and X. Cui, *J. Inorg. Biol. Chem.*, *1*, 67–72 (1996).
18. D. M. Kurtz and B. C. Prickril, *Biochem. Biophys. Res. Commun.*, *181*, 337–341 (1991).
19. P. Arosio, T. G. Adelman, and J. W. Drysdale, *J. Biol. Chem.*, *253*, 4451–4458 (1978).
20. G. R. Moore, F. H. A. Kadir, F. K. Al-Massad, N. E. LeBrun, A. J. Thomson, C. Greenwood, J. N. Keen, and J. B. C. Findlay, *Biochem. J.*, *304*, 493–497 (1994).
21. M. Von Darl, P. M. Harrison, and W. Bottke, *Eur. J. Biochem.*, *222*, 353–366 (1994).
22. A. Treffry, P. J. Lee, and P. M. Harrison, *Biochim. Biophys. Acta*, *785*, 22–29 (1984).

23. D. W. Rice, G. C. Ford, J. L. White, J. M. A. Smith, and P. M. Harrison, *Adv. Inorg. Biochem.*, *5*, 39–50 (1983).

24. G. C. Ford, P. M. Harrison, D. W. Rice, J. M. A. Smith, A. Treffry, J. L. White, and J. Yariv, *Phil. Trans. R. Soc. Lond.*, *B304*, 551–565 (1984).

25. G. Precigoux, J. Yariv, B. Gallois, A. Dautant, C. Courseille, and B. L. D'Estaintot, *Acta Cryst.*, *D50*, 739–743 (1994).

26. D. M. Lawson, PhD thesis, X-ray structure determination of recombinant ferritins, University of Sheffield, Sheffield, UK, 1990.

27. J. Trikha, E. C. Theil, and N. M. Allewell, *J. Mol. Biol.*, *248*, 949–967 (1995).

28. R. J. Hoare, P. M. Harrison, and T. G. Hoy, *Nature*, *255*, 653–654 (1975).

29. S. H. Banyard, D. K. Stammers, and P. M. Harrison, *Nature*, *271*, 282–284 (1978).

30. M.-A. Michaux, A. Dautant, B. Gallois, T. Granier, B. L. D'Estaintot, and G. Precigoux, *Proteins*, *24*, 314–321 (1996).

31. P. D. Hempstead, S. J. Yewdall, A. R. Fernie, D. M. Lawson, P. J. Artymiuk, D. W. Rice, G. C. Ford, and P. M. Harrison, *J. Mol. Biol.*, *268*, 424–448 (1997).

32. F. H. Kadir and G. R. Moore, *FEBS Lett.*, *276*, 81–84 (1990).

33. D. M. Lawson, P. J. Artymiuk, S. J. Yewdall, J. M. A. Smith, J. C. Livingstone, A. Treffry, A. Luzzago, S. Levi, P. Arosio, G. Cesareni, C. D. Thomas, W. V. Shaw, and P. M. Harrison, *Nature*, *349*, 541–544 (1991).

34. J. Trikha, G. S. Waldo, F. A. Lewandowski, Y. Ha, E. C. Theil, P. C. Weber, and N. M. Allewell, *Proteins*, *18*, 107–118 (1994).

35. P. Santambrogio, P. Pinto, S. Levi, A. Cozzi, E. Rovida, P. J. Artymiuk, P. M. Harrison, and P. Arosio, *Biochem. J.*, *322*, 461–468 (1997).

36. J. Dietzel, J. Hirzmann, D. Preis, P. Symmons, and W. Kunz, *Mol. Biochem. Parasitol.*, *50*, 245–254 (1992).

37. P. D. Hempstead, A. J. Hudson, P. J. Artymiuk, S. C. Andrews, M. J. Banfield, J. R. Guest, and P. M. Harrison, *FEBS Lett.*, *350*, 258–262 (1994).

38. F. Frolow, A. J. Kalb (Gilboa), and J. Yariv, *Nature Struct. Biol.*, *1*, 453–460 (1994).

39. P. M. Harrison and D. W. Gregory, *Nature*, *220*, 578–580 (1968).

40. R. R. Crichton and C. F. A. Bryce, *Biochem. J.*, *133*, 289–299 (1973).
41. S. Otsuka, H. Maruyama, and I. Listowsky, *Biochemistry*, *20*, 5226–5232 (1981).
42. M. Gerl and R. Jaenicke, *Biochemistry*, *27*, 4089–4096 (1988).
43. S. Stefanini, P. Vecchini, and E. Chiancone, *Biochemistry*, *26*, 1831–1837 (1987).
44. A. Luzzago and G. Cesareni, *EMBO J.*, *8*, 569–576 (1989).
45. R. Japelli, A. Luzzago, P. Tataseo, I. Pernice, and G. Cesareni, *J. Mol. Biol.*, *227*, 432–543 (1992).
46. P. M. Harrison, P. J. Artymiuk, G. C. Ford, D. M. Lawson, J. M. A. Smith, A. Treffry, and L. White, in *Biomineralization Chemical and Biochemical Perspectives* (S. Mann, J. Webb, and R. J. P. Williams, eds.). VCH, Weinheim, Germany, 1989, pp. 257–294.
47. T. Granier, B. Gallois, A. Dautant, D. Langlois, B. D'Estaintot, and G. Precigoux, *Acta Cryst.*, *D52*, 594–596 (1996).
48. A. Treffry, J. Hirzmann, S. J. Yewdall, and P. M. Harrison, *FEBS Lett.*, *302*, 109–112 (1992).
49. A. Treffry, Z. Zhao, M. A. Quail, J. R. Guest, and P. M. Harrison, *Biochemistry*, *34*, 15204–15213 (1995).
50. A. Treffry, Z. Zhao, M. A. Quail, J. R. Guest, and P. M. Harrison, *Biochemistry*, *36*, 432–441 (1997).
51. E. R. Bauminger, P. M. Harrison, D. Hechel, I. Nowik, and A. Treffry, *Biochim. Biophys. Acta*, *1118*, 48–58 (1991).
52. E. R. Bauminger, P. M. Harrison, D. Hechel, N. W. Hodson, I. Nowik, A. Treffry, and S. J. Yewdall, *Biochem. J.*, *296*, 709–719 (1993).
53. N. E. Le Brun, S. C. Andrews, J. R. Guest, P. M. Harrison, G. R. Moore, and A. J. Thomson, *Biochem. J.*, *312*, 385–392 (1995).
54. A. J. Thomson, N. E. LeBrun, A. Keech, S. C. Andrews and G. R. Moore. *Biochem. Soc. Trans.*, *25*, 96–101 (1997).
55. G. S. Waldo, J. Ling, J. Sanders-Loehr, and E. C. Theil, *Science*, *259*, 796–798 (1993).
56. K. E. Liu, A. M. Valentine, D. Qiu, D. E. Edmondson, E. H. Appelman, T. G. Spiro, and S. J. Lippard, *J. Am. Chem. Soc.*, *117*, 4997–4998 (1995).
57. P. M. Harrison, E. R. Bauminger, D. Hechel, N. W. Hodson, I. Nowik, A. Treffry, and S. J. Yewdall, in *Progress in Iron Research* (C. Hershko, A. Konijn, and P. Aisen, eds.), Plenum Press, New York, 1994, pp. 1–12.

58. P. Santambrogio, S. Levi, A. Cozzi, B. Corsi, and P. Arosio, *Biochem. J.*, *B14*, 139–144 (1996).

59. S. C. Andrews, N. E. LeBrun, V. Barynin, A. J. Thomson, G. R. Moore, J. R. Guest, and P. M. Harrison, *J. Biol. Chem.*, *270*, 23268–23274 (1995).

60. G. D. Watt, J. W. McDonald, C. H. Chiu, and K. L. M. Reddy, *J. Inorg. Biochem.*, *31*, 1–6 (1993).

61. S. Levi, S. J. Yewdall, P. M. Harrison, P. Santambrogio, A. Cozzi, E. Rovida, A. Albertini, and P. Arosio, *Biochem. J.*, *288*, 591–596 (1992).

62. R. Crichton, A. Herbas, O. Chavez-Alba, and F. Roland, *J. Biol. Inorg. Chem.*, *1*, 567–574 (1996).

63. S. Levi, P. Santambrogio, B. Corsi, A. Cozzi, and P. Arosio, *Biochem. J.*, *317*, 467–473 (1996).

64. S. Stefanini, A. Desideri, P. Vecchini, T. Drakenberg, and E. Chiancone, *Biochemistry*, *28*, 378–382 (1989).

65. A. Desideri, S. Stefanini, F. Polizio, R. Perruzelli, and E. Chiancone, *FEBS Lett.*, *287*, 10–14 (1991).

66. C. Hawkins, A. Treffry, J. B. Mackey, J. M. Williams, S. C. Andrews, J. R. Guest, and P. M. Harrison, *Nuovo Cimento*, *18*, 347–352 (1996).

67. E. R. Bauminger, P. M. Harrison, I. Nowik, and A. Treffry, *Biochemistry*, *28*, 5486–5493 (1989).

68. E. R. Bauminger, A. Treffry, A. J. Hudson, D. Hechel, N. W. Hodson, S. C. Andrews, S. Levi, I. Nowik, P. Arosio, J. R. Guest, and P. M. Harrison, *Biochem. J.*, *302*, 813–820 (1994).

69. E. R. Bauminger, P. M. Harrison, D. Hechel, I. Nowik, and A. Treffry, *Proc. Roy. Soc. Lond.*, *B244*, 211–217 (1991).

70. A. J. Thomson, N. E. LeBrun, A. M. Keech, G. R. Moore, and S. C. Andrews, *Twelfth John Innes Symp.*, Sept 9–12, Abs-30, 1996.

71. P. Nordlund, B.-M. Sjoberg, and H. Eklund, *Nature*, *345*, 593–598 (1990).

72. P. Nordlund and H. Eklund, *Curr. Opin. Struct. Biol.*, *5*, 758–766 (1995).

73. D. T. Logan, X.-D. Su, A. Aberg, K. Regnstrom, J. Hajdu, H. Eklund, and P. Nordlund, *Structure*, *4*, 1053–1064 (1996).

74. A. C. Rosenzweig, C. A. Frederick, S. J. Lippard, and P. Nordlund, *Nature*, *366*, 537–543 (1993).

75. A. C. Rosenzweig, P. Nordlund, P. M. Takahara, C. A. Frederick, and S. J. Lippard, *Chem. Biol.*, *2*, 409–418 (1995).

76. G. R. Bakker and R. F. Boyer, *J. Biol. Chem.*, *261*, 13182–13185 (1986).

77. P. J. Kraulis, *J. Appl. Cryst.*, *24*, 946–950 (1991).

78. J. Devereux (1989). The GCG Sequence Analysis Software Package, Version 6.0, Genetics Computer Group, Inc., University Research Park, 575 Science Drive, Suite B, Madison, Wisconsin, 53711, USA.

79. J. Felsenstein (1993). PHYLIP (Phylogeny Inference Package) version 3.5c. Distributed by the author. Department of Genetics, University of Washington, Seattle.

80. J. Felsenstein (1989). PHYLIP—Phylogeny Inference Package (Version 3.2). *Cladistics*, *5*, 164–166.

12

Ferritin. Uptake, Storage, and Release of Iron

N. Dennis Chasteen

Department of Chemistry, University of New Hampshire, Parsons Hall, Durham, NH 03824-3598, USA

1. INTRODUCTION 480

2. IRON DEPOSITION 481
   - 2.1. Mammalian Ferritins 481
     - 2.1.1. Iron Pathways into Ferritin 481
     - 2.1.2. Binding of Ferrous Ion 482
     - 2.1.3. Role of the H and L Subunits 483
     - 2.1.4. The Net Stoichiometric Equations 485
     - 2.1.5. The Stoichiometry of Oxidative Electron Transfer 487
     - 2.1.6. Steady-State Ferroxidase Kinetics 489
     - 2.1.7. Rapid Kinetics and Formation of "Intermediates" 490
     - 2.1.8. Radical Production and Protein Degradation 492
   - 2.2. Bacterioferritins 494
   - 2.3. Phytoferritins 496

3. IRON MOBILIZATION 498

3.1. Mammalian Ferritins 498
3.1.1. Redox Chemistry 498
3.1.2. Penetration of the Protein Shell by Reductants 499
3.1.3. Iron Removal from Ferritin 500
3.2. Bacterioferritins 501
3.3. Phytoferritins 502
4. CONCLUSIONS 503
ACKNOWLEDGMENT 504
ABBREVIATIONS 504
REFERENCES 505

1. INTRODUCTION

The ferritins form a class of iron-mineralizing proteins found widely distributed among plants, animals, and microbes [1–8]. Ferritins store iron in a biologically accessible form for the synthesis of heme and limit the participation of iron in radical-producing reactions which are deleterious to cells. Iron is stored as a crystalline, hydrous ferric oxide mineral core within the clathrate protein shell which sequesters and "solubilizes" the $FeOOH_{(s)}\downarrow$ iron particle. Up to 4500 atoms of iron can be stored in the hollow protein shell which, in mammals, is composed of 24 similar subunits of two types, H and L.

The mechanisms by which iron is acquired and released by the ferritins have been the subject of intense investigation for over three decades. During formation of the core, Fe^{2+} is thought to enter the protein through one or more channels and to be initially oxidized to iron(III) at ferroxidase sites on the H chains. Hydrolysis and mineral formation ensues at nucleation sites located principally on the L chains. Conversely, dissolution of the core occurs by reduction of the Fe^{3+} by a suitable reductant in the presence of a chelator to remove the iron(II).

Most studies have been done on mammalian ferritins, primarily horse spleen, human liver, and human recombinant ferritins, although bacterial, amphibian, insect, and plant ferritins have received increased attention. Owing to the availability of the x-ray structures of a number of ferritins [3,9–12,165] and recombinant H-chain and L-chain homopolymer proteins as well as site-directed variants, major strides have been made in understanding the structure-function relationships of the ferritins (see Chap. 11). In this chapter, we focus on the salient features of the mechanisms of iron deposition and mobilization in ferritin, principally covering the literature of the past 5 years. For discussions of the earlier literature as well as coverage of topics not presented here, the reader should consult a number of excellent review articles [1–7,13,14].

2. IRON DEPOSITION

2.1. Mammalian Ferritins

2.1.1. Iron Pathways into Ferritin

The avenues by which $Fe^{2+/3+}$, oxidants, reductants, and chelators gain access to the interior of ferritin where biomineralization/demineralization occur are of considerable importance. The four/three/two symmetry of the ferritin shell (see Fig. 1 in Chap. 11) presents a number of possible pathways for small molecule penetration into the protein. These pathways include the six hydrophobic channels located on the four-fold rotation axes [9], the eight hydrophilic channels on the three-fold axes [15–19], the two-fold axes [20], and the one-fold channel on the H subunit [10].

The hydrophobic character of the four-fold channels (Fig. 11 in Chap. 11), which are lined with the isobutyl groups of 12 Leu residues in the L chain and 8 Leu and 4 His in the H chain [9], argues against their involvement in the transport of ionic species such as $Fe^{2+/3+}$. The channels are 3–4 Å across in the L-subunit-rich horse spleen protein, HoSF, [9] but are much narrower (~1.5 Å) in the L-chain bullfrog red cell ferritin, BfLF [21], and human H-chain ferritin, HuHF [14]. The proposal that the four-fold channels may be important for the transport of molecular oxygen to the interior of the protein seems unlikely, since fluorescence-quenching studies of aromatic residues in proteins indicate that O_2 readily diffuses through protein matrices at rates approaching the diffusion control limit [22], eliminating the need of a specific mechanism of O_2 delivery to the ferroxidase site.

The hydrophilic character of the three-fold channels with their six carboxylate groups from symmetry-related Asp-131,131′,131″ and Glu-134,134′,134″ residues (human H-chain numbering) makes them the most reasonable pathways for iron entry into ferritin [9] (see Fig. 10 in Chap. 11). Recent spectroscopic, chemical modification, metal binding, and site-directed mutagenesis work [15–18,23] suggests a role for the channels in transport of Fe^{2+} to the central cavity (reviewed in [16,17]). For example, the channel mutations D131A-E134A and D131H-E134H reduce the rate of iron oxidation by nine- and five-fold, respectively, relative to the H-chain homopolymer [17]. Mutation of Cys130 and His118, both of which face the outer opening of the three-fold channel, likewise reduces the capacity of the protein to bind metals and incorporate iron [16]. The fact that such channel mutations and chemical modifications do not *completely* abolish the ferroxidase activity of the protein suggests that iron might gain access to the ferroxidase site by other, less efficient, pathways. One possible pathway is the one-fold intrasubunit channel connecting the ferroxidase site in the center of the H subunit to the outer protein surface [10]. This channel is only 1 Å in diameter in the static structure [10] but is perhaps enlarged as the protein flexes.

There are other possible pathways of entry. The porous character of the two-fold axis of the subunit dimer is revealed by the observation that demetallated protoporphyrin IX binds on this axis in horse L-chain ferritin, HoLF, to a depth of about 10 Å [20]. A dynamic structural model, whereby large molecules gain access to the interior of the protein through subunit dissociation, has also been proposed [6,24]. In this connection, moderately sized reporter molecules, such as neutral red, can be trapped inside the ferritin shell by dissociating the protein into its constituent subunits at pH 2 in the presence of the dye followed by raising the pH to 7, where the protein reassembles [25].

2.1.2. *Binding of Ferrous Ion*

The first step in the oxidative deposition of iron in ferritin involves the initial binding of Fe^{2+} to the apoprotein. Numerous studies point toward binding of a variety of metals in the channels (reviewed in [17,26]). In HuHF, Fe^{2+} forms iron nitrosyl complexes at Cys130 and His128 which are located at the outer and inner openings of the three-fold channels, respectively [26]. Binding studies with HoSF under anaerobic conditions employing size-exclusion chromatography and Mössbauer spectroscopy reveal that approximately 8 Fe^{2+} bind per protein molecule; this stoichiometry suggests binding within the eight hydrophilic channels [27]. Magnetic susceptibility data indicate that the individual S = 2 Fe^{2+} ions are remote from one another and from Fe^{3+} clusters (which begin to form when as few as 8 Fe^{3+} are present in the protein), a finding consistent with the binding of solitary Fe^{2+} ions inside the channels [28]. The protein, after repeated cycles of Fe^{2+} oxidation and Fe^{2+} addition, also shows an 8:1 Fe^{2+} binding stoichiometry, indicating that iron migration occurs on Fe^{2+} oxidation, thus freeing up the putative channel sites for the binding of additional Fe^{2+}. Evidence of iron translocation and rearrangement within the protein has also been observed in electron paramagnetic resonance (EPR) and Mössbauer spectroscopic studies of HoSF and other ferritins [17,29–35]. The presence of phosphate facilitates translocation of Fe^{3+} within the protein [31].

The sequential binding of two Fe^{2+} to the ferroxidase site prior to oxidation has been recently observed by multimixing stopped-flow kinetics [36]. Simultaneous binding of both Fe^{2+} ions is necessary for rapid oxidation of the iron.

Fe^{2+} also binds to the surface of the ferric mineral core at stoichiometries up to 50 Fe^{2+}/protein provided that the core contains surface-associated orthophosphate [37]. Reconstituted ferritins in the absence of P_i also bind Fe^{2+} but with the 8:1 stoichiometry of the apoferritin shell [27,37]. The presence of P_i in the core greatly accelerates the rate of iron oxidation at the mineral surface [37].

2.1.3. *Role of H and L Subunits*

The role of the H and L chains in mammalian ferritin structure and function has been the subject of considerable study since their discovery some 20 years ago [38]. The H and L chains are encoded on different chromosomes and consist of 182 and 174 amino acids, respectively, sharing approximately 55% homology in sequence [4,5,7]. Comparisons of amino acid sequences of a number of H- and L-chain ferritins from a variety of animal, plant, and microbial species are given elsewhere (see Fig. 2 in Chap. 11) [1,3,7,14]. A major distinction between the two types of subunits is that the H chain contains a cluster of residues, Glu27,

Glu61, Glu62, His65, and Glu107 (HuHF numbering), located in the center of the four-helix bundle and which facilitate Fe^{2+} oxidation by molecular oxygen [10,39]. The more basic L chain lacks this ferroxidase center but contains Glu residues not found in the H chain and that appear to be important for nucleation of the mineral core [40,41]. Glu53, Glu56 and Glu57 have been implicated in the mechanism of iron incorporation into L ferritins [42].

A variety of in vitro studies with recombinant proteins and site-directed variants convincingly demonstrate that the H-chain ferroxidase site is important for iron incorporation into ferritin [10,12,34,36, 39,43–48]. The mutation of the ferroxidase site ligands (Glu62,His65 → Lys,Gly) results in substantial loss of oxidation activity, being completely abolished when the putative H-chain nucleation site ligands (Glu61, Glu64, and Glu67) are also mutated [44].

Mössbauer and ultraviolet (UV)-difference spectroscopic studies show the presence of Fe^{3+} dimers at the ferroxidase site of HuHF [34,49], which is consistent with the observation of dinuclear Tb^{3+} [10] and Fe^{3+} [12] complexes by x-ray crystallography (see Fig. 9 in Chap. 11). Furthermore, a ferroxidase site has been constructed in the L chain by site-directed mutagenesis. This L-chain variant has activity comparable to that of the H chain when incorporated into a heteropolymer of 90% L-chain and 10% L-chain variant [46]. The L-chain variant cannot reassemble as a homopolymer but only when native L chain is present.

The mutation of Leu169 → Arg in HuHF produces a stable, but nonassembled form, of the H-chain protein which retains some ferroxidase activity when used in conjunction with apotransferrin or an L-chain homopolymer to accept the oxidized iron [47]. Migration of Fe^{3+} from iron-poor to iron-rich ferritin molecules has been demonstrated by Mössbauer spectroscopy [50,51].

The major role of the L chain appears to be that of facilitating iron hydrolysis and nucleation of the core [40]. The higher negative charge across the inner surface of the protein cavity of the L subunit is probably important in facilitating mineralization [44]. Although the L-chain homopolymer is capable of forming a core, it does so only at pH 7 and higher where autoxidation of iron(II) takes place at appreciable rates [40]. In contrast, the H-chain protein with inactivated ferroxidase sites does not readily form a core at this pH [45]. At pH 5.5, the L-chain homopolymer forms a core only when the H-chain homopolymer is present to facilitate iron oxidation and transfer of Fe^{3+} to the L homopolymer [45]. Thus, the H and L chains appear to have cooperative roles in iron deposition, with the H chain being important for iron oxidation and the L chain for core nucleation.

Other experiments lend support to this idea. For example, when large increments of Fe^{2+} (3000 Fe/protein) are added to apoferritin at pH 7 preloaded with 1000 Fe, conditions favoring spontaneous iron hydrolysis, H-chain homopolymers, with and without mutated ferroxidase centers, form large protein aggregates, whereas L-chain homopolymers do so only to a limited extent [52]. When the putative nucleation site ligands Glu57 and Glu60 on the inner surface of the protein cavity of the L homopolymer are mutated to His, the L chain also becomes sensitive to iron-induced protein aggregation [41]. Heteropolymers of 20% H and 80% L subunits prepared in vitro appear to resist protein aggregation better than either homopolymer and incorporate three to four times as much iron [41]. The most efficient iron-acquiring heteropolymer on an H-chain basis contains about 35% H chain [53]. The ferroxidase activity of apoferritins as measured by electrode oximetry follows the order HuHF (100% H) > ShSF (66% H) > HoSF (16% H) [54]. On an H-chain basis, however, HoSF exhibits the highest activity, which is a further indication that the more efficient proteins at harvesting iron are those with high L/H ratios [54]. They perhaps do so by facilitating iron turnover at the ferroxidase site in addition to promoting core formation at nucleation sites [54].

In contrast to H chains, the L chain lacks the ferroxidase site, being replaced by a salt-bridge between Lys62 (Glu62 in the H chain) and Glu107 (in both H and L chains) [55]. This salt-bridge is thought to be responsible for much of the physical stability of the L chain over the H chain toward denaturation by guanidine HCl at low pH 2.0–3.5 [55]. Recent thermal stability measurements employing differential scanning calorimetry of human H-chain subunit dimers and horse L-chain dimers at pH 3.5 have shown the H-chain dimer to be more stable, suggesting that the salt-bridge does not impart greater thermodynamic stability to the L-chain dimer but rather renders the L chain less susceptible to denaturation by guanidine in acidic solution [56].

2.1.4. *The Net Stoichiometric Equations*

The stoichiometry of the reaction between Fe^{2+}, O_2, and apoferritin or holoferritin has been studied in a number of laboratories [48,57–60]. It is now generally agreed that, at low iron loading (<50 Fe/protein) of H chain containing proteins, two Fe(II) are oxidized per molecule of O_2 consumed, 2 Fe/O_2. The stoichiometric reaction is given by Eq. (1) [58]:

$$2Fe^{2+} + O_2 + 4H_2O \longrightarrow 2FeOOH_{core} + H_2O_2 + 4H^+ \qquad (1)$$

Here hydrogen peroxide is the principal product of dioxygen reduction [58]. Eq. (1) is the sum of oxidation and hydrolysis reactions (2) and (3), respectively:

$$2Fe^{2+} + O_2 + 2H^+ \longrightarrow 2Fe^{3+} + H_2O_2 \qquad (2)$$

$$2Fe^{3+} + 4H_2O \longrightarrow 2FeOOH_{core} + 6H^+ \qquad (3)$$

On increasing the increment of added iron (>250 Fe/protein), the reaction stoichiometry changes to 4 Fe/O_2 (Fig. 1) according to Eq. (4) [58]:

$$4Fe^{2+} + O_2 + 6H_2O \longrightarrow 4FeOOH_{core} + 8H^+ \qquad (4)$$

Eq. (4) is the sum of the oxidation and hydrolysis reactions (5) and (6) where water is the product of dioxygen reduction:

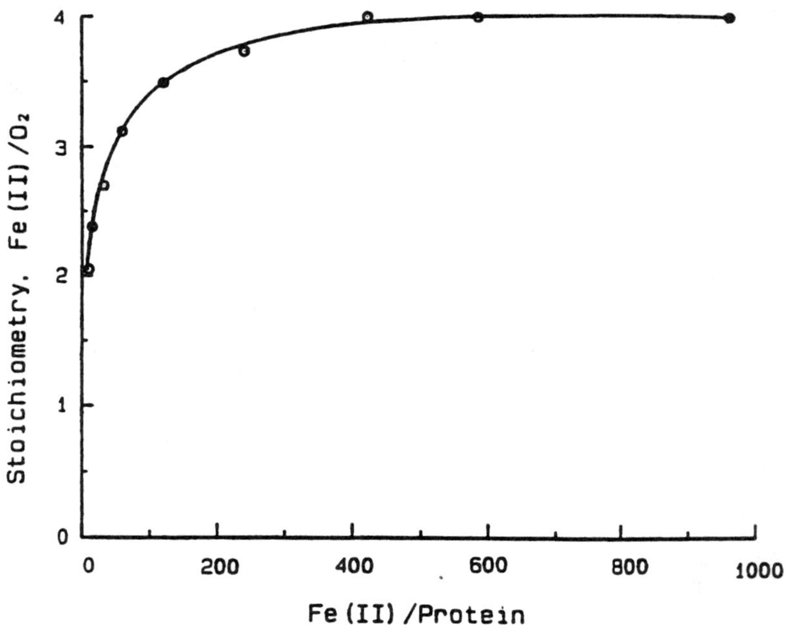

FIG. 1. Stoichiometry of iron oxidation in HoSF as a function of the increment of added iron. (Reproduced with permission from [54].)

$$4Fe^{2+} + O_2 + 4H^+ \longrightarrow 4Fe^{3+} + 2H_2O \tag{5}$$

$$4Fe^{3+} + 8H_2O \longrightarrow 4FeOOH_{core} + 12H^+ \tag{6}$$

Increasing the increment of iron added to apoferritin causes the protein ferroxidase site to become kinetically saturated and changes the dominant mechanism of iron oxidation from a protein site to a mineral surface catalyzed mechanism. With HoSF, the mechanism of iron oxidation occurs via the protein ferroxidase site independent of the amount of core already present provided that the Fe^{2+} is introduced to the protein in small increments separated by 24 h (<50 Fe/protein) [61].

The H_2O_2 produced in Eq. (1) slowly disproportionates according to reaction (7) [48,58,61]. Ferritin itself does not have catalase activity to promote this reaction.

$$H_2O_2 \longrightarrow 1/2\ O_2 + H_2O \tag{7}$$

It has been proposed that ceruloplasmin, a known ferroxidase, may be the physiological catalyst for Fe^{2+} oxidation, the resultant Fe^{3+} then being incorporated into ferritin [60,62–65]. In the presence of ceruloplasmin, an Fe/O_2 stoichiometry near 4.0 is obtained [60], which contrasts with Eq. (1). Data have recently been presented, however, showing that much or all of the iron oxidized by ceruloplasmin is not incorporated into ferritin but instead binds to the ceruloplasmin in the form of hydroxypolymers [66]. Based on this recent evidence, it remains unclear whether ceruloplasmin plays a significant role in iron uptake by ferritin.

It is also claimed that the ferroxidase activity observed in ferritin is simply an artifact of the use of Good's buffers, because no appreciable oxidation is observed when the assay is done in unbuffered saline [60,63]. However, under the conditions of the reported experiments [60,63], the protons produced from iron hydrolysis in Eq. (3) cause the pH in unbuffered saline to drop below 3 and iron oxidation immediately to cease. A buffer or pH stat setup is required to offset this effect, in which case full ferroxidase activity of the protein is observed as expected [67].

2.1.5. *The Stoichiometry of Oxidative Electron Transfer*

Since H_2O_2 is the product of oxidation of two Fe^{2+} (Eq. (1)) [58], and the ferroxidase site is a binuclear center [10,12] (see Fig. 9 in Chap. 11), iron oxidation is expected to occur by a concerted two-electron oxidation of the di-iron(II) center to produce H_2O_2 and the di-iron(III) complex seen by Mössbauer [34,35,49]; for example, reaction (8):

$$\text{P-Fe}^{2+}\text{-OH}^{-}\text{-Fe}^{2+} + O_2 + H^{+} \longrightarrow \text{P-Fe}^{3+}\text{-O}^{2-}\text{-Fe}^{3+} + H_2O_2 \qquad (8)$$

Here P represents the protein. In the early phase of iron oxidation (<0.2 s) measured by stopped-flow, the rate is second order in Fe^{2+} concentration [68] and achieves a maximum at 2 Fe^{2+}/subunit [36], which is consistent with the simultaneous oxidation of two Fe^{2+} at the ferroxidase site.

Steady-state kinetics data (>10 s) measured by electrode oximetry indicates, however, that iron oxidation is first order in Fe^{2+} concentration [61], implying that Fe^{2+} oxidation occurs by a one-electron transfer step in this slower phase of the reaction; for example, Eqs. (9) and (10):

$$\text{P-Fe}^{2+} + O_2 \longrightarrow \text{P-Fe}^{3+}\text{-O}_2^{-} \qquad (9)$$

$$\text{P-Fe}^{3+}\text{-O}_2^{-} + Fe^{2+} + H_2O \longrightarrow \text{Fe}^{3+}\text{-O}^{2-}\text{-Fe}^{3+} + H_2O_2 \qquad (10)$$

Attempts to observe the production of superoxide, $O_2^{\bullet -}$ (expected from the one-electron reduction of O_2) by spin trapping and rapid-freeze-quench EPR spectroscopy have been singularly unsuccessful [69,70], either because $O_2^{\bullet -}$ is not produced (i.e., the oxidation does not occur by a one-electron transfer) or because it remains bound to the protein. Ferritin also possesses weak superoxide dismutase (SOD) activity which may make it difficult to spin trap $O_2^{\bullet -}$ [61,71].

It should be noted that the stopped-flow and oxygen electrode experiments measure different iron oxidation processes. For example, once the rapid oxidation phase is complete, a 24-h wait is required for it to be observed again when more Fe^{2+} is added. In contrast, the slower steady-state oxidation seen by oximetry is continuous. The two types of iron oxidation may well involve different electron transfer mechanisms as their kinetics suggest. Perhaps the key to ultimately understanding the details of electron transfer in the ferritins as well as the differences between stopped-flow and oximetry results lies in the numerous radicals that have been observed in a number of these proteins [69,70,72] and in verifying that O_2 is the primary oxidant in the very early phase of the reaction. Redox centers on the protein may also be involved in the observed kinetics (see below).

2.1.6. Steady-State Ferroxidase Kinetics

Iron oxidation in HoSF, human liver ferritin (HLF), HuHF, and a number of site-directed variants measured by electrode oximetry [48,61] displays saturation kinetics with respect to Fe^{2+} and O_2 concentrations and first-order kinetics with respect to protein concentration, which are consistent with an enzyme catalysis mechanism [48,61]. The apparent $K_{M,Fe}$, K_{M,O_2}, and k_{cat} values fall in the ranges 50–300 μM, 6–140 μM, and 31–201 min^{-1}, respectively. The k_{cat} values for the various proteins follow the order HuHF (201 min^{-1}) > HoSF (80 min^{-1}) > HLF (31 min^{-1}) with increasing H-chain content as expected [48]. However, the opposite sequence is observed on an H-chain basis, providing further evidence that L chains facilitate oxidation at the H chains. The relatively large $K_{M,Fe}$ values indicate that the protein is kinetically saturated only at high iron concentrations, a property which perhaps affords protection to the cell during periods of rapid influx of iron prior to significant iron induced ferritin biosynthesis.

Zn^{2+} and Tb^{3+} are either competitive, noncompetitive, or mixed inhibitors of iron oxidation depending on the protein [48,61]. Such complex behavior can be attributed to multiple binding sites on ferritin for these metals [73]. The inhibitor constants, K_I, are around 70 μM, which is consistent with carboxylate coordination.

The kinetics of iron oxidation are more complex than the enzyme catalysis mechanism implies [48,61]. Complex mechanisms can often be modeled by simple Michaelis-Menten kinetics, masking more complex underlying behavior. In the case of bullfrog H-chain ferritin (BfHF), Michaelis-Menten kinetics is not observed [59]. Also, some iron and radical species generated during core formation in ferritin undoubtedly arise from side reactions (see below). More than one mechanism of iron oxidation, for example protein ferroxidase, mineral surface, and autoxidation mechanisms, may be operating simultaneously. As more insight into the kinetics of core formation is obtained, the previously proposed enzyme mechanism [48,61] will undoubtedly need to be modified and refined. For the moment, the apparent Michaelis-Menten kinetic parameters serve as useful benchmarks for comparing the activities of the various ferritins.

2.1.7. Rapid Kinetics and Formation of "Intermediates"

UV-Vis stopped-flow kinetics and EPR rapid-freeze-quench measurements have revealed aspects of the iron-deposition mechanism in ferritin not previously apparent in slower steady-state kinetics studies [35,59,68,70,74]. In the case of H-chain bullfrog ferritin (BfHF), a transient purple complex (λ_{max} = 550 nm) is observed and, based on its color and resonance Raman spectra, attributed to an iron(III)-tyrosinate complex [59,74]. The maximum rate of formation of this complex is achieved near 50 Fe/protein [59], a value suggestive of dimer formation (i.e., 2 Fe/subunit = 48 Fe/protein). However, the actual stoichiometry of the complex (number of Fe^{3+} and tyrosine residues) remains to be established. Its function is unknown.

A blue intermediate complex with a broad absorption centered around 650 nm has been observed with HuHF during the initial rapid oxidation of the first 48 Fe^{2+} by the protein (Fig. 2) [35,68]. It decays more rapidly than the species seen in BfHF [59,74]. The blue complex is not seen when certain ligands of the ferroxidase site or when Tyr34, which is hydrogen bonded to ligand Glu107, are mutated [68]. Although the blue color may also arise from an iron(III) tyrosinate complex, other explanations are possible such as formation of a diferric-peroxo complex [68]. Significantly, variant Y34F retains ferroxidase activity similar to that of native HuHF [35], indicating that Tyr34, although perhaps involved in the formation of the blue complex, is not essential for oxidation of the iron. In this connection, a tyrosine radical attributable to Tyr34 has been observed during iron oxidation in HuHF [72] but likewise is probably not an essential component of the mechanism.

EPR freeze-quench experiments with horse spleen apoferritin have revealed the formation of a mononuclear Fe^{3+} (g' = 4.3), a mixed-valent Fe^{2+}-Fe^{3+} (g' = 1.87), and a radical with an axial g-tensor ($g_{\parallel}$ = 2.042, $g_{\perp}$ = 2.0033) within the first 100 ms of the reaction between Fe^{2+} and O_2 (Fig. 3) [70]. The mononuclear complex is formed at a faster rate than either the mixed-valent species or the radical. The mixed-valent species appears to be generated from Fe^{2+} binding near a preformed mononuclear Fe^{3+} complex and not from the half-oxidation of a Fe^{2+}-Fe^{2+} dimer. It is not a ferroxidase site species [75]. The axial radical shows rapid EPR relaxation properties suggesting that it is located in the vicinity of paramagnetic iron. Its role, if any, in iron oxidation

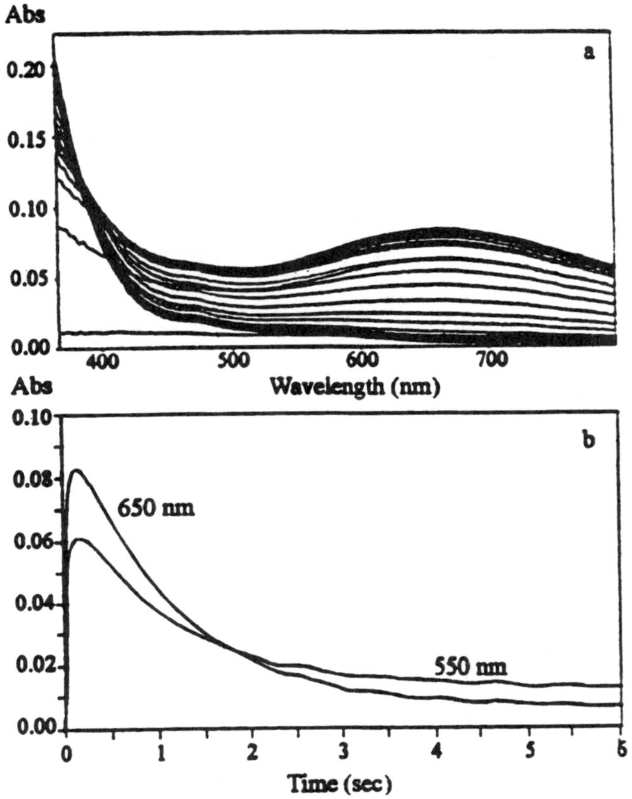

FIG. 2. (a) Spectrum development of intermediate "blue species" over a period of 50 s. (b) Progress curves at 550 and 650 nm. (Reproduced with permission from [68].)

remains to be determined. It may be simply a consequence of a radical side reaction since its concentration never exceeds a few percent of the total iron added [70]. It lacks hyperfine structure when $^{17}O_2$ is employed, indicating that it is not an oxy radical.

Mössbauer spectroscopy on the time scale of minutes has convincingly shown that a dimeric Fe^{3+}-Fe^{3+} center is formed at the ferroxidase site followed by mononuclear Fe^{3+} and then clusters [17,33–35,76]. A mononuclear Fe^{3+} complex is generated from iron migration to the three-fold channels [17]. However, there appears to be more than one mononuclear Fe^{3+} species on apoferritin (reviewed in [70]).

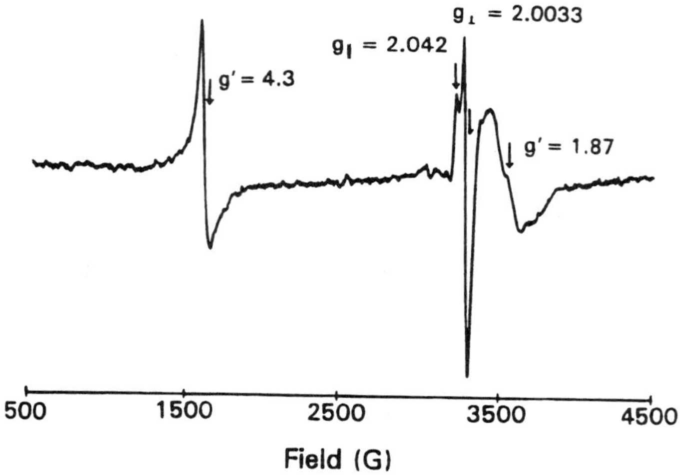

FIG. 3. Rapid-mixing freeze-quench EPR spectrum of partially oxidized HoSF at an Fe/protein ratio of 48/1. (Reproduced with permission from [70].)

A reasonable model for core formation consistent with most available data involves the binding and oxidation of two Fe^{2+} at the H chain ferroxidase site to form an Fe^{3+}-Fe^{3+} complex and the generation of the transient blue complex in the process. The rapid initial phase of iron oxidation may "prime" the protein for the slower enzymatic catalysis seen by oximetry and UV-Vis absorption changes traditionally employed to monitor iron oxidation. Iron migration to nucleation sites, principally on the L subunit, then occurs where crystallization of the core ensues. Subsequent Fe^{2+} may be oxidized at the dimeric ferroxidase sites or on the nucleated iron to give the observed kinetic behavior of the protein.

2.1.8. *Radical Production and Protein Degradation*

In addition to the formation of the tyrosine and axial radicals [70,72] alluded to earlier, other radicals have been observed in ferritin under various conditions of iron loading of the protein [69,70,72,77]. The production of H_2O_2 from the ferroxidase site-catalyzed oxidation of iron (Eq. [2]) can lead to Fenton chemistry and the production of hydroxyl radical according to equation (11):

$$Fe^{2+} + H_2O_2 \longrightarrow Fe^{3+} + HO^{\bullet} + HO^{-} \quad (11)$$

The hydroxyl radical thus produced can react with the protein at diffusion-controlled rates to generate protein radicals and other oxidation products [78–81]. Such chemistry has been observed in HoSF, which is composed of 84% L subunit and 16% H subunit [69,72]. Time-dependent EPR spectra with fine structure are obtained during the oxidative deposition of iron in HoSF (Fig. 4) [72]. Spectra of this type are attenuated when hydroxyl radical scavengers, such as mannitol and dimethyl sulfoxide, are present in the solution and are enhanced by the direct addition of H_2O_2 and Fe^{2+}, confirming the involvement of hydroxyl radical in their formation [69,72].

Interestingly, the signal in Fig. 4 continues to grow long after iron oxidation is complete (<2 min) and then ultimately decays away at 40 h. This phenomenon is postulated to be due to the diffusion of paramagnetic Fe^{3+} away from the radical center, increasing its EPR signal, followed by slow decay of the radical. The radical(s) show rapid electron spin relaxation behavior consistent with iron being bound nearby [72]. Radical formation requires an intact ferroxidase site for the production of H_2O_2, since the radicals are not observed in the L-chain homopolymer

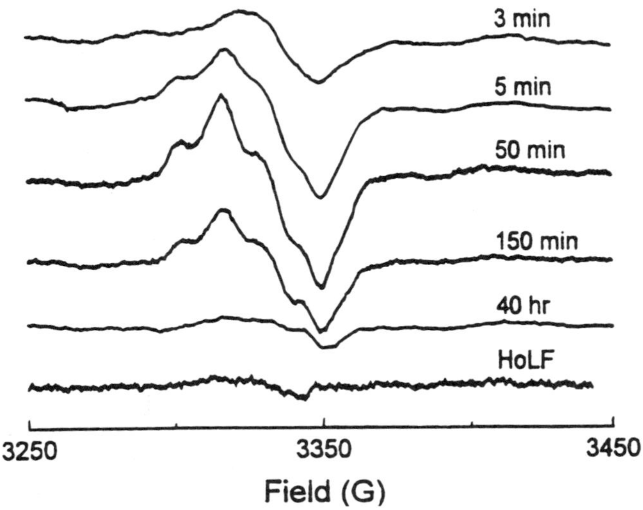

FIG. 4. Time dependence of EPR signal at 77K from Fenton chemistry-derived radicals in HoSF. (Reproduced with permission from [72].)

of HoSF (see Fig. 4). In the absence of a ferroxidase site, H_2O is the product of iron autoxidation (Eq. [5]).

The radical signals of Fig. 4 are not seen with the human H-chain homopolymer, and thus appear to be located on the L chain of the horse protein [70]. Consistent with this view is the observation that the L-chain content of HoSF as measured by SDS-PAGE progressively decreases on repeated additions of 50 Fe^{2+}/protein [82]. Other more reactive protein-based radicals have been observed in spin-trapping experiments [69,72]. Reduction in the histidine and lysine content of HoSF during the oxidative deposition of iron in the protein [63] and the formation of covalently-linked subunit dimers in sheep spleen ferritin [83] are further evidence of protein degradation. It has been suggested that the ferritin shell compartmentalizes the radical chemistry arising from Fenton chemistry and thereby serves as a sink for hydroxyl radical, minimizing damage to other biomolecules of the cell [69]. The probable role of radicals in the conversion of ferritin to hemosiderin, a degradation product of ferritin, has been discussed elsewhere [84].

2.2. Bacterioferritins

Bacterioferritins have been identified and studied in a variety of microorganisms, including *Azotobacter vinelandii* [85,86], *Pseudomonas aeruginosa* [86–88], *Escherichia coli* [86,89–94], *Cyanobacterium synechocystis* P.C.C. [95], *Mycobacterium avium* [96], *Mycobacterium leprae* [97], *Streptomyces olivaceus* [98], the fungus *Absidia spinosa* [99], *Rhodobacer capsulatus* [100], and *Helicobacter pylori*, *Anabaena variabilis*, *Bacillus subtilis*, and *Treponema palladium* [101], among others [14]. Bacterioferritins contain only one type of subunit which more closely resembles H chains than L chains in both homology and the presence of key residues [4]. Two types of ferritins have been found in *E. coli*, an eukaryote-like ferritin (FTN) which more closely resembles mammalian ferritins, and a prokaryote bacterioferritin (BFR) which is distinguished by the presence of heme in its structure [102].

Bacterioferritins, as isolated, contain variable amounts of heme (typically 8–12). Generally, 12 hemes bind per 24 mer at the interface between two subunits with bis-methionine ligation of Met52 to the heme as found for ferritins from *E. coli*, *P. aeruginosa*, *R. capsulatus*, and *A. vinelandii* [87,100,103]. The Met52 variants M52L and M52H of EcBFR form protein shells but do not bind heme, demonstrating that heme ligation is not essential for assembly of the protein [104]. Heme has also been shown to bind to HoSF [105] and, based on modeling and spectroscopic studies, appears to have bis-histidyl coordination involving His114 and His124 near the three-fold channels. HoSF lacks the coordinating Met residue found in bacterioferritins. HoLF crystallized with Sn protoporphyrin IX results in demetallation of the porphyrin with its binding at a location between subunits similar to that found in the bacterioferritins [20]. However, in HoLF, Arg59 is above and below the metal-free porphyrin ring rather than Met59 as in bacterioferritins (Met52 in BFR). In addition to heme, bacterioferritins from *P. aeruginosa* and *A. vinelandii* also contain quinones of unknown structure which may be the result of oxidative damage to the protein [106]. Whether these quinones are important for iron uptake or release is unknown.

The key residues of a ferroxidase site are also found in BFRs; namely, Glu18, Glu51, His54, Glu94, Glu127, and His130 in *R. capsulatus* BFR [14,97,100,107]. However, a key difference between the ferroxidase sites of mammalian and bacterial ferritins is the coordination to the iron of a second histidyl (His130) in place of a glutamyl residue [14]. Furthermore, the ferroxidase sites of mammalian and bacterial ferritins show a striking resemblance to the di-iron sites of ribonucleotide reductase, methane monoxygenase, and hemerythrin [12] (see Fig. 12 in Chap. 11). In the three-fold channel of *E. coli* bacterioferritin, the glutamate residues of mammalian ferritins are replaced by isoleucines [92].

The kinetics of iron uptake by EcBFR follows three phases, binding of two Fe^{2+} to the ferroxidase site, Fe^{2+} oxidation, and core formation [108]. The second phase is kinetically saturated at approximately 50 Fe/protein (~2Fe/subunit). The rapid phase is not quickly regenerated, as also found for mammalian and bullfrog ferritins [59,68]. The spectroscopic properties of the heme are perturbed by the binding of Fe^{2+} to the protein [108]; however, the presence or absence of heme does not affect the rate of iron uptake [104]. Mössbauer spectroscopy has shown the same pattern of formation of iron species as for HuHF; namely, dimer $\rightarrow$ monomer $\rightarrow$ cluster [109]. Nitric oxide, an analogue of O_2, has been used as a probe of the iron-oxygen species formed during iron deposition in EcBFR [94,108]. One type of signal has been postulated to arise from an Fe^{2+}-Fe^{2+}-NO species formed at the ferroxidase site [108]. Both Co^{2+} and Zn^{2+} (an inhibitor of iron oxidation) release nominally 2 H^+/metal on binding at the ferroxidase site and produce charge neutral dimeric complexes [110]. Co^{2+} binds with positive cooperativity and exhibits weak magnetic coupling between metal ions of the dimer [111]. In contrast to BfHF [59], an Fe^{3+}-tyrosinate species is not seen during iron deposition in EcBFR [108].

A distinguishing feature of the cores of the bacterioferritins is their high phosphate content, $Fe/P_i = 1.4/1 - 1.7/1$ in *A. vinelandii* BFR [85,112], compared with mammalian ferritins where typically $Fe/P_i \sim 8/1$. The high phosphate composition results in relatively amorphous cores unlike the crystalline ferrihydrite cores of mammalian ferritins [88,91,112]. The cores of reconstituted HoSF in the presence of sufficient P_i are structurally similar to those of bacterioferritin as measured by EXAFS [112]. Incorporation of ^{57}Fe (the Mössbauer active nucleus) into *P. aeruginosa* bacterioferritin (PaBFR) when cells are grown in ferric chloride, ferric citrate, or ferripyroverdine has been demonstrated [113], implying an iron-storage function for the protein. The Mössbauer properties of the cores of PaBFR depend on the growth conditions of the organism [88].

2.3. Phytoferritins

Phytoferritins store and detoxify iron but also play an important role in plant development and response to disease [8,114]. Ferritin is largely found in the nongreen plastids and in lesser amounts in the chloroplasts. Being developmentally regulated, phytoferritin is particularly abundant in seeds, roots, and shoots. During germination, seed ferritin is degraded and the iron released. Ferritin also disappears from buds on sprouting.

Plant ferritins consist of only one type of subunit having characteristics of both mammalian H and L subunits. However, the plant subunit is more like the H chain. Pea seed ferritin has 49% homology with the human H chain compared with 39% homology with the L chain [115]. Recently, three genes encoding the sequence of three different mature ferritin subunits have been found in developing leaves of cowpeas (*Vigna unguiculata*) [116].

Unlike mammalian and bacterioferritins, phytoferritins have two additional peptide sequences, TP and EP, on the N-terminal, resulting in a larger subunit molecular weight (~32 kDa) compared with other ferritins (~19 000–21 000) [4,115,117]. One of these, the transit peptide (TP), consists of 47 residues [115] and is involved in targeting and transporting of the protein to plastids [117]. Cleavage of the TP is a necessary step for protein shell assembly [115,118]. The mature 28-kDa subunit contains an N-terminal extension peptide (EP) of 24 amino acid residues which is predicted to extend outside of the protein shell [115]. Otherwise the structure of pea seed ferritin is thought to be analogous to that of mammalian ferritins [115].

During seed germination and subsequent iron mobilization, the 28-kDa subunit is cleaved at the N-terminal to form two shortened subunits of 25.0 and 26.5 kDa in size [119]. A similar phenomenon has been observed in vitro with iron-exchange experiments where the 28-kDa subunit is reduced to 26.5 kDa from cleavage near residue 21 [120]. This phenomenon appears to be due to radical damage from Fenton chemistry. A phytosiderin, analogous to hemosiderin in mammals, is thought to be derived in part from radical damage to phytoferritin [120].

The key residues of the ferroxidase site are conserved in plant ferritins except that Glu62 is replaced by His in pea seed ferritin. The putative nucleation site residues Glu57 and Glu60 found only in L chains are also present. The four-fold channels are hydrophilic compared with the hydrophobic nature of the channels found for mammalian ferritins [115]. The three-fold channels are hydrophilic in both plant and mammalian ferritins.

The ferroxidase activities of clover and pea seed ferritins are intermediate between those of HoSF and HuHF when Fe^{2+} and O_2 are employed as substrates [121,122]. Ferric citrate, a possible source of iron in plant tissues, can also donate iron to phytoferritin when ascorbate, another component of plants, is present [123,124]. The Fe^{3+} in the citrate complex is reduced by ascorbate and subsequently reoxidized by O_2 as it is deposited in the protein. Ferric citrate can also be photoreduced and the resultant Fe^{2+} oxidatively deposited into pea seed ferritin; however, photoreduction of the core Fe^{3+} may also occur followed by reoxidation and redeposition of the iron [123].

Cores of phytoferritins typically contain 1300–1800 Fe atoms as isolated [121,122]. These cores are rich in phosphate, Fe/P_i = 2.8–4.0

[121,122] compared with mammalian ferritins (Fe/P_i ~ 8) and may play a role in the storage of phosphate as well as iron [122]. An Fe–P distance of 3.26 Å is observed by EXAFS in the core of pea seed ferritin [125], but no Fe–Fe distance of 3.5 Å is seen, which probably reflects the high phosphate level of the core.

3. IRON MOBILIZATION

3.1. Mammalian Ferritins

3.1.1. Redox Chemistry

The redox chemistry of the ferritins has been studied by microcoulometry measurements in conjunction with Mössbauer spectroscopy [85, 126–131]. The one-electron reduction of the core of HoSF (2300 Fe/protein) shows Nernstian behavior with midpoint potentials $E_{1/2}$ of −190, −310, and −416 mV (n = 1) at pH 7, 8, and 9, respectively, relative to NHE [129]. The pH dependence of the measured $E_{1/2}$ corresponds to the uptake of 2 H^+ per Fe^{3+} according to Eq. (12) [128]:

$$FeOOH_{core} + e^- + 2H^+ = FeOH^+ + H_2O \qquad (12)$$

In order to maintain electroneutrality, transport of chloride to the interior of the protein occurs during reduction, forming a chlorohydroxy iron(II) mineral phase within the protein [28].

The protein shell of HoSF has also been shown to exhibit reversible electrochemistry (Fig. 5), likewise with $E_{1/2}$ = −310 mV (n = 1) at pH 8, the same as that of the iron core [126,130]. In this instance, however, the redox potential is pH independent. Quantitative coulometric measurements show the presence of approximately six redox centers per apoprotein shell which can exist in either oxidized $HoSF_{ox}$ or reduced $HoSF_{red}$ forms. These centers can be cycled between redox states using oxidants such as O_2, $Fe(CN)_6^{3-}$ and Fe^{3+}-cytochrome-*c* or reductants such as methylviologen and dithionite. The following reactions have been identified:

$$O_2 + HoSF_{red} \longrightarrow HoSF_{ox} \qquad (13)$$

$$Fe^{2+} + HoSF_{ox} \longrightarrow Fe^{3+}\text{-}HoSF_{red} \qquad (14)$$

$$Fe^{3+}\text{-}HoSF_{red} + O_2 \longrightarrow Fe^{3+}\text{-}HoSF_{ox} \qquad (15)$$

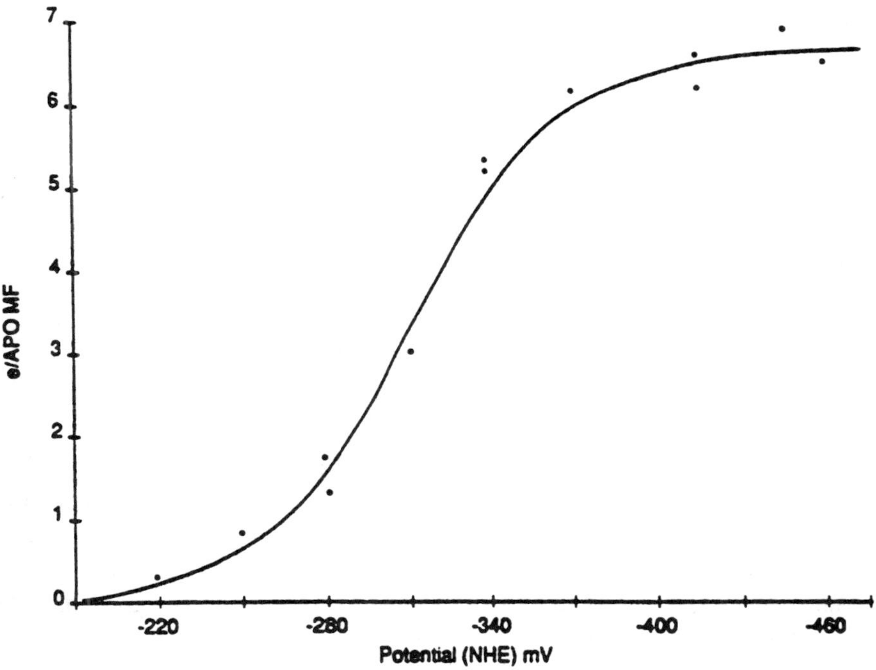

FIG. 5. Microcoulometric reduction of apoHoSF as a function of electrode potential. (Reproduced with permission from [130].)

In addition to the Fe^{3+}-$HoSF_{red}$ and Fe^{3+}-$HoSF_{ox}$ species, a stable Fe^{2+}-$HoSF_{red}$ species can be formed. Iron oxidation occurs in the absence of oxygen (Eq. (14)) to form a mononuclear Fe^{3+} complex [130]. The redox centers presumably involve aromatic amino acid residues or some tightly bound cofactor [130], perhaps quinones [106]. Whether they are related to the various radicals seen during the oxidative deposition of iron in ferritin [69,70,72,77,84] is unknown. However, they do not appear to involve the Tyr34 radical [72].

3.1.2. Penetration of the Protein Shell by Reductants

The mobilization of iron from ferritin at appreciable rates requires the use of a suitable reductant to chemically reduce the core. In principle, any reductant with a reduction potential more negative than −190 mV at pH 7 [129] should be capable of reducing Fe^{3+} in the core. Indeed, a variety of reducing agents with potentials ranging from −215 mV (FMN) to −660 mV (dithionite) at pH 7 have been shown to be suitable reductants [128,132,133]. The correlation of reduction rates with redox potentials rather than with molecular size of the reductants supports the idea that reducing agents do not have to come in direct contact with the core for reaction to occur [128,132]. The rate-limiting step of reductive removal of iron from HoSF by various diphenols also appears to be electron transfer across the protein shell [134]. Studies with large flavoproteins and ferredoxins as reductants further suggest that reduction takes place by long-range electron transfer [127]. Reverse electron transfer between reduced HoSF and ferrihemeproteins also appears to occur by such a mechanism [135].

A key issue is whether biological reductants can gain access to the interior of the protein at rates sufficiently fast to reduce the core at physiologically meaningful rates. The literature in this area is conflicting with reported times for penetration of molecules of nominally 7 Å in cross section ranging from milliseconds to days (reviewed in [136]). A recent detailed kinetics study using nitroxide spin probes indicates that diffusion into HoSF by molecules of 7–9 Å in diameter is a relatively slow process at room temperature ($t_{1/2} \sim$ 21–36 min) [136]. However, recent studies of the lability of the dinuclear iron centers of the ferroxidase site suggest that a variety of chelators readily gain access to the protein interior [137].

3.1.3. Iron Removal from Ferritin

A wide variety of reductants have been shown to assist in the mobilization of iron(III) from ferritin in vitro, including superoxide [138–140], plant phenolics [141,142], thioglycolic acid [143], 6-hydroxydopamine [144], anthracycline, bipyridyl, nitroaromatic and semiquinone radicals [145], nitric oxide [146], polyhydroxypyrimidines [147], xenobiotics [148], *DL*-dihydrolipoate [149], light [123,150,151], polyhydroxybenzenes [152,153], dithionite [128], $FMNH_2$ [128,132], tetravalent vanadium [154], and ascorbate [155–157], among others. Ascorbate has also been shown by EPR to reduce the iron core of ferritin in vivo in neuroblastoma SK-N-SH cells [158].

In the majority of the in vitro studies, a chromophoric ferrous ion chelator such as 2,2′-bipyridyl, orthophenanthroline sulfonate, bathophenanthroline sulfonate, or ferrozine has been employed to complex and remove the iron from the protein. Complexation of the Fe^{2+} can shift the redox potential of the core to a more positive value, facilitating its reduction. Thus, reductants with redox potentials as high as +950 mV can reduce the core iron in the ferrozine assay system at pH 5.6 [159,160]. The kinetics of ferrous ion generation can depend on the chelator used, bathophenanthroline sulfonate or ferrozine [161]; also, ferritins from different commercial sources can show different kinetics of iron release [162].

In some instances, electron transfer across the protein shell of HoSF appears to be the rate limiting step [159]; that is,

$$\text{Core-}(Fe^{3+})_n + e^- \xrightarrow{\text{slow}} \text{Core-}(Fe^{3+})_{n-1} + Fe^{2+} \tag{16}$$

$$Fe^{2+} + 3\ \text{ferrozine} \xrightarrow{\text{fast}} [Fe(\text{ferrozine})_3]^{2+} \tag{17}$$

Dihydroxybenzene reductants having lower redox potentials show the fastest rates as expected from electron transfer theory.

Interestingly, on binding of heme to the protein, all the reductants show the same (increased) rate, suggesting that heme facilitates electron transfer across the protein shell, in which case reaction (17) becomes the slow step [134,159,160].

Measurement of core reduction rates (as opposed to iron removal) using chromophoric ferrous ion chelators such as ferrozine thus can be misleading if the complexation reaction (17) is rate limiting. For example, the rate of reduction of the HoSF core by thioglycolic acid has been shown to be six-fold faster when studied using dispersive x-ray absorption spectroscopy (DXAS) than the rate measured from the production of the ferrous bipyridyl complex [143]. Moreover, a rapid initial phase of reduction not seen in previous work was evident [143].

3.2. Bacterioferritins

The redox chemistry of bacterial ferritins is quite different from that of mammalian ferritins. The b-type heme of AvBFR has reported $E_{1/2}$ values of −430 to −475 mV for the holoprotein, whereas it becomes significantly more positive in the apoprotein, $E_{1/2}$ = −210 to −225 mV, indicating a significant interaction between the core and the heme [85,163]. Moreover, unlike HoSF, the $E_{1/2} = -420$ mV of the core of AvBFR is pH independent in the pH range 7–9 even though 2 H^+ are taken up per Fe^{2+} formed (Eq. (12)) [85,163]. This pH independence is postulated to arise from a two-step process whereby the heme is first reduced in a pH-independent rate-limiting step followed by rapid electron transfer from the heme to the core [163]. The presence of phosphate in the core shifts the core redox potential from -345 mV for cores reconstituted without P_i to -445 mV for the holoprotein (220 Fe/protein). The high P_i content of the core of AvBFR facilitates higher levels of Fe^{2+} binding to the core than seen in mammalian ferritin [164].

The reductive mobilization of iron from *P. aeruginosa* BFR, like HoSF, is independent of the redox potential of the reductant when heme is present [159,160]. The presence of heme accelerates the rate of iron reduction, which is consistent with its role in electron transfer to the interior of the protein. In the reductive release of iron from AvBFR using dithionite, $S_2O_4^{2-}$, the kinetics is half-order with respect to $S_2O_4^{2-}$ concentration, a result implying that the radical $SO_2^{\bullet-}$, formed from the equilibrium $S_2O_4^{2-} \rightleftharpoons SO_2^{\bullet-}$, is the actual species responsible for the reduction [163]. Stopped-flow measurements indicate that the heme is first reduced followed by the core, providing further evidence that electron transfer occurs via the heme according to the following reaction sequence [163]:

$$SO_2^{\bullet-} + Fe^{3+}\text{-heme} \xrightarrow{\text{fast}} Fe^{2+}\text{-heme} + SO_3^{2-} \tag{18}$$

$$Fe^{2+}\text{-heme} + Fe^{3+}\text{-core} \xrightarrow{\text{slow}} Fe^{3+}\text{-heme} + Fe^{2+}\text{-core} \tag{19}$$

Thus, unlike mammalian ferritins, bacterioferritins have heme, a clearly identifiable electron transfer component. Although an electron transfer component is probably present in mammalian ferritins, it remains to be identified.

3.3. Phytoferritins

Studies of iron mobilization from phytoferritins have been rather limited. The iron core of pea seed ferritin can be photoreduced by visible light using fluorescent bulbs which may be one mechanism for releasing stored iron during plant growth [123]. Experiments with pea seed ferritin (PeSdF) and HoSF both suggest that the protein shell itself is not being oxidized during photoreduction of the core [123,151]. Inhibitors of Fenton chemistry slow the kinetics. The photoreduction of the core of HoSF by UV-A radiation (365 nm, $\phi = 2 \times 10^{-3}$) results in the release of about 80 Fe^{2+} per protein, a finding which may be important in skin photobiology [151]. Reoxidation of the iron by O_2 and deposition into the 2000 Fe atom core seem to occur. Other studies of HoSF, using longer wavelength fluorescent and incandescent light sources with leaves as optical filters, have likewise demonstrated significant photoproduction of Fe^{2+} which is enhanced at pH 4.6 relative to pH 7.1 [150].

The cores of phytoferritins can also be reduced by ascorbate in the presence of citrate [123,124]. Plant phenolics such as caffeic acid, dihydroxybenzoic acids, and their analogues reduce the iron core of HoSF at rates which correlate with their reduction potentials as expected for electron transfer over some distance [141,142].

4. CONCLUSIONS

Iron oxidation in ferritin proceeds by an H-chain dinuclear ferroxidase site mechanism in which dioxygen is reduced to hydrogen peroxide. At large fluxes of iron into the protein, oxidation occurs directly on the mineral surface of the growing core, in which case H_2O is the product of dioxygen reduction. There are at least two phases in the ferroxidase reaction: a rapid phase in which colored complexes are produced and a slower steady-state phase corresponding to the progressive uptake of O_2 and production of the mineral core. Since ferritin lacks catalase activity, the H_2O_2 produced during the oxidative deposition of iron in ferritin leads to Fenton chemistry-derived radicals and some degradation of the protein shell; in effect the protein itself acts as an antioxidant. Other radicals not derived from Fenton chemistry are also produced but their roles, if any, in the mechanism of iron oxidation are unknown.

Redox centers located on the apoferritin shell have been shown to oxidize iron in the absence of dioxygen and may play some role in iron accumulation. Several intermediate iron and radical species have been observed by UV-Vis, Mössbauer, and EPR spectroscopies in the early phase of iron oxidation, but a detailed mechanism which can account for the formation of all of the species has yet to be formulated. The situation is complicated by the presence of more than one pathway for iron oxidation and by various side reactions. Iron mobilization from ferritin involves reduction of the core followed by chelation and removal of the iron(II). When large reducing agents are employed, reduction of the core appears to occur by long-range electron transfer across the protein shell. Plant and bacterial ferritins share many similarities with mammalian ferritins. In the case of bacterioferritins, the heme of these proteins facilitates electron transfer during reduction of the core. The mechanism of long-range electron transfer in mammalian ferritins is unknown.

ACKNOWLEDGMENT

This work was supported by NIH grant R37 GM20194.

ABBREVIATIONS

apoHoSF	apo horse spleen ferritin
Arg	arginine
AvBFR	*Azotobacter vinelandii* bacterioferritin
BfHF	bullfrog H-chain ferritin
BfLF	bullfrog L-chain ferritin
BFR	bacterioferritin
CE	capillary electrophoresis
Cys	cysteine
D131A	aspartate-131 mutated to alanine
D131H	aspartate-131 mutated to histidine
DXAS	dispersive x-ray absorption spectroscopy
E62K	glutamate-62 mutated to lysine
E134A	glutamate-134 mutated to alanine
E134H	glutamate-134 mutated to histidine
EcBFR	*Escherichia coli* bacterioferritin
EP	extension peptide

EPR	electron paramagnetic resonance
EXAFS	extended x-ray absorption fine structure
FMN	flavin mononucleotide
$FMNH_2$	reduced flavin mononucleotide
FTN	eukaryote-like ferritin from bacteria
Glu	glutamate
Gly	glycine
H65G	histidine-65 mutated to glycine
His	histidine
HLF	human liver ferritin
HoLF	horse L-chain ferritin
HoSF	horse spleen ferritin
HuHF	human H-chain ferritin
Lys	lysine
M52H	methionine-52 mutated to histidine
M52L	methionine-52 mutated to leucine
NHE	normal hydrogen electrode
PaBFR	*Pseudomonas aeruginosa* bacterioferritin
PeSdF	pea seed ferritin
P_i	inorganic phosphate
SDS-PAGE	sodium dodecylsulfate polyacrylamide gel electrophoresis
ShSF	sheep spleen ferritin
SOD	superoxide dismutase
TP	transit peptide
Tyr	tyrosine
UV	ultraviolet
Y34F	tyrosine-34 mutated to phenylalanine

REFERENCES

1. E. C. Theil, *Adv. Enzymol. Relat. Areas Mol. Biol.*, *63*, 421 (1989).
2. M. Worwood, *Blood Rev.*, *4*, 259 (1990).
3. P. M. Harrison, S. C. Andrews, P. J. Artymiuk, G. C. Ford, J. R. Guest, J. Hirzmann, D. M. Lawson, J. C. Livingstone, J. M. A. Smith, and A. Treffry, *Adv. Inorg. Chem.*, *36*, 449 (1991).
4. S. C. Andrews, P. Arosio, W. Bottke, J. F. Briat, M. Von Darl, P. M.

Harrison, J. P. Laulhere, S. Levi, S. Lobreaux, and S. J. Yewdall, *J. Inorg. Biochem.*, *47*, 161 (1992).

5. M. J. Grossman, S. M. Hinton, V. Minak-Bernero, C. Slaughter, and E. I. Stiefel, *Proc. Natl. Acad. Sci. USA*, *89*, 2419 (1992).
6. W. H. Massover, *Micron*, *24*(4), 389 (1993).
7. G. S. Waldo and E. C. Theil, in *Comprehensive Supramolecular Chemistry* (J. L. Atwood, J. E. D. Davies, D. D. MacNicol, and F. Vogtle, eds.), Pergamon Press, New York, 1996, p. 65.
8. J. F. Briat, *J. Plant Nutr.*, *19*, 1331 (1996).
9. G. C. Ford, P. M. Harrison, D. W. Rice, J. M. A. Smith, A. Treffry, J. L. White, and J. Yariv, *Phil. Trans. R. Soc. Lond.*, *304*, 551 (1984).
10. D. M. Lawson, P. J. Artymiuk, S. J. Yewdall, J. M. A. Smith, J. C. Livingstone, A. Treffry, A. Luzzago, S. Levi, P. Arosio, G. Cesareni, C. D. Thomas, W. V. Shaw, and P. M. Harrison, *Nature*, *349*, 541 (1991).
11. J. F. Briat, I. Fobis-Loisy, N. Grignon, S. Lobréaux, N. Pascal, G. Savino, S. Thoiron, N. Von Wiren, and O. Van Wuytswinkel, *Biol. Cell*, *84*, 69 (1995).
12. P. D. Hempstead, A. J. Hudson, P. J. Artymiuk, S. C. Andrews, M. J. Banfield, J. R. Guest, and P. M. Harrison, *FEBS Lett.*, *350*, 258 (1994).
13. P. M. Harrison, S. C. Andrews, G. C. Ford, J. M. A. Smith, A. Treffry, and J. L. White, in *Iron Transport in Microbes, Plants and Animals* (G. Winkelmann, D. van der Helm, and J. B. Neilands, eds.) VCH, Weinheim, Germany, 1987, p. 445.
14. P. M. Harrison and P. Arosio, *Biochim. Biophys. Acta Bio-Energetics*, *1275*, 161 (1996).
15. A. Desideri, S. Stefanini, F. Polizio, R. Petruzzelli, and E. Chiancone, *FEBS Lett.*, *287*, 10 (1991).
16. S. Levi, P. Santambrogio, B. Corsi, A. Cozzi, and P. Arosio, *Biochem. J.*, *317*, 467 (1996).
17. A. Treffry, E. R. Bauminger, D. Hechel, N. W. Hodson, I. Nowik, S. J. Yewdall, and P. M. Harrison, *Biochem. J.*, *296*, 721 (1993).
18. S. Stefanini, A. Desideri, P. Vecchini, T. Drakenberg, and E. Chiancone, *Biochemistry*, *28*, 378 (1989).
19. S. Levi, A. Luzzago, F. Franceschinelli, P. Santambrogio, G. Cesareni, and P. Arosio, *Biochem. J.*, *264*, 381 (1989).

20. M. A. Michaux, A. Dautant, B. Gallois, T. Granier, B. L. D'Estaintot, and G. Precigoux, *Proteins*, *24*, 314 (1996).
21. J. Trikha, G. S. Waldo, F. A. Lewandowski, Y. Ha, E. C. Theil, P. C. Weber, and N. M. Allewell, *Proteins*, *18*, 107 (1994).
22. J. K. Lakowicz and B. P. Maliwal, *J. Biol. Chem.*, *258*, 4794 (1983).
23. M. J. Yablonski and E. C. Theil, *Biochemistry*, *31*, 9680 (1992).
24. W. H. Massover, in *Iron Biominerals* (R. B. Frankel and R. P. Blakemore, eds.), Plenum Press, New York, 1990, p. 349.
25. B. Webb, J. Frame, Z. Zhao, M. L. Lee, and G. D. Watt, *Arch. Biochem. Biophys.*, *309*, 178 (1994).
26. M. Lee, P. Arosio, A. Cozzi, and N. D. Chasteen, *Biochemistry*, *33*, 3679 (1994).
27. D. Jacobs, G. D. Watt, R. B. Frankel, and G. C. Papaefthymiou, *Biochemistry*, *28*, 9216 (1989).
28. S. Hilty, B. Webb, R. B. Frankel, and G. D. Watt, *J. Inorg. Biochem.*, *56*, 173 (1994).
29. N. D. Chasteen and E. C. Theil, *J. Biol. Chem.*, *257*, 7672 (1982).
30. J. G. Wardeska, B. Viglione, and N. D. Chasteen, *J. Biol. Chem.*, *261*, 6677 (1986).
31. Y. G. Cheng and N. D. Chasteen, *Biochemistry*, *30*, 2947 (1991).
32. E. R. Bauminger, P. M. Harrison, I. Nowik, and A. Treffry, *Hyperfine Interactions*, *42*, 873 (1988).
33. E. R. Bauminger, P. M. Harrison, I. Nowik, and A. Treffry, *Biochemistry*, *28*, 5486 (1989).
34. E. R. Bauminger, P. M. Harrison, D. Hechel, I. Nowik, and A. Treffry, *Biochim. Biophys. Acta*, *1118*, 48 (1991).
35. E. R. Bauminger, P. M. Harrison, D. Hechel, N. W. Hodson, I. Nowik, A. Treffry, and S. J. Yewdall, *Biochem. J.*, *296*, 709 (1993).
36. A. Treffry, A. Zhao, M. A. Quail, J. R. Guest, and P. M. Harrison, *Biochemistry*, *36*, 432 (1997).
37. H. Heqing, R. K. Watt, R. B. Frankel, and G. D. Watt, *Biochemistry*, *32*, 1681 (1993).
38. P. Arosio, T. G. Adelman, and J. W. Drysdale, *J. Biol. Chem.*, *253*, 4451 (1978).
39. S. Levi, A. Luzzago, G. Cesareni, A. Cozzi, F. Franceschinelli, A. Albertini, and P. Arosio, *J. Biol. Chem.*, *263*, 18086 (1988).

40. S. Levi, J. Salfeld, F. Franceschinelli, A. Cozzi, M. H. Dorner, and P. Arosio, *Biochemistry*, *28*, 5179 (1989).
41. P. Santambrogio, S. Levi, A. Cozzi, B. Corsi, and P. Arosio, *Biochem. J.*, *314*, 139 (1996).
42. R. R. Crichton, A. Herbas, O. Chavez-Alba, and F. Roland, *J. Biol. Inorg. Chem.*, *1*, 567 (1996).
43. D. M. Lawson, A. Treffry, P. J. Artymiuk, P. M. Harrison, S. J. Yewdall, A. Luzzago, G. Cesareni, S. Levi, and P. Arosio, *FEBS Lett.*, *254*, 207 (1989).
44. V. J. Wade, S. Levi, P. Arosio, A. Treffry, P. M. Harrison, and S. Mann, *J. Mol. Biol.*, *221*, 1443 (1991).
45. S. Levi, S. J. Yewdall, P. M. Harrison, P. Santambrogio, A. Cozzi, E. Rovida, A. Albertini, and P. Arosio, *Biochem. J.*, *288*, 591 (1992).
46. S. Levi, B. Corsi, E. Rovida, A. Cozzi, P. Santambrogio, A. Albertini, and P. Arosio, *J. Biol. Chem.*, *269*, 30334 (1994).
47. S. Levi, P. Santambrogio, A. Albertini, and P. Arosio, *FEBS Lett.*, *336*, 309 (1993).
48. S. Sun, P. Arosio, S. Levi, and N. D. Chasteen, *Biochemistry*, *32*, 9362 (1993).
49. A. Treffry, J. Hirzmann, S. J. Yewdall, and P. M. Harrison, *FEBS Lett.*, *302*, 108 (1992).
50. E. R. Bauminger, P. M. Harrison, D. Hechel, I. Nowik, and A. Treffry, *Proc. R. Soc. Lond. B*, *244*, 211 (1991).
51. E. R. Bauminger, A. Treffry, A. J. Hudson, D. Hechel, N. W. Hodson, S. C. Andrews, S. Levi, I. Nowik, P. Arosio, J. R. Guest, and P. M. Harrison, *Biochem. J.*, *302*, 813 (1994).
52. S. Levi, P. Santambrogio, A. Cozzi, E. Rovida, B. Corsi, E. Tamborini, S. Spada, A. Albertini, and P. Arosio, *J. Mol. Biol.*, *238*, 649 (1994).
53. P. Santambrogio, S. Levi, A. Cozzi, E. Rovida, A. Albertini, and P. Arosio, *J. Biol. Chem.*, *268*, 12744 (1993).
54. N. D. Chasteen, S. Sun, S. Levi, and P. Arosio, in *Progress in Iron Research* (C. Hershko, A. M. Konijn, and P. Aisen, eds.), Plenum Press, New York, 1994, p. 23.
55. P. Santambrogio, S. Levi, P. Arosio, L. Palagi, G. Vecchio, D. M. Lawson, S. J. Yewdall, P. J. Artymiuk, P. M. Harrison, R. Jappelli, and G. Cesareni, *J. Biol. Chem.*, *267*, 14077 (1992).

56. S. Stefanini, S. Cavallo, C. Q. Wang, P. Tataseo, P. Vecchini, A. Giartosio, and E. Chiancone, *Arch. Biochem. Biophys.*, *325*, 58 (1996).
57. A. Treffry, J. M. Sowerby, and P. M. Harrison, *FEBS Lett.*, *95*, 221 (1978).
58. B. Xu and N. D. Chasteen, *J. Biol. Chem.*, *266*, 19965 (1991).
59. G. S. Waldo and E. C. Theil, *Biochemistry*, *32*, 13262 (1993).
60. D. De Silva and S. D. Aust, *Arch. Biochem. Biophys.*, *298*, 259 (1992).
61. S. Sun and N. D. Chasteen, *J. Biol. Chem.*, *267*(35), 25160 (1992).
62. V. M. Samokyszyn, D. W. Reif, D. M. Miller, and S. D. Aust, *Free Radical Res. Commun.*, *12-3*, 153 (1991).
63. D. De Silva, D. M. Miller, D. W. Reif, and S. D. Aust, *Arch. Biochem. Biophys.*, *293*, 409 (1992).
64. D. M. De Silva, and S. D. Aust, *Can. J. Physiol. Pharmacol.*, *71*, 715 (1993).
65. J. H. Guo, M. Abedi, and S. D. Aust, *Arch. Biochem. Biophys.*, *335*, 197 (1996).
66. A. Treffry, D. Gelvan, A. M. Konijn, and P. M. Harrison, *Biochem. J.*, *305*, 21 (1995).
67. Y. C. Barrett, *Characterization of Radical Formation and Iron Oxidation/Hydrolysis During Iron Deposition in Ferritin*, Ph.D. thesis, University of New Hampshire, Durham, 1994.
68. A. Treffry, Z. Zhao, M. A. Quail, J. R. Guest, and P. M. Harrison, *Biochemistry*, *34*, 15204 (1995).
69. J. K. Grady, Y. Chen, N. D. Chasteen, and D. C. Harris, *J. Biol. Chem.*, *264*, 20224 (1989).
70. S. Sun and N. D. Chasteen, *Biochemistry*, *33*, 15095 (1994).
71. B. J. Bolann and R. J. Ulvik, *FEBS Lett.*, *318*, 149 (1993).
72. Y. Chen-Barrett, P. M. Harrison, A. Treffry, M. A. Quail, P. Arosio, P. Santambrogio, and N. D. Chasteen, *Biochemistry*, *34*, 7847 (1995).
73. P. M. Harrison, G. C. Ford, D. W. Rice, J. M. A. Smith, A. Treffry, and J. L. White, in *Frontiers in Bioinorganic Chemistry* (A. V. Xavier, ed.), VCH, Weinheim, Germany, 1986, p. 268.
74. G. S. Waldo, J. Ling, J. Sanders-Loehr, and E. Theil, *Science*, *259*, 796 (1993).

75. S. Sun, *Kinetics and Mechanism of Iron Oxidation in Apoferritin*, Ph.D. thesis, University of New Hampshire, Durham, 1993.

76. E. R. Bauminger, P. M. Harrison, D. Hechel, I. Nowik, and A. Treffry, *Nucl. Instrum. Meth. Phys. Res. B*, *76*, 1 (1993).

77. N. D. Chasteen, B. C. Antanaitis, and P. Aisen, *J. Biol. Chem.*, *260*, 2926 (1985).

78. K. J. A. Davies and M. E. Delsignore, *J. Biol. Chem.*, *262*, 9908 (1987).

79. K. J. A. Davies, S. W. Lin, and R. E. Pacifici, *J. Biol. Chem.*, *262*, 9914 (1987).

80. K. J. A. Davies, *J. Biol. Chem.*, *262*, 9895 (1987).

81. K. J. A. Davies, M. E. Delsignore, and S. W. Lin, *J. Biol. Chem.*, *262*, 9902 (1987).

82. J. K. Grady and N. D. Chasteen, Radical Chemistry of Ferritin During Iron Deposition. Functional Significance?, *3rd Winter Research Conference on Free Radicals*, Les Arcs, France, March 9–14, 1997.

83. J. R. Mertz and E. C. Theil, *J. Biol. Chem.*, *258*, 11719 (1983).

84. J. K. Grady and N. D. Chasteen, in *Iron Biominerals* (R. B. Frankel and R. P. Blakemore, eds.), Plenum Press, New York, 1990, p. 315.

85. G. D. Watt, R. B. Frankel, G. C. Papaefthymiou, K. Spartalian, and E. I. Stiefel, *Biochemistry*, *25*, 4330 (1986).

86. S. C. Andrews, J. B. C. Findlay, J. R. Guest, P. M. Harrison, J. N. Keen, and J. M. A. Smith, *Biochim. Biophys. Acta*, *1078*, 111 (1991).

87. M. R. Cheesman, A. J. Thomson, C. Greenwood, G. R. Moore, and F. H. A. Kadir, *Nature*, *346*, 771 (1990).

88. N. M. K. Reid, D. P. E. Dickson, C. Greenwood, A. Thompson, F. H. A. Kadir, and G. R. Moore, *Biochem. J.*, *272*, 263 (1990).

89. S. C. Andrews, J. M. A. Smith, J. R. Guest, and P. M. Harrison, *Biochem. Biophys. Res. Commun.*, *158*, 489 (1989).

90. C. W. Hawkins, *J. Magn. Magn. Mater.*, *104*, 1549 (1992).

91. S. C. Andrews, J. M. A. Smith, C. Hawkins, J. M. Williams, P. M. Harrison, and J. R. Guest, *Eur. J. Biochem.*, *213*, 329 (1993).

92. M. R. Cheesman, N. E. Le Brun, F. H. A. Kadir, A. J. Thomson, G. R. Moore, S. C. Andrews, J. R. Guest, P. M. Harrison, J. M. A. Smith, and S. J. Yewdall, *Biochem. J.*, *292*, 47 (1993).

93. F. Frolow, A. J. Kalb, and J. Yariv, *Acta Crystallogr. D-Biol. Cryst.*, *49*, 597 (1993).

94. N. E. Le Brun, M. R. Cheesman, A. J. Thomson, G. R. Moore, S. C. Andrews, J. R. Guest, and P. M. Harrison, *FEBS Lett.*, *323*, 261 (1993).

95. J. P. Laulhere, A. M. Laboure, O. Van Wuytswinkel, J. Gagnon, and J. F. Briat, *Biochem. J.*, *281*, 785 (1992).

96. N. F. Inglis, K. Stevenson, A. H. F. Hosie, and J. M. Sharp, *Gene*, *150*, 205 (1994).

97. M. C. V. Pessolani, D. R. Smith, B. Rivoire, J. McCormick, S. A. Hefta, S. T. Cole, and P. J. Brennan, *J. Exp. Med.*, *180*, 319 (1994).

98. H. Winkler, W. Meyer, A. X. Trautwein, and B. F. Matzanke, *Hyperfine Interactions*, *91*, 841 (1994).

99. C. J. Carrano, R. Boehnke, and B. F. Matzanke, *FEBS Lett.*, *390*, 261 (1996).

100. C. N. Penfold, P. L. Ringeling, S. L. Davy, G. R. Moore, A. G. McEwan, and S. Spiro, *FEMS Microbiol. Lett.*, *139*, 143 (1996).

101. D. J. Evans, Jr., D. G. Evans, H. C. Lampert, and H. Nakano, *Gene*, *153*, 123 (1995).

102. A. J. Hudson, S. C. Andrews, C. Hawkins, J. M. Williams, M. Izuhara, F. C. Meldrum, S. Mann, P. M. Harrison, and J. R. Guest, *Eur. J. Biochem.*, *218*, 985 (1993).

103. F. Frolow, A. S. Kalb (Gilboa), and J. Yariv, *Nature Struct. Biol.*, *1*, 453 (1994).

104. S. C. Andrews, N. E. Le Brun, V. Barynin, A. J. Thomson, G. R. Moore, J. R. Guest, and P. M. Harrison, *J. Biol. Chem.*, *270*, 23268 (1995).

105. F. H. A. Kadir and G. R. Moore, *FEBS Lett.*, *276*, 81 (1990).

106. F. K. Al-Massad, F. H. A. Kadir, and G. R. Moore, *Biochem. J.*, *283*, 177 (1992).

107. S. C. Andrews, J. M. A. Smith, S. J. Yewdall, J. R. Guest, and P. M. Harrison, *FEBS Lett.*, *293*, 164 (1991).

108. N. E. Le Brun, M. T. Wilson, S. C. Andrews, J. R. Guest, P. M. Harrison, A. J. Thomson, and G. R. Moore, *FEBS Lett.*, *333*, 197 (1993).

109. C. Hawkins, A. Treffry, J. B. Mackey, J. M. Williams, S. C. Andrews, J. R. Guest, and P. M. Harrison, *Nuovo Cimento*, *18*, 347 (1996).

110. N. E. Le Brun, A. M. Keech, M. R. Mauk, A. G. Mauk, S. C. Andrews, A. J. Thomson, and G. R. Moore, *FEBS Lett.*, *397*, 159 (1996).
111. A. M. Keech, N. E. Le Brun, M. T. Wilson, S. C. Andrews, G. R. Moore, and A. J. Thomson, *J. Biol. Chem.*, *272*, 422 (1997).
112. J. S. Rohrer, Q. T. Islam, G. D. Watt, D. E. Sayers, and E. C. Theil, *Biochemistry*, *29*, 259 (1990).
113. E. V. Mielczarek, S. C. Andrews, and E. R. Bauminger, *Biol. Metals*, *5*, 87 (1992).
114. J. F. Briat, A. M. Laboure, J. P. Laulhere, A. M. Lescure, S. Lobreaux, H. Pesey, D. Proudhon, and O. Van Wuytswinkel, *J. Abadia*, *1*, 265 (1995).
115. S. Lobreaux, S. J. Yewdall, J. F. Briat, and P. M. Harrison, *Biochem. J.*, *288*, 931 (1992).
116. R. E. Wicks and B. Entsch, *Biochem. Biophys. Res. Commun.*, *192*, 813 (1993).
117. M. Ragland, J. F. Briat, J. Gagnon, J. P. Laulhere, O. Massenet, and E. C. Theil, *J. Biol. Chem.*, *265*, 18339 (1990).
118. O. Van Wuytswinkel, G. Savino, and J. F. Briat, *Biochem. J.*, *305*, 253 (1995).
119. S. Lobreaux and J. F. Briat, *Biochem. J.*, *274*, 601 (1991).
120. J. P. Laulhere, A. M. Laboure, and J. F. Briat, *J. Biol. Chem.*, *264*, 3629 (1989).
121. F. Barcelo, C. O. Arean, and G. R. Moore, *BioMetals*, *8*, 47 (1995).
122. V. J. Wade, A. Treffry, J. P. Laulhere, E. R. Bauminger, M. I. Cleton, S. Mann, J. F. Briat, and P. M. Harrison, *BBA-Protein Struct. Mol. Enzym.*, *1161*, 91 (1993).
123. J. P. Laulhere, A. M. Labouré, and J. F. Briat, *Biochem. J.*, *269*, 79 (1990).
124. J. P. Laulhere and J. F. Briat, *Biochem. J.*, *290*, 693 (1993).
125. G. S. Waldo, E. Wright, Z. H. Whang, J. F. Briat, E. C. Theil, and D. E. Sayers, *Plant Physiol.*, *109*, 797 (1995).
126. G. D. Watt and R. B. Frankel, in *Iron Biominerals* (R. B. Frankel and R. P. Blakemore, eds.), Plenum Press, New York, 1990, p. 307.
127. G. D. Watt, D. Jacobs, and R. B. Frankel, *Proc. Natl. Acad. Sci. USA*, *85*, 7457 (1988).
128. D. L. Jacobs, G. D. Watt, R. B. Frankel, and G. C. Papaefthymiou, *Biochemistry*, *28*, 1650 (1989).

129. G. D. Watt, R. B. Frankel, and G. C. Papaefthymiou, *Proc. Natl. Acad. Sci. USA*, *82*, 3640 (1985).

130. R. K. Watt, R. B. Frankel, and G. D. Watt, *Biochemistry*, *31*, 9673 (1992).

131. G. D. Watt, J. W. McDonald, C. H. Chiu, and K. R. N. Reddy, *J. Inorg. Biochem.*, *51*, 745 (1993).

132. N. D. Chasteen, I. M. Ritchie, and J. Webb, *Anal. Biochem.*, *195*, 296 (1991).

133. H. Durliat, A. Couteix, A. Bergel, and M. Comtat, *J. Chim. Phys.*, *90*, 1113 (1993).

134. F. H. A. Kadir, F. K. Al-Massad, and G. R. Moore, *Biochem. J.*, *282*, 867 (1992).

135. F. H. A. Kadir, F. K. Al-Massad, S. J. A. Fatemi, H. K. Singh, M. T. Wilson, and G. R. Moore, *Biochem. J.*, *278*, 817 (1991).

136. X. Yang and N. D. Chasteen, *Biophys. J.*, *71*, 1587 (1996).

137. A. Treffry, C. Hawkins, J. M. Williams, J. R. Guest, and P. M. Harrison, *J. Biol. Inorg. Chem.*, *1*, 49 (1996).

138. Y. Bando and K. Aki, *Biochem. Biophys. Res. Commun.*, *168*, 389 (1990).

139. B. J. Bolann and R. Ulvik, *Eur. J. Biochem.*, *193*, 899 (1990).

140. P. Biemond, A. J. G. Swaak, H. G. Van Eijk, and J. F. Koster, *Free Radical Biol. Med.*, *4*, 185 (1988).

141. R. F. Boyer, J. S. McArthur, and T. M. Cary, *Phytochemistry*, *29*, 3717 (1990).

142. R. F. Boyer, H. M. Clark, and A. P. LaRoche, *J. Inorg. Biochem.*, *32*, 171 (1988).

143. M. S. Joo, G. Tourillon, D. E. Sayers, and E. C. Theil, *Biol. Metals*, *3*, 171 (1990).

144. H. P. Monteiro and C. C. Winterbourn, *Biochem. Pharmacol.*, *38*, 4177 (1989).

145. H. P. Monteiro, G. F. Vile, and C. C. Winterbourn, *Free Radical Biol. Med.*, *6*, 587 (1989).

146. D. W. Reif and R. D. Simmons, *Arch. Biochem. Biophys.*, *283*, 537 (1990).

147. H. P. Monteiro and C. C. Winterbourn, *Arch. Biochem. Biophys.*, *271*, 536 (1989).

148. C. C. Winterbourn, G. F. Vile, and H. P. Monteiro, *Free Radical Res. Commun.*, *12-3*, 107 (1991).

149. F. Bonomi, A. Cerioli, and S. Pagani, *Biochim. Biophys. Acta*, *994*, 180 (1989).

150. R. E. Macur, R. A. Olsen, and W. P. Inskeep, in *Iron Nutrition and Interactions in Plants* (Y. Chen and Y. Hadar, eds.), Kluwer, Dordrecht, The Netherlands, 1991, p. 89.

151. M. Aubailly, R. Santus, and S. Salmon, *Photochem. Photobiol.*, *54*, 769 (1991).

152. J. J. Moreno, M. Foroozesh, D. F. Church, and W. A. Pryor, *Chem. Res. Toxicol.*, *5*, 116 (1992).

153. S. Ahmad, V. Singh, and G. S. Rao, *Chem. Biol. Interact.*, *96*, 103 (1995).

154. H. P. Monteiro, C. C. Winterbourn, and A. Stern, *Free Radical Res. Commun.*, *12-3*, 125 (1991).

155. K. R. Bridges and K. E. Hoffman, *J. Biol. Chem.*, *261*, 14273 (1986).

156. A. G. Rieth, S. L. Baader, H. N. Lode, G. Bruchelt, and D. Niethammer, *BioMetals*, *5*, 223 (1992).

157. S. L. Baader, G. Bruchelt, T. C. Carmine, H. N. Lode, A. G. Rieth, and D. Niethammer, *J. Cancer Res. Clin. Oncol.*, *120*, 415 (1994).

158. S. L. Baader, E. Bill, A. X. Trautwein, G. Bruchelt, and B. F. Matzanke, *FEBS Lett.*, *381*, 131 (1996).

159. G. R. Moore, F. H. A. Kadir, and F. K. Al-Massad, *J. Inorg. Biochem.*, *47*, 175 (1992).

160. F. H. A. Kadir, F. K. Al-Massad, and G. R. Moore, *Biochem. J.*, *282*, 867 (1992).

161. R. E. Cowart, F. L. Singleton, and J. S. Hind, *Anal. Biochem.*, *211*, 151 (1993).

162. D. W. Reif, *Free Radical Biol. Med.*, *12*, 417 (1992).

163. T. D. Richards, K. R. Pitts, and G. D. Watt, *J. Inorg. Biochem.*, *61*, 1 (1996).

164. G. D. Watt, R. B. Frankel, D. Jacobs, H. Huang, and G. C. Papaefthymiou, *Biochemistry*, *31*, 5672 (1992).

165. P. D. Hempstead, S. J. Yewdall, A. R. Fernie, D. M. Lawson, P. J. Artymiuk, D. W. Rice, G. C. Ford, and P. M. Harrison, *J. Mol. Biol.*, *268*, 424 (1997).

13

Ferritin. Its Mineralization

Annie K. Powell

School of Chemical Sciences, University of East Anglia,
Norwich NR4 7TJ, UK

1. INTRODUCTION 516

2. BIOMINERAL FORMATION AS A MEANS OF CONTROLLING IRON HOMEOSTASIS 517

3. MINERALIZATION 521
   - 3.1. The Hydrolytic Mineralization of Iron Species 521
     - 3.1.1. Mineral Formation 522
     - 3.1.2. Some Important Iron Minerals in Nature 525
     - 3.1.3. Iron Hydrolysis 527
   - 3.2. Biomineralization in Ferritin 529
     - 3.2.1. Mechanisms of Core Formation 529
     - 3.2.2. Nucleation and Growth 530
   - 3.3. Biological Control of Iron(III) Hydrolysis and Mineralization 532

4. CORE STRUCTURES 534
   - 4.1. Physical Methods for Probing Core Structures 534

4.1.1. Diffraction Experiments 534
4.1.2. Electron Microscopy 535
4.1.3. UV/Vis Studies 536
4.1.4. Magnetic Measurements 536
4.1.5. EXAFS Spectroscopy 538
4.2. Assignments of the Mineral Types Observed In Vivo 538
4.2.1. Mammalian Ferritins 538
4.2.2. Nonmammalian Ferritins 540
4.3. The Mineral Types Observed In Vitro 540
5. A COMPARISON OF THE USE OF MINERAL AND CLUSTER AGGREGATE MODELS FOR ELUCIDATING THE CORE MINERAL STRUCTURE IN FERRITINS 543
5.1. The Problem of Determining the Structures of Ferritin Mineral Cores 543
5.2. The Model Compound Approach to Elucidating the Core Structure 544
5.2.1. Properties to Model 544
5.2.2. Defining a Mineral Phase 545
5.2.3. Proposed Models 546
5.2.4. The Suitability of the Cluster Aggregates as Scale Models for Ferritin 548
5.3. Synthetic Aspects Modeling Biomineralization of Ferritin Cores 553
6. CONCLUSIONS 555
ABBREVIATIONS 556
REFERENCES 556

1. INTRODUCTION

Iron is vital to virtually every life form on earth, and yet in the lithosphere, the vast proportion of iron is locked up in the form of highly insoluble minerals making iron uptake a significant challenge to biological systems. Many organisms have developed means for recycling their hard-won iron and utilize the specialized protein ferritin to store iron excess to immediate requirements as a mineralized iron(III) oxyhydroxide phase.

The structural motifs of ferritin proteins of plants, bacteria, and animals are essentially identical with 24 subunits forming a spherical protein shell enclosing a cavity of some 7–8 nm in diameter. This cavity is reached by three-fold and four-fold channels [1]. The channels probably allow for the passage of water to and from the cavity and possibly also iron, although these suggestions are difficult to prove. This cavity provides a spherical space for the storage of iron in the form of a mineralized deposit. The composition of this mineral appears to vary in different phyla. The most extensively studied ferritins are mammalian, which in the native state tend to have cores largely composed of an iron(III) oxyhydroxide mineral. The cores observed in native bacterioferritins appear to contain significant proportions of phosphate which are thought to be an integral part of the mineralized structure. These cores are rather amorphous and difficult to study. Interestingly, when the core is removed from bacterioferritins, it can be reconstituted as an iron oxyhydroxide mineral similar to that found in mammalian ferritins and such systems have been used to monitor core formation in vitro. Ferritins found in plants are relatively poorly studied as are those in invertebrates such as molluscs [2].

In this chapter, the mineralization of iron in the context of the natural environment and its relationship to the mineralization which takes place on forming ferritin cores are explored. The methods which can be used to elucidate the structures and properties of these cores are described and reviewed.

2. BIOMINERAL FORMATION AS A MEANS OF CONTROLLING IRON HOMEOSTASIS

When considering the underlying chemistry of iron biomineral formation by organisms, it is helpful to look first at the chemistry of terrestrial iron and its common mineral forms [3–5]. The oxygen-rich atmosphere of the earth makes both Fe(O) and Fe(II) unstable under normal conditions at the earth's surface and they are oxidized to Fe(III) to give various forms of iron(III) oxide, Fe_2O_3. The low solubility of these oxides means that iron cycling is dominated by solid transfer reactions with most of the iron in the form of suspended solids of finely divided particles. For example, only about 0.1–0.2% of the iron transported by rivers to the oceans is thought to exist in solution. However, the relative stabilities of the +II and +III oxidation states of iron are such that only small deviations from "normal" conditions can favor the oxidation of Fe(II) to Fe(III) or else the reduction of Fe(III) to Fe(II) with accompanying large changes in solubility. The underlying influences in this are the relative values of electrode (redox) potential, E, and pH, one a measure of the potential of electron activity, the other of the potential proton activity in the system. The value of E is found using the Nernst equation and thus the activities (sometimes approximated by concentrations) need to be taken into consideration. Chemically, for metal ions, these are often affected by the presence of coordinating species which can alter the electrode potential for oxidation or reduction through molecular orbital interactions (e.g., by changing the splitting of d orbitals in transition metals) or else overall pH through the release of protons into the medium as, for example, in hydrolysis reactions. For iron in both terrestrial and biological contexts, the available range of E and pH is governed by the stability range for water, so redox potentials which would either reduce or oxidize water need not be considered. A useful means of illustrating the stable forms of chemical species at various values of electrode potential, E, and pH is in a Pourbaix diagram. The combination of Pourbaix diagrams for water and iron indicates the species stable at normal temperatures and pressures in the environment (Fig. 1). In this diagram, it is assumed that the water is pure and has a constant activity value of 1. The standard reduction potential for dioxygen to give water,

$$O_2 + 4H^+ + 4e^- \longrightarrow 2H_2O$$

at 25°C, 1 atm and pH = 0 is E = 1.23 V. Substitution into the Nernst equation gives:

$$E = 1.23 + \frac{RT}{nF} \cdot \ln \frac{[O_2][H^+]^4}{1}$$

Substituting values for the two constants R and F plus the temperature T, n, the number of moles of electrons transferred, converting from natural logarithms so we can now equate the concentration of H^+ with

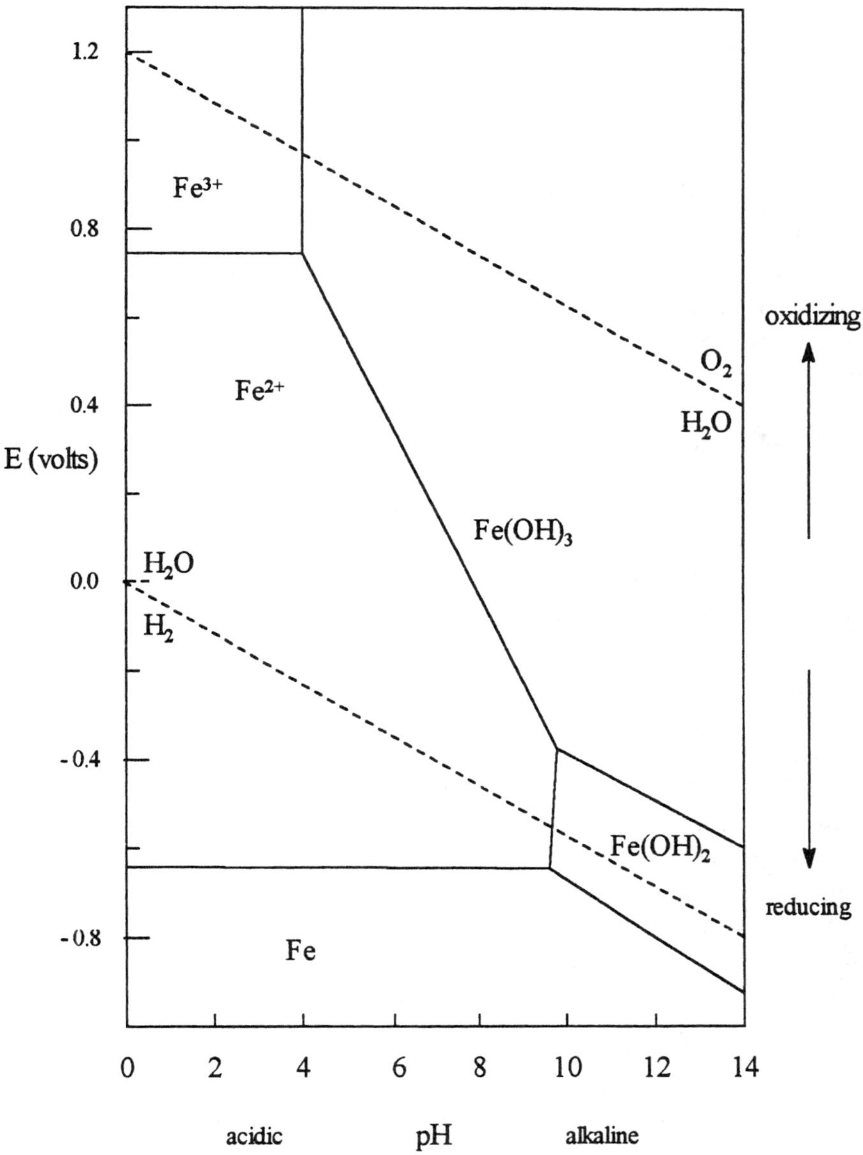

FIG. 1. The Pourbaix diagram for the iron-water system (adapted from [5]). The solid line is drawn for $[Fe^{2+}] = 10^{-7}$ M.

the negative value of pH and taking the partial pressure of oxygen into account as 0.21 atm at the earth's surface gives an electrode potential:

$$E = 1.23 + 0.0148 \log [O_2] + (4 \times 0.0148 \times \log [H^+])\ V$$

$$E = 1.23 + 0.0148 \log(0.21) - 0.059\ pH\ V$$

$$E = 1.22 - 0.059\ pH\ V$$

The diagram for the Fe/H_2O system must be drawn for each concentration of iron of interest and will also be different for other values of temperature and pressure. We can see from the diagram [5] that at low potentials and pH values, Fe(II) is quite stable giving iron in a highly soluble form, whereas at pH values around the physiological and normal environmental levels, Fe(III) predominates in the form of insoluble oxyhydroxide species unless the conditions are highly reducing. It can also be seen that, for example, in the region between pH 5 and 9, the interplay between the two oxidation states depends critically on the balance between pH and E. This has ensured that iron-dependent biology has been able to accommodate the change in the earth's atmosphere from reducing to oxidizing through the development of systems capable of tuning electrode potentials and local pH to favour the required oxidation state.

Many organisms have developed iron-storage molecules which obviate the need to take up the relatively large quantities of the metal needed for normal functioning. For example, humans contain between 2–5 g iron of which about two thirds is circulating in the red blood cells as hemoglobin and about 15–20% is in storage as ferritin and hemosiderin [2,6]. In healthy subjects, it appears that iron is available from both the storage molecules ferritin and hemosiderin. The remainder is found mostly in muscle myoglobin (~ 8%) and iron-containing enzymes. Only about 3 mg is found within the iron transport protein transferrin at any one time (although the daily exchange of iron through plasma transferrin is about 30 mg). This protein is responsible for shuttling iron between iron storage sites and sites of the biosynthesis of iron-containing proteins or the dismantling of spent iron-containing cells such as erythrocytes. Only about 1 mg iron per day is absorbed from food by healthy adults (about 10% of food iron), so overall iron balances are tightly controlled within the body, the major route for iron loss being through bleeding. Another important chemical factor in developing iron-storage molecules is the need to avoid the cellular damage which can be caused by free hydroxyl radicals produced from the Haber-Weiss reaction which is catalyzed by free iron [7].

Both iron deficiency and iron overload are serious conditions arising from defects or disease in parts of the homeostatic iron cycle. In iron-overloading syndromes such as primary and secondary hemochromatosis, the secondary iron-storage molecule, hemosiderin, becomes an important repository for the excess metal. In these conditions, the iron content of hemosiderin can increase up to 100-fold, whereas that of ferritin only increases 5- to 10-fold [8].

The normal storage of iron within the protein ferritin results in the formation of a mineralized iron(III) oxyhydroxide, or perhaps hydroxyphosphate, deposit which remains inert and "out of harm's way" until needed for the biosynthesis of iron proteins, such as hemoglobin, when the iron can be readily and easily mobilized. Given the highly insoluble nature of such iron(III) minerals encountered in the environment, as described in detail in Sec. 3, this is a remarkable achievement. The development of iron-storage proteins was a direct biological response to the huge decrease in bioavailability of iron once photosynthetic processes had led to the present oxidizing state of the atmosphere and the stabilization of largely insoluble forms of iron(III) over the more soluble iron(II). In the process, nature has achieved the means of forming an iron(III) mineral within the core of ferritin from which iron is readily accessed.

3. MINERALIZATION

3.1. The Hydrolytic Mineralization of Iron Species

In order to begin to address the question of the mineralization of ferritin cores, it is helpful to look in detail at iron mineralization processes in the natural environment. This in turn requires an understanding of the way in which chemical species form mineral phases and the physical factors which can affect the nature of the phase formed.

Biomineralization in general can be said to take place in aqueous environments; therefore the precipitation of mineralized deposits from water is of relevance to us. In the case of hydrolyzing species such as metal ions, the water can become an integral part of the solid phase in the form of oxide and hydroxide bridging anionic units as well as neutral coordinated water molecules. The presence of other coordinating species can affect the nature of the phase formed, but in the specific case of animal ferritin cores, it is first and foremost the presence of water which is the dominant influence on phase morphology. The situation for native bacterioferritins which contain significant quantities of phosphate will be discussed later.

3.1.1. Mineral Formation

There are two key processes in solid-phase formation, whether it be from gaseous or liquid phases. These are nucleation and crystal growth [9–11]. The major requirement for the production of a solid phase is that its components supersaturate the medium they are in. In aqueous solutions, we would equate this with exceeding the solubility product of the solid phase, and such solutions are termed metastable. The reason that a supersaturated solution can persist without precipitation occurring is linked to the thermodynamic and kinetic influences on nucleation and crystal growth. The first requirement for the formation of a solid phase is that nucleation occurs, and this process can be regarded as the overcoming of the activation energy barrier to the spontaneous formation of the solid phase [12]. If this energy barrier is not overcome, metastable solutions persist. The relative supersaturation of a solution can be defined as the difference between the concentration of the solute at any instance, Q, and the equilibrium solubility, S, as a fraction of S; that is, $(Q - S)/S$. Supersaturation of a solution thus occurs when $Q > S$, and this can be relieved through precipitate formation. Nucleation involves the aggregation of the constituent species of the solid phase to form "nuclei" containing perhaps as few as four or five constituent species. Such nuclei can become the solid phase once some critical nucleus size has been attained. Prior to this, the rate of redissolution of the constituents will equal or be greater than the rate of aggregation and the solid phase cannot persist. A major influence on the rate of dissolution is the thermodynamic balance between the energy required to maintain the solid and the dissolution energy. This is compounded by the entropic balances which initially favor dissolution rather than lattice formation, but the key parameter is probably the surface energy term associated with the nucleus [13]. When the nucleus is small, the surface area to radius ratio is large, and it is energetically unfavorable to form the solid phase. Once the nucleus reaches a critical size, the surface to radius ratio becomes smaller and the favorable lattice energy term takes over. At this point, the solid phase begins to precipitate out of the solution. Subsequently, the solid can precipitate further through the formation of new nuclei giving very finely divided material, which may even be colloidal, or else further components of the solid phase can add to the existing nuclei to give crystal growth. It is also possible to have a combination of these. In general terms, the more highly supersaturated a solution is, the more likely that nucleation will dominate, and the less supersaturated, the more likely that crystal growth will be dominant. Studies show that in vitro the particle size of the precipitate varies inversely with the average degree of supersaturation [13]. Furthermore, the rate of nucleation increases exponentially as a function of relative supersaturation, whereas crystallization follows a linear dependence. Overall, the optimum conditions for the formation of crystalline rather than colloidal precipitates would be to attain a sufficient level of supersaturation for some nucleation to occur and then, as the relative supersaturation drops as a result of the precipitation of these nuclei, crystal growth takes over.

The formation of a solid phase can be summarized in terms of the relationship (Eq. (1)) [14]:

$$J_N = A \exp(-\Delta G_N^*/kT) \tag{1}$$

where,

J_N is the rate of nucleation

ΔG_N^* is the activation energy of nucleation

k is Boltzmann's constant

T is temperature in K

A is a preexponential factor related to the relative supersaturation

Furthermore, ΔG_N^* is dependent on the competing forces of solid phase formation and particle dissolution (Eq. [2]) [13]:

$$\Delta G_N^* = -(4\pi r^3/3)(1/V)kT \log_e(P/K_d) + (4\pi r^2)Z \tag{2}$$

where,

r is the radius of the nucleus and $4\pi r^3/3$ and $4\pi r^2$ are the volume and surface area of this nucleus, respectively

V is the molecular volume of the mineral
k is Boltzmann's constant
T is the temperature in K
P is the ionic activity product
K_d is the stability constant for the mineral
Z is the surficial, or surface, energy

The term $\{-(4\pi r^3/3)(1/V)kT \log_e (P/K_d)\}$ represents the stability of the crystal nucleus and depends on the number of interacting species. It favors the survival of the crystal nucleus and is a negative energy term. The term $\{(4\pi r^2)Z\}$ represents the forces of dissolution or ion exchange associated with the crystal surface and acts toward the disintegration of the forming nucleus.

A further variable in this is the existence of polymorphs of a given solid as well as closely related morphologies among oxide, oxyhydroxide, and hydroxide minerals. We shall see later in Sec. 3.1.3 the importance and significance of this in iron mineralization. For polymorphic phases, it is likely that differences in the nucleation steps lead to the different crystal structures. According to Lussac's law [15,16], thermodynamically, at the lowest degree of relative supersaturation, only the most insoluble form can nucleate and grow. At higher degrees of relative supersaturation, the probability of the most soluble form predominating increases. Clearly the changing relative supersaturation on solid-phase formation can result in mixtures of phases precipitating from the same medium. The preexponential term, A, in Eq. (1), which depends on relative supersaturation, is the major influence here.

Although these factors have been quantified to a certain extent in a number of books and articles, it is also clear that there are many other, perhaps more subtle, influences which can modify the formation of solid phases. A straightforward example is the observation that the presence of "innocent" anions which are not incorporated in the final solid phase can nonetheless affect the morphology in the sense that one polymorph can be favored over another in spite of expectations to the contrary. This could be the result of a modification of the ionic strength of the solution or might be a consequence of the anions associating with the nuclei in the early stages of precipitation and affecting particle growth at the surface; that is, by affecting the ionic activity product or relative supersaturation.

3.1.2. Some Important Iron Minerals in Nature

Two further parameters of special relevance to iron mineralization are the pH and the redox potential of the system. These can have a profound effect on the stability of a given mineral phase and Pourbaix plots of the type discussed in Sec. 2 can help in rationalizing the mineral morphologies observed in the environment. However, these are usually plotted for a given temperature and pressure and both of these can be further influences on the stabilized phase. These plots also assume that there are no kinetic reasons which might prevent a given phase from forming or other factors to consider (e.g., ionic strength, presence of growth inhibitors or promoters). In addition, the interaction of iron ions with the solvent medium in the process of hydrolysis is likely to produce changes in hydrogen ion concentration, and since the redox potential is also pH dependent, this can lead to different mineral phases being favored. In natural aqueous systems, there are many different dissolved species in addition to the iron and water. These can affect the speciation and oxidation state of the iron by reacting with it to form relatively insoluble compounds (e.g., sulfide or phosphate) or stable complexes (e.g., organic acids) which prevent the iron being precipitated. Pourbaix diagrams taking all the possibilities into account (including the concentration dependencies) would be too complicated to draw, but an approximation is given in Fig. 2, which includes the effect of carbon dioxide and sulfide, both common in natural environments, on mineral phase of the iron [5]. As before, the plot takes into account the limits on the redox potential imposed by water. From this, at physiological pH values soluble Fe(II), siderite ($FeCO_3$), pyrites (FeS_2), and hematite/goethite (Fe_2O_3/Fe·O·OH) are all possibilities. Although hematite is the most stable oxide form, goethite and other hydrated oxides are almost as stable and just as insoluble and will tend to form first, aging only slowly into hematite [3–5,17]. What the diagram cannot show is the events leading up to the mineral formation and the intermediate mineral forms which precede these most stable ones. In terms of the iron oxide phases, hematite and magnetite, we can see that hematite is preferred over a wide range of pH and electrode potential, whereas magnetite is only favored in a narrow range of electrode potential at high pH values. It is also clear from this diagram that in a

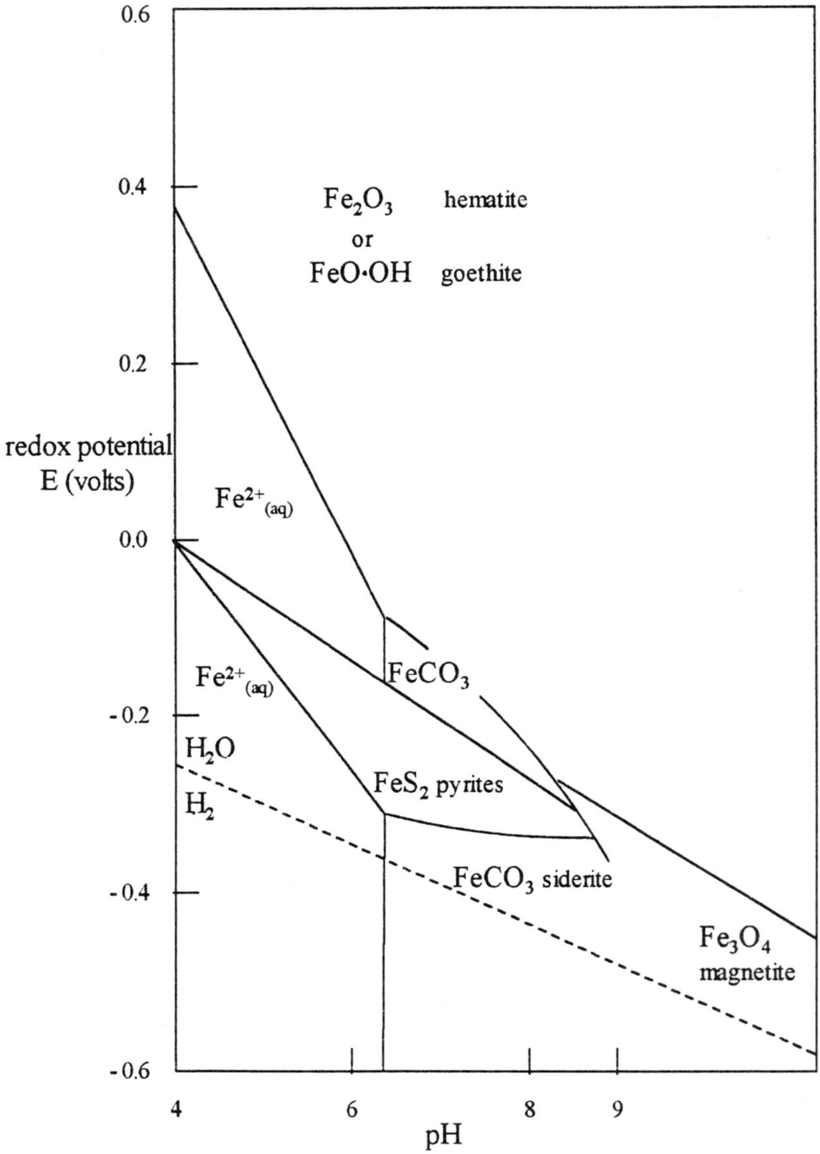

FIG. 2. The Pourbaix diagram for the Fe-H_2O-CO_2-S system (adapted from [5]). $[Fe^{2+}] = 10^{-6}$ M, $[CO_3^{2-}] = 1$ M, $[S^{2-}] = 10^{-6}$ M.

given system containing these components, it is possible to transform from one phase to another through changes in either pH, or electrode potential, or both.

3.1.3. Iron Hydrolysis

From Fig. 2 it can be seen that iron at physiological pH values is expected to be in the +III oxidation state unless the local redox potential is close to or less than zero. As noted before, iron(III) tends to exist in the form of oxides and oxyhydroxides which form through hydrolytic mineralization reactions. The likely course of these reactions gives us some insights into the way in which Nature's control of mineral phase in biological systems can be used to great advantage. Under highly acidic conditions, the hexa-aqua ion $[Fe(H_2O)_6]^{3+}$ can exist. As pH rises, the activation of O−H bonds of the coordinated water molecules becomes more important and hydroxy species are formed. These in turn can deprotonate to give oxides. In the main, these hydroxide and oxide moieties have a tendency to bridge to further metal centers unless steric or hydrogen-bonding constraints prevent this. Accompanying this is the release of water molecules (or hydroxide ions) and protons. From a different viewpoint, we can see this aggregation as a means of balancing the build-up of negative charge around the metal center. We can summarize this in a simplistic way by concentrating on the coordination environment of a single iron(III) center to follow the course of events leading to oxyhydroxide and oxide formation (Eq. (3)).

$$\begin{aligned}[Fe(H_2O)_6]^{3+} &\rightleftharpoons [Fe(H_2O)_3(OH)_3] + 3H^+ \\ &\rightleftharpoons [Fe(OH)_6]^{3-} + 6H^+ \\ &\rightleftharpoons [Fe(OH)_2(O)_3]^{6-} + 9H^+ \\ &\rightleftharpoons [Fe(O)_6]^{9-} + 12H^+ \end{aligned} \quad (3)$$

For which the overall reactions in terms of the well-defined oxyhydroxide, Fe·O·OH, and oxide minerals, Fe_2O_3 are (Eqs. (4) and (5))

$$[Fe(H_2O)_6]^{3+} \rightleftharpoons Fe{\cdot}O{\cdot}OH + 3H^+ + 4H_2O \quad (4)$$

$$[Fe(H_2O)_6]^{3+} \rightleftharpoons 0.5Fe_2O_3 + 3H^+ + 4.5H_2O \quad (5)$$

The hydrolytic reactions observed in reality result first in the formation of gelatinous precipitates of "ferric hydroxide." This is sometimes formulated as $Fe(OH)_3$ or $Fe_2O_3{\cdot}nH_2O$. Many studies have been performed following the formation and ageing of this material ([17] and references therein; [18]). A number of points emerge of relevance to the biological situation in ferritin. First, the material is probably best described as a mixture of related phases of a hydrated oxyhydroxide material which has been identified as the mineral "ferrihydrite". Second, when first formed, the material is a polydisperse colloidal precipitate of nanoscale particles. Third, depending on conditions, the material ages slowly under ambient conditions into the oxyhydroxide minerals lepidocrocite, γ-Fe·O·OH, in which the oxygen-containing layers are cubic close packed (ccp), and goethite, α-Fe·O·OH, in which the layers are hexagonally close packed (hcp). Finally, the hcp oxide mineral hematite, α-Fe_2O_3, is formed, although this can take years under normal conditions. The corresponding γ-phase oxide mineral, γ-Fe_2O_3, is maghemite and can be formed by heating lepidocrocite. In all these minerals, the iron centers occupy octahedral holes in layers of oxygen atoms. It is commonly held that the mineral phases which make up "ferrihydrite" are made up of hcp-arranged oxygen double layers with iron centers in octahedral environments. The difference from the better-defined mineral phases is that the oxygen layers derive from more hydroxide and water ligands, supporting the idea that hematite is the endproduct of the dehydration of $Fe_2O_3 \cdot nH_2O$. The overall stoichiometry of ferrihydrite is found to correspond to a mixture of Fe·O·OH and Fe_2O_3 plus water molecules and has been variously formulated as $Fe_5HO_8 \cdot 4H_2O$ by Towe and Bradley [19] and as $5Fe_2O_3 \cdot 9H_2O$ by Chukrov et al. [20], but the equivalent formula used by Russell of $Fe_2O_3 \cdot$ 2Fe·O·OH·$2.6H_2O$ underlines the presence of hydroxide as well as oxides [21]. It appears likely that the material comprises mineral arrays related to the α-phase oxides and oxyhydroxides, hematite and goethite, which also display hexagonally close-packed oxygen-containing layers with octahedrally coordinated iron centers in the interstices [22]. However, there is still a certain amount of controversy over the true structure of the material and attempts to characterize it satisfactorily have been hampered by differences in preparation procedures which can affect particle size distributions and degree of hydration, as well as in the interpretation of data in terms of a structural model. It is probably safest to describe ferrihydrite as corresponding to a mixture of related phases of hydrated ferric oxide [23–25]. Perhaps pertinent to this debate is the predominance of hcp brucite-type, $M(OH_2)$, structures in mineralized sediments such as clays which would tend to support a structural model based on hcp arrays of oxygens rather than a ccp array.

In its freshly precipitated, highly hydrated oxide form, the iron(III) is more likely to redissolve than when in the partially or fully dehydrated form. This is partly a consequence of the nanoscale proportions of the precipitated particles which lead to a high surface area to radius ratio. As we have seen above in Eq. (2), this tends to favor redissolution rather than particle growth. It would seem that in this specific case, the aging process leading to the loss of protons and water and subsequent linking of metal centers to form the better-defined goethite and hematite phases is a key element in favoring particle growth. In other words, the aging of the material and the overcoming of the unfavorable surface energy terms are intimately linked in producing the highly insoluble precipitates of oxyhydroxide or oxide iron(III) minerals which dominate in the environment and are a central problem in iron uptake by organisms. In fact, the highly insoluble nature of these iron deposits has been a key feature in the development efficient mechanisms for maintaining iron balances and of iron-storage systems such as ferritin.

3.2. Biomineralization in Ferritin

3.2.1. Mechanisms of Core Formation

The way in which the iron mineral forms is difficult to elucidate. The various models for the uptake of iron into the cavity have been reviewed recently by Chasteen in Chapter 12 of this book, Le Brun et al. [26], and Harrison and Arosio [2]. It is known that using iron(II) as a source, as is thought to be the case in vivo, an iron(III) oxyhydroxide mineral forms within the protein cavity both in the presence and absence of oxygen. The precise identity of this mineral is difficult to ascertain for a number of reasons. First, experiments performed in vivo and in vitro suggest that the mineral form can be influenced by the presence of other ions such as phosphate, as discussed in Sec. 4 of this chapter. Second, ferritins load with variable amounts of iron giving a distribution of core sizes making it difficult to perform many measurements such as x-ray diffraction, as explained in detail in Sec. 5 of this chapter. Third, the situation in vivo is much harder to follow than that in vitro, and it is by no means certain that experiments performed in vitro model the in vivo situation. Additionally, the in vitro experiments have often been inconclusive in terms of structural studies on the cores, although this must be the most promising way forward in developing means to analyze core structures of ferritins. Before discussing these difficulties in more depth, the possible course of events in the build-up of ferritin cores is addressed.

In mammals, the iron delivered by transferrin to ferritin is probably taken up into endocytic vesicles by receptor-mediated endocytosis [27]. The release of the iron is probably mediated by protons and helped by binding the receptor [28–30]. The chemical form of the iron and the intracellular donor for delivery to ferritin are not known. It has been suggested that storage involves accumulation of iron by binding to the core particle surface [31]. However, the observation that most forms of ferritin so far sequenced apart from vertebrate L chains contain residues associated with ferroxidase activity has led to the assumption that iron uptake is an oxidative process starting from iron(II) [1].

3.2.2. *Nucleation and Growth*

Once the iron ions have reached the inner cavity, it is thought that they first attach to nucleation sites on the inner surface of the protein shell. The exact identity of these nucleation sites is not known for certain, but it is possible that they are carboxylate residues. The iron mineral will then grow from such sites [1]. As we have seen above, the theories of nucleation and crystallization suggest that it is likely that only one or two crystallites will grow to any size once a critical particle size has been achieved and crystal growth can dominate over nucleation so that any new nuclei which form will redissolve, whereas atoms adding to the dominant crystallite(s) will contribute to the continued reduction in the importance of the unfavorable surficial energy term.

Since the ferritin core is composed of an iron(III) oxyhydroxide or oxide mineral, an obvious first step in attempting to elucidate the nature of the mineral core is to examine the properties of naturally occurring minerals. The experimental fact that it is hard to load more than a few thousand iron atoms into apoferritin also gives an indication of the likely density of the core. If we make the assumption that a fully loaded mammalian ferritin can hold 4500 iron centers and we take the preferred mineral model [1,19], ferrihydrite, as an example, we can calculate that, based on the formula of $5Fe_2O_3{\cdot}9H_2O$ (corresponding to $96.2\ g{\cdot}mol^{-1}$ per iron) and a cavity volume of $248\ nm^3$ (internal diameter of 7.8 nm), the density of the mineral would be 2.9 g·cm^{-3}, which is on the low side of the range observed for ferrihydrites [23]. The same calculation assuming a formula for the mineral of Fe·O·OH gives a density of 2.7 g·cm^{-3}, which is a long way from the experimentally and crystallographically determined values of 4.27 g·cm^{-3}, for goethite, α-Fe·O·OH, and 3.97 g·cm^{-3}, for lepidocrocite γ-Fe·O·OH [32]. Worse yet, the calculation for Fe_2O_3 gives a value of 2.4 g·cm^{-3}, which is less than half the value of 5.26 g·cm^{-3} observed for hematite, α-Fe_2O_3 [32]. These numbers suggest that if the cores are truly loaded to capacity with 4500 iron centers, then the mineral type is unlikely to be a pure oxide or even oxyhydroxide such as hematite, goethite, or lepidocrocite. We can see that the density of these minerals decreases as we introduce more protons into the lattice, and we can thus expect the ferritin mineral to contain relatively large proportions of these, as ferrihydrite does. The alternative is that the ferritin molecules never achieve full loading. For example, the reverse calculation for the most extreme case of Fe_2O_3 as hematite suggests that a ferritin fully loaded with this mineral would need to accommodate about 9800 iron centers.

The mineralization of iron in aqueous environments is relevant to the mineralization of the ferritin core, but it must be borne in mind that the presence of the protein shell and the local control of pH and redox potential which the protein can exert can have profound effects on the mineralization process. The fact that the cavity of ferritin can only accommodate a particle of 7–8 nm in diameter is one factor sometimes overlooked. The interactions of the surface with the mineral seem to be a key feature in the stabilization of the core. The widely accepted model of core formation has iron centers attaching to nucleation sites on this inner surface and the mineral growing to fill the cavity. It has been suggested that the cores observed in samples of mammalian ferritins correspond to the mineral phases ferrihydrite. It is noteworthy that ferrihydrite, although producing polydisperse nanoscale particles with size distributions similar to those found in ferritin cores, is likely to nucleate and grow in the opposite fashion to that suggested for ferritin, since there is no templating or nucleating effect from a boundary. Thus ferrihydrite particles, which are spherical unless growth inhibitors or promoters are present in solution [17,23,33], grow to give particles of a few nanometers in diameter from the center outward, whereas ferritin cores grow across the cavity starting from its inner surface. The formation of ferrihydrite colloids also gives insights into the chemical formula of the material as a hydrated oxide/oxyhydroxide material.

Clearly, for a spherical, bounded particle to maintain a charge balance, some means of compensating for a build-up of negative charge must be found, particularly at the surface itself, so it is likely to be the case that the surfaces of ferrihydrite particles are made up of coordinated water molecules rather than oxide or hydroxide ions. Given the likely protein interactions with the mineral core, this is unlikely to be true for ferritin. Here it might be that the protein ultimately provides a neutral container for the mineral, or, more likely, there are interactions with some negatively charged and neutral amino acid side chains such as carboxylates, alkoxides, phenolates, and imidazoles in such a way to provide the necessary charge compensation. The presence of this protein surface can also contribute to a lowering of the unfavorable surface energy term in the nucleation process. As in many other instances in metalloproteins, the presence of the protein to tune energy balances, regulate local pH, and redox potential is paramount in fitting the biomolecule to its function.

3.3. Biological Control of Iron(III) Hydrolysis and Mineralization

At this point, we can begin to appreciate that nature is able to manipulate the hydrolysis reactions of iron(III) to produce iron oxyhydroxides and oxides for specified functions and that various ligating groups can be used to act as templates to direct the structure of the mineral phase. We can summarize the general hydrolysis reaction, where some potential ligand, L^{m-}, is also present, as in Scheme 1 [34]:

$$\{FeL_n\}^{\{mn-3\}-} \rightleftharpoons \{Fe_xL_n(O)_y(OH)_z(H_2O)_p\}^{\{mn+z+2y-3x\}^- \text{ or } \{3x-mn-z-2y\}^+}$$

$$\{Fe_xL_n(O)_y(OH)_z(H_2O)_p\} \rightleftharpoons \text{"}Fe(OH)_3\text{"}$$

SCHEME 1.

Nature is able to control this hydrolysis so that, for example, the speciation of the iron can correspond to the extreme left-hand side of the scheme with the metal in a soluble and transportable form even at physiological pH; for example, as in a siderophore complex. Alternatively, it can be precipitated as a mineralized phase to perform some specific function as a biomineral. It is believed that the templating effect of the ligand is crucial in directing which phase is formed by altering the course of nucleation and relative supersaturation; for example, by manipulating terms in Eqs. (1) and (2). This corresponds to the influence of the template, L, in the initial species $\{Fe_xL_n(O)_y(OH)_z(H_2O)_p\}$. The minerals formed can include metastable minerals which would not normally be observed or persist under ambient synthetic conditions. In many such minerals, the organic matrix, L, is virtually impossible to detect, but in the specific case of ferritin mineralization, the iron biomineral now has a substantial amount of organic (matrix) material associated with it and corresponds to an identifiable molecule which we can represent as $\{Fe_xL_n(O)_y(OH)_z(H_2O)_p\}$. Here the polypeptide subunits which make up the protein coat equate to L. In fact, the ratio of the template L to mineral is so large that it is now the structure of the biomineral which is hard to detect. Thus, the details of the structure of the surrounding protein coat are long-established, but the precise nature of the mineralized core is still open to question.

In summary, the key to the structural identity of mineralized ferritin cores lies in the way in which the mineral nucleates and then grows. For the present, we have appealed to theories developed to explain equilibrium crystal growth making the assumption that for ferritin cores diffusion to the growing crystals is not rate limiting. Extracting the important features from the many complicated equations developed by crystallochemists, we can see that nucleation relies on the coming together of a critical number of individuals in order to form something we can identify as a new, solid phase. If the energy required to form this phase is greater than that of the individuals in solution, then the phase redissolves. A major driving force in the formation of the solid phase is the lattice energy of the solid, but this is countered by the surface energy at the particle/solvent interface. The larger the particle becomes, the more likely the phase is to persist, since the surface energy term becomes relatively less important. It is believed that the particular success of organisms in producing tailored mineralized species, often of unusual and metastable phase, is a result of the clever use of templates, such as protein side chains and collagen, to reduce surface energy terms and favor nucleation and crystal growth. These templates can also direct growth either by providing physical barriers or by promoting or inhibiting growth in specified crystallographic directions. Clearly the situation in ferritin is one where a metastable and mobilizable iron oxyhydroxide/phosphate mineral forms and under normal circumstances does not change into an immobilizable one.

4. CORE STRUCTURES

4.1. Physical Methods for Probing Core Structures

As explained earlier, the mineral portions formed within ferritins are constrained to achieve a maximum size of some 8 nm in diameter and therefore have the properties associated with nanoscale materials. This can lead to complications in the interpretation of some of the physical data collected from ferritin cores. In addition, the problems with achieving uniform loading of ferritin cores need to be borne in mind when constructing models for core structures on the basis of such measurements. Physical measurements which have been performed include powder x-ray and electron diffraction, electron microscopy, EXAFS measurements, and a variety of magnetic measurements, including Mössbauer, EPR, bulk susceptibility, and MCD measurements. These combined with compositional analysis can suggest possible structural types for the cores of ferritins. It should be stressed at this stage that details of core structures at the atomic resolution we would hope to achieve with single crystal x-ray diffraction, for example, are not obtainable from the measurements so far performed.

4.1.1. Diffraction Experiments

Although single-crystal x-ray diffraction experiments have been performed on many samples of both mammalian and bacterial ferritins, these have not aimed to elucidate any features of the core structures but rather the protein shell and the ferroxidase centers [35–41]. Ideally, the structural details of the core would best be found using single-crystal x-ray diffraction data on loaded ferritin as part of a complete protein structure determination. However, because of the problems of achieving uniform iron-loading and the sheer size of the task, requiring the location of the positions of several thousand iron centers, this route is not currently viable.

Another difficulty in determining the structure of the mineralized core of ferritin is a consequence of its nanoscale dimensions. Even if a ferritin is fully loaded with 4500 iron oxyhydroxide units in the core and all of these are in the same crystallite corresponding to a portion of an iron oxyhydroxide mineral, the size limitation of the cavity means that a particle of no more than 8 nm in diameter will form and about 30% of these units will be at the surface of the sphere with the remaining 70% defining the bulk. Therefore, although the iron oxyhydroxide units in the bulk will be in environments corresponding to those of the iron oxyhydroxide units in the extended mineral lattice, those at the surface will be subject to very marked boundary effects and in rather different environments. This places a limitation on diffraction experiments aimed at elucidating core structures. The physical interpretation of diffraction data requires that the assumption can be made that the lattice under study corresponds to an infinite array with no boundaries. This is reasonable even for very small single crystals of micrometer dimensions which contain very many more iron oxyhydroxide units than the ferritin core can accommodate. For these, over 99.75% of the molecules would be in the bulk with only about 0.25% on the surface giving with no discernible boundary effects. This is obviously not the case for the ferritin core.

Thus powder diffraction experiments are hampered by the limitations which small particle sizes place on the technique. It is well-recognized by mineralogists that severe line-broadening in x-ray powder traces occurs when particle sizes are in the nanoscale range [42], and this, perhaps in conjunction with the poor crystallinity of some samples, makes this technique, at best, a fingerprinting exercise for characterizing the core type under study.

4.1.2. Electron Microscopy

High-resolution electron tunneling microscopy (HRTEM) combined with electron diffraction are capable of giving the highest resolution picture of core structures available currently. The diffraction fringes observed have been used to "fingerprint" the mineral types and assign degrees of crystallinity to cores [43]. In HRTEM, the structure can be investigated by direct imaging of the crystal lattice under optimum defocus conditions. Care is needed in these studies to avoid artefacts introduced through sample preparation methods and beam damage which can lead to the loss of water and organic molecules from the system. Also, microscope aberrations, sample thickness, and the defocus conditions can all affect the appearance of the images and must be taken into account. In addition to fingerprinting the lattice type, such studies can be used to gauge core sizes in samples of loaded, or partially loaded, ferritins.

4.1.3. UV/Vis Studies

Ferritin cores have a red-brown color which results from the presence of oxide and hydroxide charge transfer transitions usually in the region of about 420 nm. The absorption band is useful in following the uptake of iron into ferritin cores [2]. Although ligand field spectra for three d-d transitions of model compounds for ferritin cores such as ferrihydrite, goethite, and lepidocrocite, have been measured in the visible region, only one transition at 900 nm has been reported for the horse spleen ferritin core [44].

4.1.4. Magnetic Measurements

The magnetic behavior of ferritin cores is secondary to their function as iron-storage sites, but it provides a further means for characterizing and categorizing core structures in terms of spectroscopic signatures and bulk magnetism. Early bulk susceptibility studies indicate an overall antiferromagnetic coupling of iron(III) centers in iron oxyhydroxide cores of ferritin [45]. More recent bulk susceptibility studies have concentrated on anomalous effects such as quantum tunneling which are a consequence of the nanoscale dimensions of the cores ([46] and references therein).

The bulk magnetic behavior of nanoparticles such as ferritin cores is a topic of interest to physicists as well as to chemists and biologists. The cores can be pictured as fragments of antiferromagnetically coupled iron oxyhydroxide minerals with uncompensated spins on their surfaces, which is, another manifestation of the boundary effects in these systems. Thus they can be regarded as "ferrimagnetic" particles (uncompensated spins) giving rise to clusters carrying overall spins which can be thought of as big paramagnets. The ground state spin should normally be trapped as "up" or "down" at low temperatures where there is insufficient thermal energy to overcome the barrier of intermediate spin levels. However, there is the possibility of tunneling through this barrier, and such quantum effects are being actively explored on a range of nanoscale superparamagnets including ferritin cores.

^{57}Fe Mössbauer spectroscopic experiments provide an indication of the crystallinity and composition of ferritin cores [43]. Under normal circumstances, isolated high-spin iron(III) centers give a single peak in the Mössbauer spectrum corresponding to transition between the $I = 1/2$ and $I = 3/2$ nuclear spin states. The presence of an electric field gradient results in the manifestation of an electric quadrupole interaction for nuclei with spin greater than 1/2, as is the case for ^{57}Fe ($I = 3/2$). In spite of a spherical distribution of d-electrons for high-spin iron(III), a quadrupole splitting can be induced through the interaction of neighboring ions which results in a partial lifting of the degeneracy of the nuclear states and can be indicative of exchange-coupled systems such as occur in ferritin cores. A particularly valuable guide to the extent of exchange coupling, core sizes, and crystallinity is the onset of superparamagnetic behavior. The generation of internal magnetic fields within the cores can lead to a splitting of the quadrupole doublet spectrum normally observed evolving into a six-line spectrum as temperature is decreased. The magnetic field completely lifts the degeneracy of the $I = 1/2$ and $I = 3/2$ states. The superparamagnetism is a characteristic of nanoscale particles of strongly coupled (antiferromagnetically or ferromagnetically) spin carriers (here the Fe(III) centers). Such particles correspond to a single magnetic domain in the sense that the energy required to create a domain wall exceeds the energy decrease attainable by having more than one domain. They behave as if they each possess one macroscopic moment, hence the term *superparamagnet*. For long superparamagnetic relaxation times, internal magnetic fields are generated giving six-line spectra. As temperature increases, the relaxation times become shorter and the spectra collapse into a quadrupole doublet. The temperature at which this occurs is termed the blocking temperature, T_B. As a general rule, the more crystalline the sample of a given mineral type, the higher the blocking temperature. Higher blocking temperatures are also expected for systems with strong magnetic coupling. It is also worth pointing out that systems with rather low magnetic ordering temperatures are unlikely to display these phenomena.

Although EPR spectroscopy would seem at first sight to be a useful further local probe of the Fe(III) centers within ferritin cores, the large degree of coupling between iron centers leading to extremely broad signals presents a significant problem in the interpretation of the spectra [47–49]. Another potentially informative magnetic spectroscopy is magnetic circular dichroism (MCD). This technique can give information on magnetization as well as optical transitions and help in the assignment of ground-state spins [50]. The study of superparamagnetic particles is a relatively unexplored area, and the independent observation of hysteresis curves which can be compared with results from bulk magnetization studies opens up new avenues in probing core structures and characteristics.

4.1.5. EXAFS Spectroscopy

Extended x-ray absorption fine structure (EXAFS) spectroscopy can provide useful information on iron environments in ferritin cores. The technique involves irradiating the Fe K-edge with high-energy x-rays (usually at a synchrotron source) and measuring the fine structure arising from interference of waves backscattered from other atomic centers in the material. This backscattering gives rise to a sinusoidal fine structure on the absorption spectrum which can be described using a Fourier transformation [51]. The quantification of such data can be problematic and is subject to overparameterization and overinterpretation. However, it is possible to propose distances and numbers of backscatterers from the iron centers with a reasonable level of confidence providing sufficient care is exercised in calibrating the EXAFS data using model systems with accurately determined metrical details.

4.2. Assignments of the Mineral Types Observed In Vivo

4.2.1. Mammalian Ferritins

Structural studies have been performed on a number of wild-type mammalian ferritins, but these have been protein crystallographic studies aimed at elucidating the structures of the protein shell and not at probing details of core structures [35–40]. Indeed, the loading of wild-type ferritins is unlikely to be particularly uniform, and this is supported by HRTEM studies which indicate a distribution of core sizes in such samples [1,6], effectively making elucidation of the core structure impossible from single crystal data. As has already been pointed out, the determination of core structures even on uniformly loaded ferritin molecules is a huge task and therefore less direct means of defining core structures have been sought.

The native cores of human, rat, and horse ferritins have been assigned a ferrihydrite-like structure largely on the basis of powder diffraction studies [1,52]. A difficulty with assigning this mineral structure to these cores is the uncertainties concerning the structure of ferrihydrite itself [23–25]. It has been found that ferrihydrite precipitates with varying particle sizes, ranging from 2.5 nm in diameter to 8 nm and larger. Clearly, these dimensions are within the same range required as ferritin cores. However, the difficulties in defining structures of nanoparticles, as explained in detail in Sec. 5.2, mean that, at best, the characteristics of ferrihydrites can only be used as a guide to the likely structure of ferritin cores. For example, x-ray powder diffraction on ferrihydrite phases reveals that they are characterized by traces with between two and six lines [19,23]. These patterns have been used as fingerprints in the study of ferritin cores.

Given the uncertainty regarding the composition of ferrihydrites, this can at best give only a crude picture of the mineral structure of native ferritin cores. Native mammalian ferritin samples are found to contain varying amounts of phosphate. It is not known how much, if any, of this phosphate is an integral part of the core structure. Experiments performed in vitro (see Sec. 4.3) show that phosphate is not a prerequisite for laying down mineral cores in mammalian ferritins. However, it may be that in vivo the phosphate plays an important role in the iron homeostasis process and is associated in some way with the core.

Mössbauer measurements performed on samples of native mammalian ferritin give spectra characteristic of superparamagnetic particles with blocking temperatures of around 40K, depending on the degree of loading [43].

EXAFS spectroscopy has been used to explore the iron environments in loaded ferritins [53–56]. In the earlier iron-edge EXAFS studies on loaded ferritin, the focus had mainly been on the Fe-O interactions, but later work suggested that the Fe-Fe interactions might provide a better guide [54]. In general, the iron centers appear to be in predominantly octahedral environments, usually interpreted as an average of six oxgen atoms at a distance of approximately 2 Å. Iron centers can be discerned beyond this first shell, and it is suggested that there are about 1.5 iron centers at 3.0 Å and about 1.1 at 3.4 Å [54]. Although it is difficult to assign numbers of scatterers from EXAFS data, the fact that there are relatively few iron centers in the split shell at 3.0/3.4 Å can be taken as further evidence of the disorder prevalent within the mineral cores of ferritins.

4.2.2. Nonmammalian Ferritins

Molluscan and plant ferritins appear to possess iron(III) oxyhydroxide mineral cores similar to those of mammalian ferritins [2], whereas most bacteria studied have high proportions of phosphate within the mineral cores [57]. It is unlikely that the high phosphate content arises solely from adventitious phosphate from biochemical purification procedures but rather that it is an important component of the core. Indeed, there have been suggestions that the role of bacterioferritins might not be as iron-storage vessels at all but rather as phosphate stores. A significant difficulty in investigating these has been their rather amorphous nature so that many of the useful signatures available from mammalian ferritin cores, such as superparamagnetic behavior at relatively high temperatures, diffraction patterns, and so on, are absent. In the case of native bacterioferritin, the magnetic coupling appears to be much weaker and the spectra do not show characteristics of superparamagnetic behavior [43]. This is supported by EPR and MCD data [50,58]. Thus all the available data suggest that native bacterioferritins possess cores containing an iron(III) hydroxyphosphate mineral.

4.3. The Mineral Types Observed In Vitro

In many ways, apoferritin can be regarded as a reaction vessel for the formation of mineralized nanoparticles. Thus it is possible to load an iron oxyhydroxide core into apobacterioferritins similar to that observed in mammalian ferritins [59–61]. In the same way, that bacterial ferritin can be used to lay down cores of oxyhydroxide minerals, mammalian ferritin can load cores of varying compositions in vitro. Studies in the laboratories of Mann, for example, show that under conditions of high pH and limited oxygen, it is possible to produce ferritin cores corresponding to the magnetic minerals magnetite (mixed valent Fe_3O_4) and maghemite (γ-Fe_2O_3, which is related to magnetite as a defect structure) [62]. It has also proved possible to produce griegite (Fe_3S_4) and MnOOH and Mn_3O_4 cores [63]. These experiments, although of interest for producing nanoparticulate grains of these mineral phases, do not shed much light on the process of mineralization itself. Indeed, the characterization of such phases is beset with the same sorts of difficulties encountered for native ferritins, and it is more useful for the purposes of this discussion to focus on experiments where the loading of ferritin with Fe(II) using at least some of the conditions operating in vivo has been explored. The basic premise that the iron is delivered to ferritin as Fe(II) has found wide acceptance, although suggestions that it might be possible to load iron as iron(III) have been made in the past [33]. Whatever the case in vivo, it is found to be necessary to use $Fe(II)SO_4$ as an iron source for loading experiments in vitro.

A particularly informative set of experiments performed on commercial horse spleen ferritin reveals that different cores can be created depending on the temperature of loading and the amount of phosphate present. St. Pierre et al. found from Mössbauer, EXAFS, HRTEM, and electron diffraction studies that cores displaying different crystallinities and compositions could be formed in a tailored fashion [64]. When cores were loaded in the absence of phosphate, it was found that at 55°C they tended to have larger core dimensions and be more crystalline than when loaded at 4°C with mean diamters of 5.14 ± 0.12 nm compared with 4.74 ± 0.10 nm. The evidence for this was from the number of diffraction lines in the electron diffraction studies and the Mössbauer spectra which had a lower blocking temperature for the low-temperature preparation suggesting either a lower degree of order or smaller particle sizes for these. In these studies, an average of 1500 iron atoms were supplied to the cores, although no estimate of the total uptake was made. The distribution of core sizes as judged from transmission electron micrographs had the wider range for the 55°C loading, and since a large number were found in both the 8- and 4-nm particle size range, it seems possible that although some of the cores must be fully loaded with about 4500 iron centers, others might be quite empty. This demonstrates the polydisperse nature of the core particle sizes and underlines the problems in interpreting physical data which are dependent on particle size such as those arising from superparamagnetic phenomena. More of the diffraction traces consistent with the ferrihydrite six-line fingerprint pattern were observed for the 55°C sample (five lines) than the 4°C sample (two diffuse peaks).

The experiments with phosphate present, from KH_2PO_4, were performed at 25°C with the phosphate added prior to supplying the iron (this time as 1000 Fe per ferritin). After dialysis to remove excess iron and phosphate, it was found from inductively coupled plasma (ICP) studies that the bulk of the phosphate supplied had been incorporated in the cores and the samples with the highest and lowest iron to phosphate ratios (7Fe:1P and 0.8Fe:1P) were investigated further. The TEM pictures revealed that with more phosphate present, the average core size was larger (6.28 ± 0.09 compared with 5.89 ± 0.08) but from electron diffraction amorphous. The cores with less phosphate had one medium and one very weak diffraction trace which fit with two peaks of the ferrihydrite fingerprint. The Mössbauer spectra on all four samples are consistent with the suggestion that the 55°C phosphate-free ferritin has the most highly ordered and crystalline structure and the 25°C ferritin with 0.8Fe:1P the most amorphous and magnetically dilute.

These results help to quantify some of the features of native ferritins. Although it is important to recognize that the polydisperse nature of the samples makes it difficult to make definitive assignments, especially with techniques such as Mössbauer spectroscopy which are local probes, there are strong parallels in this study with the situation in mammalian ferritins and bacterioferritins which are found with Fe:P ratios of about 8:1 and 1:1, respectively. Taken at face value, the results with the phosphate-containing samples are in line with many measurements made on native samples of both mammalian and bacterioferritins. Although the iron loading in mammalian ferritins is often much greater than the 1000Fe/apoferritin loaded here, this figure is consistent with that for bacterioferritins. Thus it has proved possible to use a mammalian apoferritin to reproduce a bacterioferritin-type core, providing an obverse to the experiments performed on apobacterioferritins in which mammalian-type cores have been laid down.

Furthermore, the studies at different temperatures on the phosphate-free samples show that the crystallinity of the cores can be influenced by temperature. This is perhaps not surprising in the light of the earlier discussion on factors affecting crystal growth versus nucleation, since the higher temperatures would be expected to increase the solubility of the iron(III) and thus reduce the relative supersaturation which should favor crystal growth rather than fresh nucleation. The larger average core sizes of these samples and the significant numbers found with maximum diameters of 8 nm would seem to support this. Although no experiments were performed at common physiological temperatures, the results perhaps point to the intermediate scenario of fairly crystalline cores in mammalian "ferrihydrite-like" ferritin cores. The question of the likely identity of the iron hydroxyphosphate core of bacterioferritins proves difficult to address from the available data in terms of a structural model. The added complication of phosphate anions within the structure and the much lower iron contents normally observed in bacterioferritins all point to a less well-defined structure.

5. A COMPARISON OF THE USE OF MINERAL AND CLUSTER AGGREGATE MODELS FOR ELUCIDATING THE CORE MINERAL STRUCTURE IN FERRITINS

5.1. The Problem of Determining the Structures of Ferritin Mineral Cores

The preceding sections have described the mineralization processes and methods of study in trying to classify the structural types in ferritin cores. The fact that ferritin molecules are large proteins which can accommodate about 4500 iron atoms leads to several problems in defining their core structures, as detailed in Sec. 4.1.1. The central difficulty is in performing experiments at atomic resolution which can allow us to define the core structure. Loading experiments in vitro show that ferritins can be relatively fully loaded to yield cores corresponding to spherical nanoparticles with a range of sizes up to 8 nm in diameter [64]. Studies on native samples using electron microscopy show wider distributions of size and lower levels of loading [43]. Additionally, it is hard to be certain that cores always contain one large particle and not several smaller ones. We have seen that other polydisperse nanoscale materials such as ferrihydrite phases are difficult to characterize structurally at the atomic level as a combined result of their small size leading to boundary effects in diffraction experiments and their particle size distribution and probable disorder of the lattices leading to poor definition of the structures. This situation also prevails for ferritins, and it would take a complete protein structure determination, including finding all the positions of the iron centers, to achieve a proper atomic resolution structure analysis. This in turn would require an equal loading of iron into the core within each ferritin in the crystal lattice, which implies achieving monodisperse core sizes. Even if all this were possible, the crystallographic task would be enormous. It is clear from the previous sections that this route is not currently available.

5.2. The Model Compound Approach to Elucidating the Core Structure

The use of model compounds to elucidate core structures offers the possibility of developing useful experiments to probe both cores loaded in vitro and native samples. The premise here is that well-defined model compounds can be used to calibrate experimental techniques such as EXAFS spectroscopy, magnetic measurements, and diffraction patterns on loaded ferritins in attempts to define the structural identities of their cores. These studies have concentrated almost exclusively on modeling mammalian ferritin cores, largely as a result of their better-defined spectroscopic and magnetic characteristics. In the light of this, the following discussion is restricted to systems aimed at elucidating the nature of mammalian ferritin cores.

5.2.1. *Properties to Model*

Reproducing the following features would result in a realistic model for loaded mammalian ferritin. To be of real use, the model would also need to be well defined and preferably structurally characterized by single crystal x-ray diffraction.

1. The system should contain a mineralized but bounded, nanoscale-sized core.
2. The core should be enclosed in an organic shell.
3. The approximate proportions of Fe and C, N, H, and O should be reproduced.
4. The different sorts of iron environment should be reproduced, preferably in the correct proportions.
5. The model compound should possess properties such as anomalous magnetic behaviour.

A final property to model would be that the system could reversibly assimilate and release iron. This would then produce a functional as well as structural model.

5.2.2. *Defining a Mineral Phase*

The first requirement is that the model has a nanoscale mineralized core. As yet, the philosophical question of how many nuclei are needed to define a mineral phase has not been addressed. In some cases, collections of very few centers can be clearly identified with known mineral structures, but this will also be dependent on the dimensionality of the mineral (whether it is essentially a two-dimensional sheetlike structure or a three-dimensional network). What is undoubtedly important in the early stages of mineral growth is the way in which metal centers are linked together. That is, the way in which the nucleation events proceed which in turn defines the mineral phase produced. This idea was explained in Sec. 3.1.1. For example, we know that all the common iron(III) oxide and oxyhydroxide minerals have iron centers held together by μ_2, μ_3, or μ_4, and not μ_6, oxo bridges, whereas Fe(II) minerals can also contain μ_6 bridges. This is a simple consequence of stoichiometry, with the former, for example, belonging to A_2X_3 and AX_2 minerals and the latter to AX minerals. It is also known that iron centers can occupy tetrahedral as well as octahedral holes, leading to further characteristics for identifying different mineral phases.

When the mineral phase is well defined, we only require a relatively small number of metal centers and bridging anions to be present. The requirement depends on the number of different metal and anion environments in the parent crystal structure. For example, an AX mineral (like rock salt) would be identifiable for one A atom in an octahedral environment of six X atoms, they themselves displaying octahedral coordination environments, or vice versa. However, in order to establish a true cubic close-packing lattice for the structure, we would need three layers of A or X. For hexagonally close-packed structures, we can expect to observe double strips of anions with metals in octahedral holes between them, and make the tacit assumption that the next layer up will look the same, conforming with the hcp structure. Even so, the minimum number of metals required here is six (for AX_3) or seven (for A_2X_3), both requiring 12 anions. The relevance of this to cluster aggregate models will be discussed in Sec. 5.2.4. below.

5.2.3. *Proposed Models*

The model compounds which have been suggested for ferritin cores can be placed in three categories and are considered in turn.

5.2.3.1. Iron oxyhydroxide / oxide-type minerals which reproduce the approximate elemental composition of the core. The extended iron oxyhydroxide/oxide minerals prove to be inappropriate as models, because they only reproduce certain aspects of the core and are not bounded. Naturally, they cannot be expected to model any features of the protein shell. Thus they fail to meet even the first criterion for a good model satisfactorily. In fact, it is clear from these considerations that condensed mineral phases such as goethite, lepidocrocite, hematite, and so on, fail to provide realistic models on all counts. Additionally, some of these phases, such as maghemite, are not well defined from the structural point of view. This is brought out particularly strongly in the comparison of the EXAFS spectra of horse spleen ferritin and the Fe·O·OH mineral phases goethite and lepidocrocite [65]. That the minerals fail to reproduce the range of iron environments observed in ferritin cores is hardly surprising in view of their well-ordered infinite structures. Nonetheless, these minerals have helped in understanding some aspects of the way in which mineralization occurs within ferritin and the well-defined phases can shed light on the likely structural arrangements in less well-defined phases.

5.2.3.2. Iron mineral fragments with dimensions in the region of 8 nm in diameter. The hydrolysis of iron(III) can lead initially to the production of suspensions of colloidal iron oxyhydroxide particles with dimensions of nanoscale proportions [17,23] similar to those of ferritin cores. In some cases, the addition of oxygen-containing anionic species can be used to control particle size and the resulting particles can contain 1000–4000 iron centers and may retain anionic coats derived from sugars, small organic molecules, or anions such as nitrate. The polydisperse colloidal and nanoparticulate precipitates collectively known as ferrihydrites have already been described in Sec. 3.1. These precipitates without organic coats are often in the form of ferrihydrites. They do not fulfill requirements (2) or (3) and only partially fulfill the requirements of points (1) and (4), as defined in Section 5.2.1., possessing the correct dimensions and a number of different iron environments. In addition, the nature of the mineral phase(s) in such colloidal dispersions will be hard to ascertain, since we now arrive back at the problem of structure determination on mineral portions exhibiting large boundary effects.

Other products of the hydrolysis of iron(III) salts which have been proposed as models for loaded ferritin include the "Saltmann-Spiro" balls which can be prepared from iron(III) solutions containing nitrate or citrate anions [66,67]. At first glance, the citrate polymers, in common with the ferrihydrite particles, look to provide excellent models for loaded ferritin, having dimensions of around 7 nm, as determined from electron microscopy, and apparently containing about 1500 iron centers with a coat of citrate anions. Of course, this situation is actually too close to that of ferritin: The model has essentially the same scale as the original and therefore all the corresponding problems associated with its characterization.

Therapeutic agents such as iron/dextran drugs are used to administer iron to anemics. Their sugary coat helps in the absorption of the ferric oxyhydroxide material, which probably has the composition of a ferrihydrite, and they thus possess features similar to those of loaded ferritins. This is supported by EXAFS data and to a less convincing degree by powder x-ray diffraction measurements [53,68]. Although the polysaccharide coats go some way to fulfilling criterion (2) of Sec. 5.2.1., the structures and further characteristics of these materials have not been reported, and we appear to face the same problem as with the Saltmann-Spiro balls in that the system is poorly defined and of the wrong scale.

5.2.3.3. Iron oxyhydroxide cluster aggregates as "scale models" for the whole ferritin molecule. Clearly to fulfill the first requirement for a good structural model, we need to concentrate on bounded particles, but these must also be of a size which allows for their complete characterization. Thus the use of "scale models" in the form of iron cluster aggregates can be of assistance, since compounds accessible to single crystal structure determinations can be synthesized and, in some cases, their structural and physical properties used to shed light on the chemistry and physics of loaded ferritins. A detailed analysis of the suitability of a range of structurally characterized iron oxyhydroxide aggregates with between 8–19 iron centers has been made recently [69]. To date, very few of these compounds have been subjected to the full range of characterization techniques used on ferritin (Sec. 4.1), and we would not expect size-dependent phenomena such as superparamagnetism to be reproduced with similar parameters of blocking temperature and so on. On the contrary, the fact that these "scale" models *are* structurally well characterized should allow for a much more accurate quantification of magnetic phenomena albeit at a different scale size. This in turn should aid in the understanding of the physics underlying the magnetic behavior of ferritin cores themselves. Furthermore, the actual synthetic procedures in creating these aggregates can shed light on the processes influencing the mineralization of iron.

5.2.4. *The Suitability of the Cluster Aggregates as Scale Models for Ferritin*

Although of nanoscale proportions, not all of these structurally defined cluster aggregates contain cores which can be related to mineral structures, whereas in others motifs from mineral lattices can be clearly discerned. These fall into the AX and AX_2 lattice types, the former only observed in the case of a mixed valent dodecanuclear Fe(II)/Fe(III) aggregate, $[Fe^{III}_4Fe^{II}_8(O)_2(OCH_3)_{18}(O_2CCH_3)_6(CH_3OH)_{4.67}]$, Fe_{12}, [70, 71]. The AX_2 type has been identified in a decanuclear Fe(III) aggregate, $[Fe_{10}O_4(OCH_3)_{16}(dbm)_6]$, Fe_{10}, (dbm = dibenzolylmethane) [72]. In both these cases, the cores are more accurately described as pseudo-AX and pseudo-AX_2, since some of the oxygen donors are provided by methanolate rather than hydroxide ligands. A true AX_2 core is found in the "crust" system which contains aggregates with 17 and 19 iron centers, $[Fe_{17}(heidi)_8(O)_4(OH)_{16}(H_2O)_{12}]^{3+}$ and $[Fe_{19}(heidi)_{10}(O)_6(OH)_{14}(H_2O)_{12}]^{+}$, (heidi = hydroxyethyliminodiacetate) [73,74]. These two aggregates crystallize together and have been shown to possess identifiable mineral cores, which can be thought of as corresponding to an iron(III) hydroxide lattice of the brucite ($Mg(OH)_2$) type, $\{Fe(OH)_2\}_n^{n+}$. Owing to the difficulties of nomenclature, the system has been dubbed "crust" to reflect the *c*apturing of *rust* by the heidi ligands.

The specific question of oxidation state of the iron centers has not been addressed in terms of the suitability of cluster aggregates as structural models. Two points are relevant. First, it is not known whether any iron(II) is incorporated into the core of ferritin or not and mixed valent clusters could provide useful clues. Second, we can expect the magnetic properties of mixed-valent systems to be markedly different from those of single-valent ones. This might be important in magnetostructural correlations designed to aid in the interpretation of magnetic data. The fact that some aggregates have less well-defined cores in terms of mineral structures might be helpful when considering the properties of ferritin systems with amorphous cores.

The cluster aggregates are all composed of shells of organic ligands enclosing cores of iron oxide/hydroxide units. This is in line with requirement (1) of Sec. 5.2.1 for a good model system. However, in most cases, the organic shell does not have a composition corresponding to the ferritin protein, which would require C, H, N, and O all to be present and preferably in similar forms to those in polypeptides (requirement (3)).

Thus out of all of the possible cluster aggregate models, only an aggregate with 10 iron centers as well as two sodium ions which has a ligand coat containing a mixture of carboxylate and pyridonate ligands (chp = 2-chloropyridonate), $[Fe_{10}Na_2(O)_6(OH)_4(H_2O)_2(O_2CPh)_{10}(chp)_6(Me_2CO)_2]$, [75] and the "crust" system contain all the elements of Fe, C, H, N, and O.

When the requirement (3) of Sec. 5.2.1. is further considered, it is obvious that for ferritin for the relative proportions of Fe, C, N, H, and O, the iron figure will depend in some way on the loading of individual molecules. If average figures for loaded ferritins are used an iron composition of about 25–30% is required for a good model with a ratio of about 2:3 of core:organic shell, where the core is defined as the inorganic

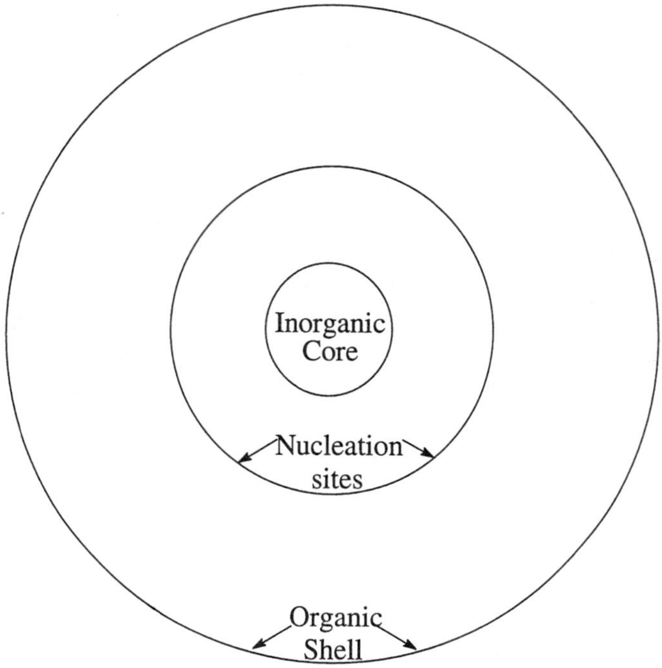

FIG. 3. Diagrammatic representation of loaded ferritin in terms of iron environments.

(hydrated) iron oxyhydroxide part and the organic shell is the protein or ligands around the core. It was found that models with AX_2 cores came closest to reproducing the figures for loaded ferritin.

The different types of iron environment which are found in loaded ferritin can be summarized as (Fig. 3):

1. The iron atoms located at nucleation sites. These have coordination spheres containing both organic (from the protein) and inorganic (from the mineral surface) ligands.
2. The iron atoms at the surface of the mineral. These are connected to the bulk mineral as well as the iron atoms within the protein shell through inorganic oxide/hydroxide linkages and there may be additional linkages to the iron atoms of the shell through dinucleating proteinaceous ligands.

3. The purely inorganic iron atoms which are in environments corresponding to those of the bulk mineral.

Any of the cluster aggregates with some purely "inorganic" iron centers will reproduce these environments to some degree, although the models with cores corresponding to mineral portions are likely to reproduce the iron environments most faithfully. Experimental evidence for this and a comparison with both hydrolytic nanoparticles and extended mineral structures as models has been provided by EXAFS analyses [53]. A recent study compared the EXAFS data for horse spleen ferritin, the minerals goethite and lepidocrocite and the cluster aggregates $[Fe_{11}O_6(OH)_6(O_2CPh)_{15}]$, Fe_{11}, [76,77] and the "crust", $Fe_{17/19}$(crust), system [65].

It was found that the $Fe_{17/19}$(crust) cluster system could be used to calibrate the parameters used in the interpretation of EXAFS data. As mentioned before in Sec. 4, the best guide in the interpretation of ferritin EXAFS data is the nature of the Fe-Fe interactions [54]. When the model used was the $Fe_{17/19}$(crust) system, this was also found to be the case [65]. The EXAFS recorded for the $Fe_{17/19}$(crust) were given an optimum fit using the known structural parameters from the single crystal structure analysis with the result that the underestimation of the Fe-O/N and Fe-Fe interactions resulting from a standard EXAFS interpretation was revised. The approach was tested by using the same parameters for the well-defined minerals goethite and lepidocrocite, where it was also found that an improved correlation between the EXAFS and crystallography was achieved. The results were also compared with previously reported data on Fe_{11} where comparative EXAFS studies on this material and horse spleen ferritin (HSF) have been performed [78]. Overall the analysis of the various models for loaded ferritin and the comparison with HSF revealed that the best fit for the HSF EXAFS was obtained using the $Fe_{17/19}$(crust) system, although the Fe_{11} system also provided a reasonable fit. In fact, the Fourier transformed EXAFS data of loaded ferritin and $Fe_{17/19}$(crust) showed a remarkable degree of congruence, with shells of oxygen and iron atoms with similar occupancies at almost identical distances, the only real difference being the better resolution of the data for the model system. The structural similarities are clear when the schematic indicating the different iron environments in loaded ferritin (see Fig. 3) and the mo-

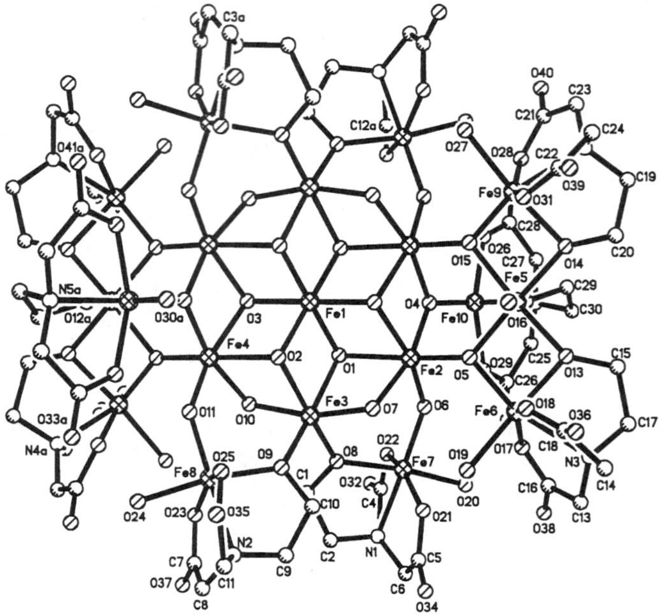

FIG. 4. The molecular structure of the Fe_{19} "crust."

lecular structure of the Fe_{19}(crust) (Fig. 4) are compared. Infinite lattices such as goethite and lepidocrocite were wholly unsatisfactory as models, underlining the requirement for model systems to be bounded particles displaying several different iron environments rather than extended arrays with perhaps just one or two highly symmetrical iron sites. The fact that the hexagonally close-packed Fe·O·OH, goethite, provided a somewhat better fit to the EXAFS than the cubic close-packed lepidocrocite was taken as support for the suggestion that loaded ferritin contains hexagonally close-packed arrays, perhaps related to hematite, goethite, or a brucite-type mineral. This is broadly in agreement with the suggestion that the core contains ferrihydrite-like arrays, although it is stressed again that the uncertainties regarding the structure of this group of phases leave much to be desired.

The magnetic properties of the cluster aggregates proposed as models for loaded ferritin indicate that in many cases the boundary effects of the particle play an important role. The elucidation of these properties is a complicated task, particularly for the larger clusters, and in some cases only very preliminary data have been reported. The most interesting magnetic properties arise in aggregates where there are some strong but unequal antiferromagnetic interactions between high-spin iron(III) ions which lead to spin frustration effects and the stabilization of intermediate ground-state spins paralleling the situation in ferritins where uncompensated spins on the surface of the core produce a net magnetic moment [45]. The highest ground-state spin reported so far is for the $Fe_{17/19}$(crust) system [74]. The interpretation of the magnetic data for this is complicated by the presence of the two cluster aggregates within the same crystal structure, but the system has been studied in depth and the conclusion drawn that at least one of the clusters stabilizes a ground-state spin of 33/2 or more. Ideally, the cluster aggregates would display degrees of superparamagnetic behavior akin to the situation found for various loadings of ferritin cores [43,64] but with correspondingly lower blocking temperatures owing to the smaller dimensions of their aggregates. Until now, no systematic study of the bulk magnetic, EPR, MCD, and Mössbauer has been undertaken, although an aggregate of 17 iron centers (1 Fe(II) and 16 Fe(III)) is reported to demonstrate superparamagnetic effects [79], and the Mössbauer data of the mixed valent Fe_{12} molecule show certain parallels with the data recorded for animal ferritins [71]. However, current work on the $Fe_{17/19}$(crust) system suggests that superparamagnetic and resonant tunneling effects may come into play at very low temperatures, paralleling the higher temperature situation for loaded ferritin [46,80]. An investigation of these properties along with a quantification of the important exchange pathways in such clusters should help in elucidating the magnetic behavior of ferritin.

5.3. Synthetic Aspects Modeling Biomineralization of Ferritin Cores

The iron oxyhydroxide cluster aggregates can only form if some water is present which can be activated on coordination to the metal to give oxide and/or hydroxides. Thus they correspond to hydrolytic oligomers of the general form of

$$\{Fe_xL_n(O)_y(OH)_z(H_2O)_p\}^{(mn+z+2y-3x)-\text{ or }(3x-mn-z-2y)+}$$

as in Scheme 1. For the majority of syntheses, nonaqueous routes have been used with adventitious water for the hydrolysis. A major exception to this is the strategy employed to produce "crust" where the only solvent present is water [74].

On examining the structures of the various cluster aggregates, it can be seen that the nature of the peripheral organic ligands can have a profound effect on the nature of the iron oxyhydroxide aggregate. Among the important factors are the charge and denticity of the ligand as well as the local pH in the reaction solution. Naturally, this last parameter is hard to define in nonaqueous systems. These effects are easiest to appreciate in the aggregates containing recognizable portions of mineral structures. The effect of a tetradentate ligand such as heidi in Fe_{17}(crust) and Fe_{19}(crust) is to block four coordination sites on the iron octahedron, favoring a layered type of structure, whereas the Fe_{12} and Fe_{10} structures both contain portions of three-dimensional arrays, probably as a consequence of the bidentate peripheral ligands (acetate or dbm) allowing for the introduction of further close-packed oxygen layers. This templating effect is obviously an important parameter to investigate in future model studies.

The formation of aggregates such as Fe_{17}(crust) and Fe_{19}(crust) can shed some light on the effect of templates on mineral phase. These molecules form under quite specific conditions of pH and pH gradient. This latter quantity is a reflection of the hydrolyzing potential of a system, and although it has been recognized as an important factor in iron oxyhydroxide speciation, its effects have only been followed qualitatively [33,34]. If the synthetic conditions are changed, dinuclear $[Fe(H_2O)(heidi)]_2$ species form. This coupled with the fact that redissolution of "crust" also results in the formation of this dimer and the release of the excess iron into solution [81] suggests that the larger aggregates are metastable and act as a transient trap for the hydrolytic iron(III) mineral. The fact that the core structures, corresponding to $Fe(OH)_2^+$, could not form a true infinite phase because of the charge imbalance underlines further the importance of the ligand/iron shell in stabilizing this phase. In this way, the "crust" system parallels the situation in ferritin not only as a miniature "scale" structural model but also by emulating features of the control of iron(III) hydrolysis utilized within ferritin in the formation of a core of a metastable iron oxyhydroxide mineral.

6. CONCLUSIONS

The mineralization of iron in ferritin cores is a difficult process to follow experimentally. Loading experiments in vitro suggest that the iron is delivered for storage within the ferritin cavity as Fe(II) and that an oxidation step occurs in forming the ferritin mineral core. Although it seems likely that some of the iron is oxidized catalytically at dinuclear ferroxidase centers within each of the 24 subunits, it is not clear if any iron reaches the growing surface of the mineral in the form of iron(II) or not. Using a knowledge of aqueous iron chemistry, it can be seen that there are several possibilities. The formation of iron(III) oxide and oxyhydroxide minerals can occur from iron(II) species, and it is possible that the growing mineral can form a catalytic surface for the oxidation of iron(II) ions joining the growing particle. Some possibilities are:

1. Some iron(II) is oxidized to iron(III) at ferroxidase sites and these interact with nucleation sites on the inner surface of the protein cavity. Subsequent mineral growth from these nucleation sites is by addition of iron(II) ions which are oxidized to iron(III) as the mineral forms
2. All the iron(II) is oxidized before forming the mineral
3. The principal function of the channels to and from the cavity is in providing pathways for the shunting of protons and electrons; that is, they provide a means of controlling the redox potential and pH to stabilize iron in a given form on the "ferritin Pourbaix diagram".

The nature of the mineral itself is also difficult to elucidate, partly as a result of the nanoscale dimensions. In native ferritins, the role of phosphate is not always clear, although it is believed that the mammalian ferritins have cores comprised largely of an iron(III) oxyhydroxide mineral. In native bacterioferritins, the cores contain a large amount of phosphate and appear to be amorphous. Interestingly, mammalian apoferritin can be used as a reaction vessel to lay down iron hydroxyphosphate cores akin to those in bacterioferritin, whereas bacterioferritin can lay down an iron oxyhydroxide core analogous to that of the mammalian case.

Using a combination of the knowledge regarding the mineralization of iron in the natural environment and the structures of such minerals with synthetic work in which the natural hydrolysis of iron is modified by the presence of templating organic ligands tends to support the suggestion that the mineral core of ferritin has a defect structure based on hcp arrays of oxygens somewhat akin to the suggested structure for ferrihydrite. The fact that model compounds, unlike ferrihydrite, can reproduce many of the features present in loaded ferritins and yet can be structurally characterized at atomic resolution offers a promising way forward in the investigation of the nature of the mineral core of ferritin.

ABBREVIATIONS

ccp	cubic close packed
chp	2-chloropyridonate
"crust"	*c*apturing of *rust*
dbm	dibenzolylmethane
EPR	electron paramagnetic resonance
EXAFS	extended x-ray absorption fine structure
hcp	hexagonally close packed
heidi	hydroxyethyliminodiacetate
HRTEM	high-resolution electron tunneling microscopy
HSF	horse spleen ferritin
ICP	inductively coupled plasma
MCD	magnetic circular dichroism
TEM	tunneling electron microscopy
UV/Vis	ultraviolet/visible

REFERENCES

1. P. M. Harrison, P. J. Artymiuk, G. C. Ford, D. M. Lawson, J. M. A. Smith, A. Treffry, and J. L. White, in *Biomineralization: Chemical and Biochemical Perspectives* (S. Mann, J. Webb, and R. J. P. Williams, eds.), VCH, Weinheim, Germany, 1989, p. 257.
2. P. M. Harrison and P. Arosio, *Biochim. Biophys. Acta*, *1275*, 161–203 (1996).

3. P. O'Neill, *Environmental Chemistry*, 2nd ed., Chapman and Hall, London, 1993.
4. K. B. Krauskopf and D. K. Bird, *Introduction to Geochemistry*, 3rd ed. McGraw-Hill, New York, 1995.
5. R. M. Garrels and L. Christ, *Solutions, Minerals and Equilibria*, Harper & Row, New York, 1965.
6. R. R. Crichton, *Inorganic Biochemistry of Iron Metabolism*, Ellis Horwood, Chichester, UK, 1991.
7. F. Haber and J. Weiss, *Proc. Roy. Soc. Ser. A*, *147*, 332–333 (1934).
8. C. Seldon, C. Seymour, and T. J. Peters, *Clin. Sci.*, *58*, 211–219 (1980).
9. A. C. Zettlemoyer (ed.), *Nucleation*, Marcel Dekker, New York, 1969.
10. G. H. Nancollas and N. Purdie, *Chem. Soc. Lond. Q. Rev.*, *18*, 1–20 (1964).
11. J. Garside, in *Biological Mineralization and Demineralization*, (G. H. Nancollas, ed.) Springer-Verlag, Berlin, 1982, pp. 37–77.
12. S. Mann, *Nature*, *332*, 119–124 (1988).
13. W. Stumm and J. J. Morgan, *Aquatic Chemistry*, Wiley-Interscience, New York, 1981.
14. S. Mann, *Struct. Bonding (Berlin)*, *54*, 125–174 (1983).
15. R. J. P. Williams, *Phil. Trans. Roy. Soc. Lond. B.*, *304*, 411–424 (1984).
16. R. J. P. Williams, in *Biomineralization: Chemical and Biochemical Perspectives*, (S. Mann, J. Webb, and R. J. P. Williams, eds.), VCH, Weinheim, Germany, 1989, p. 1.
17. C. M. Flynn, *Chem. Rev.*, *84*, 31–41 (1984).
18. A. F. Wells, *Structural Inorganic Chemistry* 3rd ed., Oxford University Press, London, 1962.
19. K. M. Towe and W. F. Bradley, *J. Colloid Inter. Sci.*, *24*, 384–392 (1967).
20. F. V. Chukrov, B. B. Zvyagin, A. I. Gorschkov, L. P. Ermilova, and V. V. Balashova, *Int. Geol. Rev.*, *16*, 1131–1143 (1973).
21. J. D. Russell, *Clay Miner.*, *14*, 109–114 (1979).
22. R. M. Cornell, R. Giovanoli, and W. Schneider, *J. Chem. Technol. Biotechnol.*, *46*, 115–134 (1989).
23. R. A. Eggleton and R. W. Fitzpatrick, *Clays Clay Miner.*, *36*, 111–124 (1988).

24. A. Manceau, J.-M. Combes, and G. Calas, *Clays Clay Miner.*, *38*, 331–334 (1990).

25. R. A. Eggleton and R. W. Fitzpatrick, *Clays Clay Miner.*, *36*, 335–336 (1990).

26. N. E. Le Brun, A. J. Thomson, and G. R. Moore, *Struct. Bonding (Berlin)*, *88*, 103–138 (1987).

27. P. K. Bali and P. Aisen, *Biochemistry*, *30*, 9947–9952 (1991).

28. P. K. Bali and P. Aisen, *Biochemistry*, *31*, 3963–3967 (1992).

29. T. J. Egan, O. Zak, and P. Aisen, *Biochemistry*, *32*, 8162–8167 (1993).

30. D. M. Sipe and R. F. Murphy, *J. Biol. Chem.*, *266*, 8002–8007 (1991).

31. A. Treffry and P. M. Harrison, *Biochem. J.*, *220*, 717–722 (1984).

32. R. A. Robie, P. M. Bethke, and K. M. Beardsley, in *CRC Handbook of Chemistry and Physics*, 61st ed. (R. C. Weast, ed.), CRC Press, Boca Raton, FL, p. B-208 (1980).

33. W. Schneider, *Chimia*, *42*, 9–20 (1988).

34. A. K. Powell and S. L. Heath, *Comm. Inorg. Chem.*, *15*, 255–296 (1994).

35. G. C. Ford, P. M. Harrison, D. W. Rice, J. M. A. Smith, A. Treffry, J. L. White, and J. Yariv, *Phil. Trans. R. Soc. Lond. B.*, *304*, 551–565 (1984).

36. D. W. Rice, G. C. Ford, J. L. White, J. M. A. Smith, and P. M. Harrison, *Adv. Inorg. Biochem.*, *5*, 39–50 (1983).

37. D. M. Lawson, P. J. Artymiuk, S. J. Yewdall, J. M. A. Smith, J. C. Livingstone, A. Treffry, A. Luzzago, S. Levi, P. Arosio, G. Cesareni, C. D. Thomas, W. V. Shaw, and P. M. Harrison, *Nature*, *349*, 541–544 (1991).

38. J. Trikha, G. S. Waldo, F. A. Lewandowski, Y. Ha, E. C. Theil, P. C. Weber, and N. M. Allewell, *Proteins*, *18*, 107–118 (1994).

39. J. Trikha, E. C. Theil, and N. M. Allewell, *J. Mol. Biol.*, *248*, 949–967 (1995).

40. F. Frolow, A. J. Kalb(Gilboa), and J. Yariv, *Nature Struct. Biol.*, *1*, 453–460 (1994).

41. P. D. Hempstead, A. J. Hudson, P. J. Artymiuk, S. C. Andrews, M. J. Banfield, J. R. Guest, and P. M. Harrison, *FEBS Lett.*, *350*, 258–262 (1994).

42. G. W. Brindley, in *Crystal Structures of Clay Minerals and Their X-ray Identification* (G. W. Brindley and G. Brown, eds.), Mineralogical Society, London, 1984, p. 222.
43. T. G. St. Pierre, J. Webb, and S. Mann, in *Biomineralization. Chemical and Biochemical Perspectives*. (S. Mann, J. Webb, and R. J. P. Williams, eds.), VCH, Weinheim, Germany, 1989, p. 295.
44. J. Webb and H. B. Gray, *Biochim. Biophys. Acta*, *351*, 224–229 (1974).
45. A. Shoden and P. Sturgeon, *Acta Haematol.*, *23*, 376–392 (1960).
46. J. Tejada, R. F. Ziolo, and X. X. Zhang, *Chem. Mater.*, *8*, 1784 (1996).
47. J. F. Boas and G. J. Troup, *Biochim. Biophys. Acta*, *229*, 68–74 (1971).
48. M. P. Weir, T. J. Peters, and J. F. Gibson, *Biochim. Biophys. Acta*, *828*, 298–305 (1985).
49. N. Deighton, A. Abu-Raqabah, I. J. Rowland, M. C. Symons, T. J. Peters, and R. J. Ward, *J. Chem. Soc.; Farad. Trans.*, *87*, 3193–3197 (1991).
50. N. E. Le Brun, G. R. Moore, and A. J. Thomson, *Mol. Phys.*, *85*, 1061–1068 (1995).
51. *EXAFS Spectroscopy: Techniques and Applications* B. K. Teo (Ed.), Plenum Press, New York, 1981.
52. P. J. Artymiuk, E. R. Bauminger, P. M. Harrison, D. M. Lawson, I. Nowik, and S. J. Yewdall, in *Iron Biominerals* (R. B. Frankel and R. P. Blakemore, eds.), Plenum Press, New York, 1990, p. 269.
53. E. C. Theil, D. E. Sayers, and M. A. Brown, *J. Biol. Chem.*, *254*, 8132–8134 (1979).
54. P. Mackle, C. D. Garner, R. J. Ward, and T. J. Peters, *Biochim. Biophys. Acta*, *1115*, 145–150 (1991).
55. S. M. Heald, E. A. Stern, B. Bunker, E. M. Holt, and S. L. Holt, *J. Am. Chem. Soc.*, *101*, 76 (1979).
56. J. S. Rohrer, T. I. Quazi, G. D. Watt, D. E. Sayers, and E. C. Theil, *Biochemistry*, *29*, 259–264 (1990).
57. J. M. A. Smith, S. C. Andrews, J. R. Guest, and P. M. Harrison, in *Iron Biominerals*. (R. B. Frankel and R. P. Blakemore, eds.), Plenum Press, New York, 1990, p. 325.
58. N. E. Le Brun, Ph.D. thesis, Studies of Iron Centres in Bacterioferritin, University of East Anglia, UK, 1993.

59. V. J. Wade, A. Treffry, J.-P. Laulhere, E. R. Bauminger, M. I. Cleton, S. Mann, J.-F. Briat, and P. M. Harrison, *Biochim. Biophys. Acta*, *1161*, 91–96 (1993).

60. S. Mann, J. M. Williams, A. Treffry, and P. M. Harrison, *J. Mol. Biol.*, *198*, 405–416 (1987).

61. P. L. Ringeling, S. L. Davy, F. A. Monkara, C. Hunt, D. P. E. Dickson, A. G. McEwan, and G. R. Moore, *Eur. J. Biochem.*, *223*, 847–855 (1994).

62. F. C. Meldrum, B. R. Heywood, and S. Mann, *Science*, *257*, 522–523 (1992).

63. F. C. Meldrum, V. J. Wade, D. L. Nimmo, B. R. Heywood, and S. Mann, *Nature*, *349*, 684–687 (1991).

64. T. G. St. Pierre, P. Chan, K. R. Bauchspiess, J. Webb, S. Betteridge, S. Walton, and D. P. E. Dickson, *Coord. Chem. Rev.*, *151*, 125–143 (1996).

65. S. L. Heath, J. M. Charnock, C. D. Garner, and A. K. Powell, *Chem. Eur. J.*, *2*, 634–639 (1996).

66. T. G. Spiro, S. E. Allerton, J. Renner, A. Terzis, R. Bils, and P. Saltman, *J. Am. Chem. Soc.*, *88*, 2721–2726 (1966).

67. T. G. Spiro, L. Pape, and P. Saltman, *J. Am. Chem. Soc.*, *89*, 5555–5559 (1967).

68. C. Yang, A. M. Bryan, E. C. Theil, D. E. Sayers, and L. H. Bowen, *J. Inorg. Biochem.*, *28*, 393–405 (1986).

69. A. K. Powell, *Struct. Bonding (Berlin)*, *88*, 1–38 (1997).

70. K. L. Taft, G. C. Papaefthymiou, and S. J. Lippard, *Science*, *259*, 1302–1305 (1993).

71. K. L. Taft, G. C. Papaefthymiou and S. J. Lippard, *Inorg. Chem.*, *33*, 1510–1520 (1994).

72. A. Caneschi, A. Cornia, A. C. Fabretti, and D. Gatteschi, *Angew. Chem. Int. Ed. Engl.*, *34*, 2716–2718 (1995).

73. S. L. Heath and A. K. Powell, *Angew. Chem. Int. Ed. Engl.*, *31*, 191–192 (1992).

74. A. K. Powell, S. L. Heath, D. Gatteschi, L. Pardi, R. Sessoli, G. Spina, F. Del Giallo, and F. Pieralli, *J. Am. Chem. Soc.*, *117*, 2491–2502 (1995).

75. C. Benilli, S. Parsons, G. A. Solan, and R. E. P. Winpenny, *Angew. Chem. Int. Ed. Engl.*, *35*, 1825–1828 (1996).

76. S. M. Gorun and S. J. Lippard, *Nature*, *319*, 666 (1986).
77. S. M. Gorun, G. C. Papaefthymiou, R. B. Frankel, and S. J. Lippard, *J. Am. Chem. Soc.*, *109*, 3337 (1987).
78. Q. T. Islam, D. E. Sayers, S. M. Gorun, and E. C. Theil, *J. Inorg. Biochem.*, *36*, 51–62 (1989).
79. W. Micklitz, V. McKee, R. L. Rardin, L. E. Pence, G. C. Papaefthymiou, S. G. Bott, and S. J. Lippard, *J. Am. Chem. Soc.*, *116*, 8061–8069 (1994).
80. R. Ballou, F. Lionti, and A. K. Powell, unpublished observations.
81. S. L. Heath, Ph.D. thesis, An Investigation of the Compounds Formed in Aqueous Solutions of Iron(III) and Chelating Carboxylate Ligands, University of East Anglia, UK, 1992.

14

Iron Storage and Ferritin in Plants

Jean-François Briat* and Stéphane Lobréaux

Biochimie et Physiologie Moléculaire des Plantes,
Centre National de la Recherche Scientifique
(URA 2133), Institut National de la Recherche
Agronomique et École Nationale Supérieure
d'Agronomie, Place Viala,
F-34060 Montpellier cedex 1, France

1. INTRODUCTION 564

2. IRON STORAGE IN THE APOPLAST AND PLANT VACUOLES 565
   2.1. Iron Storage in the Apoplast 565
   2.2. Iron Storage in Vacuoles 566

3. STRUCTURE-FUNCTION OF PLANT FERRITINS 566
   3.1. Protein 566
      3.1.1. Cellular Localization and Genetic Origin 566
      3.1.2. Plant Apoferritin 568
      3.1.3. Plant Ferritin Mineral Cores 570
   3.2. Genes 571

4. FERRITIN AND PLANT DEVELOPMENT 572

---

*Corresponding author.

4.1. Posttranscriptional Regulation of Ferritin Synthesis During Plant Development 572
4.2. RNA Stability Sequences are Present in Ferritin mRNA but Have Not Been Tested Functionally 573
4.3. Regulation of Ferritin Accumulation During Plant Development by Controlling Protein Stability 574

5. FERRITIN AND RECOVERY FROM IRON STARVATION 575

6. FERRITIN AND IRON-MEDIATED OXIDATIVE STRESS 577

7. OUTLOOK 578

ACKNOWLEDGMENTS 579

ABBREVIATIONS 579

REFERENCES 579

1. INTRODUCTION

Iron is essential for plants in many aspects, and some of them, like photosynthesis and nitrogen fixation, are specific. The basic functions involving iron in aerobic life also apply to plants, as well as the constraints of insolubility and toxicity in the use of iron [1]. However, some specificities of the plant kingdom have an important influence on iron utilization by these organisms. Among these specificities is the organization of plant cells which is characterized by three specific compartments [2]: (1) the extracellular cell wall, (2) the vacuoles which occupy most of the intracellular space in many cell types, and (3) the plastids, a class of interrelated organelles among which the chloroplasts achieve oxygen production via the photosynthetic reaction.

In addition to the active mechanisms controlling iron uptake by the roots (see Chapters 5 and 6 in this book), the three different compartments mentioned above are deeply involved in controlling iron fluxes in plants. They play key roles in buffering and storing iron in order to achieve plant iron homeostasis. The aim of this chapter is to review the respective roles of these three compartments in iron storage in plants with emphasis on the plastid compartment which harbors ferritin within plant cells.

2. IRON STORAGE IN THE APOPLAST AND PLANT VACUOLES

The ability of plants to acquire highly diluted mineral nutrients from the soil and to harvest light energy require special root and leaf surfaces. Such a morphology, and the shape of plants, can coexist because of two specific plant cell attributes: the apoplast and the vacuoles. The apoplast can be defined as the network of cell walls forming the extracellular free space of plant cells. It is composed of cellulose, hemicellulose, pectic compounds, lignin, suberin, proteins, and water [3,4], and it plays important roles in the control of water and solute movements [5]. Vacuoles are intracellular acidic compartments delimited by a membrane called the tonoplast; they are involved in many detoxification processes and participate in the control of plant cell osmotic pressure via water and ion fluxes through the tonoplast [6].

2.1. Iron Storage in the Apoplast

The apoplast plays an important role in iron metabolism, which has been well documented [7,8]. Iron resupply to iron-starved plants leads to a 10-fold increase in total root iron after 48 hs, but 90% of this iron is extracellular, trapped in the apoplastic free space [9]; possible apoplastic iron forms deposited in roots, apart from ferric ions bound to cation-exchange sites, are ferric hydroxide and ferric phosphate. Iron toxicity in flooded anaerobic soils is due to the predominance of ferrous iron which is readily taken up by roots [10]. The first event observed in rapidly detoxifying excess ferrous iron concerns the generation of an apoplastic ferric deposit, which highlights the role that the root apoplast can play in limiting iron toxicity in plants [11–13]. This apoplastic ferric iron deposit constitutes an iron reserve for use by the plant in periods of deprivation. It can be transported through the plasmalemma membrane via various transport systems activated by iron deficiency (for review, see [14], and Chapters 5 and 6 in this book). Therefore, the apoplast has to be considered as an iron-storage compartment, playing a key role in the buffering of this element. Leaf apoplast also contains iron, and in the tomato *chloronerva* mutant, which takes up too much iron resulting in an iron overload of leaf cells, iron accumulates both in the apoplast and in the symplast [15].

2.2. Iron Storage in Vacuoles

As iron has a tendency to precipitate in the presence of oxygen and to react with reduced forms of oxygen, it is sequestered by a variety of molecules, including phenolics, organic acids, amino acids and some of their derivatives, and proteins [14]. The specificity of the chelation capacity of these molecules and their iron affinity vary. Among this broad range of compounds, organic acids and mainly citrate, are used as a shuttle to transport ferric iron to the leaves by the xylem sap [16], and a link between iron metabolism and citrate concentration in plants has been clearly established [17]. In iron-overload conditions, citrate might also be involved in detoxification processes, since it is found in vacuoles which could serve as an iron store under these conditions [18]. At the molecular level, the mechanisms of iron detoxification by vacuoles is poorly documented, and the only information available has been obtained with *Saccharomyces cerevisiae*. The small phenotype observed in yeast mutants affected in the *VAM3* gene coding for the vacuolar H^+-ATPase (adenosine triphosphatase) proteolipid subunit can be suppressed by adding a large amount of iron to the growth medium [19]. Also, the use of the *vpt13* mutant deficient in yeast vacuolar acidification and of the *vpt16* mutant devoid of any vacuole structure demonstrated that vacuolar iron sequestration is not dependent on vacuolar acidification, which is in contrast to cation-proton exchangers mediating the uptake of other metal cations such as calcium, manganese, and zinc; instead, vacuolar iron uptake could be mediated through an active transport involving an ATP-dependent iron pump [20].

3. STRUCTURE-FUNCTION OF PLANT FERRITINS

3.1. Protein

3.1.1. Cellular Localization and Genetic Origin

The first characterization of plant ferritin was reported by Hyde and his collaborators [21]. Ferritin is easily visualized by classic electron microscopy methods; this explains the wealth of cytological observations concerning this protein published in the 1960s and 1970s. Ferritin was described in almost all species, and it was detected in the stroma of various plastids in many cells. Although described in chloroplasts, ferritin was mainly observed in nongreen plastids such as proplastids, etioplasts, chromoplasts, and amyloplasts. Accumulation of this protein occurs naturally in the plastids of storage tissues like seeds or shoot and root apices. Iron overload of iron-starved plants, chemical treatments with nicotine, ozone, streptomycine or ethylene, and conditions of impaired photosynthetic activity also favor ferritin accumulation within plastids (for review, see [22]).

The plastid localization of plant ferritin has been confirmed biochemically and immunologically [23–25]. At a molecular level, a transit peptide in the N-terminal extremity of plant ferritins, which is responsible for polypeptide precursor targeting to plastids (for review, see [26]), was characterized by comparing the NH_2-terminal amino acid sequence determined by microsequencing of purified mature subunits of various ferritins [9,18,27] with the amino acid sequences deduced from the nucleotide sequences of various plant ferritin cDNAs [9,25,28–30]. Synthesis of plant ferritins as precursors and their subcellular localization inside the plastids suggest they are encoded by nuclear genes. This is reinforced by the facts that (1) all the plant ferritin cDNAs cloned so far contained a polyA tail [9,25,28,29] and (2) translation of enriched polyA$^+$ mRNA fractions in eukaryotic in vitro systems gives rise to plant ferritin precursors [23,24]. It is now widely accepted that plant plastids originate from cyanobacteria and that part of the genes originally belonging to the endosymbiont were transferred during evolution in the nucleus of the host cell [31]. However, plant ferritins share an amino acid sequence identity varying between 39%–49% with animal ferritin and have no significant homology with bacterial ferritins [32], including cyanobacteria [33]. Therefore, plant ferritins, although plastid located, appear to be more related to animal ferritins than to bacterioferritins. Nevertheless, relics of a nonfunctional bacterioferritin gene are still present in the chloroplast genome [27]. It is therefore plausible that during the course of evolution, the bacterioferritin of the endosymbiont has been lost and replaced by the product of a nuclear gene, homologous to contemporary animal ferritin, and having acquired a specific sequence in its NH_2-terminal extremity responsible for its plastid subcellular localization.

3.1.2. Plant Apoferritin

Plant ferritin subunit sequences share between 39%–49% identity with mammalian ferritin sequences [32]. The seven amino acids involved in the ferroxidase center of the animal H-type ferritins, allowing an oxidation and a rapid uptake of iron inside the mineral core [34,35], are found in plant ferritin subunits. Despite this higher identity of the plant ferritins with H-type animal ferritins, it has to be noticed that carboxylic residues found at the surface of the cavity of L-type animal ferritins and responsible for an efficient iron nucleation and a better stability of the mineral core, are also conserved in plant ferritins. An additional plant-specific sequence is observed at the amino terminus when compared with animal ferritin sequences. As mentioned above, the first part of this specific amino terminus encodes a transit peptide responsible for the plastid targeting of plant ferritin precursor. The second domain of the amino terminus specific to plant ferritin belongs to the mature wild-type ferritin subunit and has been named extension peptide [27]. The 30 carboxy-terminal amino acids are highly conserved within plant ferritins, but they have diverged from their animal counterparts. In animal ferritin, conserved amino acids from this domain define the fifth or E-helix which is involved in the formation of hydrophobic channels located at the four-fold symmetry axes [36].

A very high level of similarity of the secondary structure of pea seed ferritin with mammalian ferritin secondary structure has been reported [28]. A high helical content is consistent with the adoption by residues 41–210 of a four-helix bundle conformation. A nonhelical stretch of 21 residues (positions 106–127 in pea seed ferritin [28]) showed either turn/coil or β-strand predictions and would link the B and C helices via a L-loop characteristic of the animal ferritin subunit fold [36]. Interestingly, the prediction indicated the presence of an additional three-turn α-helix in the N-terminal extension region of the protein. This helix of 11 amino acids in length would be found lying within the plant-specific sequence of 24 amino acids [27]. A further consequence of the secondary structure as predicted [28] is the conservation of the short C-terminal E-helix despite the divergent primary structures of plant and animal ferritins in this region as noted above.

Using the coordinates of the recombinant human H ferritin structure [35], a model of the three-dimensional (3-D) structure of the pea seed ferritin subunit has been proposed [28]. This model, superimposed on the known subunit structure of recombinant human H ferritin, highlights the remarkable conservation of the 3-D core structure of plants as modeled to human ferritins and the predicted regions of change. At the N-terminal extremity of the pea subunit, nearly half of the additional residues from this specific extension region would assume a helical conformation which folds back onto the surface of the subunit. This extension peptide is the site for free radical cleavages during iron exchanges in vitro [37], and possibly in vivo [38], leading to ferritin degradation. Such radical damages have been shown to confer a higher sensitivity of proteins to proteolysis [39]. Mutagenesis analysis have confirmed the role of the extension peptide in the protein stability of plant ferritin [40]. The model also supports the presence of (1) the E-helix at the same place in plant and animal ferritins and (2) a ferroxidase center in the plant ferritin subunit [28].

Ferritin subunits assemble in a compact 24-mer arrangement with 4.3.2 symmetry. The resulting packed protein shell contains many intersubunit interactions except that at the four-fold and three-fold symmetry axes, channels of about 3Å wide are formed [36]. The predicted structure of pea seed ferritin [28] has a similar quaternary structure. The plant specific N-terminal extension mentioned above would form a compact localized domain on the external surface of the 24-mer. In animals, the four-fold channels are hydrophobic, whereas the three-fold channels are hydrophilic [36]. The pea three-fold channels would remain hydrophilic, and the four-fold channels would be hydrophilic. In animals, hydrophobic channels at four-fold axes present an energy barrier to unshielded Fe^{2+} ions, whereas the hydrophilic three-fold channels have been considered as a possible entry or exit port and a site for Fe(II) oxidation [41]. The consequences of both four-fold and three-fold channels being hydrophilic, and the effects this would have on iron exchange in plant ferritins remain to be determined.

The nucleation and growth of the iron core are active biomineralization processes involving functions of the protein coat [36,42], and the demonstration of a ferroxidase activity associated with the H-type mammal ferritin has highlighted this point [35]. The predicted ferroxidase activity of plant ferritins [28] has been confirmed experimentally with ferritin purified from pea seeds [43] and with recombinant pea seed ferritin expressed in *Escherichia coli* [40]. However, this activity was found to be lower than that of the recombinant human H-ferritin; although it was much more active than the recombinant human L-ferritin [40]. Furthermore, when Ala-13 was replaced by His in recombinant pea seed ferritin, as in mammal ferritins, the ferroxidase activity was significantly reduced [44]. Ferrous iron, the substrate of the ferroxidase activity, may be formed in vitro by reduction with ascorbate of exogenous ferric chelates and/or of ferric ferritin-iron itself [45]. Photoreduction of ferric citrate and/or of ferritin-iron may also be a source of ferrous iron [46]. Depending on the concentration of ascorbate, either an overall iron release (ascorbate > 2.5 mM) or uptake (ascorbate < 2.5 mM) by pea seed ferritin may occur. In vitro studies of the different steps of these independent and simultaneous uptake and release fluxes have revealed that uptake was faster at pH 8.4 than at pH 7.0 or 6.0 and was inhibited by an excess of free iron-binding ligands. pH values and ligand/Fe ratios fit known physiological values. These free ligands were also found to be inhibitory during ferritin-iron release. The in vitro conditions mentioned above are likely to occur in vivo. Ascorbate is a natural reducer found in the plastids where iron-loaded ferritins are also located.

3.1.3. Plant Ferritin Mineral Cores

Mineral cores of animal ferritins and of bacterioferritins have significant chemical and structural differences as revealed by electron microscopic appearance and Mössbauer spectroscopy behavior [36,42]. Human and horse ferritin molecules contain a single domain crystal of ferrihydrite associated with variable amounts of inorganic phosphate. Bacterioferritin cores consist of hydrous ferric phosphate, with high amounts of phosphate, having a poor crystallinity and very low magnetic ordering temperatures. These physicochemical differences between animal and bacterial iron cores can be attributed to variations in phosphate concentration between bacterial and animal cells [47].

Plant ferritins are eukaryotic-type proteins found in a prokaryotic type environment (i.e., the plastids). Such a situation influences the composition and the structure of plant ferritin cores as determined by measurement of the Fe:P ratio and by Mössbauer spectroscopy, electron microscopy, and x-ray absorption spectroscopy analysis of native and reconstituted pea seed ferritin mineral cores [43,48]. Plant ferritin mineral cores have a high phosphate content and are amorphous. In contrast, reconstituted mineral cores using pea seed apoferritin in the absence of phosphate consist of crystalline ferrihydrite. Plant mineral cores are more related to their bacterial counterpart than to their animal one, and this is likely to be due to the high phosphate concentration found within plastids.

3.2. Genes

The complete nucleotide sequence of two maize ferritin genes [49], one soybean ferritin gene [50] and one *Arabidopsis thaliana* ferritin gene (O. Van Wuytswinkel and J.-F. Briat, unpublished observations) has been determined, making possible the comparison of ferritin gene organization in plants and animals. Although, plant and animal ferritins arose from a common ancestor and are highly conserved proteins (see above), their gene organization is remarkably different. Divergence in the genetic mechanisms for iron regulation is reflected by the absence in all plant ferritin genes of the iron-regulatory element (IRE), a conserved regulatory sequence responsible for the translational regulation of animal ferritin mRNA [51]. Moreover, fusion of an animal IRE to soybean ferritin cDNA did not result in a specific translational control of soybean ferritin mRNA [52,53], although IRE fused to reporter gene such as chloramphenicol acetyl transferase or human growth hormone led to translational regulation of the mRNA [54,55]. It has been predicted by computer modeling that the formation of the typical hairpin loop structure of the IRE is prevented in a frog IRE-soybean ferritin mRNA chimera [53]. Alternative hairpins between the crucial CAGUG loop sequence of the IRE stem loop [52] and the transit peptide sequences of soybean ferritin mRNA can form in the IRE-soybean ferritin mRNA chimera, preventing IRE function [53].

The number of introns is higher in plant (seven) than in animal (three) ferritin genes, and no intron positions are conserved between plant and animal ferritin genes; within kingdoms, the intron positions in ferritin genes are strictly conserved. In addition, the three animal ferritin introns are in frame with exon sequences, whereas four of the seven plant ferritin introns are out of frame. Finally, in animal ferritins, the position of introns has been correlated to secondary structure motifs

[56]: the first intron interrupts the sequence at the end of helix A, the second one is in the middle of the B-C loop, and the third one is at the beginning of helix D, suggesting that contemporary animal ferritin gene structure has evolved by exon shuffling. This is not the case for the gene structure of plant ferritin, since the secondary structure domains of the plant protein have no apparent relationship with the intron/exon boundaries.

4. FERRITIN AND PLANT DEVELOPMENT

4.1. Posttranscriptional Regulation of Ferritin Synthesis During Plant Development

Observation of iron-rich ferritin molecules by electron microscopy has clearly established that the presence of these molecules in various organs and tissues was regulated during the development (for review, see [22]). Ferritin naturally occurs in plant tissues such as cotyledons, root, and shoot apices, and it has also been described in cells of vessels, in vascular cambium, in reproductive cells, and in senescing cells. However, it is only recently that developmental regulation of plant ferritin has been studied at the molecular level, at defined developmental stages of leaves [57] and nodules [53,58], as well as during the life cycle of pea plants [38].

Plant ferritin accumulates in the cotyledons and embryo axis of seeds, while the amount of ferritin in vegetative organs (roots and leaves) is very low and only degraded forms of the protein can be detected [38]. In contrast, the concentration of ferritin mRNA was reported to be higher in mature soybean leaves than in young leaves [27], and this was the first indication that a posttranscriptional event could regulate ferritin synthesis during development. Release of ferritin iron during the first days of growth strongly suggests that it plays an important role in the early stage of plant development, probably owing to its contribution in building up the photosynthetic apparatus leading to the autotrophy of plants. The presence of ferritin in etioplasts and its absence from chloroplasts reinforced the idea that ferritin iron could be used for the synthesis of iron-containing enzymes during deetiolation (for review, see [59]). Using the determinate nature of leaf development in maize, it has been possible to demonstrate that only the youngest section of maize leaf had a high level of ferritin, whereas accumulation of chlorophyll and phosphoenolpyruvate carboxylase (a marker of photosynthetic activity in C4 plants) are most abundant in the middle of the leaf where the ferritin concentration reaches its minimum [57]. Thus, the leaf ferritin content correlates with the differentiation stage of chloroplasts, and this therefore reinforces the idea that this protein would be an iron source for iron-proteins involved in photosynthesis, and would allow the deposit of iron when chloroplasts degenerate. In the maize leaf model described above, an uncoordinated change in ferritin mRNA and ferritin protein concentrations was also reported, confirming that posttranscriptional control is likely to be an important level of the regulation of ferritin synthesis during plant development [57].

Besides photosynthesis and seed storage, iron is also important for nitrogen fixation, requiring iron-proteins such as leghemoglobin and nitrogenase. Ferritin accumulates in soybean nodules depending upon their age, upon interaction between the *Bradyrhizobium* strain and the host cultivar, and upon mutations affecting the bacterial strain [60]. In the effective symbiosis of *Bradyrhizobium* and soybean, ferritin has been observed only in young nodules. In contrast, in ineffective nodules produced by *Bradyrhizobium* mutants, ferritin is found in nodules of all ages. Hence, ferritin is present in nodules when they are not functional. At a molecular level, during the course of soybean nodule aging, ferritin mRNA concentrations remain elevated, whereas the protein concentration decreases four- to five-fold [58], indicating again the occurrence of a posttranscriptional control of ferritin abundance during a plant-specific developmental program.

4.2. RNA Stability Sequences are Present in Ferritin mRNA but Have Not Been Tested Functionally

There is no actual evidence that stability and/or translation of plant ferritin mRNA could be the targets for controlling ferritin synthesis during plant development. It should, however, be noted that three overlapping 5′AUUUA3′ repeats are present in the 3′ untranslated region of one of the three maize ferritin cDNA sequences [61]. Such repeats of the AUUUA sequence were first characterized in animal systems because of their role in the control of mRNA stability [62]. Generally, they are found in the 3′ untranslated regions of short half-life mRNAs such as lymphokine, cytokine, and protooncogene mRNAs, and their activity is likely to be regulated [63]. In plants, 5′AUUUA3′ repeats have already been suggested as determinants of mRNA stability [64], and a synthetic 60-base sequencing containing 11 copies of the AUUUA motif has been shown to be responsible for the destabilization of reporter genes in transgenic tobacco cell lines and tobacco plants when inserted within their 3′ untranslated regions [65]; the sequence per se rather than the A+U content was shown to be responsible. As in maize ferritin mRNA, this motif has also been found in the 3′ untranslated region of a soybean ferritin mRNA [25], although not in the same context as that of the maize one. Whether this destabilizing sequence in the 3′ untranslated region of some, but not all, plant ferritin transcripts represents a functional determinant for the control of plant ferritin synthesis during development remains to be determined.

4.3. Regulation of Ferritin Accumulation During Plant Development by Controlling Protein Stability

Posttranscriptional control of ferritin synthesis during development could be regulated, at least in part, at the protein stability level. Disappearance of ferritin protein from pea seed embryo axes during the first days following germination is accompanied by the appearance of polypeptides of lower molecular weight than the basic ferritin subunit [38]. These polypeptides react with antibodies raised against pea seed ferritin and have also been observed in samples of ferritin purified from pea seed soaked for 48 h in aerated water [37]. Also during soybean nodule aging, before ferritin disappears completely, ferritin breakdown products have been observed [58]. It is therefore likely that ferritin degradation requires processing of the ferritin subunit first. A mechanism responsible for this specific processing has been proposed based on iron-release experiments performed in vitro [37,46]. In vitro release of ferritin iron, when reduced by ascorbate or light, induces conversion of the ferritin subunit from 28.0 kDa to 26.5 and 25.0 kDa. As a consequence, pea seed ferritin has then a tendency to aggregate and to become insoluble [37]. Fenton chemistry has been involved in this process, since protein cleavage is inhibited by free radical scavengers as well as by desferrioxamine and ortho-phenanthroline, chelating ferric and ferrous iron ions, respectively [37,46]. Microsequencing of the mature 28-kDa subunit and of the processing product of 26.5 kDa has indicated that these free radical cleavages were occurring within the plant-specific amino-terminus of the pea seed ferritin subunit [37]. These observations have recently been confirmed by demonstrating that recombinant pea seed ferritin, devoid of its amino-terminus, is less soluble than recombinant wild-type pea seed ferritin [40]. A model explaining the mechanism of ferritin degradation during germination has been proposed [38]. Based on data reported on ferritin regulation during nodulation [53], this model could also apply to other plant-specific developmental processes.

5. FERRITIN AND RECOVERY FROM IRON STARVATION

Iron deficiency in plants leads to chlorosis symptoms with impaired photosynthesis and physiological disorders [14]. These symptoms are reversible when iron is added to the culture medium, and plants recover within a few days [66–68]. Iron is essential in plastids to restore photosynthesis by allowing the synthesis of iron-sulfur proteins and chlorophyll. In plants, ferritins are localized on plastids and their synthesis is induced during recovery from iron deficiency [9,24,69]. A transient accumulation of maize ferritin subunits was measured reaching its maximum 24 h after iron addition and starting to decrease 48 h after treatment [9]. It has also been reported that a large amount of iron is translocated from roots to leaves within the first 3 h during regreening of iron-deficient plantlets [70,71]. Ferritin accumulation during this recovery period is consistent with the iron-buffering function of this protein to store iron in a safe and bioavailable form. Iron resupply to iron-deficient maize plantlets leads to a transient accumulation of ferritin mRNA in both roots and leaves [9]. Ferritin transcripts are already detected 3 h after iron addition, and the maximum is reached at 6 h [72]. Therefore, regulation of plant ferritin genes in response to iron occurs at a different level in plants and mammalian cells, where a major translational control has been implicated [73]. These results are consistent with the fact that no iron regulatory elements are present in plant ferritin mRNA and genes [9,49,50]. Furthermore, it has been demonstrated in soybean cells that ferritin accumulation is regulated at the transcriptional level in response to iron [25]. Messenger RNA, known to be accumulated in response to various stresses involving the plant hormone abscisic acid (ABA), also accumulates in response to iron treatment [72]. These results suggest that ABA could be involved in the transduction pathway leading to ferritin synthesis during recovery from iron deficiency. Indeed, iron overload of maize plantlets leads to a fivefold increase in ABA concentration in roots and leaves within 3 h and ferritin mRNA accumulates in response to an exogenous ABA treatment. The proof that ABA is indeed involved in iron-induced ferritin synthesis by an iron excess came from experiments performed with an ABA-deficient mutant [72]. This maize mutant (*Vp2*) is deficient in carotenoid biosynthesis, which is a precursor for ABA synthesis. Ferritin mRNA accumulation is strongly reduced in this mutant, demonstrating that ABA is involved in a pathway leading to ferritin accumulation in response to iron. However, comparison of the maximum ferritin mRNA accumulation after supplying saturating levels of either iron or of exogenous ABA revealed that ABA cannot account for the total iron response in maize [72]. Therefore, iron resupply of maize plantlets activates a second unidentified pathway.

In maize, two subclasses of ferritin cDNAs (FM1 and FM2) have been characterized and the corresponding genes named *ZmFer1* and *ZmFer2* [9,49]. Sequence divergence between the 3′ untranslated regions of these cDNAs has allowed to use specific probes to study the accumulation of FM1 and FM2 transcripts in hydroponic cultures of maize plantlets [49]. These experiments reveal that FM1 and FM2 mRNA are differentially accumulated in response to iron treatment. The abundance of FM1 mRNA is already high 3 h after iron addition, and FM2 transcripts are only detected after 6 h. Furthermore, FM2 mRNA accumulates in response to exogenous ABA treatment and drought, which is known to increase ABA concentration in plants, although such treatments have no effect on the abundance of the FM1 transcript. It is therefore possible to conclude that, in maize plantlets during recovery from iron deficiency, an ABA-independent pathway is activated leading to FM1 mRNA accumulation and another pathway involving the plant stress hormone abscisic acid increases the abundance of the FM2 transcript. This differential expression of *ZmFer1* and *ZmFer2* maize ferritin genes could be due to the important divergence observed between the promoter sequence of these two genes [49].

6. FERRITIN AND IRON-MEDIATED OXIDATIVE STRESS

Although iron is essential for plants, its excess can be highly damaging, leading to a variety of morphological symptoms among which the one best characterized is leaf bronzing in rice [10]. In waterlogged soils, anaerobic conditions and a low pH lead to the reduction of ferric iron and stabilization of ferrous iron, which is readily taken up by plants, activating H_2O_2 to generate hydroxyl radicals via the Fenton reaction. These free radicals are among the most reactive species known, for example, through the Haber-Weiss reaction [74], and this oxidative stress can alter all the components of the cell, leading to a loss of integrity and possibly to cell death. Detoxification of reactive oxygen intermediates is an important element of the defense against oxidative stress, but controlling the concentration of free ferrous iron could be an important complementary way to prevent oxidative damages. Ferritin with its iron-storage capacity could play an important role in protecting cells against oxidative stress as observed in mammalian cells [75]. This idea is reinforced by the fact that ferritin synthesis is activated in rat cells as a direct response to oxidative stress mediated by phorone treatment [76].

As plant chloroplasts are heavily exposed to oxidative stress during photosynthesis, various antioxidant mechanisms occur in these organelles [77]. The role of ferritin in the oxidative stress response has been investigated by dipping the cut end of the stems of derooted plantlets in a solution of iron-citrate to induce a rapid iron overload of maize leaves [78], which is similar to the system used to induce leaf bronzing in rice [10]. In this system, an eightfold increase in leaf iron concentration has been measured. This high iron accumulation induces a rapid increase in ferritin mRNA abundance, and ferritin subunits can be immunodetected just 9 h after the beginning of the treatment. In these conditions, no mRNA responsive to abscisic acid accumulates in response to iron, and ferritin mRNA accumulation is still observed in the *Vp2* ABA-deficient mutant. These data show that ABA is not involved in this response. When derooted plantlets are treated with other metals (Ni, Cu, Zn, Ag, Cd), no ferritin mRNA accumulation is detected; therefore, this response is specific to iron. Since this element is tightly linked to free radical production in oxidative stress, the effect of antioxidant agents on iron-induced ferritin accumulation has also been investigated. When derooted plantlets were pretreated with glutathione or N-acetylcysteine, a precursor for glutathione synthesis, no increase in the abundance of ferritin mRNA was detected. Furthermore, hydrogen peroxide treatment performed in the presence of a low iron concentration induced an increased ferritin transcript level. These data suggest that an oxidation step is required in the pathway leading to ferritin synthesis in maize derooted plantlets, and similar results have been obtained using *Arabidopsis thaliana* plantlets [30]. In addition to an increase of ferritin concentration, iron-mediated oxidative stress also increases the concentration of dehydroascorbate and decreases the ascorbate concentration [79]. This observation is interesting, because, as already mentioned above, ascorbate concentrations lower than 2.5 mM favor iron uptake in vitro by pea seed ferritin, whereas higher concentrations promote the release of iron [45,46]. Besides ferritin accumulation, an oxidative stress in response to iron toxicity could therefore generate favorable redox conditions, increasing iron uptake by the protein and allowing iron detoxification.

7. OUTLOOK

Plant ferritins are very similar to their animal counterparts, and these proteins arise clearly from a common ancestor. Although ferritins of each kingdom have some specificities, they share a common structure responsible for the same function; namely, iron storage. However, plant ferritins are likely to play some specific roles in the biology of these organisms, and this is highlighted by their cellular localization within the plastids. It is unknown if iron from these plant-specific organelles, among which the choroplasts, where photosynthesis takes place, can be mobilized for cell functions occurring in other compartments. In contrast, it is well established that other cellular territories, such as the apoplast and the vacuoles, are also involved in the control of plant iron homeostasis. Of more general interest is the major transcriptional control of ferritin synthesis in response to iron overload, making this system unique to study iron-regulated eukaryotic gene transcription.

ACKNOWLEDGMENTS

Work of the authors quoted in this review was supported by the Centre National de la Recherche Scientifique (URA 2133 and ATIPE grant n° 93N60/0563 to J.-F.B.).

ABBREVIATIONS

ABA	abscisic acid
ATP	adenosine 5′-triphosphate
cDNA	complementary deoxyribonucleic acid
FM	maize ferritin cDNA
IRE	iron-regulatory element
mRNA	messenger ribonucleic acid
Vp2	viviparous 2
ZmFer	maize (*Zea mays*) ferritin gene

REFERENCES

1. J. F. Briat, in *Oxidative Stress in Plants* (M. Montagu and D. Inzé, eds.), Biotechnology Intelligence Unit Series, R. G. Landes Company Bioscience Publishers. In press.
2. B. Alberts, D. Bray, J. Lewis, M. Raff, K. Roberts, and J. D. Watson, in *Molecular Biology of the Cell*, Garland, New York, 1983, pp. 1099–1146.

3. G. I. Cassab and J. Varner, *Annu. Rev. Plant. Physiol. Plant Mol. Biol.*, *39*, 321–353 (1988).
4. N. C. Carpita, *Annu. Rev. Plant. Physiol. Plant Mol. Biol.*, *47*, 445–476 (1996).
5. M. J. Cany, *Annu. Rev. Plant. Physiol. Plant Mol. Biol.*, *46*, 215–236 (1995).
6. P. Matile, *Annu. Rev. Plant Physiol.*, *29*, 193–213 (1978).
7. H. F. Bienfait, W. Van Den Briel, and N. T. Mesland-Mul, *Plant Physiol.*, *78*, 596–600 (1985).
8. F. S. Zhang, V. Römheld, and H. Marschner, *Plant Physiol.*, *97*, 1302–1305 (1991).
9. S. Lobréaux, O. Massenet, and J. F. Briat, *Plant Mol. Biol.*, *19*, 563–575 (1992).
10. F. N. Ponnamperuma, R. Bradfield, and M. Peech, *Nature,* 175, 275 (1955).
11. M. S. Green and J. R. Etherington, *J. Exp. Bot.,* 28, 678–690 (1977).
12. C. D. Foy, R. L. Chaney, and M. C. White, *Annu. Rev. Plant. Physiol.*, *29*, 511–566 (1978).
13. T. Ando, S. Yoshida, and I. Nishiyama, *Plant Soil*, *72*, 57–71 (1983).
14. J. F. Briat, I. Fobis-Loisy, N. Grignon, S. Lobréaux, N. Pascal, G. Savino, S. Thoiron, N. von Wirèn, and O. Van Wuytswinkel, *Biol. Cell*, *84*, 69–81 (1995).
15. A. Pich and G. Scholz, *J. Exp. Bot.*, *42*, 1517–1523 (1991).
16. D. A. Cataldo, K. M. McFadden, T. R. Garland, and R. E. Wildung, *Plant Physiol.*, *86*, 734–739 (1988).
17. A. Pich, G. Scholz, and K. Seifert, *J. Plant Physiol.*, *137*, 323–326 (1991).
18. A. M. Lescure, O. Massenet, and J. F. Briat, *Biochem. J.*, *272*, 147–150 (1990).
19. D. J. Eide, J. T. Bridgham, Z. Zhao, and J. R. Mattoon, *Mol. Gen. Genet.*, *241*, 447–456 (1993).
20. H. P. Bode, M. Dumschat, S. Garotti, and G. F. Fuhrmann, *Eur. J. Biochem.*, *241*, 337–342 (1995).
21. B. B. Hyde, A. J. Hodge, A. Kahn, and M. L. Birnstiel, *J. Ultrastr. Res.*, *9*, 248–258 (1963).

22. J. J. Seckback, *J. Plant Nutr.*, *5*, 369–394 (1982).
23. F. Van der Mark, W. Van der Briel, and H. G. Huisman, *Biochem. J.*, *214*, 943–950 (1983).
24. D. Proudhon, J. F. Briat, and A. M. Lescure, *Plant Physiol.*, *90*, 586–590 (1989).
25. A. M. Lescure, D. Proudhon, H. Pesey, M. Ragland, E. C. Theil, and J. F. Briat, *Proc. Natl. Acad. Sci. USA*, *88*, 8222–8226 (1991).
26. G. V. Heijne, J. Steppuhn, and R. G. Herrmann, *Eur. J. Biochem.*, *180*, 535–545 (1989).
27. M. Ragland, J. F. Briat, J. Gagnon, J. P. Laulhère, O. Massenet, and E. C. Theil, *J. Biol. Chem.*, *265*, 18339–18344 (1990).
28. S. Lobréaux, S. Yewdall, J. F. Briat, and P. M. Harrison, *Biochem. J.*, *288*, 931–939 (1992).
29. M. J. Spence, M. T. Henzl, and P. J. Lammers, *Plant Mol. Biol.*, *17*, 499–504 (1991).
30. F. Gaymard, J. Boucherez, and J. F. Briat, *Biochem. J.*, *318*, 67–73 (1996).
31. M. W. Gray, *Trends Genet.*, *5*, 294–299 (1989).
32. S. C. Andrews, P. Arosio, W. Bottke, J. F. Briat, M. von Darl, P. M. Harrison, J. P. Laulhère, S. Levi, S. Lobréaux, and S. Yewdall, *J. Inorg. Biochem.*, *47*, 161–174 (1992).
33. J. P. Laulhère, A. M. Labouré, O. van Wuytswinkel, J. Gagnon, and J. F. Briat, *Biochem. J.*, *281*, 785–793 (1992).
34. E. R. Bauminger, P. M. Harrison, D. Hechel, I. Nowik, and A. Treffry, *Biochim. Biophys. Acta*, *1118*, 48–58 (1991).
35. D. M. Lawson, P. J. Artymiuk, S. J. Yewdall, J. M. A. Smith, J. C. Livingston, A. Treffry, A. Luzzago, S. Levi, P. Arosio, G. Cesareni, C. D. Thomas, W. V. Shaw, and P. M. Harrison, *Nature*, *349*, 541–544 (1991).
36. P. M. Harrison and P. Arosio, *Biochim. Biophys. Acta*, *1275*, 161–203 (1996).
37. J. P. Laulhère, A. M. Labouré, and J. F. Briat, *J. Biol. Chem.*, *264*, 3629–3635 (1989).
38. S. Lobréaux and J. F. Briat, *Biochem. J.*, *274*, 601–606 (1991).
39. K. J. A. Davies, *J. Biol. Chem.*, *262*, 9895–9901 (1987).
40. O. Van Wuytswinkel, G. Savino, and J. F. Briat, *Biochem. J.*, *305*, 253–261 (1995).

41. S. Levi, A. Luzzago, F. Franceschinelli, P. Santambrogio, G. Cesareni, and P. Arosio, *Biochem. J.*, *264*, 381–388 (1989).

42. P. M. Harrison, P. J. Artymiuk, G. C. Ford, J. R. Guest, J. Hirzman, D. M. Lawson, J. C. Livingstone, J. M. A. Smith, A. Treffry, and S. J. Yewdall, *Adv. Inorg. Chem.*, *36*, 449–486 (1991).

43. V. J. Wade, A. Treffry, J. P. Laulhère, E. R. Bauminger, M. I. Cleton, S. Mann, J. F. Briat, and P. M. Harrison, *Biochim. Biophys. Acta*, *1161*, 91–96 (1993).

44. O. Van Wuytswinkel and J. F. Briat, *Biochem. J.*, *305*, 959–965 (1995).

45. J. P. Laulhère and J. F. Briat, *Biochem. J.*, *290*, 693–699 (1993).

46. J. P. Laulhère, A. M. Labouré, and J. F. Briat, *Biochem. J.*, *269*, 79–84 (1990).

47. S. Mann, J. M. Williams, A. Treffry, and P. M. Harrison, *J. Mol. Biol.*, *198*, 405–416 (1987).

48. G. S. Waldo, E. Wright, Z. H. Wang, J. F. Briat, E. C. Theil, and D. E. Sayers, *Plant Physiol*, *109*, 797–802 (1995).

49. I. Fobis-Loisy, K. Loridon, S. Lobréaux, M. Lebrun, and J. F. Briat, *Eur. J. Biochem.*, *231*, 609–619 (1995).

50. D. Proudhon, J. Wei, J. F. Briat, and E. C. Theil, *J. Mol. Evol.*, *42*, 325–336 (1996).

51. E. C. Theil, *Biochem. J.*, *304*, 1–11 (1994).

52. D. J. Dix, P. N. Lin, A. R. Mckenzie, W. E. Walden, and E. C. Theil, *J. Mol. Biol.*, *231*, 230–240 (1993).

53. Y. Kimata and E. C. Theil, *Plant Physiol.*, *104*, 263–270 (1994).

54. N. Aziz and H. N. Munro, *Proc. Natl. Acad. Sci. USA.*, *84*, 8478–8482 (1987).

55. M. W. Hentze, S. W. Caughman, T. A. Rouault, J. G. Barriocanal, A. Dancis, J. B. Harford, and R. D. Klausner, *Science*, *238*, 1570–1572 (1987).

56. P. M. Harrison, G. C. Ford, J. M. A. Smith, and J. L. White, *Biol. Metals*, *4*, 95–99 (1991).

57. E. C. Theil and T. Hase, in *Iron Chelation in Plants and Soil Microorganisms* (L. L. Barton and B. Hemings, eds.), Academic Press, 1993, pp. 133–156.

58. M. Ragland and E. C. Theil, *Plant Mol. Biol.*, *21*, 555–560 (1993).

59. H. F. Bienfait and F. van der Mark, in *Uptake and Utilization of Metals by Plants* (D. A. Robb and W. S. Pierpont, eds.), Academic Press, 1983, pp. 111–123.

60. M. P. Ko, P. Y. Huang, J. S. Huang, and K. R. Barker, *Plant Physiol.*, *83*, 299–305 (1987).

61. J. F. Briat, A. M. Labouré, J. P. Laulhère, A. M. Lescure, S. Lobréaux, D. Proudhon, and O. Van Wuytswinkel, in *Iron in Plants and Soils* (J. Abadia, ed.), Developments in Plant and Soil Sciences series, Kluwer, Dordrecht, The Netherlands, 1995, pp. 265–276.

62. G. Shaw and R. Kamen, *Cell*, *46*, 659–667 (1986).

63. G. D. Schuler and M. D. Cole, *Cell*, *55*, 1115–1122 (1988).

64. Y. Takahashi, H. Kuroda, T. Tanaka, Y. Machida, I. Takebe, and T. Nagata, *Proc. Natl. Acad. Sci. USA*, *86*, 9279–9283 (1989).

65. M. Ohme-Takagi, C. B. Taylor, T. C. Newman, and P. J. Green, *Proc. Natl. Acad. Sci. USA*, *90*, 11811–11815 (1993).

66. N. Terry and G. Low, *J. Plant Nutr.*, *5*, 301–310 (1982).

67. K. A. Platt-Aloia, W. W. Thomson, and N. Terry, *Protoplasma*, *114*, 85–92 (1983).

68. J. N. Nishio, J. Abadia, and N. Terry, *Plant Physiol.*, *78*, 296–299 (1985).

69. F. Van Der Mark, H. F. Bienfait, and H. Van der Ende, *Biochem. Biophys. Res. Commun.*, *115*, 463–469 (1983).

70. D. Branton and L. Jacobson, *Plant Physiol.*, *37*, 539–545 (1962).

71. T. F. Young and N. Terry, *J. Plant Nutr.*, *5*, 1273–1283 (1982).

72. S. Lobréaux, T. Hardy, and J. F. Briat, *EMBO J.*, *12*, 651–657 (1993).

73. M. W. Hentze and L. C. Khun, *Proc. Natl. Acad. Sci. USA*, *93*, 8175–8182 (1996).

74. Y. Luo, Z. Han, S. M. Chin, and S. Linn, *Proc. Natl. Acad. Sci. USA*, *91*, 12438–12442 (1994).

75. G. Balla, H. S. Jacob, J. Balla, M. Rosenberg, K. Nath, F. Apple, J. W. Eaton, and G. M. Vercellotti, *J. Biol. Chem.*, *267*, 18148–18153 (1992).

76. G. Cairo, L. Tacchini, G. Popliaghi, E. Anzon, A. Tomasi, and A. Bernelli-Zazzera, *J. Biol. Chem.*, *270*, 700–703 (1995).

77. B. Halliwell, *Chem. Phys. Lipids*, *44*, 327–340 (1987).
78. S. Lobréaux, S. Thoiron, and J. F. Briat, *Plant J.*, *8*, 443–449 (1995).
79. K. Kampfenkel, M. Van Montagu, and D. Inzé, *Plant Physiol.*, *107*, 725–735 (1995).

15

Transferrin, the Transferrin Receptor, and the Uptake of Iron by Cells

Philip Aisen

Department of Physiology and Biophysics,
Albert Einstein College of Medicine,
1300 Morris Park Avenue, Bronx, NY 10461, USA

1. INTRODUCTION: EVOLUTION AND THE NEED FOR IRON-BINDING MOLECULES 586

2. TRANSFERRINS 587
   2.1. Description, Definition, and Nomenclature 587
      2.1.1. Overall Structure 588
      2.1.2. Evolutionary Origin 588
      2.1.3. Nomenclature 589
      2.1.4. Metal-Binding Properties of Transferrins 589
      2.1.5. Reversibility of Iron Binding 589
      2.1.6. The Synergistic Anion: A Defining Feature of Transferrins 591
      2.1.7. Ligand Structure of the Binding Sites 592
      2.1.8. Iron-Donating Functions of Serum Transferrin 593
      2.1.9. Atypical Transferrins 594
   2.2. Transferrin Structure: A Detailed View 595

2.2.1. Details of Ligand Structure at the Specific Sites 595
2.2.2. Overall Conformational Effects of Metal Binding 597
2.2.3. Interactions Among Nonbinding Residues 599
2.3. Spectroscopic Studies of Native and Recombinant Transferrins 600
2.3.1. Spectrophotometry 600
2.3.2. Electron Paramagnetic Resonance (EPR) Spectroscopy 601
2.3.3. Nuclear Magnetic Resonance (NMR) Spectroscopy 602
2.4. Kinetics of Iron Release from Transferrin 603
2.4.1. Physiological Background 603
2.4.2. Kinetic Studies of Iron Release 604
3. THE TRANSFERRIN RECEPTOR 606
3.1. Structure of the Receptor 606
3.2. Regulation of Transferrin Receptor Expression 609
4. UPTAKE OF IRON BY CELLS 610
4.1. Receptor-Dependent Uptake of Iron from Transferrin 610
4.2. Receptor-Independent Uptake of Iron from Transferrin 613
4.3. Uptake of Nontransferrin Iron 614
4.3.1. Ferritin 614
4.3.2. Uptake of Non-Protein-Bound Iron 615
4.4. Microbial Uptake of Iron from Transferrin 617
5. PERSPECTIVES AND PROSPECTS 618
ACKNOWLEDGMENTS 618
ABBREVIATIONS 619
REFERENCES 619

1. INTRODUCTION: EVOLUTION AND THE NEED FOR IRON-BINDING MOLECULES

When the earth was young and its atmosphere benignly anaerobic, iron-dependent organisms met no difficulty in satisfying their needs for this versatile element. In the flooded planet, the ferrous state of iron predominated, soluble, stable, and benign in the absence of atmospheric oxidants and therefore readily available to evolving creatures. On the bright day when a cyanobacterium (blue-green algae) first synthesized chlorophyll to capture the energy of sunlight, problems began. With the appearance in the atmosphere of dioxygen liberated by photosynthesis, ferrous iron was oxidized to the much less tractable ferric state. Since the solubility product of Fe^{3+} in water is only about 10^{-38} M^4, the equilibrium concentration of aquated Fe^{3+} cannot exceed 10^{-17} M at neutral pH. The iron content of the seas was deposited as insoluble iron minerals, chiefly variants of ferric hydroxide ($Fe(OH)_3$) or combinations of $Fe(OH)_3$, $Fe(OH)_2$, and $FeCO_3$, and creatures in need of iron were obliged to express iron-sequestering molecules to capture the earth's fourth most common element. The threat of shortage in the face of plenty prevailed.

Yet another problem confronted the first iron-dependent organisms seeking their way in a world made aerobic by photosynthesis. The same property underlying the versatility of iron in biological electron transfer reactions, easy one-electron interconvertibility of ferrous and ferric states in many iron complexes, enables the generation of oxygen-derived radicals and related oxidizing species when iron participates in redox chemistry. Such oxidants, which are capable of attacking almost all biomolecules, pose a threat to the viability of cells in which they arise. To survive this threat, organisms had to elaborate iron-sequestering molecules to maintain iron in soluble, bioavailable, and nontoxic form. In the unicellular universe, relatively simple iron-binding molecules, the siderophores [1], served the metabolic needs of their hosts in capturing iron from a hostile environment. With the appearance of multicellular organisms and their circulatory systems, more complex molecules—the transferrins—were expressed for seizing and transporting iron.

2. TRANSFERRINS

2.1. Description, Definition, and Nomenclature

The transferrins comprise a class of iron-binding proteins once believed to originate with the evolutionary emergence of vertebrates or prevertebrates, but now they are known to be at least as ancient as the insects [2]. Most vertebrate transferrins share structural and iron-binding characteristics.

2.1.1. Overall Structure

The typical transferrin molecule is a single-chain glycoprotein of molecular weight near 80 000 bearing two similar, but not quite identical, iron-binding sites. The polypeptide chain is disposed in two lobes, respectively representing the N-terminal and C-terminal halves of the molecule. Each lobe in turn comprises two dissimilar domains surrounding a deep hydrophilic cleft enclosing an iron-binding site. The first domain in a lobe, designated N1 or C1 depending on the lobe in which it is found, is half completed by the first 90 or so amino acid residues of the peptide chain in that lobe. The chain then crosses to the second domain, N2 or C2, which is fully formed by the next ~160 residues before returning to complete the first domain. The domains are thus joined by two connecting antiparallel β-strands, each of which contributes a ligand to the iron-binding site, with each domain also contributing a ligand. The two lobes, in turn, are connected by a short sequence which takes the form of a helix in human lactoferrin and a β-strand in rabbit serum transferrin [3].

2.1.2. Evolutionary Origin

Internal homology between N- and C-terminal lobes of transferrins [4–7], and in the organization of introns in the human transferrin gene [8], indicates that the modern transferrin molecule arose during the course of evolution by duplication and fusion of a gene specifying a single-sited precursor of modern transferrins. Each of the transferrins so far investigated is coded by a single gene, almost certainly derived from the common ancestral gene by conservative evolution. Whether the putative precursor transferrin itself still persists in any species is uncertain. A single-sited protein with iron-binding characteristics of transferrins has been identified in *Neisseria* [9], but sequence analysis, required to establish that this prokaryotic protein is a true precursor of eukaryotic transferrins, has not yet been reported (see Sec. 4.4). Another candidate is a protein with a molecular weight near 41 000, bearing a single iron-binding site with spectroscopic and iron-donating properties of true vertebrate transferrins found in the hemolymph of the sea squirts [10,11].

2.1.3. *Nomenclature*

For historical as well as functional reasons, three major classes of transferrins (once also known as siderophilins) have been distinguished: (1) serum transferrin or serotransferrin, functioning to transport iron in blood to iron-dependent cells: (2) lactoferrin or lactotransferrin, found in many extracellular fluids and in the specific granules of polymorphonuclear leukocytes; and (3) ovotransferrin (formerly called conalbumin) of avian egg white. The latter two proteins are thought to function in providing a bacteriostatic milieu by sequestering iron in physiological fluids; a basic sequence (lactoferricin) in lactoferrin may also have direct microbicidal activity [12,13]. Ovotransferrin, differing from serum transferrin of the same species only in the structure of its carbohydrate, is also capable of serving as an iron donor for immature erythrocytes in the avian embryo [14].

2.1.4. *Metal-Binding Properties of Transferrins*

Each specific site of human transferrin will accommodate a wide variety of metal ions—more than 35 have been identified, and the list continues to grow (Table 1). Of these, iron as Fe^{3+} is bound most strongly, and is the only metal definitely known to bind to transferrin in a physiologically important way. A role for transferrin in the transport of manganese has recently been reinvestigated but with conflicting results [15,16]; the ability of transferrin to bind and transport Al^{3+} may be of toxicological importance in dialysis dementia and bone disease. The capacity of serum transferrin and cellular lactoferrin to accommodate metal ions has been exploited in nuclear medicine in tumor imaging with radioactive isotopes of indium and gallium [17,18]. Metal ions with electron or nuclear paramagnetism have been especially valuable in probing the metal-binding properties of transferrins [19–24].

2.1.5. *Reversibility of Iron Binding*

The binding of iron by serum transferrin is sufficiently strong to resist hydrolysis [25], yet it is physiologically reversible. Each transferrin molecule in the human circulation experiences 100–200 cycles of iron binding, transport, and release during its lifetime [26] (see Sec. 2.4). Understanding the mechanisms of iron binding and release is therefore

TABLE 1

Metal Ions Bound at Specific Sites of Transferrins

Metal[a]	Comments	Reference[b]
Fe^{3+}	Bound to all transferrins	195
Al^{3+}	Important in dialysis-induced aluminum overload	196,197
VO^{2+}	Useful as EPR probe of binding sites	198
V^{5+}	Useful as NMR probe of binding sites	199
Cr^{3+}	Useful as EPR probe of binding sites	20
Mn^{2+}		15
Mn^{3+}	Via oxidation of Mn^{2+}	20
Fe^{2+}	Existence inferred, not directly demonstrated	200
Co^{2+}	Useful as NMR shift probe of binding sites	201
Co^{3+}	Via oxidation of Co^{2+}	20
Cu^{2+}	Useful as EPR probe of binding sites	195
Zn^{2+}		202
Ga^{3+}	^{67}Ga used in nuclear medicine imaging; may have therapeutic application	203–205
Cd^{2+}		206
In^{3+}	^{111}In used in nuclear medicine imaging; probe in perturbed angular correlation studies; may have therapeutic application	207–209
Ce^{4+}	Via oxidation of Ce^{3+}	210
Pr^{3+}	Bound to only one site	211
Nd^{3+}	Bound to only one site	211
Eu^{3+}		211
Gd^{3+}	Stoichiometry of binding uncertain	212,213
Tb^{3+}		211
Ho^{3+}		211
Er^{3+}		211
Yb^{3+}		214
Hf^{4+}	Probe in perturbed angular correlation studies	215
Pt^{2+}		216
Tl^{3+}	Useful as NMR probe of binding sites	89
Bi^{3+}		24
Th^{4+}		217

TABLE 1

Continued

Metal[a]	Comments	Reference[b]
Pa^{5+}	Oxidation state uncertain	218
Np^{4+}		219
Pu^{4+}	Only one Pu^{4+} bound	220

[a]Except for Fe^{3+}, listed in order of apperance in periodic table.
[b]Where more than one reference exists, only a typical or original report is cited.

central to understanding the function of the protein. Binding of iron by ovotransferrin is substantially similar to binding by serum transferrins, but uptake and release of iron by lactoferrin are much more sluggish than by serum transferrins. Both thermodynamic and kinetic factors may contribute to this sluggishness [27].

2.1.6. *The Synergistic Anion: A Defining Feature of Transferrins*

The binding of iron and other metals by transferrins depends on the concomitant binding of a carbonate or other "synergistic" anion, acting as a bridging ligand between protein and metal ion and completing the coordination needs of the metal ion [28]. Ordinarily, carbonate serves as the synergistic anion, but in its absence, a variety of bifunctional anions bearing at least one carboxylate group may take its place [19,29]. Just as the sites differ in their affinities for metal ions so do they differ in their affinities for synergistic anions, with carbonate by far the most tightly bound. Carbonate is anchored to transferrin by a network of electrostatic and hydrogen bonds involving, in the human sequence, Arg-124, Thr-120, and peptide-chain nitrogens of Ala-126 and Gly-127 at the N-terminus of helix 5, which bulges into the binding cleft. Cooperation between metal and anion binding is remarkable: Neither is bound strongly in the absence of the other. The anion-dependent metal binding is unique to the transferrins and may be taken as the defining feature of the proteins.

^{13}C nuclear magnetic resonance (NMR) spectroscopy of apotransferrin in the presence of excess [^{13}C]-carbonate shows two lines not evident in the spectrum of diferric transferrin taken under similar conditions. Each of these lines has an intensity corresponding to about 0.2 ^{13}C/transferrin. The lines disappear upon addition of Fe^{3+} to the protein, suggesting that they are bound at the synergistic anion sites and are broadened beyond detection by iron as it binds to the metal site. The implication is that synergistic anion binding ordinarily precedes metal binding in the capture of iron by transferrin [30]. A possible track for the synergistic anion on its way to specific binding in the N-lobe of ovotransferrin has recently been proposed based on chemical modification studies [31]. A series of lysine and histidine residues in a short sequence terminating at the anion-binding site provide a way station on this track.

The synergistic anion site in apotransferrin appears to be capable of accommodating a wide variety of anions, many of which lack the structural features required of synergistic anions [29,32] and therefore will not enable binding of iron [33,34]. Detection of anions at the specific sites of transferrin is usually based on perturbation of UV-difference spectra (see Sec. 2.3.1). The ligating residue for nonsynergistic anion binding at metal-free specific sites is probably the arginine that functions in synergistic anion binding.

2.1.7. *Ligand Structure of the Binding Sites*

The donor atoms at the specific sites of all transferrins are provided by two tyrosine residues (Tyr-95 and Tyr-188), one aspartate (Asp-63), a histidine (His-249), and the synergistic carbonate, probably in bidentate coordination to iron (designations refer to the N-terminal lobe of the human protein). Thus, each Fe^{3+} is surrounded by five donor oxygen atoms and one donor nitrogen atom (Fig. 1). The preponderance of electronegative oxygen donors substantially accounts for the low reduction potential of transferrin-bound iron; near or below -450 mV [35,36]. Ligand residues are widely dispersed in the linear sequences of each lobe, and therefore are not encoded by a single exon. The Asp has a dual role, providing an oxygen ligand to the bound metal and forming an interdomain hydrogen bond to Ser-125, thereby helping to stabilize the lobe in closed confirmation when metal is bound, but leaving it in an

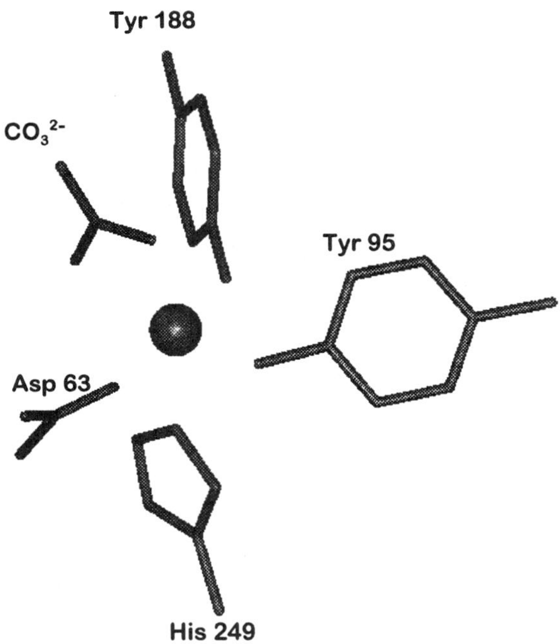

FIG. 1. Iron-binding ligands of human serum transferrin, from the crystal structure of recombinant N-lobe.

open conformation in the absence of metal [37,38] (see Sec. 2.2.2). Interlobe, as well as interdomain, interactions influence the iron-binding properties of the two sites [25,39,40] and probably contribute to differences in physicochemical properties of the binding sites despite their identical ligands.

2.1.8. *Iron-Donating Functions of Serum Transferrin*

Access of transferrin to cells is, for most cells, dependent on a specific transferrin receptor of the plasma membrane [41]. By controlling their access to transferrin, the receptor serves as the ultimate gatekeeper in regulating cellular iron uptake by these cells. Hepatocytes and malignant melanoma cells, however, also make use of receptor-independent routes for acquiring iron from transferrin [42,43]. In addition to its roles in recognizing and escorting transferrin, the human transferrin receptor functions in modulating release of iron from its target protein [44,45] (see Sec. 2.4.2).

2.1.9. *Atypical Transferrins*

2.1.9.1. Melanotransferrin. In addition to the extracellular transferrins, a soluble membrane transferrin has been identified in human malignant melanoma cells [46] and accordingly named melanotransferrin. Because of the loss of residues critical in metal and anion binding from its C-lobe, only the N-lobe of melanotransferrin possesses a functional iron-binding site [47]. Melanotransferrin provides a receptor-independent but saturable route for iron acquisition from transferrin by melanoma cells [43]. The mechanism of such acquisition appears to entail chelator-inhibitable transfer of iron from the plasma protein to the membrane protein followed by adsorptive endocytosis [43]. Circulating melanotransferrin, detected and quantified by an immunofluorescence method, has recently been reported to be a specific marker of Alzheimer's disease [48], a condition in which iron may contribute to pathogenesis [49,50]. Although probably derived from brain capillary endothelium, the source of plasma melanotransferrin is not definitely known, since the protein is also expressed by fetal intestinal cells, umbilical cord, sweat gland ducts, liver sinusoidal cells, and a wide variety of transformed cells [51].

2.1.9.2. Insect transferrins. Insect transferrins from two species, the tobacco hornworm, *Manducca sexta*, and the cockroach, *Blaberus discoidalis*, have been isolated, characterized, and cloned [52,53]. The roach protein binds two iron atoms per molecule in carbonate-dependent fashion; metal ligands in the C-lobe are identical to those of vertebrate transferrins, but the His ligand in the N-lobe is absent or displaced. Only one binding site is present in the *Manducca* protein, probably because of loss of Tyr and His ligands in the C-terminal sequence. There is no apparent evidence to indicate that the iron-transport function of transferrin in these prevertebrates differs substantially from that of vertebrates. Since the last common ancestor of vertebrates and arthropods wandered the earth some 600 million years ago [53], the progenitor transferrin sequence is probably at least that old. The evolutionary advantage conferred by the ancient emergence and subsequent persistence of a two-sited iron-binding molecule as opposed to a single-sited one is still unclear.

2.2. Transferrin Structure: A Detailed View

Binding and release of iron by transferrin involves more than the ligating residues of the protein: Interactions of amino acids remote from the specific binding sites also profoundly affect the stability and iron-binding properties of the protein. Many new insights have been garnered from comparative analyses of crystal structure of human lactoferrin [28,54], rabbit serum transferrin [55], and chicken ovotransferrin [56,57]. Critical reviews from a crystallographic perspective are particularly helpful for understanding structure-function relationships among the transferrins [3,58].

2.2.1. Details of Ligand Structure at the Specific Sites

The participation of an aspartate residue, Asp-60 in the human N-lobe lactoferrin sequence, as a ligand for metal ions came as a surprise [28], since neither chemical modification nor spectroscopic studies had previously provided any indication of carboxyl coordination. Although not essential for iron binding, the Asp ligand substantially increases the strength of iron binding as reflected in stability during urea gel electrophoresis [59]. Loss of charge neutralization by the anionic Asp, as well as of its oxygen donor atom, probably accounts for destabilization of iron binding upon mutation of the Asp. One oxygen atom (OD1) of the carboxylate group coordinates to bound metal, whereas the other (OD2) acts as a hydrogen bond acceptor from peptide NH groups at the N-termini of two helices, each in a separate domain of the N-lobe. In recombinant human transferrin N-lobe, mutation of the Asp ligand to Cys or Ser results in a protein that yields low-angle x-ray scattering profiles indicating an open or "slightly closed" conformation even when iron is bound [60]. This observation led to the suggestion that the Asp ligand provides a trigger for domain closure induced by metal binding, which is consistent with the interdomain contacts of the residue [60]. In the N-lobe of human lactoferrin, however, the Asp→Ser mutation does not abolish domain closure, but in the crystal structure, it actually brings the domains into a more closed conformation over the metal-binding site [61]. Probably, therefore, domain closure is a concerted process involving more than a single residue, and one in which the bound iron itself has a critical role in bringing the domains together.

Coordination to tyrosyl ring oxygens had been suspected since the first descriptions of transferrins because of the salmon pink color of the iron complexes and yellow color of the copper complexes. Difference UV spectrophotometry of a model complex predicted participation of two tyrosyl ligands [62], as verified by x-ray crystallography; the two tyrosyl groups are in *cis* relationship to each other. The two tyrosyl ligands may be sufficient for *weak* iron binding, since an 18-kDa fragment of duck ovotransferrin, with only these two iron ligands and the carbonate preserved, still retains iron [63].

A coordinated nitrogen atom had been identified by electron paramagnetic resonance and electron nuclear double resonance (ENDOR) spectroscopies [19,64,65], and crystallographically shown to be provided by a histidyl residue. The five-coordinate structure resulting from mutation of this His ligand is much less stable than the parent structure [66]. Thus, all four of the usual six coordination requirements of Fe^{3+} provided by the protein appear to be necessary for *stable* iron-binding. The remaining two, contributed by the essential carbonate anion, complete a distorted octahedral arrangement lacking symmetry elements.

Carbonate is anchored to the protein by a hydrogen-bonding network. In the N-lobe of human lactoferrin (2-Å resolution), the principal anchor is Arg-121, which provides two nitrogens for bonding to carbonate oxygens, and this anchor is reinforced by bonding to peptide NH groups of a helix protruding into the binding cleft. In the recombinant N-lobe of human transferrin (1.8-Å resolution) and the N-lobe of native rabbit serum transferrin [55], however, the guanidinium nitrogens of homologous Arg-124 are more than 4 Å from the nearest carbonate oxygens, substantially longer than the comparable distances in lactoferrin (2.59–3.15 Å), suggesting that the carbonate-stabilizing role of Arg-124 may be reduced in the serum proteins compared with lactoferrin. The carbonate anion is essential for iron binding, and its loss may initiate iron release within cells [67] as well as in vitro [68], so that the relative sluggishness if iron release from lactoferrin compared with transferrin may result, at least in part, from the greater stability of carbonate binding in the former. This interpretation must be qualified, however, since lactoferrin was crystallized at pH 7.8 and the serum proteins at pH 5.5–5.75 where iron binding is distinctly weaker than at the higher pH.

The critical role of arginine is further demonstrated by mutating it to a serine or lysine in the human N-lobe: Both mutants release iron to pyrophosphate with greater facility than the wild-type recombinant N-lobe [69]. Mutation of the carbonate-binding arginine in the human lactoferrin N-lobe also destabilizes the iron-protein complex as measured by loss of iron to buffers of low pH [70]. Crystal structures of R124S and R124E mutants show a severe disruption at the synergistic anion-binding site: Not only is the critical arginine link destroyed, but the carbonate is partially released from its attachment to peptide NH groups which help lock it into place in the wild-type structure [70].

The transferrin molecule is particularly well endowed with disulfide bonds. All of the 38 cysteine residues of human transferrin are coupled in disulfide bonds, with 11 linkages in the C-lobe and the remaining 8 in the N-lobe. The larger number in the C-lobe may account in part for the increased stability of metal binding (Table 2) and the relative resistance to conformational change associated with loss of metal [58] in this lobe (Table 2).

2.2.2. *Overall Conformational Effects of Metal Binding*

A seminal insight into the mechanism of iron binding and release by transferrins was provided by Bates and his colleagues, who noted a hyperbolic dependence of observed release rates on acetohydroxamate concentration when that chelator was used to capture iron from human transferrin [71]. A "rate-limiting, bond-breaking labilization of Fe^{3+}-transferrin-CO_3" to an "open conformational state" was therefore inferred. The reliability of the observation was confirmed by other kinetic investigations [72,73], and the accuracy of the prediction was dramatically verified by crystallographic studies of apolactoferrin showing an open-jaw configuration of the two domains enclosing the binding cleft in the N-lobe of the protein [74]. Surprisingly, the C-lobe was found in the closed conformation, but subsequent low-angle x-ray scattering studies demonstrated an open configuration in both lobes of apolactoferrin and of apotransferrin in solution [38]. Very likely, therefore, a minority conformation of apolactoferrin was selected by crystal lattice forces, and the Bates model remains valid. Human monoferric transferrin bearing an iron atom at the C-lobe site, and showing the closed conformation of the C-lobe and the open conformation of the N-lobe, is shown in Fig. 2.

The conformational switch consequent to metal binding or release by lactoferrin is a rigid-body 54-degree rotation of the two domains to bring about a transition between closed and open states. The major part of this rotation occurs as a "see-saw" action about an axis through the

TABLE 2

Selected Stability Constants[a] for Metal Ion Binding by Transferrins

Protein	Metal	Method	pH	log K_1	log K_{1N}	log K_{1C}	log K_2	log K_{2N}	log K_{2C}	Reference
Human transferrin	Fe^{3+}	Equil. Dial.	7.4	20.7	19.8	20.6	19.4	19.4	20.2	25
Human lactoferrin	Fe^{3+}	Equil. Dial.	6.4	19.0			18.7			221
Human transferrin	Fe^{2+}	LFER[b]	7.4	3.2			2.5			200
Human transferrin	Al^{3+}	UV Diff.	7.4	12.9			12.3			222
Human transferrin	Zn^{2+}	UV Diff.	7.4	5.7			4.3			202
Human transferrin	Cd^{2+}	UV Diff.	7.4	5.9			4.9			206
Human transferrin	Ga^{3+}	UV Diff.	7.4	20.3			19.3			203
Human lactoferrin	Ga^{3+}	UV Diff.	7.4	21.4			20.6			223
Human transferrin	In^{3+}	UV Diff.	7.4	18.5			16.6			224

[a]Values are conditional or effective constants measured in air [25].
$K_1 = K_{1N} + K_{1C}$; $K_2^{-1} = K_{2N}^{-1} + K_{2C}^{-1}$, where K_{iN} and K_{iC} (i = 1,2) are site constants [25].
[b]Linear free energy relationship.

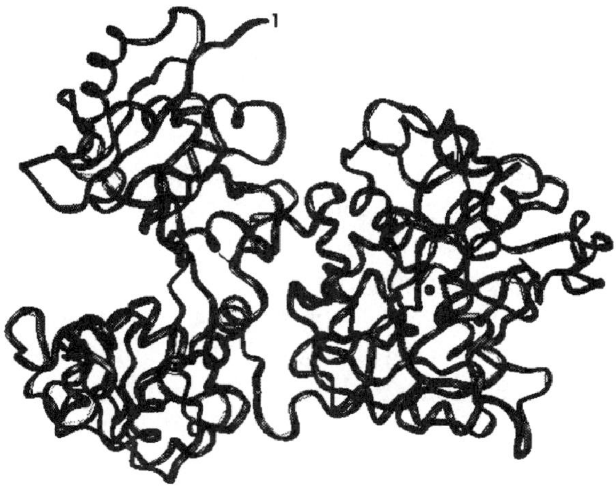

FIG. 2. Structure of human monoferric transferrin. The empty N-lobe is in the open conformation, and the C-lobe bearing an iron atom (shown as a ball) is in the closed conformation. The N-terminal valine residue is marked with a 1.

antiparallel β-strands linking the two domains [75], with a resulting change in interfacial contacts of the domains. In the larger interface on one side of the rotation axis, residues are solvent accessible in the open form but become buried in the closed form, whereas the reverse situation is obtained in the smaller interface. Interfacial contacts between domains are therefore likely to regulate the iron-binding properties of the sites enclosed by the domains.

2.2.3. *Interactions Among Nonbinding Residues*

Ligands and their arrangements at the specific sites are much the same in all transferrins with determined structures, yet the metal-binding properties of lactoferrins differ substantially from those of serum transferrins and ovotransferrins. Interactions of residues remote from the metal ligands must therefore modulate the iron-binding properties of transferrins. Several such interactions have recently been uncovered.

Perhaps the most important of these remote effects involves a dilysine pair first discovered in the N-lobe of chicken ovotransferrin [57]. In that protein's structure at a resolution of 2.3 Å, the side chain NZ atoms of two lysine residues, Lys-209 and Lys-301, are separated by

2.3 Å. Considering that the protein crystals were grown from a pH 5.9 acetate buffer, the strong likelihood is that both lysines bear a positive charge. The close proximity of these charges on residues located on opposite domains were therefore postulated to provide a pH-sensitive force for driving the domains apart, thereby switching the lobe into its open conformation from which release is facile. Dilysine arrangements, or equivalent lysine-arginine pairs, are common occurrences in other transferrin N-lobes, reinforcing the suggestion that pH-sensitive repulsive interactions in these pairs trigger the change from closed to open conformations. An interesting exception is found in the C-lobe of human lactoferrin, where an asparagine replaces lysine on domain II and so precludes a charge-repulsive interaction of the domains; again helping to explain the sluggish iron release by this protein.

Several interdomain contacts stabilized by hydrogen bonding or salt bridging have been identified in both lobes of human lactoferrin [54]. Only one of these, however, is common to both lobes and to other transferrins, made by the iron ligand Asp-60 (Asp-63 in human transferrin), already discussed (see Sec. 2.1.7). Differences in metal-binding properties of the two sites in the same transferrin, or of homologous sites of different transferrins, are at least in part ascribable to differences among stabilizing interdomain contacts. The relative abundance of interdomain contacts in lactoferrin helps explain the stability of iron binding and its resistance to low pH in lactoferrin. Creation of such contacts by mutation of the human transferrin N-lobe recombinant correspondingly retards release of iron from the protein [69]. Changing Lys-206 to Arg or His-207 to Glu creates products with interdomain hydrogen-bonding possibilities akin to those in lactoferrin. Both mutants release iron to pyrophosphate much more slowly than the wild-type protein. Similarly, the K206Q mutant is a more sluggish releaser at pH 4 to deferoxamine than its wild-type parent [59].

2.3. Spectroscopic Studies of Native and Recombinant Transferrins

2.3.1. *Spectrophotometry*

The distinctive colors of transferrin-metal complexes have long been appreciated. Iron transferrins have a characteristic salmon-pink color attributed to charge transfer from the tyrosyl ligands to the Fe^{3+}. Transferrins display a yellow hue when binding Cu^{2+}, a deep orange color with Co^{3+}, a pale blue-green tint with Cr^{3+}, and a deep inky-brown with Mn^{3+} [20], all probably resulting from ligand-to-metal charge transfer [76]. Most other complexes of transferrin and multivalent metal ions are colorless, or nearly so, probably because of unfavorable optical electronegativities of the metal ions. However, all metal-transferrin complexes studied exhibit pronounced UV absorption with peaks near 245 nm ascribed to perturbation of π to π^* transitions of the tyrosyl phenolic rings. The different molar absorptivities of these peaks with apotransferrin as a reference are typically in the range of 15 000–20 000 M^{-1} cm^{-1}, so that spectrophotometric titration provides a sensitive measure of metal binding by a transferrin. Anion binding at or near the specific sites of apotransferrin may also be detected and quantified by UV difference spectrophotometry, in this case with negative absorption bands [34].

2.3.2. *Electron Paramagnetic Resonance (EPR) Spectroscopy*

Assessing specific iron binding by transferrins has occasionally been confounded by nonspecific binding. EPR spectroscopy is particularly helpful in distinguishing iron at the specific sites from nondescript iron bound elsewhere. The X-band EPR spectrum of high-spin Fe^{3+}, a spin 5/2 species, in the transferrin low symmetry typically shows a split line centered at $g' = 4.3$, and broader lines near $g' = 9$ and $g' < 2$. Because of the high-spin multiplicity and ligand-field effects, interpretation and simulation of line shapes and splittings in Fe^{3+}-transferrin EPR spectra had been difficult or uncertain [77]. A signal advance in such interpretation was achieved with the realization that parameters of the spin Hamiltonian describing such spectra were not fixed quantities but rather represented a distribution of values [78,79] analogous to the "g-broadening" or hyperfine-broadening effects in simpler spin systems [80]. Using distributions of zero-field (also called ligand field or fine structure) parameters, near perfect simulations of Fe^{3+}-transferrin EPR spectra were achieved [79]. In these simulations, the zero-field term is predominantly rhombic but with some axial character present.

The value of EPR spectroscopy in revealing alterations in ligand fields experienced by specifically bound Fe^{3+} has long been appreciated.

Where optical spectroscopy is insensitive to replacement of carbonate by other synergistic anions, EPR spectroscopy is dramatically revealing [19,21,81]. Not surprisingly, therefore, EPR spectra respond to mutations of transferrin ligands. Replacement of the ligating Asp-63 residue by Ser or Cys results in a protein with a nearly perfectly rhombic EPR spectrum giving an unsplit line at $g' = 4.3$ [82]. When the His-249 ligand is mutated to glutamine, the spectrum generated is again nearly perfectly rhombic, but mutations to glutamate results in a spectrum with substantial axial character. Less expected is the responsiveness to mutations of nonligating residues. Changing Gly-65 to Arg produces a spectrum interpreted as representing a sum of rhombic and normal components such as might arise from a population with two ligand conformations. Mutation of K296, positioned near the interdomain connecting strands of the N-lobe and the bound Fe^{3+}, exerts surprisingly large axial effects. Detailed analysis of these changes and how they are brought about is not yet feasible, but the descriptive value of EPR spectroscopy is substantial.

The variety of multivalent metal ions bound by transferrins, many of them paramagnetic, has provided EPR probes for studying the metal-binding properties of the proteins. These studies, reviewed earlier [37], have demonstrated differences between the sites and access of sites to more than one conformation. The EPR responsiveness of Cu^{2+} and VO^{2+} to changes in ligands and their arrangements has been particularly exploited in detecting pH-induced conformational changes in host transferrins [21,83,84].

2.3.3. *Nuclear Magnetic Resonance (NMR) Spectroscopy*

Recombinant technology has made it possible to incorporate isotopically labeled amino acids in expressed transferrin, thereby enhancing the usefulness of NMR spectroscopy in exploring conformational analysis of the protein. Using $^{13}CH_3$-methionine and two-dimensional (2-D) nuclear Overhauser effect spectroscopy to probe the binding of Ga^{3+} to recombinant human holotransferrin, it was shown that the metal binds sequentially to the two sites, occupying the more strongly binding C-lobe site first when oxalate serves as the synergistic anion. After first loading the C-lobe, further addition of gallium to the N-lobe perturbs the $^{13}CH_3$ resonance of Met-382, situated in the interlobe contact region, and thereby demonstrating at the residue level interlobe communications previously inferred from kinetic and thermodynamic studies of metal binding in human serum transferrin and lactoferrin [25,39,85].

Lanthanide ions with incompletely filled f-shells have short electron relaxation times, and hence act as paramagnetic shift reagents. Sequential binding of Tb^{3+} to human transferrin was therefore exploited in corroborating conformational differences between the two lobes of the protein [86]. Nonparamagnetic metal ions with nuclear moments have also been valuable in studying conformational differences between transferrin binding sites. Al^{3+} and its congener, Sc^{3+}, have proved to be useful in defining site differences in the affinities of ovotransferrin for different metal ions and the dependence of relative site affinities on the particular synergistic anion used to enable metal ion binding [87,88]. NMR spectroscopy of diamagnetic multivalent metal ions of large atomic radii, Bi^{3+} and Tl^{3+}, verified the capacity of transferrins to accommodate large metal ions and again distinguished between the sites in their affinities for such large ions [24,89,90].

2.4. Kinetics of Iron Release from Transferrin

2.4.1. Physiological Background

In the human circulation, the half-life of transferrin is about 7.6 days, whereas that of transferrin-bound iron is close to 1.7 h [26]. On the average, therefore, a transferrin molecule experiences more than 100 cycles of iron binding, transport, and release during its lifetime. Most of the iron transported by transferrin is destined for the hemoglobin-synthesizing immature red blood cell, which gathers iron by receptor-mediated uptake from circulating transferrin [91]. Since a typical rabbit reticulocyte displays a steady-state binding of some 124 000 transferrin molecules, and takes up some 90 000 iron atoms per minute from iron-saturated transferrin (unpublished observations), the transfer of iron from protein to cell takes no more than 2–3 min [67]. Since iron uptake from transferrin is less than 100% efficient [92], the actual dwell time of a transferrin molecule with a reticulocyte may be even less than simple calculations indicate. Understanding the physiology of transferrin accordingly demands understanding of the mechanisms of iron release from the protein.

2.4.2. *Kinetic Studies of Iron Release*

Among the factors known to influence iron release from transferrin are the following:

1. The site from which release takes place
2. pH and ionic composition of the supporting buffer
3. The sequestering ligand needed to capture released iron
4. Binding of transferrin to its receptor

Each of these will be considered in turn.

2.4.2.1. The Iron-Releasing Site. Binding constants of the two sites of transferrin have been determined by correlating relative concentrations of each of the four possible species of transferrin (apotransferrin, monoferric transferrin loaded at N-lobe or C-lobe sites, and diferric transferrin), revealed by urea gel electrophoresis, with overall thermodynamic stability constants measured by equilibrium dialysis [25] (Table 2). As expected from the relative acid lability of the N-lobe site, binding of iron to this site is weaker at pH 7.4 than binding of iron to the C-lobe site and weaker still at pH 6.7. Correspondingly, iron release from the N-lobe to EDTA (ethylenediamine-*N,N,N′,N′*-tetraacetate) is more facile than from the C-lobe, but the relative lability of the sites can be dramatically reversed by moderate concentrations of simple salts, particularly perchlorate [93], or at pH above 8.2 [94].

2.4.2.2. pH and Ionic Composition. As expected, iron release rates from both lobes increase as pH is lowered. Even at pH 5, however, release of iron from diferric transferrin in a self-buffered solution without iron-sequestering agents is much slower than release within the cell, with a half-time greater than 3 h compared with the 1–3 min achieved physiologically. Clearly, simple unfolding of the protein by protonation inadequately accounts for release to the iron-dependent cell.

A considerable advance in understanding iron release was provided by measurements of release rates from monoferric human transferrins as a function of ionic strength. With a synthetic tricatechol-based iron-sequestering agent, observed release rates from both lobes were found to extrapolate nearly to zero as ionic strength approached zero, indicating that ion binding to transferrin promotes a conformational change essential for release [95]. Corroboration of this interpretation was then provided by studies focusing on iron release from the C-site to pyrophosphate. Variation of observed release rates with concentration of simple anions (Cl^-, ClO_4^-, NO_3^-, HPO_4^{2-}) conformed to a model invoking a dual-pathway for release. The model postulates iron removal from a species in which a postulated kinetically active anion-binding site must be occupied by either the simple anion or the sequestering pyrophosphate anion and a rapid equilibrium links the two protein species [96]. The profile of observed release rates as a function of ligand concentration predicted by this model is a saturation-linear function, as observed in earlier studies of release to a variety of sequestering agents [73,97]. Presumably, occupancy of the kinetic anion-binding site is required for the change from closed to open conformations of the releasing lobe, thereby accounting for the saturation part of the release profile as the conformational change approaches completion, while the iron-sequestering ligand exerts a simple linear effect.

Things are not quite so appealingly simple, however. Detailed studies of release from the N-lobe of native human transferrin, or from recombinant wild-type and mutant N-lobe half transferrins, reveal an inhibitory rather than an accelerating effect of chloride [69]. The effect of chloride resembles that of Michaelis-Menten competitive inhibition in enzyme kinetics, and can be rationalized by a model invoking a ligand-binding site where ligand-exchange of Fe^{3+} takes place, but direct evidence for such a model is not at hand. Furthermore, mutation of Lys-569 to Gln (K569Q) in the C-lobe converts the release profile of that lobe from saturation-linear to simple saturation (unpublished results), whereas replacement of Lys-569 by a similarly charged Arg leaves the release profile substantially unaltered. No explanation now apparent accounts for differences in release rates and profiles with different ligands and from different lobes.

2.4.2.3. Sequestering Ligand. For a ligand to function effectively in iron removal from transferrin, it must be kinetically as well as thermodynamically competent in sequestering iron. For example, deferoxamine, which binds Fe^{3+} much more tightly than transferrin, removes iron from the protein sluggishly, but inclusion of a kinetically competent mediating chelator accelerates removal by an order of magnitude [98,99]. Ligands that are both kinetically and thermodynamically competent release iron at varying rates, so that release mechanisms may depend on ligand interactions with transferrin as well as on conformational changes of the protein.

2.4.2.4. Effects of the Transferrin Receptor. Binding of iron-loaded transferrin to its receptor modulates release of iron from the transport protein. At the pH of the cell surface where transferrin first encounters its receptor, 7.4, the detergent-solubilized complex of receptor and transferrin releases iron to pyrophosphate at about one tenth the rate of free transferrin [44]. At the pH of the endosome where iron is released within the cell, 5.6, the complex becomes much more facile than free transferrin in releasing iron. Since the detergent (CHAPS) does not alter iron release kinetics from free transferrin, and the effect of lowering pH is seen in transferrin bound to receptor at the surface of intact cells in culture [45], the likelihood is that receptor-modulated release of iron from transferrin is a physiological phenomenon.

Modulating effects of receptor on N-lobe and C-lobe sites are different. At pH 7.4, release is retarded from both sites, but at pH 5.6, the receptor exerts little influence on the N-site while substantially accelerating release from its counterpart in the C-lobe [100]. The net effect is that the C-site becomes more facile than the N-site in surrendering iron, thereby helping to account for the preponderance of iron bound to the N-site in the human circulation despite the weaker binding to that site [101,102].

3. THE TRANSFERRIN RECEPTOR

Recognition of the central role of a cell-surface receptor in cellular uptake of iron from transferrin accompanied recognition of the role of transferrin itself in iron uptake. Treatment of reticulocytes with trypsin virtually abolished incorporation of transferrin-borne iron into hemoglobin, leading to the inference that iron transfer depends on specific receptors [91]. Precise, reliable, and reproducible characterization of the receptor then eluded investigation for over 20 years, but uncertainties finally gave way to cloning technology [103]. What had become clear from hydrodynamic and gel electrophoresis studies was that the transferrin receptor is a disulfide-linked homodimer of subunits with molecular weights near 90 000 [104,105].

3.1. Structure of the Receptor

Like the genes for transferrin melanotransferrin, the structural gene for the transferrin receptor resides on the long arm of human chromosome 3 [106]. A protein of 760 amino acid residues and a mass of 84 910 kDa is encoded in the mRNA specified by the 19 exons of the gene [103,107]. The protein lacks a leader sequence, and shows a single well-defined hydrophobic region comprising some 18 residues near its N-terminus, probably representing the transmembrane domain of the protein. This transmembrane sequence may therefore function as a noncleavable signal guiding insertion into the membrane [108]. A strongly hydrophilic sequence precedes the hydrophobic region, suggesting that this sequence would not enter the cell membrane and therefore is likely to lie on the cytoplasmic side of the membrane. The extracellular C-terminal portion of the protein bears a trypsin-sensitive site, cleavage at which generates a 70-kDa fragment and accounts for the loss of transferrin-binding activity in trypsinized cells. The 70-kDa soluble fragment retains an intact transferrin-recognition site, and crystals of a complex of this fragment with transferrin have been obtained but diffract too poorly for structural crystallography [109].

The subunits of the functional receptor homodimer are joined by two disulfide bonds at cysteines 89 and 98, within or adjacent to the membrane-spanning regions of the subunits [110] (Fig. 3). Anchorage of the receptor to the membrane may be further aided by posttranslational acylation by palmitate of another cysteine residue, Cys-62, but abolition of the acylation site by site-directed mutagenesis has little effect on receptor function [111]. The receptor experiences additional posttranslational modification by N-linked glycosylation at three asparagine residues [112] and O-linked glycosylation at a threonine site [113], all in the extracellular domain. Glycosylation is necessary for normal receptor function, since mutated receptor lacking asparagine glycosylation sites shows impaired transferrin-binding activity, reduced intersubunit disulfide bonding, and diminished surface expression [114], whereas loss of the threonine site increases susceptibility of the receptor vulnerable to proteolytic cleavage [115]. Carbohydrates of the transferrin receptor are quite heterogeneous, including biantennary and triantennary forms, oligomannosidic glycans, and complex-type oligosaccharides [116].

A phosphorylation site responsive to phorbol ester stimulation is present on Ser-24 in the cytoplasmic domain of the receptor [117], but phosphorylation appears to play little role in receptor function [118].

Only a fraction of total cellular receptors are localized to the cell surface at any time; the remainder are interiorized but are in dynamic equilibrium with surface receptors [119]. Shuttling of receptors from the interior of the cell to the plasma membrane is ongoing and largely

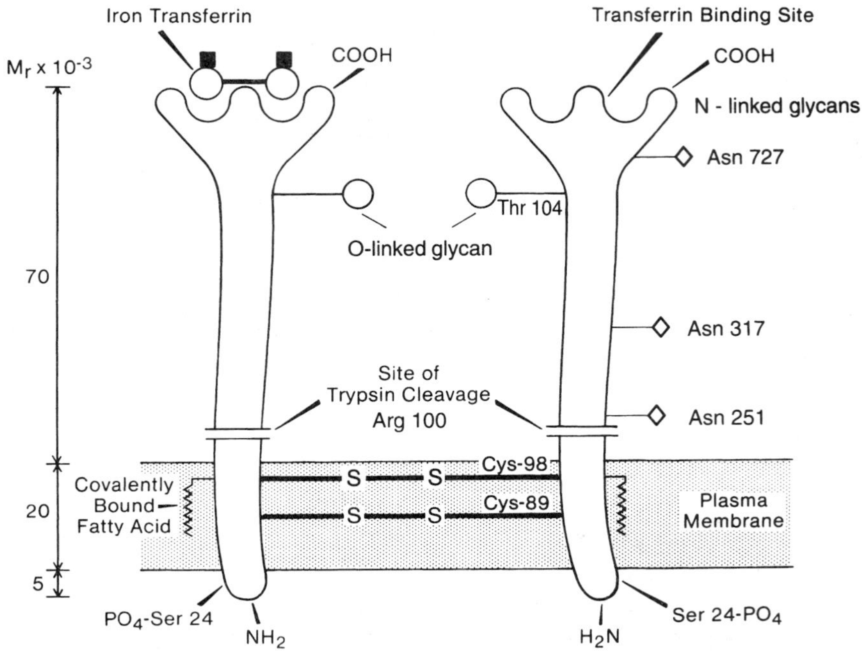

FIG. 3. Diagrammatic representation of the human transferrin receptor.

independent of receptor function in binding and internalizing transferrin. Mutational studies show a tight-turn structural motif in the cytoplasmic tail of the receptor to be the recognition signal for this continual endocytosis [120].

Perhaps the most critical unanswered question in structure-function relationships of the transferrin receptor concerns the location and architecture of the transferrin-recognition site in each subunit of the receptor. Since truncations or deletions in the receptor result in loss of detectable receptor, cDNA constructs specifying chimeras of human receptor and chicken receptor were prepared and expressed in chick embryo fibroblasts [121]. By exploiting the relative indifference of avian receptor to human transferrin, the transferrin recognition sequence could be localized to the carboxy-terminal region in the extracellular domain of the human receptor, probably in the last 192 residues.

3.2. Regulation of Transferrin Receptor Expression

The transferrin receptor is the ultimate gatekeeper for controlling access of transferrin-borne iron to most cells (the hepatocyte and the melanoma cell are the outstanding exceptions to this generalization). Cells with large requirements for iron, such as developing erythroid cells, placental trophoblasts, and rapidly proliferating cells, correspondingly express large numbers of receptors. Regulation of receptor expression is therefore critical in controlling cellular iron metabolism. A sequence of events responsive to cellular iron status underlies this regulation.

Regulation of receptor expression is largely posttranscriptional. The 3′-untranslated region of receptor mRNA contains a series of five hairpin stem-loop structures required for iron-dependent regulation [122]; deletion of this noncoding region in mouse fibroblasts stably transfected with human receptor cDNA abolishes regulation and results in simple constitutive expression of receptor [123]. The stem-loop structures, now known as iron-regulatory elements (IREs) bind a cytoplasmic iron-regulatory protein, or IRP [124], that substantially increases the half-life of receptor mRNA [125,126] by protecting it from endonuclease attack. Binding of protein to RNA is extremely strong, with a K_d of 0.09 nM [127]. Activation of binding is induced by depletion of cellular iron, and suppression of binding by iron surfeit. The question, then, is how iron sensing occurs.

A major advance in answering this question came when the iron-sensing species was identified as a form of the iron-sulfur enzyme aconitase [128]. The aconitase is enzymatically inactive but capable of binding to IREs when depleted of iron and, conversely, enzymatically active but without RNA-binding activity when replete with iron. Thus, by accepting iron in iron-replete cells, or relinquishing it in iron deficiency, the protein effectively functions as an iron sensor regulating expression of transferrin receptor. RNA binding and regulatory activity of the IRE-binding protein are activated as well by nitric oxide and induction of nitric oxide synthase [129–131]. Whether activation entails conversion of the protein to its iron-free form or interference with its 4Fe-4S cluster is not yet clear [132]. In general, molecular mechanisms by which cellular iron effects insertion or removal of the essential 4Fe-4S cluster in the regulatory protein remain uncertain.

The IRE-binding protein coordinately regulates expression of the iron-storage protein ferritin by binding to an IRE in the 5′-untranslated region of the ferritin message, thereby restraining translation of the storage protein when iron is in short supply. When the regulatory protein is replete with iron and therefore unable to bind to the ferritin message IRE, the storage protein is expressed to gather potentially toxic cellular burdens of iron. Thus, feedback regulation by iron is positive for ferritin and, possibly for transferrin [133], but negative for the transferrin receptor, poising the cell for effective response to iron surfeit and iron deficiency.

This model of cellular regulation of iron uptake and storage, appealing in its simplicity and elegance, is incomplete. A second and equally important IRE-binding protein with a regulatory function has recently been described. Unlike its predecessor, this protein does not appear to bind iron and does not exhibit aconitase activity [134] despite apparent 57% identity in amino acid sequence of the two proteins [135]. Discovery of the second protein necessitated a revision of nomenclature: the "original" IRE-binding protein has been renamed from IRE-BP to IRP (iron-regulatory protein)-1, whereas the newcomer is designated IRP-2. Both proteins appear to regulate expression of transferrin receptor and ferritin by binding to IREs, although the specificity of binding may differ [136]. An important distinction is that cellular levels of IRP-1 do not change appreciably in response to change in iron status—only its iron content does—but IRP-2 declines to undetectable levels in response to iron loading [134]. Increased degradation, rather than decreased synthesis, of IRP-2 causes the decline, possibly mediated by a labile protein (an iron sensor in IRP-2 function?) marking IRP-2 for degradation in the multicatalytic endopeptidase complex or proteasome [129,137]. Just how the cascade of iron-induced degradation is brought about remains unclear, and the last word about regulation of cellular iron metabolism is yet to be spoken.

4. UPTAKE OF IRON BY CELLS

4.1. Receptor-Dependent Uptake of Iron from Transferrin

Iron-dependent cells in general, and hemoglobin-synthesizing immature erythroid cell in particular, depend on receptor-mediated uptake of iron from transferrin to meet their needs for the essential metal. In the human circulation, transferrin is normally only 30% saturated with iron, so that the predominant forms of transferrin lack iron at one or both sites. At the pH of the cell surface, where the transferrin receptor encounters its target, the receptor must discriminate in favor of iron-carrying transferrin to assure that competition does not overwhelm cell surface receptors with iron-free protein. Such assurance appears to be provided by preference of the receptor for monoferric and diferric transferrins over apotransferrin; in the rabbit, apotransferrin is bound with only about one twenty-fifth the affinity of iron-loaded transferrin [138]. Whether the same holds true for the interaction of human transferrin with its receptor is not completely clear, however. The primary receptor recognition site in the human protein has been localized to the C-lobe [139], so it might therefore be expected that monoferric transferrin loaded in the N-lobe and with its C-lobe in the open state might not be an effective iron donor for human cells. That this is indeed the case has not been experimentally tested. In any event, rabbit diferric transferrin is bound six to seven times more tightly by its receptor than either monoferric transferrin, so that the effectiveness of iron transfer is likely to be greater from the fully loaded than from either half-loaded protein [140,141].

Binding of transferrin by its receptor on the surface of iron-dependent cells is a simple chemical event not dependent upon metabolic energy. Binding initiates an elaborate series of events culminating in iron transfer from protein to cell and return to the circulation of iron-depleted protein for another cycle of iron transport (Fig. 4). The complex of receptor and transferrin first invaginates to a clathrin-coated pit, which buds from the plasma membrane to internalize in a clathrin-coated, membrane-bound vesicle or endosome. Rapid sorting of the endosome bearing the complex, to deflect it from lysosomal degradation, occurs almost immediately after internalization. Upon maturation and loss of the clathrin coat, the endosome becomes competent to pump protons in a process energized by adenosine triphosphate (ATP), thereby lowering its pH to about 5.6 [142–144]. Transferrin remains bound to receptor throughout this process, but at the low pH of the endosome, binding of apotransferrin becomes stronger than binding of iron transferrin [145], thereby helping account for acceleration of iron release from receptor-bound transferrin at endosomal pH [44,45]. Acidification of the endosome is essential for efficient release of iron from transferrin,

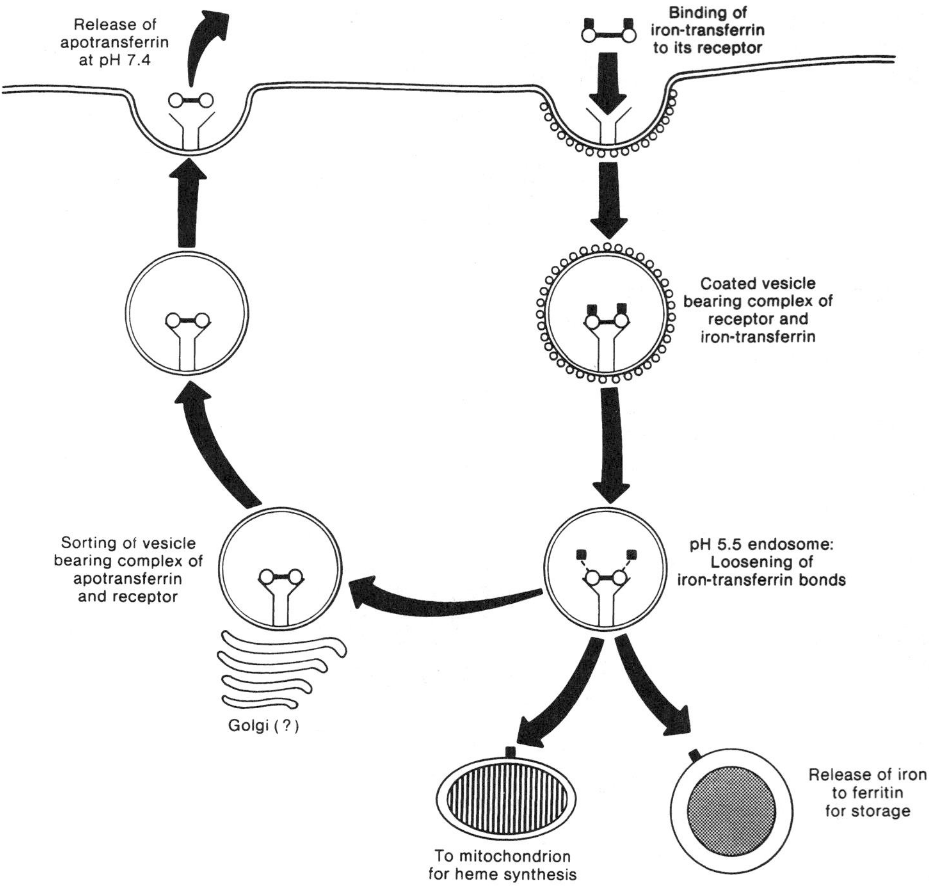

FIG. 4. The transferrin-to-cell cycle in iron metabolism. (Reproduced with permission from [225].)

since weak bases substantially inhibit the net gain of iron by cells [146]. Iron is now freed from its bonds to transferrin, but the actual mechanism of iron release and the chemical form of released iron are not known. Released iron must then cross the endosomal membrane. This translocation in isolated reticulocyte endosomes is inhibited by blockers of the endosomal H^+-ATPase, which has therefore been postulated to function as the transmembrane iron transporter [147,148]. Further studies, using a more specific inhibitor of the endosomal ("vacuolar")

ATPase in a variety of experimental systems, point to an acidification rather than a transport function for the ATPase [149], so that the issue of how released iron escapes the endosome is still in doubt. In any event, the low-pH endosome carrying iron-depleted transferrin now finds its way back to the cell surface where it fuses with the plasma membrane. Again encountering a pH of 7.4, transferrin is freed from receptor to reenter the circulation and again capture iron released from cells where the metal is absorbed, stored, or recovered from senescent erythrocytes. Despite its complexity, the process of receptor-mediated iron uptake from transferrin is efficient and rapid, preserving both receptor and transferrin for repeated cycles of iron transport and delivery.

4.2. Receptor-Independent Uptake of Iron from Transferrin

Most cell types appear to meet their needs for iron largely, if not exclusively, on receptor-mediated uptake from transferrin, a high-affinity but saturable process. Hepatocytes, however, also make use of a low-affinity but nonsaturable pathway which is not suppressible by antibodies to transferrin receptor, although such antibodies effectively block uptake of iron from transferrin by reticulocytes [150]. Because of its high capacity, at physiological concentrations of transferrin the nonsaturable pathway may convey more iron to hepatocytes in primary culture. The importance of this pathway in the intact liver has not been clarified.

Transfection of the HuH-7 human hepatoma cell with an antisense cDNA sequence for the transferrin receptor results in a 50–60% suppression of receptor expression [151] and a corresponding decline in transferrin-inhibitable receptor-mediated iron uptake [42]. Uptake of labeled iron via a process that is not inhibitable by a 100-fold excess of transferrin bearing unlabeled iron is similar in transfected and wild-type cells, however, indicating that this second process is receptor independent. The receptor-independent pathway may also be taken by iron secured from isolated transferrin N-lobe, which is not recognized by receptor [152]. A proposed model for the receptor-independent process entails nonspecific adsorption of transferrin to proteins on the plasma membrane, endocytic internalization, and release of iron by a pH-dependent step. Receptor-mediated and receptor-independent routes for iron uptake ultimately converge to deliver iron for storage in ferritin or incorporation into heme and iron enzymes.

Another pathway, saturable but independent of transferrin receptor, is available for iron uptake from transferrin in the human melanoma cell [43,153] or Chinese hamster ovary cells lacking transferrin receptor but transfected with melanotransferrin [154]. This pathway may involve transfer of iron from transferrin or simple chelators to membrane-anchored melanotransferrin, and thence to the cell interior by an endocytic route that is less sensitive to the effect of weak bases but more accessible to hydrophilic chelators than the receptor-mediated pathway.

Finally worth noting is the nonphysiological but effective receptor-independent uptake of transferrin-borne iron by rabbit reticulocytes in conditions of low pH (5.9) and low ionic strength [155]. This uptake may be a carrier-mediated process, perhaps at some point involving reduction of iron to its ferrous state which readily crosses reticulocyte membranes [156].

4.3. Uptake of Nontransferrin Iron

4.3.1. Ferritin

Several cell types have been shown to express ferritin-binding sites, presumably ferritin receptors [157–167]. An iron-donating function for ferritin has been clearly demonstrated only in rat hepatocytes and guinea pig erythroid precursor cells, however. When rat spleen ferritin endogenously labeled with ^{59}Fe was injected into rats, the label disappeared from the circulation with a half-time of about 40 min, and almost all of the cleared iron could be accounted for in liver [168]. Hepatocytes in primary culture take up ferritin in a manner consistent with receptor-mediated endocytosis [169]. The ferritin protein is degraded, probably in lysosomes, but its iron is recovered and stored in ferritin synthesized by the cells [170,171]. A possible physiological role for ferritin uptake by liver cells in assuring adequate iron supply to the cells even in the presence of iron deficiency has been suggested but not proved [170].

Guinea pig erythrocytes also take up ferritin by a saturable, time- and temperature-dependent process characteristic of receptor-mediated endocytosis [157,172]. Up to 70% of the iron so acquired soon appears in heme, thus arguing for a physiological function of ferritin uptake. Of interest, in this respect, is an early electron microscopic observation of ferritin cores (identified by their distinctive morphology) in close apposition to pits in the plasma membranes of developing human red cells [173]. In the light of more current studies, it now seems likely that the ferritin was on its way into the cells rather than on its way out as originally suggested.

Red cell precursors and hepatocytes exhibit far fewer receptors for ferritin than for transferrin, and the rates at which ferritin receptors convey their ligands to cell interiors are far slower than corresponding rates of transferrin receptors. Nevertheless, the potential for iron delivery for ferritin may exceed that for transferrin, since the polynuclear iron core of ferritin contains up to 4500 iron atoms, whereas transferrin can carry no more than two iron atoms per molecule.

4.3.2. Uptake of Non-Protein-Bound Iron

All nonheme iron in the circulation is normally carried by transferrin. In disorders characterized by iron overload, such as transfusion-dependent thalassemia and hemochromatosis, nonheme iron may also be found in the plasma. Such iron, largely as its complex with citrate [174], may be more reactive and available than transferrin-borne iron [175] and hence a threat to virtually all cells and tissues. The behavior of non-transferrin-bound iron has accordingly been a subject of ongoing investigation.

Iron is readily taken up by Chang liver cells in culture when presented in soluble complex with transferrin, citrate, or nitrilotriacetate [176]. Iron acquired from transferrin is largely destined for ferritin, but when it is taken up from the chelators, much of the cell-associated iron becomes membrane bound, indicating the importance of cell fractionation in evaluating the physiology of iron uptake by cells in culture. In contrast, iron offered to K562 cells complexed with a 4:1 molar excess of nitrilotriacetate is assimilated into heme with an efficiency comparable to that of iron derived from transferrin, although uptake from the complex is less than one quarter that of transferrin iron [177]. Uptake of nitrilotriacetate iron is not suppressed by weak bases, indicating that the mechanisms of uptake and assimilation differ for transferrin-borne and chelate-derived iron. Hepatocytes in primary culture can also secure iron from its complex with nitrilotriacetate, with V_{max} greater than 10^8 atoms/cell/min [178]. This remarkable value is far greater than the maximum rate of iron uptake from transferrin, and emphasizes the toxic potential of nontransferrin iron uptake by the liver.

A possible pathway for nontransferrin iron uptake by intestinal and erythroid cells has recently been proposed [179,180]. A proposed model, based on classic pulse-chase methodology, invokes participation of a 56-kDa iron-binding protein named mobilferrin for the city (Mobile, Alabama) in which it was discovered. The postulated new path entails transfer of iron from its complex with nitrilotriacetate to integrin, an integral membrane protein, then from integrin to the single iron-binding site of soluble cystosolic mobilferrin, and finally to the ferritin core. Published data do not permit quantitative assessment of the kinetics of iron transfer, nor how exchange of iron from one carrier protein to another might be effected. For the time being at least, the model is probably best considered as calling for further research rather than as firmly established.

In the intact animal, non-transferrin-bound iron administered as ferric citrate is largely directed to the liver, where at least some if it is destined for biliary excretion [181]. The iron is available for chelation by intravenously administered deferoxamine and by orally effective dimethylhydroxypyridinone which enhance its excretion in bile. Relative effectiveness of the chelators is greater in iron-overloaded than in control animals; in the former, some 25–50% of the administered labeled iron could be recovered in bile. Transport mechanisms for iron uptake and iron excretion are not known.

Ferrous iron is avidly taken up by perfused rat liver, with the apparent formation of new ferritin cores as seen in electron microscopy [182]. Uptake is temperature dependent but relatively insensitive to inhibition of cellular energy metabolism and to the iron status of the liver. Unlike uptake of ferric iron, uptake of ferrous iron by liver cells is inhibitable by other divalent metal ions. The physiological significance of ferrous iron uptake is not apparent, since it is unlikely that ferrous iron is found in the oxygen-rich circulation, even in chronic iron overload. Possibly, however, membrane reductases for nontransferrin iron are involved in, and may even be essential for [183], its acquisition by cells. As already noted, ferrous iron traverses red cell membranes with facility [156], and in yeast cells involvement of an iron reductase has been shown by genetic as well as biochemical methods to be essential for iron uptake [184]. Whether a specific membrane carrier, or a membrane permease, or both are required for Fe^{2+} transport in animal cells, and whether transport mechanisms for nontransferrin iron are common to all eukaryotic cells, remain to be determined.

A cautionary note: Assay of true ferric iron reductase is not as simple as many reports suggest. The use of so-called ferrous iron-specific chelators such as FerroZine, bathophenanthroline or 2,2′-bipyridyl to report on iron reduction can be misleading. Although these chelators display intense colors only when binding Fe^{2+}, due to metal-to-ligand charge transfer, they also form colorless complexes with Fe^{3+} [185]. More importantly, the high reduction potentials of the Fe^{2+}/Fe^{3+} couple in these complexes, which may exceed + 1100 mV [186], and the extremely strong binding of Fe^{2+}, make their complexes with Fe^{3+} potent oxidants which can drive, and not simply report on, ferric iron reduction [187]. A far more rigorous approach for identifying and assaying ferric reductases, or for corroborating and validating results of simpler methods, would be to use nonperturbing techniques such as EPR and Mössbauer spectroscopies. Difficulties in their use would be more than compensated in the reliability of their results: Uncritically accepting a poor experiment is not justified by difficulty in doing a good one.

4.4. Microbial Uptake of Iron from Transferrin

Many important human pathogens, among them *Haemophilus influenzae*, *Neisseria meningitidis*, *Moraxella catarrhalis*, and *Neisseria gonorrhoeae*, are capable of securing iron from transferrin or lactoferrin. To accomplish this, they elaborate a pair of transferrin-binding proteins, Tbp1 and Tbp2 or Lbp1 and Lbp2, which act in concert to trap transferrin [188,189]. The pathogens are not known to elaborate siderophores, but effectively remove iron from transferrin by an unknown process involving these transferrin receptors [190]. Interestingly, if coincidentally, the primary site in the human transferrin molecule recognized by the bacterial receptors is in the C-lobe [191], as it is for recognition by the human transferrin receptor.

A soluble iron-binding protein in the periplasmic (intermembranous) space of pathogenic *Neisseria*, functioning in the delivery of iron from simple iron chelates to the bacteria, has been isolated and characterized [192]. In its iron-binding properties, the protein displays remarkable similarities to transferrins: a broad visible absorption band with a peak at 481 nm and intensity comparable to that of human transferrin, luminescence of its complex with terbium [193], ability to bind a variety of multivalent metal ions, and enhancement by bicarbonate of color associated with Fe^{3+} binding. Furthermore, in its mass, about 34 kDa, the microbial protein, which binds one Fe^{3+}/molecule, approximates the size of a transferrin lobe. It remains to be seen whether sequence analysis confirms this iron-binding protein of *Neisseria* as the long-sought single-sited transferrin precursor, but for the moment the speculation is irresistible.

5. PERSPECTIVES AND PROSPECTS

With the exception of the lactic acid bacteria, all known organisms are dependent upon iron, and even the lactic acid bacteria are able to secure iron from their environment [194]. Pathways exploited by cells in satisfying their needs for iron, of defending themselves against present or future iron deficiency, are remarkable for their variety and versatility. Structural answers to biological questions that once eluded analysis are now routinely provided with the tools of crystallography and molecular biology. Yet questions persist, and even increase, as new knowledge reveals new problems. Identifying questions that are at once interesting and experimentally answerable is a difficult, sometimes insurmountable, task. Despite decades of intensive investigation, disorders of iron metabolism, primary or secondary, of deficiency or of overload, remain among the most common and widespread of mankind's ills.

ACKNOWLEDGMENTS

Preparation of this chapter, and the research carried out in this laboratory, were supported in part by Grant DK 15056 from the National Institutes of Health, U.S. Public Health Service. Coordinates for human monoferric transferrin bearing iron at the C-lobe site, and for iron-loaded recombinant human N-lobe, were generously provided by Dr. Harmon Zuccola and Dr. Gary Brayer, respectively.

ABBREVIATIONS

ATP	adenosine 5′-triphosphate
BP	binding protein
cDNA	complementary deoxyribonucleic acid
CHAPS	3[(3-cholamidopropyl)dimethylammonio]-1-propane sulfonate
EDTA	ethylenediamine-*N*,*N*,*N*′,*N*′-tetraacetate
ENDOR	electron nuclear double resonance
EPR	electron paramagnetic resonance
IRE	iron-regulatory element
IRP	iron-regulatory protein
Lbp	lactoferrin-binding protein
LFER	linear free energy relationship
mRNA	messenger ribonucleic acid
NMR	nuclear magnetic resonance
NZ	ε-amino nitrogen of lysine (designation used in Protein Data Bank files)
Tbp	transferrin-binding protein
UV	ultraviolet

REFERENCES

1. K. N. Raymond and C. J. Carrano, *Coord. Chem. Rev.*, *12*, 183–190 (1979).
2. J. R. Gasdaska, J. H. Law, C. J. Bender, and P. Aisen, *J. Inorg. Biochem.*, *64*, 247–258 (1996).
3. E. N. Baker and P. F. Lindley, *J. Inorg. Biochem.*, *47*, 147–160 (1992).
4. R. T. A. MacGillivray, E. Mendez, J. G. Shewale, S. K. Sinha, J.

Lineback-Zins, and K. Brew, *J. Biol. Chem.*, *258*, 3543–3553 (1983).

5. J. Mazurier, M. H. Boutigue-Metz, J. Jolles, G. Spik, J. Montreuil, and P. and Jolles, *Experientia*, *39*, 135–141 (1983).
6. J. M. Jeltch and P. Chambon, *Eur. J. Biochem.*, *122*, 291–295 (1982).
7. G. S. Baldwin, *Comp. Biochem. Physiol. [B]*, *106B*, 203–218 (1993).
8. I. Park, E. Schaeffer, A. Sidoli, F. E. Baralle, G. N. Cohen, and M. M. Zakin, *Proc. Natl. Acad. Sci. USA*, *82*, 3149–3153 (1985).
9. C.-Y. Chen, S. A. Berish, S. A. Morse, and T. A. Mietzner, *Mol. Microbiol.*, *10*, 311–318 (1993).
10. A. W. Martin, E. Huebers, H. Huebers, J. Webb, and C. A. Finch, *Blood*, *64*, 1047–1052 (1984).
11. B. H. Bowman, F. Yang, and G. S. Adrian, *Adv. Genet.*, *25*, 1–38 (1988).
12. R. T. Ellison, III, *Adv. Exp. Med. Biol.*, *357*, 71–90 (1994).
13. E. W. Odell, R. Sarra, M. Foxworthy, D. S. Chapple, and R. W. Evans, *FEBS Lett.*, *382*, 175–178 (1996).
14. S. C. Williams and R. C. Woodworth, *J. Biol. Chem.*, *248*, 5845–5853 (1973).
15. W. R. Harris and Y. Chen, *J. Inorg. Biochem.*, *54*, 1–19 (1994).
16. T. K. Dickinson, A. G. Devenyi, and J. R. Connor, *J. Lab. Clin. Med.*, *128*, 270–278 (1996).
17. D. A. Goodwin, R. Goode, L. Brown, and C. J. Imbornone, *Radiology*, *100*, 175–179 (1971).
18. S. R. Vallabhajosula, J. F. Harwig, J. K. Siemsen, and W. Wolf, *J. Nucl. Med.*, *21*, 650–656 (1980).
19. P. Aisen, R. Aasa, B. G. Malmström, and T. Vänngård, *J. Biol. Chem.*, *242*, 2484–2490 (1967).
20. P. Aisen, R. Aasa, and A. G. Redfield, *J. Biol. Chem.*, *244*, 4628–4633 (1969).
21. J. L. Zweier and P. Aisen, *J. Biol. Chem.*, *252*, 6090–6096 (1977).
22. G. Kubal, A. B. Mason, S. U. Patel, P. J. Sadler, and R. C. Woodworth, *Biochemistry*, *32*, 3387–3395 (1993).
23. W. Kiang, P. J. Sadler, and D. G. Reid, *Magn. Reson. Chem.*, *31 Suppl.*, S110–S112 (1993).

24. H. Y. Li, P. J. Sadler, and H. Z. Sun, *J. Biol. Chem.*, *271*, 9483–9489 (1996).
25. P. Aisen, A. Leibman, and J. Zweier, *J. Biol. Chem.*, *253*, 1930–1937 (1978).
26. J. H. Katz, *J. Clin. Invest.*, *40*, 2143–2152 (1961).
27. W. R. Harris, *Adv. Exp. Med. Biol.*, *249*, 67–93 (1989).
28. B. F. Anderson, H. M. Baker, E. J. Dodson, G. E. Norris, S. V. Rumball, J. M. Waters, and E. N. Baker, *Proc. Natl. Acad. Sci. USA*, *84*, 1769–1773 (1987).
29. M. R. Schlabach and G. W. Bates, *J. Biol. Chem.*, *250*, 2182–2188 (1975).
30. J. L. Zweier, J. B. Wooten, and J. S. Cohen, *J. Biol. Chem.*, *20*, 3505–3510 (1981).
31. O. W. Nadeau, A. M. Falick, and R. C. Woodworth, *Biochemistry*, *35*, 14294–14303 (1996).
32. D. A. Schneider, A. L. Roe, R. J. Mayer, and L. Que, Jr., *J. Biol. Chem.*, *259*, 9699–9703 (1984).
33. W. R. Harris and D. Nesset-Tollefson, *Biochemistry*, *30*, 6930–6936 (1991).
34. W. R. Harris, *Biochemistry*, *24*, 7412–7418 (1985).
35. S. A. Kretchmar, Z. E. Reyes, and K. N. Raymond, *Biochim. Biophys. Acta*, *956*, 85–94 (1988).
36. D. C. Harris, A. L. Rinehart, D. Hereld, R. W. Schwartz, F. P. Burke, and A. P. Salvador, *Biochim. Biophys. Acta*, *838*, 295–301 (1985).
37. D. C. Harris, and P. Aisen, in Iron Carriers and Iron Proteins (T. M. Loehr, H. B. Gray and A. B. P. Lever, eds.), VCH, Weinheim, Germany, 1989, pp. 239–351.
38. J. G. Grossmann, M. Neu, R. W. Evans, P. F. Lindley, H. Appel, and S. S. Hasnain, *J. Mol. Biol.*, *229*, 585–590 (1993).
39. C. L. Day, K. M. Stowell, E. N. Baker, and J. W. Tweedie, *J. Biol. Chem.*, *267*, 13857–13862 (1992).
40. E. J. Beatty, M. C. Cox, T. A. Frenkiel, B. M. Tam, A. B. Mason, R. T. A. MacGillivray, P. J. Sadler, and R. C. Woodworth, *Biochemistry*, *35*, 7635–7642 (1996).
41. P. Aisen, *Adv. Exp. Med. Biol.*, *356*, 31–40 (1994).
42. D. Trinder, O. Zak, and P. Aisen, *Hepatology*, *23*, 1512–1520 (1996).

43. D. R. Richardson and E. Baker, *J. Cell. Physiol.*, *161*, 160–168 (1994).
44. P. K. Bali, O. Zak, and P. Aisen, *Biochemistry*, *30*, 324–328 (1991).
45. D. M. Sipe and R. F. Murphy, *J. Biol. Chem.*, *266*, 8002–8007 (1991).
46. J. P. Brown, R. M. Hewick, I. Hellstrom, K. E. Hellstrom, R. F. Doolittle, and W. J. Dreyer, *Nature*, *296*, 171–173 (1982).
47. E. N. Baker, H. M. Baker, C. A. Smith, M. R. Stebbins, M. Kahn, K. E. Hellström, and I. Hellström, *FEBS Lett.*, *298*, 215–218 (1992).
48. M. L. Kennard, H. Feldmann, T. Yamada, and W. A. Jefferies, *Nature Med.*, *2*, 1230–1235 (1996).
49. S. V. Kala, B. B. Hasinoff, and J. S. Richardson, *Int. J. Neurosci.*, *86*, 263–269 (1996).
50. M. A. Smith and G. Perry, *J. Neurol. Sci.*, *134 Suppl.*, 92–94 (1995).
51. R. Alemany, M. R. Vilá, C. Francí, G. Egea, F. X. Real, and T. M. Thomson, *J. Cell Sci.*, *104*, 1155–1162 (1993).
52. N. S. Bartfeld and J. H. Law, *J. Biol. Chem.*, *265*, 21684–21691 (1990).
53. R. C. Jamroz, J. R. Gasdaska, J. Y. Bradfield, and J. H. Law, *Proc. Natl. Acad. Sci. USA*, *90*, 1320–1324 (1993).
54. M. Haridas, B. F. Anderson, and E. N. Baker, *Acta Cryst.*, *D51*, 629–646 (1995).
55. S. Bailey, R. W. Evans, R. C. Garratt, B. Gorinsky, S. Hasnain, C. Horsburgh, H. Jhoti, P. F. Lindley, A. Mydin, and R. Sarra, *Biochemistry*, *27*, 5804–5812 (1988).
56. H. Kurokawa, B. Mikami, and M. Hirose, *J. Mol. Biol.*, *254*, 196–207 (1995).
57. J. Dewan, B. Mikami, M. Hirose, and J. C. Sacchettini, *Biochemistry*, *32*, 11963–11968 (1993).
58. E. N. Baker, *Adv. Inorg. Chem.*, *41*, 389–461 (1994).
59. R. C. Woodworth, A. B. Mason, W. D. Funk, and R. T. A. MacGillivray, *Biochemistry*, *30*, 10824–10829 (1991).
60. J. G. Grossmann, A. B. Mason, R. C. Woodworth, M. Neu, P. F. Lindley, and S. S. Hasnain, *J. Mol. Biol.*, *231*, 554–558 (1993).

61. H. R. Faber, T. Bland, C. L. Day, G. E. Norris, J. W. Tweedie, and E. N. Baker, *J. Mol. Biol.*, *256*, 352–363 (1996).

62. V. L. Pecoraro, W. R. Harris, C. J. Carrano, and K. N. Raymond, *Biochemistry*, *20*, 7033–7039 (1981).

63. P. F. Lindley, M. Bajaj, R. W. Evans, R. C. Garratt, S. S. Hasnain, H. Jhoti, P. Kuser, M. Neu, K. Patel, and R. Sarra, *Acta Cryst.*, *D49*, 292–304 (1993).

64. J. E. Roberts, T. G. Brown, B. F. Hoffman, and P. Aisen, *Biochim. Biophys. Acta*, *747*, 49–54 (1983).

65. W. Froncisz and P. Aisen, *Biochim. Biophys. Acta*, *700*, 55–58 (1982).

66. H. Nicholson, B. F. Anderson, T. Bland, S. C. Shewry, J. W. Tweedie, and E. N. Baker, *Biochemistry*, *36*, 341–346 (1997).

67. P. Aisen and A. Leibman, *Biochim. Biophys. Acta*, *304*, 797–804 (1973).

68. J. M. El Hage Chahine and R. Pakdaman, *Eur. J. Biochem.*, *230*, 1102–1110 (1995).

69. O. Zak, P. Aisen, J. B. Crawley, C. L. Joannou, K. J. Patel, M. Rafiq, and R. W. Evans, *Biochemistry*, *34*, 14428–14434 (1995).

70. H. R. Faber, C. J. Baker, C. L. Day, J. W. Tweedie, and E. N. Baker, *Biochemistry*, *35*, 14473–14479 (1996).

71. R. E. Cowart, N. Kojima, and G. W. Bates, *J. Biol. Chem.*, *257*, 7560–7566 (1982).

72. S. A. K. Nguyen, A. Craig, and K. N. Raymond, *J. Am. Chem. Soc.*, *115*, 6758–6764 (1993).

73. P. K. Bali, W. R. Harris, and D. Nesset-Tollefson, *Inorg. Chem.*, *30*, 502–508 (1991).

74. B. F. Anderson, H. M. Baker, G. E. Norris, S. V. Rumball, and E. N. Baker, *Nature*, *344*, 784–787 (1990).

75. M. Gerstein, B. F. Anderson, G. E. Norris, E. N. Baker, A. M. Lesk, and C. Chothia, *J. Mol. Biol.*, *234*, 357–372 (1993).

76. M. G. Patch and C. J. Carrano, *Inorg. Chim. Acta*, *56*, L71–L73 (1981).

77. M. I. Scullane, L. K. White, and N. D. Chasteen, *J. Mag. Res.*, *47*, 383–397 (1982).

78. A. S. Yang and B. J. Gaffney, *Biophys. J.*, *51*, 55–67 (1987).

79. B. J. Gaffney and H. J. Silverstone, in EMR of Paramagnetic Molecules. (L. J. Berliner and J. Reuben, eds.) Plenum Press, New York, 1993, pp. 1–57.
80. W. Froncisz and J. S. Hyde, *J. Chem. Phys.*, *73*, 3123–3131 (1980).
81. K. S. Doctor, B. J. Gaffney, G. Alvarez, and H. J. Silverstone, *J. Phys. Chem.*, *97*, 3028–3033 (1993).
82. J. K. Grady, A. B. Mason, R. C. Woodworth, and N. D. Chasteen, *Biochem. J.*, *309*, 403–410 (1995).
83. N. D. Chasteen, L. K. White, and R. F. Campbell, *Biochemistry*, *16*, 363–368 (1977).
84. J. L. Zweier, *J. Biol. Chem.*, *253*, 7616–7621 (1978).
85. P. K. Bali and W. R. Harris, *J. Am. Chem. Soc.*, *111*, 4457–4461 (1989).
86. L. Messori and M. Piccioli, *J. Inorg. Biochem.*, *42*, 185–190 (1991).
87. J. M. Aramini and H. J. Vogel, *J. Am. Chem. Soc.*, *115*, 245–252 (1993).
88. J. M. Aramini and H. J. Vogel, *J. Am. Chem. Soc.*, *116*, 1988–1993 (1994).
89. I. Bertini, C. Luchinat, and L. Messori, *J. Am. Chem. Soc.*, *105*, 1347–1350 (1983).
90. J. M. Aramini, P. H. Krygsman, and H. J. Vogel, *Biochemistry*, *33*, 3304–3311 (1994).
91. J. H. Jandl, J. K. Inman, R. L. Simmons, and D. W. Allen, *J. Clin. Invest.*, *38*, 161–185 (1959).
92. E. Baker and E. H. Morgan, *Biochemistry, 8*, 2954–2958 (1969).
93. D. A. Baldwin and D. M. R. de Sousa, *Biochem. Biophys. Res. Commun.*, *99*, 1101–1107 (1981).
94. D. A. Baldwin, D. M. R. de Sousa, and R. von Wandruszka, *Biochim. Biophys. Acta*, *719*, 140–146 (1982).
95. S. A. Kretchmar and K. N. Raymond, *Inorg. Chem.*, *27*, 1436–1441 (1988).
96. T. J. Egan, D. C. Ross, L. R. Purves, and P. A. Adams, *Inorg. Chem.*, *31*, 1994–1998 (1992).
97. W. R. Harris, *J. Inorg. Biochem.*, *21*, 263–276 (1984).
98. S. Pollack, P. Aisen, F. D. Lasky, and G. Vanderhoff, *Br. J. Haematol.*, *34*, 231–235 (1976).

99. S. Pollack, G. A. Vanderhoff, and F. Lasky, *Biochim. Biophys. Acta*, *497*, 481–487 (1977).
100. P. K. Bali and P. Aisen, *Biochemistry*, *30*, 9947–9952 (1991).
101. O. Zak and P. Aisen, *Blood*, *68*, 157–161 (1986).
102. J. Williams and K. Moreton, *Biochem. J.*, *185*, 483–488 (1980).
103. A. McClelland, L. C. Kühn, and F. H. Ruddle, *Cell*, *39*, 267–274 (1984).
104. H.-Y. Y. Hu and P. Aisen, *J. Supramol. Struct.*, *8*, 349–360 (1978).
105. P. A. Seligman, R. B. Schleicher, and R. H. Allen, *J. Biol. Chem.*, *254*, 9943–9946 (1979).
106. M. Rabin, A. McClelland, L. C. Kühn, and F. H. Ruddle, *Am. J. Hum. Genet.*, *37*, 1112–1116 (1985).
107. C. Schneider, M. J. Owen, D. Banville, and J. G. Williams, *Nature*, *311*, 675–678 (1984).
108. M. Zerial, P. Melancon, C. Schneider, and H. Garoff, *EMBO J.*, *5*, 1543–1550 (1997).
109. D. W. Borhani and S. C. Harrison, *J. Mol. Biol.*, *218*, 685–689 (1991).
110. S. Jing and I. S. Trowbridge, *EMBO J.*, *6*, 327–331 (1987).
111. S. Jing and I. S. Trowbridge, *J. Biol. Chem.*, *265*, 11555–11559 (1990).
112. M. B. Omary and I. S. Trowbridge, *J. Biol. Chem.*, *256*, 12888–12892 (1981).
113. G. R. Hayes, C. A. Enns, and J. J. Lucas, *Glycobiology*, *2*, 355–359 (1992).
114. A. M. Williams and C. A. Enns, *J. Biol. Chem.*, *266*, 17648–17654 (1991).
115. E. A. Rutledge, B. J. Root, J. J. Lucas, and C. A. Enns, *Blood*, *83*, 580–586 (1994).
116. G. Orberger, R. Geyer, S. Stirm, and R. Tauber, *Eur. J. Biochem.*, *205*, 257–267 (1992).
117. R. J. Davis, G. L. Johnson, D. J. Kelleher, J. K. Anderson, J. E. Mole, and M. P. Czech, *J. Biol. Chem.*, *261*, 9034–9041 (1986).
118. S. Rothenberger, B. J. Iacopetta, and L. C. Kuhn, *Cell*, *49*, 423–431 (1987).
119. J. L. Frazier, J. H. Caskey, M. Yoffe, and P. A. Seligman, *J. Clin. Invest.*, *69*, 853–865 (1982).

120. J. F. Collawn, A. Lai, D. Domingo, M. Fitch, S. Hatton, and I. S. Trowbridge, *J. Biol. Chem.*, *268*, 21686–21692 (1993).

121. F. Buchegger, I. S. Trowbridge, L. F. Liu, S. White, and J. F. Collawn, *Eur. J. Biochem.*, *235*, 9–17 (1996).

122. J. L. Casey, M. W. Hentze, D. M. Koeller, S. W. Caughman, T. A. Rouault, R. D. Klausner, and J. B. Harford, *Science*, *240*, 924–928 (1988).

123. D. Owen and L. C. Kühn, *EMBO J.*, *6*, 1287–1293 (1987).

124. E. A. Leibold and H. N. Munro, *Proc. Natl. Acad. Sci. USA*, *85*, 2171–2175 (1988).

125. E. W. Müllner and L. C. Kühn, *Cell*, *53*, 815–825 (1988).

126. K. Rao, J. B. Harford, T. Rouault, A. McClelland, F. H. Ruddle, and R. D. Klausner, *Mol. Cell. Biol.*, *6*, 236–240 (1986).

127. H. A. Barton, R. S. Eisenstein, A. Bomford, and H. N. Munro, *J. Biol. Chem.*, *265*, 7000–7008 (1990).

128. S. Kaptain, W. E. Downey, C. Tang, C. Philpott, D. Haile, D. G. Orloff, J. B. Harford, T. A. Rouault, and R. D. Klausner, *Proc. Natl. Acad. Sci. USA*, *88*, 10109–10113 (1991).

129. B. R. Henderson and L. C. Kühn, *J. Biol. Chem.*, *270*, 20509–20515 (1995).

130. J.-C. Drapier, H. Hirling, J. Wietzerbin, P. Kaldy, and L. C. Kühn, *EMBO J.*, *12*, 3643–3649 (1993).

131. G. Weiss, B. Goossen, W. Doppler, D. Fuchs, K. Pantopoulos, G. Werner-Felmayer, H. Wachter, and M. W. Hentze, *EMBO J.*, *12*, 3651–3657 (1993).

132. M. W. Hentze and L. C. Kühn, *Proc. Natl. Acad. Sci. USA*, *93*, 8175–8182 (1996).

133. L. A. Cox, M. C. Kennedy, and G. S. Adrian, *Biochem. Biophys. Res. Commun.*, *212*, 925–932 (1995).

134. B. Guo, Y. Yu, and E. A. Leibold, *J. Biol. Chem.*, *269*, 24252–24260 (1994).

135. T. A. Rouault, C. K. Tang, S. Kaptain, W. H. Burgess, D. J. Haile, F. Samaniego, O. W. McBride, J. B. Harford, and R. D. Klausner, *Proc. Natl. Acad. Sci. USA*, *87*, 7958–7962 (1990).

136. E. R. Henderson, E. Menotti, and L. C. Kühn, *J. Biol. Chem.*, *271*, 4900–4908 (1996).

137. B. Guo, J. D. Phillips, Y. Yu, and E. A. Leibold, *J. Biol. Chem.*, *270*, 21645–21651 (1995).

138. S. P. Young, A. Bomford, and R. Williams, *Biochem. J.*, *219*, 505–510 (1984).

139. O. Zak, D. Trinder, and P. Aisen, *J. Biol. Chem.*, *269*, 7110–7114 (1994).

140. M. Cazzola, H. A. Huebers, M. H. Sayers, A. P. MacPhail, M. Eng, and C. A. Finch, *Blood*, *66*, 935–939 (1985).

141. G. Bergamaschi, M. J. Eng, H. A. Huebers, and C. A. Finch, *Proc. Soc. Exp. Biol. Med.*, *183*, 66–73 (1986).

142. J. vanRenswoude, K. R. Bridges, J. B. Harford, and R. D. Klausner, *Proc. Natl. Acad. Sci. USA*, *79*, 6186–6190 (1982).

143. R. D. Klausner, J. V. Ashwell, J. B. VanRenswoude, J. Harford, and K. Bridges, *Proc. Natl. Acad. Sci. USA*, *80*, 2263–2266 (1983).

144. A. Dautry-Varsat, A. Ciechanover, and H. F. Lodish, *Proc. Natl. Acad. Sci. USA*, *80*, 2258–2262 (1983).

145. B. Ecarot-Charrier, V. L. Grey, A. Wilczynska, and H. M. Schulman, *Can. J. Biochem.*, *58*, 418–426 (1980).

146. E. H. Morgan, *Biochim. Biophys. Acta*, *642*, 119–134 (1981).

147. C.-Y. Li, J. A. Watkins, and J. Glass, *J. Biol. Chem.*, *269*, 10242–10246 (1994).

148. C.-Y. Li, J. A. Watkins, S. Hamazaki, J. D. Altazan, and J. Glass, *Biochemistry*, *34*, 5130–5136 (1995).

149. D. L. Savigni and E. H. Morgan, *Biochem. Pharmacol.*, *52*, 371–377 (1996).

150. D. Trinder, E. H. Morgan, and E. Baker, *Biochim. Biophys. Acta*, *943*, 440–446 (1988).

151. K. Sasaki, O. Zak, and P. Aisen, *Am. J. Hematol.*, *42*, 74–80 (1993).

152. K. Thorstensen, D. Trinder, O. Zak, and P. Aisen, *Eur. J. Biochem.*, *232*, 129–133 (1995).

153. D. R. Richardson and E. Baker, *Biochim. Biophys. Acta Mol. Cell Res.*, *1053*, 1–12 (1990).

154. M. L. Kennard, D. R. Richardson, R. Gabathuler, P. Ponka, and W. A. Jefferies, *EMBO J.*, *14*, 4178–4186 (1995).

155. L. I. Hodgson, E. A. Quail, and E. H. Morgan, *Arch. Biochem. Biophys.*, *308*, 318–326 (1994).

156. A. Egyed, *Br. J. Haematol.*, *68*, 483–486 (1988).

157. S. Pollack and T. Campana, *Biochem. Biophys. Res. Commun.*, *100*, 1667–1672 (1981).

158. U. Mack, E. L. Storey, L. W. Powell, and J. W. Halliday, *Biochim. Biophys. Acta*, *843*, 164–170 (1985).

159. M. S. Bretscher and J. N. Thomson, *EMBO J.*, *2*, 599–603 (1983).

160. K. Pattanapanyasat, T. G. Hoy, A. Jacobs, S. Courtney, and D. J. T. Webster, *Br. J. Cancer*, *57*, 193–197 (1988).

161. S. Fargion, P. Arosio, A. L. Fracanzani, S. Cislaghi, S. Levi, A. Cozzi, A. Piperno, and G. Fiorelli, *Blood*, *71*, 753–757 (1988).

162. A. M. Covell and J. D. Cook, *Br. J. Haematol.*, *69*, 559–563 (1988).

163. A. M. Konijn, E. G. Meyron-Holtz, R. Levy, H. Ben-Bassat, and Y. Matzner, *FEBS Lett.*, *263*, 229–232 (1990).

164. S. Fargion, A. L. Fracanzani, V. Cislaghi, S. Levi, M. D. Cappellini, and G. Fiorelli, *Curr. Stud. Hematol. Blood Transfus.*, *58*, 164–170 (1991).

165. D. Moss, S. Fargion, A. L. Fracanzani, S. Levi, M. D. Cappellini, P. Arosio, L. W. Powell, and J. W. Halliday, *J. Inorg. Biochem.*, *47*, 219–227 (1992).

166. G. A. Ramm, R. S. Britton, R. O'Neill, and B. R. Bacon, *J. Clin. Invest.*, *94*, 9–15 (1994).

167. A. M. Konijn, E. G. Meyron-Holtz, E. Fibach, and D. Gelvan, *Adv. Exp. Med. Biol.*, *356*, 189–198 (1994).

168. A. Unger and C. Hershko, *Br. J. Haematol.*, *28*, 169–179 (1974).

169. K. Osterloh and P. Aisen, *Biochim. Biophys. Acta*, *1011*, 40–45 (1989).

170. J.-C. Sibille, H. Kondo, and P. Aisen, *Hepatology*, *8*, 296–301 (1988).

171. J.-C. Sibille, M. Ciriolo, H. Kondo, R. R. Crichton, and P. Aisen, *Biochem. J.*, *262*, 685–688 (1989).

172. G. D. Blight and E. H. Morgan, *Br. J. Haematol.*, *55*, 59–71 (1983).

173. M. C. Bessis and J. Breton-Gorius, *Blood*, *19*, 635–663 (1962).

174. M. Grootveld, J. D. Bell, B. Halliwell, O. I. Aruoma, A. Bomford, and P. J. Sadler, *J. Biol. Chem.*, *264*, 4417–4422 (1989).
175. C. Hershko, G. Graham, G. W. Bates, and E. A. Rachmilewitz, *Br. J. Haematol.*, *40*, 255–263 (1978).
176. A. Jacobs, T. Hoy, J. Humphrys, and P. Perera, *Br. J. Exp. Path.*, *59*, 489–498 (1978).
177. R. S. Inman and M. Wessling-Resnick, *J. Biol. Chem.*, *268*, 8521–8528 (1993).
178. D. Barisani, C. L. Berg, M. Wessling-Resnick, and J. L. Gollan, *Am. J. Physiol. Gastrointest. Liver Physiol.*, *269*, G570–G576 (1995).
179. M. E. Conrad and J. N. Umbreit, *Am. J. Hematol.*, *42*, 67–73 (1993).
180. M. E. Conrad, J. N. Umbreit, E. G. Moore, C. Uzel, and M. R. Berry, *J. Biol. Chem.*, *269*, 7169–7173 (1994).
181. P. Brissot, G. Zanninelli, D. Guyader, J. Zeind, and J. Gollan, *Am. J. Physiol. Gastrointest. Liver Physiol.*, *267*, G135–G142 (1994).
182. T. L. Wright, P. Brissot, W.-L. Ma, and R. A. Weisiger, *J. Biol. Chem.*, *261*, 10909–10914 (1986).
183. E. W. Randell, J. G. Parkes, N. F. Olivieri, and D. M. Templeton, *J. Biol. Chem.*, *269*, 16046–16053 (1994).
184. G. J. Anderson, A. Dancis, D. G. Roman, and R. D. Klausner, *Adv. Exp. Med. Biol.*, *356*, 81–90 (1994).
185. R. M. Smith, and A. E. Martell, *Critical Stability Constants*, Vol. 2, New York, Plenum Press, 1975.
186. G. R. Buettner, *Arch. Biochem. Biophys.*, *300*, 535–543 (1993).
187. K. Thorstensen and P. Aisen, *Biochim. Biophys. Acta*, *1052*, 29–35 (1990).
188. R. A. Bonnah, R. H. Yu, and A. B. Schryvers, *Microb. Pathog.*, *19*, 285–297 (1995).
189. C. N. Cornelissen and P. F. Sparling, *Mol. Microbiol.*, *14*, 843–850 (1994).
190. S. D. Gray-Owen and A. B. Schryvers, *Microb. Pathog.*, *14*, 389–398 (1993).
191. J. Alcantara, R. H. Yu, and A. B. Schryvers, *Mol. Microbiol.*, *8*, 1135–1143 (1993).

192. A. J. Nowalk, S. B. Tencza, and T. A. Mietzner, *Biochemistry*, *33*, 12769–12775 (1994).
193. P. O'Hara, S. M. Yeh, C. F. Meares, and R. Bersohn, *Biochemistry*, *20*, 4704–4708 (1981).
194. E. Kot, S. Furmanov, and A. Bezkorovainy, *J. Food Sci.*, *60*, 547–550 (1995).
195. R. Aasa, B. G. Malmström, P. Saltman, and T. Vänngård, *Biochim. Biophys. Acta*, *75*, 203–222 (1963).
196. M. Cochran, J. Coates, and S. Neoh, *FEBS Lett.*, *176*, 129–132 (1984).
197. G. Kubal, P. J. Sadler, and R. W. Evans, *J. Am. Chem. Soc.*, *114*, 1117–1118 (1992).
198. J. C. Cannon and N. D. Chasteen, *Biochemistry*, *14*, 4573–4577 (1975).
199. A. Butler and H. Eckert, *J. Am. Chem. Soc.*, *111*, 2802–2809 (1989).
200. W. R. Harris, *J. Inorg. Biochem.*, *27*, 41–52 (1986).
201. I. Bertini, C. Luchinat, L. Messori, and A. Scozzafava, *Eur. J. Biochem.*, *141*, 375–378 (1984).
202. W. R. Harris, *Biochemistry*, *22*, 3920–3926 (1983).
203. W. R. Harris and V. L. Pecoraro, *Biochemistry*, *22*, 292–299 (1983).
204. C. R. Chitambar and D. Sax, *Blood*, *80*, 505–511 (1992).
205. P. A. Seligman, P. L. Moran, R. B. Schleicher, and E. D. Crawford, *Am. J. Hematol.*, *41*, 232–240 (1992).
206. W. R. Harris and L. J. Madsen, *Biochemistry*, *27*, 284–288 (1988).
207. S. Kulprathipanja, D. J. Hnatowich, R. Beh, and D. Elmaleh, *Int. J. Nucl. Med. Biol.*, *6*, 138–141 (1979).
208. P. W. Martin and K. Skov, *J. Nucl. Med.*, *19*, 1171–1174 (1978).
209. P. L. Moran and P. A. Seligman, *Cancer Res.*, *49*, 4237–4241 (1989).
210. C. A. Smith, E. W. Ainscough, H. M. Baker, A. M. Brodie, and E. N. Baker, *J. Am. Chem. Soc.*, *116*, 7889–7890 (1994).
211. C. K. Luk, *Biochemistry*, *10*, 2838–2843 (1971).
212. P. B. O'Hara and S. H. Koenig, *Biochem. J.*, *25*, 1445–1550 (1986).
213. O. Zak and P. Aisen, *Biochemistry*, *27*, 1075–1080 (1988).

214. L. Messori, R. Monnani, and A. Scozzafava, *Inorg. Chim. Acta*, *124*, L15–L17 (1986).
215. G. M. Then, H. Appel, J. Duffield, D. M. Taylor, and W.-G. Thies, *J. Inorg. Biochem.*, *27*, 255–270 (1986).
216. R. Stjernholm, F. W. Warner, J. W. Robinson, E. Ezekiel, and N. Katayama, *Bioinorg. Chem.*, *9*, 277–280 (1978).
217. W. R. Harris, C. J. Carrano, V. J. Pecoraro, and K. N. Raymond, *J. Am. Chem. Soc.*, *103*, 2231–2237 (1981).
218. D. M. Taylor and L. C. Farrow, *Nucl. Med. Biol.*, *14*, 27–31 (1987).
219. R. Wirth, D. M. Taylor, and J. Duffield, *Int. J. Nucl. Med. Biol.*, *12*, 327–330 (1985).
220. M. Lehmann, H. Culig, and D. M. Taylor, *Int. J. Radiat. Biol. Relat. Stud. Phys. Chem. Med.*, *44*, 65–74 (1983).
221. A. Leibman and P. Aisen, *Biochim. Biophys. Acta*, *257*, 314–323 (1972).
222. R. B. Martin, J. Savory, S. Brown, R. L. Bertholf, and M. R. Wills, *Clin. Chem.*, *33*, 405–407 (1987).
223. W. R. Harris, *Biochemistry*, *25*, 803–808 (1986).
224. W. R. Harris, Y. Chen, and K. Wein, *Inorg. Chem.*, *33*, 4991–4998 (1997).
225. E. C. Theil and P. Aisen, in *Iron Transport in Microbes, Plants and Animals* (G. Winkelmann, D. van der Helm, and J. B. Neilands, eds.), VCH, Weinheim, Germany, 1987, p. 503.

16

Iron Homeostasis

***Robert R. Crichton*[1] *and Roberta J. Ward*[2]**

[1]Unité de Biochimie, Université Catholique de Louvain, Place Louis Pasteur 1, B-1348 Louvain-la-Neuve, Belgium

[2]Unité de Biologie Animale, Université Catholique de Louvain, Place Croix du Sud 1, B-1348 Louvain-la-Neuve, Belgium

1.	INTRODUCTION	634
2.	IRON HOMEOSTASIS IN BACTERIA	638
3.	IRON HOMEOSTASIS IN PLANTS AND YEAST	639
4.	CELLULAR IRON HOMEOSTASIS IN HUMANS	643
	4.1. Generalities	643
	4.2. Iron Homeostasis in Various Tissues	645
	4.2.1. Brain	645
	4.2.2. Gastrointestinal Tract	647
	4.2.3. Liver	648
	4.3. Relationship Between Reactive Oxygen Intermediates, Hydrogen Peroxide, Nitric Oxide, and Iron Homeostasis	649
5.	IRON TRANSPORT, STORAGE, AND HOMEOSTASIS IN HUMANS	650
	5.1. Iron Transport and Uptake Proteins	653

5.2. Iron-Storage Proteins 655
6. CONCLUSIONS AND UNRESOLVED PROBLEMS 657
ACKNOWLEDGMENT 660
ABBREVIATIONS 660
REFERENCES 661

1. INTRODUCTION

Homeostasis is defined, in physiological terms, as the maintenance of normal, internal stability in an organism by the coordinated responses of the organ systems that automatically compensate for environmental changes. As far as the essential element iron is concerned, this implies the control of iron levels at the cellular, and in the case of multicellular organisms, at the level of the whole organism, of appropriate amounts of iron to ensure the correct functioning of the system while avoiding the potentially toxic effects of the accumulation of excess amounts of the metal. This is particularly important as far as iron is concerned, because, in tiny amounts, iron, like copper, can catalyse hydroxyl radical production by the redox cycling indicated in reactions (1) and (2):

$$Fe^{3+} + O_2^- \rightarrow Fe^{2+} + O_2 \tag{1}$$

$$Fe^{2+} + H_2O_2 \rightarrow Fe^{3+} + OH^- + OH^\bullet \tag{2}$$

Hydroxyl radicals can cause all sorts of oxidative damage to proteins, lipids and nucleic acids, and although there are substantial antioxidative protective mechanisms, iron-derived oxidative stress has been implicated in a growing number of human diseases. We shall return to this point later in the chapter.

In the case of iron, there are two further complications; namely, the tendency of ferrous iron to be oxidized to the ferric state under physiological conditions, and the inexorable hydrolysis and polymerization of ferric iron in aqueous solution, to form insoluble hydroxides and oxyhydroxides. This explains why it is absolutely essential that iron is kept out of harm's way (by its close association with proteins). In other words, the only way to prevent *free* iron from being either potentially toxic, by catalyzing Fenton chemistry leading to the production of the hydroxyl radical, or becoming unavailable to living organisms on account of its insolubility, is to maintain the levels of free iron (the low molecular weight iron pool) within biological systems at extremely low levels, while maintaining the most important amounts of iron within the cell relatively tightly bound to proteins (for a detailed review of iron biochemistry, see [1]).

For unicellular organisms, and indeed for individual cells in a multicellular organism, iron homeostasis represents the equilibrium between uptake, intracellular utilization and, where possible, storage, and excretion from the cell (Fig. 1). Iron homeostasis can be regulated entirely at the level of uptake, and it seems likely that this is the case in *Escherichia coli* and probably in many other bacteria. It could also be achieved by balancing uptake in a reciprocal way with intracellular storage and utilization, and this is thought to be the case in mammalian cells. The importance of excretion in iron homeostasis is not well understood, although one example could be the secretion of iron in the form of ferritin by macrophages which have acquired excess iron from the phagocytosis of effete red blood cells.

Metal ion homeostasis may be regulated at various levels (Table 1). In the case of bacteria such as *E. coli*, when sufficient iron is present, the Fe^{2+}-Fur complex acts directly as a repressor on RNA transcription from iron-uptake operons. Indirect effects of iron on transcription in-

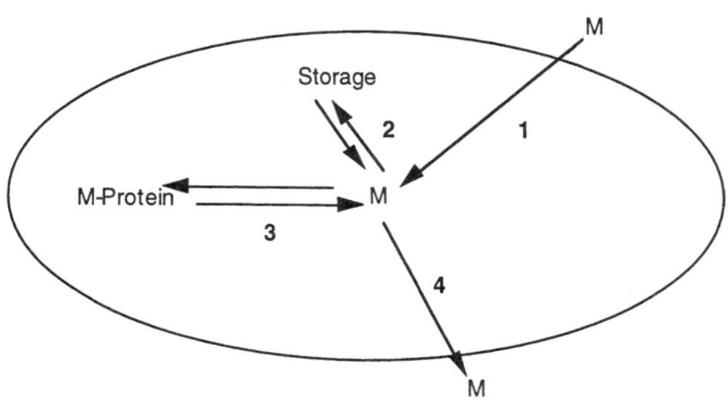

FIG. 1. Metal ion homeostasis depends on the balance between (1) uptake of the metal ion from the extracellular medium, (2) intracellular storage, (3) intracellular utilization, and (4) export.

TABLE 1

Levels of Metal Ion Homeostasis

Transcription
directly (protein binding to DNA)
indirectly (via activation of nuclear transcription factors)
Translation
via regulatory elements on mRNAs affecting either ribosome binding or mRNA stability
Protein Turnover
ubiquinylation of intracellular proteins
desalyalation of extracellular proteins

clude the oxidative stress-induced activation of a number of nuclear transcription factors in many different biological systems. Regulation at the level of mRNA translation by ribosomes into protein can be of two types. Iron-responsive elements on mRNAs can be targeted by specific proteins which bind to the iron-responsive elements affecting either ribosomal binding or mRNA stability to nuclease digestion. This is the case for example in many mammalian cells.

A third factor in cellular iron homeostasis is the rate at which proteins involved in iron storage, transport, and utilization are turned over. In eukaryotic cells, for intracellular proteins, there are two mechanisms, one lysosomal and the other cytosolic. The latter is dependent on adenosine triphosphate (ATP), involving ubiquitinylation (illustrated in Fig. 2) and subsequent degradation in large multiprotein complexes called proteosomes. This mechanism is probably involved in turnover of the mammalian iron-regulatory protein 2 (IRP-2). Extracellular proteins are also subject to turnover.

Finally, as we will develop in greater detail in Section 5 below, humans represent not only the complexity of a multicellular organism but also a remarkable lack of capacity both to absorb dietary iron and to excrete iron when compared with other mammals. This has the consequence of placing humans in grave danger not only of iron deficiency (anemia affects one third of the world's population) but also of iron

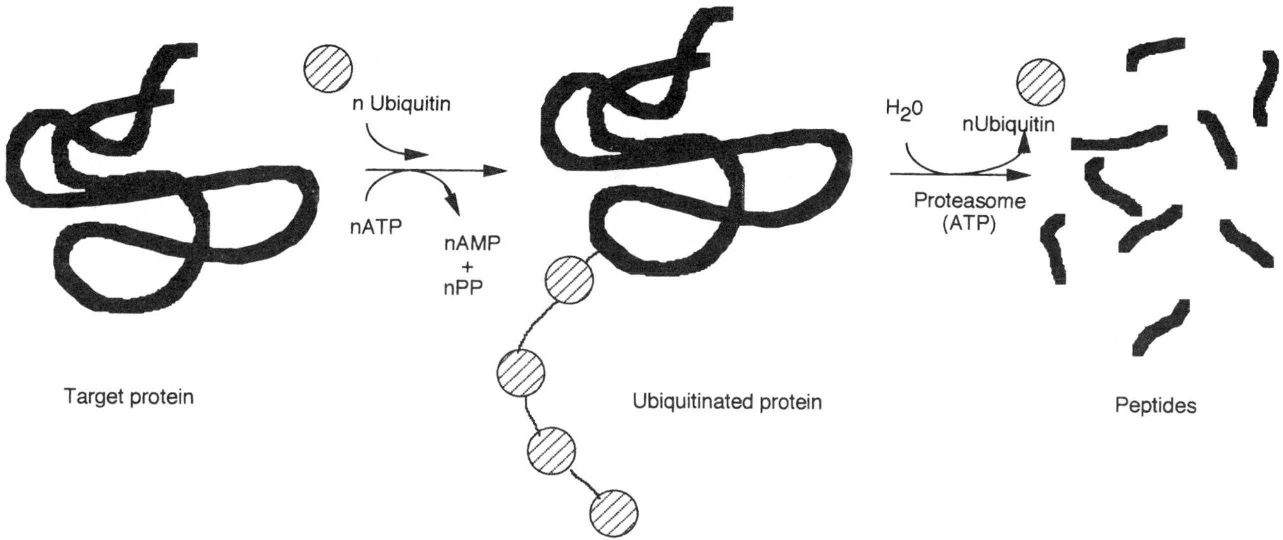

FIG. 2. The reactions involved in the ubiquitination and hydrolysis of a protein. Ubiquitination enzymes catalyse the attachment of numerous molecules of ubiquitin through Lys amino groups to the protein targeted for degradation. This marks out the protein for ATP-dependent hydrolysis by the 26S proteasome, releasing peptides and ubiquitin.

overload (between 1 in 200 to 1 in 400 individuals of Northern European descent have hereditary hemochromatosis).

2. IRON HOMEOSTASIS IN BACTERIA

In the course of evolution, bacteria have developed highly effective systems for iron assimilation, secreting into the medium strongly iron-chelating molecules, called siderophores, which have stability constants for Fe^{3+} typically in excess of 10^{30} (for a discussion of this point, see [2]). We consider here iron homeostasis in *E. coli*, which has by far the best characterized systems of bacterial iron uptake [3]. In *E. coli*, there are at least eight iron-uptake systems [3], six of which require siderophores (the organism also has the capacity to take up iron from ferric citrate and has a specific ferrous iron-transport system). In conditions of iron stress, the siderophore is synthesized and secreted into the extracellular medium, where it binds iron tightly. The ferric-siderophore complex is then recognized and bound by a specific outer-membrane receptor protein, which acts as an energy-dependent gated channel, allowing the ferric siderophore to pass into the periplasmic space. The ferric siderophore is then transported across the periplasmic space to the cytoplasmic membrane and bound to a periplasmic binding protein, which as has been previously pointed out [1], probably has close structural similarities to the two lobes of the transferrin molecule. There, in an energy-dependent mechanism involving a cytoplasmic membrane receptor complex, the ferric siderophore is released into the cytoplasm. The iron is then released by reduction to Fe(II) by an NAD[P]H-dependent ferrireductase (in the case of ferric enterobactin, enterobactin is first hydrolyzed by an esterase, Fes). A detailed account of iron transport in bacteria is presented in Chapters 2 and 3 of this volume [4,3].

The genes of *E. coli* which are involved in iron uptake (there are about 50) are regulated by the Fur (ferric uptake regulation) repressor protein. Fur is a 17-kDa protein, rich in histidine, which can be converted to a transcriptional repressor by binding Fe^{2+} forming a dimeric metalloprotein. This dimer binds to specific DNA sequences in the promotor regions of *fur*-controlled genes, acting as a repressor of transcription. When the cellular iron concentration decreases below a critical level, repression is released and the synthesis of the iron uptake systems can recommence (as reviewed in [3]). Systems analogous to Fur have been reported in a great many other bacteria [3]. A similar iron-dependent regulation of diptheria toxin and siderophore expression has been described in *Corynebacterium diphtheriae* involving the dimeric Fe^{2+}-DtxR (diphtheria toxin repressor) protein [5]; the structures of both the apo- and the metal-activated forms of DtxR have been recently reported [6]. Systems analogous to DtxR have been reported in *Streptococcus pilosus* [7] and in *Mycobacterium tuberculosis* [8]. We conclude that Fur or Fur-related systems are involved in the regulation of iron uptake in both gram-positive and gram-negative bacteria.

What might be the role of iron-storage proteins in regulating bacterial iron homeostasis? As has been pointed out in a previous chapter [9] *E. coli* has two bacterioferritins, one heme containing (BFR) and one nonheme containing (FTN). However, even in conditions where the two bacterioferritins have been overproduced in *E. coli*, their respective iron contents are 5–20 atoms/holomer for FTN and 25–75 for BFR [10,11]. Taken together with the important regulation exerted at the level of repression of transcription of iron-uptake operons by the Fe^{2+}-Fur dimer, we would suggest that iron homeostasis in *E. coli* is essentially regulated at the level of the siderophore-based and other iron uptake systems. The role of bacterioferritins of the BFR and FTN type may therefore not be to function as iron-storage proteins but rather as an electron-storage molecule [12], as has been suggested for BFR—the heme-containing BFR was after all first discovered as a contaminant of the electron-consuming nitrogenase—or, for the nonheme-containing FTN in *E. coli*, as a protection against oxidative stress. Recent unpublished studies that gene knockout of *Bfr* in *Fur* mutants renders the cells much more sensitive to oxygen-derived free radicals (D. Touati, personal communication) support this idea.

3. IRON HOMEOSTASIS IN PLANTS AND YEAST

Mineral nutrition is extremely important in plants for both growth and development, and iron is no exception. Plants must control the uptake of iron by the roots, transport the iron from the roots to the rest of the plant, and somehow communicate to the roots the iron status in the leaves. The mechanism of iron uptake by plant roots [13–15] involves ferric reductases, which use NAD[P]H as electron donor, coupled to proton extrusion, and a specific Fe(II) transport system localized in the root plasma membrane. In grasses, there is also iron-inducible synthesis of phytosiderophores from methionine [14]. Transport of iron in the xylem sap (which delivers water and solutes to the leaves) appears to involve ferric iron chelated to organic acids, notably citrate. Iron uptake from sap into the leaves again involves reduction. In plants, iron can be stored in vacuoles or in the iron-storage protein, ferritin. However, plant ferritin is found in the chloroplasts, whereas in animal cells, it is present in the cytosol, and its synthesis is developmentally regulated. In noniron stress conditions, it is virtually undetectable in the plastids of roots and leaves, whereas at the time of fecondation, phytoferritin accumulates in flowers and developing seeds. Iron metabolism in plants has been dealt with in Chapters 5, 6, and 14 of this volume [13–15], and some discussion of reductive pathways for iron uptake in yeast can be found in Chapter 4 [16]. As has been stressed in an earlier review of iron metabolism in plants [17], on account of the great differences in environmental iron bioavailability, plants manifest physiological disorders of both iron deficiency (chlorosis) and iron excess. Iron availability in the environment can induce stress-dependent signals which trigger the control of the expression of genes involved in the control of iron uptake by roots, long-distance iron transport between roots and shoots, in sensing the iron status in the leaves and signaling it to the roots, in intracellular iron distribution, and in detoxifying and buffering iron in excess. However, most of these signals are as yet not well established. Following the resupply of iron to iron-deficient plants, transcripts of the chlorophyll a/b binding protein and the small and large subunits of ribulose bis-phosphate carboxylase/oxygenase were found to be about fivefold more abundant [18]. A transitory activation of phytoferritin synthesis in plastids, albeit of limited duration, is also observed, underlining the role of phytoferritin as a cellular iron buffer, which can supply iron for the synthesis of iron-containing proteins required for photosynthesis, as was suggested by Bienfait and van der Mark [19]. Phytoferritin mRNA accumulation in roots and leaves is mediated by two pathways. In maize, one is dependent on the plant hormone abscisic acid, and the other on a second maize ferritin gene (designated ZmFer1). The expression of the latter gene seems to be rapidly induced in response to iron supply in iron-deficient plants [20], whereas the activation of the former seems to be the result of a general stress response involving abscisic acid [21]. Induction of phytoferritin synthesis also seems likely to be a characteristic response in the protection of plastids from oxidative stress [22]. This may be provoked by the potential for Fenton chemistry resultant on an important influx of iron.

The common yeast, *Saccharomyces cerevisiae*, on account of its short generation time and the ease with which it can be grown (not to forget that its complete genome has now been completely sequenced), is an extremely attractive eukaryotic model system. Some years ago, we showed that iron uptake in yeast requires (1) reduction of Fe(III) by a NADPH-dependent enzyme; (2) transport of Fe(II) into the cell, which can be inhibited by other divalent cations; and (3) storage of iron, not in ferritin but in polyphosphate-rich vacuoles, from which the iron can be mobilized for the synthesis of mitochondrial heme [23,24]. More recent studies using the powerful modern techniques of molecular genetics by the groups of Klausner and Eide have confirmed these proposals and led to the model [25] presented in Fig. 3. Extracellular iron is reduced by plasma-membrane ferric reductases, encoded by the genes *FRE1* and *FRE2*. Iron enters the cell via a recently characterized high-affinity iron transporter, *FTR1* [26]. It has been found that iron uptake also requires the involvement of *CTR1* and of *FET3*, a copper-containing oxidase, which has sequence homologies with the family of blue copper oxidases, including ceruloplasmin. Ceruloplasmin seems to play an important role in human iron homeostasis (see Sec. 6). It is postulated that such multicopper oxidases are required for iron transport across membranes, although their precise role is unclear.

In the course of selecting mutants of *Saccharomyces cerevisiae* with abnormalities of iron metabolism, a gene, designated *AFT1*, has been identified [27] which is involved in the control of iron uptake. The gene encodes a 78-kDa protein, which has a highly basic amino-terminal region and a C-terminal domain rich in glutamine, which is similar to many other transcriptional activators. The N-terminal and C-terminal regions contain 10% His residues. A dominant mutant allele of *AFT1* expresses high levels of ferric reductase and Fe^{2+} uptake which are not repressed by exogenous iron, associated with the failure of iron to

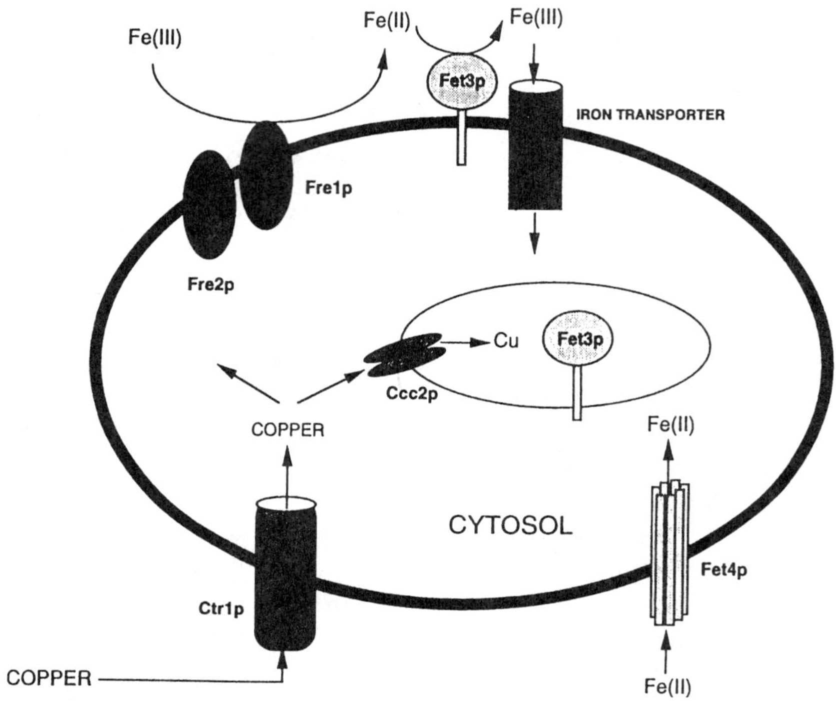

FIG. 3. Relationship between iron and copper uptake in *Saccharomyces cerevisiae*. (Reproduced with permission from [72].)

repress transcription of *FRE1*, *FRE2* and *FET3*. In contrast, a strain with interruption of the *AFT1* gene shows low ferric reductase activity and low Fe^{2+} uptake, and is susceptible to iron deprivation on account of deficient expression of *FRE1* and negligible expression of *FRE2* and *FET3*. Thus *AFT1* seems to function as a transcriptional regulator of iron uptake systems in yeast, but in a manner which is different to that exerted by Fur in bacterial systems.

The role of systems related to those found in mammalian systems (see Sec. 4) in the regulation of iron homeostasis in plants and in yeast is at present negative. There is no evidence of the presence of IREs in the 5′UTR of plant ferritin messenger RNAs [20], nor of IRP activity as measured by gel band shift assay with mammalian IRE in either plant or yeast extracts [28]. Although in *Arabidopsis thaliana*, an aconitase gene has been found which shows significant structural homology with mammalian IRP [29], no evidence has so far been found for IRP activity. So the IRP/IRE system, which functions in mammalian cells does not seem to be present in either plants or yeast. Nor for that matter does the Fur system of transcriptional regulation found in microbial systems appear to be present in plants and yeast, although the *AFT1* gene product does seem to operate globally in a somewhat similar way to Fur with the difference that, although it functions at the level of transcription, its effect is to switch on transcription of iron uptake systems in contrast to Fe^{2+}-Fur, which acts as a repressor of iron uptake.

4. CELLULAR IRON HOMEOSTASIS IN HUMANS

4.1. Generalities

We now consider how iron homeostasis in mammalian cell is regulated by balancing uptake with intracellular iron storage and utilization. As has been pointed out in Chapter 10 of this volume [30] and in recent reviews [31,32], this is largely at the level of protein synthesis (translation) rather than as in plants and microbes at the level of transcription (mRNA synthesis). Highly conserved stem-loop sequences designated iron-responsive elements (IREs) are present at the 5′-untranslated region of ferritin and erythroid 5-aminolevulinate synthase (ALA) mRNAs and at the 3′-untranslated region of transferrin receptor mRNAs (the latter in five copies). Two IRE-binding proteins (designated iron-regulatory protein, IRP1 and IRP2) act as iron sensors within the cytosol. They exist in two conformations: when iron is in short supply, the apo-IRPs can bind the IREs with high affinity; when iron levels within the cell are high, IRP-1 incorporates a cubane iron-sulfur cluster present which prevents IRE binding and confers cytoplasmic aconitase activity. It is unresolved whether IRP-2 coordinates an Fe-S cluster. At low iron concentrations, IRPs bind to the IREs in the 3′-untranslated region of transferrin receptor mRNA and protect it against nuclease digestion, thereby maintaining synthesis of transferrin receptors, and cellular iron uptake. IRP binding to the IREs in the 5′-untranslated regions of ferritin and ALA mRNAs blocks the formation of an initiation complex with the small ribosomal subunit and prevents their translation, thus decreasing intracellular iron storage and utilization. At high iron concentrations, lack of binding to the IREs allows transferrin receptor mRNA to be degraded, resulting in low iron uptake, whereas the mRNAs for ferritin and ALA are translated leading to enhanced iron storage and utilization. The amino acid sequence of IRP-1 shows extensive homology to mitochondrial aconitase, an enzyme with a 4Fe-4S cluster involved in the conversion of citrate to isocitrate [33], including three Cys residues which are thought to anchor the 4Fe-4S cluster in the mitochondrial enzyme. On the basis of the structure of mitochondrial aconitase, it is proposed that apo-IRP-1 (without its Fe-S cluster) is able to bind RNA (Fig. 4) with high affinity, whereas the holo-IRP-1 (with the 4Fe-4S cluster) is unable to bind RNA but has aconitase activity. Mutagenesis of the three Cys residues thought to anchor the iron-sulfur cluster abolishes aconitase activity and leads to constitutive RNA binding [33]. The decrease in IRP-1 RNA-binding activity in iron-repleted cells occurs by a switch between apoprotein and the 4Fe-4S forms without changes in IRP-1 levels. However, the decrease in IRP-2 RNA-binding activity in iron repletion reflects a reduction in IRP-2 levels [34]. Pulse-chase experiments demonstrate that iron results in a threefold increase in the degradation rate of IRP-2, which is prevented by inhibitors that block proteasome function but not by inhibitors of lysosomal proteases or calpain. It seems, therefore, highly probable that the proteasome (and also ubiquinylation) is involved in iron-mediated IRP-2 proteolysis [34].

The molecular mechanism of cellular iron homeostasis, namely, that in iron penury IRPs respond to intracellular iron levels to down-regulate iron storage and utilization and to favor iron uptake, whereas when iron is available, intracellular pathways of iron storage and utilization are upregulated while the uptake of iron is blocked, has been well established in vitro, using for the greater part permanent cell lines, which have the characteristic of all tumor cells of expressing very large numbers of transferrin receptors. The IRP-based model of cellular homeostasis has also been shown to be operational in animal models of both iron deficiency and iron overload [35], particularly in the liver when combined with altered transferrin saturation and serum iron levels. In both the spleen and the heart, the affinities of iron-regulatory factor changed in parallel with both altered iron status and hematological markers. However, in the brain and duodenum, there was no consistent changes with iron loading or depletion, possibly suggesting that

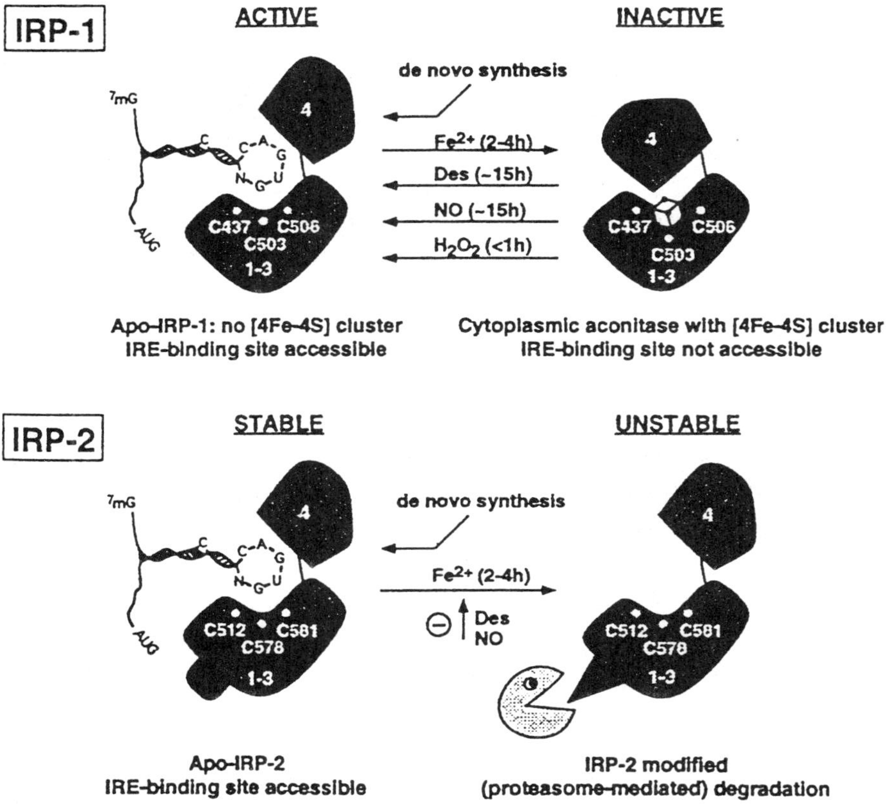

FIG. 4. Regulation of IRP-1 and IRP-2. The two IRPs are shown as homologous proteins which bind IREs. In iron-replete cells, IRP-1 assembles a cubane Fe-S cluster and is stable and active as a cytoplasmic aconitase, whereas IRP-2 is targeted for proteasome-mediated degradation. (Reproduced with permission from [31].)

such a regulatory system for iron homeostasis within these tissues may not exist.

4.2. Iron Homeostasis in Various Tissues

4.2.1. Brain

The specific functions for which iron is required in the brain are not well delineated. Neurons and glia all require iron for many aspects of their cell physiology, including the cytochrome oxidase system, NADPH reductase activity, and as a cofactor for a number of enzymes, including ribonuclecleotide reductase, an essential enzyme for DNA synthesis. Iron is also involved in the function and synthesis of various neurotransmitters, such as dopamine, serotonin, and GABA.

The blood-brain barrier presents a formidable obstacle for iron to traverse. The uptake of iron into the brain appears to be by a two-stage process involving initial deposition of iron in the blood capillary endothelium by serum transferrin (Tf) and the subsequent transfer of iron to brain-derived transferrin and then transfer within the brain to sites with a high transferrin receptor density [36]. Transferrin is synthesized within the brain by oligodendrocytes, transferrin mRNA occurring in both gray and white matter of oligodendrocytes in the brain, as well as by the choroid plexus. Exactly what the stimulus for transferrin synthesis within the brain is remains undefined, as well as the process by which iron exchanges from the serum transferrin to the glycoprotein synthesized by the brain. The level of serum transferrin saturation does not appear to be a determining factor, since genetic hemochromatosis patients who have a high serum transferrin saturation do not normally exhibit either increased brain iron content or neurological symptoms, although one exceptional case of hereditary hemochromatosis with iron accumulation in basal ganglia associated with a parkinsonian syndrome has been recently reported [37]. Neuronal and glia cells receive their iron via a transferrin-mediated process. The distribution and morphological appearance of Tf-containing cells in the gray matter are similar to those reported for iron and ferritin in studies of normal brain postmortem.

The mechanism by which the brain maintains iron homeostasis at either the whole organ or cellular level is unknown. Connor et al. [38] showed that subsets of neurons in subhuman primates immunolabel specifically for the H-ferritin protein, whereas cells resembling microglia showed immunolabelling with L-chain ferritin antibodies. Only one cell type immunostains for both H- and L-ferritin; these are cells which are morphologically similar and have the same distribution patterns as oligodendrocytes. Clearly different cells play an important role in maintaining general brain cell homeostasis.

Hemopexin is present in the brain with neurons in all regions of the brain showing immunostaining. However, oligodendrocytes and choroid plexus epithelial cells lack hemopexin immunoreactivity; such that it has been suggested that this protein may play an important role in heme transport for neuronal iron homeostasis [39].

Studies of postmortem brains have indicated that there is disruption of iron homeostasis in the aging brain, with alterations in the iron-regulatory proteins, with transferrin being consistently reduced in the white matter of various cerebral cortical regions, whereas there are no significant alterations in either iron or ferritin [40]. In Alzheimer's disease, transferrin is consistently decreased, particularly in the white matter of various cerebral cortical regions, suggesting that alterations in iron regulatory proteins could exacerbate the disease [40]. From these results, it was suggested that the diminution in transferrin levels could lead to decreases in both the mobilization and utilization of iron in the brain causing degeneration and increased peroxidation of brain lipids.

It may well be that neurotransmitters themselves can modulate brain iron homeostasis. In one study [41], lesions of the neostriatum/globus pallidus, induced by unilateral ibotenic acid infusions, showed progressive decrease in both the volume and iron concentration in the substantia nigra, indicating that GABAergic transmission/metabolism may exert an effect on iron homeostasis.

4.2.2. Gastrointestinal Tract

Despite exhaustive investigations, the manner in which iron is regulated by means of its uptake, transfer, and release across the enterocyte of the duodenum remains unresolved. Ferricyanide reductase activity has been reported on the mucosal surface of the mouse duodenum [42]; it may be an important enzyme controlling iron uptake from the duodenum, increasing its activity when additional amounts of iron are required by the tissues, such as hypoxia and iron deficiency. Furthermore, recent studies [43] have shown that the rates of iron reduction and uptake by duodenal mucosal biopsy specimens from patients with genetic hemochromatosis (GH) were significantly increased by comparison to control biopsy samples indicating the important role played by this enzyme in controlling iron homeostasis within the body. Earlier studies have suggested that gut mucosal iron concentration regulates the uptake and transfer of iron in the intestine [44]. Within the enterocytes of the intestinal mucosa, it is unknown whether transferrin actually exists, there being little evidence that transferrin is synthesized in intestinal cells. Pietrangelo et al. [45] were unable to detect transferrin mRNA in human gastric and duodenal biopsies, confirming earlier findings in the rat. Transferrin receptors have been located to the basolateral membrane of the enterocyte, while their role in the transport of iron either into or away from the enterocyte into the plasma remains undefined.

Several groups have examined ferritin protein and mRNAs in the gut, first in animals with iron overload and subsequently in human subjects with hemochromatosis. Studies of human duodenal specimens from normal and untreated hemochromatosis showed no significant differences in the ratio of H- and L-ferritin. Higher levels of transferrin receptor (TfR) mRNA were detected in the duodenums of subjects with idiopathic hemochromatosis, whereas both H- and L-ferritin mRNA were significantly lower in these patients. Recently, Pietrangelo et al. [46] have studied the regulation of ferritin synthesis in the duodenum of normal, anemic, and siderotic subjects. In patients with GH, ferritin gene transcription is preserved in intestinal cells. The abnormally low level of ferritin transcripts is not caused by a defective control of ferritin synthesis but by low expression of ferritin mRNA and sustained activity of duodenal IRP. Thus, both TfR mRNA and ferritin mRNA are regulated in concert in hemochromatosis patients.

4.2.3. Liver

The liver has considerable capacity for storing excess iron in both the hepatocytes and reticuloendothelial cells. Although many studies have examined the regulation of transferrin, transferrin receptor and ferritin subunit gene expression in experimental systems there have been few studies in human subjects. In one study of liver samples from normal subjects and from patients with genetic hemochromatosis and secondary hemochromatosis (SH) patients, it was clear that transferrin mRNA was not depressed by iron overload, but it was actually increased in patients with SH [47], indicating that iron maintains the ability to induce transferrin gene expression even when the cellular iron content is significantly increased. A lower signal for transferrin receptor mRNA was consistently detected in iron-overloaded patients in both GH and

SH versus control subjects. Ferritin, L-type mRNA was significantly increased in patients with severe siderosis, whereas no significant changes were found in hepatic ferritin H gene expression, confirming preferential production of L-subunit-enriched ferritins in long-term iron overload.

It is interesting to note that after partial hepatectomy, the expression of transferrin receptors is induced independent of total liver iron concentration [48], and it was concluded that IRP mediated this post-transcriptional control of the transferrin receptor mRNA and could be caused either by a drop of iron levels in the regulatory pool or by an iron-independent mechanism [48].

4.3. Relationship Between Reactive Oxygen Intermediates, Hydrogen Peroxide, Nitric Oxide, and Iron Homeostasis

The generation of reactive oxygen species (ROS) is an inevitable by-product of aerobic metabolism and their concentration will be maintained at a certain threshold level by the vast armory of cytoprotective enzymes and antioxidants present within the cell [49]. ROS will be required both as mediators of the antimicrobial respiratory burst of phagocytic cells, such as macrophages, as well as at lower levels reactive oxygen intermediates (ROIs) [50,51], for example, H_2O_2, to act as intracellular signaling molecules to activate transcription factors such as NFκB and APT-1 [52]. The toxicity of such ROIs is exacerbated in the presence of ferrous iron, which will react with H_2O_2 to yield the hydroxyl radical, $OH^{\bullet}$.

Exposure of cell lines to H_2O_2 leads to a reduction in the synthesis of the intracellular iron-storage protein ferritin and stimulation of transferrin receptor mRNA expression [53]. Both responses are post-transcriptional and result from the induction of iron-regulatory protein binding to iron-responsive elements in ferritin and TfR mRNAs. The induction of IRP by H_2O_2 involves the disassembly of the 4Fe-4S cluster, with inactivation of aconitase activity, and the rapid stimulation of IRE binding leading to increased TfR mRNA levels and repressed ferritin synthesis. In addition to altering iron homeostasis, H_2O_2 can also activate transcriptional factors within the cytoplasm, possibly through the formation of the hydroxyl radical, with their translocation to the nucleus to enhance expression of certain inflammatory responses such as cytokines [54] through the transcription of the NFκB-binding site. This has been shown to be inhibited when the iron chelator desferrioxamine is added to the incubation media [52], indicating an important role for iron in the exacerbation of the inflammatory response. Overall, it is apparent that the immune system, through the production of cytokines, plays a regulatory role in iron homeostasis, inducing the anemia of chronic disease [55], whereby there is increased iron retention in tissues and increased ferritin synthesis. The regulation of TfR expression by inflammatory cytokines is more controversial and no data are available about its regulation in vivo. Studies of experimental inflammation indicated that IRP is activated by an iron-independent pathway, and they furthermore suggest that nitric oxide (NO) may be the trigger candidate [56]. Hepatic NO synthesis is stimulated by inflammatory substances in vivo and by specific cytokines in cultured cells. Cytokine tumor necrosis factor α specifically induces synthesis of ferritin H-subunit [57] which involves a selective increase in ferritin H transcription mediated by transcription factors of the NFκB family. The physiological explanation for this increase in ferritin H-chain transcription is unclear.

As well as being an important mediator of the antimicrobial respiratory burst of macrophages, NO will also activate IRE binding by IRP-1 and IRP-2 accompanied by a loss of aconitase activity of IRP-1. Stimulation of macrophages with interferon and lipolysaccharides induces the inducible form of nitric oxide synthase, with the release of NO. Our recent studies of in vivo iron-loaded macrophages has shown that there is decreased release of NO from such cells after stimulation with γ-interferon (IFN) and lipopolysaccarides [58] which would minimize the activation of IRE binding and uptake of further iron, although possibly compromise its antimicrobial activity.

5. IRON TRANSPORT, STORAGE, AND HOMEOSTASIS IN HUMANS

To achieve global homeostasis within the whole organism, iron absorption must balance iron excretion. In humans, iron metabolism is essentially controlled at the level of dietary iron absorption [59]. There is no physiological mechanism in humans for excretion of excess iron: In this regard, iron excretion in humans is very limited when compared with other mammals [60]. Loss of iron, mostly by sloughing of cells from the skin, the gastrointestinal, and the genitourinary tracts, is limited to about 1 mg/day in the adult male and about twice this amount in premenopausal women. Normal iron balance is maintained by regulating iron absorption which takes place essentially in the duodenum and the upper jejunum. The most important factor in the physiological regulation of mucosal iron absorption is *total body iron content*. Although the principal control of intestinal iron absorption lies within the intestinal mucosal cell itself, other elements, such as the extent of erythropoeisis, and perhaps the presence of circulating humoral factors must be considered [61]. Hypoxia increases iron absorption, independent of changes in the rate of erythropoeisis; the effects of hypoxia are observed before a response from the erythroid bone marrow can be observed. Iron absorption is also high during the neonatal period, although the mechanism of high neonatal absorption is unknown.

In genetic hemochromatosis, which affects some 1 in 400 individuals of Northern European descent, iron absorption is assumed to be set at a slightly higher level than normal, such that over a period of decades, the body accumulates excess iron, which finally results in toxic levels of tissue iron. The recent discovery of a novel major histocompatibility class I-like gene, as one at least of the candidate genes for genetic hemochromatosis [62], (see Sec. 6) may help us to understand how this increased iron absorption takes place. Recently [63], a similar, but apparently distinctive, genetic condition also characterized by increased dietary iron absorption has been reported in a central African population.

As to the factors that intervene to regulate dietary iron uptake from the gastrointestinal tract, several have been proposed. Transferrin saturation clearly is not the determining factor, since in genetic hemochromatosis saturation levels are high and yet "excess" iron is still absorbed, nor is the level of serum ferritin. The levels of transferrin receptors, or fragments thereof, circulating in the plasma have been proposed to reflect the rate of erythropoeisis, and by implication of iron absorption [64], but more recent investigations indicate that there is no correlation between circulating levels of transferrin receptors in serum and iron absorption [65] nor between transferrin receptors in the intestinal mucosa and the absorption of iron [66,67].

In humans [1], iron is absorbed from the gastrointestinal tract, mostly in the duodenum. Iron absorption can be divided into three distinct phases (Fig. 5). First, the delivery of iron to the brush border of the intestinal cell is determined by the type of food ingested and by its interactions with gastrointestinal secretions. Heme iron is better absorbed than nonheme iron and ferrous iron better than ferric, suggesting but not proving, the involvement of a reductive step. The presence of strong iron-complexing agents, such as phytates and phosphates, will decrease the absorption of iron from the gastrointestinal tract. The

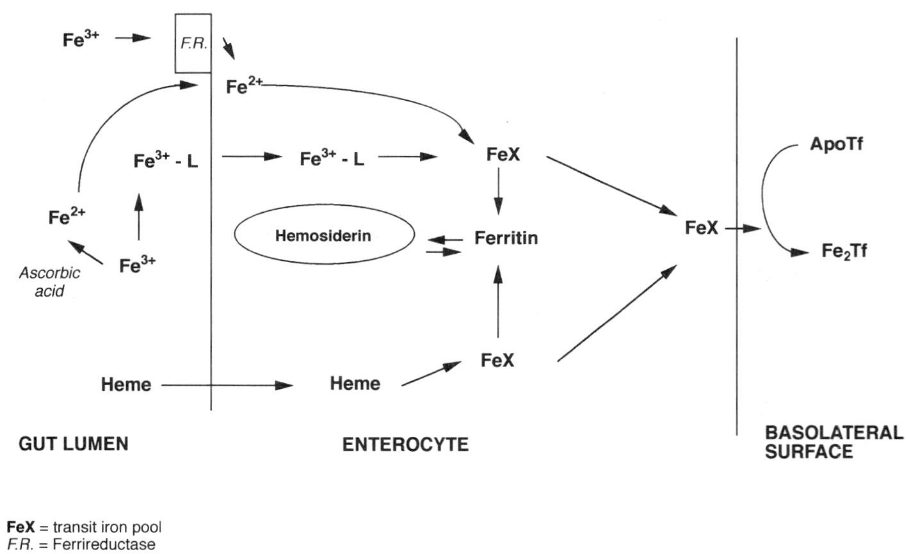

FIG. 5. Iron uptake from the gastrointestinal tract can be divided into three distinct phases: (1) uptake from the lumen of the gut across the brush-border membrane of the mucosal epithelial cell, involving both heme-bound (probably via a receptor) and nonheme iron; the latter can be absorbed as ferrous or ferric, chelated or not. (2) Transport within the enterocyte from the brush-border membrane to the basolateral membrane. (3) Finally, either its transfer out of the mucosal cell and its release to plasma apotransferrin, or its sequestration within the mucosal cell, and its eventual loss back to the gastrointestinal tract, when the cell exfoliates. (Reproduced with permission from [1].)

second phase in iron absorption, the mucosal phase, in which the iron which has been absorbed into the mucosal epithelial cells is processed; some will be transferred to the circulation and some will be incorporated into mucosal cell ferritin, which will be lost when the mucosal epithelial cell is sloughed off into the gastrointestinal tract at the end of its relatively short life span. Finally, there is the systemic phase, where the iron assimilated by the mucosal cell is transferred to the circulation, through mechanisms as yet uncharted, but which, in the normal subject, take into account the global iron requirements of the whole organism.

Many years ago, we showed that anemic rats have a transferrin-like protein in their gut [68]. Recently, a similar protein has been described linked to a GPI (glycosyl-phosphatidylinositol) anchor on the apical surface of fetal porcine intestinal epithelial cells [69]; once again raising the possibility that a mucosal transferrin is involved in iron absorption. The possible role of a reductive mechanism in iron absorption has been mentioned earlier: Mouse duodenum seems to possess mucosal surface transplasma membrane ferricyanide reductase activity which correlates with intestinal iron absorption [42]. Rates of iron reduction and uptake were significantly increased in biopsy samples of both untreated GH patients and in GH patients in whom iron stores were normal after venesection treatment compared with controls, suggesting that there is a persistent abnormality in both reduction and uptake of iron by the intestine in GH [43].

5.1. Iron Transport and Uptake Proteins

Whatever their involvement in iron absorption, transferrins are a class of iron-binding proteins found in the physiological fluids of many vertebrates (reviewed in [70]). Serotransferrrin, found in serum and in other extracellular secretions, transports iron between cells. The structure of transferrin consists of two identical lobes each of which tightly binds one atom of ferric iron. The iron-binding sites are buried at the inner end of a deep interdomain cleft within each of the two lobes of the molecule. In the absence of any bound metal, one of the two iron-binding lobes of human lactotransferrin lies wide open, with its molecular jaws poised to close firmly around an incoming iron cation—transferrin can indeed be likened to a Venus fly trap. In view of the involvement of periplasmic binding proteins in the transport of ferric siderophores between the outer and the cytoplasmic membranes in bacteria, it comes as no surprise that bacterial periplasmic binding proteins have the same Venus fly trap structure as transferrin.

From very early studies with double-labeled transferrin, it became clear that whereas the iron was rapidly cleared from the circulation, the protein recycles many times. Many mammalian cells have transferrin receptors located on their plasma membranes. These are disulfide-linked dimers, each of molecular weight 90–95 kDa. The receptor is anchored in the plasma membrane by a transmembrane helix with a short intracellular N-terminal domain and a large extracellular diferric-transferrin-binding domain.

The transferrin to cell cycle [70] has been well established in erythroid cells and in many transformed cell lines. After binding to its receptor, the transferrin-receptor complex is internalized within an endosomal vesicle in which an ATP-dependent proton pump establishes a pH of 5.0–5.5. Iron release from diferri-transferrin at mildly acidic pH values is substantially increased when it is bound to its receptor compared with release from free diferri-transferrin: thus within the endosome, the transferrin receptor facilitates iron release from transferrin. The apotransferrin has a high affinity for its receptor, and it is then recycled back to the plasma membrane, where at the slightly alkaline extracellular pH, it dissociates from the receptor and is released into the circulation in search of iron.

Where and from what source circulating apotransferrin acquires its iron remains yet another of the enigmas of mammalian iron metabolism. This has been dealt with elsewhere [70], and although we know that apotransferrin can readily take up Fe^{3+} in vitro, the source of the iron in the circulation remains unknown. One possibility is that iron bound to ferritin which has been released from reticuloendothelial cells might be a source of transferrin iron, and once again from in vitro studies, it has been shown that ferritin can transfer its iron to apotransferrin with the help of a number of low molecular weight ligands present in plasma [71].

It also seems clear that mammalian cells can obtain iron by mechanisms which do not involve the transferrin-transferrin receptor pathway—such transferrin-independent systems have been recently reviewed [72]. In two genetic defects, congenital hypotransferrinemia and genetic hemochromatosis, nontransferrin iron is rapidly cleared from plasma and accumulated by a variety of parenchymal tissues. The principal characteristics of the transferrin independent iron transport system are the following: (1) the pathway requires a cell-surface ferri-reductase. (2) It does not require endocytosis, as indicated by the observation that agents which inhibit endocytosis have no effect on the transferrin-independent pathway. (3) It is dependent on extracellular calcium. (4) It is inhibited by other transition metals, unlike the transferrin-dependent route. (5) It is principally regulated by iron overload, increasing the activity of the uptake of iron, and unlike the transferrin-dependent pathway, it does not respond to iron deprivation. One additional source of nontransferrin iron could be circulating ferritin, derived in part from iron-replete macrophages [70]. Since some parenchymal cells have plasma membrane receptors for ferritin, subsequent uptake of ferritin by such cells, hepatocytes, for example, could provide a pathway for the exchange of iron from the reticuloendothelial to the parenchymal system. This could be a probable cause of parenchymal cell damage in secondary iron-loading disorders, like the thalassemias.

5.2. Iron-Storage Proteins

Intracellular iron storage involves two proteins, ferritin, and hemosiderin. Hemosiderin is a water-insoluble protein complex characterized by a higher iron to protein ratio than ferritin, and it is typically found in lysosomes, where it is thought to be formed from ferritin by aggregation and proteolytic degradation. For more information see [73].

Ferritin is a cytosolic protein consisting of a soluble, hollow protein shell of external diameter 12–13 nm which encloses an internal cavity of some 7–8 nm diameter within which a core of variable iron content is deposited. Apoferritin, the iron-free protein, is made up of 24 subunits of two types, H and L [74]. The central iron-rich core of mammalian ferritins consists of a ferrihydrite biomineral phase, containing up to a maximum of 4500 atoms of ferric oxyhydroxide together with additional variable amounts of phosphate.

Access to the interior of the molecule for iron is probably via the eight three-fold channels [75] which are extremely polar, with three Glu residues situated toward the wider end of the channel on the outer surface of the protein shell and three Asp residues toward the narrower inner surface. Iron deposition in ferritin [75] involves the protein itself, at least in the early stages. Initially Fe^{2+} is oxidized by the protein to Fe^{3+}, which then migrates to nucleation sites where the biomineralization to ferrihydrite proceeds. The H-subunits have a much greater

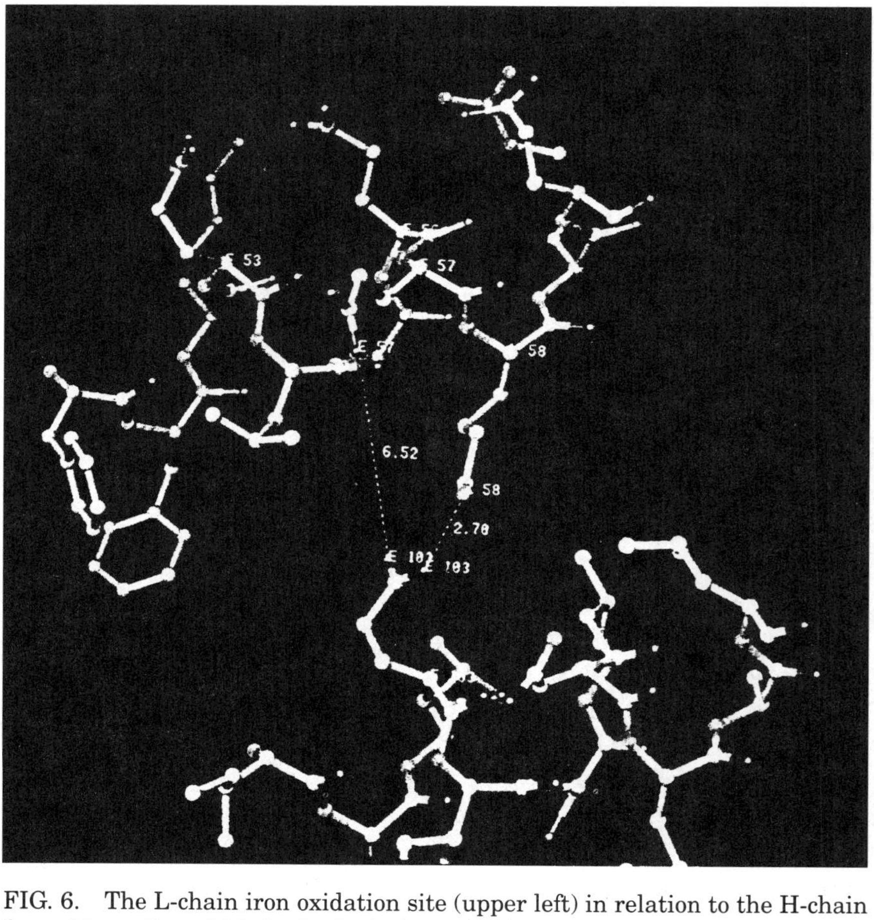

FIG. 6. The L-chain iron oxidation site (upper left) in relation to the H-chain ferroxidase site, which in the L-chain apoferritin is blocked by the salt-bridge between Glu-103 and Lys-58 (which replaces Glu-58 in H). (Reproduced with permission from [76].)

ferroxidase activity than the L-subunits, although the latter predominate in most tissues where large amounts of iron are stored. The putative ferroxidase site of *E. coli* ferritin has been identified directly by x-ray crystallography and is a dinuclear iron site in which the two iron atoms are linked by a bridging carboxyl group and perhaps also by a μ-oxo bridge [74]. This geometry is reminiscent of a number of other proteins which have dinuclear μ-oxo-bridged iron centers, often with a carboxyl group bridging the two iron atoms—they include hemerythrin, methane monooxygenase, and ribonucleotide reductase.

In mammalian L-subunits, the ferroxidase site of the H-subunit is blocked by a salt-bridge between Glu-103 and Lys-58 (Fig. 6). In recent studies, we have identified the carboxyl groups which constitute the iron oxidation and/or nucleation site in horse spleen apoferritin (containing more than 90% of L chains) and in recombinant horse L-chain apoferritin. The cluster of three Glu residues (53, 56, and 57) in the upper part of the figure, pointing into the interior of the protein shell constitutes this site [76].

An important difference between ferritins from bacteria and ferritins from plants and animals is that bacterial ferritins have heme bound in the interior of the protein shell at an interface between subunit dimers. Two Met residues form the fifth and sixth ligands bound to the heme iron [74]. However, when heme is diffused into crystals of horse spleen apoferritin, or to recombinant horse L-apoferritin, the porphyrin binds in roughly the same interdimer site, but the metal ion has been removed [77,78]. A cluster of carboxylates is situated at the entrance to the pocket in which the protoporphyrin IX molecule is bound [78], and this turns out to be the same Glu-53, Glu-56, and Glu-57 which had been previously identified as the site of iron incorporation into L-chain apoferritins [76].

6. CONCLUSIONS AND UNRESOLVED PROBLEMS

We can now draw together the main conclusions concerning iron homeostasis, and, since we have seen that there are important differences between prokaryotes, lower eukryotes such as yeasts, plants, and mammals, we will deal with each in turn.

Clearly, our understanding of bacterial iron homeostasis draws largely on the Fe^{2+} Fur system in *E. coli*. But as we have pointed out, not only are there many more examples of Fur or Fur-related systems in other bacteria, considerable uncertainty still remains over the real role of bacterioferritins—we really need more extensive studies in other organisms. And there is also the interesting question of just how many other enzymes with key roles in important metabolic pathways, are also under control by iron [3].

When we consider the situation in yeast and in higher plants, there are essentially two conclusions. First, once again, iron seems to play a key role in the regulation of proteins not directly involved in iron metabolism. Second, there may be some evidence for transcriptional regulation of iron homeostasis. However, at present, there is no evidence for the finely tuned translational regulation represented by the IRE/IRP system in either plants or yeast, leaving us with the rather unsatisfactory conclusion that yeast may not be a very good model for understanding iron metabolism and homeostasis in mammals, particularly in humans. The link between iron and copper uptake in yeast has helped to draw our attention once again to the important interaction between copper and iron [26]. This has been recently underlined by the observation that a congenital defect in a Japanese family, in which the C-terminal 60 residues of the serum copper-bearing protein ceruloplasmin have been deleted, results in the clinical observation of hepatic and brain iron loading [79].

The role played by T cells in the regulation of iron metabolism is another unresolved factor. Studies by Maria de Sousa and her colleagues has identified a relationship between high CD4-CD8 ratios, the two main T-cell subsets, and the degree of iron overload in hemochromatosis patients [80,81]. Moreover, the high CD4-CD8 ratios seemed to precede the development of severe iron overload, indicating that such ratios can be related to the disregulation of iron absorption and tissue iron accumulation. The investigators speculated that in hemochromatosis patients the defect in the major histocompatibility complex class I product could result in failure to expand the $CD8^+$ T cells and consequently or concomitantly fail to switch off iron absorption appropriately [82].

We then come to the problems of cellular iron homeostasis in mammalian cells, and address the question of the universal applicability of iron homeostasis via IRPs. This has been discussed in Section 4. Our conclusions are that, however attractive the IRE/IRP regulatory system may appear, its universal applicability to all mammalian tissues and cell types remains to be established. In addition, the effects of both transcriptional regulation of iron transport and storage proteins, and the effects of intracellular iron pools in the production of reactive oxygen species, notably the hydroxyl radical, with their effects on both the activation of nuclear transcription factors and the inducible form of NO synthase, should not be underestimated. One recent observation, that is, in humans, single-point mutations in the IRP-binding stem loop of IREs in L-chain ferritin mRNA lead to constitutive synthesis of L-ferritin in members of families affected with dominantly inherited hyperferritinemia and cataract [83,84], can only underline the extent to which our increasing understanding at the molecular and cellular level of iron homeostasis still leave us perplexed when we are confronted by their clinical manifestations.

As we have pointed out earlier, there are still many uncertainties concerning iron absorption and its regulation. We might have hoped that the discovery of a gene for GH [63] would represent the beginning of the identification of the protein involved in regulating iron uptake from the gastrointestinal tract. The gene, related to the major histocompatibility antigen class I family, contained two missense alterations, one of which is predicted to inactivate this class of proteins and was found homozygous in 83% of the GH patients examined. It had been reported previously that progressive hepatic iron overload, indistinguishable from that observed in GH, was found in β_2-microglobulin-deficient mice [85], and subsequent studies confirmed that mice which have altered expression of class I gene products, the β_2-microglobulin knockout mice do indeed develop iron overload [86]. It now remains to be established exactly how β_2-microglobulin-associated proteins are involved in the control of intestinal iron absorption. As the cutting edge of our newly acquired tools of molecular biology (and we understand by that not only the unidimensional transfer of information but also the three-dimensional knowledge of protein structures and their functions) begin to make their incisions into the whole fabric of iron homeostasis, we are still left with the great holistic dilemma of how these cellular and molecular discoveries translate into the physiology of the functioning of multicellular organisms as complex as humans.

ACKNOWLEDGMENT

R. R. C. thanks the Fonds du Development Scientifique of the Université Catholique de Louvain for financial support.

ABBREVIATIONS

AFT1	a mediator of iron-regulated transcriptional control in yeast
ALA	5-aminolevulinate synthase
AMP	adenosine 5′-monophosphate
APT	activator protein 1
ATP	adenosine 5′-triphosphate
BFR	heme containing bacterioferritin
CTR	copper transport
DtxR	diphtheria toxin repressor protein
Fes	Fe enterobactin esterase
FET	ferrous transport
FRE	ferric reductase
FTN	nonheme containing bacterioferritin
FTR	Fe transporter
Fur	ferric uptake regulation protein
GABA	γ-aminobutyric acid
GH	genetic hemochromatosis
GPI	glycosylphosphatidylinositol
IFN	γ-interferon
IRE	iron-responsive element
IRP	iron-regulatory protein
mRNA	messenger ribonucleic acid
NAD	nicotinamide adenine dinucleotide
NADPH	nicotinamide adenine dinucleotide phosphate (reduced)
NFκB	nuclear factor κB
ROI	reactive oxygen intermediate
ROS	reactive oxygen species
SH	secondary hemochromatosis
Tf	transferrin
TfR	transferrin receptor

5′UTR	5′-untranslated region
ZmFer	maize ferritin

REFERENCES

1. R. R. Crichton, in *Inorganic Biochemistry of Iron Metabolism*, Ellis Horwood, Chichester, UK, 1991.
2. A. L. Crumbliss and A.-M. Albrecht-Gary, this volume, Chap. 7.
3. V. Braun, K. Hantke and W. Köster, this volume, Chap. 3.
4. B. R. Byers and J. E. L. Arcenaux, this volume, Chap. 2.
5. X. Tao, H.-Y. Zeng, and J. R. Murphy, *Proc. Natl. Acad. Sci. USA*, *92*, 6803–6807 (1995).
6. N. Schiering, X. Tao, H. Zeng, J. R. Murphy, G. A. Petsko, and D. Ringe, *Proc. Natl. Acad. Sci. USA*, *92*, 9843–9850 (1995).
7. K. Gunter, C. Toupet, and T. Schupp, *J. Bacteriol.*, *175*, 3295–3302 (1993).
8. M. P. Schmitt, M. Predich, L. Doukhan, I. Smith, and R. K. Holmes, *Infect. Immunol.*, *63*, 4284–4289 (1995).
9. P. M. Harrison, P. C. Hempstead, P. J. Artymiuk, and S. C. Andrews, this volume, Chap. 11.
10. A. J. Hudson, S. C. Andrews, C. Hawkins, J. M. Williams, M. Izuhara, F. C. Meldrum, S. Mann, P. M. Harrison, and J. R. Guest, *Eur. J. Biochem.*, *218*, 985–995 (1993).
11. S. C. Andrews, J. M. A. Smith, C. Hawkins, J. M. Williams, P. M. Harrison, and J. R. Guest, *Eur. J. Biochem.*, *213*, 329–339 (1993).
12. E. I. Stiefel and G. D. Watt, *Nature*, *279*, 81–83 (1979).
13. J. Fett, K. LeVier, and M. L. Guerinot, this volume, Chap. 5.
14. S. Mori, this volume, Chap. 6.
15. J. F. Briat and S. Lobréaux, this volume, Chap. 14.
16. S. A. Leong and G. Winkelmann, this volume, Chap. 4.
17. J.-F. Briat, I. Fobis-Loisy, N. Grignon, S. Lobréaux, N. Pascal, G. Savino, S. Thoiron, N. von Wirén and O. Van Wuytswinkel, *Biol. Cell*, *84*, 69–81 (1995).
18. S. C. Spiller, L. S. Kaufmann, W. F. Thomson, and W. R. Briggs, *Plant Physiol.*, *84*, 409–414 (1987).

19. H. F. Bienfait and F. van der Mark, in *Uptake and Utilisation of Iron by Plants* (D. A. Robb and W. S. Pierpont, eds.), Academic Press, New York, 1983, pp. 111–123.
20. I. Fobis-Loisy, K. Loridon, S. Lobréaux, and J.-F. Briat, *Eur. J. Biochem.*, *231*, 609–619 (1995).
21. S. Lobréaux, T. Hardy, and J.-F. Briat, *EMBO J.*, *12*, 651–657 (1993).
22. S. Lobréaux, S. Thoiron, and J.-F. Briat, *Plant J.*, *8*, 443–449 (1995).
23. E. Lesuisse, F. Raguzzi, and R. R. Crichton, *J. Gen. Microbiol.*, *133*, 3229–3236 (1987).
24. F. Raguzzi, E. Lesuisse, and R. R. Crichton, *FEBS Lett.*, *231*, 253–258 (1988).
25. A. Dancis, D. S. Yuan, D. Haile, C. Askwith, D. Eide, C. Moehle, J. Kaplan, and R. D. Klausner, *Cell*, *76*, 393–402 (1994).
26. R. Stearman, D. S. Yuan, Y. Yamaguchi-Iwai, R. D. Klausner, and A. Dancis, *Science*, *271*, 1552–1557 (1996).
27. Y. Yamaguchi-Iwai, A. Dancis, and R. D. Klausner, *EMBO J.*, *14*, 1231–1239 (1995).
28. S. Rothenberger, E. W. Müllner, and L. C. Kühn, *Nucleic Acids Res.*, *18*, 1175–1179 (1990).
29. P. Peyret, P. Perez, and M. Alric, *J. Biol. Chem.*, *270*, 8131–8137 (1995).
30. E. Theil, this volume, Chap. 10.
31. M. W. Hentze and L. C. Kühn, *Proc. Natl. Acad. Sci. USA*, *93*, 8175–8182 (1996).
32. E. Paraskeva and M. W. Hentze, *FEBS Lett.*, *389*, 40–43 (1996).
33. H. Hirling, B. R. Henderson, and L. C. Kühn, *EMBO J.*, *13*, 453–461 (1994).
34. B. Guo, J. D. Phillips, Y. Yu, and E. A. Liebold, *J. Biol. Chem.*, *270*, 21645–21651 (1995),
35. R. J. Ward, L. C. Kühn, P. Kaldy, A. Florence, T. J. Peters, and R. R. Crichton, *Eur. J. Biochem*, *220*, 927–931 (1994).
36. C. M. Morris, J. M. Candy, A. B. Keith, A. E. Oakley, G. A. Taylor, R. G. Pullen, C. A. Bloxham, A. Goucht, and J. A. Edwardson, *J. Inorg. Biochem.*, *47*, 257–265 (1992).

37. J. E. Nielsen, L. N. Jensen and K. Krabbe, *J. Neurol. Neurosurg. Psychiatry*, *59*, 318–321 (1995).

38. J. R. Connor, K. L. Boeshore, S. A. Benkovic and S. L. Menzies, *J. Neurosci. Res.*, *37*, 61–465 (1994).

39. C. M. Morris, J. M. Candy, J. A. Edwardson, C. A. Bloxham, and A. Smith, *Neurosci. Lett.*, *149*, 141–144 (1993).

40. J. R. Connor, B. S. Snyder, J. L. Beard, R. E. Fine, and E. J. Mufson, *J. Neurosci. Res.*, *31*, 327–335 (1992).

41. S. Sastry and G. W. Arendash, *Neuroscience*, *67*, 649–666 (1965).

42. D. J. Pountney, K. B. Raja, M. J. Bottwood, J. M. Wrigglesworth, and R. J. Simpson, *Biometals*, *9*, 15–20 (1996).

43. K. B. Raja, D. Pountney, A. Bomford, R. Przemioslo, D. Sherman, R. J. Simpson, R. Williams, and T. J. Peters, *Gut*, *38*, 765–769 (1996).

44. P. C. Adams, R. Zhong, J. Haist, P. R. Flanagan, and D. R. Grant, *Gastroenterology*, *100*, 370–374 (1991).

45. A. Pietrangelo, E. Rocchi, G. Casalgrandi, G. Rigo, A. Ferrari, M. Perini, E. Ventura, and G. Cairo, *Gastroenterology*, *102*, 802–809 (1992).

46. A. Pietrangelo, G. Casalgrandi, D. Quaglino, R. Gualdi, D. Conte, S. Milani, G. Montosi, L. Cesarini, G. Ventura, and G. Cairo, *Gastroenterology*, *108*, 208–217 (1995).

47. A. Pietrangelo, E. Rocchi, A. Ferrari, E. Ventura, and G. Cairo, *Hepatology*, *14*, 1083–1089 (1991).

48. G. Cairo and A. Pietrangelo, *J. Biol. Chem.*, *269*, 6405–6409 (1994).

49. R. J. Ward, C. Abiaka, and T. J. Peters, *J. Nephrol.*, *7*, 89–96 (1994).

50. J. Remacle, M. Raes, O. Toussaint, P. Renard, and G. Rao, *Mutation Res.*, *316*, 103–122 (1995).

51. R. Schreck and P. A. Baeuerle, *Trends Cell Biol.*, *1*, 39–42 (1991).

52. C. Sappey, J. R. Boelaert, S. Legrand-Poels, C. Forceille, A. Favier, and J. Piette, *AIDS Res. Hum. Retroviruses*, *11*, 1049–1061 (1995).

53. K. Pantopoulos and M. W. Hentze, *EMBO J.*, *14*, 2917–2924 (1995).

54. R. J. Ward, Y. Zhang, R. R. Crichton, B. Piret, J. Piette and Ph. de Witte, *FEBS Lett.*, *389*, 119–122 (1996).

55. L. L. Moldawer, M. A. Marano, H. Wai, Y. Fong, M. L. Silen, and G. Kuo, *FASEB J.*, *3*, 637–643 (1989).
56. G. Cairo and A. Pietrangelo, *Eur. J. Biochem.*, *232*, 358–363 (1995).
57. E. L. Kwak, D. A. Larochelle, C. Beaumont, S. V. Torti, and F. M. Torti, *J. Biol. Chem.*, *270*, 15285–15293 (1995).
58. Y. Zang, R. R. Crichton, J. R. Boelaert, P. G. Jorens, A. G. Herman, R. J. Ward, F. Lallemand, and Ph. de Witte, *Biochem. Pharmacol.* In press.
59. R. A. McCance and E. M. Widdowson, *Lancet*, *233*, 680–684 (1937).
60. C. A. Finch, H. A. Ragan, I. A. Dyer, and J. D. Cook, *Proc. Soc. Exp. Biol. Med.*, *159*, 335–338 (1978).
61. T. H. Bothwell, R. W. Charlton, J. D. Cook, and C. A. Finch, in *Iron Metabolism in Man*, Blackwell, Oxford, UK, 1979.
62. J. N. Feder, A. Gnirke, W. Thomas, Z. Tsuchihashi, D. A. Ruddy, A. Basava, F. Dormishian, R. Domingo, Jr., M. C. Ellis, A. Fullan, L. M. Hinton, N. L. Jones, B. E. Kimmel, G. S. Krnmal, P. Lauer, V. K. Lee, D. B. Loeb, F. A. Mapa, E. McClelland, N. C. Meyer, G. A. Mintier, N. Moeller, T. Moore, E. Morikang, C. E. Prass, L. Quintana, S. M. Starnes, R. C. Schatzman, K. J. Brunke, D. T. Drayna, N. J. Risch, B. R. Bacon, and R. K. Wolff, *Nature Genet.*, *13*, 399–408 (1996).
63. V. Gordeuk, J. Mukibi, S. J. Haastedt, W. Samowitz, C. Q. Edwards, G. West, S. Ndamira, J. Emmanuel, N. Nkanza, Z. Chapanduka, M. Randall, P. Boone, P. Romano, R. W. Martell, T. Yamashita, P. Effler, and G. Brittenham, *N. Engl. J. Med.*, *326*, 95–100 (1992).
64. H. A. Huebers, Y. Beguin, P. Pootrakul, D. Einspahr, and C. A. Finch, *Blood*, *75*, 102–107 (1990).
65. J. D. Cook, S. Dassenko, and B. S. Skikne, *Br. J. Haematol.*, *75*, 603–609 (1990).
66. M. Lombard, A. B. Bomford, R. J. Polson, A. J. Bellingham, and R. Williams, *Gastroenterology*, *98*, 976–984 (1990).
67. G. J. Anderson, L. W. Powell, and J. W. Halliday, *Gastroenterology*, *98*, 576–585 (1990).
68. H. Huebers, E. Huebers, W. Rummel, and R. R. Crichton, *Eur. J. Biochem.*, *66*, 447–455 (1976).
69. E. M. Danielsen and B. van Deurs, *J. Cell Biol.*, *131*, 939–950 (1995).

70. P. Aisen, this volume, Chap. 15.
71. Y. Jin and R. R. Crichton, *FEBS Lett.*, *217*, 41–46 (1987).
72. D. M. de Silva, C. VC. Askwith, and J. Kaplan, *Physiol. Rev.*, *76*, 31–47 (1996).
73. R. R. Crichton and R. J. Ward, *Biochemistry*, *31*, 11255–264 (1992).
74. P. M. Harrison, P. C. Hempstead, P. J. Artymiuk, and S. C. Andrews, this volume, Chap. 11.
75. N. D. Chasteen, this volume, Chap. 12.
76. R. R. Crichton, A. Herbas, O. Chavez-Alba, and F. Roland, *J. Biol. Inorg. Chem.*, *1*, 567–574 (1996).
77. G. Précigoux, J. Yariv, B. Gallois, A. Dautant, C. Courseille, and B. Langlois d'Estaintot, *Acta Cryst.*, *D50*, 739–743 (1994).
78. B. Gallois, B. Langlois d'Estaintot, M.-A. Michaux, A. Dautant, T. Granier, G. Précigoux, J. A. Soruco, F. Roland, O. Chavez-Alba, A. Herbas, and R. R. Crichton, *J. Biol. Inorg. Chem.*, *2*, 360–367 (1997).
79. K. Yoshida, K. Furihata, S. Takeda, A. Nakamura, K. Yamamoto, H. Morita, S. Hiyamuta, S. Ikeda, N. Shiizu, and N. Yanagisawa, *Nature Genet.*, *9*, 267–272 (1995).
80. R. Reimao, G. Porto, and M. De Sousa, *CR Acad. Sci. Paris*, *313*, 481–483 (1991).
81. G. Porto, R. Reimao, C. Goncalves, C. Vicente, B. Justica, and M. De Sousa, *Eur. J. Haematol.*, *52*, 283–290 (1994).
82. G. Porto, C. Vicente, M. Anjos Teixeira, O. Martins, J. Manuel Cabeda, R. Lacerda, C. Goncalves, J. Fraga, G. Macedo, B. M. Silva, H. Alves, B. Justica, and M. De Sousa, *Hepatology*, *25*, 397–402 (1997).
83. C. Beaumont, P. Leneuve, I. Devaux, Y. Y. Scoazec, M. Berthier, M. N. Loiseau, B. Grandchamp, and D. Bonneau, *Nature Genet.*, *11*, 444–446 (1995).
84. D. Girelli, R. Corrocher, L. Bisceglia, O. Olivieri, L. De-Francheschi, L. Zelante, and P. Gasparini, *Blood*, *86*, 4050–4053 (1995).
85. M. de Sousa, R. Reimao, R. Lacerda, P. Hugo, S. H. Kaufmann, and G. Porto, *Immunol. Lett.*, *39*, 105–111 (1994).
86. B. E. Rothenberg and J. R. Voland, *Proc. Natl. Acad. Sci. USA*, *93*, 1529–1534 (1996).

17

The Role of Other Metals in Iron Transport

Barry Chiswell,[1] *Kelvin O'Halloran,*[2] *and Jarrod Wall*[1]

[1]Department of Chemistry, The University of Queensland, Brisbane Qld 4072, Australia

[2]Centre for Water and Waste Technology, University of NSW, Sydney NSW 2052, Australia

1. INTRODUCTION 668
   1.1. Iron Transport in Biological Systems 668
   1.2. Modes of Iron Transport 669
   1.3. Processes Associated with Iron Transport 670
   1.4. Iron Transport and Other Metals 670
2. OTHER METALS AND IRON-BINDING PROTEINS IN PROKARYOTES AND LOWER EUKARYOTES 672
   2.1. Bacteria 672
   2.2. Fungi and Phytoplankton 674
   2.3. Plants 674
3. OTHER METALS AND IRON-BINDING PROTEINS IN HIGHER EUKARYOTES 675
   3.1. Animals 675
   3.2. Humans 679
4. NITROGENASE AND IRON TRANSPORT 682

5. CONCLUSIONS 683
ABBREVIATIONS 684
REFERENCES 684

1. INTRODUCTION

1.1. Iron Transport in Biological Systems

Although iron is the fourth most abundant element and after aluminum the most abundant metal in the earth's crust, and although present in almost all biological systems, it is not present in readily available forms in the lithosphere. Iron is normally found in the oxidized forms of Fe(III) of hematite (Fe_2O_3), magnetite (Fe_3O_4), limonite (approx. $2Fe_2O_3{\cdot}3H_2O$), and pyrite (FeS_2), all of which are particularly insoluble in normally encountered aerobic, aqueous environments, with equilibrium concentrations of the order of 10^{-17} M [1].

Solubilization of iron(III) for biological growth to a roughly optimal level of 10^{-7} M is required via the chemical processes of chelation and reduction (including protonation). Such processes may of course be the same, as it is rather arbitrary to attempt to classify oxidation states for systems whose redox potentials may vary from +300 mV in some cytochromes to −490 mV in some iron-sulfur proteins [2]. The best-known microbially produced chelators are the siderophores, which are low molecular weight (virtually) iron-specific compounds, and whose role in iron(III) uptake and transport is well documented. Uptake and transport of iron(II) appears to be much less common, although there is evidence of some systems being able to utilize iron(II) rather than iron(III) [3]. Indeed, on the basis of the fact that it is known that several microorganisms can produce more than one siderophore, with increasing chelating power being used if a previous siderophore fails to obtain the required iron [4], one may expect that many organisms may be eventually shown to be facultative with respect to their use of iron in various oxidation states.

The complex nature of the selectivity of the transport system and iron availability is well illustrated by reference to work on *Saccharomyces cerevisiae*, which appears to have a high-affinity transportation system operating above 5 μM and a low-affinity system operating at less than 5 μM of iron(II) [5]. This latter type of system appears to be nonspecific in operation for iron polymers and does not require carriers, binding such compounds to the surface of cells and dissolving them by the release of hydroxy acids [6].

1.2. Modes of Iron Transport

The topic of iron transport in biological systems has a long and illustrious history. From the exploratory review of Worwood [7] in 1974, through the review of Aisen and Listowsky [8] and that of Crichton and Charloteaux-Wauters [9] of the 1980s, up to the recent work of Guerinot [3].

We can now understand that distinctions must be made between systems operating on land against those in aquatic environments, between systems used by prokaryotes and those used by animals, and between those which operate in aerobic situations compared with those present in an anaerobic environment.

Thus, whereas adsorption of siderophores to inert surfaces poses a problem in terrestrial environments, in water-based environments, the major problem is siderophore diffusion from the producing organism [10].

Although we know that, in animals, much iron is present intercellularly in heme, in iron-sulfur proteins, and in ferritin, and extracellularly that iron is complexed by glycoproteins [11], we have much less knowledge of how microbial siderophores in soils and on plant (rhizosphere) surfaces operate [12]. Indeed, questions of the type posed by De Angelis et al. [13], about whether microorganisms choosing a particular siderophore go about such activity to obtain carbon or iron (or perhaps some other required element), indicates the lack of detailed knowledge in this area.

It is argued that anaerobic systems are much more likely to have access to iron(II), which is appreciably more soluble than iron(III), than do aerobic systems. Consequently, much of the work done on anaerobic systems has been aimed at elucidating the role of iron oxides as electron acceptors [14]; thus the question of whether iron(II) has specific transport systems is largely unresolved.

1.3. Processes Associated with Iron Transport

Siderophores, which are responsible for extracting iron from the environment surrounding the organism, are chelates with either hydroxamic acid or catechol-binding sites which are very strong iron(III) binders [15]. These complexes are then assimilated by cells at a receptor specific for the complex [16], and once inside the cell or its membrane, undergo dissociation [17,18] or reduction to iron(II) [19] to release intercellular iron. Once inside the cell, the iron is retained in a biologically usable form via complexation with a storage protein such as ferritin [20].

The transferrins responsible for extracellular iron transport (originally known as siderophilins) have been widely studied since as early as 1900 [21], whereas work on the ferritins is much more recent, dating from the 1960s [22]. In the period of almost a century, much detailed work has been undertaken as exemplified in the excellent review of Crichton and Charloteaux-Wauters [9]; but what of work that clarifies the role of other metals in the operation of iron-transport systems?

1.4. Iron Transport and Other Metals

Commencing with the review of Worwood [7], now some two decades old, we have combed the literature seeking references dealing with the role played by other metals in, and upon, iron transport; the results are given in the following sections. However, it is worth noting now the overall approaches which were adopted in the literature of the 1960s and 1970s to the influence exerted by other transition metals upon iron transport in biological systems.

There appear to have been two main approaches which were followed in the earlier literature.

1. To study the variation in iron assimilation by biological systems, or in specific compounds isolated from them, by the presence of other transition metals. Both enhancement and reduction in iron uptake was found depending (mainly) upon the actual metal ion used, concentrations, and pH of experiment.
2. To study the effect of replacing iron-transport moieties by

various transition metal ions. Transport of these other metals was often studied, but also structural elucidation, including attempts to discern ligand binding sites, was commonly attempted. An excellent example of such work can be seen, for example, in the use of chromium(III)-substituted siderophores instead of the iron(III) analogues by Raymond and coworkers [23]. Here, the use of the inert d^3 chromium analogue allows solution of an isomer problem not easily solved for the far more labile d^5 iron(III) siderophore.

Much of the above work appears to have been driven by the ethos that transition metals may often act as biological "poisons", and thus that their presence may be deleterious to biological systems. This explains why studies were undertaken not only on the more common trace metals such as chromium, manganese, cobalt, copper, zinc, cadmium, lead, and molybdenum [24] but also on metals such as indium [25], plutonium, and americium [26] and the lanthanide elements [27].

As early as 1960, Pollack et al. [28] had shown that the intestinal absorption of both cobalt and manganese was greater in iron-deficient rats than in iron-replete rats. Similar absorption of zinc and nickel was also reported for iron-deficient rats [29]. The results of this work were interpreted in terms of competition between iron and the added transition metal for uptake sites on the iron-containing entity under consideration. In particular, competition between cobalt and iron was studied in the work of Schade et al. [30], who concluded that when oral doses of cobalt and iron were given to rats, the cobalt diminished the mucosal uptake of iron from the lumen of the intestine but did not primarily reduce iron transfer from the intestinal cells to the blood. Later work by Thomson and Valberg [31] concluded, similarly, that cobalt, iron, and manganese all competed for uptake binding sites upon the iron-chelating siderophore. Conversely, Schade et al. [32] were able to show that cobalt is not incorporated into ferritin.

There is much evidence that transferrin will bind many metals [33], and it has been found that it binds copper(II), cobalt(II), chromium(III), and manganese(II) in a similar manner to its binding to iron(III), which involves a stoichiometry of two atoms of metal per ferritin moiety and the presence of a bicarbonate anion [34]. Larger metal ions, such as the lanthanides on the other hand, appear to bind only one metal atom per transferrin [27].

Interference by other metals in iron transport is obviously of importance in any review of this topic, and thus will be taken up in due course. However, the possibility that other metals may be intimately involved in the naturally occurring iron-transport systems is also an approach that should be kept in mind; as we will see, there appears to be very little current knowledge in this area.

2. OTHER METALS AND IRON-BINDING PROTEINS IN PROKARYOTES AND LOWER EUKARYOTES

2.1. Bacteria

Nearly all work on siderophore systems has concentrated on gram-negative bacteria. However, recent evidence that *Bacillus subtilis* uses a catechol siderophore similar to that used by enterobactin [35], and also uses the siderophores ferrichrome and ferrioxamine [36], suggests that both gram-positive and gram-negative bacteria may use similar iron-uptake systems.

In situations of iron deficiency, *Escherichia coli* coderepresses many iron-uptake processes, including outer membrane, periplasmic, and several inner membrane proteins. Silver and Walderhaug [37] have written an excellent review of the siderophores for enterobactin, aerobactin, and iron(III) citrate of *E. coli*. All of the genes associated with iron uptake are negatively regulated by a gene, identified by chemical mutagenesis, known as "Fur" (ferric uptake regulator) [38]. It is proposed that Fur binds iron as a corepresssor and then acts as a negative regulator via sequence specific protein-DNA interactions [39].

Of interest is the requirement of the presence of a divalent first-row transition metal; that is, Mn(II), Fe(II), Co(II), Cu(II) or Zn(II) or of Cd(II) to activate the specific binding of Fur to DNA. Of these metals, it is believed that Fe(II) is the natural activator because of the relative abundance of iron [40].

In the operation of Fur, the so-called "iron box" or "Fur box" sequence in the DNA operator consists of an amino acid array which is (with some small variation) common in the regulatory DNA of iron-affected systems of a number of microbial entities [41]. Recent work [42] using electron microscopy, suggests that Fur polymerizes around the operator to bind to it. Other work involving electron paramagnetic resonance spectroscopy [43], and a site mutagenesis study [44], have not been able to define the metal-binding sites, although a recent report [45] suggests that each monomer of Fur possesses both an octahedral (iron?) binding site together with a tetrahedral metal-binding site. The mode of operation of Fur is still very unclear, as the Fur regulator also contains many genes which appear unrelated to iron uptake [46].

The fact that both Fur and ArcA, the latter of which is the repressor for the *sod*A gene coding for manganese-superoxide dismutase, bind at the same site [47], presents the possibility of some association here between iron and manganese; a not uncommon association in the natural environment.

Iron transport across the cell membrane (outer and inner) in bacterial systems is covered in great detail in the review by Guerinot [3]. A broad spectrum of microbial systems is covered, but no evidence of the involvement of metals other than iron is indicated.

Rhizobia (*Azorhizobium*, *Rhizobium*, and *Bradyrhizobium*) engage in nitrogen-fixing symbiosis with a suitable legume. The importance of iron-containing proteins such as nitrogenase has led to very extensive literature on how rhizobia acquire iron [48]. In general, where siderophores have been isolated in such systems, they appear to be system specific, with no one siderophore being produced by all rhizobia.

The interest in nitrogenase in the context of this Chapter lies in the fact that nitrogenase uses not only iron but also molybdenum; further details will be given in Section 4.

Recently, a new class of siderophores, which do not appear to use either hydroxamate or catecholate groups, has been isolated from the Zygomycetes group of fungi. These siderophores bind iron strongly, apparently through carboxylate and/or hydroxyl groups [49,50]. The archetype of these siderophores, rhizoferrin, has been shown to chelate not only to iron(III) but to gallium(III), cobalt(III), chromium(III), and rhodium(III); there is, however, no suggestion that any of these metals other than iron are involved in the iron transport system of rhizoferrin.

Many marine bacteria and cyanobacteria produce siderophores, often with very high affinity for iron [51]. There appears to be no published evidence of the involvement of other metals in the operation of such siderophores, and indeed under conditions of iron limitation, which may readily occur in aqueous solution via diffusion of the siderophore from the cell surface, marine organisms appear to have substituted non-iron-containing compounds for siderophores; thus flavodoxin has replaced ferredoxin in the operation of various cyanobacteria [52] and phytoplankton [53].

2.2. Fungi and Phytoplankton

The above reference to marine bacteria is a logical step to consideration of phytoplankton production in the sea. Work in this area stems from the pioneering paper of Martin and Fitzwater in 1988 [54], which provided evidence that iron limitation had an important effect upon phytoplankton growth in the North Pacific Ocean. The generality of this finding has since been confirmed [55]. Once again, although there is not necessarily a direct connection between iron availability and regulation of phytoplankton growth, it must be kept in mind that phytoplankton are most important in the geochemical cycles of not only iron, but of manganese, copper, and zinc as well [56].

Recent evidence [57] suggests that iron additions may also stimulate phytoplankton productivity in a low-nitrate, low-chlorophyll marine environment, which indicates that magnesium and iron may be acting in a closer relationship than presently conceived.

The connection between iron transport and the presence of a copper-containing oxidase has been described for *S. cerevisiae* [58]. The evidence is that a single gene abrogates both high-affinity copper uptake and high-affinity iron uptake. The two metals do not, however, share a common transport system, as iron does not compete for high-affinity copper uptake and vice versa.

At this stage, no *S. cerevisiae* siderophores have been isolated, and yeasts, like plants, rely upon the reduction of various iron(III) compounds in the soil via a plasma membrane-bound iron(III) reductase for their iron requirements.

2.3. Plants

This is not an area of study that appears to have yielded a great wealth of new information. Workers seem to have concentrated their studies on iron transport, particularly on simpler organisms or on the higher animals and humans.

Although work, as described above, had been undertaken on the iron-transport procedures connected with the operations of rhizobia, little appears to be known about metal iron transport within plants and there is very little published regarding the involvement of other metals. Two typical studies are noted below; the first of these most likely has much more to do with iron transport by external (to the plant) microorganisms rather than with any actual plant cell procedures.

The interference effects of copper, zinc, and cobalt upon the uptake of a mugineic acid-iron(III) complex have been studied for barley (*Hordeum vulgare*) grown in nutrient solution [59]. Copper was most effective in decreasing iron uptake when added in an ionic form. The inhibition order (at higher metal ion concentrations), which was consistent with the stability constants of the metals with mugineic acid, was Cu(II) $>$ Zn(II) $\geqslant$ Co(II) $>$ Co(III). It was proposed that competition by the metal ions for mugineic acid complexation led to the reduction in iron uptake. This approach will be discussed in more detail in Section 3, in which it will become clear how often it has been used.

Typical of metal ion replacement experiments that have been undertaken by many workers on iron proteins is that reported by Iametti et al. [60] on spinach ferredoxin (and bovine adrenodoxin). Several metals, including Cd(II), Zn(II), V(IV), Mn(II), Co(II), and Ni(II), were introduced into the proteins from which the iron had been removed. Visible absorption and electron paramagnetic resonance spectroscopy were measured to elucidate structure and biological activity assessed; Cd(II) and Zn(II) appear to give the most reactive species. All of the metal-substituted proteins could be reconverted into the iron-containing form by incubation with iron and an external source of sulfide. The workers concluded that eukaryotic ferredoxins may well have a “sensing” function for intercellular levels of potentially toxic metals.

3. OTHER METALS AND IRON-BINDING PROTEINS IN HIGHER EUKARYOTES

3.1. Animals

A brief mention must first be made here about the excellent review by Locke and Nichol on “Iron Economy in Insects” [61]. Apart from the material in this work, little has been found about iron transport in insects and nothing about associated metals with such transport. As these investigators point out: "study of the physiological role that iron plays in insect biology has been neglected".

As might be expected, work on rats has been a major area of study on the topic of iron transport. Much work has been reported on the extent and mode of interference with iron transport by a large number of transition metal ions. The most widely studied (but by no means exclusively) have been cobalt, manganese, copper, and cadmium.

At the basic level, early experiments aimed at assessing the effect of perfusing increasing concentrations of metal ions, other than iron, into rats which had iron deficiency. Work by Thomson et al. [62] suggested that the affinity of the transport system for cobalt was greater than its affinity for iron, that is, cobalt reduced iron absorption more effectively than iron reduced cobalt absorption. Manganese, on the other hand, had to be added in very high concentrations to inhibit iron uptake. Thomson and Valberg [31] concluded that cobalt, iron, and manganese all competed for the same binding sites; however, in the epithelial cell, if the iron is not transferred into the blood, it is incorporated into ferritin. Such incorporation cannot occur for cobalt [32].

In a similar study, using oral doses of iron and cobalt, Schade et al. [30] concluded that cobalt diminished the mucosal uptake of iron from the lumen of the intestine of rats but did not primarily reduce iron transfer from the intestinal cells to the blood.

To demonstrate the continuity of this line of research, we can look at a recent publication dealing with iron uptake by cells of rabbit reticulocytes [63]. Reticulocytes were suspended in low ionic strength media and the ability to take up non-transferrin-bound iron and utilize it for heme synthesis was measured under conditions of increased transmembrane potential difference. Very potent inhibition of the highly efficient iron uptake was found for the cations Mn(II), Co(II), Ni(II), and Zn(II); this was attributed to direct action by these ions in competition for the binding sites of the uptake system. Quail and Morgan [63] concluded that high negative surface potential on the reticulocytes promoted the approach of iron (as well as the other metal cations) to the transport site on the cell membrane. Work using mixtures of solutions of radiolabeled iron(II) sulfate and iron(III) chloride indicated that uptake of the iron appeared to be via Fe(II) rather than Fe(III).

As a metal often found in the environment in close proximity to iron, manganese is commonly used to assess its interference with iron uptake for various systems. As early as the 1950s, Hartman et al. [64] found that in lambs and Matrone et al. [65] found that in rabbits and pigs feeding large amounts of dietary manganese interfered with hemoglobin regeneration. This antagonism could be overcome by increasing dietary iron input, suggesting that there was a direct competition for uptake sites between iron and manganese.

This ability of manganese to bind in sites which are normally occupied by iron is demonstrated in a study by Tan and Woodworth [24], who assessed the efficiency of binding of manganese(II) (among other metals) to hen's egg conalbumin. Such conalbumin normally binds 2 Mol of iron(III) per mole of protein, and the relative stability of the iron- and metal-replacement complexes of the protein were found to be in the order Fe(III) > Cr(III), Cu(II) > Mn(II), Co(II), Cd(II) > Zn(II) > Ni(II).

Recent work on the axonal transport of manganese and its relevance to selective neurotoxicity in the rat basal ganglia [66] indicates that manganese toxicity and iron transport are closely connected. Although it has been established [67] that the distribution of manganese after overexposure is predominately confined to brain areas containing high levels of nonheme iron, further details of binding and mechanism of action of the manganese are little understood.

There are, however, several pieces of evidence indicating that accumulation and transport of manganese occurs in a similar manner to that of iron [68,69]. High levels of nonheme iron in the form of ferritin are found in the basal ganglia [67]. The regional distribution of brain iron and transferrin receptors is uneven, but frequently iron accumulation corresponds to efferent areas, which possess high transferrin receptor density, suggesting that basal ganglia reception of iron is via neuronal transport [70]. In addition, astrocytes have been shown to take up manganese by a specific high-affinity transport system [71]. Furthermore, in catecholamine-containing neuroblastoma cells, manganese is present as a manganese transferrin complex which is subsequently bound to ferritin [72].

This work on the axonal transport of manganese [66] proposes that the injected manganese is bound to ferritin associated with glial cells, whereas the axonally transported manganese could be bound to transferrin. To summarize: "Iron therefore determines the selective accumulation of manganese in and within the basal ganglia."

The reference above to the fact that chromium(III) binds tightly to iron sites in hen's egg conalbumin [25] raises the question of competitive binding of chromium(III) in the iron-transport system.

Chromium forms a stable complex with transferrin [25], and it has been logical to study its influence upon iron uptake by, for example, mucosal transferrin. Although the hexaquachromium(III) ion has poor absorption, the chromate [Cr(VI)] ion is absorbed strongly; it is claimed that such absorption is followed by reduction of the chromium to a cationic form [73]. When the chromium is injected as Cr(VI), iron is released into the blood of rats, particularly if the rats are iron deficient [74].

Of the other earlier transition metals, vanadium interactions with iron-transport systems have been the most widely studied. This is due to the usefulness of the VO^{2+} ion as a spin probe in electron paramagnetic resonance (EPR) spectroscopy. Binding site and EPR studies of VO^{2+} complexes of rat ferritin have been undertaken [75], and the results appear to indicate that in the binding of the iron to the protein vanadium(IV) will compete for at least a histidine site.

For higher animals and humans, it is claimed that nickel toxicity is associated with interference in iron metabolism [76]. Certainly, nickel-deficient rats exhibit poor iron absorption and develop anemia [77]. It appears that Ni(II) enhances the intestinal absorption of Fe(III) but not of Fe(II) [78]. Studies on the effect of nickel supplements, fed to a variety of iron-deficient animals, yielded a variety of effects. Hepatic and/or lung and/or liver iron concentrations were assessed in such studies, which were undertaken on rats, pigs, calves, and young lambs [79]. No discernible pattern of effects has been summarized for this work.

It has been known for several years that copper-deficient animals develop anemia. Despite adequate liver iron, copper deficiency results in low plasma ceruloplasmin and reduces iron mobilization. As early as 1966, Frieden proposed that ceruloplasmin, a copper transporter, was the molecular link between iron and copper metabolism for iron mobilization [80]. And, indeed the role played by ceruloplasmin in the release of liver iron to the plasma has been fully confirmed [81]. Such mobilization occurs via release of iron from ferritin by a specific ferritin reductive step which yields iron(II). Ceruloplasmin then controls the rate at which this reduced iron is converted to circulating iron(III) transferrin. This mode of operation distributes iron to all cells, particularly those involved in heme biosynthesis.

Recent work [82] on the influence of copper upon iron transport has looked at the effect of coupled iron and copper loading on rat brain, liver, kidney, femur, blood, and plasma concentrations of the two metals and iron transport into the organs during young rat growth. Iron-loaded diets together with copper-loaded diets were used and resulted in large increases in the iron content of the liver compared with controls with no added copper; liver, kidney, and plasma copper levels were conversely lowered. Dual iron and copper loading increased concentrations of brain nonheme iron and copper compared with iron-loading only experiments. These results indicate clearly the effect of copper upon the brain iron-uptake mechanisms; no details of such a mechanism, however, were proposed.

The effect of the heavier transition metals upon iron transport has also been studied. In particular, cadmium, which is well-known for its toxicity to humans, has been studied using mice. It was shown [83] that the addition of cadmium(II) to the diet of mice inhibited iron uptake through the stomach. Iron-deficient mice appeared to take up more cadmium than those with replete iron stores [84]. Binding site competition is again implied as an explanation of these results.

Lead also has an effect upon iron transport; it is well known to induce anemia via impairment of heme synthesis [85]. It has been proposed that this occurs via inhibition of δ-aminolevulinic acid dehydrase, resulting in a negative exponential relationship between the dehydratase and blood lead [86]. Lead also decreases ferrochelatase activity, which catalyses the incorporation of Fe(II) ion into the porphyrin structure of heme. Failure to insert iron into protoporphyrin leads to depressed heme formation; excess protoporphyrin takes the place of heme in hemoglobin, and zinc replaces iron in the center of the molecule [87].

3.2. Humans

In the previous section (3.1), appreciable detail was given concerning work on the effect of copper upon iron transport in rats, in which it was claimed that the linkage was through the copper transporter ceruloplasmin [81]; it appears that copper is required to maintain the flow of iron across the hepatocyte. This work has been extended into the study of human cells.

To review: Iron in plasma is mostly carried by transferrin, a glycoprotein, carrying two iron atoms per molecule bound in high-affinity sites, one in the N-terminal and the other in the C-terminal domain (lobe) [33]. Although transferrin is the major iron transporter, human cells can utilize the iron in serum ferritin [88,89].

Nearly all plasma copper is also protein bound and associated with ceruloplasmin, an α-globulin; the copper is incorporated into the protein in the liver with six or seven copper atoms per molecule [90].

In a detailed review of the transport of both copper and iron across the cell membrane, McArdle [91] points to the differences of how the two metals are transferred, but claims that there are important similarities, that is, both metals are carried in chelating molecules which are important in the recognition system; both metals are taken up by a carrier system, although the system may be different; both metals are possibly reduced during the uptake process. McArdle reiterates his conviction that our knowledge of the iron-transport process will be invaluable in developing an understanding of the processes involved in copper uptake.

Work on the antioxidant properties of ceruloplasmin toward iron- and copper-dependent oxygen radical formation represents a further aspect of the connection between copper and iron transport. Fractionation of human serum has shown that the copper-containing α_2-globulin ceruloplasmin and the β-globulin transferrin contribute significantly to important antioxidant activities of extracellular fluids [92]. Gutteridge [93] has proposed a coordinated role for both the iron and the copper proteins in protection against catalytic iron damage, and has further shown that ceruloplasmin is very potent when coupled with transferrin in antioxidant activity of normal human plasma [94]. Proposals that ceruloplasmin operates as an extracellular "superoxide dismutase" [95], or that it might decompose peroxides, or acts as a scavenger for organic oxide radicals [96] were discounted by this work.

Gutteridge [93] also refers to the unusual antioxidant property of ceruloplasmin as perhaps having important implications for Wilson's disease, in which patients are known to have a deficiency of ceruloplasmin synthesis. This connection, which has been documented over a long period [7], is still not clearly understood.

Aluminum has also been observed to bind to iron-transport moieties. Early observations on the binding of aluminum, as well as of gallium and zinc, were made by Tominatsu and Donovan [97], who proposed from spectral studies that Al(III) and Ga(III) binding to ovotransferrin and human serum transferrin involved perturbation of tryptophan and tyrosine; Zn(II) on the other hand, was believed to bind via ionization of tyrosine.

Such work has been followed by observations that intestinal absorption of aluminum may involve either the iron- or the calcium-transport mechanism [98]. Aluminum toxicity in renal patients, who have high total aluminum loads, has been addressed by administration of the microbial iron-transport agent desferrioxamine, which consists of basic repeating units of three 1-amino-5-hydroxyaminopentanes and two linking succinic acids that give rise to three hydroxamate groups which strongly bind iron(III) [99]. Indeed, thermodynamically, desferrioxamine should draw Al(III) from transferrin in the plasma, to which Al(III) binds very strongly. In such binding, Al(III) does not displace Fe(III) but binds to the unoccupied transferrin binding sites. As transferrin is the leading plasma protein for aluminum binding, and as some 70% of the transferrin-binding sites are empty, both Fe(III) and Al(III) can bind at the same time to this iron-transport moiety [99].

The site of binding of aluminum in human serum transferrin has been studied using ^{1}H-NMR (nuclear magnetic resonance) by Kubal et al. [100]. Given that the transferrin glycoprotein contains two structurally similar lobes, each of approximately 40 kDa with a high-affinity metal-binding site [101,102], these workers [100] showed that human apotransferrin can preferentially bind Al(III) in the N-lobe. Hongyan et al. [103] have suggested that Bi(III) may also bind strongly to both the N- and C-lobes of human serum transferrin.

The fact of accumulation of aluminum in the human bone during hemodialysis is well known, but how it occurs is not. Work by McGregor et al. [104] has shown that transferrin may act as the physiological carrier of aluminum to the bone, and that aluminum may affect iron uptake and metabolism in bone.

Given that aluminum binds strongly to transferrin, it is not surprising that gallium also binds to it, and Ga-67 has been widely used as an imaging agent for tumors and inflammatory abscesses. Ga(III) travels through the circulatory system bound to the serum iron-transport protein transferrin, whose binding is an essential step in tumor localization. The binding of gallium to transferrin has been found to be about

300 times weaker than that of Fe(III) [105]. Unlike Al(III), Ga(III) appears from ^{1}H-NMR evidence to bind to the C-lobe of human serum transferrin first and then to the N-lobe of the transferrin [106].

It has been shown that through its binding to transferrin, gallium is incorporated into cells through transferrin receptors and inhibits iron-dependent DNA synthesis. In human myeloid H60 cells, gallium regulates transferrin expression at the posttranscriptional level by mechanisms which require de novo protein synthesis and involve interaction with iron [107]. It was also noted that gallium-transferrin had different effects on the transferrin receptor mRNA in human myeloid H60 and lymphoid CCRF-CEM cells.

4. NITROGENASE AND IRON TRANSPORT

The importance of iron for living organisms is underlined by its role in a large number of proteins for their activity. Some of the most important of these involve reactions with the major gases in the earth's atmosphere, oxygen and nitrogen. Enzymes catalyzing oxygenase and nitrogenase activity are very prominent among other electron-transfer reactions.

Although neither hemoglobins nor myoglobins are considered as iron-transport systems, being usually classified as oxygen transporters, in both cases, iron actually undergoes transportation. Similarly, nitrogenase is seen as a nitrogen-related transporter rather than one which transports iron [108].

Nitrogenase depends upon the presence of both iron and molybdenum in the same protein for its operation, and is thus an example of a naturally occurring biological system in which another metal is essential to the iron in the enzymatic reaction. For this reason, we have chosen to include a description of the mechanism of operation of the nitrogenase iron proteins as a special case, and we offer a brief description of the state of knowledge in this area.

The catalytic operation of nitrogenase can be described overall as an acidic reductive process whereby molecular nitrogen is reduced to ammonia with concomitant hydrolysis of magnesium-ATP to magnesium-ADP.

The molybdenum-iron protein component of the overall action of the metalloenzyme nitrogenase is an $\alpha_2\beta_2$ tetramer containing two unique molybdenum-iron-sulfur cofactors [109], which are substrate binding and reduction sites, and two iron-sulfur protein clusters [110, 111], which are believed initially to accept the electron before transferring it to the iron-molybdenum cofactor [112].

The current model for the nitrogenase mechanism proposes that the reduced iron protein, together with two bound Mg-ATP molecules, binds to the Mo-Fe-protein; the Fe-protein is oxidized by transfer of an electron to the Mo-Fe-protein, and the two associated Mg-ATP molecules are hydrolyzed [113,114]. The oxidized iron-protein with its two bound Mg-ATP molecules now releases the Mo-Fe-protein, and another reduced Fe-protein, and its associated Mg-ATP entities, bind to the partially reduced Mo-Fe-protein with further hydrolysis of Mg-ATP [115].

The dissociation of the Fe-protein-Mo-Fe-protein complex is the rate-determining step for nitrogenase catalysis, and the necessary presence of molybdenum together with iron in the one cluster is one of the few reasonably well-elucidated symbiotic trace metal-iron protein systems known.

5. CONCLUSIONS

Whereas there is an excess of material available on the gross effects on iron transport of importing various trace metals into all manner of organisms, particularly in situations of iron deficiency, there is a very limited literature on the detailed understanding of any necessary presence of metals other than iron upon biological iron transport processes. One of the few situations in which it appears that another metal may be essentially involved in iron transport is that in which copper in ceruloplasmin appears to play a role in transferrin operation [96].

One is almost obliged to conclude this chapter by amplifying the orchestral metaphor so eloquently penned, some 10 years, by Crichton and Chaloteaux-Wauters [9], in which it is suggested that “we may compare intracellular iron with a symphony orchestra: we can, in the absence of the conductor, identify the different instruments, but not the role they play”. Within this context, our knowledge of any essential role played by other metals in iron transport can be likened to that played in the orchestra by the triangle; probably necessary, but at this stage essentially overlooked by the research audience.

ABBREVIATIONS

ADP	adenosine 5′-diphosphate
ArcA	repressor for the *sod*A gene coding for manganese-superoxide dismutase
ATP	adenosine 5′-triphosphate
CCRF-CEM	a lymphoid cell line, obtained from the Americal Type Culture Collection, Rockville, MD, USA
DNA	deoxyribonucleic acid
EPR	electron paramagnetic resonance
Fur	ferric uptake regulator
NMR	nuclear magnetic resonance
mRNA	messenger ribonucleic acid

REFERENCES

1. N. N. Greenwood and A. Earnshaw, in *Chemistry of the Elements*, Pergamon Press, Oxford, UK 1984, p. 1243.
2. S. M. Payne, *CRC Microbiol*, *16*, 81–111 (1988).
3. M. L. Guerinot, *Annu. Rev. Microbiol.*, *48*, 743–772 (1994).
4. M. S. Sevine and W. J. Page, *J. Gen. Microbiol.*, *138*, 587–596 (1992).
5. D. Eide, S. Davis-Kaplan, I. Jordan, D. Sipe, and J. Kaplin, *J. Biol. Chem.*, *267*, 2074–2081 (1992).
6. T. J. Beveridge, *Annu. Rev. Microbiol.*, *43*, 147–171 (1989).
7. M. Worwood, in *Iron in Biochemistry and Medicine* (A. Jacobs and M. Worwood, eds.), Academic Press, London, 1974, pp. 335–367.
8. P. Aisen and I. Listowsky, *Annu. Rev. Microbiol.*, *49*, 357–393 (1980).

9. R. R. Crichton and M. Charloteaux-Wauters, *J. Biochem.*, *164*, 485–506 (1987).
10. D. A. Hutchins, J. G. Reuter, and W. Fish, *Limnol. Oceanogr.*, *36*, 1–12 (1991).
11. S. M. Payne, *Trends Microbiol.*, *1*, 66–69 (1993).
12. J. E. Lager and J. S. Bayer, *Mel. Plant Microbe Interact.*, *4*, 5–13 (1991).
13. R. De Angelis, M. Forsyth, and D. Castignetti, *BioMetals*, *6*, 234–238 (1993).
14. K. H. Nealson and C. R. Myers, *Appl. Environ. Microbiol.*, *58*, 439–443 (1992).
15. K. N. Raymond, G. Muller, and B. F. Matzanke, *Top. Curr. Chem.*, *123*, 50–100 (1984).
16. V. Braun, in *The Enzymes of Biological Membranes* (A. N. Martenosi, ed.), Plenum Press, New York, 1985, pp. 617–652.
17. C. Ratledge, P. V. Patel, and J. Mundy, *J. Gen. Microbiol.*, *128*, 1559–1565 (1982).
18. T. P. Fleming, M. S. Wahlik, J. B. Neilands, and M. A. McIntosh, *Gene, 34*, 47–54 (1985).
19. J. F. Ernst and G. Winkelmann, *Biochim. Biophys. Acta*, *500*, 27–41 (1977).
20. E. C. Theil, in *Advances in Inorganic Biochemistry*, Vol. 5 (E. C. Theil, G. L. Eichhorn and L. G. Marzilli, eds.), Elsevier, New York, 1983, pp. 1–38.
21. T. B. Osborne and G. F. Cambell, *J. Am. Chem. Soc.*, *22*, 422–426 (1900).
22. M. Awai and E. B. Brown, *J. Lab. Clin. Med.*, *61*, 363–368 (1963).
23. K. N. Raymond and T. P. Tufano, in *The Biological Chemistry of Iron*, (H. B. Dunford, D. Dolphin, K. N. Raymond, and L. Sieher, eds.). D. Reidel, Dordrecht, The Netherlands, 1982, pp. 85–105.
24. A. T. Tan and R. C. Woodworth, *Biochemistry*, *8*, 3711–3716 (1969).
25. F. Hosain, P. A. McIntyre, K. Poulose, H. S. Stern, and M. N. Wagner, *Clin. Chim. Acta*, *24*, 69–78 (1969).
26. D. M. Taylor, *Health Phys.*, *22*, 575–581 (1972).
27. C. K. Luk, *Biochemistry*, *10*, 2838–2843 (1971).

28. S. Pollack, J. N. George, R.C. Reba, R. M. Kautman, and W. H. Crosby, *J. Clin. Invest.*, *44*, 1470–1476 (1965).
29. W. Forth, in *Trace Element Metabolism in Animals* (C. F. Mills, ed.), Livingstone, London, 1970, p. 298.
30. S. G. Schade, B. F. Felsher, B. E. Glaler, and M. E. Conrad, *Proc. Soc. Exp. Biol. Med.*, *134*, 741–746 (1970).
31. A. B. R. Thomson and L. S. Valberg, *Am. J. Physiol.*, *223*, 1327–1331 (1972).
32. S. G. Schade, B. F. Felsher, G. M. Bernie, and M. E. Conrad, *J. Lab. Clin. Med.*, *75*, 435–438 (1970).
33. G. de Jong, J. P. van Dijk, and H. G. van Eijk, *Clin. Chim. Acta*, *190*, 1–46 (1990).
34. P. Aisen, R. Aasa, and A. G. Redfield, *J. Biol. Chem.*, *244*, 4628–4633 (1969).
35. T. H. Grossman, M. Tuckman, S. Ellestad, and M. S. Osburne, *J. Bacteriol.*, *175*, 6203–6211 (1993).
36. R. Schneider and K. Hantke, *Mol. Microbiol.*, *8*, 111–121 (1993).
37. S. Silver and M. Walderhaug, *Microbiol. Rev.*, *56*, 195–228 (1992).
38. A. Bagg and J. B. Neilands, *Microbiol. Rev.*, *51*, 509–518 (1987).
39. A. Bagg and J. B. Neilands, *Biochemistry*, *26*, 5471–5477 (1987).
40. V. de Lorenzo, S. Wee, M. Herrero, and J. B. Neilands, *J. Bacteriol.*, *169*, 2624–2629 (1987).
41. V. de Lorenzo, F. Giovannini, M. Herrero, and J. B. Neilands, *J. Mol. Biol.*, *197*, 875–884 (1988).
42. E. Le Cam, D. Frechon, M. Barray, A. Fourcade, and E. Delain, *Proc. Natl. Acad. Sci. USA*, *91*, 11816–11820 (1994).
43. M. Y. Hamed and J. B. Neilands, *J. Inorg. Biochem.*, *53*, 235–248 (1994).
44. M. Coy, C. Doyle, J. Besser, and J. B. Neilands, *BioMetals*, *7*, 292–298 (1994).
45. I. Michaud-Soret, A. Adrait, N. Debaecker, J.-M. Latour, and D. Touati, *J. Inorg. Biochem.*, *59*, 99 (1995).
46. V. de Lorenzo, M. Herrero, F. Giovannini, and J. B. Neilands, *Eur. J. Biochem.*, *173*, 537–546 (1988).
47. B. Tardat and D. Tourti, *Mol. Microbiol.*, *9*, 53–56 (1993).

48. D. Expert and P. R. J. Gill, in *Molecular Signals in Plant Microbe Communications* (D. P. S. Verma, ed.), Boca Raton, FL: CRC Press, 1991, pp. 229–245.

49. H. Drechsel, J. Metzger, S. Freund, G. Jung, J. R. Boelaert, and G. Winkelmann, *BioMetals*, *4*, 238–243 (1991).

50. H. Drechsel, G. Jung, and G. Winkelmann, *BioMetals*, *5*, 141–148 (1992).

51. R. T. Reid, D. H. Live, D. J. Faulkner, and A. Butler, *Nature*, *366*, 455–458 (1993).

52. A. Bovy, G. de Vrieze, L. Lugones, P. van Horssen, and C. van den Berg, *Mol. Microbiol.*, *7*, 429–439 (1993).

53. J. La Roche, R. J. Geider, L. M. Graziano, H. Murray, and K. Lewis, *J. Phycol.*, *29*, 767–777 (1993).

54. J. H. Martin and S. E. Fitzwater, *Nature*, *331*, 341–343 (1988).

55. J. H. Martin, S. E. Fitzwater, R. M. Gordon, C. N. Hunter, and S. J. Tanner, *Deep-Sea Res.*, *40*, 115–134 (1993).

56. K. W. Bruland, J. R. Donat, and D. A. Hutchins, *Limnol. Oceanogr.*, *36*, 1555–1577 (1991).

57. G. R. D. Tullio, D. A. Hutchins, and H. W. Bruland, *Limnol. Oceanogr.*, *38*, 495–508 (1993).

58. A. Dancis, D. S. Yuan, D. Halle, C. Aswith, and D. Eide, *Cell*, *76*, 393–402 (1994).

59. J. F. Ma and K. Nomoto, *Physiol. Plant.*, *89*, 331–334 (1993).

60. S. Iametti, H. Uhlmann, N. Sala, R. Bernhardt, E. Ragg, and F. Bonomi, *Eur. J. Biochem.*, *239*, 818–826 (1996).

61. M. Locke and H. Nichol, *Annu. Rev. Entomol.*, *37*, 195–215 (1992).

62. A. B. R. Thomson, L. S. Valberg, and D. G. Sinclair, *J. Clin. Invest.*, *50*, 2384–2398 (1971).

63. E. A. Quail and E. H. Morgan, *J. Cell. Physiol.*, *159*, 238–244 (1994).

64. R. H. Hartman, G. Matrone, and G. H. Wise, *J. Nutr.*, *57*, 429–435 (1955).

65. G. Matrone, R. H. Hartman, and A. J. Clawson, *J. Nutr.*, *67*, 309–316 (1959).

66. W. N. Sloot and J.-B. P. Gramsbergen, *Brain Res.*, *657*, 124–132 (1994).

67. J. M. Hill and R. C. Switzer, *Neuroscience*, *11*, 595–603 (1984).
68. M. Aschner and J. L. Aschner, *Brain Res. Bull.*, *24*, 857–860 (1990).
69. M. Aschner and J. L. Aschner, *Neurosci. Biobehav. Rev.*, *15*, 333–340 (1991).
70. J. M. Hill, M. R. Ruff, R. J. Weber, and C. B. Pert, *Proc. Natl. Acad. Sci. USA*, *82*, 4553–4557 (1985).
71. M. Aschner, M. Gannon, and H. K. Kimelberg, *J. Neurochem.*, *58*, 730–735 (1992).
72. N. Suárcz and H. Eriksson, *J. Neurochem.*, *61*, 127–131 (1993).
73. R. D. McKenzie, R. A. Anwar, R. U. Byerrum, and C. A. Hoppert, *Arch. Biochem.*, *79*, 200–202 (1959).
74. P. Wollenberg, H. Huebers, and W. Rummel, in *The Biochemistry and Physiology of Iron* (P. Saltmann and J. Hegenauer, eds.), Elsevier, New York, 1982, pp. 287–290.
75. G. J. Gerfen, P. M. Hanna, N. D. Chasteen, and D. J. Singel, *J. Am. Chem. Soc.*, *113*, 9513–9519 (1991).
76. P. Stokes, in *Metal Ions in Biological Systems*, Vol. 23, *Nickel and Its Role in Biology* (H. Sigel and A. Sigel, eds.), Marcel Dekker, New York, 1988, p. 33.
77. M. Anke, B. Groppel, H. Krunemann, and M. Grün, in *Nickel in the Human Environment* (F. W. Sunderman, Jr., ed.), IARC, Oxford University Press, Oxford, UK, 1984, pp. 339–365.
78. F. H. Nielson, *Annu. Rev. Nutr.*, *4*, 21–41 (1984).
79. J. W. Spears, R. W. Harvey, and L. J. Samsell, *J. Nutr.*, *116*, 1873–1882 (1986).
80. E. Frieden, in *Metal Ions in Biological Systems*, Vol. 13, *Copper Proteins* (H. Sigel, ed.), Marcel Dekker, New York, 1981, p. 133.
81. E. Frieden, *Advan. Chem. Ser. Bioinorg. Chem.*, *100*, 293–321 (1971).
82. A. Crowe and E. H. Morgan, *J. Nutr.*, *126*, 183–190 (1996).
83. D. L. Hamilton and L. S. Valberg, *Am. J. Physiol.*, *227*, 1033–1037 (1974).
84. L. S. Valberg, J. Sorbie, and D. L. Hamilton, *Am. J. Physiol.*, *231*, 462–467 (1976).
85. R. A. Goyer, in *Handbook on Toxicity of Inorganic Compounds*

(H. G. Seiler, H. Sigel, and A. Sigel, eds.), Marcel Dekker, New York, 1988, pp. 369–370.

86. R. A. Goyer, in *Casarett and Doull's Toxicology: The Basic Science of Poisons*, 3rd ed. (J. Doull, C. D. Klaasen, and M. D. Amdur, eds.), Macmillan, New York, 1986, p. 582.

87. M. D. Bassis and W. N. Jensen, *Br. J. Haematol.*, *11*, 49–56 (1965).

88. G. D. Blight and E. H. Morgan, *Eur. J. Cell Biol.*, *43*, 260–265 (1987).

89. R. D. Camparelli, B. M. Friedman, A. P. MacPhail, T. H. Bothwell, J. I. Phillips, and R. D. Baynes, *Br. J. Haematol.*, *72*, 100–105 (1989).

90. A. Messerschmidt and R. Huber, *Eur. J. Biochem.*, *187*, 341–352 (1990).

91. H. J. McArdle, *Proc. Nutr. Soc.*, *51*, 199–209 (1992).

92. J. Stocks, J. McGutteridge, R. J. Sharp, and T. L. Dormandy, *Clin. Sci. Mol. Med.*, *47*, 223–233 (1974).

93. J. M. C. Gutteridge, *Biochem. Soc. Trans.*, *10*, 72–73 (1982).

94. J. M. C. Gutteridge, *FEBS Lett.*, *157*, 37–40 (1983).

95. I. M. Goldstein, H. B. Kaplan, K. S. Edelson, and G. Weismann, *J. Biol. Chem.*, *254*, 4040–4045 (1979).

96. J. M. C. Gutteridge, *Ann. Clin. Biochem.*, *15*, 293–296 (1978).

97. Y. Tomimatsu and J. W. Donovan, in *Proteins of Iron Metabolism* (E. B. Brown, P. Airen, J. Fielding, and R. R. Crichton, eds.), Grune & Stratton, New York, 1977, pp. 221–226.

98. M. R. Wills and J. Savory, in *Metal Ions in Biological Systems*, Vol. 24, *Aluminum and Its Role in Biology* (H. Sigel and A. Sigel, eds.), Marcel Dekker, New York, 1988, p. 322.

99. R. B. Martin in *Metal Ions in Biological Systems*, Vol. 24, *Aluminum and Its Role in Biology* (H. Sigel and A. Sigel, eds.), Marcel Dekker, New York, 1988, pp. 42–45.

100. G. Kubal, A. B. Mason, P. J. Sadler, A. Tucker, and R. C. Woodworth, *Biochem. J.*, *285*, 711–714 (1992).

101. R. R. Crichton, *Adv. Protein Chem.*, *40*, 281–363 (1990).

102. K. N. Raymond, T. D. Y. Chung, V. L. Pecoraro, and C. J. Carrano, in *The Biochemistry and Physiology of Iron* (P. Saltmann and J. Hegenauer, eds.), Elsevier, New York, 1982, pp. 649–662.

103. L. Hongyan, S. Hongzhe, and P. J. Sadler, *J. Inorg. Biochem.*, *59*, 105 (1995).

104. S. J. McGregor, M. J. F. Menendez, M. L. Naves, R. Elloriaga, J. H. Brock, and J. B. Cannata, *Trace Elements Med.*, *11*, 187–191 (1994).

105. W. R. Harris and V. L. Pecoraro, *Biochemistry*, *22*, 292–299 (1983).

106. G. Kubal, A. B. Mason, S. V. Patel, P. J. Sadler, and R. C. Woodworth, *Biochemistry*, *32*, 3387–3385 (1993).

107. R. Ul-Haq and C. R. Chitambar, *Biochem. J.*, *294*, 873–877 (1993).

108. J. B. Neilands, *J. Biol. Chem.*, *270*, 26723–26726 (1995).

109. J. Kim and D. C. Rees, *Nature*, *360*, 553–560 (1992).

110. T. R. Hawkes, P. A. McLean and B. E. Smith, *Biochem. J.*, *217*, 317–321 (1984).

111. J. T. Bolin, A. Ronco, T. V. Morgan, L. E. Mortenson, and N. H. Xuong, *Proc. Natl. Acad. Sci. USA*, *90*, 1078–1082 (1993).

112. J. W. Peters, K. Fisher, W. E. Newton, and D. R. Dean, *J. Biol. Chem.*, *270*, 27007–27013 (1995).

113. R. N. F. Thorneley and D. J. Lowe, *Biochem. J.*, *215*, 393–403 (1983).

114. J. B. Howard and D. C. Rees, *Annu. Rev. Biochem.*, *63*, 235–264 (1994).

115. R. V. Hageman and R. H. Borris, *Proc. Natl. Acad. Sci. USA*, *75*, 2699–2702 (1978).

18

Iron Chelators for Clinical Use

Gary S. Tilbrook and Robert C. Hider

Department of Pharmacy, King's College London,
University of London, Manresa Road,
London SW3 6LX, UK

1. INTRODUCTION 692

2. DESIGN FEATURES OF Fe(III) LIGANDS 695
   2.1. Choice of Denticity 695
   2.2. Choice of Ligand 699
   2.3. Influence of Molecular Weight 701
   2.4. Influence of Lipophilicity 703

3. IRON CHELATOR TOXICITY 704
   3.1. Inhibition of Iron-Dependent Enzymes 704
   3.2. Distribution of Chelators 706
   3.3. Toxicity of Iron Complexes 707
      3.3.1. Redox Cycling 707
      3.3.2. Partially Complexed Iron 708

4. CLINICALLY USEFUL BIDENTATE AND TRIDENTATE LIGANDS 709
   4.1. Bidentate Chelators 709

4.2. Tridentate Chelators 713

5. CLINICALLY USEFUL HEXADENTATE LIGANDS 714
5.1. Desferrioxamine 714
5.2. DTPA 716
5.3. HBED 716
5.4. DBED 717
5.5. Razoxane 718

6. THERAPEUTIC APPLICATIONS OF Fe(III)-SELECTIVE LIGANDS 719
6.1. Treatment of Systemic Iron Overload 719
6.2. Elevated Iron Levels in the Brain 720
6.3. Temporary Elevated Intracellular Iron Levels in Localized Tissue 721
6.3.1. Ischemic Tissue 721
6.3.2. Inflamed Tissue 722
6.3.3. Protection Against Redox-Cycling Chemicals 722
6.4. Temporary Elevated Extracellular Iron Levels 722
6.5. Selective Inhibition of Nonheme-Containing Enzymes 723
6.6. Iron-Transport-Mediated Drug Delivery 724

7. CONCLUSIONS 725

ABBREVIATIONS 726

REFERENCES 727

1. INTRODUCTION

Iron is an essential element for life and plays a central role in biochemical processes by virtue of its facile redox chemistry and the high affinity of both redox states (Fe(II) and Fe(III)) for oxygen. The 2+ and 3+ oxidation states of this transition metal have found a critical position in biology, since through appropriate ligation and pH control, a range of redox potentials can be achieved. Thus, iron serves as a prosthetic group in such vital enzymes as oxidases, hydroxylases, ribonucleotide reductase, and lipoxygenases.

Iron is nontoxic when appropriately ligated, but loosely bound iron is able to redox-cycle, thereby generating oxygen-derived free radicals which are toxic to biological material [1]. As a direct consequence of these properties, biological systems have evolved complex mechanisms for strict compartmental control of the metal. Iron metabolism in humans is highly conservative; there exists no specific excretory mechanism. Thus, homeostasis is wholly reliant on dietary absorption and storage regulation. Aberrations in the homeostatic control mechanisms (i.e., the common genetic disorder hereditary hemochromatosis [2]) or circumvention (i.e., chronic blood transfusion therapy for hemoglobinopathies [3]) will lead to an elevation in the total body burden of iron and result ultimately in tissue/organ damage and eventual mortality. Although venesection is a treatment option in diseases where adequate erythropoietic reserve exists (i.e., hereditary hemochromatosis), clearly iron removal by alternative mechanisms is required in dyserythropoietic diseases such as β-thalassemia major. In addition, focal iron imbalances have been associated with the neurodegenerative disorders, Parkinson's disease, and Alzheimer's disease [4] and in conditions related to ischemia-reperfusion damage; that is, myocardial infarction and cerebral vascular accident [5].

Iron plays a key role in the de novo biosynthesis of DNA, since it is intimately involved in the catalytic conversion of nucleotides to 2′-deoxynucleotides via the action of the metalloenzyme ribonucleotide reductase. Hence, development of selective inhibitors of this enzyme may prove to be of useful therapeutic benefit in conditions characterized by uncontrolled cellular proliferation such as psoriasis and neoplasia. Parasitic infections such as malaria have been demonstrated to be sensitive to iron-chelating agents using compounds such as desferrioxamine. This finding offers an alternative strategy for the development of therapeutic agents to address the increasing problem of drug resistance to current therapies. Also, the issue of increased microbial resistance to antibiotics requires urgent attention. Examples of current lines of research include fusion of β-lactam derivatives to siderophore-type units in an attempt to exploit active iron-complex uptake of the conjugate, thereby increasing drug localization within the organism and hence potentiating toxicity. Clearly, iron-chelating agents for clinical use are required to be selectively directed to specific organ/cellular/subcellular targets in order to optimize their efficacy and to minimize exposure of nontarget sites. These issues will be addressed in Section 4, 5, and 6.

In designing potential therapeutic iron-chelating agents, a range of specifications must be considered and can be broadly subdivided into two categories: (1) *chemical factors* such as solubility, lipophilicity, metal affinity and selectivity, together with ligand and metal-complex speciation; and (2) *biological factors* such as metabolism, pharmacokinetics, and toxicity of both the free ligand and the metal complex.

In designing iron chelators for therapeutic application, the properties of metal selectivity and resultant ligand-metal complex stability are paramount. Chelating agents can be designed either for the ferrous or ferric oxidation states, although in practice the ferric form provides the best target for complexation in vivo. Ligands that prefer Fe(II) contain "soft" donor atoms, exemplified by the unsaturated 1-4 nitrogen arrays found in phenathrolines and bipyridyl derivatives. Although these compounds are selective for Fe(II) over Fe(III), ligands of this type retain appreciable affinity for other biologically important bivalent metals such as copper and zinc (Table 1), and thus the design of a nontoxic Fe(II)-selective ligand is extremely difficult and may not be possible. Ligands selective for Fe(III) over Fe(II) typically contain "hard" donors such as oxyanions found prevalently in siderophore structures, notably hydroxamates and catecholates. Thus, in principle, it is possible to design a ligand which possesses considerable selectivity

TABLE 1

Metal Affinity Constants for Selected Ligands[a]

	Log cummulative stability constant		
Ligand	Fe(III)	Cu(II)	Zn(II)
1,10-Phenanthroline	14.1	21.4	17.5
DFT (14)	29.6	20.9	16.7
DFO (4)	30.6	14.1	11.1
EDTA	25.1	18.8	16.5
8-Hydroxyquinoline	36.9	22.9	15.8

[a]Data sourced from Martell and Smith (1974–1989) [6a].

for Fe(III) under most biological conditions. An additional advantage of high-affinity Fe(III) ligands is that, under aerobic conditions, they will chelate Fe(II) cations and autoxidize them to the Fe(III) species [7]. Thus, high-affinity Fe(III)-selective chelators bind both Fe(III) and Fe(II) under most physiological conditions.

2. DESIGN FEATURES OF Fe(III) LIGANDS

2.1. Choice of Denticity

The coordination requirements of Fe(III) are best satisfied by six donor atoms ligating in an octahedral fashion to the metal center. In order to maximize the thermodynamic stability of the Fe(III) complex, it is necessary to incorporate all six donors into a singular molecular structure, thereby creating a hexadentate ligand (Fig. 1). An increment in the overall stability constant for the metal complex can be readily observed when comparing ligands of differing denticities that contain the same ligating units, for instance, bidentate and tridentate ligands (Fig. 1). Thus, the connecting structure to which the coordinating units are appended contributes to the overall free energy change associated with the complexation. This increase in thermodynamic chelate stability is largely associated with entropic changes that occur going from solvated free ligand and solvated free metal to solvated ligand-metal complex. The overall stability constant trends for bidentate and hexadentate

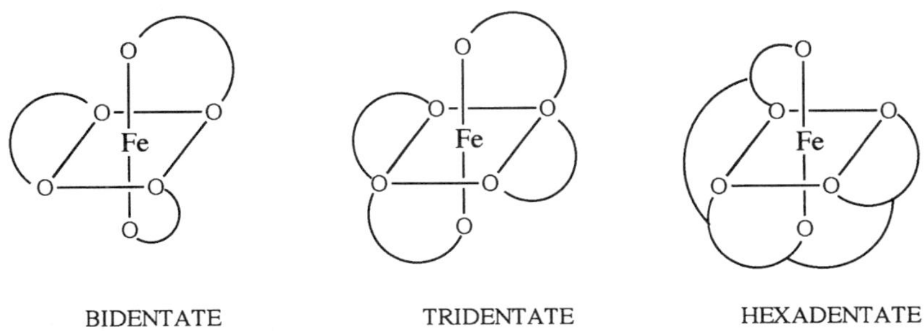

FIG. 1. Octahedral coordination of Fe(III) by bidentate, tridentate and hexadentate ligands.

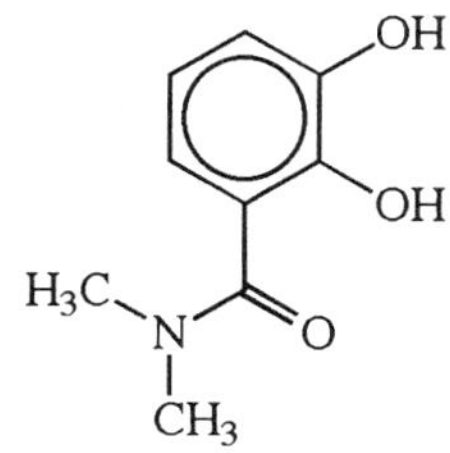

(1)

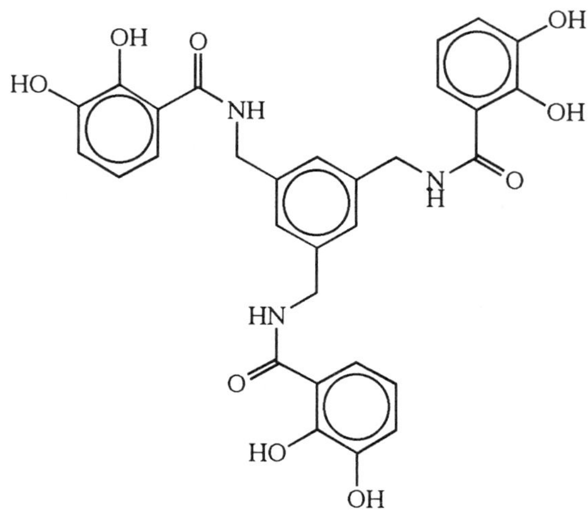

(2)

ligands are typified by the bidentate ligand *N,N*-dimethyl-2,3-dihydroxybenzamide (**1**) and the hexadentate congener MECAM (**2**) a differential of almost 6 log units in stability is observed. Similarly, comparison of the bidentate ligand acetohydroxamic acid (**3**) with a linear hexaden-

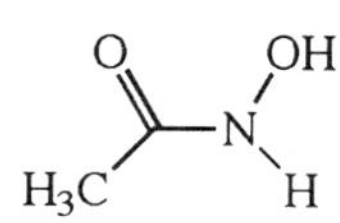

(3)

tate ligand, desferrioxamine B (**4**) a smaller but nevertheless positive increment of 2.3 log units is observed. In contrast, the incorporation of this hydroxamate group into tripodal hexadentate ligands gives rise to a reduction in stability relative to acetohydroxamic acid by 2 log units.

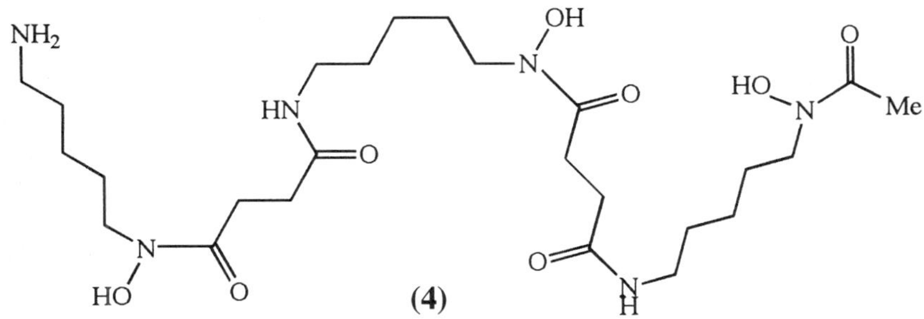

(4)

Thus, changes in ligand basicity aside, the architecture of the connecting backbone of hexadentate ligands can detrimentally influence the overall chelate stability if not correctly designed.

In extreme cases, certain ligands can adopt low-energy conformations in solution which closely match the preferred conformation of the metal chelate. Under these circumstances, rotational and translational entropic changes associated with the ligand upon complexation are minimized. Ligands with these characteristics are either termed preorganized or predisposed toward chelation.

Although the magnitude of the Fe(III)-ligand formation constants can be exceptionally large (e.g., enterobactin (**5**), log K = 49), it is nevertheless a fact that, at pH 7.4 protons will compete effectively with

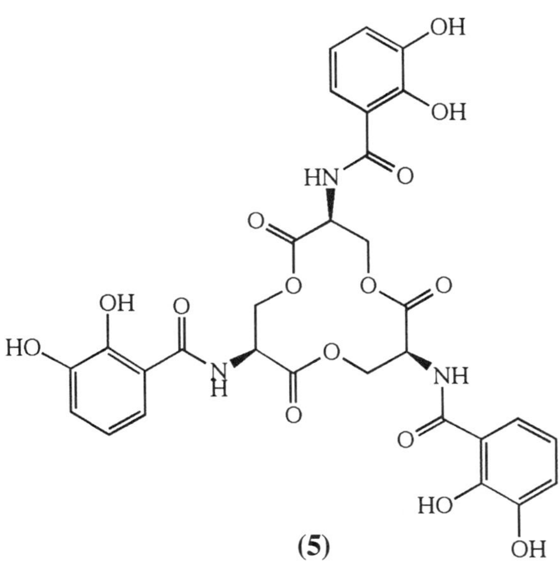

(5)

Fe(III) for bonding to many donor groups. This leads to the presence of other protonated metal-ligand species, and hence log K or log β_3 is not a suitable standard by which to gauge the effectiveness of a potential iron chelating agent under in vivo conditions. The concentration of free Fe(III) in solution provides a superior measure and is best expressed in terms of pFe^{3+} (where $pFe^{3+} = -\log[Fe^{3+}]$). Typically, pFe^{3+} values are calculated for total [ligand] = 10^{-5} M, total [iron] = 10^{-6} M at pH 7.4. The comparison of ligands under these conditions is now more meaningful, since pFe^{3+}, unlike log K, takes into account the effects of ligand basicity, ligand protonation, and hydrolysis as well as differences in metal-ligand stoichiometries. Comparison of the pFe^{3+} values for the ligands in Table 2 reveals that hexadentate ligands are far superior to their bidentate counterparts at efficient chelation under in vivo conditions. This is a direct consequence of the fact that dissociation of the complex is dependent on $[\text{ligand}]^3$ for a bidentate ligand. The degree of dissociation for a hexadentate ligand-metal complex is only dependent on [ligand]. Hence, the dilution sensitivity to chelate dissociation follows the order hexadentate < tridentate < bidentate < monodentate. It is for this reason that the majority of siderophores are hexadentate [8]. The kinetic lability of the resultant metal complex can be minimized by increasing the effective denticity of the ligand so as to satisfy fully the coordination requirements for Fe(III). Hence, hexadentate ligands exchange chelated metal much more slowly with competing ligands when

TABLE 2

A Comparison of the pFe^{3+} Values for Bidentate and Hexadentate Ligands[a]

Ligand	pFe^{3+}
N,N-dimethyl-2,3-dihydroxybenzamide (**1**)	15
MECAM (**2**)	28
Acetohydroxamic acid (**3**)	13
DFO (**4**)	26

[a]Data sourced from Motekaitis, Sun, and Martell [6b]. If the pFe^{3+} value is less than 16, the ligand is unable to compete with the hydrolysis of iron.

compared with the corresponding tris-bidentate metal complexes. Low kinetic lability is important so as to limit possible metal redistribution in vivo [9] (see Sec. 3).

2.2. Choice of Ligand

High-spin Fe(III), a spherically symmetrical tripositive cation of radius 0.65 Å, is classified as a hard Lewis acid by virtue of its high charge density and hence forms most stable bonds with ligands containing weakly polarizable atoms, such as oxygen. The bonding in such complexes is largely electrostatic in nature. This high affinity for oxyanions offers an explanation for both the extreme insolubility of Fe(III) in aqueous solution at neutral pH, and the observation that dioxobidentate ligands, namely, catecholate and hydroxamate, are generally preferred in siderophore structures. The obvious method of choice is to model prototype compounds on natural catechol and hydroxamate siderophores (see Chap. 7). However, both ligands are unsuitable when considered for clinical application. Catechols are susceptible to oxidation in the gut, are poorly absorbed orally, and form charged Fe(III) complexes resulting in intracellular trapping. This is a consequence of the resulting large charged species exhibiting poor membrane permeability properties. Hydroxamates, in contrast to catecholates, form neutral Fe(III) complexes, since they are monoprotic ligands; however, many are metabolically labile and are only poorly absorbed via the oral route.

The 3-hydroxypyridin-4-ones (3,4-HPOs) (**6**) can be considered to embody the desired chelation properties of both catechol and hydroxamate groups. Like both catechol and hydroxamate the HPOs form five-membered chelate rings in which the metal is coordinated by two vicinal oxygen atoms. Like hydroxamates, the hydroxypyridinones are monoprotic acids and thus form neutral tris Fe(III) complexes. The deprotonated form of the hydroxypyridinones have a zwitterionic aromatic resonance form which is isoelectronic and isostructural with the catecholate dianion [10]. In fact, three classes of HPO ligands exist (**6**, **7**, and **8**) and possess an intermediate chelating effectiveness for Fe(III), which is between that of a phenolate donor group and a catecholate group. This is due to the delocalization of the nitrogen electron lone pair with the carbonyl oxygen to afford a resonance structure whereby two

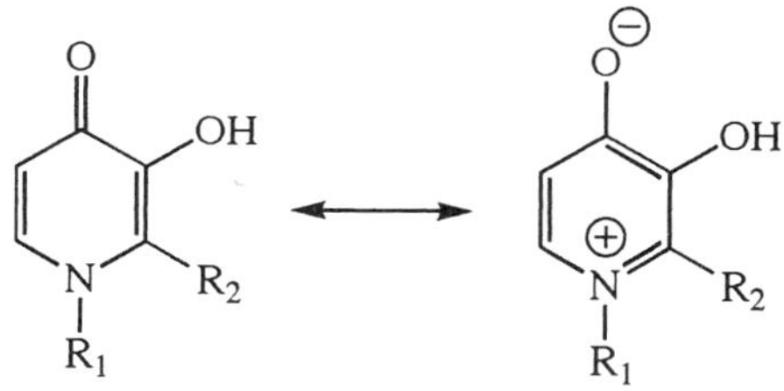

(6)

R_1	R_2	Ligand
CH_3	CH_3	CP20
CH_2CH_2OH	CH_3	CP40
CH_2CH_3	CH_2CH_3	CP94
CH_2CH_2OH	CH_2CH_3	CP102
$CH_2CH_2CH_2OH$	CH_2CH_3	CP106
$CH_2CH_2CO_2H$	CH_2CH_3	CP110

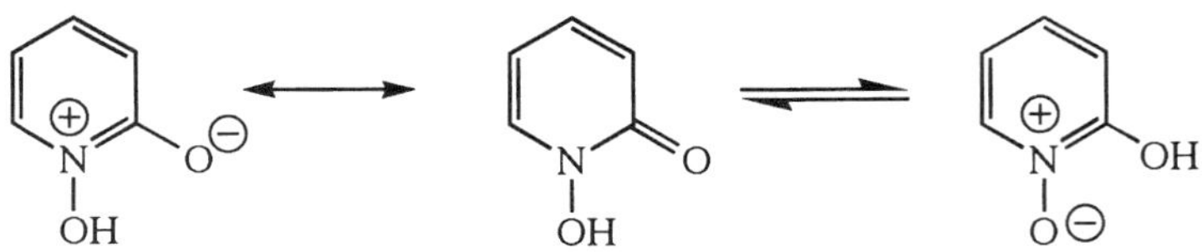

(7)

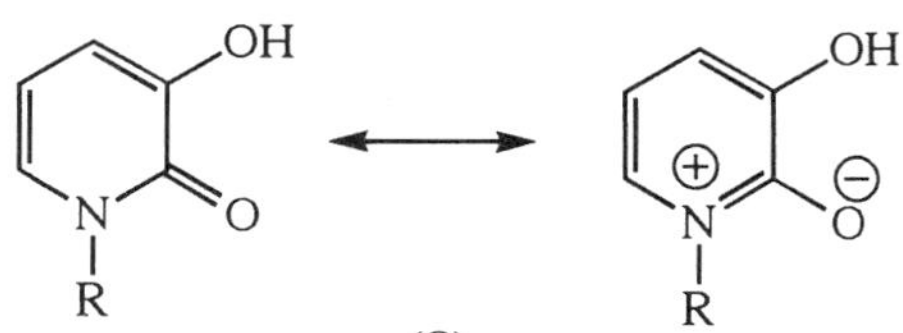

(8)

negative charges are distributed between the two donor oxygens. The resulting charge associated with these oxygen atoms is therefore greater than one but less than two. The extent of electron-pair delocalization from the nitrogen onto the carbonyl oxygen will dictate the hardness of that donor atom and hence influence overall chelate stability. Of the three classes of HPO ligands, the 3-hydroxypyridin-4-one class possesses the highest affinity for Fe(III). This is a direct consequence of the elevated pK_a value associated with the 4-oxo group compared with the 2-oxo congeners (Table 3). Hider and coworkers have pursued all three classes of HPO ligands since 1979, culminating in several patents in the early 1980s [11]. The 3,4-HPO ligands are of particular medicinal interest, because the cation discrimination for Fe(III) over that of biologically important divalent cations is excellent. Two bidentate 3,4-HPO ligands, namely, CP20 and CP94 (**6**), have progressed to preliminary assessment in humans with limited success [12,13].

TABLE 3

pK_a Values and Fe(III) Affinity Constants for Bidentate Ligands

Ligand	pK_{a1}	pK_{a2}	$\log \beta_3$ (Fe^{3+})	pFe^{3+}
N,N-Dimethyl-2,3-dihydroxybenzamide (**1**)	8.4	12.1	40	15
3-Hydroxypyridin-4-one (**6**)	3.6	9.9	37	20
3-Hydroxypyridin-2-one (**7**)	0.2	8.6	32	16
1-Hydroxypyridin-2-one (**8**)	—	5.8	27	16
Acetohydroxamic acid (**3**)	—	9.4	28	13
8-Hydroxyquinoline	4.9	9.7	37	21

2.3. Influence of Molecular Weight

In order for a chelating agent to exert its pharmacological effect, it must be able to reach the target sites at sufficient concentrations. Hence, following oral administration, the ligand must be efficiently absorbed from the gastrointestinal tract and transverse many biological membranes to access the desired target sites. Three major factors influence a compound's ability to permeate lipid membranes freely; that is, lipophilicity, ionization, and molecular size. A major distinguishing feature between bidentate and hexadentate ligands is molecular size. Bidentate ligands typically fall in the molecular weight range of 100–250 Da, whereas the larger hexadentate ligands fall in the range of 400–1000 Da. Most siderophores, including desferrioxamine B (**4**), have molecular weights between 600–900 Da. Although EDTA has a molecular weight of only 292, it is too small to encompass the chelated iron fully, thereby facilitating the potential toxicity of this metal (see Sec. 3.3). DTPA (**9**) (MW 393) and HBED (**10a**) (MW 388) are probably close to the minimum size possible for effective hexadentate Fe(III) ligands. The estimated orally absorbed fraction for a range of octadentate and hexadentate ligands, including DTPA, were found to be of the order between 2–5%; for example, the fraction of the oral dose absorbed for the hexadentate TREN-(3,2-HPO) (**11**) was found to be 3.5% in mice [14]. In contrast, the

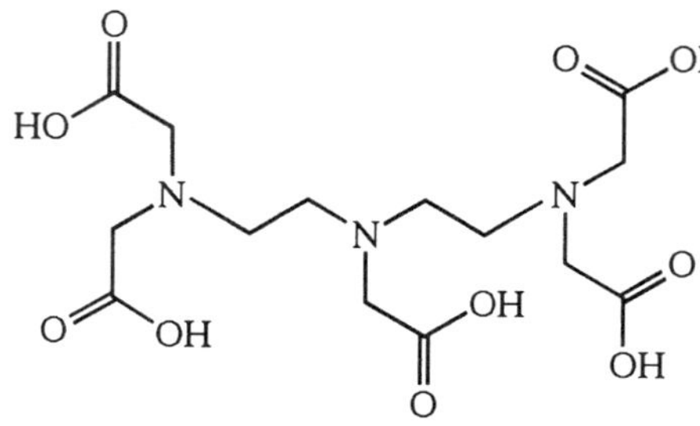

O
OH
HO
N
N
N
O
O
OH
OH
O
OH
O

(9)

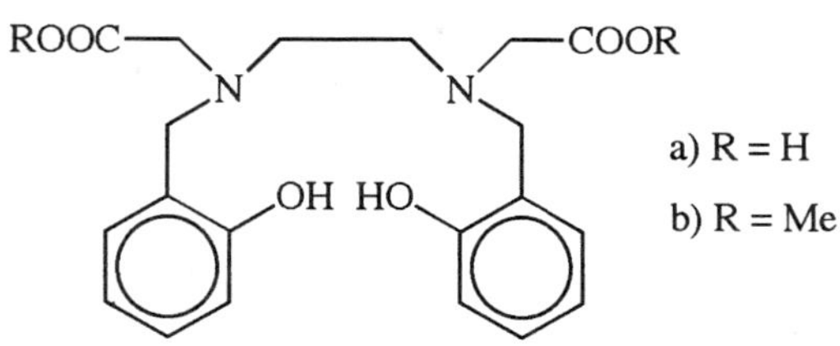

ROOC
COOR
N
N
OH
HO
a) R = H
b) R = Me

(10)

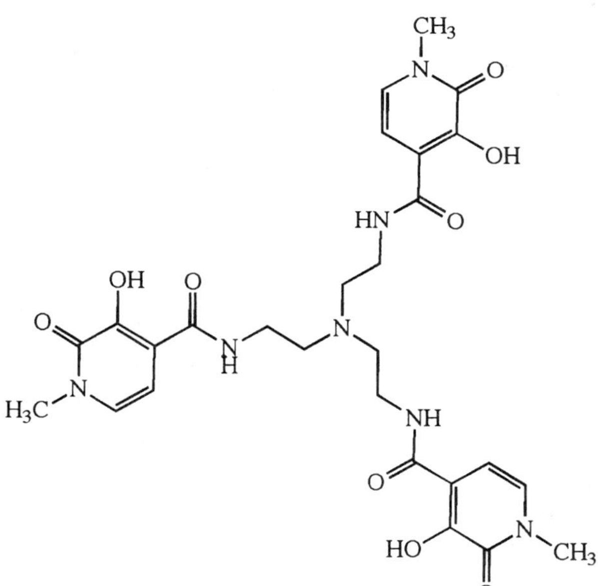

CH3
N
O
OH
HN
O
OH
O
O
N
H
N
H3C
N
NH
O
HO
N
CH3
O

(11)

bidentate ligands such as the 3,4-HPOs, by virtue of their lower molecular mass are efficiently absorbed when dosed orally. The fraction of the absorbed dose for a range of bidentate 3,4-HPOs has been found to be typically between 50–70% as assessed in the rabbit [15]. Indeed, the ligands CP20 and CP94 are both efficiently absorbed in humans [12,13].

Although efficient oral absorption is desirable, so too is the selective accumulation of the ligand in target tissues such as liver and spleen. Exposure of nontarget sites can give rise to unwanted toxicity. One such site is the brain. Access to the central nervous system via the blood-brain barrier (BBB) is sensitive to molecular size. In a systematic study by Levin, an excellent correlation between logP $(MW)^{1/2}$ and the permeability for molecules with MW <300 was shown, although no relationship held with larger ligands (MW >500) [16]. Thus, most hexadentate ligands are predicted to permeate the BBB inefficiently at best. In contrast, the smaller bidentate ligands such as the 3,4-HPOs permeate the BBB with relative ease, which is an undesirable feature in most therapies. Such ligands have the opportunity to interfere with metalloenzymes in the central nervous system that are critical for neurotransmitter activity thereby inducing potential toxicity (see Sec. 3).

2.4. Influence of Lipophilicity

Lipophilicity is a measure of the affinity of the molecule for a lipophilic environment and is commonly quantified by measuring the distribution ratio of the solute between two phases, typically *n*-octanol and water. Expressed as logP, the lipophilicity of a molecule can profoundly influence the pharmacokinetic and metabolic behaviour in vivo. Iron chelation efficiency has been found to correlate well with the lipophilicity of both the free ligand and Fe(III) complex. The partition coefficient of the free ligand of a range of 1,2-disubstituted-3,4-HPOs critically influenced their ability to mobilize iron from rat hepatocytes displaying a familiar parabolic relationship. The optimal $D_{7.4}$ value in the system was found to be close to unity. The efficacy of a range of tridentate PIH analogues (**12**) in mobilizing iron from reticulocytes also lead to the conclusion that a distribution coefficient close to unity for both free ligand and Fe(III) complex gave maximal results. Although lipophilic compounds tend to be orally absorbed to a greater extent, the fraction

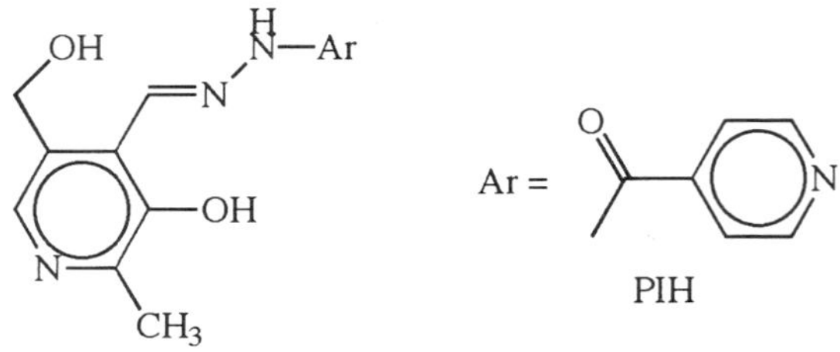

(12)

orally absorbed (F) for a range of bidentate 3,4-HPOs in the rabbit was found to be insensitive to lipophilicity, since, for example, CP40 ($D_{7.4}$ = 0.08) and CP94 (**6**) ($D_{7.4}$ = 1.78) possess F values of 73 and 62%, respectively, suggesting that molecular weight also plays a dominant role in influencing gastrointestinal absorption of compounds of this type.

Penetration of the BBB by small molecular mass solutes (MW <300) is critically influenced by lipophilicity. In principle, hydrophilic chelators ($D_{7.4}$ <0.01) are predicted to permeate the BBB inefficiently, and thus entry to the central nervous system is predicted to be low. In contrast, lipophilic 3,4-HPO ligands have been demonstrated to be toxic, with the LD_{50} in mice sharply rising for ligands with partition coefficients greater than 1 [17].

3. IRON CHELATOR TOXICITY

Although bidentate and tridentate ligands possess a clear advantage over hexadentate ligands with respect to oral bioavailability, they are potentially more toxic, as is indicated in Table 4. By virtue of their lower molecular weight, for a given distribution coefficient ($D_{7.4}$), they penetrate membranes more easily than hexadentate analogues and therefore gain access to the cytoplasm of all cells more readily.

3.1. Inhibition of Iron-Dependent Enzymes

In principle, chelating agents can inhibit metalloenzymes via three different mechanisms: removal of the essential metallic cofactor, formation of a stable ternary complex, or deprivation of the apoenzyme of its normal source of metal ion. Fe(III)-chelating agents generally do not

TABLE 4

Comparative Toxicities of Fe(III) Chelators

Bidentate and tridentate ligands	Hexadentate ligands
Penetration of BBB is dependent on $D_{7.4}$ values	Generally low penetration of BBB
Kinetically labile—iron redistribution is possible	Kinetically inert thus iron redistribution is unlikely
Affinity for iron is concentration dependent—redistribution of iron facilitated	Affinity for iron is concentration independent—redistribution of iron not facilitated
Form 2:1 and 1:1 complexes which are potentially toxic due to incompletely complexed iron	Only form 1:1 complexes, which are generally nontoxic

directly inhibit heme-containing enzymes or iron-sulfur cluster proteins. They do, however, interfere with mono-iron and bi-iron centers coordinated to oxygen ligands. Examples of such enzymes are given in Table 5. In general, lipoxygenases are inhibited by hydrophobic chelators [18], and therefore the introduction of hydrophilic characteristics into a chelator tends to minimize inhibitory potential. Concerning post-translational modification of proteins, hydroxylation of specific protein residues is achieved by a closely related group of proteins, each of which being selective for the -GlyXYGly motif. High chelator concentrations are generally required to cause appreciable inhibition; the IC_{50} values for the hydroxypyridinone CP20 (**6**), for instance, are prolyl-4-hydroxylase (150 μM) and deoxyhyposyl hydroxylase (90 μM) [19]. Thus, if the clinical levels of chelator can be limited to ≤25 μM, inhibition of these enzymes should not present serious problems. The aromatic amino acid hydroxylases are particularly susceptible to inhibition by bidentate aromatic ligands, for instance, catechols and hydroxypyridinones, owing to the similar molecular characteristics of the inhibitor and the substrate. However, the introduction of an alkyl function at the 2 position of hydroxypyridinones markedly increases the K_I value to the millimolar range, presumably due to steric interference of the chelation process at the enzyme active site [20]. Thus, by careful design, iron chelators can

TABLE 5

Nonheme Fe-Containing Enzymes
Which Are Inhibited by Iron Chelators

Lipoxygenase family (5-, 12-, and 15-lipoxygenases)
Enzymes associated with post translational modification of proteins
prolyl-4-hydroxylase
lysine-4-hydroxylase
deoxyhyposyl hydroxylase
Aromatic hydroxylase family
phenylalanine hydroxylase
tyrosine hydroxylase
tryptophan hydroxylase
Ribonucleotide reductase

be synthesized which exert minimal inhibitory influence on most cytoplasmic metalloenzymes.

In contrast to other enzymes discussed above, ribonucleotide reductase presents a problem, as it possesses a rapid turnover and hence if the concentration of iron in the precursor pool is reduced by a chelator, then the production of active enzyme is adversely effected [21]. This can present a critical problem as inhibition of ribonucleotide reductase leads to the depletion of deoxynucleotide pools, thereby inhibiting DNA synthesis. Inhibition of ribonucleotide reductase therefore interferes with the G1→S transition, blocking the cell cycle. It is likely that the bone marrow toxicity induced by certain iron chelators results from the inhibition of this enzyme [22]. Agranulocytosis and neutropenia have both been associated with high doses (> 80 mg kg^{-1}) of CP20 [23]. This problem may prove to be impossible to avoid, and therefore the distribution of iron chelators should be limited to nonproliferating cells if at all possible.

3.2. Distribution of Chelators

As there are a number of enzymes and cell responses (for instance, those dependent on the iron response protein) [24] which are dependent on intracellular iron levels, a practical strategy to minimize toxicity is to limit the distribution of the chelator to the region of elevated iron levels. Thus, with thalassemia major, the liver is one of the major sites of iron overload, with neurodegenerative diseases, it is the brain, and more specifically with Parkinson's disease, it is the substantia nigra.

One of the reasons why desferrioxamine is relatively nontoxic is that it is extremely hydrophilic and possesses a relatively large molecular weight. As a result, this chelator enters cells by simple diffusion at an extremely slow rate, whereas being rapidly cleared by the kidneys. Thus, bone marrow toxicity is not a frequently observed side effect of desferrioxamine even though it is supplied to some patients at doses close to 2 g 24 h^{-1} for extended periods. Clearly, hydrophilic chelators are likely to be less toxic than more hydrophobic analogues, with an optimal $D_{7.4}$ value being close to 0.01 [25].

3.3. Toxicity of Iron Complexes

3.3.1. Redox Cycling

As indicated in Section 3.1, high selectivity for Fe(III) can be achieved by utilizing charged oxyanions; for instance, catecholate, hydroxamate, and pyridonate. Complexes derived from such interaction possess high redox potentials and are thus difficult to reduce [26]. Chelators which utilize nitrogen as well as oxyanions as ligands possess lower redox potentials and the coordinated iron can be reduced enzymatically under biological conditions (Table 6). Such ligands include 8-hydroxyquinoline, EDTA, NTA (**13**), aryl hydrazones (**12**), and desferrithiocin (**14**).

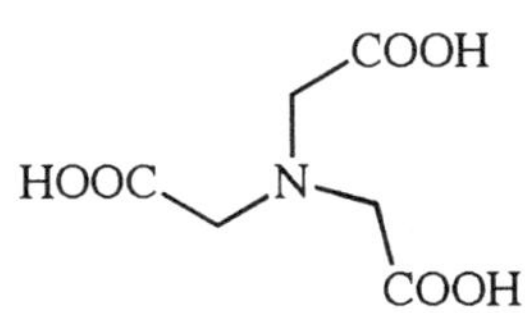

(13)

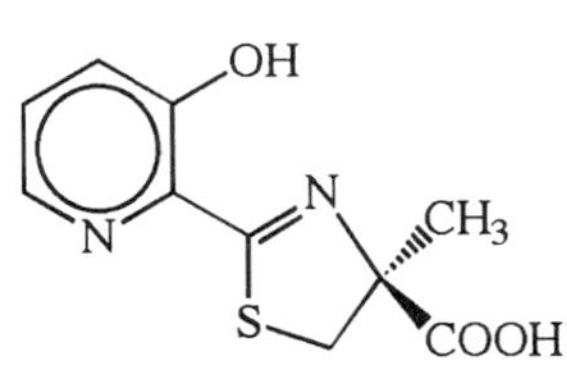

(14)

TABLE 6

The Hydroxyl Radical-Generating Ability of a Range of Iron Complexes: Comparison with EDTA[a]

Chelator	Hydroxyl radical production (% of EDTA effect)
Control	8.1 ± 4.9
EDTA	100
NTA (**13**)	39.6 ± 9.2
DTPA (**9**)	75.6 ± 11.7
HBED (**10a**)	10.8 ± 4.5
DFO (**4**)	7.9 ± 3.9
DFT (**14**)	43.7 ± 9.4
CP20 (3:1)[b] (**6**)	16.9 ± 4.1
CP20 (5:1)[b] (**6**)	1.6 ± 2.0

[a]Xanthine oxidase was used as a source of superoxide and hydrogen peroxide and *N,N'*-(5-nitro-1,3-phenylene) bisglutaramide as a trapping agent. Iron complexes (0.5×10^{-3} M) were incubated at 37°C for 30 min [27].
[b]Molar ratio ligand:iron.

In principle, therefore, these complexes can redox cycle under aerobic conditions, producing oxygen radicals. Significantly, ferrithiocin has been reported to be toxic under cell culture conditions [28].

3.3.2. Partially Complexed Iron

Iron should be coordinated by a therapeutically useful ligand in such a manner as to prevent direct access to oxygen or hydrogen peroxide. If either of these molecules binds directly to the iron atom, hydroxyl radical (•OH) production is possible. Most hexadentate ligands, for instance, desferrioxamine B, reduce •OH production to a minimum by completely masking the surface of the iron. This is not the situation with EDTA, where the chelator is not sufficiently large to envelope the iron completely, a so-called basket complex resulting. Direct access by small ligands to the iron atom is possible with such structures. Many bidentate ligands, for instance, catechol and hydroxamates, are efficient •OH generators in the presence of iron under aerobic conditions. This is due to the partial dissociation of the 3:1 complex to the 2:1 complex (see Sec. 4) (see Table 5). In contrast, 3-hydroxypyridin-4-ones behave more like hexadentate ligands in that the 3:1 complex is the dominant species at pH 7, and the iron atom is completely protected from attack by oxygen in this form (see Fig. 1). Thus, as long as the neutral 3:1 pyridinone complex (ML_3) dominates the speciation of the low molecular weight forms of iron, then minimal hydroxyl radical production should occur. Indeed, the viability of cultured hepatocytes, as judged by the release of lactate dehydrogenase, is enhanced by the presence of 3-hydroxypyridin-4-ones in the concentration range of 20–100 μM [29]. A major reason for the marked difference in behavior between the abilities of catechol and 3-hydroxypyridin-4-one to generate •OH is the percentage of the 2:1 (ligand:iron) complex present at pH 7. Although the catecholate anion possesses a high affinity for Fe(III), it also binds two protons tightly (pK_a 13.0 and 9.2), the 3:1 complex only forming in appreciable quantities above pH 8 [9]. In contrast, the corresponding 2:1 (hydroxypyridin-4-one:iron) (ML_2) complex only makes a minor contribution to iron speciation at pH 7.4 (Fig. 2). Plasma non-transferrin-bound iron (NTBI) levels in transfused thalassemic patients typically fall in the range of 2–15 μM [30]. If this level can be maintained at less than 2 μM during chelation therapy (and preliminary evidence indicates that this is indeed possible), then maintaining a 3-hydroxypyridinone plasma concentration of 20 μM would limit the concentration of the potentially toxic 2:1 species $\leqslant 10^{-8}$ M. At this concentration, the hydroxyl radical flux would be negligible.

4. CLINICALLY USEFUL BIDENTATE AND TRIDENTATE LIGANDS

4.1. Bidentate Chelators

Ideally, monobasic bidentate ligands should be adopted for clinical use as they result in the formation of neutral Fe(III) complexes, which will

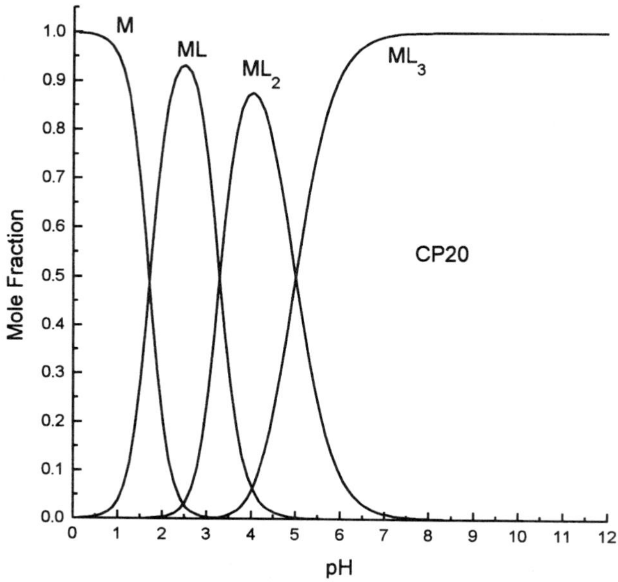

FIG. 2. Speciation plot of a CP20 (**6**) and Fe(III). Concentration of ligand = 1×10^{-5}M, concentration of Fe(III) = 1×10^{-6} M. The following parameters were incorporated into the model $\log K_1 = 15.1$; $\log K_2 = 12.1$; $\log K_3 = 9.8$; $pK_{a1} = 3.58$; $pK_{a2} = 9.92$; $\log K$ (MOH/M·OH) = 11.81; $\log K$ ($M(OH)_2$/MOH·OH) = 10.5; $\log K$ ($M_2(OH)_2$)/$M^2 \cdot OH^2$) = 25.1.

penetrate membranes if the $D_{7.4}$ value falls in the range of 0.5–5.0. Clearly, the dominant species under most physiological conditions should be the neutral 3:1 species, and the pFe^{3+} value of the ligand should be greater than that of the hydroxide anion—at pH 7.4, this value is 16. If the pFe^{3+} value is less than 16, then the ligand at a concentration of 10 μM will not compete efficiently with the hydroxide anion for Fe(III). Only two bidentate ligand classes fit this category, 8-hydroxyquinoline and 3-hydroxypyridin-4-one (see Table 3). Owing to the nitrogen donor atom in 8-hydroxyquinoline, this compound is much less selective than the hydroxypyridinone and hence is toxic. Speciation plots of 3-hydroxypyridin-4-one with Fe(III) demonstrate that at a concentration of 10 μM, the neutral 3:1 complex (ML_3) dominates over the pH range of 6–9 (see Fig. 2). The reason for the marked difference between the pyridin-4-one and pyridin-2-one classes is the greater delocalization of the nitrogen lone pair on the 4-carbonyl oxygen in the former class.

The limited introduction of Deferiprone (CP20) and CP94 into humans has established that it is feasible to undertake effective iron chelation via the oral route using this class of molecule [12,13]. However, both molecules have limited efficacy in humans, although in rodents, CP94 is highly effective at removing iron. In both the rat and human, the major metabolite of CP20 is the glucuronide which, as a result of conjugation of the 3-hydroxyl function, is unable to bind iron [31]. This critical dependence of chelator efficacy on metabolic behavior has led to a concept of ligand design which prevents extensive phase II metabolism; that is, conjugation with glucuronide and sulfate. In principle, such a modification would permit the use of lower pyridinone doses, since in contrast to both CP20 and CP94, the bulk of the absorbed dose would retain its chelating ability. A further refinement of the concept is to design pyridinones which are only metabolized by phase I processes, thereby leaving the essential bidentate chelating function of the pyridinone intact (Fig. 3). Such a strategy renders it possible to control the distribution of the chelator. Thus, if the metabolite is much more hydrophilic than the parent pyridinone, the majority of extracellular chela-

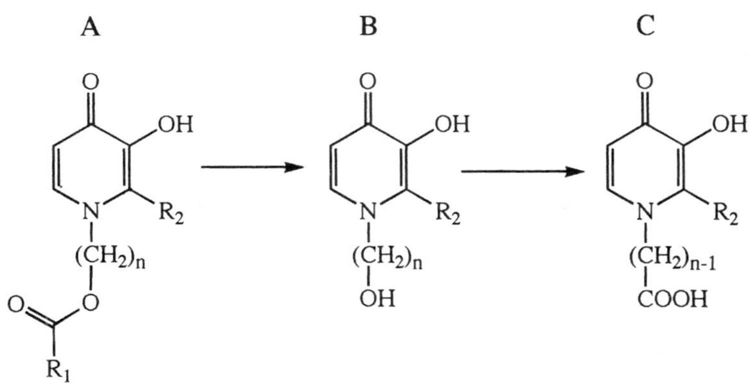

FIG. 3. Phase I metabolism of 3,4-HPOs. By controlling the value of n, it is possible to design molecules which predominately undergo one of the three following transformations: A → B; A → C; B → C. By suitable choice of the R_2 substituent, undesirable phase II conjugation reactions involving the 3-hydroxyl function are minimized. R_1 can either be aliphatic or aromatic.

tion would be achieved by the hydrophilic metabolite and thus access to the brain and other critically sensitive organs would be minimized. Pyridinones possessing both these properties have recently been identified [32]. Thus, the hydroxyethyl-substituted compound (CP102 (**6**)) is not extensively metabolized, and whereas the hydroxypropyl-substituted compound (CP106 (**6**)) undergoes extensive phase I metabolism to the corresponding carboxyethyl-substituted pyridinone (CP110), neither CP106 nor CP110 undergo appreciable conjugation with glucuronic acid [32]. CP110 is much more hydrophilic than CP106 ($D_{7.4}$ values being <0.001 and 0.48, respectively).

In practice, a clinically useful iron chelator for the treatment of body iron overload, requires contradictory partitioning properties (Table 7). To ensure good oral absorption and extraction by the liver, a $D_{7.4}$ value of 1–5.0 is ideal. However, in order to minimize penetration of the bone marrow, brain, and placenta, a much lower value is required; typically < 0.001. Clearly, these parameters cannot be the property of a single molecule but can be achieved by "prodrug" design. One such possibility involves the selective hydrolysis of hydrophobic esters, to hydrophilic hydroxypyridinones (see Fig. 3). These esters are relatively stable in both the lumen of the small intestine and in plasma, but are rapidly hydrolyzed in hepatocytes by cytoplasmic carboxyesterases. By adopting this strategy, it is possible to design chelators with distribu-

TABLE 7

Lipophilicity Paradox: $D_{7.4}$ Requirements for Efficient Absorption and Selective Distribution

Desired property	$D_{7.4}$
Good absorption from the gastrointestinal tract	>0.2
Efficient liver extraction	>1.0
Poor entry into peripheral cells (thymus, muscle, heart, bone marrow)	<0.001
Poor ability to penetrate the blood-brain and maternal/placental barriers	<0.001

tion properties similar to the ideal ranges outlined in Table 7. The concept works extremely well in mammals, with hydrophobic esters enhancing iron excretion in several iron-overload models [33].

4.2. Tridentate Chelators

Tridentate ligands fall into two major classes (**15** and **16**). The fundamental difference between these two types is that in type **16**, one of the

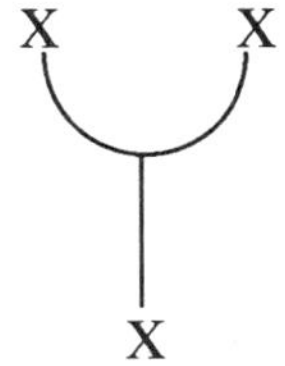

(15)

(16)

donor atoms (Y) is also part of the linking chain. This renders it possible to use smaller chelating rings and therefore to decrease the adverse changes in entropy on chelation of iron. However, for optimal activity and selectivity with Fe(III), the three ligands should be negatively charged oxygen, which excludes type **16**. Thus, although desferrithiocin (**14**), which belongs to this class, does form a stable 2:1 complex with Fe(III) [34], it also possesses appreciable affinity for Fe(II) and thus has the potential to redox cycle in the presence of intracellular reductases (see Sec. 3).

A series of compounds based on type **15** have also been synthesized (**17** and **18**). Unfortunately, the entropy factor is such that these molecules behave as bidentate pyridinones and the potential ligating side arm containing either the phenolic or carboxylic function plays no demonstrable role in the coordination of Fe(III) [35]. The aryl hydrazones (**12**) also bind Fe(III) in a tridentate manner and have been demon-

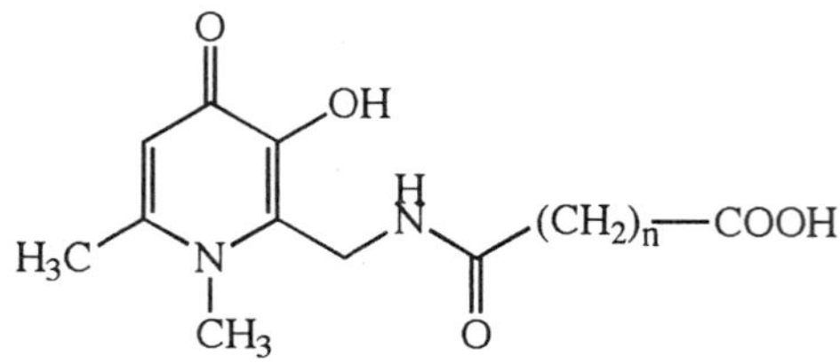

(17)

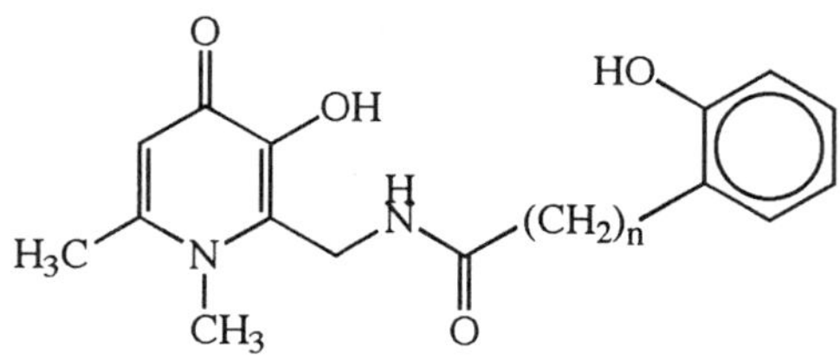

(18)

strated to be effective under in vivo conditions [36]. Again, the presence of "linking/coordinating" nitrogen endows this molecular class with an appreciable affinity for Fe(II) [37].

A serious potential problem associated with tridentate ligands is that, unlike bidentate compounds, there is a possibility of forming polymeric structures. Such structures would be difficult to clear via the kidney.

5. CLINICALLY USEFUL HEXADENTATE LIGANDS

5.1. Desferrioxamine

The naturally occurring microbial siderophore desferrioxamine B (**4**), a hexadentate tris-hydroxamic acid derivative produced by *Streptomyces pilosus*, is regarded as the therapeutic agent of choice for the treatment of transfusional iron overload. Unfortunately, it has to be parenterally administered, since its oral bioavailability is poor, as it is inefficiently absorbed from the gastrointestinal tract. Following intravenous bolus injection, desferrioxamine (DFO) is rapidly cleared from the plasma with a half-life of only 5–10 min. Formulated as the mesylate salt (Desferal®, Ciba Pharmaceuticals) desferrioxamine is able to maintain iron balance in patients receiving chronic blood transfusional regimens only when administered via intravenously or subcutaneous ambulatory infusion 8–12 h a day, 5–7 days a week. Consequently, patient compliance with this expensive and cumbersome regimen is often poor. Although desferrioxamine has been successfully employed clinically since the mid 1970s, its use has associated limitations. Since desferrioxamine is a natural siderophore, several microorganisms, notably *Yersinia enterocolitica*, are able to utilize the ferric chelate giving rise to opportunistic infections. Other undesirable properties include visual and auditory abnormalities associated with aggressive high-dose therapy [38]. The in vivo chelation efficiency of desferrioxamine has been calculated to be at best 10%.

Hence, although DFO has been of proven therapeutic benefit, considerable effort has been expended in the pursuit of orally active derivatives. The preparation of a range of DFO prodrugs obtained via esterification of the labile hydroxamate functions, thereby increasing lipophilicity, has been investigated by Ciba Pharmaceuticals as a potential means of improving the oral bioavailability of the drug. Improvements in oral activity were, however, only found to be marginal with a range of tetra-acyl derivatives. This prodrug approach, however, only addresses the bioavailability problem and not the in vivo iron chelation efficiency limitation [39]. To address this latter issue, several strategies centered on molecular modifications of the backbone of desferrioxamine have been pursued. A reductionist approach whereby the *N*-propanoyl-*N*-pentyl moieties were replaced with smaller C-5 or C-8 methylene units yielded novel "amide-less" DFO analogues of lower molecular weight and increased lipophilicity. In vivo assessment in a rodent model failed to identify a superior ligand to DFO. However, changes in the relative amounts of iron excreted via the urinary and biliary routes were observed together with protracted biliary iron clearance [40]. Modifications at the terminals of DFO by means of accessing neutral lipophilic polyether derivatives has also been pursued. When assessed in a rodent model, both analogues performed better than DFO when administered subcutaneously. Disappointedly, when compared at the same dose orally, the polyether analogues could only attain 0.4–0.5 efficiency compared with that of DFO, suggesting that lipophilicity is not the only determinant in oral activity [41].

By coupling DFO to a variety of biocompatible polymers such as dextran and hydroxyethyl-starch, improvements can be afforded in reducing the rapid plasma clearance of the free ligand which currently limits DFO administration. Also, the high molecular weight forms of DFO have been demonstrated to be less toxic when compared with the free ligand, since they are largely confined to the extracellular compartment [42].

5.2. DTPA

Diethylenetriaminepentaacetic acid, DTPA (**9**), was first clinically evaluated in 1961. DTPA is largely confined to the extracellular compartment in vivo and is excreted within 24 h of parenteral administration and almost exclusively in the urine. The lack of oral activity may be rationalized by consideration of the extent of ionization at pH 7.4. An additional disadvantage of the carboxylate function is its relatively poor metal selectivity. This lack of selectivity leads to zinc depletion in patients receiving DTPA therapy. Consequently, zinc supplementation is required to prevent the toxic sequelae of such depletion. DTPA has been successfully administered to patients withdrawn from desferrioxamine therapy for up to 18 months without toxicity provided zinc supplementation was administered concurrently [43]. Currently, the clinical use of DTPA is reserved for iron-overloaded patients who develop toxicity from treatment with DFO.

5.3. HBED

N,N′-bis(2-hydroxybenzyl)ethylenediamine-*N,N′*-diacetic acid, HBED (**10a**), was designed by Martell and coworkers [64] some 30 years ago in an attempt to improve the poor selectivity of the aminocarboxylate ligands EDTA and DTPA (**9**) for Fe(III) over other bivalent and trivalent cations. HBED is structurally related to EDTA in which two carboxylate units are replaced with *ortho*-phenolates. As a result, cation selectivity is markedly improved and the undesirable “basket” nature of EDTA is lost. Efficacy of HBED in iron chelation was first demonstrated in a hypertransfused rat model of iron overload in which HBED dosed intraperitoneally produced three times total iron excretion compared with that of DFO given intraperitoneally at equivalent doses [65]. Also, in the same study, HBED given orally produced 0.7 times iron excretion compared with that of DFO given intraperitoneally. The lipophilic methyl diester dimethyl-HBED, DmHBED (**10b**), also yielded encouraging results producing six times elevation and four times elevation in total iron excretion when given intraperitoneally and orally, respectively, when compared with DFO given via the intraperitoneal route [44]. In a limited phase I clinical trial, HBED was compared with DFO (60 mg/kg infused subcutaneously over 8 h) in four hypertransfused patients at either 40 or 80 mg/kg divided into three equal doses/day given orally. Blood profiles and serum chemistries were unremarkable. The total iron excretion relative to that of DFO was, however, disappointing. At 40 mg/kg, HBED was able only to achieve 11.4% net relative iron excretion, and at 80 mg/kg, this only improved to 16.6% that of DFO. A two- to threefold increase in zinc excretion was also observed [45]. The low iron excretion observed with HBED can possibly be attributed to poor bioavailability associated with low gastrointestinal absorption. Improving the lipophilicity such as by the use of ester derivatives may afford improvement in iron excretion efficiency. However, no human data are currently available for the lipophilic prodrug DmHBED. HBED and DmHBED have, however, been concurrently assessed in a validated primate model of iron overload. Using the *Cebus* monkey, HBED (**10a**) and DmHBED (**10b**) dosed at 150 μmol/kg orally were compared with DFO at the same dose (as subcutaneous bolus injection). The efficiency of HBED was found to be 11% that of DFO, and the dimethyl ester DmHBED approximately 30% that of DFO [46].

5.4. DBED

Structurally related to HBED, *N,N′*-bis(di(3,4,5-trimethoxybenzyl))-ethylenediamine-*N,N′*-diacetic acid, DBED (**19**), has been developed as a site-specific hydroxyl radical scavenger. The ligand undergoes reaction with a hydroxyl radical to afford a pentadentate chelating product able to complex Fe(III), $\log K_1 = 28$. It is hoped that in vivo activation of the ligand would occur in cellular environments characterized by abnormal levels of oxidative stress resulting in the production of oxygen-derived free radicals [47].

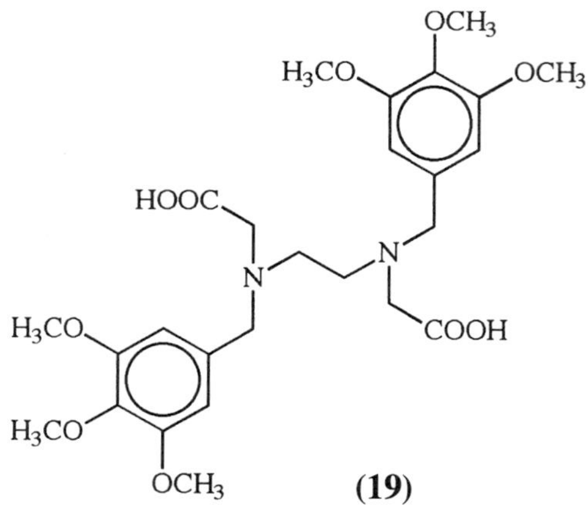

(19)

5.5. Razoxane

The *bis*-dioxopiperazine compound, razoxane (ICRF-159, Razoxin®, Zeneca Pharma) (20), is a racemic mixture of levrazoxane (ICRF-186) and

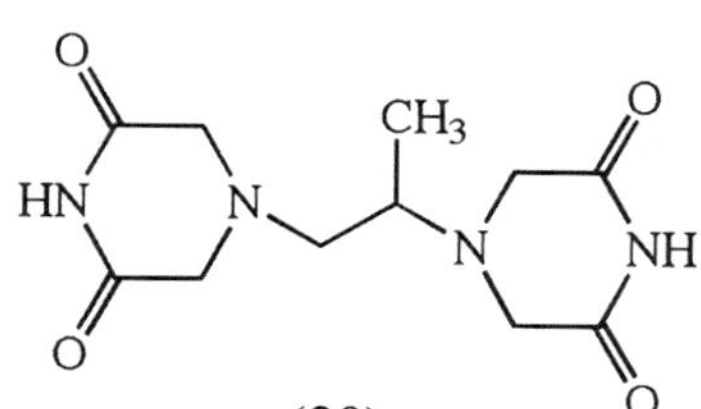

(20)

dexrazoxane (ICRF-187) and can be considered as a diimide derivative of EDTA. This compound initially found successful therapeutic utility in the systemic treatment of cutaneous psoriasis and psoriatic arthropathy. However, the development of acute myeloid leukemias and other malignancies made it contraindicated in nonmalignant diseases [48].

Recently, the (+)-(*S*)-enantiomer, dexrazoxane (ICRF-187, ADR 529), has shown great promise in phase III clinical trials being conducted in order to assess its efficacy at reducing the cumulative dose-limiting cardiotoxicity associated with the use of the antineoplastic agent doxorubicin. Dexrazoxane has not only been shown to be effective at offering cardioprotection against doxorubicin but also ameliorates the toxicity associated with other anthracyclines, notably daunorubicin and epirubicin. Considerable evidence has now amassed to implicate iron-based oxidative stress resulting in the generation of oxygen-derived free radical-mediated damage in anthracycline-induced cardiomyopathy. Razoxane (**20**) is considered a prodrug, as it requires in vivo conversion via hydrolysis to a two ring-opened active metabolite. The neutral lipophilic form is believed to diffuse intracellularly where both imide rings are hydrolytically cleaved to afford the diacid diamide product ADR 925 (*S*-enantiomer of ICRF-198) via one of two one-ring opened intermediates. This conversion can be effected both enzymatically in such organs as liver and kidney as well as nonenzymatic iron-promoted hydrolysis in the case of the one-ring opened intermediates. Both the two one-ring opened intermediates and the final product, ADR 925, are able to chelate Fe(III) from the doxorubicin-iron complex [49]. Since the production rates of the one-ring open intermediates are greater than for the second hydrolysis step, intracellular iron chelation may occur via the intermediates followed by rapid iron-promoted conversion to the stronger chelating derivative, ADR 925, where it is either able to chelate "free" iron or interact directly with the anthracycline-metal complex directly. This two-ring opened product, ADR 925, although able to chelate Fe(III), is anticipated to retain appreciable affinities for biologically important divalent cations such as Cu(II), Zn(II), and Mg(II) by virtue of the relatively nonselective aminocarboxylate ligands.

6. THERAPEUTIC APPLICATIONS OF Fe(III)-SELECTIVE LIGANDS

6.1. Treatment of Systemic Iron Overload

The adult human has a total of 4–5 g of body iron, a large portion of which is in the form of hemoglobin. Several blood diseases require the regular transfusion of red cells. One of the most important of these on a worldwide basis is thalassemia. This disease is associated with an abnormality of red blood cells, arising from an unbalanced production of the globin chains of hemoglobin. This leads to a shortened red cell survival in the circulation and consequently to anemia. In the severe form of the disease, thalassaemia major, life-long blood transfusion is necessary, starting from 2–6 months after birth [3]. As each unit of blood (500 mL) contains approximately 250 mg of iron, repeated blood transfusion will inevitably lead to an excess accumulation of iron in the body. Some patients may accumulate a total body iron load in excess of 60 g from repeated blood transfusions. This excess iron is distributed throughout the body but is found in highest concentrations within the liver. The excess iron is also present in a low molecular weight form within cells and in plasma leading to free radical-mediated damage to the heart, endocrine glands, and the liver.

Over 40 000 cases with thalassemia major occur worldwide annually, particularly in the Mediterranean and Far East Regions, most of whom will die before the third decade of life without treatment to remove the excess iron with the iron chelator desferrioxamine. Clearly, an orally active iron chelator would be much preferred.

The treatment of sickle cell crises can also be effectively managed by blood transfusion, but many clinicians prefer not to adopt such treatment on a regular basis owing to the associated problems of iron overload. If a nontoxic orally active iron chelator were available, this form of disease management would undoubtedly increase. Iron absorption may be increased secondary to chronic hemolysis, as in patients with thalassaemia intermedia. With these patients, the anemia is not usually sufficiently severe to necessitate regular blood transfusion, as in thalassemia major, but iron accumulates over many years from increased absorption via the gut, and again there is an urgent need for an orally active iron chelator [50]. The World Health Organization (WHO) estimates that in the year 2000, 7% of the world's population will be carriers of one form of a hemoglobinopathy. The reasons for the common occurrence of this inherited condition is that such changes provide resistance from malaria [51].

6.2. Elevated Iron Levels in the Brain

There is considerable overlap between the iron-rich areas of the brain and those regions where GABA neurons terminate. In these regions, approximately 30% of the iron is bound to ferritin, the remainder being associated with enzymes, respiratory proteins, and transferrin. A large proportion of this iron is located within neurons and is probably associated with neurotransmitter metabolism; brain tyrosine hydroxylase, for instance, is an iron-dependent enzyme. The presence of such high concentrations of iron in certain nuclei of the brain is a potential hazard, particularly as dopamine, an iron chelator, is present in dopaminergic neurons at a concentration of 50 mM. Not only is dopamine a chelator of iron, but it also enters an internal redox reaction with the metal, thereby generating free radicals [52]. Thus, dopamine-rich areas of the brain are predicted to be particularly sensitive to imbalance of iron levels. Significantly, an increased nigral iron content has been reported for postmortem parkinsonian brain [53]. This elevated level of iron could be, at least partially, responsible for the observed enhancement of lipid peroxidation that occurs in the substantia nigra of patients dying with Parkinson's disease. Elevated levels of iron have also been associated with many neurodegenarative disorders of the brain, including Alzheimer's disease. Clearly, such elevated iron levels are undesirable and will almost certainly enhance the progression of each of these disease states. Iron-specific chelators which are capable of crossing the blood-brain barrier could, in principle, minimize the oxidative damage induced by iron and thereby slow down the progression of the disease.

6.3. Temporary Elevated Intracellular Iron Levels in Localized Tissue

6.3.1. *Ischemic Tissue*

Over the last decade, overwhelming evidence has been gathered which suggests that oxygen-derived free radicals play the major role in producing microvascular and parenchymal damage associated with reperfusion of ischemic tissue. Tissue damage is thought to be related to the production of oxygen radicals either within the affected tissues or the associated extracellular fluid via white blood cells and in particular neutrophils. On reperfusion, however, the rapid increase in oxygen tension results in the production of a large pulse of hydroxyl radicals due to the presence of the higher concentration of the labile iron pool. The origin of the redox active iron is likely to be ferritin, which is mobilized as a result of an increased cellular reducing potential under the anaerobic conditions associated with ischemia [54]. Such pathogenic mechanisms may well hold for ischemic injury to the small bowel, kidney, heart, and central nervous system. There is some evidence for the beneficial influence of the presence of iron chelators during reperfusion of ischemic tissue [55].

6.3.2. *Inflamed Tissue*

Disturbance of iron metabolism is a prominent feature of rheumatoid disease; many rheumatoid arthritis patients have raised levels of synovial fluid ferritin and iron deposits within their synovial tissue [56]. Although iron chelators undoubtedly possess anti-inflammatory properties, high concentrations are necessary and selective direction of these molecules to the site of inflammation presents a major problem.

6.3.3. *Protection Against Redox-Cycling Chemicals*

Paraquat is selectively accumulated by lung tissue via the polyamine-transport system. Once located intracellularly, it is capable of undergoing a one-electron redox cycling mechanism with cytochrome P_{450} reductase. The resulting paraquat radical is capable of mobilizing iron from ferritin, thereby elevating intracellular levels of "low molecular weight iron" which can lead to hydroxy radical-induced damage. Such damage is minimized by treatment with selective iron chelators [57]. Doxorubicin (and related drugs) are capable of interacting with DNA probably via intercalation, thereby leading to the inhibition of tropoisomerase. It is an extremely successful chemotherapeutic agent, being a first-line choice for the treatment of acute leukemias, lymphomas, and a variety of solid tumors. Doxorubicin possesses an iron-chelating site and is able to redox cycle and thereby mobilize iron from ferritin. This undesirable effect has been associated with cardiac toxicity, with doxorubicin being accumulated by heart tissue.

Desferrioxamine B (**4**) has been shown to protect tissues against doxorubicin-induced damage, and currently, a prodrug of a chelator, razoxane (**20**), has been introduced for cardiac protection from doxorubicin-induced damage [48].

6.4. Temporary Elevated Extracellular Iron Levels

Under normal conditions, transferrin is not fully saturated, and thus any low molecular weight form of iron that enters the blood supply, for instance, iron-citrate, is rapidly transferred to apotransferrin. Thus, in the normal state of health, the concentration of non-transferrin-bound iron (NTBI) is vanishingly small. However, under certain physiological states, transferrin becomes saturated with iron and NTBI levels rise to levels of 10–15 μM. NTBI is removed from the circulation by several tissues, including the heart, and therefore is potentially toxic. The presence of NTBI in the systemic circulation also renders the organism susceptible to bacterial and fungal infections. Iron chelators can be used to reduce NTBI levels [30] and thus, in principle, protect against radical damage and infection by microorganisms. A number of conditions lead to the presence of NTBI.

Thus, premature infants born earlier than 35 weeks of gestation frequently possess NTBI levels of up to 10 μM [58]. This probably results from the fetus being well supplied with iron from the maternal blood stream, while possessing relatively low levels of transferrin. This high level drops to zero within a period of 1–2 days; however, during this critical period, irreversible damage can occur to the lungs. Chemotherapy using a wide range of regimens associated with the treatment of cancer or bone marrow transplantation leads to saturation of transferrin and the consequential build up of NTBI levels [59]. These changes result from the "shutdown" or "partial shutdown" of the bone marrow in response to the toxic influence of the chemotherapeutic agent selected for clinical treatment. NTBI can remain at an elevated state for periods of up to 2 weeks and during this time be responsible for free radical-induced damage to critical organs and, more specifically, to increased infection by microorganisms normally controlled by the presence of apotransferrin. Liver disease is also frequently associated with the presence of NTBI. It is possible that iron chelation could prove to be beneficial in all these circumstances.

6.5. Selective Inhibition of Noheme-Containing Enzymes

Iron-containing metalloenzymes can be divided into the heme and nonheme types. In general, chelating agents are not effective inhibitors of heme-containing enzymes, but in contrast many nonheme iron-containing enzymes are susceptible to such inhibition. Such iron centers are found in ribonucleotide reductase and the lipoxygenase family.

Ribonucleotide reductase catalyzes an essential metabolic step in DNA synthesis, and a wide range of iron chelators have been shown to inhibit this enzyme, and this is undoubtedly the reason for the cytotoxic properties of many such molecules (see Sec. 3). Such agents hold the cell cycle in the S-stage, because the synthesis of DNA is inhibited. The ability of iron chelators to inhibit ribonucleotide reductase has led to several proposals for therapeutic application; for instance, the treatment of neoplastic disease [55]. Iron ligands, and in particular oligodentate catecholates, have been demonstrated to be potent inhibitors of DNA synthesis in a murine leukemia cell line L1210 [60]. Unfortunately, many such potent iron chelators are also toxic to bone marrow and they are only likely to become useful antineoplastic agents if they can be specifically directed to these cell types. Synchronization of cell cycling can be achieved by the use of iron chelators, thereby reversibly inhibiting proliferating cells in the S-phase. Removal of the chelator leads to the complete recovery of the cells, but now the cells are all in a similar stage of the cell cycle [22]. This synchronization has potential for enhancing the efficiency of certain chemotherapeutic agents.

The lipoxygenase family of enzymes catalyze stereospecific oxygenation reactions of fatty acid substrates and are inhibited by a range of iron chelators. The inhibition of 5-lipoxygenase decreases leukotriene production without adversely influencing cyclooxygenase-based prostaglandin production. A wide range of iron chelators have been shown to inhibit this enzyme without inhibiting cyclooxygenase. In particular, hydroxamates have been developed as potent, selective 5-lipoxygenase inhibitors [61], some of which are orally active. Leukotrienes have been suggested to be important in a wide range of diseases, including asthma, arthritis, and psoriasis, and consequently chelators with a highly selective action may prove to possess therapeutic potential.

6.6. Iron-Transport-Mediated Drug Delivery

Some natural antibiotics contain a siderophore structure, for instance, δ_1-albomycin (**21**), which is produced by *Streptomyces subtropicus*. The linear tripeptide portion chelates Fe(III) and thereby is able to utilize the iron-transport system of a range of microorganisms. Subsequent to uptake, it is believed that peptidases localized in the cytoplasmic membrane hydrolytically release the toxic thioribosyl moiety. In principle, this property can be used for selected drug delivery. Preliminary studies

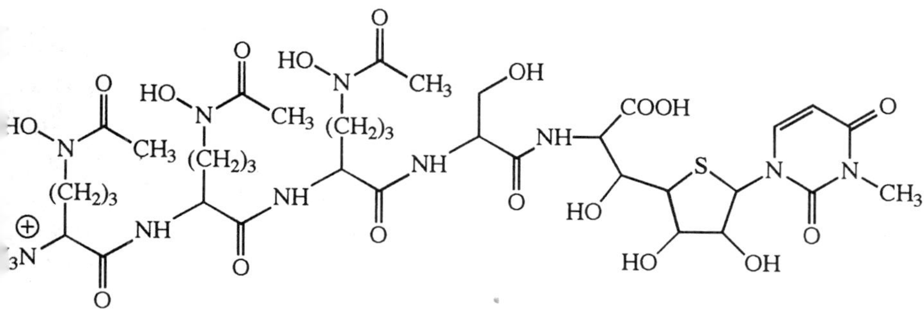

(21)

have indicated that substantial modification of the siderophore framework can be tolerated by microbial iron-transport systems [62].

Surprisingly, much simpler modifications can also be made to cephalosporin molecules, which endow them with the ability to interact with the microorganism iron-transport mechanisms; for instance, the simple incorporation of a catechol moiety as in GR69153 (22). This

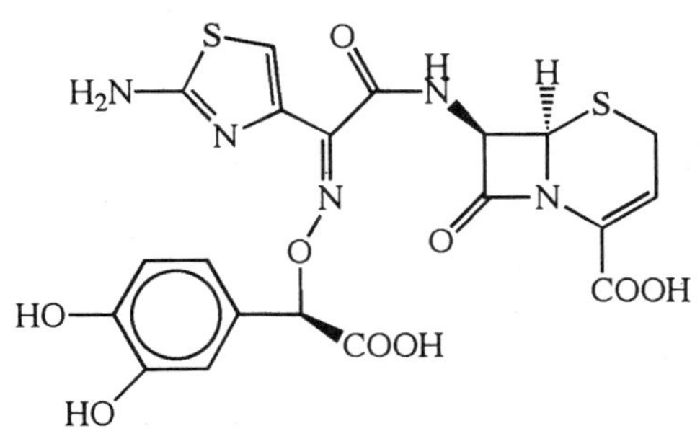

(22)

molecule has enhanced activity against *Pseudomonas aeruginosa* when compared with other cephalosporins [63].

7. CONCLUSIONS

Over the past 30 years it has generally been the accepted view that it would be difficult to produce an orally active iron chelator for therapeutic application. To date, several chemical leads have been identified and are under active development for the treatment of transfusional iron overload. The successful introduction of such compounds will impact considerably on the therapeutic outcome and quality of life for the thalassemic population. There clearly exists the potential for the application of iron chelation in a wide range of clinical situations (see Sec. 6). Once a chelator has been clinically proven in thalassemic patients, such compounds will almost certainly find application in other clinical areas such as sickle cell anemia, cancer chemotherapy, and neurodegenerative disorders.

Orally active ligands are most likely to be small molecular weight entities and thus will fall into the category of either bidentate or tridentate ligands. However, for parenteral application, such as in the case for cancer chemotherapy, the larger hexadentate ligands are more likely to enjoy successful application for reasons outlined in Section 2.

ABBREVIATIONS

BBB	blood-brain barrier
$D_{7.4}$	distribution coefficient at pH 7.4
DBED	*N,N'*-bis(di(3,4,5-trimethoxybenzyl))ethylenediamine-*N,N'*-diacetic acid (Structure **19**)
DFO	desferrioxamine B (Structure **4**)
DFT	desferrithiocin (Structure **14s)**
DmHBED	dimethyl-*N,N'*-bis(2-hydroxylbenzyl)ethylenediamine-diacetic acid (Structure **10b**)
DTPA	diethylenetriaminepentaacetic acid (Structure **9**; here and elsewhere, for simplicity, the structure of aminocarboxylates were drawn in uncharged rather than zwitterion form)
EDTA	ethylenediamine-*N,N,N',N'*-tetraacetic acid
HBED	*N,N'*-bis(2-hydroxylbenzyl)ethylenediaminediacetic acid (Structure **10a**)
HPO	hydroxypyridinone (Structure **6**)
IC_{50}	concentration of the inhibitor which leads to a 50% inhibition of the enzyme activity

LD_{50}	lethal single dose of a substance which leads to the death of 50% of a population
logP	logarithm of the *n*-octanol/H_2O partition coefficient
MECAM	1,3,5-tris(2,3-dihydroxybenzoyl)triaminomethylbenzene (Structure **2**)
NTA	nitrilotriacetic acid (Structure **13**)
NTBI	non-transferrin-bound iron
PIH	pyridoxal isonicontoyl hydrazone
WHO	World Health Organization

REFERENCES

1. B. Halliwell and J. M. C. Gutteridge, in *Free Radicals in Biology and Medicine*, 2nd ed., Clarendon Press, Oxford, UK, 1989.
2. J. C. Barton and L. F. Bertoli, *Nature Med.*, *2*, 394–395 (1996).
3. D. J. Weatherall and J. B. Clegg, in *The Thalassaemia Syndromes*, 3rd ed., Blackwell Scientific Publications, Oxford, UK, 1981.
4. P. Jenner, *Lancet*, *344*, 796–798 (1994).
5. B. Halliwell and J. M. C. Gutteridge, *Trends Neurosci.*, *8*, 22–26 (1985).
6. (a) A. E. Martell and R. M. Smith, in *Critical Stability Constants*, Vols. 1–6, Plenum Press, London, 1974–1989. (b) R. J. Motekaitis, Y. Sun, and A. E. Martell, *Inorg. Chem.*, *30*, 1554–1556 (1991).
7. D. C. Harris and P. Aisen, *Biochim. Biophys. Acta*, *329*, 156–158 (1973).
8. R. C. Hider, *Structure Bonding*, *58*, 25–87 (1984).
9. R. C. Hider and A. D. Hall, *Prog. Med. Chem.*, *28*, 41–173 (1991).
10. P. S. Dobbin, R. C. Hider, A. D. Hall, P. D. Taylor, P. Sarpong, J. B. Porter, G. Xiao, and D. van der Helm, *J. Med. Chem.*, *36*, 2448–2458 (1993).
11. R. C. Hider, G. Kontoghiorghes, J. Silver, and M. A. Stockham, UK Patent 2117766 (1982).
12. N. F. Olivieri, G. M. Brittenham, D. Matsui, M. Berkovitch, L. M. Blendis, R. G. Cameron, R. A. McClelland, P. P. Liu, D. M. Templeton, and G. Koren, *N. Engl. J. Med.*, *332*, 918–922 (1995).

13. J. B. Porter, S. Singh, K. P. Hoyes, O. Epemolu, R. D. Abeysinghe, and R. C. Hider, *Adv. Exp. Med. Biol.*, *356*, 361–370 (1994).
14. J. Xu, B. Kullgren, P. W. Durbin, and K. N. Raymond, *J. Med. Chem.*, *38*, 2606–2614 (1995).
15. R. A. Yokel, A. M. Fredenburg, K. A. Meurer, and T. L. Skinner, *Drug Met. Dispos.*, *23*, 1178–1180 (1995).
16. V. A. Levin, *J. Med. Chem.*, *23*, 682–684 (1980).
17. J. B. Porter, J. Morgan, K. P. Hoyes, L. C. Burke, E. R. Huehns, and R. C. Hider, *Blood*, *76*, 2389–2396 (1990).
18. R. D. Abeysinghe, P. J. Roberts, C. E. Cooper, K. H. MacLean, R. C. Hider, and J. B. Porter, *J. Biol. Chem.*, *271*, 7965–7972 (1996).
19. R. W. Grady, R. Newfield, K. I. Kivirikko, M. H. Pork, J. E. Folk, T. McCaffrey, P. J. Giardina, and H. M. Hanauske-Abel, *5th Conf. Int. Ass. Study of Disorders of Iron Metabolism*, Boston. Abstract A23, 1995.
20. R. C. Hider and K. Lerch, *Biochem. J.*, *257*, 289–290 (1989).
21. C. E. Cooper, G. R. Lynagh, K. P. Hoyes, R. C. Hider, R. Cammack, and J. B. Porter, *J. Biol. Chem.*, *271*, 20291–20299 (1996).
22. K. P. Hoyes, H. M. Jones, R. D. Abeysinghe, R. C. Hider, and J. B. Porter, *Exp. Haematol.*, *21*, 86–92 (1993).
23. G. M. Brittenham, *Blood*, *80*, 569–574 (1992).
24. M. W. Hautze and L. Kühn, *Proc. Natl. Acad. Sci. USA*, *93*, 8175–8182 (1996).
25. R. C. Hider, J. B. Porter, and S. Singh, in *The Development of Iron Chelators for Clinical Use* (R. J. Bergeron and G. M. Brittenham, eds.), CRC Press, London, 1994, pp. 353–371.
26. T. M. Garrett, T. J. McMurray, M. W. Hosseini, Z. E. Reyes, F. E. Hahn and K. N. Raymond, *J. Am. Chem. Soc.*, *113*, 2965–2977 (1991).
27. S. Singh, H. Khodr, M. I. Taylor, and R. C. Hider, *Biochem. Soc. Symp.*, *61*, 127–137 (1995).
28. E. Baker, W. H. Peter, and A. Jacobs, *Br. J. Haematol.*, *81*, 424–431 (1992).
29. J. B. Porter, M. Gyparaki, L. C. Burke, E. R. Huehns, P. Sarpong, V. Saez, and R. C. Hider, *Blood*, *72*, 1497–1503 (1988).
30. J. B. Porter, R. D. Abeysinghe, L. Marshall, R. C. Hider and S. Singh, *Blood*, *88*, 705–713 (1996).
31. S. Singh, R. O. Epemolu, P. S. Dobbin, G. S. Tilbrook, B. L. Ellis,

L. A. Damani, and R. C. Hider, *Drug Met. Dispos.*, *20*, 256–261 (1992).

32. S. Singh, R. Choudhury, R. O. Epemolu, and R. C. Hider, *Eur. J. Drug. Metab. Pharmacokinet.*, *21*, 33–41 (1996).
33. R. C. Hider, R. Choudhury, B. L. Rai, L. S. Dehkordi, and S. Singh, *Acta Haematol.*, *95*, 6–12 (1996).
34. G. Anderegg and M. Raber, *J. Chem. Soc. Chem. Commun.*, 1194–1196 (1990).
35. G. S. Tilbrook, M. Y. Moridani, H. Khodr, and R. C. Hider, Manuscript in preparation.
36. S. Avramovici-Grisaru, S. Sarel, G. Link, and C. Hershko, *J. Med. Chem.*, *26*, 298–302 (1983).
37. S. Avramovici-Grisaru, S. Sarel, S. Cohen, and R. E. Bauminger, *Israel J. Chem.*, *25*, 288–292 (1985).
38. J. B. Porter and E. R. Huehns, in *Balliére's Clinical Haematology*, Vol. 2 (C. Hershko, ed.), Baillière-Tindal, London, 1989, pp. 459–474.
39. B. Fechtig and H. A. Peter, Patent WO 8603745 (1984).
40. R. J. Bergeron, Z. R. Liu, J. S. McManis, and J. Wiegand, *J. Med. Chem.*, *35*, 4739–4744 (1992).
41. R. J. Bergeron, J. Wiegand, J. S. McManis, and P. R. Perumal, *J. Med. Chem.*, *34*, 3182–3187 (1991).
42. P. E. Hallaway, J. W. Eaton, S. S. Panter, and B. E. Hedlund, *Proc. Natl. Acad. Sci. USA*, *86*, 10108–10112 (1989).
43. M. J. Pippard, M. J. Jackson, K. Hoffman, M. Petrou, and C. B. Modell, *Scand. J. Haematol.*, *36*, 466–472 (1986).
44. R. W. Grady and C. Hershko, *Semin. Haematol.*, *27*, 105–111 (1990).
45. R. W. Grady, A. D. Salbe, M. W. Hilgartner, and P. J. Giardina, in *The Development of Iron Chelators for Clinical Use* (R. J. Bergeron and G. M. Brittenham, eds.), CRC Press, London, 1994, pp. 395–406.
46. H. H. Peter, R. J. Bergeron, R. R. Streiff, and J. Wiegand, in *The Development of Iron Chelators for Clinical Use* (R. J. Bergeron and G. M. Brittenham, eds.), CRC Press, London, 1994, pp. 373–394.
47. J. B. Galey, J. Dumats, I. Beck, B. Fernandez, and M. Hocquaux, *Free Rad. Res.*, *22*, 67–87 (1995).
48. K. L. Malisza and B. B. Hasinoff, *Arch. Biochem. Biophys.*, *316*, 680–688 (1995).

49. J. L. Buss and B. B. Hasinoff, *Agents Actions*, *40*, 86–95 (1993).
50. M. J. Pippard, S. J. Callender, G. T. Warner, and D. J. Weatherall, *Lancet*, *2*, 819–821 (1979).
51. R. C. Hider, M. Habgood, J. Abbot, and Z. Liu, *J. Pharm. Pharmacol.*, *49*, 54, 106 (1997).
52. R. C. Hider, A. R. Mohd-Nor, J. Silver, I. E. G. Morrison, and L. V. C. Rees, *J. Chem. Soc. Dalton Trans.*, 609–622 (1981).
53. D. T. Dexter, F. R. Wells, A. J. Lees, F. Agid, Y. Agid, P. Jenner, and C. D. Marsden, *J. Neurochem.*, *52*, 1830–1836 (1989).
54. S. D. Aust, and B. C. White, *Adv. Free Rad. Biol. Med.*, *1*, 1–17 (1985).
55. R. C. Hider, S. Singh, and J. B. Porter, *Proc. Royal Soc. Edin.*, *99B*, 137–168 (1992).
56. D. R. Blake and P. A. Bacon, *Br. Med. J.*, *282*, 189–197 (1981).
57. N. A. A. van der Waal, L. L. Smith, F. L. M. van Oirschot, and B. S. van Asbeck, *Am. Rev. Respir. Dis.*, *145*, 180–186 (1992).
58. R. Kime, A. Gibson, R. C. Hider, and H. Powers, *Clin. Sci.* In press, 1997.
59. S. J. Bradley, I. Gosriwatana, S. Srichairatanakool, R. C. Hider, and J. B. Porter, *Br. J. Haematol.* In press.
60. R. J. Bergeron, P. F. Gavanaugh, S. J. Kline, R. G. Hughes, G. T. Elliot, and C. W. Porter, *Biochem. Biophys. Res. Commun.*, *121*, 848–854 (1984).
61. J. B. Summers, K. H. Kim, H. Mazdiyasni, J. H. Holms, J. D. Ratajczyk, A. O. Stewart, R. D. Dyer, and G. W. Carter, *J. Med. Chem.*, *33*, 992–998 (1990).
62. M. S. Diarra, M. C. Lavoie, M. Jacques, I. Darwish, E. K. Dolence, J. A. Dolence, A. Ghosh, M. Ghosh, M. J. Miller, and F. Malonin, *Antimicrob. Agents Chemother.*, *40*, 2610–2617 (1996).
63. P. Silley, J. W. Griffiths, D. Monsey, and A. M. Harris, *Antimicrob. Agents Chemother.*, *34*, 1806–1808 (1990).
64. F. L'Eplattenier, I. Murase, and A. E. Martell, *J. Am. Chem. Soc.*, *89*, 837–843 (1967).
65. R. W. Grady and A. Jacobs, in *Development of Iron Chelators for Clinical Use* (A. E. Martell, W. F. Anderson, and D. G. Badman, eds.), Elsevier-North Holland, New York, 1981, pp. 133–164.

Subject Index

A

A.
eutrophus, see Alcaligenes
flavescens, see Arthrobacter and Aureobacter
fumigatus, see Aspergillus
haemolyticus, see Acinetobacter
hydrophila, see Aeromonas
luteoviolacea, see Alteromonas
pleuropneumoniae, see Actinobacillus
quadricinctus, see Aspergillus
spinosa, see Absidia
thaliana, see Arabidopsis
tumefaciens, see Agrobacterium
variabilis, see Anabaena
Abscisic acid, 423, 576–578, 640, 641
Absidia spinosa, 98, 444, 494
Absorption bands and spectra (*see also* Spectrophotometry *and* UV absorption)
visible, 248–253, 344, 490, 492, 504, 536, 618, 675
Aceruloplasminemia, 175
Acetate (or acetic acid), 548, 554
buffer, *see* Buffer
iodo-, 459
N-Acetylcysteine, 578
O-Acetylserine, 101
Achromobactin, 105, 199
Acidity constants (*see also* Equilibrium constants), 253, 269–271, 273, 285, 292, 303, 310, 700, 701, 709
Acinetobacter haemolyticus, 267
Acinetoferrin, 267

Aconitase, 100, 406, 418, 419, 425, 609, 610, 642, 643, 649, 650
  c-, 418, 645
  m-, 405, 407, 408, 412–415, 418, 644
Actinobacillus sp., 110
  pleuropneumoniae, 53, 95, 103, 113
Active sites (*see also* Coordination spheres)
  enzymes, *see* Enzymes
Adenosine diphosphate, *see* ADP
Adenosine triphosphate, *see* ATP
ADP, 218, 682
Adrenodoxin
  bovine, 675
Aedes aegytia, 439, 442, 443
Aerobactins, 19, 59, 71, 87, 88, 90, 103, 114, 192, 243, 244, 272, 274, 357, 361, 366, 373, 374, 388, 672
Aeromonas sp., 105
  hydrophila, 51
Aerosols (*see also* Atmosphere)
  iron in, 4
Affinity constants, *see* Stability constants
Agaricus
  bisporus, 149
  ochraceus, 149
Agarose column
  nickel-nitrilotriacetate, 83, 377
Agranulocytosis, 706
Agrobacterium tumefaciens, 198
Agrobactins, 198, 251, 252, 362
Akaganite, 15
Alanine (and residues), 82, 100, 101, 149, 251, 255, 257, 280, 281, 300, 339, 343, 348, 364, 369, 385, 453, 455–457, 461, 467, 570, 591
Albomycin, 75, 76, 79, 86–88, 90, 94, 101, 103, 361, 373, 383, 384, 724
Albumin, 110
Alcaligen E, 363, 366
Alcaligenes sp., 539
  eutrophus, 363, 366, 439, 443
Alcaligin, 123, 247, 248, 271, 272, 284
Alfafuran, 216
Alfalfa, 216
Algae (*see also* individual names), 21, 25, 27
  blue-green, 587
  green, 15
  nutrients, 5, 17
Alginate, 125
Alkaline earth ions, *see* individual names
Alkoxide, 532
Alterobactin
  A, 20, 270, 272, 273
  B, 20
Alteromonas sp., 19
  luteoviolacea, 20
Aluminum(III), 11, 26, 40, 241, 242, 279, 380, 589, 590, 598, 603, 668, 680–682
  silicate, *see* Silicate
  toxicity, 681
Alzheimer's disease, 594, 647, 693, 721
Americium, 671
Amide (*see also* individual names)
  groups, 335, 336, 349, 454, 595, 596
  linkage, 162, 281, 290, 292, 337, 340, 349, 384, 385
Amines (*see also* individual names), 337
  benz-, 377
Amino acids (*see also* individual names), 72
  sequences, 49, 78, 89, 93, 96, 97, 114, 162, 176, 358, 367, 370, 439–441, 451, 454, 483, 567, 568, 610, 644
  substitutions, 459–464

γ-Aminobutyric acid, 646, 647, 720
Aminolevulinate
  dehydratase, 679
  synthase, 405, 407, 408, 412–415, 643, 644
Ammonium, 15, 16, 21, 23, 24
Amonabactin, 50, 51
  synthesis, *see* Synthesis
Amphibia(n) (*see also* individual names), 413
  ferritin, *see* Ferritins
  red cells, 445
Anabaena sp., 19, 20
  variabilis, 494
Anemia, 636, 648, 650, 679, 719, 720
  rat, 653, 671, 676, 678
Anguibactin, 50, 85, 103, 104, 123, 126, 360, 366
Anhydromevalonic acid, 384
Animal (*see also* Mammal *and* individual names and species), 669, 674–679
  ferritins, *see* Ferritins
  hemoglobin, *see* Hemoglobin
  iron studies, 644, 647, 648
  iron uptake, 403–421
  mRNA (*see* mRNA)
  pathogens, 87, 189, 198
Antagonistic action, *see* Interdependencies
Antarctic
  Ocean, 5
  Vostoc ice core, 26
  Weddell Sea, 5
Anthracene fluorescence, 255, 345
Anthracyclines, 500, 718, 719
Anthranilic acid, 176, 257
Antibacterial activity of lactoferrin, 57
Antibiotic(s) (*see also* individual names), 75, 79, 86, 88, 90, 101, 126, 196, 361, 383, 390, 693, 724
[Antibiotic(s)]
  peptide, 50, 51
  resistance, 693
Antibodies, 15, 74, 574, 613, 646
  monoclonal, 78, 79, 84, 202, 357, 369, 380
Antineoplastic agents, 718, 724
Antioxidant(s), 578, 649, 680
  defense, 41
Apoplasm, 227, 233
Apoplasts, 234, 565, 579
Arabidopsis sp., 207, 422
  thaliana, 200, 201, 217, 235, 439, 443, 571, 573, 578, 642
Archaebacteria, 87
Arginine (and residues), 83, 116, 166, 167, 364, 369, 385, 419, 446, 449, 452, 453, 455, 458, 459, 464, 467, 484, 495, 591, 592, 596, 597, 600, 602, 605, 608
Arsenate, 81
Artemia sp., 439, 442, 443
Artemin, 438–441
Arthritis, 724
  psoriasis, 718
  rheumatoid, 722
Arthrobacter flavescens, 284, 339
Arthropods, 594
Ascomycetes, 149, 152, 195
Ascorbate (or ascorbic acid), 497, 500, 503, 570, 574, 578, 652, 656
Asparagine (and residues), 453, 455, 459, 462, 464, 466, 600, 607, 608
Aspartic acid (and residues), 74, 78, 83, 114, 116, 251, 369, 449, 453, 455, 457, 461, 462, 464, 466, 467, 469, 470, 482, 592, 593, 595, 600, 602
Asperchromes, 149, 151, 154
Aspergillus sp., 149, 153, 195
  fumigatus, 156, 178
  quadricinctus, 162

Association constants (*see also* Dissociation constants *and* Stability constants), 380
Astrocytes, 677
Atmosphere, 682
  carbondioxide in, 25–28, 332
  dust, *see* Dust
  iron in, 2, 5, 9, 21
Atomic emission spectrometry
  inductively coupled plasma, 542
ATP, 18, 49, 72, 81, 85–87, 93, 94, 103, 157, 162, 218, 388, 611, 636, 654, 683
  [^{32}P]8-azido-, 106
  dependent iron pump, 566
  dependent proton pump, 654
  hydrolase, *see* Hydrolases
  hydrolysis, 85, 86, 95, 106, 637, 682
ATPase(s), 84, 85, 91, 92, 95, 104, 107, 113, 155, 216, 217, 363, 372, 566, 613
  H^+-, 612
  P-type, 174, 175
Aureobacterium flavescens, 281, 312
Avenic acid, 219
Avian (*see also* Monkey)
  egg white, 589
  mRNA, *see* mRNA
  transferrin, *see* Transferrins
  transferrin receptor, 608
Azorhizobium sp., 189, 673
Azotobacter, 96, 253
  vinelandii, 100, 101, 267, 424, 439, 443, 494–496, 501, 502
Azotobactin δ, 243, 245, 253, 255, 259, 270, 272, 302

B

B.
  abortus, see Brucella
  bronchiseptica, see Bordetella
[*B.*]
  fragilis, see Bacterioides
  intermedius, see Bacillus
  japonicum, see Bradyrhizobium
  lactofermentum, see Brevibacterium
  melitensis, see Brucella
  pertussis, see Bordetella
  subtilis, see Bacillus
Bacillus
  intermedius, 88
  subtilis, 88, 89, 103, 115, 277, 363, 494, 672
Bacitracin, 74, 79
Bacteria(l) (*see also* individual names), 42, 50–61, 163, 191, 196, 199, 202, 206, 422, 424, 635, 642, 723
  anaerobic, 86, 442
  archae-, 87
  cyano-, *see* Cyanobacteria
  enteric, *see* Enterobacteria
  ferritin, *see* Ferritins
  gram-negative, 52–54, 70–73, 85, 122, 127, 155, 192, 356, 357, 386, 388, 639, 672
  gram-positive, 54, 72, 74, 85, 87, 105, 115, 116, 127, 363, 639, 672
  heterotrophic, 9–12, 18–20
  iron homeostasis, 638, 639, 658
  iron metabolism, 38, 113, 122, 123
  lactic acid, 618
  marine, 8, 20, 673, 674
  myco-, *see* Mycobacteria
  pathogenic (*see also* Pathogens), 61, 113
  phototrophic, 19
  siderophore receptor proteins, 355–391
  siderophores, *see* Siderophores

Bacterioferritin (*see also* Ferritins), 41, 96-98, 153, 438–444, 449, 453, 461, 462, 464–466, 480, 494–497, 501, 502, 517, 522, 534, 540–543, 555, 567, 570, 571, 639, 657, 658
Bacterioides fragilis, 108, 111
Bacteriophage T1, 374
Bacteroids, 193, 194
Barium ion, 27
Barley (*see also Hordeum vulgare*), 201, 202, 205, 219, 227, 233, 235, 675
  leaves, 219–225, 229–231, 234
  roots, 219–224, 229, 230, 234
  tiller, 219, 221, 225
Base pairs, 409, 411, 412, 414
  AU, 407, 417
  CG, 411
  GC, 414–416
  UA, 411
Basidiomycetes, 149, 152, 156
Bathophenanthroline, 219, 233, 617
Beans, 198
  leaves, *see* Leaves
  roots, *see* Roots
  soy-, *see* Soybean
Benzamine, 377
Beryllium(III)
  ^{10}Be, 27
  Betaine, 452
Bicarbonate, *see* Hydrogen carbonate
Bicinchoninic acid, 377
Bile, 73
  excretion of iron, 616, 715
Binding constants, *see* Stability constants
Biomineralization, *see* Mineralization
Biosynthesis (*see also* Synthesis)
  anguibactin, 123
  chlorophyll, 227
[Biosynthesis]
  chrysobactin, 105
  Fe-S clusters, 100
  isoleucine, 101
  proteins, 404–407
  pseudobactin, 122, 123, 125
  pyochelin, 123
  pyoverdin, 122
  rhodotorulic acid, 157
  siderophore, 50–52, 116, 122–126, 128, 156–171, 193, 197, 199
  valine, 101
  yersiniabactin, 123
Biota, 2, 9, 27
Biotin, 74
2,2'-Bipyridyl, 95, 500, 501, 617, 694
Birds, 413
Bismuth(III), 590, 603, 681
Bisucaberin, 247
Blaberus discoidalis, 594
Black Sea, 6
Bleomycin
  iron, 411
Blood, 646, 671, 676, 678, 717, 722, 723
  diseases (*see also* individual names), 719
  mouse, 58
  plasma (*see also* Plasma), 311
  rat, 678, 679
  transfusion, 422, 693, 714, 715, 719, 720
Blood-brain barrier, 646, 703–705, 712, 721
Boltzmann constant, 523, 524
Bone
  disease, 589
  iron metabolism, 681
  marrow, 651, 706, 707, 712, 723, 724
Bordetella
  bronchiseptica, 123, 126
  pertussis, 360, 362, 366

Bovine (*see also* Calf)
  adrenodoxin, 675
  c-aconitase, 418
  mRNA, 408, 414
Bradyrhizobia, 191, 194
Bradyrhizobium sp., 189, 573, 673
  japonicum, 191, 192, 194
Brain, 644–647, 658, 677, 703, 707, 712
  elevated iron levels, 720, 721
  human, 445, 594, 721
  iron homeostasis, 645–647, 658
Brassica napus, 439, 443
Brevibacterium lactofermentum, 54
Brucella
  abortus, 59
  melitensis, 96, 439, 443, 466
Brucite, 528, 549, 552
Buffer, 382
  acetate, 259, 600
  Good's, 487
  oxalate, 94
  phosphate, 94
Bullfrog (*see also Rana catesbeiana*) ferritin, *see* Ferritins

C

C.
  coli, see Campylobacter
  diphtheria, see Corynebacterium
  glutamicum, see Corynebacterium
  jejuni, see Campylobacter
  paradoxa, see Cyanophore
  perfringens, see Clostridium
  synechocystis, see Cyanobacterium
Cadmium(II), 2, 170, 459, 462, 466, 578, 590, 598, 671, 672, 675, 677, 679
  interdependency with other elements, *see* Interdependencies
Caffeic acid, 216
Calcium(II), 90, 241, 376, 379, 381–383, 387, 389, 391, 452, 456, 459, 462, 566, 655, 681
Calorimetry
  differential scanning, 485
Calpain, 644
Calf (*see also* Bovine), 678
Cambium, 200
Campylobacter
  coli, 88, 89, 105
  jejuni, 42, 88, 89, 98, 439, 443
Carbohydrates (*see also* Sugars *and* individual names), 589, 607, 716
  acidic, 99
Carbon, 17, 28
  assimilation, 18, 45
  ^{13}C NMR, *see* NMR
  fixation rates, 24
  global cycle, 25
Carbonate(s)
  ^{13}C, 592
  hydrogen, 95
  iron, 12, 525, 526, 591, 593, 594, 596, 597, 618, 671
  synergistic, 592, 597, 602
Carbon dioxide, 525
  biological pump, 25, 27
  in atmosphere, *see* Atmosphere
Carbon monoxide, 454
Carbonyl oxygen, 699, 700, 711
Carboxylases
  phosphoenolpyruvate, 573
  ribulose-1,5-diphosphate, 640
Carboxylate(s) (or carboxylic acids) (*see also* individual names), 106, 107, 114, 116, 290, 341, 424, 462, 482, 489, 530, 532, 549, 568, 591, 595, 657, 713, 716
  α-hydroxy-, 242, 243, 255, 272–275

Cardiac
dysfunction, 177
muscle, *see* Heart
Cardiomyopathy, 719
Cardiotoxicity, 718, 722
Carotenoids, 576
Catalase, 41, 116, 121
Catalytic sites, *see* Active sites
Cataract, 659
Catecholamides, 268
chiral analogues, 261, 263
siderophores, *see* Siderophores
Catechol(ates) (and residues), 19, 71, 104, 105, 127, 248–253, 261, 264, 288–290, 292, 295, 302, 304, 306–308, 361, 377, 380, 501, 677, 705, 707, 709, 724, 725
chiral derivatives, 262
ferric, *see* Iron(III)
fluorescent, 362
siderophores, *see* Siderophores
CD, *see* Circular dichroism
Cell(s)
channels (*see also* Channels), 73
Chinese hamster ovary, 614
damage, 42
epidermal, 216
epithelial, 107
erythroid, 407, 413, 609, 610, 614, 616, 654
eukaryotic, 636
extracellular iron levels, 722, 723
intestinal, 616
malignant, 445, 593, 594
mammalian, 576, 577, 635, 636, 654
melanoma, 609, 614
membrane, 200, 204, 217, 218, 273, 276, 332, 346, 347, 422, 564, 594, 607, 670, 680
metabolism, 100, 124
reticuloendothelial, 648, 654
[Cell(s)]
surface, 16, 296, 369, 425, 613, 669
T, 658
tumor, 644
Cellulose, 565
Central nervous system, 703, 704, 721
Cephalosporins, 105, 726
Cerebral vascular accident, 693
Cerium(III), 590
Cerium(IV), 590
Ceruloplasmin, 174, 175, 487, 641, 658, 678–680, 683
synthesis, *see* Synthesis
Channels, 390
diffusion, 371
four-fold, 464–467, 481, 497, 517, 569
gated, 372
hydrophilic, 481–483, 497, 569
hydrophobic, 481, 497, 568, 569
three-fold, 462–464, 467, 482, 491, 495, 497, 517, 569, 655
Charge transfer, 462, 536
ligand-to-metal, 248, 249, 253, 259, 262, 601
metal-to-ligand, 617
Chelating agents and therapy (*see also* Drugs, Therapeutic agents *and* individual names), 295, 709
iron, *see* Iron chelators
Chemoluminescence assay
iron, 15
Chemotherapy (*see also* Drugs *and* Therapeutic agents), 347, 724
Chicken, 416
embryo, 608
ferritin, 408, 412, 414, 439, 443
fibroblasts, 608
ovotransferrin, *see* Ovotransferrin

Chloride, 311, 605
ferric, *see* Iron(III)
Chlorophylls, 205, 573, 575, 587, 640, 674
a, 21, 23, 24, 198
b, 198
biosynthesis, *see* Biosynthesis
Chloroplasts, 227, 231, 423, 424, 426, 496, 567, 572, 573, 577, 579, 640
light-dependent iron transport, 231–235
maturation, 405
Chlorosis, 201, 202, 205, 233, 575, 640
Chromatin, 167, 168
Chromatography
chelate affinity, 114
column, 377
fast-protein liquid, 377, 378
gel, *see* Gel chromatography
high performance liquid, *see* High performance liquid chromatography
ion exchange, *see* Ion exchange chromatography
size exclusion, 378, 483
Chromium(III), 341, 380, 590, 601, 671, 673, 677
interdependency with other elements, *see* Interdependencies)
Chromium(VI), 678
Chromophores, 6, 252, 283
Chromosomes, 83, 90, 94, 101, 103, 104, 112, 114, 115, 173, 191, 376, 483, 606, 607
Chrysobactin, 103, 105, 199, 272, 363
Ciliates, 12
Circular dichroism, 259–266, 378, 458
Cotton effect, 262
magnetic, *see* Magnetic circular dichroism
Citrate(s) (or citric acid), 55, 170, 172, 191, 192, 267, 296, 348, 357, 360, 361, 363, 373, 380, 503, 547, 566, 615, 644
ferric, *see* Iron(III)
iso-, 644
synthase, *see* Synthases
Clay, 529
Cloacin, 373
Clostridium perfringens, 439, 443
Clover, 497
Cobalamins (*see also* Vitamin B_{12}), 366, 381, 382
Cobalt
interdependency with other elements, *see* Interdependencies
Cobalt(II), 86, 114, 170, 496, 590, 671, 672, 675–677
Cobalt(III), 590, 601, 673, 675
Coccolithophores, 27
Cockroach, 594
Coenzymes (*see also* individual names)
A, 157, 163
Colicins, 83, 105, 382, 390, 391
B, 78, 79, 82, 360, 366, 373, 377
D, 78, 79, 360, 366, 373
M, 75, 76, 79, 80, 361, 362, 369, 373, 383
Collagen, 534
Colorimetry, *see* Absorption bands and spectra *and* Spectrophotometry
Conalbumin (*see also* Ovotransferrin), 200, 677, 678
Conifers, 195
Coordination spheres, 116, 247, 296–313, 469, 592, 593
Copepod, 13
Copper (different oxidation states), 174, 333, 422, 578, 634, 642, 671, 674, 680
deficiency, 678

[Copper (different oxidation states)]
dietary, 679
geochemical cycle, 674
interdependency with other elements, *see* Interdependencies
metabolism, *see* Metabolism
transport proteins, 172, 174, 175
Copper(II), 15, 241, 590, 596, 601, 602 671, 672, 675, 677, 694, 719
Coprogens, 70, 71, 87, 88, 103, 151–154, 268, 270, 274, 331, 334, 361, 373, 374, 384
B, 342
labeled, 153, 154
neo-, *see* Neocoprogen
Corynebacterium
diphtheriae, 52, 54, 113, 115, 116, 122, 639
glutamicum, 116
Cotton, 203
Coulometry
micro-, 498, 499
Crayfish (*see also* individual names), 462
Cricetulus griseus, 439, 443
Cucumber, 203
Crystal structure, *see* X-ray
Cyanide
hydrogen, 196
Cyanobacteria, 11–13, 16, 17, 19, 20, 87, 567, 587, 673, 674
Cyanobacterium synechocystis, 494
Cyanocobalamin, *see* Vitamin B_{12}
Cyanophora paradoxa, 439, 442, 443
Cyclic voltammetry, 285, 294
Cycling of
carbon, 25
dioxygen, 18
iron, 1–29
Cyclodextrin, 281
β-Cystathionase, 101
Cysteine (and residues), 74, 78, 100, 101, 114–116, 169, 368, 419, 442, 459, 462, 466, 467, 482, 595, 597, 602, 607, 608, 644
N-acetyl-, 578
Cystic fibrosis, 125
Cytochrome(s) (*see also* individual names), 99, 193, 668
oxidase, *see* Oxidases
Cytochrome b, 49
b_1, 96
b_{557}, 96
reductase, 625
Cytochrome c, 193, 498
c_{553}, 18
Cytochrome P_{450} reductase, *see* Reductases
Cytokine(s), 650
mRNA, 405
Cytolysins, 46, 47, 52, 57
Cytoplasm (*see also* Plasma), 49, 55, 99, 170–172, 174, 193, 194, 204, 330, 331, 372, 418, 419, 423, 608, 609, 643, 645, 650, 704, 706, 712
bacterial, 70-95, 654
eukaryotes, 17
in plants, 216-218, 233
membrane, 53, 70–72, 74, 76, 80–95, 103, 106, 107, 109, 111, 113, 117–120, 127, 374, 388–390, 607, 638, 654, 724
Cytosine, 416
Cytosol, 15, 47, 59, 85, 92, 93, 95, 204, 636, 640, 642, 643
Cytotoxicity (*see also* Toxicity), 332, 724

D

D.
vulgaris, see Desulfovibrio
Daunomycin, 303

Daunorubicin, 718
Deferriferrichromes, *see* Ferrichromes
Deferrioxamines, 575, 600, 605, 616, 650, 681, 693, 707, 720
  B, 15, 244–246, 249, 255, 260, 273, 278, 299, 305–307, 311, 422, 694, 696, 698, 701, 708, 714–717, 722
  E, 246, 247, 270, 279
Deferiprone, 700, 703, 705, 706, 710, 711
Dehydratase
  dehydroxyacid, 101
Dehydrogenases
  glyceraldehyde, 161
  lactate, 709
  succinic, 408, 413
Dementia
  dialysis, 589
2'-Deoxymugineic acid, 201, 219
Deoxyribonuclease, *see* DNase
Deoxyribonucleic acid, *see* DNA
Deprotonation constants, *see* Acidity constants
Dermatophytes, 156
Desferal, 9, 177, 219, 278, 311, 422, 714
Desferrioxamine, *see* Deferrioxamines
Desferrithiocin, 694, 707, 708, 713
Desulfovibrio vulgaris, 439, 440, 442–444
Detergents (*see also* individual names), 376–379
  CHAPS, 606
Detoxification (*see also* Toxicity)
  iron, 41, 42, 56, 98, 565, 566, 578, 640
  reactive oxygen, 577
Detritus
  iron in, 9–12
Dexrazoxane, 718
Dextran
  iron drugs, 547, 716
Dialysis
  dementia, 589
  equilibrium, *see* Equilibrium dialysis
  hemo-, 681
  induced aluminum overload, 590
1,2-Diaminoethane, *see* Ethylenediamine
2,6-Diaminohexanoate, *see* Lysine
2,5-Diaminopentanoate, *see* Ornithine
Diatoms (*see also* individual names), 15, 24
Dibenzolylmethane, 548, 554
Dicyclohexylcarbodiimide, 232, 235
Diet
  copper, 679
  iron, *see* Iron
Diethylenetriamine-*N,N,N',N'',N''*-pentaacetate, 15, 701, 703, 708, 716
1,2-Dihydroxybenzene-3,5-disulfonate, 299, 302–304, 307
2,3-Dihydroxybenzoic acid (and residues), 51, 52, 69, 105, 503
N-[N^2-(2,3-Dihydroxybenzoyl)-D-lysyl]-L-serine, *see* Chrysobactin
2,3-Dihydroxy-*N*-benzoylserine, 104, 105, 275, 277, 366, 373
N,N-Dimethyl-2,3-dihydroxybenzamide, 696, 698, 701
Dimethylhydroxypyridinone, 616
Dimethyl sulfate, 408
Dimethylsulfoxide, 493
Dinitrogen (*see also* Nitrogen), 682
Dinitrogen oxide (N_2O), 482
Dioxygen (*see also* Oxygen), 6, 7, 157, 424, 464, 470, 481, 484–490, 495, 497–499, 503, 518–520, 587, 634
  ^{16}O, 27

[Dioxygen]
^{17}O, 491
^{18}O, 27
assimilation, 12
carrier proteins, 39
electrode, *see* Electrode, oximetry
radicals, *see* Radicals
in seawater, *see* Seawater
intermediates, 649
Dioxygenases
2-oxoglutarate-dependent, 203
Diphosphate, 157, 162, 597, 600, 604–606
Diphtheria toxin, *see* Toxins
Diseases (*see also* individual names)
blood, 719
iron overload, 278, 287, 311, 340, 422, 615, 638
Menkes, 175
Parkinson's, *see* Parkinson's disease
Wilson's, *see* Wilson's disease
Dissociation constants (*see also* Association constants, Equilibrium constants *and* Stability constants), 73, 88, 90, 110, 374, 378–382
Distichonic acid, 219
Disulfide
bonds, 597, 607
bridge, 74
iron, 525, 526
links, 101, 606, 654
Dithionite, 498, 500, 502
Dithiothreitol, 100, 101
DNA (*see also* Nucleic acids), 54, 74, 77, 80, 94, 98, 114, 160, 165, 167, 169, 174, 198, 235, 404, 407, 426, 636, 638, 722
binding site, 116
[DNA]
c, 161, 176, 202, 217, 410, 567, 571, 573, 608, 609, 613
damage, 40, 56, 176, 177
footprinting, 113–116, 175
human, 609
mobility band shift experiments, 118, 169
nucleotide sequences, 567, 573
protein interactions, *see* Proteins
recognition, 167
repair, 124, 177
sequence analysis, 160, 161, 164, 175
synthesis, 646, 682, 693, 706, 723, 724
transcription, 405, 418
DNase I, 113
Dodecyldimethylamine oxide, 377
Dodecyl maltoside, 378
Dopamine, 646, 721
6-hydroxy-, 500
Doxorubicin, 718, 719, 722
Drosophila, 413
Drugs (*see also* individual names), 177, 332, 340
antimalarial, 347–349
iron/dextran, 547
pro-, 717, 719
resistance, 347, 693
selected delivery, 724, 725
DTPA, *see* Diethylenetriamine-*N,N,N',N'',N''*-pentaacetate
Duck ovotransferrin, 596
Duodenum, 644
human, 648, 651, 652
mouse, 647, 653
mucosa, 647
rat, 648
Dust
iron in, 2, 4, 5, 26–28
zinc in 28

E

E.
chrysanthemi, see Erwinia
coli, see Escherichia
herbicola, see Erwinia
Earth's
crust, 2, 668
surface, 241, 520
Ecosystem
iron-limited, 24
EDTA, *see* Ethylenediamine-*N,N,N',N'*-tetraacetate
Egg
white, 200
yolk, 200
EGTA, 381
Electrochemistry, 270, 273, 276, 285
Electrode
normal hydrogen, 275, 498
oximetry, 485, 488, 489, 492
Electrode potentials, *see* Redox potentials
Electron density, 306, 445, 451, 453, 459
Electron diffraction studies, 534, 535, 541
Electronegativity, 241
Electronic spectra, *see* Absorption bands and spectra
Electron nuclear double resonance, *see* ENDOR
Electron microscopy, 77, 115, 387–389, 534, 543, 547, 566, 570, 572, 615, 616, 672
scanning, 219
tunneling, *see* Electron tunneling microscopy
Electron paramagnetic resonance, *see* EPR
Electron spin resonance, *see* EPR
Electron transfer, 14, 100, 501, 502, 587, 682
long-range, 500, 503
oxidative, 487–489
Electron tunneling microscopy, 542
high-resolution, 535, 536, 539, 541
Electrophoresis
gel, *see* Gel electrophoresis
mobility shift assay, 165, 169, 175
SDS-PAGE, 494
Electrospray mass spectrometry, 267, 284
Electrostatic interactions, 242, 591, 699
Enantioselectivity, 335
Endocytosis, 425, 530, 594, 608, 614, 615, 655
Endonucleases, 609
Endopeptidase, 610
Endophytes, 195
Endoproteases, 371
ENDOR spectroscopy, 596
Endosomes, 606, 611–613, 654
Enterobacteria, 87, 105, 242
Enterobacteriaceae, 86, 104
Enterobactin, 19, 50, 55, 59, 69, 78, 79, 84, 94, 103–105, 123–125, 158, 163, 192, 251, 254, 259, 261–263, 265, 270, 272, 274, 277, 285–287, 334-338, 341, 342, 349, 357, 359–362, 366, 368, 369, 371, 373, 374, 378–382, 672, 697
chiral analogues, 265
enantio-, 385
ferric, *see* Iron(III)
models, 286
synthesis, *see* Synthesis
V(IV)-, 337, 385
Enterochelin, *see* Enterobactin
Enterocytes, 648, 652
Enteromophthorales, 152
Enzyme(s) (*see also* individual names), 42, 100, 101
active sites, 100, 705
adenylate-forming, 50

[Enzyme(s)]
antioxidant defense, 41
heme-containing, 705
iron-dependent, 704–706
-linked immunosorbent assay, 202
nonheme containing, 706, 723, 724
3-Epihydroxymugineic acid, 219–221, 225–227, 229, 232, 235
Epirubicin, 719
EPR, 368, 378, 461, 483, 488, 493, 500, 504, 534, 538, 540, 553, 590, 596, 601, 602, 617, 673, 675, 678
rapid freeze-quench, 488, 490, 492
spin trapping, 488, 494
Equilibrium constants (*see also* Acidity constants *and* Stability constants), 269, 298, 305, 308
Equilibrium dialysis, 598, 604
Equus caballus, *see* Horse
Erbium(III) (*see also* Lanthanides), 590
Erwinia
chrysanthemi, 88, 89, 103, 105, 198–200, 363
herbicola (*see also Pantoea agglomerans*), 342
Erythrin, 442–444
Erythrocytes, 107, 347, 349, 520, 589, 603, 613, 635
guinea pig, 614
human, 615, 719
lysis, 52, 109
Erythroid
5-aminolevulinate synthase, 643
bone marrow, 651
cells, *see* Cells
Erythromycin, 79
Erythropoeisis, 651, 693
Escherichia coli, 42, 48, 50, 54, 55, 59, 69–71, 73, 74, 76–78, 81, 83, 85–89, 94, 96–111, 113–115, 117, 120–122, 124–128, 155, 158, 160, 163, 165, 176, 192, 276, 277, 284, 357, 360, 361, 363, 366, 372–376, 381, 385, 388, 424, 439, 440, 443, 444, 458, 569, 635, 638, 658, 672
ferritin, *see* Ferritins
receptor proteins, 360, 361, 366
ESR, *see* EPR
Esters (*see also* individual names)
hydrolysis, 712
hydrophobic, 712, 713
linkage, 340
phorbol, 424, 607
Estuaries
iron in, 3
Ethanol
mercapto-, 47, 419
Ethylene, 567
Ethylenebis(oxyethylenenitrilo)tetraacetate, *see* EGTA
Ethylenediamine, 513
di-(*o*-hydroxyphenylacetic acid), 205
Ethylenediamine-*N,N'*-diacetic acid
N,N'-bis(dibenzyl)-, 717, 718
N,N'-bis(*O*-hydroxybenzyl)-, 701, 702, 708, 716, 717
Ethylenediamine-*N,N,N',N'*-tetraacetate, 15, 17, 169, 205, 219, 270, 281–283, 287, 290, 296, 312, 313, 348, 376, 382, 408, 411, 604, 694, 701, 707, 708, 716, 718
derivatives, 291, 297
Ethyleneglycol-bis-(β-aminoethylether) *N,N,N',N'*-tetraacetate, *see* EGTA
Eubacteria, 438, 442, 443

Eukaryotes (*see also* individual names), 9, 14, 52, 53, 100, 148, 149, 173, 405, 424, 567, 570, 579, 617, 641, 657, 672
cells, *see* Cells
cytoplasm, *see* Cytoplasm
ferredoxins, 675
ferritin, *see* Ferritins
fungi, 148, 149
mRNA, 415
phototrophic, 9, 12
phytoplankton, 7, 11, 13, 16, 17, 19
transferrin, *see* Transferrins
yeast, *see* Yeast
Europium(III) (*see also* Lanthanides), 590
Evolution, 586-588, 594
EXAFS, 496, 498, 534, 538–541, 544, 546, 547, 551, 552
Excretion of
iron, *see* Iron
zinc, 717
Exons, 161, 571, 592, 607
Exotoxin A, 122, 124
Extended absorption fine structure spectroscopy, *see* EXAFS

F

FAB-MS, *see* Fast atom bombardment mass spectrometry
Fast atom bombardment mass spectrometry, 267, 268
Fenton
chemistry, 39, 492–494, 497, 503, 575, 635, 641
reaction, 42, 577
Ferredoxins, 18, 121, 500, 674
eukaryotic, 675
spinach, 675
Ferrichromes (*see also* individual names), 71, 74–81, 83, 84, 87, 88, 90, 101, 103, 149–151, 155, 158, 162–164, 169–171, 177, 192, 204, 263, 266, 268, 276, 283, 284, 331, 334, 341, 342, 357, 359–361, 366, 369, 371, 373, 374, 377–381, 383, 384, 387, 390, 672
A, 88, 149, 155, 170, 171, 204, 272, 274, 276, 359, 373, 379, 380
analogues, 338–340, 343, 344, 36-349
C, 149
deferri-, 246, 247, 257, 260, 380
enantio-, 155, 385
models, 281–285
radiolabeled, 153
stability constants, 282
synthetic, 347
tetraglycyl, 149
Ferrichrysin, 149, 150, 154, 155, 272, 359
Ferricrocin, 149, 150, 178, 195, 268, 272, 274, 357, 380
Ferricyanide, *see* Hexacyanoferrate
Ferrihydrite, 96, 97, 438, 467, 468, 496, 528, 531, 532, 536, 539, 542–544, 547, 552, 556, 570, 571, 655, 656
Ferrioxamine(s), 672
analogues, 340–342
B, 88, 103, 104, 116, 148, 172, 192, 195, 205, 246, 251, 263, 266, 268, 271–274, 276, 279, 280, 299, 306–308, 310, 312, 331, 334, 340–342, 348, 349, 357, 360–362, 366, 384
D, 272
de-, *see* Deferrioxamines

[Ferrioxamine(s)]
E, 88, 272, 281, 380
H, 247
models, 279–282
stability constants, 280
Ferrirhodin, 149, 151, 268, 359, 380, 384
des(diserylglycyl)-, 149
Ferrirubin, 149, 150, 154, 359, 380, 384
Ferrithiocin, 272, 275, 708
des-, *see* Desferrithiocin
nor-, 273, 274
Ferritins, 42, 47, 57, 73, 97–100, 194, 257, 296, 363, 403–426, 610, 614–616, 635, 640, 641, 643, 644, 646, 647, 652–655, 669–671, 676, 677, 680
amphibian, 462, 480
animal, 408, 412, 413, 453, 467, 480, 483, 517, 522, 567, 569, 571, 572, 657
apo-, 457, 483, 485, 487, 490, 491, 498, 499, 530, 540, 568–570, 655–657
bacterial, *see* Bacterioferritin
bullfrog H chain, 489, 490, 496
bullfrog L chain, 451, 453, 454, 463, 481
cDNA, 571, 576
channels, *see* Channels
E. coli, 439, 443–451, 456, 457, 459–461,463–465, 467–470, 494–496, 639, 657, 658
eukaryotic, 438–444, 455, 456, 466, 494
evolution, 438–444
frog, 408, 412, 414, 571
fungal, 152, 153, 438
genes, *see* Genes
H chain, 97, 98, 646, 648–650, 656
holo-, 485
horse, 570
[Ferritins]
horse L chain, 448, 450–455, 462, 465, 466
horse spleen, 449–452, 457–459, 462, 466, 481–483, 485–487, 489, 492–503, 536, 539, 541, 546, 551, 657
human, 56, 69, 408, 412, 414, 439, 443, 444, 480, 539, 570, 720–722
human H chain, 439, 440, 448, 450–462, 464–468, 470, 471, 481, 482, 484, 485, 489, 490, 494, 495, 497
human L chain, 452, 454, 455, 462, 466
human liver, 489
invertebrate, 456, 517
iso-, 444, 445
L chain, 97–99, 646, 648, 649, 656, 659
maize, 571, 573–575, 578, 641
mammalian, 99, 444, 457–459, 462, 465, 480–502, 517, 530, 531, 534, 538–544, 555, 568, 570, 655
microbial, *see* Bacterioferritin
mineral cores, 466, 467, 486, 498, 501, 529, 530, 534–554
mineralization, *see* Mineralization
molluscan, 540
mRNA, 407–410, 425, 571–578, 640, 642, 644, 648, 649, 659
myco-, 153
pea seed, 462, 464, 497, 498, 502, 503, 568–571, 574, 575, 578
phosphate content, 438
phylogenetic tree, 442, 443
phyto-, *see* Phytoferritins

[Ferritins]
plant (*see also* Phytoferritins), 438, 442, 443, 453, 455, 462, 464, 466, 467, 480, 483, 496–498, 517, 540, 563–579, 657
prokaryotic, 424, 438–443, 455, 466, 494
protein concentrations, 573
rat, 408, 412, 414, 439, 678
rat L chain, 453, 465
rat spleen, 614
recombinant bullfrog L chain, 447, 451, 452, 454, 462, 464
recombinant horse L chain, 437, 445–454, 456, 462, 464, 466, 467, 657
recombinant human H chain, 450, 452–456, 459, 461, 462, 464, 466, 469–471, 568–570
recombinant human L chain, 467, 570
recombinant pea seed, 569, 570, 575
recombinant rat L chain, 451, 452, 464
recombinant *Schistosoma mansoni* "type 1," 451, 456, 463, 464
redox chemistry, 498, 499
sequences, 568
sheep spleen, 485, 494
soybean, 571, 573, 574, 576
structure-function relationships, 435–471, 566–572
subunits, 437, 438, 444, 452–460, 480, 568, 575
synthesis in plants, 572–574, 576, 577, 579
three-dimensional structures, 444–459
vertebrate, 41, 96, 99, 438, 442–445, 467, 530

[Ferritins]
x-ray structures, 480
zygo-, 152, 153
Ferroxidase(s), 96–98
activity of ferritin, 438, 439, 442, 453, 456, 457, 461, 462, 466, 467, 470, 480–485, 487, 488, 490–497, 500, 503, 530, 534, 555, 568–570, 656, 657
kinetics, 489
Ferrozine, 501
Fetus, 594, 723
porcine, 653
Fibroblasts
chicken embryo, 608
mouse, 609
"Fingerprint" studies, 536–543
Fish pathogens, 104
Flagellates, 12
Flavin reductase, 48, 99
Flavin mononucleotide, *see* FMN
Flavin mononucleotide (reduced), *see* $FMNH_2$
Flavocytochromes, 49
Flavodoxins, 18, 121, 674
Flavoproteins, 500
Fluorescein, 346
Fluorescence
anthracene, 255, 345
probe, 77, 88
pseudomonads, *see* Pseudomonads
pyrene, 257
quenching, 343, 481
spectroscopy, 458
titration, 344
Fluorescence emission spectroscopy, 253–259, 283
FMN, 500
$FMNH_2$, 500
Food, 520, 652
chain, 18
Footprinting, 411
of DNA, *see* DNA

Formation constants, *see* Equilibrium constants *and* Stability constants
Fourier transformation, 538, 551
Frog ferritin, *see* Ferritins
Fungi (*see also* individual names), 50, 57, 70, 73, 98, 127, 242, 494, 673, 674, 723
  eukaryotic, *see* Eukaryotes
  ferritins, *see* Ferritins
  iron transport, 147–178
  metabolism, 153
  mycorrhizal, 195, 196
  siderophores, *see* Siderophores
  unicellular, 14
  virulence factors, 177, 178
Fusarinines, 151, 152, 195
Fusarium sp., 151
  oxysporum, 196
Fusigens, *see* Fusarinines

G

Gadolinium(III) (*see also* Lanthanides), 590
Galactose, 85, 165, 166, 176
β-Galactosidase, 113, 166, 458
Galapagos Islands, 24
Gallichrome, 381
Gallium(III), 268, 278, 279, 380, 381, 384, 390, 589, 590, 598, 602, 673, 680, 682
  ^{67}Ga, 589, 590, 681
Gas chromatography-mass spectrometry, 267
Gastrointestinal tract (*see also* Intestine), 651, 701, 704, 712, 714, 717
  iron homeostasis, 647, 648, 651–653, 659
GC-MS, *see* Gas chromatography-mass spectrometry
Gel
  chromatography, 378, 379
  filtration, 162, 379
Gel electrophoresis, 458, 459, 606
  polyacrylamide, 155
  sodium dodecyl sulfate poly-acrylamide, *see* SDS-PAGE
  urea, 595, 604
Gene(s), 49, 74, 496, 566, 567, 571, 572, 576, 606, 607, 641–643, 659, 673, 674
  afu, 113
  alg, 125
  bfr, 98, 639
  btu, 103
  cbs, 105, 199
  cft, 98
  cluster, 102
  cor, 86
  css, 191
  cyt, 193
  dtxR, 54, 116
  ent, 104, 105, 199, 376
  exb, 82, 83, 94, 117
  expression, 53–56, 103, 404, 405, 425
  fat, 103, 104, 126
  fbp, 112, 113
  fct, 105
  fec, 103, 117, 118, 120
  feg, 192
  feo, 58, 59, 86
  fep, 104, 374, 380
  ferritin, 423, 424, 426, 571, 572, 575–577, 641, 648
  fet, 176
  feu, 193
  fhu, 80, 82, 90, 101–103, 164
  fre, 172, 173
  frg, 59
  froh, 217
  frp, 173
  ftn, 97, 98
  fur, 42, 43, 54, 98, 113–116, 121, 124, 176, 192, 197, 376, 638, 639, 672
  gac, 126
  glt, 191

[Gene(s)]
has, 110
hem, 107, 109, 110
his, 103, 174
hit, 113
hpu, 109
hup, 111
ide, 42, 43, 116
ids, 202, 203
ina, 197
irg, 126
lac, 113, 165, 166, 173, 193, 199
lem, 126
nif, 100
omp, 377
operons, *see* Operons
organization of iron transport, 101–113, 119
pcs, 191
pfe, 125
pst, 103
pub, 119, 197
pvd, 163
rec, 98, 124
reg, 124
rep, 126
rsg, 97
rum, 107
sfu, 112
sid, 158, 160–165, 167–170, 177
siderophore biosynthesis, 156–163
siderophore regulation, 163–171
sod, 122, 673
tbp, 111, 112
tol, 82, 117
tonB, 58, 59, 80-82, 94, 108, 117
tox, 116, 124
transcription, 54, 55, 102, 104, 113–127, 165, 422, 642, 648
transferrin, 588
[Gene(s)]
urbs, 164–166, 170, 171, 177
viu, 105
yec, 97
yfu, 113
Geochemical cycle of
copper, 674
iron, 674
manganese, 674
zinc, 674
Glia, 645, 646, 677
Globins (*see also* Hemoglobin), 407
hapto-, *see* Haptoglobin
Globulins
α-, 680
β -, 680
β_2-micro-, 659
Glomus mosseae, 196
Glucopyranoside
n-octyl-β-D-, 376
Glucose (*see also* Sugars), 85, 165
Glucuronic acid, 712
β-Glucuronidases, 163–165, 167, 169
Glutamic acid (and residues), 78, 82, 97, 114, 116, 251, 263, 342, 385, 446, 449, 451–457, 459, 461, 462, 464, 466, 467, 469, 470, 482–485, 490, 495, 497, 600, 656, 657
Glutamine (and residues), 175, 364, 453, 455, 456, 461, 466, 469, 602, 605, 641
γ-L-Glutamyl-L-cysteinylglycine, *see* Glutathione
Glutarate
2-oxo-, 203
Glutathione, 578
Glycine (and residues), 51, 93, 149, 162, 257, 280, 339, 343, 364, 375, 455, 457, 458, 462, 591, 602, 705
Glycine max. (*see also* Soybean), 439, 443

Glycolic acid
thio-, 500, 501
Glycoproteins, 53, 588, 646, 669, 680, 681
Goethite, 15, 97, 525, 526, 528, 529, 531, 536, 546, 551, 552
Good's buffer, *see* Buffer
Gramicidin synthase, 51
Gramineae, 202
Grass (*see also* individual names), 189, 200, 201, 216, 640
Greenhouse effect, 332
Griegite, 541
Growth
factors, 425
hormone, *see* Hormones
inhibition, 332, 339, 346, 347
promotion, 332, 334, 338–342, 346, 362, 371
Guanidine, 166
Guinea pig, 614
Gut, 73, 124, 647, 648, 652, 653, 699, 720

H

H.
ducreyi, see Haemophilus
gonorrhoeae, 108
influenzae, see Haemophilus
paragallinarum, see Haemophilus
pylori, see Helicobacter
Haber-Weiss reaction, 521, 577
Haemophilus sp., 53, 109–111, 127, 362
ducreyi, 108, 109
influenzae, 83, 88, 89, 95, 96, 103, 108–110, 112, 113, 439, 443, 444, 617
paragallinarum, 112
Hafnium(IV), 590
Hamster
Chinese, 614
Haptoglobin, 47, 57, 108–110
Heart, 644, 712, 720–723
human, 445
Helicobacter pylori, 97, 439, 443, 494
Hematite, 7, 15, 525, 526, 528, 529, 531, 546, 552, 668
Hematoporphyrin
IX, 97
platinum, 97
Heme(s), 8, 41, 46, 47, 52, 57–61, 69, 70, 81, 84, 87–94, 96–98, 102, 103, 127, 193, 363, 406, 419, 438, 449, 450, 456, 457, 466, 494, 495, 500–502, 504, 614, 615, 641, 647, 652, 657, 669
oxygenase, *see* Oxygenases
synthesis, *see* Synthesis
transport systems, 107–111
Hemerythrins, 97, 495, 657
Hemicellulose, 565
Hemisphere
Northern, 4
Southern, 4
Hemochromatosis, 177, 422, 521, 615, 658
genetic, 638, 646–649, 651, 653, 655, 659, 693
secondary, 648, 649
Hemodialysis (*see also* Dialysis), 681
Hemoglobin(s), 56, 57, 70, 107–110, 127, 363, 606, 610, 677, 679, 682
animal, 109
bacterial, 108
human, 109, 170, 520, 521, 603, 719
leg-, 573
vertebrate, 107
Hemoglobinopathy, 693, 720
Hemolysins, 52, 110, 363

Hemopexin, 47, 57, 70, 103, 108, 110, 127, 646, 647
Hemosiderin, 494, 497, 520, 521, 652, 655
Hepatocytes, 593, 609, 613, 615, 616, 648, 655, 709, 712
  human, 679
  rat, 614, 703
Hexacyanoferrate, 15, 498
High-performance liquid chromatography, 195, 377
Histidine (and residues), 72, 82, 85, 90, 95–97, 114, 116, 128, 164, 165, 169, 174, 446, 449, 451–453, 455, 457, 461, 462, 464, 466, 469, 470, 481, 482, 484, 485, 494, 495, 497, 592–594, 596, 602, 638, 641, 678
  kinase, *see* Kinases
Holmium, 590
Homeostasis (*see also* Metabolism)
  iron, *see* Iron
Hordeum vulgare (*see also* Barley), 675
Hormones (*see also* individual names), 332, 423, 425
  human growth, 571
  plant, *see* Plants
  thyroid, 424
Horse, 439
  ferritin, *see* Ferritins
  spleen, 96, 444, 445, 480, 490
Horseradish peroxidase, 111
HPLC, *see* High-performance liquid chromatography
Human(s) (*see also* Mammal), 177, 201, 311, 332, 416, 418, 422, 520, 636, 674, 678–682, 693, 700, 703, 711, 719
  brain, *see* Brain
  diseases, *see* Diseases
  duodenum, *see* Duodenum
[Human(s)]
  erythrocytes, *see* Erythrocytes
  ferritin, *see* Ferritins
  genetic diseases (*see also* individual names), 176
  growth hormone, *see* Hormones
  heart, *see* Heart
  hemoglobin, *see* Hemoglobin
  hepatocytes, *see* Hepatocytes
  iron homeostasis, 641, 643–657, 693
  lactoferrin, *see* Lactoferrin
  liver, *see* Liver
  melanoma, 593, 594
  mRNA, *see* mRNA
  pathogens, 38, 87, 177, 617
  serum, *see* Serum
  total body iron content, 651, 693
  transferrin, *see* Transferrins
Humic matter (*see also* Organic matter), 6, 8
Hydrazones
  aryl, 707, 713
  carbonylcyanide-*m*-chlorophenyl-, 81, 118, 232, 235
  pyridoxal isonicotinoyl-, 703, 704
Hydrogen bonds, 295, 300, 336–340, 349, 385, 386, 446, 447, 452, 453, 455, 490, 527, 591, 592, 595, 596, 600
Hydrogen carbonate, 95
Hydrogen peroxide, 6, 7, 40–45, 60, 98, 486-488, 492, 493, 503, 577, 578, 634, 645, 649, 650, 708
Hydrogen sulfide (*see also* Sulfide), 100
Hydrolases (*see also* Proteases *and* individual names)
  ATP, 92, 103
Hydrolysis (of)
  ATP, *see* ATP
  hydrophobic esters, 712

[Hydrolysis (of)]
iron, 284, 310, 480, 484–486, 518, 525, 527–529, 532–534, 546, 547, 554, 556, 589, 634, 698
Hydrophilicity, 387, 614, 704, 705, 707, 711, 712
region, 446, 447, 451–457, 466, 467, 588, 607
Hydrophobic
contacts, 446, 449
interactions (*see also* Lipophilicity), 308, 346–348, 705, 707, 712, 713
pocket, 97, 449
protein, 91, 113, 372, 374
region, 85, 95, 104, 607
residues, 364, 455–457, 464, 467
Hydroxamates (or hydroxamic acids, and residues) (*see also* individual names), 19, 71, 80, 85, 87, 88, 90, 94, 101–104, 121, 248–253, 263, 267, 285, 288, 295, 299, 302–304, 307–310, 312, 349, 363, 380, 383, 696, 707, 709, 715, 724
aceto-, 244–246, 248, 287, 288, 298, 306, 307, 597, 696, 698, 701
ferric, *see* Iron(III)
linkage, 341
N-methylaceto-, 302, 305, 307, 308
siderophores, *see* Siderophores
thio-, 248
Hydroxide(s)
ion, 241
iron, *see* Iron(III)
oxy-, *see* Oxyhydroxides
Hydroxo
bridges, 285, 301
complexes, 298, 301, 303, 304, 307
Hydroxyl
groups (*see also* Phenolates, Polysaccharides *and* Sugars), 673
radicals, *see* Radicals
Hydroxylases, 692, 705, 706
methane monooxygenase, 97
tyrosine, 720
3-Hydroxymugineic acid, 219
Hydroxypyridinones, 279, 287–290, 311, 616, 699–701, 703–705, 709, 711
8-Hydroxyquinoline, 251, 252, 295, 694, 701, 707, 711
5-Hydroxytryptamine, *see* Serotonin
Hymenoschyphus ericae, 195
Hyperferritinemia, 659
Hyphomycetes, 195
Hypotransferrinemia, 655
Hypoxia, 647, 651

I

Ice age, 26–28
ICPAES, *see* Inductively coupled plasma atomic emission spectrometry
Imaging, 589, 590
Imidazole (*see also* Histidine), 532
Iminodiacetate
hydroxyethyl-, 549, 554
Immune system, 650
Immunostaining, 646
Indium (including +III state), 589, 598, 671
^{111}In, 589, 590
Inductively coupled plasma atomic emission spectrometry, 542
Infections
bacterial, 723
by microorganisms, 723
fungal, 723
role of iron, 45–61

Inflammation, 650, 681, 722
Inositol, 287
Insect(s) (*see also* individual names and species), 480, 588, 594, 675, 676
  transferrin, *see* Transferrins
Integrin, 616
Interdependencies
  cadmium-iron, 676, 679
  chromium-iron, 678
  cobalt-iron, 671, 675, 676
  copper-iron, 175, 422, 423, 642, 658, 675, 676, 678–680
  iron-lead, 679
  iron-magnesium, 674
  iron-manganese, 673, 676, 677
  iron-molybdenum, 683
  iron-nickel, 671, 678
  iron-other metal ions, 667–684
  iron-vanadium, 678
  iron-zinc, 671, 675
Interferon, 650
Interleukins, 416
Intestine, 671, 681, 712
  cells, 616
  mouse, 58
  rat, 676
Introns, 168, 571, 572, 588
Invertebrates (*see also* individual names), 413
  ferritin, *see* Ferritins
Iodoacetic acid, 459
Ion exchange chromatography
  DEAE, 377, 378
Ion pairs (*see also* Salt bridges), 448, 449, 452, 453, 455
Ionic interactions, *see* Electrostatic interactions
Ionization constants, *see* Acidity constants *and* Equilibrium constants
Ionophores (*see also* individual names), 271
  ferric, *see* Siderophores
Iron (different oxidation states)
  ^{53}Fe, 268
  ^{55}Fe, 154, 170, 195, 206, 345
  ^{57}Fe, 496, 537
  ^{59}Fe, 219–221, 225–227, 229, 232, 235, 614
  absorption, 650–653, 659, 701
  abundance, 2
  acquisition, 13–15, 37–61
  assimilation, 41, 45
  binuclear centers, 468–471, 484, 487, 488, 491, 492, 495, 496, 500, 503
  biofulcrum, 56-60
  chemistry in seawater, 5-9
  cycling in the ocean, 1–29
  deficiency, 15, 18, 42, 155, 172, 176, 178, 188, 191–193, 196–199, 201, 202, 206, 216, 217, 219, 222, 224, 225, 227, 233, 347, 349, 358, 421, 422, 521, 565, 575–577, 609, 610, 614, 636, 640, 644, 647, 671, 672, 676, 678, 679, 683
  detoxification, *see* Detoxification
  detrital, *see* Detritus
  dextran drugs, 547
  di-iron carboxylate proteins, 468–471
  dietary, 636, 651, 677, 679, 693
  excretion, 650, 651, 713, 715–717
  finger, 170
  flux, *see* Iron flux
  homeostasis, *see* Iron homeostasis
  in plants, *see* Plants
  in storage cavities, *see* Ferritins, mineral cores
  interdependency with other elements, *see* Interdependencies
  limitation hypothesis in the sea, 20–24, 674

[Iron]
lithogenic, 10–12
metabolism, *see* Metabolism
minerals, 525–527
mixed valent, 548
non-protein bound, 615–617
normal balance, 651
overload, 42, 521, 644, 648, 649, 651, 655, 659, 707, 714, 716, 719–723
overload diseases, *see* Diseases *and* individual names
oxidase, *see* Oxidases
oxyhydroxide, *see* Oxyhxdroxides
proteins, *see* Proteins
role in infections, 39, 45–60
speciation, 19
toxicity, 73, 86, 98, 564, 565, 578, 634, 635, 651, 701, 723
turnover rates, 13
Iron(III)
$^{55}Fe^{3+}$, 81, 88, 334
chloride, 95, 496, 676
citrate, 69–72, 94, 95, 103, 106, 117–120, 122, 128, 163, 194, 235, 310, 311, 361, 366, 422, 496, 497, 566, 570, 577, 616, 638, 640, 672, 722
colloidal, 6, 7, 15
coprogen, 374
cyanide, 647, 653
design features of ligands, 695–704
dimeric, 491, 492
enterobactin, 357, 361, 369, 371, 374, 378–380, 383, 385, 386, 388, 390, 638
hydrolysis, *see* Hydrolysis
hydroxamate, 87, 127, 268, 363, 681
hydroxide, 69, 189, 203, 217, 269, 296–304, 309, 310, 480, 486, 519, 524, 527, 532, 536, 549, 565, 587, 634, 711
[Iron(III)]
hydroxyphosphate mıneral, 540, 543
oxides, 188, 517, 527–532, 545, 546, 555, 669
oxyhydroxide, *see* Oxyhydroxides
parabactin, 379, 380, 385
phosphate, 565, 570
pseudobactin, 362, 366, 379, 381–383
reductase, *see* Reductases
selective ligands, 719–725
Iron chelators (*see also* individual names)
for clinical use, 691–726
hydrophilic, 705
hydrophobic, 705
toxicity, 703–709, 716, 724
Iron flux, 6
atmosphere to ocean, 4
deepwater to surface, 5
Iron homeostasis (in), 153, 176, 517–521, 539, 564, 579, 633–660
bacteria, 638, 639, 658
brain, 645–647, 658
gastrointestinal tract, 647, 648, 651–653, 659
humans, 641, 643–658, 693
liver, 648, 649, 658
plants, 639–641, 658
yeast, 641–643, 658
Iron regulatory element, 100, 113–127, 168, 403–426, 571, 576, 609, 610, 636, 642–645, 650, 658, 659
binding protein, *see* Iron regulatory proteins
iso-, 406, 413–416
Iron regulatory proteins (*see also* Proteins), 42, 43, 119, 403–426, 609, 642, 644, 645, 647–650, 658, 659, 706

[Iron regulatory proteins]
apo-, 644
footprinting, 411
holo-, 644
mammalian, 642, 643
Iron-sulfur clusters, 8, 18, 69, 99–101, 406, 643, 683, 705
[Fe-S], 100, 419, 425, 645
[2Fe-2S], 98
[4Fe-4S], 101, 609, 644, 645, 650
Iron-sulfur proteins (*see also* Ferredoxins *and* individual names), 18, 99, 575, 668, 669
Ischemia, 693, 721
Isocitrate, 644
Isoleucine (and residues), 348, 459, 464, 495
biosynthesis, 101

J

Jejunum, 651

K

Keto groups, *see* Carbonyl oxygen
Kidney, 707, 714, 719, 721
rat, 679
Kinases
histidine, 125, 127
protein kinase c, 418

L

L.
innocua, see Listeria
monocytogenes, see Listeria
pneumophila, see Legionella
β-Lactam, 693
β-Lactamase, 91, 92
Lactate (or lactic acid)
bacteria, 618
dehydrogenase, *see* Dehydrogenases
Lactobacillus sp., 39, 69
Lactoferricin, 589
Lactoferrins, 47, 52, 53, 56, 57, 127, 200, 363, 589, 591, 603
antibacterial activity, 57
apo-, 597
bacterial, 70, 71, 73, 94, 95, 111, 112, 617
human, 69, 71, 588, 595–600, 653
receptor, 57, 109
Lactose, 113, 114
Lamb, 677, 678
Lanthanides (*see also* individual elements), 603, 671
LD_{50}
infections, 59
iron chelators, 704
Lead, 671, 679
interdependency with other elements, *see* Interdependencies
Leaves, 199, 200, 226, 228, 233, 565, 566, 572, 640
barley, *see* Barley
bean, 233
bronzing, 577
iron concentrations, 196
maize, 573, 575–577
rice, 577
Legionella pneumophila, 59
Legumes, 191
Lepidocrocite, 97, 528, 531, 536, 546, 551, 552
Lethal concentration, *see* LD_{50}
Leucine (and residues), 78, 83, 96, 167, 255, 257, 261, 266, 348, 458, 459, 464, 467, 481, 485
iso-, *see* Isoleucine
Leukemias
acute, 722
acute myeloid, 718
murine L1210, 724

Leukocytes, 69
  polymorphonuclear, 56, 589
Leukotrienes, 724
Levrazoxane, 718
Ligand exchange, 205, 206, 273, 296, 297, 299
  kinetics, 281
  reactions, 310–313
Lignin, 565
Lime, 205
Limonite, 668
Lipid(s), 127, 378, 387, 566, 634, 647
  bilayer, 74, 78, 92, 128
  membrane, *see* Membranes
  peroxidation, 176, 721
  vesicles, 85
Lipoate
  D,L-dihydro-, 500
Lipophilicity (*see also* Hydrophobic effects), 288, 346, 694, 703, 704, 715, 717, 719
Lipopolysaccharides, 104, 377, 650
Lipoproteins, 74, 85, 103, 109, 110, 112, 127
Liposomes, 79
Lipoxygenase, 692, 705, 706, 723, 724
Listeria
  innocua, 99
  monocytogenes, 48, 59, 88, 89
Liver, 613, 614, 616, 644, 659, 678, 703, 707, 712, 719, 720, 723
  cirrhosis, 177
  ferritins, *see* Ferritins
  human, 444, 445, 480, 489, 594, 680
  iron homeostasis, 648, 649, 658
  mouse, 58, 59
  rat, 445, 616, 679
Lungs, 678
  human, 178, 722, 723
  mouse, 58
Lymnea stagnalis, 414, 439, 442, 444, 464
Lymph nodes, 59
Lymphomas (*see also* Leukemias *and* Tumors), 722
Lysine (and residues), 50, 51, 78, 84, 251, 261, 287, 341, 371, 452, 455, 457, 462, 484, 485, 494, 592, 599, 600, 605, 637, 656, 657
Lysosomes, 614, 636, 655

M

M.
  avium, see Mycobacterium
  barkeri, see Methanosarcina
  catarrhalis, see Moraxella
  genitalium, see Mycoplasma
  jannaschii, see Methanococcus
  leprae, see Mycobacterium
  morganii, see Morganella
  pneumoniae, see Mycobacterium
  smegmatis, see Mycobacterium
  tuberculosis, see Mycobacterium
Macrophages, 59, 422, 635, 650, 655
Maghemite, 528, 541, 546
Magnesium ion, 86, 121, 382, 415, 682, 683, 719
Magnetic circular dichroism, 534, 538, 540, 553
Magnetic susceptibility measurements, 483, 534, 536, 544, 553
Magnetite, 525, 526, 541, 668
Maize, 196, 201, 202, 205, 296, 406, 573, 640, 641
  ferritin, *see* Ferritins
  leaves, *see* Leaves
  roots, *see* Roots
Malaria, 693, 720
  drugs, 347–349

Maltodextrins, 75, 79, 80
Maltose, 72, 85, 128
Mammal(ian) (*see also* individual names and species), 413, 530, 651, 657, 713
  cells, *see* Cells
  ferritin, *see* Ferritins
  mRNA, *see* mRNA
Man, *see* Human
Manduca, 413
  sexta, 594
Manganese (different oxidation states), 41, 42, 45, 116, 589, 671, 674
  geochemical cycle, 674
  interdependency with other elements, *see* Interdependencies
  toxicity, 677
Manganese(II), 86, 114, 121, 460–462, 464, 468, 541, 566, 590, 671, 672, 675–677
Manganese(III), 590, 601
Mannitol, 493
Marine bacteria, *see* Bacteria
Mass spectrometry, 267, 268
  electrospray, *see* Electrospray mass spectrometry
  fast atom bombardment, *see* Fast atom bombardment mass spectrometry
  inductively coupled plasma atomic emission spectrometry, 542
MCD, *see* Magnetic circular dichroism
MECAM, 251, 254, 268, 285–287, 379, 696, 698
Melanomas (*see also* Tumors), 609
  human malignant, 593, 594, 614
Melanotransferrin, 594, 606, 614
Membrane(s), 234, 612, 648, 652, 711
  brush-border, 652
  cell, *see* Cells
[Membrane(s)]
  cytoplasmic, *see* Cytoplasm
  damage, 57
  inner, 192, 387, 388, 672, 673
  lipid, 701
  outer, *see* Outer membrane
  peribacteroid, 194
  plasma, *see* Plasma
  potential, 155, 676
  proteins, 155, 192
  reductase, *see* Reductases
  surface, 276
  transport, 73–95, 331
Menkes disease, 175
Mercaptoethanol, 74, 419
Mesophyll, 227, 231, 233, 234
Metabolism (of) (*see also* Homeostasis)
  bacterial, *see* Bacteria
  cell, *see* Cells
  copper, 422, 423, 678
  fungal, *see* Fungi
  iron, 38, 43, 72, 96–100, 113, 116, 122, 123, 422, 423, 565, 566, 612, 654, 678, 681, 693
  microbial, 38, 39
  neurotransmitter, 720
Metallothioneins, 203
Metazoa, 13
  heterotrophic, 10, 11
Methane monooxygenase, *see* Monooxygenases
Methanococcus jannaschii, 86, 88, 89, 439, 443
Methanosarcina barkeri, 88, 89
Methionine (and residues), 90, 96, 116, 201, 218, 219, 419, 457, 464, 494, 495, 602, 640, 657
Methylviologen, 498
Mice, 659, 679, 701, 704
  blood, 58
  ferritin, 408, 412, 414, 439, 443
  fibroblasts, 609
  intestine, *see* Intestine
  LD_{50}, 704
  liver, *see* Liver

[Mice]
lung, *see* Lungs
spleen, *see* Spleen
Micelles (*see also* Vesicles), 308
Michaelis-Menten
constants, 90, 489
kinetics, 14, 488–490, 605
Microbes (*see also* individual names), 17, 19, 69, 278, 330, 335, 480, 589, 643, 681, 693
ferritin, *see* Bacterioferritins
growth promoters, 340–342
iron acquisition, 37–61
iron metabolism, 38, 39
iron transport, 37–61, 257, 277, 296, 331, 726
iron uptake from transferrin, 617, 618
siderophores, *see* Siderophores
Microcin 25, 75, 76, 384
Microorganisms (*see also* individual names and species), 113, 287, 332, 333, 668, 669, 715, 723, 724
aerobic, 40, 50
iron acquisition systems, 46, 47
iron uptake, 421–426
pathogenic, 37–61, 332
plant, *see* Plant
soil, *see* Soil
Microscopy
atomic force, 115
electron, *see* Electron microscopy
Mineralization
ferritin, 481, 484, 485, 492, 515–556, 656
hydrolytic, 321–329
re-, *see* Remineralization
Mitochondria, 218, 418, 641, 644
Mixed ligand complexes, *see* Ternary complexes
Mobilferrin, 616
Molecular recognition, 332, 384, 386
Molluscs, 517
Molybdenum (different oxidation states), 95, 100, 377, 671, 673, 682
interdependency with other elements, *see* Interdependencies
Molybdopterin, 95
Monkeys
Cebus, 717
Monooxygenases
methane, 97, 462, 468–470, 495, 657
Moraxella catarrhalis, 53, 617
Morganella morganii, 88, 103, 106, 363, 366
Mosquito, 464
Mössbauer spectroscopy, 69, 152, 467, 468, 470, 471, 483, 484, 488, 491, 495, 496, 498, 504, 534, 537, 539, 541, 542, 553, 570, 617
Mouse, *see* Mice
mRNA, 100, 164, 201, 218, 567, 609, 636, 682
animal, 408, 413, 414, 574
avian, 416
cytokine, 405, 574
degradation, 420
eukaryotic, *see* Eukaryotes
ferritin, *see* Ferritins
human, 408, 412, 607
lymphokine, 574
mammalian, 416
nucleotide sequences, 412
oncogene, 405, 574
protooncogene, 574
regulation, 403-426
synthesis, 405, 643
transferrin receptor, *see* Transferrin receptor
transferrin, *see* Transferrins
translation, 406, 407, 413–416, 636
vertebrate, 413, 414
viral, 405, 425

Mucorales, 152
Mucosa, 56, 107, 112, 647, 648, 651–653, 671, 676, 678
Mugineic acids, 201–203, 216, 218, 219, 223, 234, 235, 243, 245, 272, 274, 296, 675
  2-deoxy-, *see* 2-Deoxymugineic acid
  3-epihydroxy, *see* 3-Epihydroxymugineic acid
  3-hydroxy, 219
  radiolabeled, 206, 219–221, 225–227, 229, 232
Murein acid, 74, 127
Mus musculus, *see* Mice
Muscle, 520, 712
  cardiac, *see* Heart
Mushrooms (*see also* Fungi *and* individual names), 149
Mutagenesis, 158, 164, 177
  analysis, 569
  site-directed, 97, 100, 162, 166, 167, 457, 458, 461, 466, 482, 484, 489, 607, 673
Mycobacteria, 87, 96, 116
Mycobacterium
  avium, 439, 443, 466, 494
  leprae, 439, 443, 466, 494
  smegmatis, 42–45, 116, 122
  tuberculosis, 54, 58–60, 88, 89, 639
Mycoplasma
  genitalium, 88, 444
  pneumoniae, 88
Mycorrhiza, 156
Myocardial infarction, 693
Myoglobin, 520, 682

N

N.
  crassa, *see Neurospora*
  gonorrhoeae, *see Neisseria*
  meningitidis, *see Neisseria*
  winogradsky, *see Nitrobacter*
NADH, 7, 40, 100, 155, 638, 640
NADPH, 156, 173, 204, 217, 638, 640, 641
  reductase, *see* Reductases
Neisseria sp., 53, 127, 588, 618
  gonorrhoeae, 94, 96, 103, 110, 111, 113, 121, 197, 439, 443, 444, 457, 617
  meningitidis, 83, 88, 94, 108, 109, 111, 113, 197, 617
Neisseriaceae, 110, 112
Nematodes, 201
Neocalanus plumchrus, 13
Neocoprogen, 151, 154
Neodymium(III) (*see also* Lanthanides), 590
Neoplasia, 693
Neovossia indica, 149
Neptunium(IV), 591
Nernst equation, 518
Neuroblastomas, 500, 677
Neurodegenerative diseases, *see also* individual names, 693, 707, 721
Neurons, 645–647, 677, 720
  dopaminergic, 721
Neurospora, 195
  crassa, 154–156
Neurotoxicity of manganese, 677
Neurotransmitters, 646, 647, 703
  metabolism, 720
Neutropenia, 706
Nickel (oxidation state not defined), 2, 377, 578
  deficiency, 678
  interdependency with other elements, *see* Interdependencies
  toxicity, 678
Nickel(II), 83, 86, 377, 675–678
Nicotianamine, 202, 218
Nicotinamide adenine dinucleotide (reduced), *see* NADH
Nicotinamide adenine dinucleotide phosphate (reduced), *see* NADPH

Nicotine, 567
Nitrate(s), 2, 5, 16, 21–25, 547, 605
assimilation, 24
reductase, *see* Reductases
uptake, 24
Nitric oxide, *see* Nitrogen monoxide
Nitrilotriacetate, 377, 615, 616, 707, 708
Nitrite, 23
Nitrobacter winogradsky, 96
Nitrogen (*see also* Dinitrogen)
assimilation, 45, 682
fixation, 189, 192, 193, 207, 424, 564, 573, 673
Nitrogen monoxide, 419, 425, 495, 496, 500, 609, 645, 649, 650
hepatic synthesis, 650
synthase, 609, 650, 659
Nitrogenases, 100, 573, 639, 673, 682, 683
Nitrosyl complexes
iron, 482
Nitrous oxide, *see* Dinitrogen oxide
Nitroxide spin labels, 500
NMR, 267, 268, 375, 381, 408, 411, 590, 602, 603
^{1}H, 114, 116, 267, 268, 681, 682
^{13}C, 267, 268, 592, 602
NOESY, 602
NTA, *see* Nitrilotriacetate
Nuclear magnetic resonance,*see* NMR
Nuclear Overhauser effect spectroscopy
two-dimensional, *see* NOESY
Nucleases, 408, 411, 420, 425, 636, 643
Nucleic acids (*see also* DNA *and* RNA), 634
-protein complexes, *see* Proteins
Nucleoside phosphates, *see* Nucleotides *and* individual names
Nucleotides (*see also* individual names), 415, 693
oligo-, *see* Oligonucleotides
sequences, 412, 567, 573
Nutrients, 220
algal, *see* Algae
essential, 188

O

Oats, 202
Ocean (*see also* Seawater *and* individual names)
cycling of iron, 1–29, 518
fertilization, 24, 28
iron enrichment, 21-24
iron limitation, 20-24
Oidiodendron griseum, 195
Oligonucleotides (*see also* Nucleotides), 165, 170
Oligosaccharides, 607
Oncogene mRNA, 405, 574
Oncorhynchus myleiss, 439
Operons (*see also* Gene), 101, 107, 110–112, 125, 193, 199, 635, 639
2,3-dihydroxybenzoic acid, 51, 52
siderophore, 54
transport, 103–105, 107, 113
Optical spectra, *see* Absorption bands and spectra
Organic matter (*see also* Humic matter), 27
particulate, 5, 25
Ornithine (and residues), 50, 149, 154, 157, 160, 385
δ-*N*-hydroxy-, 157–159, 162, 177
oxygenase, *see* Oxygenases
Oryctolagus cuniculus, *see* Rabbit
Outer membrane, 70, 71, 192, 372, 379, 387–389, 654

[Outer membrane]
bacterial, 73–84, 87, 94, 95, 101, 104, 106, 110–112, 117, 118, 120, 125–127, 155, 331, 638, 672, 673
receptors, 19, 101, 105, 107, 109, 164, 197, 199, 355–391, 638
Outer sphere association constants, 298
Ovotransferrins, 589, 591, 592, 681
chicken, 595, 599, 677, 678
duck, 596
Oxalate
buffer, *see* Buffer
synergistic, 602
Oxidases
blue copper, 641
copper, 174, 674
cytochrome, 646
iron, 49, 172, 175, 692
xanthine, 708
Oxidation-reduction potentials, *see* Redox potentials
Oxidative
damage, 41, 42, 60, 98, 191, 419, 495, 634, 721
stress, 43, 45, 116, 122, 423, 577, 578, 636, 639, 641, 717, 719
Oxides
iron, 6, 15, 188
Oxidoreductases (*see also* Peroxidases), 173, 277
NAD$^+$-dependent, 204
μ-Oxo-bridges, 285, 468, 470, 471, 488, 545, 657
Oxygen (*see also* Dioxygen)
^{16}O, 268
electrode experiments, *see* Electrode, oximetry
Oxygenases, 640, 682, 724
di-, *see* Dioxygenases
heme, 47, 52, 57
[Oxygenases]
lip-, *see* Lipoxygenases
ornithine, 158, 160, 163
Oxyhydroxide
iron, 3, 5, 7, 8, 517, 520, 521, 524, 526–532, 534–536, 540, 541, 545–550, 552–555, 634, 655
Ozone, 567

P

P.
aeruginosa, see Pseudomonas
agglomerans, see Pantoea
denitrificans, see Paracoccus
fluorescens, see Pseudomonas
putida, see Pseudomonas
syringae, see Pseudomonas
tinctorius, see Pisolithus
viridiflava, see Pseudomonas
Pacifastacus leniusculus, 439
Pacific Ocean
Antarctic, 5, 21
equatorial, 5, 19, 21, 23–28
iron in, 2, 5, 674
North, 2, 5, 674
South, 28
subarctic, 3, 9, 11, 17, 21
Palmitate, 607
Pancreas
human, 177
Pantoea agglomerans (*see also Erwinia herbicola*), 77, 78
Parabactin, 251, 272, 275, 277, 379, 381
A, 251, 252, 259, 261, 274, 277
enantio-, 385
ferric, *see* Iron(III)
L-, 252, 259, 261, 380
Paracoccus denitrificans, 259, 277, 380, 381
Paramagnetic resonance, *see* EPR *and* NMR
Paraphysemonas sp., 16, 18

Paraquat, 98, 722
  radical, *see* Radicals
Parasites, 111
  growth, 346
  infections, 693
  malaria-causing, 349
Parkinson's
  disease, 693, 707, 721
  syndrome, 646
Pasteurellaceae, 110
Pathogens (*see also* Bacteria), 46, 47, 53, 57–60, 107, 177, 617, 618
  animal, *see* Animals
  bacterial, *see* Bacteria
  fish, 104
  human, *see* Human
  plant, *see* Plants
Pea(s), 198, 572
  cow-, 496
  ferritin, *see* Ferritins
  pigeon, 216
  seed, 496
Peanuts, 205
Pectic compounds, 565
Pectinase, 199
Penicillium sp., 149, 153
Peptide(s) (*see also* individual names), 78, 89, 90, 92–94, 103, 281, 368, 574, 637
  antibiotics, *see* Antibiotics
  bond, 162, 384, 385
  cleavage, 497
  mapping, 75, 80
  nitrogen, *see* Amide groups
  synthases, *see* Synthases
  synthesis, 157
  synthetic, 75, 77
Periplasm, 70–72, 75, 76, 78, 81, 87, 90, 91, 105, 112, 117, 118, 120, 193, 331, 368, 369, 371, 379, 387, 390, 638, 672
  binding protein, 84, 85, 90–95, 103–106, 112, 192, 363, 372–374, 388, 389, 617, 618
Permeases, 49, 53, 85, 93, 617
Peroxidase
  heme-containing, 41
  horseradish, 111
Peroxide(s), 116, 419, 680
  hydrogen, *see* Hydrogen peroxide
Pesticin, 362
Phages, 79, 82, 382
  bacterio-, *see* Bacteriophages
  Φ80, 75–77, 80, 361, 362, 373, 375
  T1, 75–77, 80, 361, 362, 373, 383
  T5, 75–80, 83, 361, 362, 372, 373, 378, 383, 387
  UC-1, 76
Phagocytes, 60, 61, 649
Phagocytosis, 635
Pharmacokinetics, 694, 703
Phaseolus vulgaris, 439, 443
9,10-Phenanthrenequinone diimine
  rhodium complexes, 410
1,10-Phenanthroline (and derivatives), 501, 575, 694
  batho-, *see* Bathophenanthroline
  copper complexes, 408, 410
  rhodium complexes, 410
Phenolates (or phenols, and phenolic groups) (*see also* individual names), 104, 217, 290, 292, 295, 302, 309, 310, 500, 503, 532, 566, 601, 699, 713, 716
  poly-, 200, 500
Phenylalanine (and residues), 51, 348, 364, 371, 387, 459, 464
Phloem, 200, 222, 233, 234
Phorbol esters, 424, 607
Phosphate(s) (including hydrogen and dihydrogen phosphate), 2, 5, 21, 25, 40, 96, 438, 483, 496–498, 502, 517, 522, 525, 529, 534, 539–543, 555, 570, 571, 605, 641, 652, 655

[Phosphate(s) (including hydrogen and dihydrogen phosphate)]
buffer, *see* Buffer
deficiency, 170
di-, *see* Diphosphate
pyro-, *see* Diphosphate
transport, 220, 222, 224, 227–231
Phosphoenolpyruvate carboxylase, 573
Phosphorus, 195
^{32}P label, 106, 224, 227, 230, 410
Phosphatidylinositol
glycosyl-, 653
Phosphorylation, 119, 406, 418, 419, 607
Photoaffinity labeling, 153
Photoautotrophs, 17
Photoreduction of iron, 6-8, 13, 497, 502, 503, 570, 574
Photosynthesis, 18, 21, 24, 25, 521, 564, 567, 572, 573, 575, 577, 579, 587, 640
Photosystems
I, 17
II, 18, 24
Phyllosphere, 196
Phylogenetic
sequence conservation, 408, 416
tree, 364–366, 442–444
Phytates, 652
Phytoferritins (*see also* Ferritins), 496–498, 502, 503, 640, 641
Phytoplankton, 332
blooming, 21
eukaryotic, *see* Eukaryotes
growth rates, 16, 17
iron in, 5, 9, 12–18, 21, 23–28, 674
Phytosiderin, 497
Phytosiderophores, 189, 190, 201–206, 275, 640
Pig, 408, 414, 653, 677, 678
Pigments, 196
Pine, 195
Pisolithus tinctorius, 195
Pistidic acid, 216
Pisum sativum, 439
Placenta(l), 609, 712
barrier, 712
Plankton, 15
phyto-, *see* Phytoplankton
Plants (*see also* individual names and species)
apoferritin, 568–570
(beneficial) microorganisms, 189–197, 675
cotyledons, 204, 572
dicots, 201, 216, 217
ferritin, *see* Ferritins
graminaceous, 215–235
higher, 14, 188–208
hormones, 576, 577, 640
iron homeostasis, 564, 639–641, 657, 658
iron homeostasis, 639–641
iron in vacuoles, 565, 566, 579
iron overload, 576, 577, 640
iron regulation of ferritin, 405, 407
iron-starved, 565, 567, 575–577, 641
iron storage, 563–579
iron toxicity, 565
iron transport, 215–235, 674, 675
iron uptake, 188–208, 277, 296, 421–426
leaves, *see* Leaves
mineral nutrition, 195, 196, 295, 565
monocotyledons, 148, 216
nodules, *see* Roots
nongraminaceous, 217
pathogens, 38, 87, 156, 189, 191, 196, 198–200, 206, 363
phosphorus uptake, 195
roots, *see* Roots
seedlings, *see* Seed
shoot, 198, 201

[Plants]
strategy I, 200–203, 216, 217, 233
strategy II, 201–205, 216, 218, 219, 233, 235
transgenic, 198
utilization of microbial siderophores, 203–205, 669
Plasma (*see also* Serum), 520, 594, 651, 652, 679–681, 712, 714, 716, 720
membrane, 204, 216, 593, 594, 608, 611, 613, 615, 640, 641, 654, 655, 674
Plasmalemma, 14, 565
Plasmids, 71, 86, 94, 98, 103, 114, 117, 126, 166, 191, 197, 361, 363, 374, 376, 377, 380, 424
Plasmodium falciparum, 346–349
Plastids, 496, 497, 564, 567, 570, 571, 575, 578, 641
Plastocyanin, 18
Platinum (different oxidation states), 97, 449, 462, 590
Plutonium (including +IV state), 591, 671
PMR, *see* NMR
Polar
effects, *see* Electrostatic interactions
ice, 26, 27
Polyethylene glycol, 377
Polymerases
chain reaction, 357
RNA, *see* RNA polymerases
Polysaccharides, 547
lipo-, *see* Lipopolysaccharides
Porins, 73, 74, 77, 79, 117, 363–365, 368, 371, 372, 377, 386, 387
Porphyrins (*see also* Hemes *and* individual names), 97, 109, 657, 679
proto-, *see* Protoporphyrins
Porphyromonas gingivalis, 53
Potentials, *see* Redox potentials
membrane, *see* Membrane
Potentiometric titrations, 249, 270
Pourbaix diagram, 518–520, 525, 526, 555
Praseodymium(III) (*see also* Lanthanides), 590
Prodrugs, 717, 719
Prokaryotes, 9, 148, 149, 163, 169, 405, 422, 570, 588, 657, 669, 672
ferritin, *see* Ferritins
Proline (and residues), 348, 452, 455, 458
Prostaglandins, 724
Protactinium
^{231}Pa, 27
(V), 591
Proteases (*see also* individual names), 46, 47, 57, 74, 82, 369, 644
endo-, *see* Endoproteases
K, 88, 90, 373
staphylococcal, 95
Proteasomes, 636, 637, 644, 645
Proteins (*see also* individual names)
Aft, 175
Afu, 95, 103
Ale, 358, 363-366
Alg, 123, 125
amino acid sequences, *see* Amino acids
Ang, 126
Ara, 55, 123, 125, 127
Arc, 673
ATP-dependent hydrolysis, 637
Bar, 123, 126
Bfd, 98
Bfe, 358, 360, 362, 365, 366
biosynthesis, *see* Biosynthesis
Btu, 109, 361, 366, 381, 382, 388
Bvg, 123, 126
catabolic activator, 121

[Proteins]
Cbr, 103
Ccc, 172, 174
Ceu, 103
channels (*see also* Channels), 74–84, 93
Chu, 108
Cir, 105, 358, 361, 365, 366, 373, 384
Cld, 104
Cor, 86
di-iron carboxylate, 468–471
-DNA interactions, 417, 672
Dps, 98, 99
DtxR, 54, 72, 115, 116, 122, 128, 639
Exb, 70, 71, 76, 82, 83, 118, 374, 389
Fat, 103, 104, 358, 360, 363, 366
Fbp, 94, 95, 103, 112
Fct, 199, 358, 363–365
Fcu, 357, 358, 360, 362, 366
Fec, 94, 103, 106, 117–120, 122, 124, 125, 235, 358, 360, 361, 363, 365, 366, 372, 373, 377, 379, 422
Feo, 70, 86, 363
Fep, 78–80, 84, 103, 104, 126, 357, 358, 360, 363, 365–369, 371–380, 382, 383, 385, 386, 388, 390
Fes, 105
Fet, 172, 174–176
Fhu, 74–84, 88, 90–94, 103, 104, 331, 357, 358, 360–363, 365, 366, 369–373, 375, 377, 378, 380–385, 387, 388, 390
Fiu, 105, 361, 373, 384
flavo-, 500
Fox, 357, 358, 360, 362, 365, 366, 371, 387
Fpt, 125, 357, 358, 360, 361, 365
Fpv, 357, 358, 360, 361, 366

[Proteins]
Fre, 174
Ftn, 98
Ftr, 172, 174, 175
Fur, 43, 54, 55, 59, 72, 73, 98, 101, 103, 104, 107, 112–116, 118, 121–128, 169, 170, 192, 361, 635, 638, 639, 642, 643, 658, 673
Fyu, 357, 358, 362, 365, 366
glyco-, *see* Glycoproteins
Has, 103, 108, 110, 111
Hem, 103, 107–109, 362
Hgb, 108, 109
Hhu, 108, 109
Hit, 95, 103
Hmb, 108, 109
Hmu, 108
holo-, 644
Hpu, 108, 110
Hup, 108, 111
Hut, 108, 109
Hxu, 103, 108, 110
IdeR, 43, 54, 116, 122
Irg, 123, 126, 362
iron-regulatory, *see* Iron-regulatory proteins
iron-sulfur, *see* Iron-sulfur proteins
isoelectric point, 378
Iut, 357, 358, 361, 365, 366, 373
Lac, 199
Lam, 75, 80, 364, 386
Lbp, 109, 112, 617
Lem, 126
Lex, 115
lipo-, *see* Lipoproteins
Lys, 55, 126, 127
membrane, *see* Membrane
Nif, 100, 101
-nucleic acid complexes, 417
Omp, 377
Pbr, 122, 124
Pbu, 358, 360, 362, 365, 366
Pch, 123, 125

[Proteins]
Pfe, 105, 123, 125, 357, 360, 361, 365, 366
Pfr, 122–125, 197
Pho, 364, 374
-protein interactions, 417
Psn, 125
Pup, 119, 122, 197, 357, 358, 360, 362, 366, 371, 377, 379, 381–384
Pvd, 122, 124
radicals, *see* Radicals
receptor, *see* Receptor proteins
recognition, 417–421
Rep, 125
-RNA interactions, 418, 425
Rum, 103, 105, 107, 358, 363, 365, 366
sequence identity, 406, 412, 414, 416
Sfu, 94, 95, 103, 113
Shl, 110
Shu, 107–109
Sid, 162
SodB, 124
SoxR, 100
synthesis, *see* Sythesis
Tbp, 109–112, 617
Tol, 81, 110
TonB, 70, 71, 76, 79, 80, 82, 83, 107–112, 118–120, 126, 192, 361, 364, 365, 368, 371, 372, 374–376, 380, 382, 384, 387–390
toxic, 75
Urbs, 164–170
Uvr, 123
Viu, 105, 358, 362, 365, 366
Ybt, 123, 125
Yfu, 95, 103
Proton magnetic resonance, *see* NMR
Protonation constants, *see* Acidity constants
Protoporphyrins, 445, 452, 457, 679
[Protoporphyrins]
IX, 97, 98, 449, 450, 453, 462, 482, 495, 657
Protozoa
heterotrophic, 10–13, 16, 18
Pseudobactins (*see also* Pyoverdins), 117–120, 122, 123, 196, 197, 206, 253, 357, 360, 362, 365, 366, 379, 381–384, 386
biosynthesis, *see* Biosynthesis
ferric, *see* Iron(III)
gallium, 381
Pseudomonads, 125, 192, 365
fluorescent, 196, 197, 206
receptor proteins, 361, 362
Pseudomonas sp., 55, 205, 253, 360–362
aeruginosa, 43, 54, 55, 96, 105, 121–125, 160, 163, 178, 192, 197, 252, 266, 357, 360, 361, 366, 374, 444, 494–496, 502, 725
fluorescens, 122–124, 126, 193, 196, 197, 343
putida, 83, 119, 122–125, 128, 196, 197, 206, 257, 339, 342, 343, 345, 346, 357, 362, 366, 374, 375, 439, 443, 457
syringae, 126, 196
viridiflava, 123, 125
Psoriasis, 693, 724
arthritis, 718
cutaneous, 718
Putrebactin, 247
Pyochelin, 125, 196, 357, 360, 361, 363, 366
biosynthesis, *see* Biosynthesis
Pyoverdins (*see also* Pseudobactins), 122, 125, 193, 196, 197, 253, 266, 280, 357, 360–362, 496
biosynthesis, *see* Biosynthesis
PaA, 243, 244, 251–253, 258, 266, 270, 272, 274, 299, 302, 310

Pyrene
  fluorescence, 257
Pyridine
  hydroxy-, 280, 290
Pyridinones, 711–713
  hydroxy-, *see* Hydroxy-pyridinones
Pyridoxal phosphate, 100, 101
Pyrite, 525, 526, 668
Pythium sp., 196

Q

Quinine, 347
Quinizarin-2-sulfonate, 303
Quinolines, 256, 303
  dihydroxy-, 248, 253
  hydroxy-, *see* Hydroxy-quinoline
Quinones (*see also* individual names), 410, 495, 499
  semi-, 500

R

R.
  capsulatus, see Rhodobacter
  leguminosarum, see Rhizobium
  meliloti, see Rhizobium
  phaseoli, see Rhizobium
Rabbit, 418, 677, 703, 704
  ferritin, 408, 412, 439
  reticulocytes, *see* Reticulocytes
  serum transferrin, *see* Transferrins
Radicals, 419, 488–494, 499, 503
  cleavage, 569
  free, 41, 578, 720, 721, 723
  hydroxyl, 40, 492–494, 521, 577, 634, 635, 649, 650, 659, 708, 709, 717, 721, 722
[Radicals]
  mediated damage, 497, 521, 569, 720, 723
  oxy-, 42, 44, 45, 49, 56, 60, 61, 73, 176, 442, 587, 639, 680, 693, 708, 717, 719, 721
  paraquat, 722
  protein, 493
  scavengers, 575
  semiquinone, 500
  superoxide, 488, 502
  tyrosyl, 470, 490, 492, 499
Radish, 197
Raman spectroscopy, 490
Rana catesbeiana, 439–443, 445
Rat, 416, 439, 539, 577, 676–679, 711, 716
  anemia, *see* Anemia
  blood, *see* Blood
  ferritin, *see* Ferritins
  hepatocytes, *see* Hepatocytes
  kidney, *see* Kidney
  liver, *see* Liver
  spleen, 614
Rate constants (*see also* Kinetic studies), 12, 14, 268, 269, 296–313
  chelate ring formation, 299
  first order, 298
  pseudo-first order, 6, 282, 300, 310
  second order, 298, 299, 303, 308, 309
Rattus norvegicus (*see also* Rat), 439
Razoxane, 718, 719, 722
Reaction rates, *see* Rate constants
Receptor proteins (*see also* Proteins), 73–127, 331, 355–391
  models, 365–371
  purification, 376
  transferrin, *see* Transferrin receptor
  with known sequences, 357–363

Redox potentials, 69, 71, 80, 81, 83, 106, 118, 156, 269, 273–277, 285, 295, 498–503, 518–520, 525–527, 531, 532, 555, 592, 617, 668, 692, 707
Reductases, 7, 15, 18, 49, 50, 171, 172, 194, 713
  cytochrome P_{450}, 722
  ferric, 173–175, 188, 198, 200, 206, 217, 233, 235, 617, 638, 640–642, 647, 653, 655, 674
  flavin, 48, 99
  membrane, 48, 616
  NADPH, 646
  nitrate, 15, 204
  oxido-, *see* Oxidoreductases
  ribonucleotide, *see* Ribonucleotide reductases
  siderophore, 72, 99, 204
Reduction potentials, *see* Redox potentials
Remineralization, 4, 5, 17, 25
Resonance Raman spectroscopy, *see* Raman spectroscopy
Reticulocytes, 606, 612, 613, 703
  rabbit, 407, 603, 614, 676
Rheumatoid arthritis, 722
Rhizobactin, 192
Rhizobia, 189–194, 673
Rhizobium sp., 189, 242, 673
  leguminosarum, 191, 193
  meliloti, 191, 192
  phaseoli, 191
Rhizoferrin, 87, 103, 106, 107, 151, 152, 205, 243, 245, 272, 274, 363, 366, 673
Rhizopus sp., 242
  microsporus, 152
Rhizosphere, 188, 191, 196, 197, 201–207, 216–219, 669
Rhodamin, 346
Rhodium(III), 673
  phenanthrenequinonediimine complexes, 410
Rhodobacter capsulatus, 96, 439, 443, 466, 494, 495
Rhodothamnus chamaecistus, 195
Rhodotorula, 152
  pilimanae, 156-158, 276
Rhodotorulic acid, 88, 103, 151, 152, 156–158, 172, 192, 204, 247, 267, 271, 272, 274, 276, 284, 285, 361, 380, 384
  enantio-, 385
Riboflavin 5'-monophosphate, *see* FMN
Ribonucleic acid, *see* RNA
Ribonucleotide reductases, 97, 468–470, 495, 646, 657, 692, 693, 706, 723, 724
Ribosomes, 405–407, 415, 418, 420, 421, 425, 636, 643
Rice, 201, 219, 223, 577
  shoots, 225, 227
Rifampin, 79
Rifamycins, 384
River (*see also* Water)
  iron in, 2–5, 518
RNA (*see also* Nucleic acids), 170, 635, 644
  antisense, 56, 126, 127
  double-stranded, 408
  m-, *see* mRNA
  -protein interactions, *see* Proteins
  single-stranded, 408
  t-, 415
RNA polymerases, 118, 120
  II, 167
Rodents (*see also* individual species), 711, 715
Roots, 193–196, 198, 200, 202–206, 216–225, 275, 421, 496, 565, 567, 572, 640
  barley, *see* Barley
  bean, 197
  grapevine, 196
  maize, 575, 576

[Roots]
  nodules, 193, 194, 423, 572–574
Rubredoxin (*see also* Iron-sulfur proteins), 442, 470
Rubrerythrin, 439, 442–444, 451, 460, 468–470

S

S.
  cerevisiae, see Saccharomyces
  dysenteriae, see Shigella
  marcescens, see Serratia
  mutans, see Streptococcus
  olivaceus, see Streptomyces
  paratyphi, see Salmonella
  pilosus, see Streptococcus and Streptomyces
  subtropicus, see Streptomyces
  tropici, see Sinorhizobium
  typhimurium, see Salmonella
Saccharides (*see also* Sugars *and* individual names)
  lipopoly-, *see* Lipopolysaccharides
  oligo-, *see* Oligosaccharides
Saccharomyces cerevisiae (*see also* Yeast), 49, 156, 171, 200, 201, 217, 363, 422, 424, 444, 566, 641, 642, 668, 674
Saintpaulia, 199
L-Salicylaldoxime, 8
Salicylate (or salicylic acid), 196, 268, 290
Salmo salmar, 439, 443
Salmonella sp., 124
  paratyphi, 77, 78
  typhimurium, 58, 59, 77, 78, 85, 86, 113, 114, 122, 158, 160, 164
Salt bridges (*see also* Electrostatic interactions), 446, 447, 449, 451–453, 455, 462, 485, 600, 656, 657
Scandium(III), 603
Scanning electron microscopy, 219
Schistosoma mansoni, 439–443, 451
  ferritin, *see* Ferritins
Schizokinen, 19, 20, 88, 103, 104, 204
Schizosaccharomyces pombe, 171
SDS-PAGE, *see* Electrophoresis
Sea squirts, 588
Seawater (*see also* Ocean *and* Water)
  iron in, 1–29
Sedimentation studies, 378
  velocity measurements, 458
Sediments
  anoxic, 2, 3
  iron in, 6, 26, 27, 529
  mineralized, 529
Seedling
  diseases, 196
  soybean, 204
Seeds, 423, 496, 567, 572, 573, 640
Serine (and residues), 101, 149, 385, 452, 453, 455, 456, 462, 592, 595, 602, 607, 608
  D-, 385
  2,3-dihydroxy-*N*-benzoyl-, *see* 2,3-Dihydroxy-*N*-benzoylserine
  N-[N^2-(2,3-dihydroxybenzoyl)-D-lysyl]-L-, *see* Chrysobactin
Serotonin, 646
Serotransferrin (*see also* Transferrins), 653
Serratia marcescens, 94, 103, 108, 110–112, 363
Serum (*see also* Plasma), 109, 110, 349, 651, 680
  human, 680
  rabbit, *see* Rabbit
  transferrin, *see* Transferrins
Sheep spleen ferritin, *see* Ferritins

Shift reagents
  paramagnetic, 603
Shigella sp., 59
  dysenteriae, 107–109
Shoots, 567, 572
  iron transport, 219–223, 496, 640
Sickle cell anemia, 287, 720
Siderite, 525, 526
Siderophilins, *see* Transferrins
Siderophore(s) (*see also* individual names), 8, 9, 17–20, 38, 49–57, 59, 61, 69–74, 81–84, 87–94, 99, 100, 102, 104, 105, 107, 109, 113, 116, 117, 121–128, 188, 191–194, 202, 533, 587, 617, 638, 639, 654, 668–674, 693, 698, 701, 715, 724, 725
  bacterial/microbial, 189–194, 200, 203–205, 268, 422, 669, 672, 714
  binding groups, 242, 243
  biomimetic, 257, 277, 329–350
  biosynthesis, *see* Biosynthesis
  carboxylate, 152, 191, 242, 243, 247, 255, 272, 274, 338, 673
  catecholamide, 248, 252, 261, 263
  catecholate, 51, 105, 198, 199, 242–244, 259, 268, 270, 272–276, 279, 285–287, 299, 301, 310, 334, 338, 362, 363, 373, 379, 381, 670, 672, 673, 694, 699
  complex dissociation reactions, 306–309
  endogenous, 46, 50, 55
  exogenous, 46, 50, 55
  ^{55}Fe-labeled, 154, 373
  fluorescent, 123–125, 253, 343–346
  fungal, 106, 148–170, 192, 196, 205, 360, 363
[Siderophore(s)]
  hydroxamate, 152, 155–158, 191, 192, 195, 199, 242–244, 247 –253, 255–257, 268, 270–276, 278–285, 290, 299, 301, 310, 335, 338, 340, 341, 357, 361, 362, 373, 380, 384, 670, 673, 694, 699, 714
  metal selectivity, 241, 242
  phenolate, 51, 52, 362
  phyto-, *see* Phytosiderophores
  receptors, *see* Receptor proteins
  redox potentials, 273–277
  reductase, *see* Reductases
  reversed, 346–349
  rhizobial, 191–193
  synthetic models, 277–296
  thermodynamics and kinetics, 241, 242
  transport, 153–156
  uptake and release mechanisms, 296–313, 421
Siderosis, 649
Silicate, 4, 5
  aluminum, 4–6, 9
Silver, 578
Sinorhizobium sp., 189
  tropici, 191
Skin, 503, 651
Snails (*see also* individual names), 414, 438, 464
Sodium dodecyl sulfate, 74, 75, 79, 155
  gel electrophoresis, *see* Electrophoresis
  micelles, 308
Soil (*see also* Lithosphere)
  anaerobic, 565, 577
  calcareous, 201, 205, 216
  microorganisms, 188–208, 669
Sorghum, 201
Soybeans, 193, 194, 216, 406, 423, 439, 464, 573, 575
  ferritin, *see* Ferritins
  seedlings, 204

Spectrophotometry (*see also* Absorption bands and spectra *and* UV absorption), 270, 600–602
Spermidine, 105
Spheroplasts, 369
Spinach ferredoxin, 675
Spin-labeling
  site directed, 78, 79
Spleen, 644, 703
  horse, *see* Horse
  mouse, 58, 59
  rat, *see* Rat
Spliceosome, 418
Squaric acid, 303
Squash, 204
Stability constants (*see also* Equilibrium constants, Association constants, Dissociation constants, *and* Michaelis-Menten constants), 49, 72, 152, 170, 172, 206, 217, 241–245, 249, 268–296, 309, 312, 523, 524, 598, 604, 609, 675, 694–698, 701, 705, 711, 717
  conditional, 8, 9
  of ferric siderophores, 270–272, 638
Stachyose, 79
Standard potentials, *see* Redox potentials
Staphylococcus aureus, 42
Steady-state
  balance of Fe(II) and Fe(III), 7, 13
  kinetics, *see* Michaelis-Menten kinetics
Stomach, 124
Stopped flow kinetics (*see also* Kinetic studies), 483, 488, 490, 502
Streptococcus
  mutans, 45, 48
  pilosus, 639
  pyogenes, 444
Streptomyces sp., 105, 115, 116, 242, 422
  olivaceus, 279, 494
  pilosus, 54, 714
  subtropicus, 724
Streptomycin, 567
Structure-function relationships, 75, 153, 277, 595, 608
Suberin, 565
Succinate, 121, 681
  dehydrogenase, *see* Dehydrogenases
Sugars (*see also* individual names), 72, 75, 547
Sulfate
  iron, 24, 541, 676, 711
  thio-, 95
Sulfide, 101, 418, 525, 675
  di-, *see* Disulfide
  hydrogen, 100
  iron clusters, *see* Iron-sulfur clusters
  per-, 100
Sulfur
  ^{35}S, 90
  S^0, 101
  iron clusters, *see* Iron-sulfur clusters
Superoxide, 6, 7, 40, 41, 488, 500, 708
  radical, *see* Radicals
  stress, 100
Superoxide dismutase, 41, 42, 45, 442, 488, 680
  iron, 42, 54, 121
  manganese, 42, 116, 673
Supramolecular assemblies, 271
Susceptibility measurements, *see* Magnetic susceptibility measurements
Sweat gland, 594
Symbiosis, 194
  nitrogen-fixing, 189, 573
Synchrotron radiation, 470
Synechococcus sp., 16, 17, 19, 89, 98, 121

Synechocystis sp., 88, 96, 235, 439, 442–444, 457
Synthases
- aerobactin, 59
- aminolevulinic acid, *see* Aminolevulinic acid
- citrate, 191
- gramicidin, 51
- nitric oxide, *see* Nitrogen monoxide
- peptide, 50, 51, 61, 162
- tyrocidine, 51

Synthesis (of)
- amonabactin, 50
- bio-, *see* Biosynthesis
- ceruloplasmin, 680
- DNA, *see* DNA
- enterobactin, 50
- ferritin, 420, 423, 641, 648, 650
- heme, 405, 480, 676, 678, 679
- mRNA, *see* mRNA
- photo-, *see* Photosynthesis
- protein, 643

T

T.
- *palladium, see Treponema*
- *saginata, see Taenia*

Taenia saginata, 439, 443
Tannins, 200
Teichoic acid, 74
Terbium(III), 460–462, 484, 489, 590, 603, 618
Ternary complexes, 308, 310–313, 704
2,2',2''-Terpyridyl, 410
Thalassemia, 287, 615, 655, 693, 707, 709, 720
Thalassiosira sp., 7
- *oceanica*, 15

Thallium(III), 590, 603
Therapeutic agents (*see also* Drugs, Chelating agents and therapy *and* individual names), 278, 287, 311, 547, 691–726
- bacteriostatic effects, 278
- iron, *see* Iron chelators

Thiazolidine ring, 361
Thioglycolic acid, 500, 501
Thiosulfate, 95
Thorium
- ^{230}Th, 27
- (IV), 590

Threonine (and residues), 457, 464, 591, 607, 608
Thymus, 712
Thyroid hormone, 424
Tiller, 222
- barley, *see* Barley

Tin (different oxidation states), 449, 462, 495
Tiron, *see* 1,2-Dihydroxybenzene-3,5-disulfonate
Tissue(s) (*see also* individual names), 645
- damage, 107, 175, 693, 721
- inflamed, 722
- ischemic, 721
- mammalian, 659

Tobacco, 574
- hornworm, 594

Tomato, 216, 565
Toxicity (of)
- aluminum, 681
- cardio-, *see* Cardiotoxicity
- cyto-, *see* Cytotoxicity
- iron chelators, *see* Iron chelators
- iron, *see* Iron
- ligands, 694
- manganese, 677
- metal ions, 675, 694
- neuro-, *see* Neurotoxicity
- nickel, 678
- retinal, 716

Toxins (*see also* individual names), 122, 126, 198, 405
  diphtheria, 113, 115, 116, 639
Transacetylase
  δ-*N*-hydroxyornithine:acetyl-CoA-*N*-, 157, 163
Transcription factors, 175, 636, 649, 650, 659
  Fnr, 100
  GATA, 164-170
Transfer
  electron, *see* Electron transfer
Transferases
  chloramphenicol acetyl, 571
  phosphopantethiene, 163
Transferrin(s), 52, 53, 58–61, 69–71, 73, 94, 95, 100, 110–112, 127, 200, 257, 310, 311, 363, 422, 425, 470, 530, 587–618, 638, 647, 648, 651, 653, 654, 670, 671, 677, 678, 680, 683
  apo-, 484, 592, 597, 601, 604, 611, 652, 654, 681, 722, 723
  atypical, 594
  avian, 112
  coordination spheres, 592, 593
  definitions, 587–594
  eukaryotic, 588
  evolutionary origin, 588
  half-life, 603
  human, 56, 69, 71, 111, 112, 520, 589, 593, 597–599, 603, 605, 611, 617, 618, 681, 682, 720, 722, 723
  insect, 594
  kinetics of iron release, 603–606
  lacto-, *see* Lactoferrins
  list of metal ions bound, 590
  list of stability constants, 598
  melano-, 594
  microbial uptake, 617, 618
  mRNA, 646, 648
  nomenclature, 587–594
[Transferrin(s)]
  ovo-, *see* Conalbumin *and* Ovotransferrin
  rabbit apo-, 611
  rabbit serum, 588, 595, 596, 611
  receptor, *see* Transferrin receptor
  receptor-dependent iron uptake, 610–613
  receptor-independent iron uptake, 613, 614
  recombinant human, 595, 602
  serum, 589, 591, 593, 594, 599, 603, 646, 653, 681, 682
  spectroscopic studies, 600–603
  structure, 588, 592, 593, 595–606
  synergistic anion, 591, 592
  vertebrate, 111, 112, 588, 594, 653
Transferrin receptor, 57, 100, 109, 404–409, 414–417, 421, 425, 593–595, 605–618, 643, 646, 648–651, 654, 677, 682
  avian, 608
  diagram, 608
  human, 608, 617
  mRNA, 643, 644, 648, 649
  regulation, 609, 610
  structure, 606-608
Transposons, 42, 121
Trees
  deciduous, 195
TREN, 283, 285, 290, 292–295, 312, 701, 702
TRENCAM, 285, 286, 290, 295
TRENDROX, 251, 256, 282, 283, 290, 294, 295
TRENSOX, 251, 256, 294–296
Treponema palladium, 494
Triacetylfusarinine C, 268
Tris, 348
Tris(2-aminoethyl)amine, *see* TREN

Tris[2-aminoethyl-(2,3-dihydroxybenzoyl)]-amine, *see* TRENCAM
Trisbenzamide, 335–337
N,N',N''-Tris(2,3-dihydroxybenzoyl)-1,3,5-tris(aminomethyl)-benzene, *see* MECAM
Tris(hydroxymethyl)methyl-amine, *see* Tris
Triton X-100, 376, 378
Tropoisomerase, 722
Trypsin, 84, 88, 90, 371, 606–608
Tryptophan, 51, 364, 371, 387, 681
Tumor(s) (*see also* individual names)
  imaging, 589, 681
  necrosis factor a, 424, 650
  solid, 722
Typhoid
  mouse, 58
Tyrocidine synthase, *see* Synthases
Tyrosine (and residues), 78, 95, 97, 387, 439, 451–453, 456, 461, 462, 469, 490, 496, 592–594, 596, 601, 681
  hydroxylase, *see* Hydroxylases
  radical, *see* Radicals

U

Ubiquitin, 637
Ultraviolet, *see* UV
Umbilical cord, 594
Urea
  dichlorophenyldimethyl-, 232
  gel electrophoresis, *see* Gel electrophoresis
Urine
  iron excretion, 715, 716
Ustilago sp., 163–171, 177
[*Ustilago* sp.]
  maydis, 156, 158, 160–163, 165, 166
  sphaerogena, 149, 155, 171, 276
UV
  absorption or spectra (*see also* Absorption bands and spectra), 348, 458, 468, 484, 601
  crosslinking experiments, 419
  difference spectra, 592, 596, 598, 601
  radiation, 503

V

V.
  anguillarum, *see Vibrio*
  cholerae, *see Vibrio*
Vacuoles, 47
  iron in, 424, 564, 640, 641
Valine (and residues), 82, 101, 457, 464
Vanadium (different oxidation states), 333, 385, 590
Vanadium(IV) and/or vanadylion, 500, 675, 678
  enterobactin, 337, 385
  interdependency with other elements, *see* Interdependencies
Van der Waals forces, 338, 339
Vertebrates (*see also* individual names), 53, 56–58, 60, 587
  ferritin, *see* Ferritins
  hemoglobin, *see* Hemoglobins
  iron-withholding defenses, 60
  mRNA, *see* mRNA
  transferrin, *see* Transferrins
Vesicles (*see also* Micelles), 611
  endocytic, 530
  endosomal, 654
  lipid, 78

Vibrio
anguillarum, 50, 56, 85, 88, 89, 103, 104, 123, 126, 197, 360, 363, 366
cholerae, 55, 105, 108, 109, 123, 126, 362, 366
Vibriobactin, 105, 251, 252, 362, 366
Vigna unguiculata, 496
Virulence factor, 125, 126, 191, 198
fungi, *see* Fungi
Viruses, 74
mRNA, *see* mRNA
polio, 420
Visible spectra, *see* Absorption bands and spectra
Vitamin B_{12}, (*see also* Cobalamin), 82, 87, 92, 109, 361, 374, 388
receptor protein, 381, 382, 388
Vitamin C, *see* Ascorbate
Volcanoes
eruption, 4, 5, 28
Mauna Loa, 28
Mount Pinatubo, 28
Voltammetry (*see also* Electrochemistry), 284, 285
stripping, 8

W

Water
coastal, 2–4, 6, 15, 16, 21
estuaries, 3
ocean, *see* Oceans
river, *see* River
sea, *see* Seawater
Wheat germ, 407, 420
WHO, *see* World Health Organization
Wilcoxina sp., 195
Wilson's disease, 175, 680
Wilt, 196
World Health Organization, 720

X

X-ray
crystal structures, 85, 96, 116, 271, 278, 284, 288, 290, 368, 371, 378, 385, 386, 449, 452, 456, 457, 461, 468, 470, 480, 484, 593, 596, 597, 657
absorption spectroscopy, 570
dispersive, 501
diffraction studies, 334, 335, 337, 529, 534, 535
low-angle scattering profiles, 595, 597
powder diffraction, 534, 535, 539, 547
single crystal diffraction, 534, 535, 539, 544, 548, 551
Xenobiotics, 500
Xenopus laevis, 439–443
Xylem, 200, 203, 227, 234, 566, 640

Y

Y.
pestis, see Yersinia
enterolitica, see Yersinia
Yeasts (*see also* individual names), 49, 57, 152, 156, 161, 165, 166, 170–173, 207, 421–424, 617, 657, 658, 674
eukaryotic, 49
iron homeostasis, 641–643
iron transport, 171–176, 200, 201
mutants, 566
Yersinia
enterolitica, 50, 88, 90, 95, 103, 107–109, 113, 357, 360, 362, 366, 715
pestis, 50, 55, 88, 89, 108, 123, 125
Yersiniabactin, 50, 123, 125, 357, 362, 366

Ytterbium(III) (*see also* Lanthanides), 590

Z

Zea mays, 439, 443
Zinc(II), 2, 28, 170, 241, 333, 461, 489, 496, 566, 578, 590, 598, 671, 672, 674–677, 679, 681, 694, 719
  depletion, 716
[Zinc(II)]
  excretion, 717
  geochemical cycle, 674
  interdependency with other elements, *see* Interdependencies
  phytosiderophore complexes, 202
  transporter, 201
Zinc fingers, 164, 165, 169, 170
Zooplankton, 24
Zygomycetes, 152, 177, 673